Marine Propellers and Propulsion

Marine Propellers and Propulsion

Third Edition

J S Carlton FREng
Professor of Marine Engineering, City University London
President of the Institute of Marine Engineering,
Science and Technology 2010/11

AMSTERDAM • BOSTON • HEIDELBERG • LONDON • NEW YORK • OXFORD
PARIS • SAN DIEGO • SAN FRANCISCO • SINGAPORE • SYDNEY • TOKYO

Butterworth-Heinemann is an imprint of Elsevier

Butterworth-Heinemann is an imprint of Elsevier
The Boulevard, Langford Lane, Kidlington, Oxford OX5 1GB UK
225 Wyman Street, Waltham, MA 02451, USA

First edition 1994
Second edition 2007
Third edition 2012

Copyright © 2012 John Carlton. Published by Elsevier Ltd. All right reserved.

The right of John Carlton to be identified as the author of this work has been asserted in accordance with the Copyright, Designs and Patents Act 1988

No part of this publication may be reproduced, stored in a retrieval system or transmitted in any form or by any means electronic, mechanical, photocopying, recording or otherwise without the prior written permission of the publisher

Permissions may be sought directly from Elsevier's Science & Technology Rights Department in Oxford, UK: phone (+44) (0) 1865 843830; fax (+44) (0) 1865 853333; email: permissions@elsevier.com. Alternatively you can submit your request online by visiting the Elsevier web site at http://elsevier.com/locate/permissions, and selecting *Obtaining permission to use Elsevier material*

Notice
No responsibility is assumed by the publisher for any injury and/or damage to persons or property as a matter of products liability, negligence or otherwise, or from any use or operation of any methods, products, instructions or ideas contained in the material herein. Because of rapid advances in the medical sciences, in particular, independent verification of diagnoses and drug dosages should be made

British Library Cataloguing-in-Publication Data
A catalogue record for this book is available from the British Library

Library of Congress Cataloging-in-Publication Data
A catalog record for this book is available from the Library of Congress

ISBN: 978-0-08-097123-0

For information on all Butterworth-Heinemann publications
visit our web site at http://books.elsevier.com

Transferred to Digital Printing in 2013

Working together to grow
libraries in developing countries

www.elsevier.com | www.bookaid.org | www.sabre.org

ELSEVIER   BOOK AID International   Sabre Foundation

To Jane and Caroline

Contents

Preface to the Third Edition   xi
Preface to the Second Edition   xiii
Preface to the First Edition   xv
General Nomenclature   xvii

1. The Early Development of the Screw Propeller   1
References and Further Reading   9

2. Propulsion Systems   11
2.1. Fixed Pitch Propellers   11
2.2. Ducted Propellers   13
2.3. Podded and Azimuthing Propulsors   16
2.4. Contra-Rotating Propellers   16
2.5. Overlapping Propellers   17
2.6. Tandem Propellers   18
2.7. Controllable Pitch Propellers   19
2.8. Surface Piercing Propellers   21
2.9. Waterjet Propulsion   21
2.10. Cycloidal Propellers   22
2.11. Paddle Wheels   23
2.12. Magnetohydrodynamic Propulsion   24
2.13. Whale-Tail Propulsion   27
References and Further Reading   27

3. Propeller Geometry   29
3.1. Frames of Reference   29
3.2. Propeller Reference Lines   30
3.3. Pitch   31
3.4. Rake and Skew   33
3.5. Propeller Outlines and Area   35
3.6. Propeller Drawing Methods   38
3.7. Section Geometry and Definition   38
3.8. Blade Thickness Distribution and Thickness Fraction   42
3.9. Blade Interference Limits for Controllable Pitch Propellers   43
3.10. Controllable Pitch Propeller Off-Design Section Geometry   43
3.11. Miscellaneous Conventional Propeller Geometry Terminology   46
References and Further Reading   46

4. The Propeller Environment   47
4.1. Density of Water   47
4.2. Salinity   48
4.3. Water Temperature   49
4.4. Viscosity   49
4.5. Vapor Pressure   50
4.6. Dissolved Gases in Sea Water   50
4.7. Surface Tension   51
4.8. Weather   51
4.9. Silt and Marine Organisms   55
References and Further Reading   56

5. The Ship Wake Field   57
5.1. General Wake Field Characteristics   57
5.2. Wake Field Definition   59
5.3. The Nominal Wake Field   61
5.4. Estimation of Wake Field Parameters   62
5.5. Effective Wake Field   65
5.6. Wake Field Scaling   67
5.7. Wake Quality Assessment   70
5.8. Wake Field Measurement   72
References and Further Reading   77

6. Propeller Performance Characteristics   79
6.1. General Open Water Characteristics   79
6.2. The Effect of Cavitation on Open Water Characteristics   85
6.3. Propeller Scale Effects   87
6.4. Specific Propeller Open Water Characteristics   89
6.5. Standard Series Data   93
6.6. Multi-Quadrant Series Data   112
6.7. Slipstream Contraction and Flow Velocities in the Wake   119
6.8. Behind-Hull Propeller Characteristics   133
6.9. Propeller Ventilation   134
References and Further Reading   136

7. Theoretical Methods — Basic Concepts   137
7.1. Basic Aerofoil Section Characteristics   139
7.2. Vortex Filaments and Sheets   141

vii

7.3.	Field Point Velocities	142	
7.4.	The Kutta Condition	144	
7.5.	The Starting Vortex	145	
7.6.	Thin Aerofoil Theory	146	
7.7.	Pressure Distribution Calculations	149	
7.8.	Boundary Layer Growth Over an Aerofoil	154	
7.9.	The Finite Wing	158	
7.10.	Models of Propeller Action	160	
7.11.	Source and Vortex Panel Methods	163	
7.12.	Euler, Lagrangian and Navier–Stokes Methods	164	
References and Further Reading		166	

8. Theoretical and Analytical Methods Relating to Propeller Action — 169

- 8.1. Momentum Theory – Rankine (1865); R.E. Froude (1887) — 169
- 8.2. Blade Element Theory – W. Froude (1878) — 171
- 8.3. Propeller Theoretical Development (1900–1930) — 172
- 8.4. Burrill's Analysis Procedure (1944) — 174
- 8.5. Lerbs Analysis Method (1952) — 180
- 8.6. Eckhardt and Morgan's Design Method (1955) — 183
- 8.7. Lifting Surface Correction Factors – Morgan et al. — 188
- 8.8. Lifting Surface Models — 193
- 8.9. Lifting Line–Lifting Surface Hybrid Models — 194
- 8.10. Vortex Lattice Methods — 194
- 8.11. Boundary Element Methods — 198
- 8.12. Methods for Specialist Propulsors — 200
- 8.13. Computational Fluid Dynamics Analysis — 202
- References and Further Reading — 204

9. Cavitation — 209

- 9.1. The Basic Physics of Cavitation — 210
- 9.2. Types of Cavitation Experienced by Propellers — 214
- 9.3. Cavitation Considerations in Design — 221
- 9.4. Cavitation Inception — 229
- 9.5. Cavitation-Induced Damage — 234
- 9.6. Cavitation Testing of Propellers — 239
- 9.7. Analysis of Measured Pressure Data from a Cavitating Propeller — 243
- 9.8. The CFD Prediction of Cavitation — 245
- References and Further Reading — 248

10. Propeller Noise — 251

- 10.1. Physics of Underwater Sound — 251
- 10.2. Nature of Propeller Noise — 255
- 10.3. Noise Scaling Relationships — 261
- 10.4. Noise Prediction and Control — 262
- 10.5. Transverse Propulsion Unit Noise — 263
- 10.6. Measurement of Radiated Noise — 263
- 10.7. Noise in Relation to Marine Mammals — 264
- References and Further Reading — 268

11. Propeller, Ship and Rudder Interaction — 271

- 11.1. Bearing Forces and Moments — 271
- 11.2. Hydrodynamic Interaction — 288
- 11.3. Propeller–Rudder Interaction — 293
- References and Further Reading — 297

12. Ship Resistance and Propulsion — 299

- 12.1. Froude's Analysis Procedure — 300
- 12.2. Components of Calm Water Resistance — 301
- 12.3. Methods of Resistance Evaluation — 311
- 12.4. Propulsive Coefficients — 323
- 12.5. The Influence of Rough Water — 325
- 12.6. Restricted Water Effects — 327
- 12.7. High-Speed Hull form Resistance — 328
- 12.8. Air Resistance — 330
- References and Further Reading — 330

13. Thrust Augmentation Devices — 333

- 13.1. Devices Before the Propeller — 334
- 13.2. Devices at the Propeller — 337
- 13.3. Devices Behind the Propeller — 341
- 13.4. Combinations of Systems — 342
- References and Further Reading — 342

14. Transverse Thrusters — 343

- 14.1. Transverse Thrusters — 343
- 14.2. Steerable Internal Duct Thrusters — 350
- References and Further Reading — 352

15. Azimuthing and Podded Propulsors — 353

- 15.1. Azimuthing Thrusters — 353
- 15.2. Podded Propulsors — 356
- References and Further Reading — 362

16. Waterjet Propulsion — 363

- 16.1. Basic Principle of Waterjet Propulsion — 364
- 16.2. Impeller Types — 366
- 16.3. Maneuvering Aspects of Waterjets — 367
- 16.4. Waterjet Component Design — 367
- References and Further Reading — 371

17. Full-Scale Trials — 373

- 17.1. Power Absorption Measurements and Trials — 373
- 17.2. Bollard Pull Trials — 379

Contents

17.3. Propeller-Induced Hull Surface
Pressure Measurements   381
17.4. Cavitation Observations   382
References and Further Reading   383

18. Propeller Materials   385
18.1. General Properties of Propeller
Materials   385
18.2. Specific Properties of Propeller
Materials   388
18.3. Mechanical Properties   393
18.4. Test Procedures   394
References and Further Reading   396

19. Propeller Blade Strength   397
19.1. Cantilever Beam Method   397
19.2. Numerical Blade Stress
Computational Methods   402
19.3. Detailed Strength Design
Considerations   405
19.4. Propeller Backing Stresses   408
19.5. Blade Root Fillet Design   408
19.6. Residual Blade Stresses   409
19.7. Allowable Design Stresses   410
19.8. Full-Scale Blade Strain
Measurement   413
References and Further Reading   414

20. Propeller Manufacture   415
20.1. Traditional Manufacturing Method   415
20.2. Changes to the Traditional Technique
of Manufacture   419
References and Further Reading   420

21. Propeller Blade Vibration   421
21.1. Flat-Plate Blade Vibration in Air   421
21.2. Vibration of Propeller Blades in Air   422
21.3. The Effect of Immersion in Water   422
21.4. Simple Estimation Methods   425
21.5. Finite Element Analysis   426
21.6. Propeller Blade Damping   427
21.7. Propeller Singing   428
References and Further Reading   429

22. Propeller Design   431
22.1. The Design and Analysis Loop   431
22.2. Design Constraints   433

22.3. The Energy Efficiency Design Index   433
22.4. The Choice of Propeller Type   435
22.5. The Propeller Design Basis   438
22.6. The Use of Standard Series Data
in Design   442
22.7. Design Considerations   445
22.8. The Design Process   452
References and Further Reading   458

23. Operational Problems   459
23.1. Performance Related Problems   459
23.2. Propeller Integrity Related Problems   465
23.3. Impact or Grounding   467
References and Further Reading   467

24. Service Performance and Analysis   469
24.1. Effects of Weather   469
24.2. Hull Roughness and Fouling   469
24.3. Hull Drag Reduction   478
24.4. Propeller Roughness and Fouling   478
24.5. Generalized Equations for the
Roughness-Induced Power Penalties
in Ship Operation   482
24.6. Monitoring of Ship Performance   485
References and Further Reading   492

25. Propeller Tolerances and
Inspection   495
25.1. Propeller Tolerances   495
25.2. Propeller Inspection   495
References and Further Reading   500

26. Propeller Maintenance and Repair   501
26.1. Causes of Propeller Damage   501
26.2. Propeller Repair   503
26.3. Welding and the Extent of Weld
Repairs   505
26.4. Stress Relief   507
References and Further Reading   508

Bibliography   509
Index   511

Preface

Preface to the Third Edition

Since the second edition was published in 2007 a number of important changes have and are taking place which relate to ship propulsion. Among these are the developments arising from the IMO's initiatives on the Energy Efficiency Design Index and the bringing to fruition of their earlier work on hull coatings stemming from their resolution on the subject in 2001. In addition to these legislative initiatives is a greater understanding of the physics associated with cavitation development and collapse and its subsequent erosive effects on propeller materials. Morover, some new types of propulsor are being developed while in other technical fields there is a greater awareness, for example, of the effects of shipping activity on the behavior of marine mammals. It is the intention of this third edition to capture these and other developments that have occurred. In addition a certain amount of rearrangement of the subject matter has also taken place.

As with previous editions thanks are once again due to many colleagues around the world who have made very valuable suggestions and comments as well as providing me with further material for inclusion from their own libraries and archives. As in previous years I would like to particularly acknowledge Mr P.A. Fitzsimmons, Mr J. Th. Ligtelijn, Dr D. Radosavljevic and Prof. Dr T. van Terwisga who have continued in their support as well as particular contributions for this addition from Mr T. Veitkomeno, Mr J. Gonzalez-Adalid, Mr P. van Terwisga and Dr S. Whitworth. Finally, thanks are again due to Jane, my wife, for her encouragement, support and proof-reading activities in undertaking this new edition of the book.

J.S. Carlton
Battle, East Sussex
July 2012

Preface

Preface to the Second Edition

It is now rather over a decade since much of the material was written for the first edition of this book. During that time advances have been made in the understanding of several branches of the subject and it is now time to incorporate much of that material into the text. These advances in understanding, together with the natural progression of the subject, relate particularly to cavitation dynamics, theoretical methods including the growing development of computational fluid dynamics in many parts of the subject and the use of carbon fiber materials for certain propeller types. Moreover, podded propulsors have emerged in the intervening years since the first edition was written and have become a propulsion option for certain types of ship, particularly cruise ships and ice breakers but with a potential to embrace other ship types in the future.

Some other aspects of the subject were not included in the original publication for a number of reasons. In this new edition I have attempted to rectify some of these omissions by the inclusion of material on high-speed propellers, propeller—rudder interaction as well as a new chapter dealing with azimuthing and podded propulsors and a substantial revision to the chapter on cavitation. These additions, together with a reasonably extensive updating of the material and the removal of the inevitable typographical errors, in the first edition form the basis of this new

addition. Furthermore, experience in using the book over the last 10 years or so has shown that the arrangement of some of the material could be improved. As a consequence it will be seen that a certain amount of re-grouping of the subject matter has taken place in the hope that this will make the text easier to use.

Finally, thanks are once again due to many colleagues around the world who have made very valuable suggestions and comments as well as providing me with further material for inclusion from their own libraries and archives. Furthermore, the normal day-to-day discussions that are held on various aspects of the subject frequently trigger thought processes which have found their way into various parts of the narrative. In particular, my thanks are due to Mrs W. Ball, Mr P.A. Fitzsimmons, Mr M. Johansen, Mr J. Th. Ligtelijn, Dr D. Radosavljevic, Prof. Dr T. van Terwisga and Mr J. Wiltshire. Thanks are also due to Dr P. Helmore who, having read the book some 10 years ago, kindly supplied me with a list of errata for this edition. Finally, thanks are also due to Jane, my wife, for her encouragement and support in undertaking this revision to the book in a relatively short-time frame.

J.S. Carlton
Hythe, Kent
December 2006

Preface

Preface to the First Edition

Although the propeller normally lies well submerged out of sight and therefore, to some extent, also out of mind, it is a deceptively complex component in both the hydrodynamic and the structural sense. The subject of propulsion technology embraces many disciplines: for example, those of mathematics, physics, metallurgy, naval architecture and mechanical and marine engineering. Clearly, the dependence of the subject on such a wide set of basic disciplines introduces the possibility of conflicting requirements within the design process, necessitating some degree of compromise between opposing constraints. It is the attainment of this compromise that typifies good propeller design.

The foundations of the subject were laid during the latter part of the last century and the early years of this century. Since that time much has been written and published in the form of technical papers, but the number of books which attempt to draw together all of these works on the subject from around the world is small. A brief study of the bibliography shows that, with the exception of Gerr's recent book dealing with the practical aspects of the design of small craft propellers, little has been published dealing with the subject as an entity since the early 1960s. Over the last 30 or so years an immense amount of work, both theoretical and empirical, has been undertaken and published, probably more than in any preceding period. The principal aim, therefore, of this book is to collect together the work that has been done in the field of propeller technology up to the present time in each of the areas of hydrodynamics, strength, manufacture and design, so as to present an overall view of the subject and the current levels of knowledge.

The book is mainly directed towards practising marine engineers and naval architects, principally within the marine industry but also in academic and research institutions. In particular when writing this book I have kept in mind the range of questions about propeller technology that are frequently posed by designers, ship operators and surveyors and I have attempted to provide answers to these questions. Furthermore, the book is based on the currently accepted body of knowledge of use to practical design and

analysis; current research issues are addressed in a less extensive manner. For example, recent developments in surface panel techniques and Navier–Stokes solutions are dealt with in less detail than the currently more widely used lifting line, lifting surface and vortex lattice techniques of propeller analysis. As a consequence knowledge of mathematics, fluid mechanics and engineering science is assumed commensurate with these premises. Notwithstanding this, it is to be hoped that students at both undergraduate and post-graduate levels will find the book of value to their studies.

The first two chapters of the book are essentially an introduction to the subject: first, a brief history of the early development of propellers and, second, an introduction to the different propeller types that are either of topical interest or, alternatively, will not be considered further in the book; for example, paddle wheels or superconducting electric propulsion. Chapter 3 considers propeller geometry and, consequently, this chapter can be viewed as a foundation upon which the rest of the book is built. Without a thorough knowledge of propeller geometry, the subject will not be fully understood. Chapters 4 and 5 concern themselves with the environment in which the propeller operates and the wake field in particular. The wake field and its various methods of prediction and transformation, particularly from nominal to effective, are again fundamental to the understanding of the design and analysis of propellers.

Chapters 6–15 deal with propulsion hydrodynamics, first in the context of model results and theoretical methods relating to propellers fixed to line shafting, then moving on to ship resistance and propulsion, including the important subjects of propeller–hull interaction and thrust augmentation devices, and finally to consideration of the specific aspects of fixed and rotable thrusters and waterjets. Chapter 17 addresses the all-important subject of sea trials in terms of the conditions necessary for a valid trial, instrumentation and analysis.

Chapters 18–20 deal with the mechanical aspects of propellers. Materials, manufacture, blade strength and vibration are the principal subjects of these four chapters,

xv

and the techniques discussed are generally applicable to all types of propulsors. The final five chapters, 21–25, discuss various practical aspects of propeller technology, starting with design, then continuing to operational problems, service performance and, finally, to propeller inspection, repair and maintenance.

In each of the chapters of the book the attainment of a fair balance between theoretical and practical considerations has been attempted, so that the information presented will be of value to the practitioner in marine science. For more advanced studies, particularly of a theoretical nature, the data presented here will act as a starting point for further research: in the case of the theoretical hydrodynamic aspects of the subjects, some of the references contained in the bibliography will be found to be of value.

This book, representing as it does a gathering together of the subject of propulsion technology, is based upon the research of many scientists and engineers throughout the world. Indeed, it must be remembered that without these people, many of whom have devoted considerable portions of their lives to the development of the subject, this book could not have been written and, indeed, the subject of propeller technology could not have developed so far. I hope that I have done justice to their efforts in this book. At the end of each chapter a series of references is given so that, if necessary, the reader may refer to the original work, which will contain full details of the specific research topic under consideration. I am also considerably indebted to my colleagues, both within Lloyd's Register and in the marine industry, for many discussions on various aspects of the subject over the years, all of which have helped to provide a greater insight into, and understanding of, the subject. Particularly, in this respect, thanks are given to Mr C.M.R. Wills, Mr P.A. Fitzsimmons and Mr D.J. Howarth who, as specialists in particular branches of the subject, have also read several of the chapters and made many useful comments concerning their content. I would also like to thank Mr A.W.O. Webb who, as a specialist in propeller materials technology and colleague, has given much helpful advice over the years in solving propeller problems and this together with his many technical papers has influenced much of the text of Chapters 17 and 25. Also, I am particularly grateful to Mr J.Th. Ligtelijn of MARIN and to Dr G. Patience of Stone Manganese Marine Ltd, who have supplied me with several photographs for inclusion in the text and with whom many stimulating discussions on the subject have been had over the years. Thanks are also due to the many kind ladies who have so painstakingly typed the text of this book over the years and without whom the book would not have been produced.

J.S. Carlton
London
May 1993

General Nomenclature

Unless otherwise stated in the text the following general nomenclature apply.

Upper case

A	Cross-sectional area
A_C	Admiralty coefficient
A_D	Developed area
A_E	Expanded area
A_M	Mid-ship section area
A_O	Disc area
A_P	Projected area
AR	Aspect ratio
B	Moulded breadth of ship
B_P	Propeller power coefficient
BAR	Blade area ratio
C_A	Correlation factor
	Section area coefficient
C_b	Ship block coefficient
C_D	Drag coefficient
C_F	Frictional resistance coefficient
C_L	Lift coefficient
C_M	Moment coefficient
	Section modulus coefficient
C_P	Pressure coefficient
	Ship prismatic coefficient
	Propeller power coefficient
C_T	Thrust loading coefficient
	Total resistance coefficient
C_W	Wave-making resistance coefficient
D	Drag force
	Propeller diameter
D_b	Behind diameter
D_o	Diameter of slipstream far upstream
D_s	Shaft diameter
F	Force
	Fetch of the sea
F_B	Bollard pull
F_n	Froude number
G	Boundary layer unique shape function
	Non-dimensional circulation coefficient
H	Hydraulic head
H_p	Pump head
I	Dry inertia
I_e	Polar entrained inertia
IVR	Inlet velocity ratio
J	Advance coefficient
J_p	Ship polar moment of inertia
K	Prandtl or Goldstein factor
K_n	Knapp's similarity parameter
K_p	Pressure coefficient
K_Q	Propeller torque coefficient
K_{QS}	Spindle torque coefficient
K_T	Thrust coefficient
K_{TN}, K_{TD}	Duct thrust coefficient
K_{TP}	Propeller thrust coefficient
K_Y	Side force coefficient
L	Length of ship or duct
	Lift force
	Section centrifugal bending moment arm
L_P	Sound pressure level
L_{PP}	Length of ship between perpendiculars

xvii

L_R	Length of run	T	Temperature
L_{WL}	Length of ship along waterline		Draught of ship
M	Moment of force		Propulsor thrust
M_a	Mach number	T_A	Draught aft
N	Rotational speed (RPM)	T_F	Draught forward
	Number of cycles	T_N, T_D	Duct thrust
	Number of fatigue cycles	T_p	Propeller thrust
N_S	Specific speed	U_T	Propeller tip speed
P	Propeller pitch	V	Volume velocity
P_B	Brake power	V_a	Speed of advance
P_D	Delivered power	V_s	Ship speed
P_E	Effective power	X	Distance along co-ordinate axis
P_G	Generator power	Y	Distance along co-ordinate axis
P_S	Shaft power	W	Resultant velocity
Q	Flow quantity		Width of channel
	Propeller torque	W_e	Weber number
QPC	Quasi-propulsive coefficient	Z	Blade number
Q_S	Total spindle torque		Distance along co-ordinate axis
Q_{SC}	Centrifugal spindle torque	Z_m	Section modulus
Q_{SF}	Frictional spindle torque	**Lower case**	
Q_{SH}	Hydrodynamic spindle torque	a	Propeller axial inflow factor
R	Radius of propeller, paddle wheel or bubble	a_1	Propeller tangential inflow factor
		a_c	Crack length
	Specific gas constant	a_r	Resistance augmentation factor
R_{AIR}	Air resistance of ship	b	Span of wing
R_{APP}	Appendage resistance	c	Wake contraction factor
R_e	Real part		Section chord length
R_F	Frictional resistance	c_d	Section drag coefficient
R_n	Reynolds number	c_l	Section lift coefficient
R_T	Total resistance	c_{li}	Ideal section lift coefficient
R_V	Viscous resistance	c_m	Section moment coefficient
R_W	Wave-making resistance	c_{max}	Limiting chord length
S	Surface tension	f	Frequency
	Ship wetted surface area		Function of . . .
S_A	Additional load scale factor	g	Acceleration due to gravity
S_a	Apparent slip		Function of . . .
SBF	Solid boundary factor	h	Fluid enthalpy
S_C	Camber scale factor		Height

General Nomenclature

	Hydraulic head
h_b	Height of bulbous bow centroid from base line in transverse plane
i	Counter
i_G	Section generator line rake
i_P	Propeller rake
i_S	Section skew-induced rake
i_T	Total rake of propeller section
j	Counter
k	Counter
k_c	Lifting surface camber correction factor
k_s	Mean apparent amplitude of surface roughness
k_t	Lifting surface thickness correction factor
k_x	Lifting surface ideal angle of attack correction factor
$(1 + k)$	Frictional form factor
l	Counter
	Length
lcb	Longitudinal center of buoyancy
m	Mass counter
$\dot{m}$	Specific mass flow
n	Rotational speed (rps)
p	Section pitch
	Pressure
p_c	Cavity variation-induced pressure
p_H	Propeller-induced pressure
p_o	Reference pressure
	Non-cavitating pressure
	Pitch of reference section
p_v	Hull-induced vibratory pressure
	Vapor pressure
p^1	Apparent-induced pressure
q	Dynamic flow pressure
r	Radius of a propeller section
r_h	Hub or boss radius
s	Length parameter

t	Time
	Thrust deduction factor
	Section thickness
t_F	Thickness fraction
t_{max}	Maximum thickness
t_o	Notional blade thickness at shaft center line
u	Local velocity
v	Local velocity
v_a	Axial velocity
v_r	Radial velocity
V_t	Tangential velocity
v_T	Tide speed
w	Downwash velocity
	Mean wake fraction
w_F	Froude wake fraction
w_{max}	Maximum value of wake fraction in propeller disc
w_n	Nominal wake fraction
w_p	Potential wake fraction
w_T	Taylor wake fraction
w_v	Viscous wake fraction
w_w	Wave-induced wake fraction
x	Distance along a co-ordinate axis
	Non-dimensional radius (r/R)
x_c	Distance along chord
	Radial position of centroid
x_{cp}	Centre of pressure measured along chord
x_o	Reference section
y	Distance along co-ordinate axis
y_c	Camber ordinate
y_L	Section lower surface ordinate
y_t	Thickness ordinate
y_U	Section upper surface ordinate
z	Distance along co-ordinate axis

Suffixes

m	Model
s	Ship
U	Upper

L	Lower	θ_{fp}	Face pitch angle
b	Bound, behind	θ_{ip}	Propeller rake angle
F	Free	θ_{nt}	Nose–tail pitch angle
O	Reference value	θ_o	Effective pitch angle
x	Reference radius	θ_s	Section skew angle

Greek and other symbols

		θ_{sp}	Propeller skew angle
α	Angle of attack Gas content	θ_w	Angular position of transition wake roll-up point
α_d	Cavitation bucket width	Λ	Frequency reduction ratio
α_i	Ideal angle of attack	λ	Wavelength
α_K	Air content ratio		Source–sink strength
α_o	Zero lift angle		Ship–model scale factor
β	Advance angle	μ	Coefficient of dynamic viscosity
β_ε	Hydrodynamic pitch in the ultimate wake	ρ	Density of water
β_i	Hydrodynamic pitch angle	ρ_a	Density of air
Γ	Circulation	ρ_L	Leading edge radius
γ	Local vortex strength	ρ_m	Density of blade material
	Length parameter		
	Ratio of drag to lift coefficient (C_d/C_e)	σ	Cavitation number Stress on section
γ_g	Correction to angle of attack due to cascade effects	σ_a	Alternating stress
		σ_F	Corrosion fatigue strength
Δ	Change in parameter Displacement of ship	σ_i	Inception cavitation number
		σ_L	Local cavitation number
δ	Boundary layer thickness Linear displacement Propeller speed coefficient	σ_{MD}	Mean design stress
		σ_n	Cavitation number based on rotational speed Relative shaft angle
ε	Thrust eccentricity Transformation parameter	σ_o	Free steam cavitation number
ζ	Bendemann static thrust factor Damping factor Transformation parameter	σ_R	Residual stress
		σ_s	Blade solidity factor
η_b	Propeller behind hull efficiency	σ_x	Blade stress at location on blade
η_h	Hull efficiency	τ	Shear stress
η_i	Ideal efficiency	τ_C	Thrust loading coefficient
η_m	Mechanical efficiency	υ	Coefficient of kinematic viscosity
η_o	Propeller open water efficiency	ϕ	Angle of rotation in propeller plane Hull-form parameter Velocity potential Angular displacement Flow coefficient Shaft alignment angle relative to flow
η_p	Pump efficiency		
η_r	Relative rotative efficiency		
θ	Pitch angle Transformation parameter Momentum thickness of boundary layer		

General Nomenclature

ψ	Transformation parameter		ITTC	International Towing Tank Conference
	Gas content number		LDV	Laser Dopple Velocimetry
	Energy transfer coefficient		LE	Leading edge
Ω	Angular velocity		LES	Large Eddy Simulation
ω	Angular velocity		LNG	Liquid Natural Gas
∇	Volumetric displacement		MARIN	Maritime Research Institute of the Netherlands, formerly NSMB

Abbreviations

a.c.	Alternating current		MEPC	Marine Environment Protection Committee of IMO
AEW	Admiralty Experiment Works, Haslar		MCR	Maximum Continuous Rating
AP	After Perpendicular		mph	Miles per hour
ATTC	American Towing Tank Conference		NACA	National Advisory Council for Aeronautics
BHP	Brake Horse Power		NC	Numerically Controlled
BS	British Standard		NCR	Normal Continuous Rating
CAD	Computer Aided Design		OD	Oil Distribution
CAM	Computer Aided Manufacture		PIV	Particle Image Velocimetry
cwt	Hundred weight (1 cwt =112 lbf = 50.8 kgf)		PHV	Propulsor Hull Vortex
			qrs	Quarters (4 qrs = 1 cwt; 1 cwt = 50.8 kgf)
DES	Design		RANS	Reynolds Averaged Navier Stokes
DHP	Delivered Horse Power		RH	Right Handed
DTNSRD	David Taylor Naval Ship Research and Design Centre		rpm	Revolutions per minute
			shp	Shaft horsepower
EEDI	Energy Efficiency Design Index		SM	Simpson's Multiplier
EHP	Effective Horse Power		SPA	Self Polishing Anti-fouling
ft	Feet		SSPA	Statens Skeppsprovningsanstalt, Göteborg
HMS	Her Majesty's Ship		TE	Trailing Edge
hp	Horsepower		THP	Thrust Horse Power
HSVA	Hamburg Ship Model BasinTnQ		VLCC	Very Large Crude Carrier
IMO	International Maritime Organization		VTOL	Vertical Take−Off and Landing
ISO	International Standards Organization			

Chapter 1

The Early Development of the Screw Propeller

Both Archimedes (*c*. 250 BC) and Leonardo da Vinci (*c*.1500) can be credited with having considered designs and ideas which would subsequently be explored by ship propulsion engineers many years later. In the case of Archimedes, his thinking centered on the application of the screw pump which bears his name and this provided considerable inspiration to the nineteenth-century engineers involved in marine propulsion. Unfortunately, however, it also gave rise to several subsequent misconceptions about the basis of propeller action by comparing it to that of a screw thread. In contrast Leonardo da Vinci, in his sketchbooks which were produced some 1700 years after Archimedes, shows an alternative form of screw propulsion based on the idea of using fan blades having a similar appearance to those used for cooling purposes today.

The development of screw propulsion as we recognize it today can be traced back to the work of Robert Hooke, who is perhaps better remembered for his work on the elasticity of materials. Hooke in his *Philosophical Collections*, presented to the Royal Society in 1681, explained the design of a horizontal watermill which was remarkably similar in its principle of operation to the Kirsten-Boeing vertical axis propeller developed two and a half centuries later. Returning however to Hooke's watermill, it comprised six wooden vanes, geared to a central shaft and pinned vertically to a horizontal circular rotor. The gearing constrained the vanes to rotate through 180° about their own spindle axes for each complete revolution of the rotor.

During his life Hooke was also interested in the subject of metrology and in the course of his work he developed an air flow meter based on the principle of a windmill. He successfully modified this instrument in 1683 to measure water currents and then foresaw the potential of this invention to drive ships through the water if provided with a suitable means of motive power. As seen in Figure 1.1 the instrument comprises four, flat rectangular blades located on radial arms with the blades inclined to the plane of rotation.

Some years later in 1752, the Académie des Sciences in Paris offered a series of prizes for research into theoretical methods leading to significant developments in naval architecture. As might be expected, the famous mathematicians and scientists of Europe were attracted by this offer and names such as d'Alembert, Euler and Bernoulli appear in the contributions. Bernoulli's contribution, for which he won a prize, introduced the propeller wheel, shown in Figure 1.2, which he intended to be driven by a Newcomen steam engine. With this arrangement he calculated that a particular ship could be propelled at just under 2½ knots by the application of some 20–25 hp. Opinion, however, was still divided as to the most suitable propulsor configuration, as indeed it was to be for many years to come. For example, the French mathematician Paucton, working at about the same time as Bernoulli, suggested a different approach, illustrated in Figure 1.3, which was based on the Archimedean screw.

Thirty-three years after the Paris invitation Joseph Bramah in England proposed an arrangement for a screw propeller located at the stern of a vessel which, as may be seen from Figure 1.4, contains most of the features that we associate with screw propulsion today. It comprises a propeller with a small number of blades driven by a horizontal shaft which passes into the hull below the

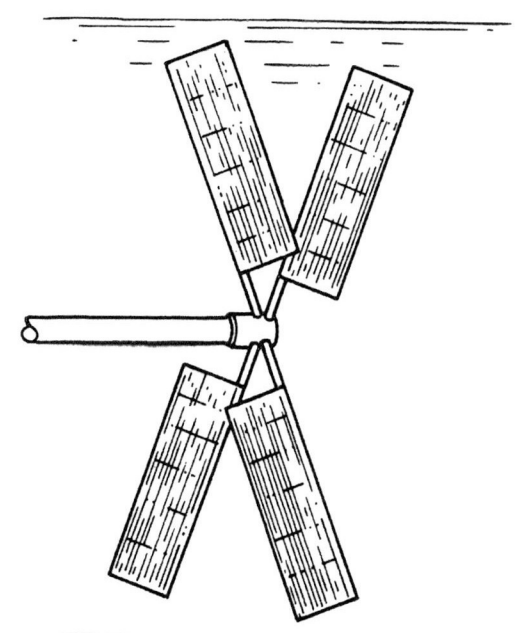

FIGURE 1.1 Hooke's screw propeller (1683).

Marine Propellers and Propulsion, Third Edition.
Copyright © 2012 John Carlton. Published by Elsevier Ltd. All rights reserved.

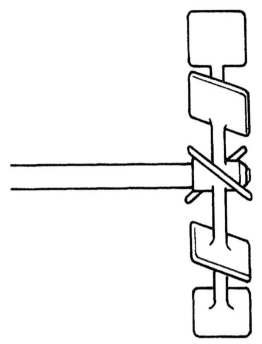

FIGURE 1.2　Bernoulli's propeller wheel (1752).

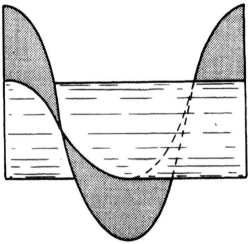

FIGURE 1.3　Archimedean screw of Paucton.

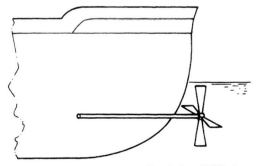

FIGURE 1.4　Bramah's screw propeller design (1785). *Reproduced with permission*[3].

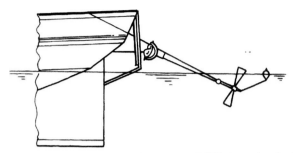

FIGURE 1.5　Shorter's propulsion system (1802). *Reproduced with permission*[3].

waterline. There appears, however, to be no evidence of any trials of a propeller of this kind being fitted to a ship and driven by a steam engine. Subsequently, in 1802 Edward Shorter used a variation of Bramah's idea to assist sailing vessels that were becalmed to make some headway. In Shorter's proposal, Figure 1.5, the shaft was designed to pass into the vessel's hull above the waterline and consequently eliminated the need for seals; the motive power for this propulsion arrangement was provided by eight men at a capstan. Using this technique Shorter managed to propel the transport ship *Doncaster* in Gibraltar and again at Malta at a speed of 1.5 mph in calm conditions: perhaps understandably, in view of the means of providing power, no further application of Shorter's propeller was recorded, but he recognized that this propulsion concept could be driven by a steam engine. Nevertheless, it is interesting to note the enthusiasm with which this propeller was received by Admiral Sir Richard Rickerton and his Captains (Figure 1.6).

Colonel John Stevens, who was a lawyer in the USA and a man of substantial financial means, experimented with screw propulsion in the year following Shorter's proposal. As a basis for his work he built a 25 ft long boat into which he installed a rotary steam engine and coupled this directly to a four-bladed propeller. The blades of this propeller were flat iron plates riveted to forgings which formed a 'spider-like' boss attachment to the shaft. Stevens later replaced the rotary engine with a steam engine of the Watt type and managed to attain a steady cruising speed of 4 mph with some occasional surges of up to 8 mph. However, he was not impressed with the overall performance of his craft and decided to turn his attention and energies to other means of marine propulsion.

In 1824 contra-rotating propellers made their appearance in France in a design produced by Monsieur Dollman. He used a two-bladed set of windmill type propellers rotating in opposite directions on the same shaft axis to propel a small craft. Following on from this French development the scene turned once again to England, where John Ericsson, a former Swedish army officer residing at that time in London, designed and patented in

The Early Development of the Screw Propeller

CERTIFICATES

From

ADMIRAL SIR RICHARD RICKERTON

and the

CAPTAINS of His MAJESTY'S Ships DRAGON, SUPERB, &c.&c.&c.

GIBRALTER BAY, July 4, 1802

Sir,

I Arrived here on the 1st after a Passage of Ten Days from England, and at the Time of my Arrival had a fresh breeze S.W. in consequence of which, had not an Opportunity of making use of the PROPELLER, but Yesterday being Calm, I got the DONCASTER under-way by Desire of some Captains in the Navy and several others when it was exhibited to the great Surprise and Satisfaction of every Spectator, at the same Time the Log was hove, and found the Ship, although deep loaded, went one-Knot and a Half through the Water, entirely by the Use of your new invented PROPELLER.

The inclosed Certificate I have received from the Captains of His Majesty's Ships DRAGON, and SUPERB in order that the Utility of the grand Machine may be made known to all Persons concerned in Shipping especially Ships coming up the Mediterranean where we are so much subject to Calms.

I have received Orders to go to Malta, and shall sail this Afternoon, if Wind permits.

Your's JOHN SHOUT,

Master of the DONCASTER Transport.

To Mr. SHORTER,

No.83, Wapping-Wall.

We, the under-mentioned Captains of His Majesty's ships DRAGON and SUPERB, have seen the DONCASTER moved in a Calm, the Distance of Two Miles, in GIBRALTER BAY, and with sufficient Velocity, by the sole Use of Mr. SHORTER's PROPELLER, to give her Steerage-way.

GIBRALTER BAY, July 4, 1802.

S. AYLMER, Captain of H.M.S. DRAGON

R. KEATS, Captain of H.M.S. SUPERB

FIGURE 1.6 **Certificate of performance for Mr Shorter's propeller arrangement.** *Courtesy: Mr J. Wiltshire, Qinetiq.*

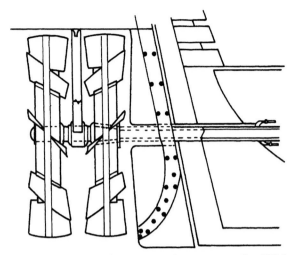

FIGURE 1.7 Ericsson's contra-rotating screw propeller (1836).

1836 a propulsion system comprising two contra-rotating propeller wheels. His design is shown in Figure 1.7, from which it can be seen that the individual wheels were not dissimilar in outline to Bernoulli's earlier proposal. Each wheel comprised eight short, wide blades of a helical configuration mounted on a blade ring with the blades tied at their tips by a peripheral strap. In this arrangement the two wheels were allowed to rotate at different speeds, probably to overcome the problem of the different flow configurations induced in the forward and after wheels. Ericsson conducted his early trials on a 3 ft model, and the results proved successful enough to encourage him to construct a 45 ft vessel which he named the *Francis B. Ogden*. This vessel was fitted with his propulsion system and had blade wheels with a diameter of 5 ft 2 in. Trials were conducted on the Thames in the presence of representatives from the Admiralty and the vessel was observed to be capable of a speed of some 10 mph. However, in his first design Ericsson placed the propeller astern of the rudder and this had an adverse effect both on the steerability of the ship and also on the flow into the propeller. The Admiralty Board expressed disappointment with the trial although the propulsion results were good when judged by the standards of the day. However, it was said that one reason was their concern over a vessel's ability to steer reliably when propelled from the stern. Following this rebuff Ericsson left England for the USA and in 1843 designed the US Navy's first screw-propelled vessel, the *Princeton*. It has been suggested that by around this time the US merchant marine had some forty-one screw-propelled vessels in operation.

The development of the screw propeller depended not only on technical development but also upon the availability of finance, politics and the likely return on the investment made by the inventor or his backers. Smith was rather more successful in these respects than his contemporary Ericsson. Francis Petit Smith took out a patent in which a different form of propeller was used, more akin to an Archimedean screw, but, more importantly, based on a different location of the propeller with respect to the rudder. This happened just a few weeks prior to Ericsson establishing his patent and the British Admiralty modified their view of screw propulsion shortly after Ericsson's trials due to Smith's work. Smith, who despite being frequently referred to as a farmer had a sound classical education, explored the concepts of marine propulsion by making model boats and testing them on a pond. From one such model, which was propelled by an Archimedean screw, he was sufficiently encouraged to build a six tonne prototype boat, the *F P Smith*, powered by a 6 hp steam engine to which he fitted a wooden Archimedean screw of two turns. The vessel underwent trials on the Paddington Canal in 1837; however, by one of those fortunate accidents which sometimes occur in the history of science and technology, the propeller was damaged during the trials and about half of it broke off, whereupon the vessel immediately increased its speed. Smith recognized the implications of this accident and modified the propeller accordingly. After completing the calm water trials he took the vessel on a voyage down the River Thames from Blackwall in a series of stages to Folkestone and eventually on to Hythe on the Kentish coast: between these last two ports the vessel averaged a speed of some 7 mph. On the return voyage to London, Smith encountered a storm in the Thames Estuary and the little craft apparently performed excellently in these adverse conditions. In March 1830 Smith and his backers, Wright and the Rennie brothers, made an approach to the Admiralty, who then requested a special trial for their inspection. The Navy's response to these trials was sufficiently encouraging to motivate Smith and his backers into constructing a larger ship of 237 tonnes displacement which he called *Archimedes*. This vessel, which was laid down by Henry Wilmshurst and engined by George Rennie, was completed in 1839. It had a length of 125 ft and was rigged as a three-masted schooner. The *Archimedes* was completed just as the ill-fated Screw Propeller Company was incorporated as a joint stock company. The objectives of this company were to purchase Smith's patents, transfer the financial interest to the company and sell licenses to use the location for the propeller within the deadwood of a ship as suggested by Smith, but not the propeller design itself. The *Archimedes* was powered by two 45 hp engines and finally fitted with a single turn Archimedean screw which had a diameter of 5 ft 9 in., a pitch of 10 ft and was about 5 ft in length. This propeller was the last of a series tried on the ship, the first having a diameter of 7 ft with a pitch of 8 ft and a helix making one complete turn. This propeller was subsequently replaced by a modification in which double-threaded screws, each of half a turn, were employed in

Chapter | 1   The Early Development of the Screw Propeller

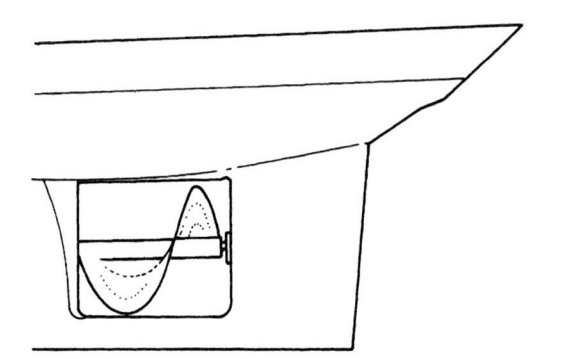

FIGURE 1.8   Propeller fitted to the *Archimedes* (1839).

accordance with Smith's amended patent of 1839. The propeller is shown in Figure 1.8. After undergoing a series of proving trials in which the speed achieved was in excess of nine knots the ship arrived at Dover in 1840 to undertake a series of races against the cross-channel packets, which at that time were operated by the Royal Navy. The Admiralty was duly impressed with the results of these races and agreed to the adoption of screw propulsion in the Navy. In the meantime, the *Archimedes* was lent to Brunel, who fitted her with a series of propellers having different forms.

Concurrent with these developments other inventors had introduced novel features into propeller design. In 1838 Lowes patented a propeller comprising one or more blades where each blade was a portion of a curve which if continued would produce a screw. The arrangement was equivalent to a pair of tandem propellers on a single shaft with each blade being mounted on a separate boss. Subsequently, the *SS Novelty* was built at Blackwall by Mr Wilmshurst between 1839 and 1840 to test the principle of screw propulsion. Indeed, this ship can be considered to be the first screw-propelled cargo ship. Also in 1839 Rennie patented a conoidal design in which he proposed increases in pitch from forward to aft of the blade; three-bladed helices and the use of skewback in the design. Taylor and Napier, a year later, experimented with tandem propellers, some of which were partially submerged. Also by 1842 the 'windmill' propeller, as opposed to the Archimedean screw, had developed to a fairly advanced state as witnessed by Figure 1.9, which depicts the propeller fitted to the

Napoleon, a ship having a displacement of 376 tonnes. This propeller is particularly interesting since it was developed to its final form from a series of model tests in which diameter, pitch, blade area and blade number were all varied. The first propeller in the series was designed with three blades each having a length of a third of a turn of a screw thread, thereby giving a high blade area ratio. Nevertheless, as the design evolved better results were achieved with shorter-length blades of around 22 per cent of a full thread turn. The ship was built by Augustin Normand at Le Havre and the propellers were designed and manufactured in Manchester by John Barnes who also built the engines. Although the ship was originally destined for postal service duties on the Mediterranean Sea, she was later acquired by the French Navy and deployed as a dispatch boat. The eventual propeller was manufactured from cast iron and rotated at 126 rpm giving the ship a speed of 10–12 knots.

The result of Brunel's trials with the *Archimedes* was that the design of the *Great Britain*, which is now preserved at Bristol in England and was originally intended for paddle propulsion, was adapted for screw propulsion. It is, however, interesting to note that the general form of the propeller adopted by Brunel for the *Great Britain* did not follow the type of propellers used by Smith but was similar to that proposed by Ericsson, except that in the case of the *Great Britain* the propeller was not of the contra-rotating type (Figure 1.10). Indeed, the original propeller designed by Brunel was subsequently modified since it had a tendency to break in service. Nevertheless, the pitch

FIGURE 1.10   Replica of Brunel's propeller for the *Great Britain*.

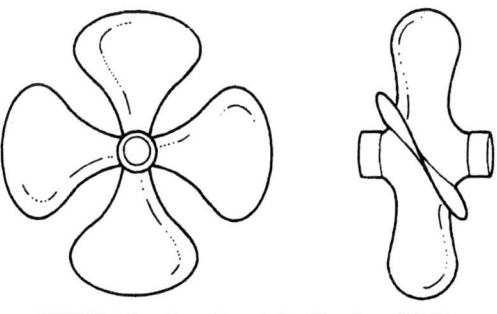

FIGURE 1.9   Propeller of the *Napoleon* (1842).

chosen was not dissimilar, in effective pitch terms, from that which would have been chosen today. Although the original propeller was 16 ft in diameter, had six blades and was made from a single casting, the propeller which was finally adopted was a built-up wrought iron propeller, also with six blades but having a diameter of 15.5 ft and a pitch of 25 ft.

As a direct result of the Royal Navy's commitment to screw propulsion *HMS Rattler* was laid down in 1841 at Sheerness Dockyard and underwent initial sea trials in the latter part of 1843 when she achieved a speed of some $8^{3}/_{4}$ knots. *HMS Rattler* was a sloop of approximately 800 tonnes and was powered by a steam engine of about 200 hp. Subsequently she ran a race against her paddle half-sister, *HMS Polyphemus*. A design study was commissioned in an attempt to study the various facets of propeller design and also to optimize a propeller design for *Rattler*; by January 1845 some thirty-two different propeller designs had been tested. The best of these propellers was designed by Smith and propelled the ship at a speed of about nine knots. This propeller was a two-bladed design with a diameter of 10 ft 1 in., a pitch of 11 ft and weighed 26 cwt 2 qrs (1.68 tonnes). During the spring of 1845 the *Rattler* ran a series of competitive trials against the paddle steamer *Alecto*. These trials embraced both free-running and towing exercises and also a series of separate sail, steam and combined sail and steam propulsion trials. By March 1845 the Admiralty was so convinced of the advantages of screw propulsion that they had ordered seven screw-propelled frigates together with a number of lesser ships. In April 1845 the famous 'tug of war' between the *Rattler* and the *Alecto* was held; however, this appears to have been more of a public relations exercise than a scientific trial.

In 1846 Joseph Maudsley patented a two-bladed propeller design in which the propeller could be lifted by a rope and tackle connected to a cross-head and which permitted the propeller to be raised to deck level. One year later *HMS Blenheim*, which had been built in 1813, was fitted with a similar arrangement to that proposed by Maudsley when she was converted from sail to screw propulsion. The following year, 1848, he patented a further design in which the blades of a two-bladed propeller, when not working, could be turned into the plane of the shaft to reduce sailing resistance. This theme of raising the propeller and thereby reducing the resistance of the ship when under sail was continued by Seaward who, also in 1848, developed a folding propeller in which the blades were cut into five radial segments which could be folded so as to be contained within the projection of the ship's deadwood. Indeed, the configuration of the propeller blades resembled, to some extent, a lady's hand-held fan in its form and operation. Later, in 1865, the Rev. P.A. Fothergill patented a self-feathering propeller which removed the need to raise the propeller when under sail. In this design the blades were so arranged as to take up a position of least resistance when not being rotated.

In 1853 John Fisher patented a two-bladed design with perforated blades. These perforations were in the form of slots to disperse any air that may have been entrained on the blades. A year later Walduck patented a design which was intended to attenuate the centrifugal motion of water over the blade surfaces by introducing a series of terraces, concentric with the shaft, but each being greater in pitch than its inner neighbor. This theme was returned to many times during the subsequent development of the propeller, one of the later developments being in 1924 where chordal plates were introduced into the blade design.

Peacock, in 1855, patented an auxiliary propeller in which each blade was built from iron plate and supported by a stay rod projecting radially from the boss. Interestingly, each blade was shaped to correspond to the general form of a bee's wing and the working surfaces of the blade given a parabolic form.

Although accepted by the Navy, screw propulsion had not been universally accepted for seagoing ships in preference to paddle propulsion, as witnessed by the relatively late general introduction of screw propulsion by the North Atlantic Steamship companies. However, the latter part of the nineteenth century saw a considerable amount of work being undertaken by a great number of people to explore the effects of radial pitch distribution, adjustable blades, blade arrangement and outline and cavitation. For example, in 1860 Hirsch patented a propeller having both variable chordal pitch, which we know today as camber, and variable radial pitch; as an additional feature this propeller also possessed a considerable amount of forward skew on the blades.

A type of propeller known as the Common Screw emerged and this was the most successful type of propeller in use before 1860. The working surfaces of the blades were portions of helices cut-off by parallel lines about an eighth of the pitch apart and located on a small cylindrical boss. With these propellers the blade chord lengths increase from root to tip, however, Robert Griffiths modified a blade of this type to have rounded tips and this was particularly successful. Indeed, the Admiralty, which had a number of Common Screws, reduced their broad tips by cutting away the leading corners and this resulted in significant reductions in vibration.

During this period of rapid development the competition between rival designers was great. In 1865 Hirsch designed a four-bladed propeller for the *SS Périere* which had originally been fitted with a Griffiths propeller design. In this case a one knot improvement was recorded on trial and similar results were noted when Hirsch propellers replaced other designs. At the same time a four-bladed, 22.8 ft diameter, 21.37 ft pitch, 11.7 ft long propeller was constructed for *HMS Lord Warden*. This propeller was

The Early Development of the Screw Propeller

a built-up design with the blades bolted through slots to permit adjustments to the blade pitch. The ship attained a speed of 13.5 knots during trials. In the 1890s Hirsch also introduced the idea of bolted-on blades, thereby providing another early example of built-up propellers which achieved considerable popularity in the first half of the twentieth century.

Thornycroft in 1873 designed a propeller with restricted camber in the mid-span regions of the blade and also combined this with a backward curvature of the blades in an attempt to suppress tangential flow. Zeise carried the ideas of the development of the radial pitch distribution a stage further in 1886 when he increased the pitch of the inner sections of the blade in an attempt to make better use of the inner part of the blades.

In parallel with the development of what might be termed fixed pitch propeller designs in the period 1844 through to about 1911 a number of inventors turned their attention to the potential for controllable pitch propellers. In reality, however, a number of these designs would be better termed adjustable rather than controllable pitch propellers. Bennett Woodcroft in 1844 patented a design with adjustable blades and this design had blades with increasing pitch from forward to the after edge in keeping with his earlier patent of 1832. Later in 1844 he patented a further modification where short links to the blade stems replaced his earlier idea of grooves on a collar to actuate the blades. In order to fix the blades in the desired position a similar collar was provided aft with two wedge-shaped arms that acted on small sliding pieces. In 1868 Mr H.B. Young patented a method of altering pitch by which the shanks of two blades are inserted in a hollow boss and extend through it. These shanks were then retained in position by arms projecting from them and the arms were controlled by a nut on a screwed rod which extended through the main boss and was turned by a key which was manipulated from within the ship. In the same year R. Griffiths introduced his concept of an adjustable pitch propeller. In this design, within the boss the shank of a blade was provided with an arm, connected to a link with a collar, and screwed on a sleeve that was loose on the propeller shaft. The after end of the sleeve turned in a groove, in which fitted a brake, and could be tightened on a collar by actuation from on deck of a tab on a screw. The movement for this pitch actuation was achieved by slowly rotating the propeller shaft, thereby screwing the collar along and in the process setting the blades to the desired position or indeed feathering them. In a second arrangement, patented in 1858, Young designed a system with a cotter which passed through the stem of the blades and rested in a sector-shaped recess in the boss. The alteration of pitch was effected by varying a number of packing pieces. The Bevis–Gibson reversible propeller was patented in 1911 and was a development of an earlier feathering propeller patent in 1869 by Mr R.R. Bevis which had been

used extensively. This new patent provided a means of reversing a small vessel driven by an internal combustion engine. In this design the roots of two blades were provided with toothed pinions which mated with a rack. The racks were yoked together and were actuated parallel to the shaft by means of a central rod which passed through a hollow propeller shaft. This rod was then operated by a lever at the control position in the craft so as to adjust the pitch into a forward, astern or feathering position.

The contra-rotating propeller received further attention in 1876 when Mr C.S. de Bay designed a propulsion system for the steam yacht *Iolair*, a 40.4 gross ton schooner rigged vessel having a length of 81.5 ft. His design, a model of which is located in the Science Museum in London and is shown in Figure 1.11, comprised two propellers of equal and opposite pitch mounted on the same shaft but revolving in opposite directions. The diameters of the propellers differed slightly with the larger having three blades and the smaller four blades. The blade shapes were of considerable complexity with portions of the blades being cut out so that the remainder of the blades could revolve in an interlocking manner. This complexity was introduced to try and prevent energy losses caused by the centrifugal and other motions of the water. Comparative trials in 1879 were made between this propeller and a Griffiths design, of a similar type to that used on *HMS Lord Warden*, and it was stated that the de Bays design achieved an efficiency at least 40 per cent greater than that of the competing design which represented a speed increase of around one knot.

In 1878 Col. W.H. Mallory in the USA introduced the concept of the azimuthing propeller. In this design the propeller was carried in a frame which rather resembled a rudder and was rotated by a bevel gear driven by an engine mounted on the deck.

FIGURE 1.11   de Bay's contra-rotating propeller design.

Other developments worthy of note in the context of this introductory review are those by Mangin, Zeise and Taylor. Mangin in 1851 attempted to increase the thrust of a propeller by dividing the blades radially into two portions. Griffiths also used this idea in 1871 but he used only a partial division of the blades in their center regions. Zeise in 1901 experimented with the idea of flexible blades, in which the trailing part of the blade was constructed from lamellae, and Taylor some six years later introduced air injection on the blade suction surface in order to control the erosive effects of cavitation.

Figure 1.12 shows a collage of some of these propellers together with their novel features during the period 1838–1907.

The latter part of the nineteenth century also saw the introduction of theoretical methods which attempted to explain the action of the screw propeller. Notable among these theoretical treatments were the works of Rankin and Froude; these, together with subsequent developments which occurred during the twentieth century, will, however, be introduced in the appropriate later chapters, notably Chapter 8.

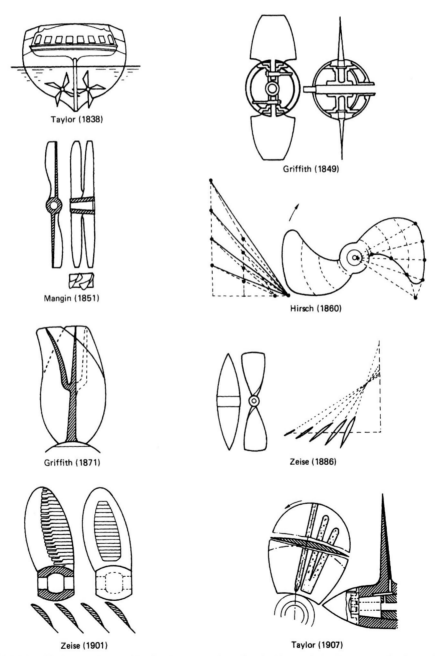

FIGURE 1.12 **Various early propeller developments.** *Reproduced with permission from parts of reference 2 and 3.*

These, therefore, were some of the activities and developments in the early years of propeller application, which paved the way for the advancement of marine propeller technology during the twentieth century and the subject that we practice today. With the exception of ducted propellers, propeller design after the turn of the nineteenth century advanced principally in matters of detail aimed at improving efficiency, maneuverability and controlling cavitation in the context of either vibration or erosion. For example, in 1907, just ten years after Sir Charles Parsons had introduced the steam turbine into marine practice at the fleet review on the 26 June 1897 with the 2000 hp *Turbinia*, the steam turbine driven liner *Mauritania* absorbed 70 000 shp on four propellers rotating 180 rpm and achieved a speed of 26.3 knots. These propellers, weighing 18.7 tonnes, had a diameter of 16.75 ft, a pitch of 15.5 ft and a blade area ratio of 0.467. This ship held the Blue Riband for the North Atlantic from 1910–1929. Then some years later the *Queen Mary*, powered by four single reduction geared turbine sets aggregating 160 000 shp on four propellers, achieved speeds of 30–32 knots. In her case the propeller blade area ratios had increased significantly from those of the *Mauritania* and the propellers weighed around 35 tonnes each. Her first series of propellers suffered from cavitation erosion and gave rise to cavitation excited vibration in the ship. This was cured by a redesign of the propellers, particularly with respect to blade shape and section form. This is again in contrast to a 380 000 dwt VLCC (Very Large Crude Carrier) in the early 1970s which was propelled by a six-bladed, single propeller.

This required the casting of 93 tonnes of nickel–aluminum bronze to yield a propeller of 70 tonnes finished weight. Today some of the major propulsion challenges are container and LNG (Liquid Natural Gas) ships. In the former case a 25 knot 12 500 teu container ship will absorb on a single, six-bladed propeller some 67.3 MW at a rotational speed of 90 rpm[7]. This propeller will have a diameter of 9600 mm, a pitch ratio of 1.04, a blade area ratio of 0.85 and will weigh around 128 tonnes.

REFERENCES

1. Geissler R. *Der Schraubenpropeller: Eine Darstellung seiner Entwicklung nach dem Inhalt der deutschen, amerikanischen und englischen Patentliteratur.* Berlin: Krayn; 1921.
2. Dirkzwager JM. *Some aspects on the development of screw-propulsion in the 19th and early 20th century.* 4th Lips Propeller Symposium; October 1969.
3. Brown DK. *A Century of Naval Construction.* London: Conway; 1983.

FURTHER READING

Taggart R. *Marine Propulsion: Principles and Evolution.* Texas: Gulf Publishing; 1969.

Lambert AD. The Royal Navy and the introduction of the screw propeller. History of Technology 1999;**21**.

John Ericsson 1803–1899. *RINA Affairs*; October 2004.

Carlton JS. The propulsion of a 12500 ton container ship. Trans.I.Mar.EST; April 2006.

Chapter 2

Propulsion Systems

Chapter Outline

2.1 Fixed Pitch Propellers	11	2.8 Surface Piercing Propellers	21
2.2 Ducted Propellers	13	2.9 Waterjet Propulsion	21
2.3 Podded and Azimuthing Propulsors	16	2.10 Cycloidal Propellers	22
2.4 Contra-Rotating Propellers	16	2.11 Paddle Wheels	23
2.5 Overlapping Propellers	17	2.12 Magnetohydrodynamic Propulsion	24
2.6 Tandem Propellers	18	2.13 Whale-Tail Propulsion	27
2.7 Controllable Pitch Propellers	19	References and Further Reading	27

The previous chapter gave an outline of the early development of the propeller up to around 1900 together with a few insights into its subsequent progress. In this chapter we move forward to the present day and consider, again in outline, the range of propulsion systems that are either currently in use or have been under development. The majority of the topical concepts and systems discussed in this chapter are considered in greater detail in later chapters; however, it is important to gain an overview of the subject prior to discussing the various facets of propulsion technology in more depth. Accordingly, the principal propeller types are briefly reviewed by outlining their major features and characteristics together with their general areas of application.

2.1 FIXED PITCH PROPELLERS

The fixed pitch propeller has traditionally formed the basis of propeller production over the years in either its mono-block or built-up forms. Whilst the mono-block propeller is commonly used today the built-up propeller, whose blades are cast separately from the boss and then bolted to it after machining, is now rarely used. This was not always the case since in the early years of the last century built-up propellers were very common, partly due to the inability to achieve good quality large castings at that time and partly to difficulties in defining the correct blade pitch. In both these respects the built-up propeller has obvious advantages. Nevertheless, built-up propellers generally have larger boss radii than their fixed pitch counterparts and this can cause difficulty with cavitation problems in the blade root section regions in some cases.

Mono-block propellers cover a broad spectrum of design types and sizes, ranging from those weighing only a few kilograms for use on small power-boats to those, for example, destined for large container ships which can weigh around 130 tonnes and require the simultaneous casting of significantly more metal in order to produce the casting. Figure 2.1 shows a collage of various types of fixed pitch propeller in use today. These range from a large four-bladed propeller fitted to a bulk carrier, seen in the figure in contrast to a man standing on the dock bottom, through highly skewed propellers for merchant and naval applications, to small high-speed patrol craft and surface piercing propellers.

As might be expected, the materials of manufacture vary considerably over such a wide range of designs and sizes. For the larger propellers, over 300 mm in diameter, the non-ferrous materials predominate: high-tensile brass together with the manganese and nickel—aluminum bronzes are the most favored types of materials. However, stainless steel has also gained limited use. Cast iron, once a favorite material for the production of spare propellers, has now virtually disappeared from use. Alternatively, for small propellers use is frequently made of materials such as the polymers, aluminum, nylon and more recently carbon fiber composites.

For fixed pitch propellers the choice of blade number, notwithstanding considerations of blade-to-blade clearances at the blade root to boss interface, is largely an independent variable and is normally chosen to give a mismatch to the range of hull, superstructure and machinery vibration frequencies which are considered

Marine Propellers and Propulsion, Third Edition.
Copyright © 2012 John Carlton. Published by Elsevier Ltd. All rights reserved.

11

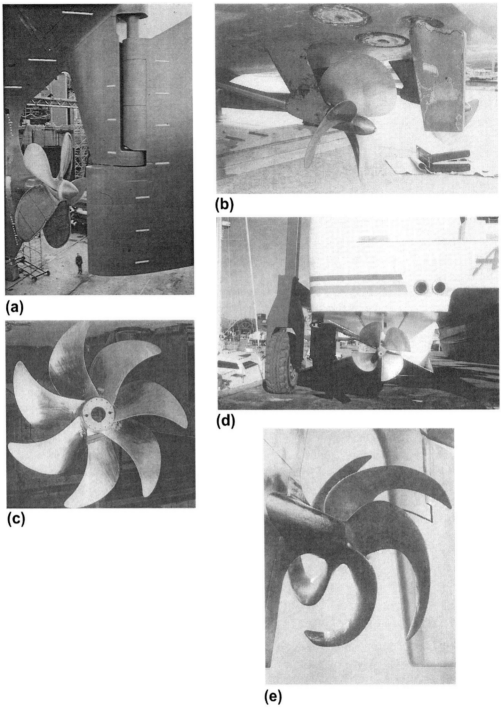

FIGURE 2.1 Typical fixed pitch propellers: (a) large four-bladed propeller for a bulk carrier; (b) high-speed patrol craft propeller; (c) seven-bladed balanced high-screw design; (d) surface piercing propeller and (e) biased high-skew, low-blade-area ratio propeller.

likely to cause concern. Additionally, blade number is also a useful parameter in controlling unwelcome cavitation characteristics. Blade numbers generally range from two to seven, although in some naval applications, where considerations of radiated noise become important, blade numbers greater than these have been researched and used to solve a variety of propulsion problems. For merchant vessels, however, four, five and six blades are generally favored, although many tugs and fishing vessels frequently use three-bladed designs. In the case of small work or pleasure power-boats two- and three-bladed propellers tend to predominate.

Chapter | 2 Propulsion Systems

The early propeller design philosophies centered on the optimization of the efficiency from the propeller. Whilst today this aspect is no less important, and, in some respects associated with energy conservation, has assumed a greater importance, other constraints on design have emerged. These are in response to calls for the reduction of vibration excitation and radiated noise from the propeller. This latter aspect has of course been a prime concern of naval ship and torpedo propeller designers for many years; however, pressure to introduce these constraints, albeit in a generally less stringent form, into merchant ship design practice has grown in recent years. This has been brought about by the increases in power transmitted per shaft; the use of after deckhouses; the maximization of the cargo carrying capacity, which imposes constraints on the hull lines; ship structural failure and international legislation. Moreover, in recent years there has been a growing awareness of the effects of underwater radiated noise on marine mammals and fish.

For the majority of vessels of over 100 tonnes displacement it is possible to design propellers on whose blades it is possible to control, although not eliminate, the effects of cavitation in terms of its erosive effect on the material, its ability to impair hydrodynamic performance and it being the source of vibration excitation. In this latter context it must be remembered that there are very few propellers which are free from cavitation since the greater majority experience cavitation at some position in the propeller disc: submarine propellers when operating at depth, the propellers of towed array frigates and research vessels when operating under part load conditions are notable exceptions, since these propellers are normally designed to be subcavitating to meet stringent noise emission requirements to minimize either detection or interference with their own instruments. Additionally, in the case of propellers operating at significant water depths such as in the case of a submarine, due account must be taken of the additional hydrostatic pressure-induced thrust which will have to be reacted by the ship's thrust block.

For some small, high-speed vessels where both the propeller advance and rotational speeds are high and the immersion low, a point is reached where it is not possible to control the effects of cavitation acceptably within the other constraints of the propeller design. To overcome this problem, all or some of the blade sections are permitted to fully cavitate, so that the cavity developed on the back of the blade extends beyond the trailing edge and collapses into the wake of the blades in the slipstream. Such propellers are termed super-cavitating propellers and frequently find application on high-speed naval and pleasure craft. Figure 2.2(c) illustrates schematically this design philosophy in contrast to non-cavitating and partially cavitating propeller sections, shown in Figure 2.2(a) and (b), respectively.

When design conditions dictate a specific hydrodynamic loading together with a very susceptible cavitation

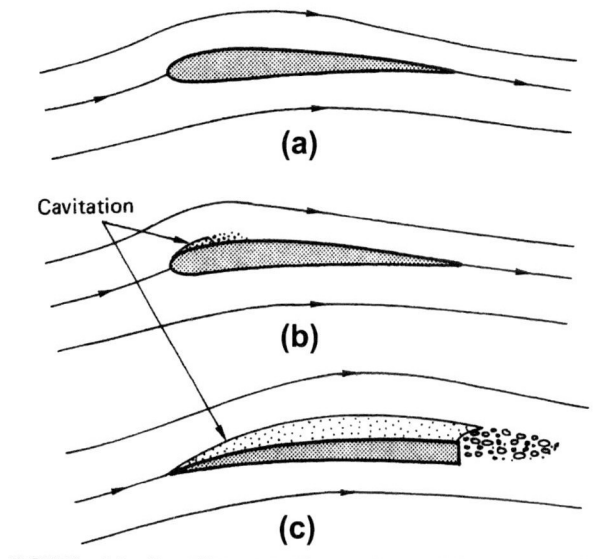

FIGURE 2.2 Propeller operating regimes: (a) non-cavitating; (b) partially cavitating and (c) super-cavitating.

environment, typified by a low cavitation number, there comes a point when even the super-cavitating propeller will not perform satisfactorily: for example, if the propeller tip immersion becomes so small that the propeller tends to draw air from the surface, termed ventilation, along some convenient path such as along the hull surface or down a shaft bracket. Eventually, if the immersion is reduced sufficiently by either the design or operational constraints the propeller tips will break the surface. Although this condition is well known on cargo vessels when operating in ballast conditions and may, in these cases, lead to certain disadvantages from the point of view of material fatigue and induced vibration, the surface breaking concept can be an effective means of propelling relatively small high-speed craft. Such propellers are termed surface piercing propellers and their design immersion, measured from the free surface to the shaft center line, can be reduced to zero; that is, the propeller operates half in and half out of the water. In these partially immersed conditions the propeller blades are commonly designed to operate such that the pressure face of the blade remains fully wetted and the suction side is fully ventilated or dry. This is an analogous operating regime to the super-cavitating propeller, but in this case the blade surface suction pressure is at atmospheric conditions and not the vapor pressure of water.

2.2 DUCTED PROPELLERS

Ducted propellers, as their name implies, generally comprise two principal components: the first is an annular duct having an aerofoil cross-section which may be either of uniform shape around the duct and, therefore, symmetric with respect to the shaft center line, or have certain asymmetric features to accommodate the wake field flow variations. However, due to the cost of wake adapted ducts it is normally axisymmetric ducts that predominate. The second component, the propeller, is a special case of a non-ducted propeller in which the design of the blades has been modified to take account of the flow interactions caused by the presence of the duct in its flow field. The propeller for these units can be either of the fixed or controllable pitch type and in some special applications, such as torpedo propulsion, may be a contra-rotating pair. Ducted propellers, sometimes referred to as Kort nozzles by way of recognition of the Kort Propulsion Company's initial patents and long association with this type of propeller, have found application for many years where high thrust at low speed is required; typically in towing and trawling situations. In such cases, the duct generally contributes some 50 per cent of the propulsor's total thrust at zero ship speed: termed the bollard pull condition. However, this relative contribution of the duct falls to more modest amounts with increasing ship speed and it is also possible for a duct to give a negative contribution to the propulsor thrust at high advance speeds. This latter situation would nevertheless be a most unusual design condition to encounter.

There are nominally two principal types of duct form, the accelerating and decelerating duct, and these are shown in Figure 2.3(a), (b), (c) and (d), respectively. The underlying reason for this somewhat artificial designation can be appreciated, in global terms, by considering their general form in relation to the continuity equation of fluid mechanics. This can be expressed for incompressible flow in a closed conduit between two stations a-a and b-b as,

$$\rho A_a v_a = \rho A_b v_b \qquad (2.1)$$

where v_a is the velocity at station a-a; v_b is the velocity at station b-b; A_a is the cross-section area at station a-a; A_b is the cross-section area at station b-b and ρ is the density of the fluid.

In this context station b-b can be chosen in way of the propeller disc while a-a is some way forward, although not necessarily at the leading edge. In the case of Figure 2.3(a), which shows the accelerating duct, it can be seen that A_a is greater than A_b since the internal diameter of the duct is greater at station a-a. Hence, from equation (2.1) and since water is incompressible, v_a must be less than v_b which implies an acceleration of the water between stations a-a and b-b; that is, up to the propeller location. The converse situation is true in the case of the decelerating duct shown in Figure 2.3(d). To determine precisely which form the duct actually is, if indeed this is important, the induced velocities of the propeller also need to be taken into account in the velocity distribution throughout the duct.

By undertaking a detailed hydrodynamic analysis it is possible to design complex duct forms intended for specific application and duties. Indeed, attempts at producing non-symmetric duct forms to suit varying wake field conditions have been made which result in a duct with both varying aerofoil section shape and incidence, relative to the shaft center line, around its circumference. However, with duct forms it must be appreciated that the hydrodynamic desirability for a particular form must be balanced against the practical manufacturing problem of producing the desired

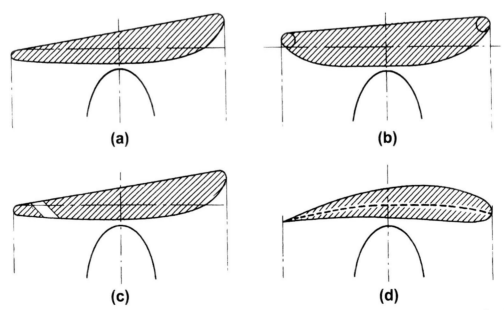

FIGURE 2.3 Duct types: (a) accelerating duct; (b) 'pull–push' duct; (c) Hannan slotted duct and (d) decelerating duct.

Propulsion Systems

shape if an economic, structurally sound and competitive duct is to result. This tenet is firmly underlined by appreciating that ducts have been produced for a range of propeller diameters from 0.5 m or less up to around 8.0 m. For these larger sizes, fabrication problems can be difficult, not least in maintaining the circularity of the duct and providing reasonable engineering clearances between the blade tips and the duct: recognizing that from the hydrodynamic viewpoint the clearance should be as small as possible.

Many standard duct forms are in use today but those most commonly used are shown in Figure 2.3. While the duct shown in Figure 2.3(a), the Wageningen 19A form, is probably the most widely used because it has a good ahead performance, its astern performance is less good due to the aerofoil form of the duct having to work in reverse: that is, the trailing edge effectively becomes the leading edge in astern operations. This is of relatively minor importance in, say, a trawler or tanker, since for the majority of their operating lives they are essentially unidirectional ships. However, this is not true for all vessels since some, such as tugs, are expected to have broadly equal capabilities in both directions. In cases where a bidirectional capability is required a duct form of the type illustrated in Figure 2.3(b), the Wageningen No. 37 form, might be selected since its trailing edge represents a compromise between a conventional trailing edge and leading edges of, for example, the 19A form. For this type of duct the astern performance is improved but at the expense of the ahead performance, thereby introducing an element of compromise in the design process. Several other methods of overcoming the disadvantages of the classical accelerating duct form in astern operations have been patented over the years. One such method is the 'Hannan slot', shown in Figure 2.3(c). This approach, whilst attempting to preserve the aerodynamic form of the duct in the ahead condition allows water when backing to enter the duct both in the conventional manner and also through the slots at the trailing edge in an attempt to improve the astern efficiency of the unit.

When the control of cavitation and more particularly the noise resulting from cavitation is of importance, use can be made of the decelerating duct form. A duct form of this type, Figure 2.3(d), effectively improves the local cavitation conditions by slowing the water before it passes through the propeller. Most applications of this duct form are found in naval situations, for example, with submarines and torpedoes. Nevertheless, some specialist research ships also have needs which can be partially satisfied by the use of this type of duct in the appropriate circumstances.

An interesting development of the classical ducted propeller form is found in the pump jet, Figure 2.4. Early pump jets sometimes comprise a row of inlet guide vanes, which double as duct supports, followed by a row of rotor

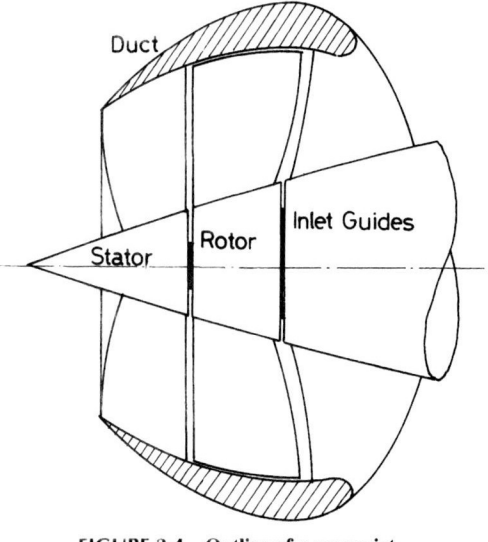

FIGURE 2.4 Outline of a pump jet.

blades which are finally followed by a stator blade row. Typically, rotor and stator blade numbers might lay between 15 and 20, respectively, each row having a different blade number. Naturally there are variants of this basic design in which the blade numbers may be reduced or the inlet guide vanes dispensed with. Indeed, some later designs comprise only a leading stator ring followed by the rotor. The efficiency achievable from the unit is dependent upon the design of the rotor; the rotor–stator interaction; the final stator row in reducing the swirl component of the flow and the reduction of the guide vane size in order to limit skin friction losses: hence, the desirability of not using guide vanes if possible. The pump jet in this form is largely restricted to military applications and should not be confused with a type of directional thruster, referred to as a pump-jet, which is relatively widely used for providing directional thrust for ships.

The ducts of ducted propellers, in addition to being fixed structures rigidly attached to the hull, are in some cases found to be steerable. The steerable duct, which obviates the need for a rudder, is mounted on pintles whose axes lie on the vertical diameter of the propeller disc. This then allows the duct to be rotated about the pintle axes by an inboard steering motor and consequently the thrust of the propeller can be directed towards a desired direction for navigation purposes. Clearly, however, the arc through which the thrust can be directed is limited by geometric constraints. Applications of this type can range from small craft, such as harbor tugs, to comparatively large commercial vessels as shown by Figure 2.5. A further application of the steerable ducted propeller which has gained considerable popularity in recent years, particularly in the offshore field, is the azimuthing thruster; in many cases these units can be trained around a full 360°.

FIGURE 2.5 Steerable ducted propeller.

2.3 PODDED AND AZIMUTHING PROPULSORS

Azimuthing thrusters have been in common use for many years and can have either non-ducted or ducted propeller arrangements. They can be further classified into pusher or tractor units as seen in Figure 2.6. The essential difference between azimuthing and podded propellers lies in where the engine or motor driving the propeller is sited. If the engine or motor is sited in the ship's hull then the system would be termed an azimuthing propulsor and most commonly the mechanical drive would be of a Z or L type to the propeller shaft. Frequently, the drive between the vertical and horizontal shafts is via spiral bevel gears.

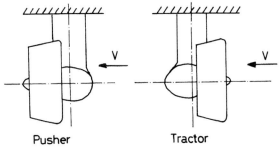

FIGURE 2.6 Pusher and tractor thruster units.

In the case of a podded propulsor the drive system normally comprises an electric motor directly coupled to a propeller shaft which is supported on two rolling element bearing systems: one frequently being a radial bearing closest to the propeller while the other is a spherical roller bearing at the opposite end of the shaft line. Nevertheless, variants of this arrangement do exist and designs incorporating conventional journal and thrust bearings in addition to rolling element CARB bearings have been proposed. The propellers associated with these propulsors have been of the fixed pitch type and are commonly built-up although their size is not particularly large. Currently, the largest size of unit is around the 23 MW capacity and the use of podded propulsors has been mainly in the context of cruise ships and ice breakers where their maneuvering potentials have been fully realized. Clearly, however, there are a number of other ship types which might benefit (and have benefitted) from their application. Figure 2.7 shows a typical example of a large podded propulsor unit being maintained in dry dock.

Tractor arrangements of podded and azimuthing propulsors generally have an improved inflow velocity field since they do not have a shafting and A-bracket system ahead of them to cause a disturbance to the inflow. This tends to help suppress the blade rate harmonic pressures since the relatively undisturbed wake field close to zero azimuthing angles is more conducive to maintaining low rates of growth and collapse of cavities. However, there can be a tendency for these propellers to exhibit broadband excitation characteristics and during the design process care has to be exercised to minimize these effects. At high azimuthing angles, however, the flow field is clearly more disturbed.

Azimuthing or podded propulsors offer significant maneuverability advantages; however, when used in combinations of two or more care has to be exercised in preventing the existence of sets of azimuthing angle where the propulsors can mutually interfere with each other. If this occurs large fluctuating forces and moments can be induced on the shaft system and significant vibration can be encountered.

2.4 CONTRA-ROTATING PROPELLERS

The contra-rotating propeller principle, comprising two coaxial propellers sited one behind the other and rotating in opposite directions, has traditionally been associated with the propulsion of aircraft, although Ericsson's original proposal of 1836, Figure 1.7, used this method as did de Bay's design for the *Iolair* featured in Figure 1.11.

Contra-rotating propulsion systems have the hydrodynamic advantage of recovering part of the slipstream rotational energy which would otherwise be lost to a conventional single-screw system. In marine applications of contra-rotating propulsion it is normal for the aftermost propeller to have a smaller diameter than the forward propeller and, in this way, accommodate the slipstream

Chapter | 2 Propulsion Systems

FIGURE 2.7 Typical podded propulsor unit.

contraction effects. Similarly, the blade numbers of the forward and aft propellers are usually different; typically, four and five for the forward and aft propellers, respectively. Furthermore, because of the two propeller configuration, contra-rotating propellers possess a capability for balancing the torque reaction from the propulsor which is an important matter for torpedo and other similar propulsion problems, Figure 2.8.

Contra-rotating propeller systems have been the subject of considerable theoretical and experimental research as well as some practical development exercises. Whilst they have found a significant number of applications, particularly in small high-speed outboard units, operating for example at around 1500−2000 rpm, the mechanical problems associated with the longer line shafting systems of larger vessels have generally precluded them from use on merchant ships. Interest in the concept has had a cyclic nature: interest growing and then waning. An upsurge in interest in 1988, however, has resulted in a system being fitted to a 37 000 dwt bulk carrier[1] and subsequently to a 258 000 dwt VLCC in 1993.

More recently, however, an interesting variant of the traditional contra-rotating propulsor has been proposed and fitted to some ships. This comprises a combination of a traditional propeller, driven from a conventional line shaft, as the forward member of the pair, with a podded propulsor acting as the astern component of the propulsor pair. Figure 2.9 demonstrates this concept. Such an arrangement also has the potential benefit of dispensing with the rudder since the azimuthing podded propulsor of the propulsor pair provides this feature. Clearly, for these arrangements when at high podded propulsor angles it will be seen that the after propulsor of the pair operates obliquely in the helical flow generated by the forward propulsor.

2.5 OVERLAPPING PROPELLERS

This again is a two-propeller concept. In this case the propellers are not mounted coaxially but are each located on separate shaft systems with the distance between the shaft center lines being less than the diameter of the propellers. Figure 2.10 shows a typical arrangement of such a system; again this is not a recent idea and references may be found dating back over a hundred years: for example, Figure 1.12 showing Taylor's design of 1830.

As in the case of the contra-rotating propeller principle, recent work on this concept has been largely confined to research and development, and the system has rarely been used in practice. Research has largely centered on the effects of the shaft spacing to propeller diameter ratio on the overall propulsion efficiency in the context of particular hull forms.[2−3] The principal aim of this type of propulsion arrangement is to gain as much benefit as possible from the

FIGURE 2.8 Contra-rotating torpedo propeller.

FIGURE 2.9 Contra-rotating pair comprising a conventional and podded propulsor. *Courtesy ABB.*

low-velocity portion of the wake field and, thereby, increase propulsion efficiency. Consequently, the benefits derived from this propulsion concept are intimately related to the propeller and hull propulsion coefficients.

Despite one propeller working partially in the wake of the other, cavitation problems are not currently thought to pose insurmountable design problems. However, significant increases in the levels of fluctuating thrust and torque have been identified when compared to single-screw applications. In comparison to the twin-screw alternative, research has indicated that the overlapping arrangement may be associated with lower building costs, and this is portrayed as one further advantage for the concept.

When designing this type of propulsion system several additional variables are presented to the designer. These are the direction of propeller rotation, the distance between the shafts, the longitudinal clearance between the propellers and the stern shape. At the present time there are only partial answers to these questions. Research tends to suggest that the best direction of rotation is outward, relative to the top dead center position and that the optimum distance between the shafts lies below 0.8 D. In addition there are indications that the principal effect of the longitudinal spacing of the propellers is to be found in vibration excitation and that propulsion efficiency is comparatively insensitive to this variable.

2.6 TANDEM PROPELLERS

Tandem propeller arrangements are again not a new propulsion concept. Perhaps the best-known example is that of Parson's *Turbinia* where eventually three propellers were mounted on each of the three propulsion shafts in order to overcome the effects of cavitation-induced thrust breakdown, Figure 2.11. Indeed, the principal reason for the employment of tandem propellers has been to ease difficult propeller loading situations; however, these occasions have been relatively few. The disadvantage of the tandem propeller arrangement when applied to conventional single- and twin-screw ships is that the weights and axial distribution of the propellers create large bending moments which have to be reacted, principally by the stern tube bearings.

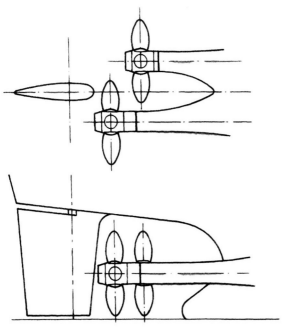

FIGURE 2.10 Overlapping propellers.

Propulsion Systems

FIGURE 2.11   Tandem propeller arrangement on a shaft line of *Turbinia*.

Some azimuthing and podded propulsor arrangements, however, employ this arrangement by having a propeller located at each end of the propulsion shaft, either side of the pod body. In this way the load is shared by the tractor and pusher propellers and the weight-induced shaft moments controlled.

2.7 CONTROLLABLE PITCH PROPELLERS

Unlike fixed pitch propellers whose only operational variable is rotational speed, the controllable pitch propeller provides an extra degree of freedom in its ability to change blade pitch. However, for some propulsion applications, particularly those involving shaft-driven generators, the shaft speed is held constant, thus reducing the number of operating variables again to one. While this latter arrangement is very convenient for electrical power generation it can cause difficulties in terms of the cavitation characteristics of the propeller by inducing back and face cavitation at different propulsion conditions.

The controllable pitch propeller has found application in the majority of the propeller types and applications so far discussed in this chapter, with the possible exception of the podded propulsors, contra-rotating and tandem propellers, although even in these extreme examples of mechanical complexity some development work has been undertaken for certain specialist propulsion problems. In the last forty years the controllable pitch propeller has grown in popularity from representing a small proportion of the propellers produced to its current position of having a very substantial market share. Currently the controllable pitch propeller has about a 35 per cent market share when compared to fixed pitch propulsion systems. The controllable pitch propeller tends to be most favored in the passenger ship and ferry, general cargo, tug and trawling markets.

The controllable pitch propeller, although of necessity possessing a greater degree of complexity than the fixed pitch alternative, does possess a number of important advantages. Clearly, maneuvering is one such advantage, in that fine thrust control can be achieved without necessarily the need to accelerate and decelerate the propulsion machinery. Furthermore, fine control of thrust is particularly important in certain cases: for example, in dynamic positioning situations or where frequent berthing maneuvers are required, such as in short sea route ferry operations. Moreover, the basic controllable pitch propeller hub design can in many instances be modified to accommodate the feathering of the propeller blades. The feathering position is the position where the blades are aligned approximately fore and aft and in the position in which they present least resistance to forward motion when not rotating. Such arrangements find applications on double-ended ferries or in small warships. In this latter application, the vessel could, typically, have three propellers; the two wing screws being used when cruising with the center screw not rotating, implying, therefore, that it would benefit from being feathered in order to produce minimum resistance to forward motion in this condition. Then, when the sprint condition is required all three propellers could be used at their appropriate pitch settings to develop maximum speed.

The details and design of controllable pitch propeller hub mechanisms are outside the scope of this book since this text is primarily concerned with the hydrodynamic aspects of ship propulsion. It will suffice to say, therefore, that each manufacturer has an individual design of pitch actuating mechanism, but that these designs can be broadly grouped into two principal types; those with inboard and those with outboard hydraulic actuation. Figure 2.12 shows these principal types in schematic form. For further discussion and development of these matters reference can be made to the works of Plumb, Smith and Brownlie,[4–6] which provide introductions to this subject. Alternatively, propeller manufacturers' catalogues frequently provide a source of outline information on this aspect of controllable pitch propeller design.

The hub boss, in addition to providing housing for the blade actuation mechanism, must also be sufficiently strong to withstand the propulsive forces supplied to and transmitted from the propeller blades to the shaft. In general, therefore, controllable pitch propellers tend to have larger hub diameters than those for equivalent fixed pitch propellers. Typically, the controllable pitch propeller hub has a diameter in the range 0.24–0.32 D, but for some applications this may rise to as high as 0.4 or even 0.5 D. In contrast, fixed pitch propeller boss diameters are generally within the range 0.16–0.25 D. The large boss diameters may give rise to complex hydrodynamic problems, often cavitation related, but for the majority of normal applications the larger diameter of the controllable pitch propeller hub does not generally pose problems that cannot be either directly or indirectly solved by known design practices.

Certain specialist types of controllable pitch propeller have been designed and patented in the past. Two examples are the self-pitching propeller and the Pinnate propeller, both of which are modern versions of much earlier designs. Self-pitching propellers are a modern development of Griffiths' work in 1849. The blades are sited on an external crank which is pinned to the hub and they are free to take up any pitch position. The actual blade pitch position taken up in service depends on a balance of the blade loading and

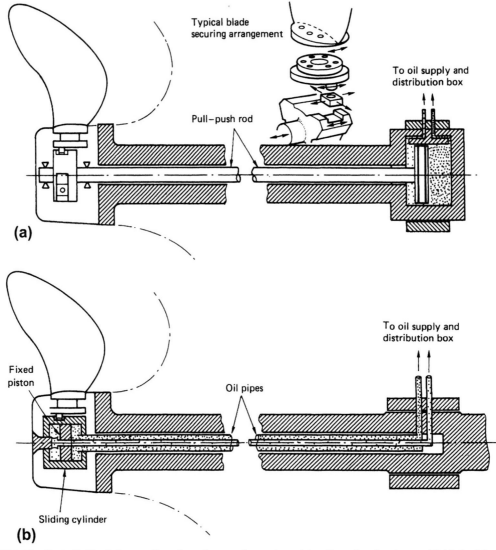

FIGURE 2.12 Controllable pitch propeller schematic operating systems: (a) pull–push rod system and (b) hub piston system.

spindle torque components which are variables depending on, amongst other parameters, rotational speed: at zero shaft speed but with a finite ship speed the blades are designed to feather. At the present time these propellers have only been used on relatively small craft.

The Pinnate design is to some extent a controllable pitch–fixed pitch propeller hybrid. It has a blade activation mechanism which allows the blades to change pitch about a mean position by varying angular amounts during one revolution of the propeller. The purpose of the concept is to reduce both the magnitude of the blade cyclical forces and cavitation by attempting to adjust the blades for the varying inflow velocity conditions around the propeller disc. Trials of these types of propeller have been undertaken on small naval craft and Simonsson describes these applications.[7]

2.8 SURFACE PIERCING PROPELLERS

Surface piercing propellers, Figure 2.1(d), are sometimes referred to as ventilated or partially submerged propellers. They are normally used in special cases of high-speed propulsion: in some cases in the region of 100 knots. These types of propeller provide a means of maintaining a reasonable propulsion efficiency when operating under difficult hydrodynamic conditions. With these types of propeller and with the ship at rest the propeller is usually, although not in all cases, fully submerged. Then as the vessel accelerates to high-speed and the hull starts to plane the propeller takes up a partially submerged attitude: in these conditions the degree of partial submergence may be up to 0.5 D.

The blade chordal section forms of surface piercing propellers differ considerably from more conventional propellers: typically they might take a form not dissimilar to that shown in Figure 2.2(c). During operation in the fully ventilated design condition, the backs, or suction surfaces, of the propeller blades should be surrounded by an air film which extends to the free surface and only the pressure faces remain wetted. As such, these types of propellers have specific design and analysis methods which are applicable to their mode of operation in order to achieve the correct performance and power absorption characteristics.

In the case of surface piercing propellers three principal operating regimes may be identified. These are the partially ventilated, the transition and the fully ventilated conditions. In the former case of the partially ventilated condition, air cavities start near the blunt trailing edges of the blade sections and vent towards the free surface. In this condition the extent and volume of the air cavities are frequently seen to increase as the propeller advance coefficient decreases: similarly with the time-averaged thrust and torque coefficients developed by the propeller. In the final condition, the

fully ventilated design condition, when the advance coefficient of the propeller is further reduced the flow over the propeller suction surfaces has cavities which start near the leading edge and extend over the blade surfaces and eventually vent to atmosphere. This fully developed flow regime is relatively stable and the blade trailing edges remain continuously ventilated. The intermediate transition region between these two operating conditions is very unstable and considerable vibratory forces are frequently developed. This is because during this transition regime the air cavities on the blade surfaces begin to spread towards the leading edges but in doing so suffer significant fluctuations both in shape and size.

Surface piercing propellers, in this context, should be distinguished from a conventional merchant ship propeller which is driving a ship in a light draft condition and in so doing is not fully submerged: the two propellers and their operating regimes are quite different.

2.9 WATERJET PROPULSION

The origin of the waterjet principle can be traced back to 1661, when Toogood and Hayes produced a description of a ship having a central water channel in which either a plunger or centrifugal pump was installed to provide the motive power. In more recent times waterjet propulsion has found considerable application on a wide range of small high-speed craft while its application to larger craft is growing with tunnel diameters of upwards of 2m being considered.

The principle of operation of the present-day waterjet is that in which water is drawn through a ducting system by an internal pump which adds energy, after which the water is expelled aft at high velocity. The unit's thrust is primarily generated as a result of the momentum increase imparted to the water. Figure 2.13 shows, in outline form, the main features of the waterjet system and this method of propulsion is further discussed in Chapter 16.

The pump configuration adopted for use with a waterjet system depends on the specific speed of the pump; specific speed N_s being defined in normal hydraulic terms as

$$N_\mathrm{s} = \frac{(N)Q^{1/2}}{H^{3/4}} \qquad (2.2)$$

where Q is the quantity of fluid discharged, N is the rotational speed and H is the head.

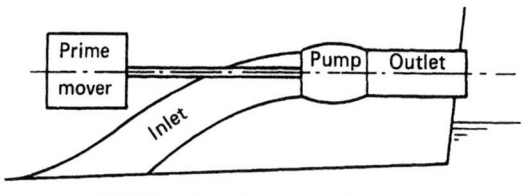

FIGURE 2.13 Waterjet configuration.

For low values of specific speed centrifugal pumps are usually adopted, whereas for intermediate and high values of N_s axial pumps and inducers are normally used, respectively. The prime movers usually associated with these various pumps are either gas turbines or high-speed diesel engines.

Waterjet propulsion offers a further dimension to the range of propulsion alternatives and tends to be used where other propulsion forms are rejected for some reason: typically for reasons of efficiency, cavitation extent, noise or immersion and draught. For example, in the case of a small vessel traveling at say 45 knots one might expect that a conventional propeller would be fully cavitating, whereas in the corresponding waterjet unit the pump should not cavitate. However, waterjet propulsors are not always cavitation free as there are operating conditions where cavitation problems may be experienced. Consequently, the potential for waterjet application, neglecting any small special purpose craft with particular requirements, is where conventional, trans-cavitating and super-cavitating propeller performance is beginning to fall off. Indeed surface-piercing propellers and waterjet systems are to some extent competitors for some similar applications. Waterjet units, however, tend to be heavier than conventional propeller-based systems and, therefore, might be expected to find favor with larger craft; for example, large wave-piercing ferries.

In terms of maneuverability the waterjet system is potentially very good, since deflector units are normally fitted to the jet outlet pipe which then direct the water flow and hence introduce turning forces by changing the direction of the jet momentum. Similarly, for stopping maneuvers, flaps or a 'bucket' can be introduced over the jet outlet to redirect the flow forward and hence apply an effective reactive retarding force to the vessel.

2.10 CYCLOIDAL PROPELLERS

Cycloidal propeller development started in the 1920s, initially with the Kirsten−Boeing and subsequently the Voith−Schneider designs. As discussed in Chapter 1, it is interesting to note that the Kirsten−Boeing design was very similar in its hydrodynamic action to the horizontal waterwheel developed by Robert Hooke some two and half centuries earlier in 1681.

The cycloidal or vertical axis propellers basically comprise a set of vertically mounted vanes, six or eight in number, which rotate on a disc mounted in a horizontal or near horizontal plane. The vanes are constrained to move about their spindle axis relative to the rotating disc in a predetermined way by a governing mechanical linkage. Figure 2.14(a) illustrates schematically the Kirsten−Boeing principle. It can be seen from the figure that the vanes' relative attitude to the circumference of the

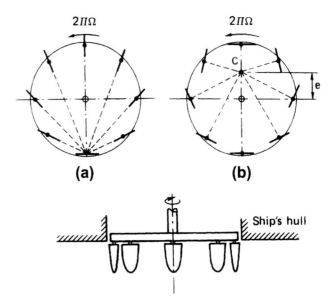

FIGURE 2.14 Vertical axis propeller principle: (a) Kirsten−Boeing propeller and (b) Voith−Schneider propeller.

circle, which governs their tracking path, is determined by referring the motion of the vanes to a particular point on that circumference. As such, it can be deduced that each vane makes half a revolution about its own pintle axis during one revolution of the entire propeller disc. The thrust magnitude developed by this propeller design is governed by rotational speed alone and the direction of the resulting thrust by the position of the reference point on the circumference of the vane-tracking circle.

The design of the Voith−Schneider propeller is rather more complex since it comprises a series of linkages which enable the individual vane motions to be controlled from points other than on the circumference of the vane-tracking circle. Figure 2.14(b) demonstrates this for a particular value of the eccentricity (e) of the vane-control center point from the center of the disc. By controlling the eccentricity, which in turn governs the vane-pitch angles, both the thrust magnitude and direction can be controlled independently of rotational speed. In the case of the Voith−Schneider design, in contrast to the Kirsten−Boeing propeller, the individual vanes make one complete revolution about their pintle axes for each complete revolution of the propeller disc. In many cases the units are provided with guards to help protect the propulsor blades from damage from external sources.

Vertical axis propellers do have considerable advantages when maneuverability or station keeping is a high priority and this is an important factor in the ship design, since the resultant thrust can be readily directed along any navigational bearing and have variable magnitude. Indeed, this type of propeller avoids the necessity for a separate rudder installation on the vessel. Despite the relative mechanical complexity, these propellers have shown themselves to be reliable in operation over many years of service.

2.11 PADDLE WHEELS

Paddle propulsion, as is well known, predates screw propulsion; however, this form of propulsion has almost completely disappeared except for very few specialized applications. These are to be found largely on lakes and river services, either as tourist or nostalgic attractions, or alternatively, where limited draughts are encountered. Nevertheless, the Royal Navy, until a few years ago, also favored their use on certain classes of harbor tug, where they were found to be exceptionally maneuverable. The last example of a seagoing paddle steamer, the *Waverley*, is seen in Figure 2.15.

The principal reason for the demise of the paddle wheel was its intolerance of large changes of draught and the complementary problem of variable immersion in seaways. Once they were superseded by screw propulsion for ocean-going vessels their use was largely confined through the first half of the twentieth century to river steamers and tugs. Paddle wheels, however, also suffered from damage caused by flotsam in rivers and were relatively expensive to produce when compared to the equivalent fixed pitch propeller.

Paddle design progressed over the years from the original simple fixed float designs to the feathering float system which then featured throughout much of its life. Figure 2.16 shows a typical feathering float paddle wheel design from which it can be seen that the float attitude is governed from a point just slightly off-center of the wheel axis. Feathering floats is essential for good efficiency on relatively small diameter and deeply immersed wheels. However, on the larger wheels, which are not so deeply immersed, feathering floats are not essential and fixed float designs were normally adopted. This led to the practice of adopting feathered wheels in side-mounted wheel applications, such as were found on the Clyde or Thames excursion steamers, because of the consequent wheel diameter restriction imposed by the draught of the vessel. In contrast, on the stern wheel propelled vessels, such as those designed for the Mississippi services, the use of fixed floats was preferred since the wheel diameter restriction did not apply.

The design of paddle wheels is considerably more empirical than that of screw propellers today; nevertheless, high propulsion efficiencies were achieved and these were of similar orders to equivalent screw-propelled steamers. Ideally, each float of the paddle wheel should enter the water 'edgeways' and without shock, having taken due account of the relative velocity of the float to the water. Relative velocity in still water has two components: the angular speed due to the rotation of the wheel and the speed of the vessel V_a. From Figure 2.17 it can be seen that at the point of entry A, a resultant vector $\bar{a}$ is produced from the combination of advance speed V_a and the rotational vector ωR. This resultant vector represents the absolute velocity at the point of entry and to avoid shock at entry, that is a vertical thrusting action of the float, the float should be aligned parallel to this vector along the line YY. However, this is not possible practically and the best that can be achieved is to align the floats to the point B and this is achieved by a linkage EFG which is introduced into the system. Furthermore, from Figure 2.17 it is obvious that the less the immersion of the wheel (h), the less is the advantage to be gained from adopting a feathering float system. This explains why the fixed float principle is adopted for large, shallowly immersed wheels.

FIGURE 2.15 *P.S. Waverley*: **Example of a side wheel paddle steamer.**

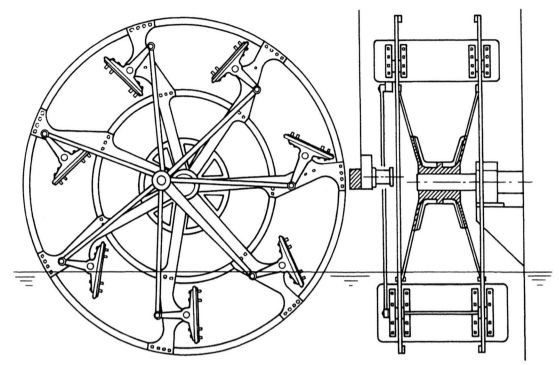

FIGURE 2.16 Paddle wheel. *Reproduced from Hamilton.*[8]

With regard to the overall design parameters, based on experience it was found that the number of fixed floats on a wheel should be about one for every foot of diameter of the wheel and for feathering designs this number was reduced to around 60 or 70 per cent of the fixed float 'rule'. The width of the floats used in a particular design was of the order of 25–40 per cent of the float length for feathering designs, but this figure was reduced for the fixed float paddle wheel to between 20 and 25 per cent. A further constraint on the immersion of the floats was that the peripheral speed at the top of the floats should not exceed the ship speed and, in general, feathering floats were immersed in the water up to about half a float width, whilst with sternwheelers, the tops of the floats were never far from the water surface.

The empirical nature of paddle design was recognized as being unsatisfactory and in the mid 1950s Volpich and Bridge[9–11] conducted systematic experiments on paddle wheel performance at the Denny tank in Dumbarton. Unfortunately, this work came at the end of the time when paddle wheels were in use as a common form of propulsion and, therefore, never achieved its full potential.

2.12 MAGNETOHYDRODYNAMIC PROPULSION

Magnetohydrodynamic propulsion potentially provides a means of ship propulsion without the aid of either propellers or paddles. The laws governing magnetohydrodynamic propulsion were known in the nineteenth century and apart from a few isolated experiments, such as that by Faraday when he attempted to measure the voltage across the Thames induced by its motion through the earth's magnetic field and the work of Hartmann on electromagnetic pumps in 1918, the subject had largely to wait for engineering development until the 1960s.

The idea of electromagnetic thrusters was first patented in the USA by Rice in 1961.[12] Following this patent the USA took a leading role in both theoretical and experimental studies, culminating in a report from the Westinghouse Research Laboratory in 1966. This report showed that greater magnetic field densities were required before

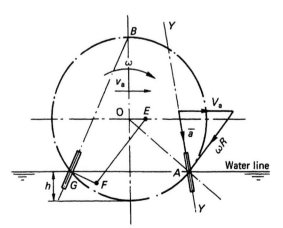

FIGURE 2.17 Paddle wheel float relative velocities.

the idea could become practicable in terms of providing a realistic alternative for ship propulsion. In the 1970s superconducting coils enabled further progress to be made with this concept.

The fundamental principle of electromagnetic propulsion is based upon the interaction of a magnetic field B produced by a fixed coil placed inside the ship and an electric current passed through the sea water from electrodes in the bottom of the ship or across a duct, as shown diagrammatically in Figure 2.18. Since the magnetic field and the current are in mutually orthogonal directions, then the resulting Lorentz force provides the necessary pumping action. The Lorentz force is $J \times B$ where J is the induced current density. Iwata *et al.*[13,14] present an interesting description of the state of the art of superconducting propulsion.

In theory the electrical field can be generated either internally or externally, in the latter case by positioning a system of electrodes in the bottom of the ship. This, however, is a relatively inefficient method for ship propulsion and the environmental impact of the internal system

is considerably reduced due to the containment of the electromagnetic fields. Most work, therefore, has concentrated on systems using internal magnetic fields and the principle of this type of system is shown in Figure 2.19(a) in which a duct, through which sea water flows, is surrounded by superconducting magnetic coils which are immersed in a cryostat. Two electrodes are placed inside the duct, which create the electric field necessary to interact with the magnetic field in order to create the Lorentz forces for propulsion. Nevertheless, the efficiency of a unit is low due to the losses caused by the low conductivity of sea water. The efficiency, however, is proportional to the square of the magnetic flux intensity and to the flow speed, which is a function of ship speed. Consequently, in order to arrive at a reasonable efficiency it is necessary to create a strong magnetic flux intensity by using powerful magnets. In order to investigate the full potential of these systems at prototype scale a small craft, *Yamato 1*, was built for trial purposes by the Japanese and Figure 2.19(b) shows a cross section through one of the prototype propulsion units, indicating the arrangement of the six dipole propulsion

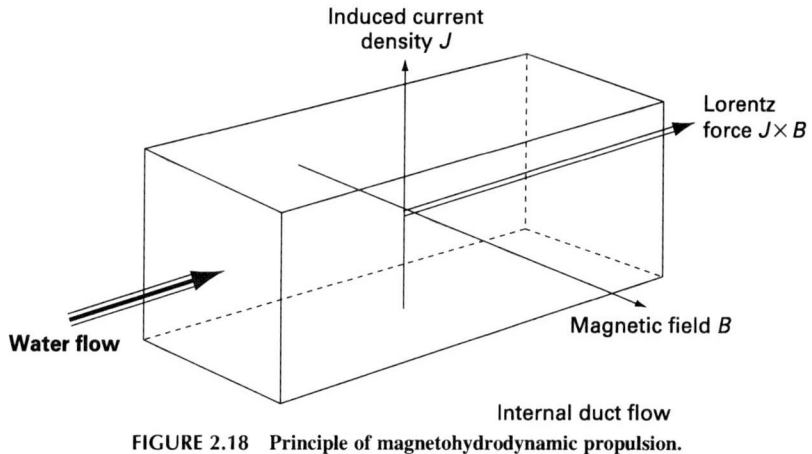

FIGURE 2.18   Principle of magnetohydrodynamic propulsion.

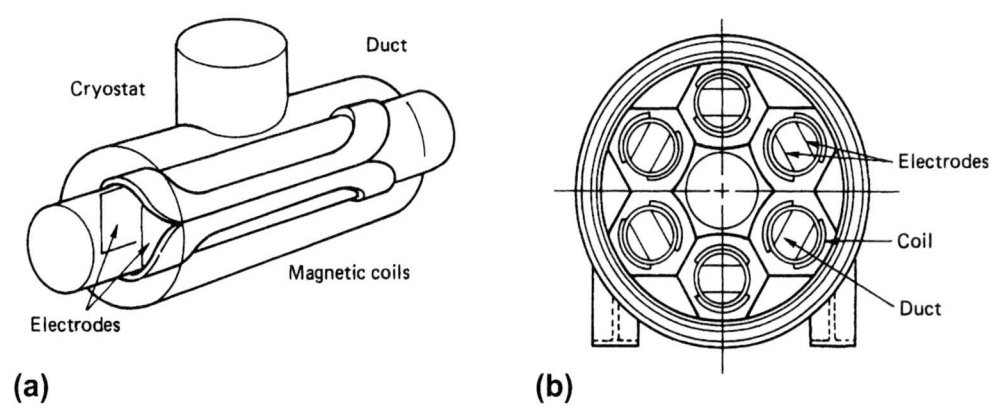

(a)   **(b)**

FIGURE 2.19   Internal magnetic field electromagnetic propulsion unit: (a) the dipole propulsion unit with internal magnetic field and (b) a cross-section through a prototype propulsion unit.

ducts within the unit. Figure 2.20 shows the experimental craft, *Yamato 1*.

Electromagnetic propulsion does have certain potential advantages in terms of providing a basis for noise and vibration-free hydrodynamic propulsion. However, a major obstacle to the development of electromagnetic propulsion until relatively recently was that the superconducting coil, in order to maintain its zero-resistance property, needed to be kept at the temperature of liquid helium, 4.2 K ($-268°$). This clearly requires the use of thermally well-insulated vessels in which the superconducting coil can be placed in order to maintain these conditions. The criticality of this thermal condition can be seen from Figure 2.21, which indicates how the resistance of a superconductor behaves with temperature and eventually reaches a critical temperature when the resistance falls rapidly to zero. Superconductors are also sensitive to current and magnetic fields; if either become too high then the superconductor will fail in the manner shown in Figure 2.22.

Superconductivity began with the work of Kamerlingh Onnes at Leiden University in 1911 when he established the superconducting property for mercury in liquid helium; for

FIGURE 2.20 *Yamato 1*: Experimental magnetohydrodynamic propulsion craft.

this work he won a Nobel Prize. Work continued on superconductivity; however, progress was slow in finding metals which would perform at temperatures as high as that of liquid nitrogen, $-196°$ C. By 1973 the best achievable temperature was 23 K. However, in 1986 Muller and Bednorz in Zurich turned their attention to ceramic oxides which had hitherto been considered as insulators. The result of this shift of emphasis was to immediately increase the critical temperature to 35 K by the use of a lanthanum, barium, copper oxide compound: this discovery led to Muller and Bednorz also being awarded a Nobel Prize for their work. Consequent on this discovery, work in the USA,

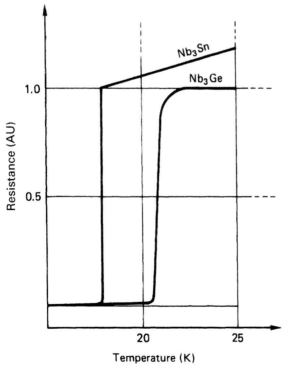

FIGURE 2.21 Superconducting effect.

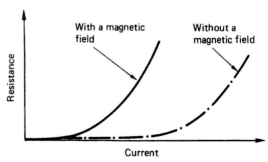

FIGURE 2.22 Effect of a magnetic field on a superconductor.

Propulsion Systems

TABLE 2.1 Development of Superconducting Ceramic Oxides

Date	Ceramic Oxide	Superconducting Temperatures (K)
September 1986	La—Ba—Cu—O	35
January 1987	Y—Ba—Cu—O	93
January 1988	Bi—Sr—Ca—Cu—O	118
February 1988	Tl—Ba—Ca—Cu—O	125

China, India and Japan intensified, leading to the series of rapid developments depicted in Table 2.1.

While these advances are clearly encouraging since they make the use of superconducting coils easier from the thermal insulation viewpoint, many ceramic oxides are comparatively difficult to produce. Consequently, while this form of propulsion clearly has potential and significant advances have been made, both in the basic research and application, much work still has to be done before this type of propulsion can become a reality on a commercial scale or before the concept is even fully tested.

Notwithstanding the problems for magnetohydrodynamic propulsion, superconductivity has in the last few years shown its potential for the production of marine propulsion motors using the high-temperature superconductors of Bi-2223 material [$(Bi,Pb)_2Sr_2Ca_2Cu_3O_x$] which have a T_c of 110 K but operate at a temperature of 35–40 K. This material has, at the present time, been demonstrated to be the most technically viable material for propulsion motors. In the USA a 5 MW demonstrator machine has proved satisfactory and a 25 MW demonstrator is being constructed to demonstrate the potential for marine propulsion purposes. In addition to other marine propulsion applications the relatively small diameter of these machines, if finally proved satisfactory, may have implications for podded propulsors since the hub diameter may then be reduced given that this diameter is principally governed by the electric motor size.

2.13 WHALE-TAIL PROPULSION

Throughout the ages designers have endeavored to mimic marine mammals and fish in trying to develop enhanced and more efficient solutions for ship propulsion. It is clear that when comparing conventional ship propulsion to that, for example, of whales, in the case of the ship the ratio of the propulsor swept area to that of the ship's mid-ship section area is low, while the converse is the case for the marine mammal. Such an observation suggests relatively light loading of the tail of the mammal, particularly when swimming at high-speed.

During the late 1980s work at both Glasgow and Memorial Universities and elsewhere bore testament to this quest to learn from the marine mammals and fish. Subsequently, van Manen *et. al.*[15] developed a whale-tail propulsion concept. This was based, in part, on his earlier studies of vertical axis and trichoidal propulsors where he noted, albeit in a vertical plane, that the blade angular orientations, as the propulsor passed through one complete revolution, were not dissimilar to the tail movements of a whale: Figure 2.14. When the normal vertical axis of these propulsors was turned through 90 degrees to become a horizontal axis then the motion of the blades became a reasonable approximation of the action of a whale's tail movements. This led to the whale-tail propulsor concept which comprised a system in which the blades were supported at both ends, unlike the Voith—Schneider or Kirsten—Boeing propulsors, and typically was designed to accommodate between four and seven blades located around the periphery of the wheel.

Two advantages which accrue from this concept are the possibility of large propulsor sizes and flow regimes over the blades which are almost two-dimensional. Moreover, since the blades are lightly loaded in terms of thrust loading per unit envelope area this implies the possibility of achieving high efficiency and the development of a good cavitation environment. Indeed, simplified quasi-steady computational studies supported this in that at low propulsor loading coefficients, high equivalent open water efficiencies, approaching the ideal efficiency of propulsors, could be a possibility. Clearly more detailed estimates attenuated this potential to some extent but the concept was sufficiently attractive to be taken to a full-scale demonstrator.

REFERENCES AND FURTHER READING

1. IHI. *CRP system for large merchant ships*. Ship Technology International '93, SPG; 1993.
2. Kerlen H, Esveldt J, Wereldsman R. *Propulsion, Cavitation and Vibration Characteristics of Overlapping Propellers for a Container Ship*. Berlin: Schiffbautechnische Gesellschaft; November 1970.
3. Restad K, Volcy GG, Garnier H, Masson JC. *Investigation on free and forced vibrations of an LNG tanker with overlapping propeller arrangement*. Trans. SNAME; 1973.
4. Plumb CM. *Warship Propulsion System Selection*. I. Mar. EST; 1987.
5. Smith DW. *Marine Auxiliary Machinery*. 6th ed. Oxford: Butterworth-Heinemann; 1983.
6. Brownlie, K. *Controllable Pitch Propellers*. I. Mar. EST. ISBN 1-902536-01-X.
7. Simonsson P. *The Pinnate Propeller. Department of Mechanics*. Stockholm: Royal Institute of Technology; 1983.
8. Hamilton FC. *Famous Paddle Steamers*. London: Marshall; 1948.

9. Volpich, Bridge. *Preliminary model experiments*. Trans. IESS, **98**:1954–55
10. Volpich, Bridge. *Systemic model experiments*. Trans. IESS, **99**:1955–56
11. Volpich, Bridge. *Further model experiments and ship model correlation*. Trans. IESS, **100**:1956–57
12. Rice WA. *US Patent 2997013* August 1961;**22**.
13. Iwata A, Tada E, Saji Y. *Experimental and theoretical study of superconducting electromagnetic ship propulsion*. Paper No. 2. 5th Lips Propeller Symposium; May 1983.
14. Iwata A. *Superconducting electromagnetic propulsion system*. Bull. of Mar. Eng. Soc. 1990;**18**(1).
15. Manen JD, van Terwisga, van T. *A new way of simulating whale tail propulsion*. ONR Symposium; 1996.

Chapter 3

Propeller Geometry

Chapter Outline

3.1 Frames of Reference	29	3.9 Blade Interference Limits for Controllable Pitch Propellers	43
3.2 Propeller Reference Lines	30	3.10 Controllable Pitch Propeller Off-Design Section Geometry	43
3.3 Pitch	31		
3.4 Rake and Skew	33		
3.5 Propeller Outlines and Area	35	3.11 Miscellaneous Conventional Propeller Geometry Terminology	46
3.6 Propeller Drawing Methods	38	References and Further Reading	46
3.7 Section Geometry and Definition	38		
3.8 Blade Thickness Distribution and Thickness Fraction	42		

To appreciate fully propeller hydrodynamic action from either the empirical or theoretical standpoint, it is essential to have a thorough understanding of basic propeller geometry and the corresponding definitions used. Whilst each propeller manufacturer, consultant or test tank has proprietary ways of presenting propeller geometric data on drawings or in dimension books produced either by hand or with the aid of a computer, these differences are most commonly in matters of detail rather than in fundamental changes of definition. Consequently, this chapter will not generally concern itself with a detailed account of each of the different ways of representing propeller geometric information: instead it will present a general account of propeller geometry which will act as an adequate basis for any particular applications with which the reader will be concerned.

3.1 FRAMES OF REFERENCE

A prerequisite for the discussion of the geometric features of any object or concept is the definition of a suitable reference frame. In the case of propeller geometry and hydrodynamic analysis many reference frames are encountered in the literature, each, no doubt, chosen for some particular advantage or preference of the author concerned. However, at the 10th International Towing Tank Conference (ITTC) in 1963 the preparation of a dictionary and nomenclature of ship hydrodynamic terms was initiated; this work was completed in 1975 and the compiled version presented in 1978.[1] The global reference frame proposed by the ITTC is that shown in Figure 3.1(a) which is a right-handed, rectangular Cartesian system. The X-axis

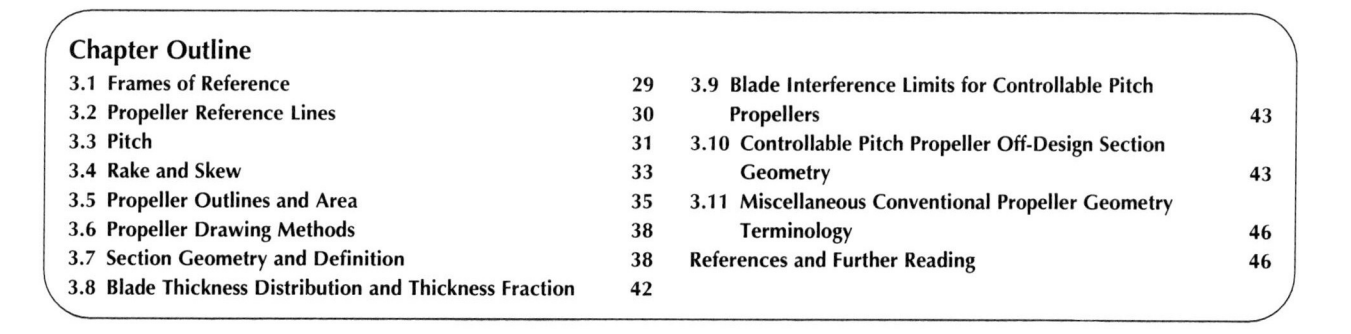
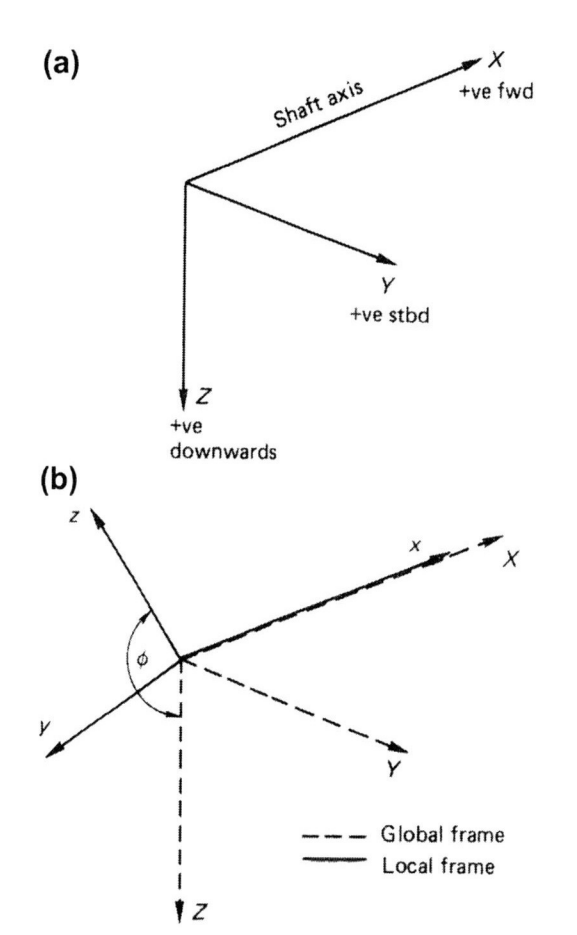

FIGURE 3.1 Reference frames: (a) global reference frame and (b) local reference frame.

Marine Propellers and Propulsion, Third Edition.
Copyright © 2012 John Carlton. Published by Elsevier Ltd. All rights reserved.

is positive, forward and coincident with the shaft axis; the Y-axis is positive to starboard and the Z-axis is positive in the vertically downward direction. This system is adopted as the global reference frame for this book since no other general agreement exists in the field of propeller technology. For propeller geometry, however, it is convenient to define a local reference frame having a common axis such that OX and Ox are coincident, but allowing the mutually perpendicular axes Oy and Oz to rotate relative to the OY and OZ fixed global frame as shown in Figure 3.1(b).

3.2 PROPELLER REFERENCE LINES

The propeller blade is defined about a line normal to the shaft axis called either the 'propeller reference line' or the 'directrix': the word 'directrix' being the older term used for this line. In the case of the controllable pitch propeller the term 'spindle axis' is frequently synonymous with the reference line or directrix. However, in a few special design cases the spindle axis has been defined to lie normally to the surface of a shallow cone which is coaxial with the shaft axis and tapers towards the aft direction. In these cases the spindle axis is inclined to the reference line by a few degrees; such applications are, however, comparatively rare. For the majority of cases, therefore, the terms spindle axis, directrix and reference line relate to the same line, as can be seen in Figure 3.2. These lines are frequently, but not necessarily, defined at the origin of the Cartesian reference frame discussed in the previous section.

The aerofoil sections which together comprise the blade of a propeller are defined on the surface of cylinders whose axes are concentric with the shaft axis; hence the term

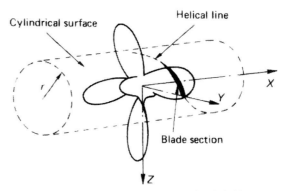

FIGURE 3.3 Cylindrical blade section definition.

'cylindrical sections' which is frequently encountered in propeller technology. Figure 3.3 shows this cylindrical definition of the section, from which it will be seen that the section lies obliquely over the surface of the cylinder and thus its nose–tail line, connecting the leading and trailing edges of the section, forms a helix over the cylinder. The point A shown in Figure 3.2 where this helix intersects the plane defined by the directrix and the x-axis is of particular interest since it forms one point, at the radius r of the section considered, on the 'generator line'. The generator line is thus the locus of all such points between the tip and root of the blade as seen in Figure 3.2. Occasionally the term 'stacking line' is encountered, this is most frequently used as a synonym for the generator line; however, there have been instances when the term has been used by designers to mean the directrix: consequently care is needed for all cases except the special case when the generator line is the same as the directrix.

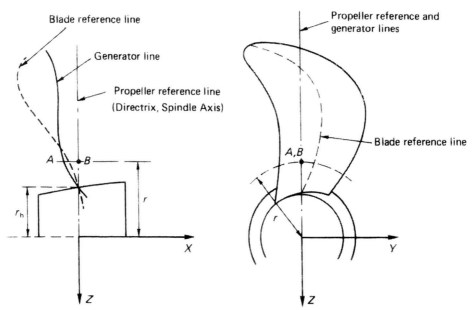

FIGURE 3.2 Blade reference lines.

Propeller Geometry   31

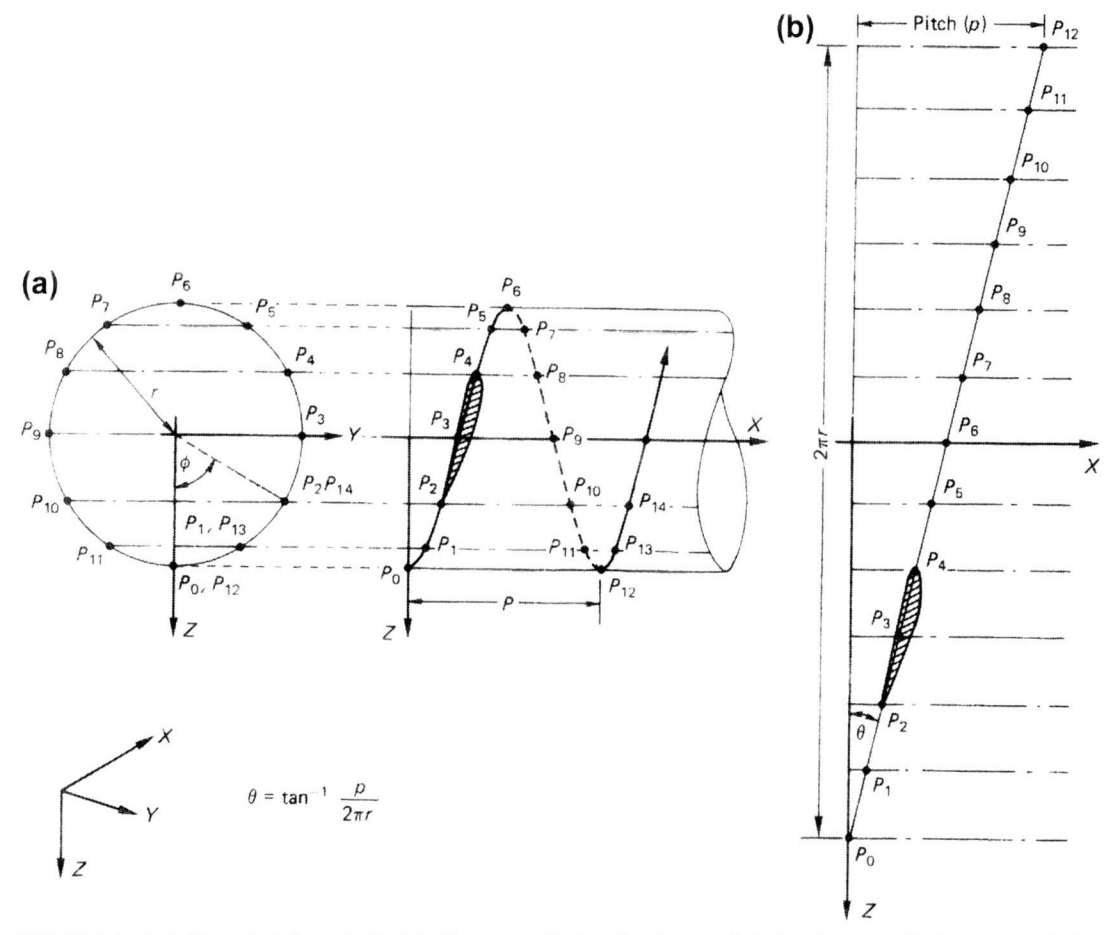

FIGURE 3.4   Definition of pitch: (a) helix definition on a cylinder of radius *r* and (b) development of helix on the cylinder.

3.3 PITCH

Consider a point P lying on the surface of a cylinder of radius r which is at some initial point P_0 and moves so as to form a helix over the surface of a cylinder. The equations governing the motion of the point P over the surface of the cylinder (points P_0, P_1, P_2, ... , P_n) in Figure 3.4(a) are as follows:

$$\left. \begin{array}{l} x = f(\phi) \\ y = r\sin(\phi) \\ z = r\cos(\phi) \end{array} \right\} \qquad (3.1)$$

where ϕ is the angle of rotation in the $Y-Z$-plane of radius arm r relative to the OZ-axis in the global reference frame. When the angle $\phi = 360°$, or 2π radians, then the helix, defined by the locus of the points P_n, has completed one complete revolution of the cylinder and again intersects the $X-Z$-plane but at a distance p measured along the OX-axis from the origin. If the cylinder is now 'opened out' as shown in Figure 3.4(b), we see that the locus of the point P, as it was rotated through 2π radians on the surface of the cylinder, lies on a straight line. In the

projection one revolution of the helix around the cylinder, measured normal to the OX direction, is equal to a distance $2\pi r$. The distance moved forward by the helical line during this revolution is p and hence the helix angle (θ) is given by

$$\theta = \tan^{-1}\left(\frac{p}{2\pi r}\right) \qquad (3.2)$$

The angle θ is termed the pitch angle and the distance p is the pitch. Hence equation (3.1), which defines a point on a helix, can be written as follows:

$$\left. \begin{array}{l} x = r\phi\tan\theta \\ y = r\sin(\phi) \\ z = r\cos(\phi) \end{array} \right\} \qquad (3.1a)$$

There are several pitch definitions that are of importance in propeller analysis and the distinction between them is of considerable importance if serious analytical mistakes are to be avoided. In all cases, however, the term pitch in propeller technology refers to the helical progress along a cylindrical surface rather than, for example, in mechanical gear design where pitch refers to the distance between

FIGURE 3.5 Pitch lines.

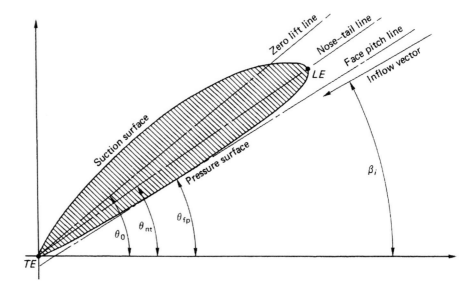

teeth. The important pitch terms with which the analyst needs to be thoroughly conversant are as follows:

1. Nose–tail pitch.
2. Face pitch.
3. Effective or 'no-lift' pitch.
4. Hydrodynamic pitch.

Figure 3.5 shows these pitch lines in association with an arbitrary aerofoil section profile. The nose–tail pitch line is today the most commonly used reference line by the principal propeller manufacturers in order to define blade sections, and it is normally defined at a pitch angle θ_{nt} to the thwart-ship direction. The nose–tail line also has a hydrodynamic significance too, since the section angles of attack are defined relative to it in the conventional aerodynamic sense.

Face pitch is now relatively rarely used by the large propeller manufacturers, but it will frequently be seen on older drawings and is still used by some smaller manufacturers. Indeed many of the older model test series, for example the Wageningen B series, use this pitch reference as a standard to present the open water characteristics. Face pitch has no hydrodynamic significance at all: it was a device invented by the manufacturers to simplify the propeller production process by obviating the need to 'hollow out' the surface of the propeller mould to accommodate that part of the section between the nose–tail and face pitch lines. The face pitch line is basically a tangent to the section's pressure side surface and, therefore, has a degree of arbitrariness about its definition since many tangents can be drawn to the aerofoil pressure surface.

The effective pitch line of the section corresponds to the conventional aerodynamic no-lift line and is the line that if the incident water flowed along it, zero lift would result from the aerofoil section. The effective pitch angle (θ_0) is greater than the nose–tail pitch angle by an amount corresponding to the three-dimensional zero lift angle of the section. As such, this is a fundamental pitch angle since it is the basis about which the hydrodynamic forces associated with the propeller section are calculated in classical analysis. Finally, the hydrodynamic pitch angle (β_i) is the angle at which the incident flow encounters the blade section and is a hydrodynamic inflow rather than a geometric property of the propeller: neither this angle nor the effective pitch angle would, however, be expected to be found on the propeller drawing in normal circumstances.

From the above discussion it can be seen that the three pitch angles, effective, nose–tail and hydrodynamic pitch, are all related by the equations:

Effective pitch angle = nose-tail pitch angle
   +3D zero lift angle
 = hydrodynamic pitch angle
   +angle of attack of section
   +3D zero lift angle.

The fuller discussion of the effective pitch, hydrodynamic pitch and zero lift angles will be left until Chapters 7 and 8; they have only been included here to underline the differences between them and thereby prevent confusion and serious analytical mistakes.

The mean pitch of a propeller blade is calculated using a moment mean principle. As such it is defined by

$$\bar{p} = \frac{\int_{x=x_h}^{1.0} px \, dx}{\int_{x=x_h}^{1.0} x \, dx} \quad (3.3)$$

The reason for adopting a moment mean is a practical expedient, which has been justified both experimentally and by calculation. As a consequence it can be used, in the context of effective pitch, to compare propellers, which may have different radial pitch distributions, from the

Chapter | 3 Propeller Geometry

viewpoint of power absorption. For continuous and fair distributions of pitch from the root to the tip it will be frequently found that the moment mean pitch corresponds in magnitude to the local pitch in the region of $0.6-0.7R$.

For practical calculation purposes of equation (3.3), because the radial pitch distribution is normally represented by a well-behaved curve without great changes in gradient (Figure 3.6), it is possible to use a lower-order numerical integration procedure. Indeed the trapezoidal rule provides a satisfactory procedure if the span of the blade is split into ten intervals giving eleven ordinates. Then the mid-points of these intervals x_j ($j = 1, 2, 3, \ldots, 10$) are defined as follows, where x is the non-dimensional radius $x = r/R$:

$$x_j = \frac{x_i + x_{i+1}}{2} \quad i,j = 1, 2, 3, \ldots, 10$$

Since the integral

$$\int_{x=x_h}^{1.0} p(x)x \ dx = \sum_{j=1}^{10} p(x_j)x_j \left(\frac{x_{TIP} - x_{HUB}}{10} \right)$$

and similarly,

$$\int_{x=x_h}^{1.0} x \ dx = \sum_{j=1}^{10} x_j \left(\frac{x_{TIP} - x_{ROOT}}{10} \right)$$

Hence,

$$\bar{p} = \frac{\sum_{j=1}^{10} p(x_j)x_j}{\sum_{j=1}^{10} x_j} \tag{3.4}$$

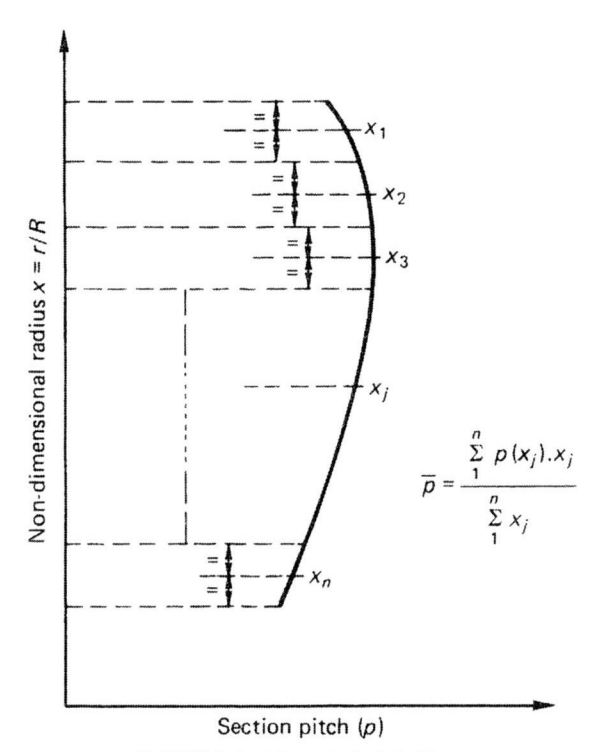

$$\bar{p} = \frac{\sum_{1}^{n} p(x_j).x_j}{\sum_{1}^{n} x_j}$$

FIGURE 3.6 Mean pitch definition.

where

$$x_j = \frac{x_i + x_{i+1}}{2} \quad i,j = 1, 2, 3, \ldots, 10$$

and

$$x_{i=1.0} = 1.0 \quad x_{i=11} = \text{root radius}$$

3.4 RAKE AND SKEW

The terms rake and skew, although defining the propeller geometry in different planes, have a cross-coupling component due to the helical nature of blade sections. As with the Cartesian reference frame, many practitioners have adopted different definitions of skew. The author prefers the following definition since, as well as following the ITTC code, it has also been adopted by several other authorities in Europe, the USA and the Far East. The skew angle $\theta_s(x)$ of a particular section, Figure 3.7, is the angle between the directrix and a line drawn through the shaft center line and the mid-chord point of a section at its non-dimensional radius (x) in the projected propeller outline; that is, looking normally, along the shaft center line, into the $y-z$-plane of Figure 3.1. Angles forward of the directrix, which is in the direction of rotation, in the projected outline are considered to be negative.

The propeller skew angle (θ_{sp}) is defined as the greatest angle, measured at the shaft center line, in the projected plane, which can be drawn between lines passing from the shaft center line through the mid-chord position of any two sections.

Propeller skew tends to be broadly classified into two types: balanced and biased skew designs. Notwithstanding this classification, a unique definition of these two types of skew is difficult because of the differing practices of the manufacturers. However, in general terms a balanced skew design is one where the locus of the mid-chord line tends initially to be thrown forward of the directrix in the inner regions of the blade span and then backwards in the outer regions. In contrast with the biased skew design, the mid-chord locus tends to predominantly move aft relative to the directrix along the span of the blade. Figure 3.7 illustrates these concepts which tend to be descriptive rather than have any intrinsic quantitative significance.

Propeller rake is divided into two components: generator line rake (i_G) and skew induced rake (i_s). The total rake of the section with respect to directrix (i_T) is given by

$$i_T(r) = i_s(r) + i_G(r) \tag{3.5}$$

The generator line rake is measured in the $x-z$-plane of Figure 3.1 and is simply the distance AB shown in Figure 3.2. That is, it is the distance, parallel to the x-axis, from the directrix to the point where the helix of the section at radius r cuts the $x-z$-plane. To understand skew induced rake consider Figure 3.8, which shows an 'unwrapping' of

FIGURE 3.7 Skew definition.

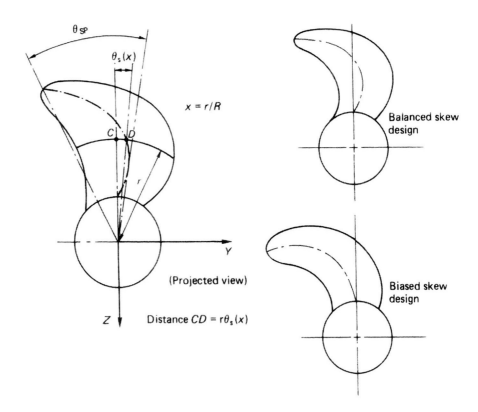

two cylindrical sections, one at the root of the propeller and the other at some radius r between the tip and root of the blade. It will be seen that skew induced rake is the component, measured in the x-direction, of the helical distance around the cylinder from the mid-chord point of the section to the projection of the directrix when viewed normally to the y–z-plane. That is,

$$i_s = r\theta_s \tan(\theta_{nt}) \qquad (3.6)$$

Consequently, it is possible then to define the locus of the mid-chord points of the propeller blade in space as follows for a rotating right-handed blade initially defined, $\phi = 0$, about the OZ-axis of the global reference frame (Figure 3.9):

$$\left.\begin{array}{l} X_{c/2} = -[i_G + r\theta_s \tan(\theta_{nt})] \\ Y_{c/2} = -r\sin(\phi - \theta_s) \\ Z_{c/2} = r\cos(\phi - \theta_s) \end{array}\right\} \qquad (3.7)$$

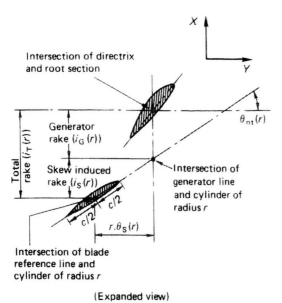

FIGURE 3.8 Definition of total rake.

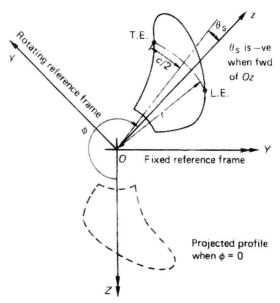

FIGURE 3.9 Blade co-ordinate definition.

Propeller Geometry

And for the leading and trailing edges of the blade equation (3.7) can be extended to give:
for the leading edge:

$$
\left.\begin{array}{l}
X_{LE} = -\left[i_G + r\theta_s \tan(\theta_{nt}) + \dfrac{c}{2}, \sin(\theta_{nt})\right] \\[2mm]
Y_{LE} = -r \sin\left[\phi - \theta_s + \dfrac{90c \cos(\theta_{nt})}{\pi r}\right] \\[2mm]
Z_{LE} = r \cos\left[\phi - \theta_s + \dfrac{90c \cos(\theta_{nt})}{\pi r}\right]
\end{array}\right.
$$

and for the trailing edge:

$$
\left.\begin{array}{l}
X_{TE} = -\left[i_G + r\theta_s \tan(\theta_{nt})\right] - \dfrac{c}{2}\sin(\theta_{nt}) \\[2mm]
Y_{TE} = -r \sin\left[\phi - \theta_s - \dfrac{90c \cos(\theta_{nt})}{\pi r}\right] \\[2mm]
Z_{TE} = r \cos\left[\phi - \theta_s - \dfrac{90c \cos(\theta_{nt})}{\pi r}\right]
\end{array}\right\} \tag{3.8}
$$

where c is the chord length of the section at radius x and ϕ and θ_s are expressed in degrees.

In cases when the generator line is a linear function of radius it is meaningful to talk in terms of either a propeller rake (i_p) or a propeller rake angle (θ_{ip}). These are measured at the propeller tip as shown in Figure 3.10, where the propeller rake is given by

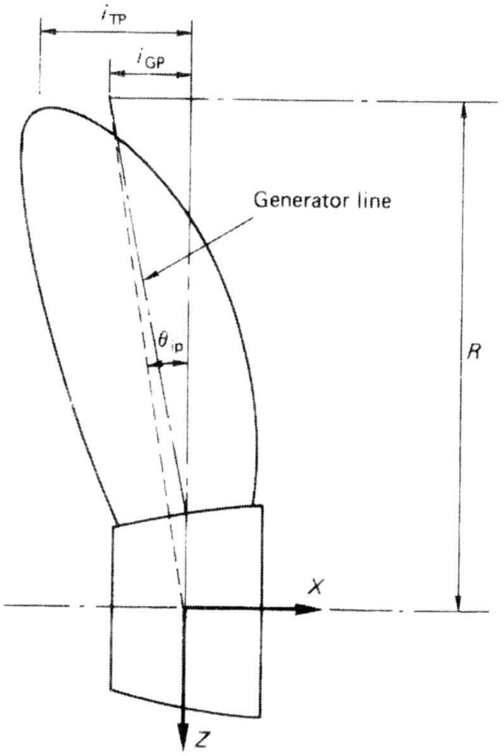

FIGURE 3.10 Tip rake definition.

$$
\left.\begin{array}{l}
i_p = i_G|_{(r/R=1.0)} \\
\text{and} \\
\theta_{ip} = \tan^{-1}\left[\dfrac{i_G|_{(r/R=1.0)}}{R}\right]
\end{array}\right\} \tag{3.9}
$$

In equation (3.9), i_p is taken as positive when the generator line at the tip is astern of the directrix, and similarly with θ_{ip}. In applying equation (3.9) it should be noted that some manufacturers adopt the alternative notation of specifying the rake angle from the root section:

$$
\theta'_{ip} = \tan^{-1}\left[\dfrac{i_G|_{(r/R=1.0)}}{(R - r_h)}\right]
$$

where r_h is the radius of the root section. Consequently some care is needed in interpreting specific propeller applications.

3.5 PROPELLER OUTLINES AND AREA

The calculation of the blade width distribution is always made with reference to the cavitation criteria to which the propeller blade will be subjected. However, having once calculated the blade section widths based on these criteria, it is necessary then to fair them into a blade outline. This can either be done by conventional drawing techniques or by the fitting of a suitable mathematical expression. One such expression which gives good results is:

$$
\begin{aligned}
\frac{c}{D} = {} & K_0(1 - x)^{1/2} + K_1 + K_2(1 - x) + K_3(1 - x)^2 \\
& + K_4(1 - x)^3 + K_5(1 - x)^4
\end{aligned}
$$

where x is the non-dimensional radius and $K_n, (n = 0, 1, \ldots, 5)$ are coefficients. There are four basic outlines in general use currently which describe the propeller blade shape:

1. The projected outline.
2. The developed outline.
3. The expanded outline.
4. The swept outline.

The projected outline is the view of the propeller blade that is actually seen when the propeller is viewed along the shaft center line; that is, normal to the y–z-plane. Convention dictates that this is the view seen when looking forward. In this view the helical sections are defined in their appropriate pitch angles and the sections are seen to lie along circular arcs whose center is the shaft axis; Figure 3.11 shows this view together with the developed and expanded views. The projected area of the propeller is the area seen when looking forward along the shaft axis. From Figure 3.11 it is clear that the projected area A_p is given by

$$
A_p = Z \int_{r_h}^{R} (\theta_{TE} - \theta_{LE}) r \, dr \tag{3.10}
$$

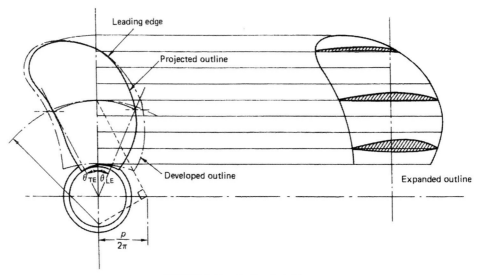

FIGURE 3.11 Outline definition.

where the same sign convention applies for θ as in the case of the skew angle and Z is the number of blades.

Projected area is of little interest today. However, in the early years of propeller technology the projected area was used extensively on a thrust loading per unit projected area basis for determining the required blade area to avoid the harmful effects of cavitation. It will be noted that the projected area is the area in the plane normal to the thrust vector.

The developed outline is related to the projected outline in so far as it is a helically based view, but the pitch of each section has been reduced to zero; that is the sections all lie in the thwart-ship plane. This view is used to give an appreciation of the true form of the blade and the distribution of chord lengths. The developed and projected views are the most commonly seen representations on propeller drawings; Figure 3.11 shows this view in relation to the projected outline.

To calculate the developed area it is necessary to integrate the area under the developed profile curve numerically if a precise value is required. For most purposes, however, it is sufficient to use the approximation for the developed area A_D as being

$$A_D \simeq A_E$$

where A_E is the expanded area of the blade.

In the past several researchers have developed empirical relationships for the estimation of the developed area; one such relationship, proposed by Burrill for non-skewed forms, is

$$A_D \simeq \frac{A_p}{(1.067 - 0.229 P/D)} \quad (3.11)$$

In general, however, the developed area is greater than the projected area and slightly less than the expanded area.

The expanded outline is not really an outline in any true geometric sense. It could more correctly be termed a plotting of the chord lengths at their correct radial stations about the directrix; no attempt in this outline is made to represent the helical nature of the blades and the pitch angle of each section is reduced to zero. This view is, however, useful in that it is sometimes used to give an idea of the blade section forms used, as these are frequently plotted on the chord lengths, as seen in Figure 3.11.

The expanded area is the most simple of the areas that can be calculated, and for this reason is the area most normally quoted, and is given by the relationship:

$$A_E = Z \int_{r_n}^{R} c \, dr \quad (3.12)$$

In order to calculate this area it is sufficient for most purposes to use a Simpson's procedure with eleven ordinates, as shown in Figure 3.12.

Blade area ratio is simply the blade area, either the projected, developed or expanded depending on the context, divided by the propeller disc area A_o:

$$\left. \begin{array}{l} \dfrac{A_p}{A_o} = \dfrac{4A_p}{\pi D^2} \\[6pt] \dfrac{A_D}{A_o} = \dfrac{4A_D}{\pi D^2} \\[6pt] \dfrac{A_E}{A_o} = \dfrac{4A_D}{\pi D^2} \end{array} \right\} \quad (3.13)$$

By way of example, the difference in the value of the projected, developed and expanded area ratio for the propeller shown in Figure 3.11 can be seen from Table 3.1. The propeller was assumed to have four blades and a constant pitch ratio for this example.

Chapter | 3 Propeller Geometry

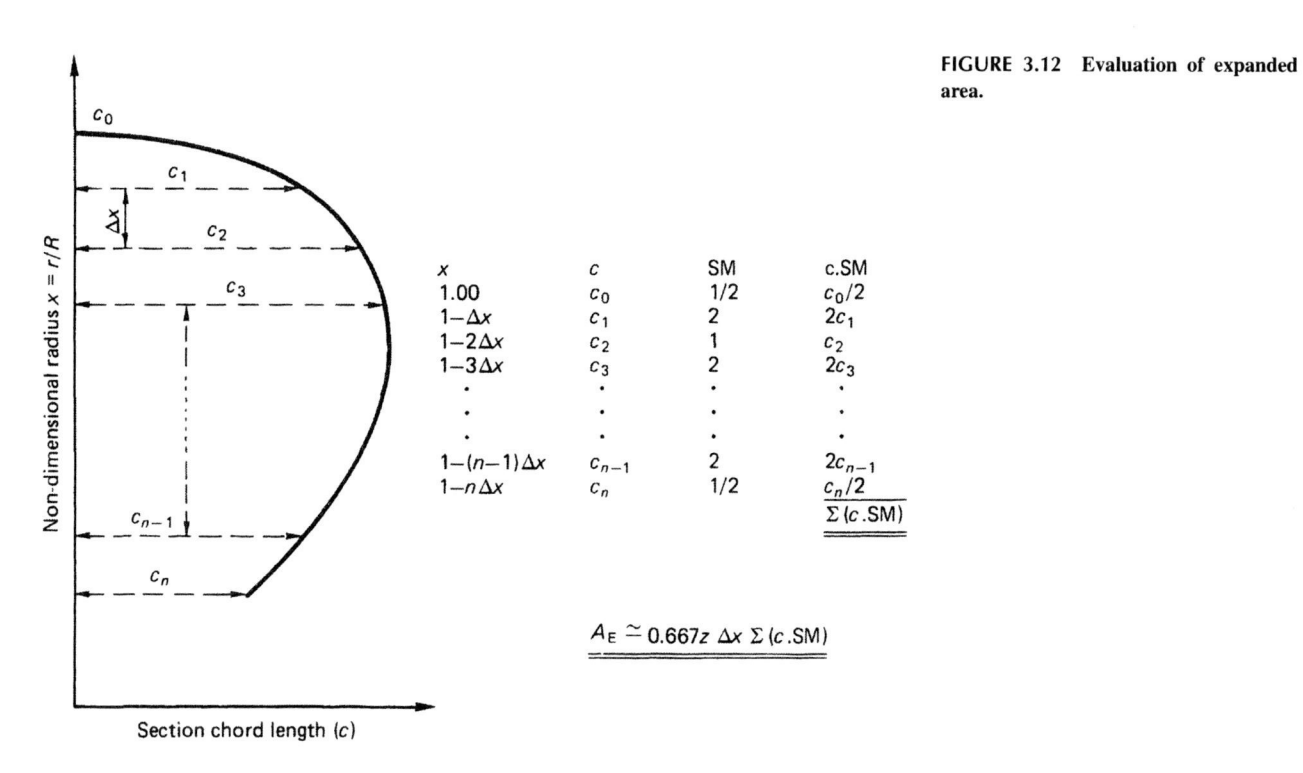

FIGURE 3.12 Evaluation of expanded area.

$$A_E \cong 0.667z\,\Delta x\,\Sigma(c.\text{SM})$$

In each of the areas discussed so far the blade has been represented by a lamina of zero thickness. The true surface area of the blade will need to take account of the blade thickness and the surface profile on the suction and pressure faces; which will be different in all cases except for the so-called 'flat plate' blades found in applications like controllable pitch transverse propulsion units. To calculate the true surface area of one of the blade surfaces the algorithm of Figure 3.13 needs to be adopted.

This algorithm is based on a linear distance — that is between the successive points on the surface. This is sufficient for most calculation purposes, but higher-order methods can be used at the expense of a considerable increase in computational complexity.

The swept outline of a propeller is precisely what is conventionally meant by a swept outline in normal engineering terms. It is normally used only to represent stern frame clearances. For the case of the highly skewed propeller a representation of the swept outline is important since the skew induced rake term, if not carefully controlled in design, can lead to considerable 'overhang' of

TABLE 3.1 Example of Comparative Blade Area Ratios			
	Projected Area	Developed Area	Expanded Area
Area ratio (A/A_0)	0.480	0.574	0.582

Generate an array of points at each radial location on the chosen surface of the blade, increasing the density of the array where the curvature of the surface is greatest. Typically this might involve calculating an array of surface offsets at, for example, 0, 0.125, 0.25, 0.50, 0.75, 1.25, 1.75, 2.5, 5.0, 7.5, 10, 15, 20, 25, 30, 35, 40, 45, 50, 55, 60, 65, 75, 80, 85, 90, 95, 97.5, 100% of the chord. At each point the Cartesian co-ordinates should be calculated

For each successive pair of points along the surface calculate the length between successive points l_{ij}. That is, for the ith radial position evaluate the distance between the jth and $(j + 1)$th point using the relationship:
$$l^2 = (x_{j+1} - x_j)^2 + (y_{j+1} - y_j)^2 + (z_{j+1} - z_j)^2$$

For each radius calculate the true length around the surface from the leading to trailing edge by
$$L_i = \sum_{j=0}^{TE} l_{ij}$$

Evaluate the true area A_T of the chosen surface by the relationship
$$A_T = Z \int_{r_h}^{R} L_i \, dr$$

FIGURE 3.13 Algorithm for calculating surface area.

the blade which, in turn, can lead to mechanical interference with the stern frame. The swept outline is derived by plotting the rotation of each of the leading and trailing edges about the shaft axis.

3.6 PROPELLER DRAWING METHODS

Prior to the introduction of computer-based methods the most commonly used method for drawing a propeller was that developed by Holst.[2] This method relies on being able to adequately represent the helical arcs along which the propeller sections are defined by circular arcs, of some radius greater than the section radius, when the helical arcs have been swung about the directrix into the zero pitch or developed view (see Figure 3.11). This drawing method is an approximation but does not lead to significant errors unless used for very wide bladed or highly skewed propellers; in these cases errors can be significant and the alternative and more rigorous method of Rosingh[3] would then be used to represent the blade drawings.

The basis of Holst's construction is shown in Figure 3.14. This figure shows the construction for only one particular radius in the interests of clarity; other radii are treated identically. A series of arcs with center on the shaft axis at O are constructed at each of the radial stations on the directrix where the blade is to be defined. A length $p/2\pi$ is then struck off along the horizontal axis for each section and the lines AB are joined for each of the sections under consideration. A right angle ABC is then constructed, which in turn defines a point C on the extension of the directrix below the shaft center line. An arc AC is then drawn with center C and radius r'. The distances from the directrix to the leading edge AA_L and the directrix to the trailing edge AA_T are measured around the circumference of the arc. Projections, normal to the directrix, through A_L and A_T meet the arc of radius r, about the shaft center line, at P_L and P_T, respectively. These latter two points form two points on the leading and trailing edges of the projected outline, whilst A_L and A_T lie on the developed outline. Consequently, it can be seen that distances measured around the arcs on the developed outline represent 'true lengths' that can be formed on the actual propeller.

The Holst drawing method was a common procedure used in propeller drawing offices years ago. However, the advent of the computer and its associated graphics capabilities have permitted the designer to plot automatically blade outlines using points calculated by analytical geometry, for example equation (3.8), together with curve-fitting routines, typically cubic splines.

3.7 SECTION GEOMETRY AND DEFINITION

The discussion so far has, with the exception of that relating to the true surface area, assumed the blade to be a thin lamina. This section redresses this assumption by discussing the blade section geometry.

In the early 1930s the National Advisory Committee for Aeronautics (NACA) in the USA — now known as NASA — embarked on a series of aerofoil experiments which were based on aerofoil geometry developed in a rational and systematic way. Some of these aerofoil shapes have been adopted for the design of marine propellers, and as such have become widely used by manufacturers all over the world. Consequently, this discussion of aerofoil geometry will take as its basis the NACA definitions whilst at the same time recognizing that with the advent of prescribed velocity distribution capabilities some designers are starting to generate their own section forms to meet specific surface pressure requirements.

Figure 3.15 shows the general definition of the aerofoil. The mean line or camber line is the locus of the mid-points between the upper and lower surfaces when measured perpendicular to the camber line. The extremities of the camber line are termed the leading and trailing edges of the aerofoil and the straight line joining these two points is termed the chord line. The distance between the leading and trailing edges when measured along the chord line is termed the chord length (c) of the section. The camber of the section is the maximum distance between the mean camber line and the chord line, measured perpendicular to

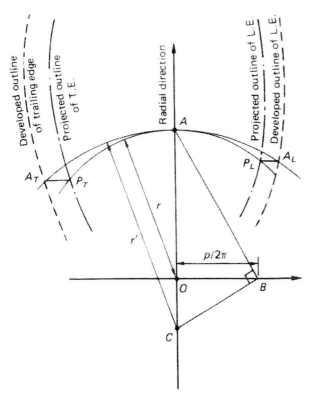

FIGURE 3.14 Holst's propeller drawing method.

Propeller Geometry

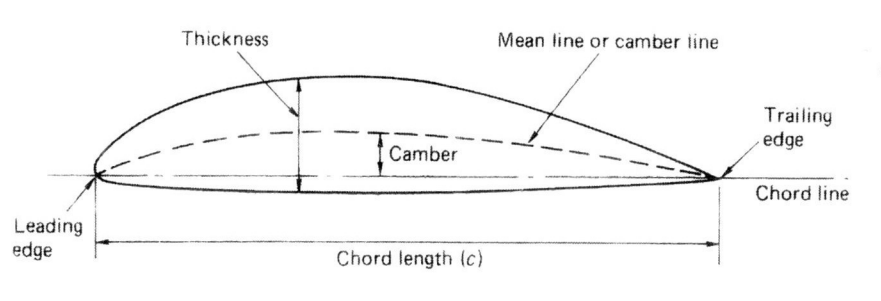

FIGURE 3.15 General definition of an aerofoil section.

the chord line. The aerofoil thickness is the distance between the upper and lower surfaces of the section, usually measured perpendicularly to the chord line although strictly this should be to the camber line. The leading edges are usually circular, having a leading edge radius defined about a point on the camber line. However, for marine propellers, leading edge definition practices vary widely from manufacturer to manufacturer and care should be taken in establishing the practice actually used for the propeller in question.

The process of combining a chosen camber line with a thickness line in order to obtain the desired section form is shown in Figure 3.16 for a given chord length c. In the figure only the leading edge is shown for the sake of clarity; however, the trailing edge situation is identical. The mean line is defined from the offsets (y_c) relating to the chosen line and these are 'laid off' perpendicularly to the chord line. The upper and lower surfaces are defined from the ordinates y_t of the chosen symmetrical thickness distribution, and these are then laid off perpendicularly to the camber line. Hence, a point P_u on the upper surface of the aerofoil is defined by

$$\left. \begin{array}{l} x_u = x_c - y_t \sin \psi \\ y_u = y_c + y_t \cos \psi \end{array} \right\} \qquad (3.14a)$$

where ψ is the slope of the camber line at the non-dimensional chordal position x_c.

Similarly for a point P_L on the lower surface of the aerofoil we have

$$\left. \begin{array}{l} x_L = x_c + y_t \sin \psi \\ y_L = y_c - y_t \cos \psi \end{array} \right\} \qquad (3.14b)$$

Although equations (3.14a and b) give the true definition of the points on the section surface, since y_c/c is usually of the order of 0.02−0.06 for marine propellers, the value of ψ is small. This implies $\sin \psi \to 0$ and $\cos \psi \to 1$. Hence, it is generally valid to make the approximation

$$\left. \begin{array}{l} x_u = x_c \\ y_u = y_c + y_t \\ x_L = x_c \\ y_L = y_c - y_t \end{array} \right\} \qquad (3.15)$$

where $y_t = t(x_c)/2$ (i.e. the local section semi-thickness) and the approximation defined by equation (3.15) is generally used in propeller definition. The errors involved in this approximation are normally small — usually less than 0.5 mm and certainly within most manufacturing tolerances.

The center for the leading edge radius is found from the NACA definition as follows. A line is drawn through the forward end of the chord at the leading edge with a slope equal to the slope of the mean line close to the leading edge. Frequently the slope at a point $x_c = 0.005$ is taken, since the slope at the leading edge is theoretically infinite. This approximation is justified by the manner in

FIGURE 3.16 Aerofoil section definition.

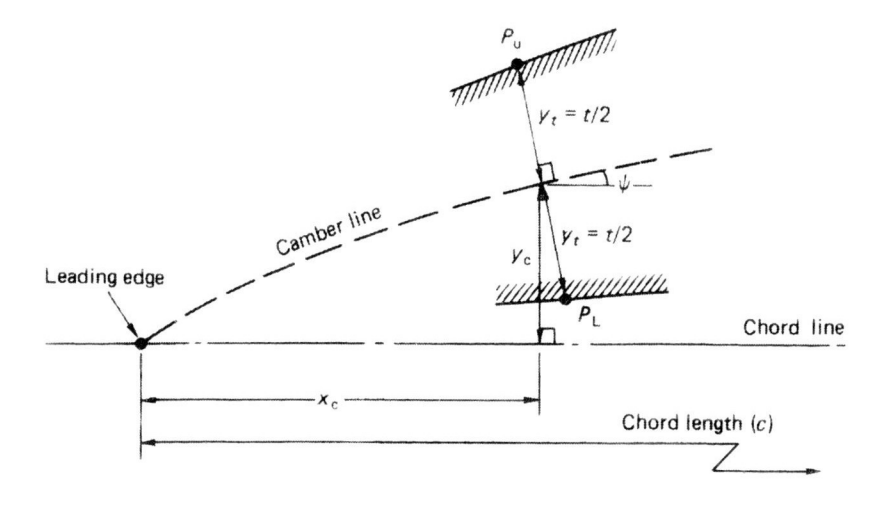

which the slope approaches infinity close to the leading edge. A distance is then laid off along this line equal to the leading edge radius and this forms the center of the leading edge radius.

Details of all of the NACA series section forms can be found in Abbott and von Doenhoff[4]; however, for convenience the more common section forms used in propeller practice are reproduced here in Tables 3.2 and 3.3. In Table 3.3 the NACA 66 (Mod) section has been taken from Brockett[5] who thickened the edge region of the parent NACA 66 section for marine use. The basic NACA 65 and 66 section forms cannot be represented in the same y/t_{max} form for all section t_{max}/c ratios, as with the NACA 16 section, and Reference 4 needs to be consulted for the ordinates for each section thickness to chord ratio. In practice, for marine propeller purposes all of the basic NACA sections need thickening at the edges, otherwise they would frequently incur mechanical damage by being too thin. Typical section edge thicknesses are shown in Table 3.4 as a proportion of the maximum section thickness for conventional free-running, non-highly skewed propellers. In the case of a highly skewed propeller, defined by the Rules of Lloyd's Register as one having a propeller skew angle in excess of 25°, the trailing edge thicknesses would be expected to be increased from those of Table 3.4 by amounts depending on the type and extent of the skew. The implication of Table 3.4 is that the leading and trailing edges have 'square' ends. This clearly is not the case: these are the thicknesses that would exist at the edges if the section thicknesses were extrapolated to the edges without rounding.

It is frequently necessary to interpolate the camber and thickness ordinates at locations away from those defined by Tables 3.2 and 3.3. For normal types of camber lines standard interpolation procedures can be used, provided they are based on either second- or third-order polynomials. This is also the case with the thickness distribution away from the rapid changes of curvature that occur close to the leading edge. To overcome this difficulty van Oossanen[6] proposed a method based on defining an equivalent ellipse having a thickness to chord ratio equal to that of the section under consideration. Figure 3.17 demonstrates the method in which a thickness ratio T_R is formed between the actual section and the equivalent elliptical section:

$$T_R = \frac{y_t}{y_{t_{max}} \sin[\cos^{-1}(1 - 2x_c/c)]}$$

This provides a smooth well-behaved function between the leading and trailing edges and having a value of unity at these points. This function can then be interpolated at any required point x'_c and the required thickness at this point derived from the relationship

$$y'_t = T'_R y_{t_{max}} \sin[\cos^{-1}(1 - 2x_c/c)] \qquad (3.16)$$

This method can be used over the entire section in order to provide a smooth interpolation procedure; however, a difficulty is incurred right at the leading edge where the thickness distribution gives way to the leading edge radius. For points between this transition point, denoted by P in Figure 3.17, and the leading edge, the value of the thickness ratio T_R is given by

$$T_R = \frac{\rho_L^2 - (x_c - \rho_L)^2}{y_{t_{max}} \sin[\cos^{-1}(1 - 2x_c/c)]}$$

At the point P it should be noted that the tangent to both the leading edge radius and the thickness form are equal.

Having, therefore, defined the basis of section geometry, it is possible to revert to equation (3.8) and define the co-ordinates for any point P on the surface of the aerofoil section. Figure 3.18 shows this definition, and the equations defining the point P about the local reference frame (Ox, Oy, Oz) are given by

$$\left.\begin{aligned} x_p &= -[i_G + r\theta_s \tan(\theta_{nt})] + (0.5c - x_c)\sin(\theta_{nt}) + y_{u,L}\cos(\theta_{nt}) \\ y_p &= r\sin\left[\theta_s - \frac{180[(0.5c - x_c)\cos(\theta_{nt}) - y_{u,L}\sin(\theta_{nt})]}{\pi r}\right] \\ z_p &= r\cos\left[\theta_s - \frac{180[(0.5c - x_c)\cos(\theta_{nt}) - y_{u,L}\sin(\theta_{nt})]}{\pi r}\right] \end{aligned}\right\} \qquad (3.17)$$

where $y_u = y_c \pm y_t \cos\psi$ as per equations (3.14a and b). To convert these to the global reference frame (OX, OY, OZ), we simply write the transformation

$$\begin{bmatrix} X_p \\ Y_p \\ Z_p \end{bmatrix} = \begin{bmatrix} 1 & 0 & 0 \\ 0 & \cos\phi & -\sin\phi \\ 0 & \sin\phi & \cos\phi \end{bmatrix} \begin{bmatrix} x_p \\ y_p \\ z_p \end{bmatrix} \qquad (3.18)$$

where ϕ is the angle between the reference frames as shown in Figure 3.9. By combining equations (3.17) and (3.18) and inserting the appropriate values for x_c and $y_{u,L}$, the expressions for the leading, trailing edges and mid-chord

Chapter | 3 Propeller Geometry

TABLE 3.2 NACA Series Camber or Mean Lines

	64 Mean Line		65 Mean Line		66 Mean Line	
x_c (% c)	y_c (% c)	dy_c/dx_c	y_c (% c)	dy_c/dx_c	y_c (% c)	dy_c/dx_c
0	0	0.30000	0	0.24000	0	0.20000
1.25	0.369	0.29062	0.296	0.23400	0.247	0.19583
2.5	0.726	0.28125	0.585	0.22800	0.490	0.19167
5.0	1.406	0.26250	1.140	0.21600	0.958	0.18333
7.5	2.039	0.24375	1.665	0.20400	1.406	0.17500
10	2.625	0.22500	2.160	0.19200	1.833	0.16667
15	3.656	0.18750	3.060	0.16800	2.625	0.15000
20	4.500	0.15000	3.840	0.14400	3.333	0.13333
25	5.156	0.11250	4.500	0.12000	3.958	0.11667
30	5.625	0.07500	5.040	0.09600	4.500	0.10000
40	6.000	0	5.760	0.04800	5.333	0.06667
50	5.833	−0.03333	6.000	0	5.833	0.03333
60	5.333	−0.06667	5.760	−0.04800	6.000	0
70	4.500	−0.10000	5.040	−0.09600	5.625	−0.07500
80	3.333	−0.13333	3.840	−0.14400	4.500	−0.15000
90	1.833	−0.16667	2.160	−0.19200	2.625	−0.22500
95	0.958	−0.18333	1.140	−0.21600	1.406	−0.26250
100	0	−0.20000	0	−0.24000	0	−0.30000

	a = 0.8 Mean Line		a = 0.8 (mod) Mean Line		a = 1.0 Mean Line	
x_c (% c)	y_c % c)	dy_c/dx_c	y_c (% c)	dy_c/dx_c	y_c (% c)	dy_c/dx_c
0	0		0		0	
0.5	0.287	0.48535	0.281	0.47539	0.250	0.42120
0.75	0.404	0.44925	0.396	0.44004	0.350	0.38875
1.25	0.616	0.40359	0.603	0.39531	0.535	0.34770
2.5	1.077	0.34104	1.055	0.33404	0.930	0.29155
5.0	1.841	0.27718	1.803	0.27149	1.580	0.23430
7.5	2.483	0.23868	2.432	0.23378	2.120	0.19995
10	3.043	0.21050	2.981	0.20618	2.585	0.17485
15	3.985	0.16892	3.903	0.16546	3.365	0.13805
20	4.748	0.13734	4.651	0.13452	3.980	0.11030
25	5.367	0.11101	5.257	0.10873	4.475	0.08745
30	5.863	0.08775	5.742	0.08595	4.860	0.06745
35	6.248	0.06634	6.120	0.06498	5.150	0.04925
40	6.528	0.04601	6.394	0.04507	5.355	0.03225
45	6.709	0.02613	6.571	0.02559	5.475	0.01595

(Continued)

TABLE 3.2 NACA Series Camber or Mean Lines—cont'd

	a = 0.8 Mean Line		a = 0.8 (mod) Mean Line		a = 1.0 Mean Line	
x_c (% c)	y_c (% c)	dy_c/dx_c	y_c (% c)	dy_c/dx_c	y_c (% c)	dy_c/dx_c
50	6.790	0.00620	6.651	0.00607	5.515	0
55	6.770	−0.01433	6.631	−0.01404	5.475	−0.01595
60	6.644	−0.03611	6.508	−0.03537	5.355	−0.03225
65	6.405	−0.06010	6.274	−0.05887	5.150	−0.04925
70	6.037	−0.08790	5.913	−0.08610	4.860	−0.06745
75	5.514	−0.12311	5.401	−0.12058	4.475	−0.08745
80	4.771	−0.18412	4.673	−0.18034	3.980	−0.11030
85	3.683	−0.23921	3.607	−0.23430	3.365	−0.13805
90	2.435	−0.25583	2.452	−0.24521	2.585	−0.17485
95	1.163	−0.24904	1.226	−0.24521	1.580	−0.23430
100	0	−0.20385	0	−0.24521	0	

points, equations (3.8) and (3.7), respectively, can be derived.

The term 'washback' is sometimes seen in older papers dealing with propeller technology and in classification society rules. It relates to the definition of the after part of the face of the section and its relation to the face pitch line as shown in Figure 3.19. From this figure it is seen that for a section to have no 'washback', the face of the blade astern of the maximum thickness position is coincident with the face pitch line. When there is a 'washback', the blade section lifts above the face pitch line.

Section edge geometry is a complex matter, since cavitation properties can be influenced greatly by the choice of the geometric configuration. In the case of the leading edge it is becoming increasingly popular to use a NACA type definition; however, some quite complex edge definitions will be found. For example, the choice of a radius defined about some arbitrary but well-defined point relative to the section chord line. These types of definition have largely been introduced from empiricism and experience of avoiding one type of cavitation or another prior to the advent of adequate flow computational procedures. Consequently, care must be exercised in interpreting drawings from different designers and manufacturers. With regard to the trailing edge, this generally receives less detailed consideration. In the absence of an anti-singing edge, see Figure 21.9, it is usual to specify either a half or quarter round trailing edge.

3.8 BLADE THICKNESS DISTRIBUTION AND THICKNESS FRACTION

Blade maximum thickness distributions are normally selected on the basis of stress analysis calculations. Sometimes this involves a calculation of the stress at some radial location, for example at the $0.25R$ radius, together with the use of a standard radial thickness line found by the designer to give satisfactory service experience. More frequently today the thickness distributions are the subject of detailed stress calculations over the entire blade using finite element techniques.

The resulting thickness distributions for large propellers are normally non-linear in form and vary considerably from one manufacturer to another. In the case of smaller propellers a linear thickness distribution is sometimes selected, and although this gives a conservative reserve of strength to the blade, it also causes an additional weight and drag penalty to the propeller. On propeller drawings it is customary to show the maximum thickness distribution of the blade in an elevation as shown in Figure 3.20. In this elevation the maximum thicknesses are shown relative to the blade generator line. The blade thickness fraction is the ratio:

$$t_F = \left(\frac{t_0}{D}\right) \quad (3.19)$$

where t_0 is the notional blade thickness defined at the shaft center line as shown in Figure 3.20. In the case of a linear

Propeller Geometry

TABLE 3.3 Typical Aerofoil Section Thickness Distributions

	x/c	NACA 16 y/t_{max}	NACA 66 (mod) y/t_{max}
LE	0	0	0
	0.005	–	0.0665
	0.0075	–	0.0812
	0.0125	0.1077	0.1044
	0.0250	0.1504	0.1466
	0.0500	0.2091	0.2066
	0.0750	0.2527	0.2525
	0.1000	0.2881	0.2907
	0.1500	0.3445	0.3521
	0.2000	0.3887	0.4000
	0.2500	–	0.4363
	0.3000	0.4514	0.4637
	0.3500	–	0.4832
	0.4000	0.4879	0.4952
	0.4500	–	0.5000
	0.5000	0.5000	0.4962
	0.5500	–	0.4846
	0.6000	0.4862	0.4653
	0.6500	–	0.4383
	0.7000	0.4391	0.4035
	0.7500	–	0.3612
	0.8000	0.3499	0.3110
	0.8500	–	0.2532
	0.9000	0.2098	0.1877
	0.9500	0.1179	0.1143
TE	1.0000	0.0100	0.0333

Section t_{max}/c 0.06 0.09 0.12 0.15 0.18 0.21
LE radius/c (%)
0.176 0.396 0.703 1.100 1.584 2.156 $\rho_L = 0.448c \left(\dfrac{t_{max}}{c} \right)^2$

thickness distribution the value of t_0 is easy to calculate since it is simply a linear extrapolation of the maximum thickness distribution to the shaft center line:

$$t_0 = t(1.0) + \frac{t(x) - t(1.0)}{(1.0 - x)}$$

where $t(x)$ is the blade maximum thickness at the non-dimensional radius x and $t(1.0)$ is the blade maximum thickness at the tip before any edge treatment. In the case of

a non-linear thickness distribution the thickness fraction is calculated by a moment mean approximation as follows:

$$t_F = \frac{1}{D} \left[\frac{\sum t(x)x/(1-x)}{\sum x} + \left[t(1.0) - \frac{t(1.0) \sum x/(1-x)}{\sum x} \right] \right]$$

where x can take a range of nine or ten values over the blade span. For example,

$$x = 0.9, 0.8, 0.7, 0.6, 0.5, 0.4, 0.3, 0.2;\ (x \neq 1.0)$$

or

$$x = 0.9375, 0.875, 0.75, 0.625, 0.5, 0.375, 0.25;\ (x \neq 1.0)$$

3.9 BLADE INTERFERENCE LIMITS FOR CONTROLLABLE PITCH PROPELLERS

In order that a controllable pitch propeller can be fully reversible, in the sense that its blades can pass through the zero pitch condition, care has to be taken that the blades will not interfere with each other. To establish the limiting conditions for full reversibility, use can either be made of equation (3.8), together with an interpolation procedure, or alternatively, the limits can be approximated using Holt's drawing method.

The latter method, as shown by Hawdon *et al.* (Reference 7), gives rise to the following set of relationships for the interference limits of three-, four- and five-bladed controllable pitch propellers:

$$\left. \begin{array}{l} \text{Three-bladed propeller} \\ c_{max}/D = [1.01x + 0.050(P/D - 1) + 0.055] \\ \text{Four-bladed propeller} \\ c_{max}/D = [0.771x + 0.025(P/D - 1) + 0.023] \\ \text{Five-bladed propeller} \\ c_{max}/D = [0.632x + 0.0125(P/D - 1) + 0.010] \end{array} \right\}$$

(3.20)

3.10 CONTROLLABLE PITCH PROPELLER OFF-DESIGN SECTION GEOMETRY

A controllable pitch propeller presents further complications in blade section geometry if rotated about its spindle axis from the design pitch conditions for which the original helical section geometry was designed. Under these conditions it is found that helical sections at any given radii are subjected to a distortion when compared to the original designed section profile. To illustrate this point further, consider a blade in the designed pitch setting together with a section denoted by a projection of the arc ABC at some given radius r (Figure 3.21). When the blade is rotated

TABLE 3.4 Typical Section Edge Thickness Ratio for Conventional Free-Running, Non-Highly Skewed Propellers

	Edge Thickness Ratios $\dfrac{t(x_c/x = 0 \text{ or } 1.0)}{t_{max}}$	
r/R	Leading Edge	Trailing Edge
0.9	0.245	0.245
0.8	0.170	0.152
0.7	0.143	0.120
0.6	0.134	0.100
0.5	0.130	0.085
0.4	0.127	0.075
0.3	0.124	0.068
0.2	0.120	0.057

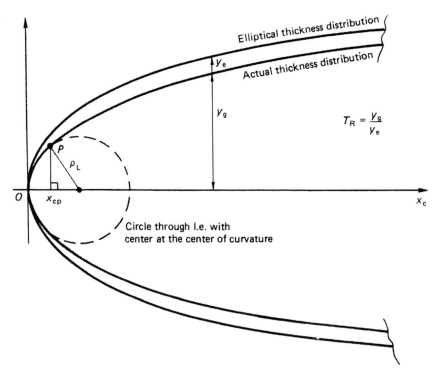

FIGURE 3.17 Van Oossanen's section thickness interpolation procedure.

about its spindle axis, through an angle $\Delta\theta$, such that the new pitch angle attained is less than the designed angle, then the blade will take up a position illustrated by the hatched line in the diagram. Therefore, at the particular radius r chosen, the helical section is now to be found as a projection of the arc $A'BC'$. However, the point A' has been derived from the point A'', which, with the blade in the design setting, was at a radius r_1 ($r_1 < r$). Similarly with the point C', since this originated from the point C'' at a radius r_2 ($r_2 < r$). Consequently, the helical section $A'BC'$ at radius r becomes a composite section containing elements of all the original design sections at radii within the range r to r_2 assuming $r_1 > r_2$. These distortions are further accentuated by the radially varying pitch angle distribution of the blade, causing an effective twisting of the leading and trailing edges of the section. A similar argument applies to the case when the pitch angle is increased from that of the design value. This latter case, however, is normally of a fairly

Chapter | 3 Propeller Geometry   **45**

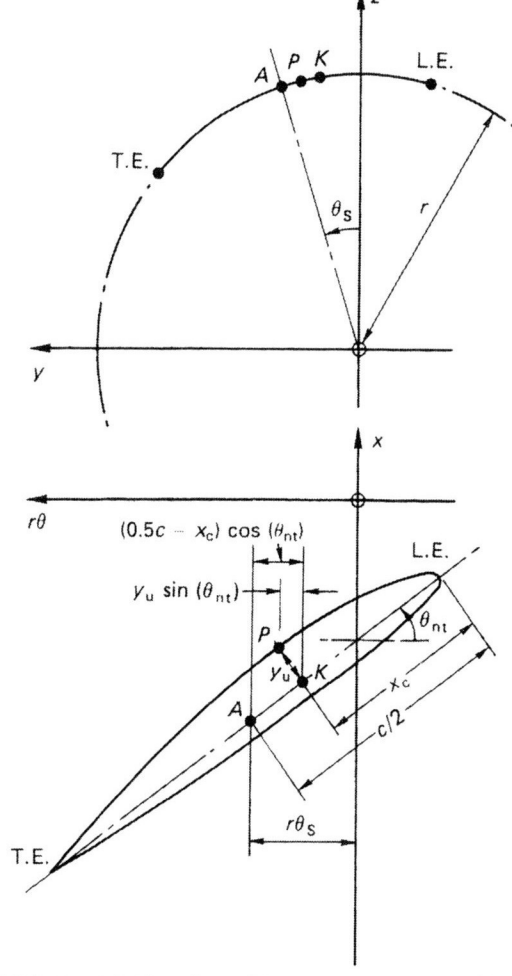

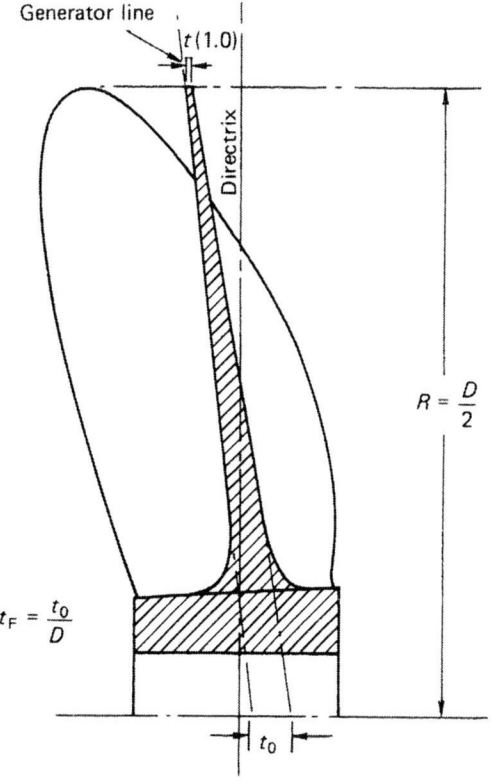

FIGURE 3.20   Typical representation of propeller maximum thickness distribution and notional thickness at shaft center line.

FIGURE 3.18   Definition of an arbitrary point p on a propeller blade surface.

trivial nature from the section of definition viewpoint, since the pitch changes in this direction are seldom in excess of $4-5°$.

The calculation of this 'distorted' section geometry at off-design pitch can be done either by draughting techniques, which is extremely laborious, or by using computer-based surface geometry software packages. The resulting section distortion can be quite significant, as seen

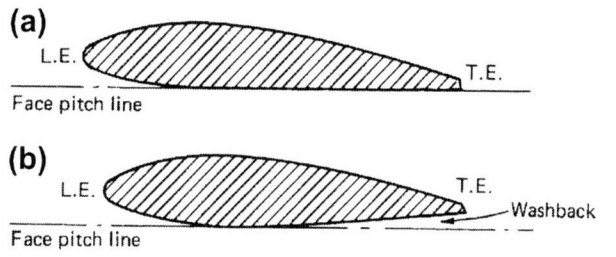

FIGURE 3.19   Section 'washback': (a) section without washback and (b) section with washback.

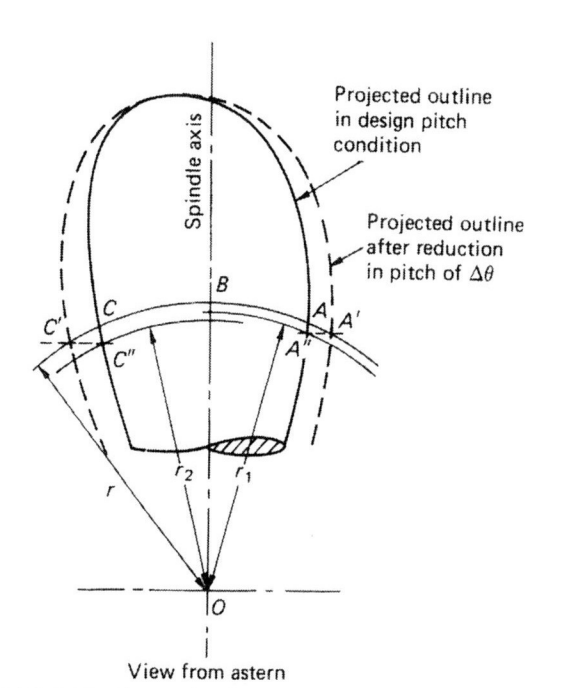

FIGURE 3.21   Geometric effects on blade section resulting from changes in pitch angle.

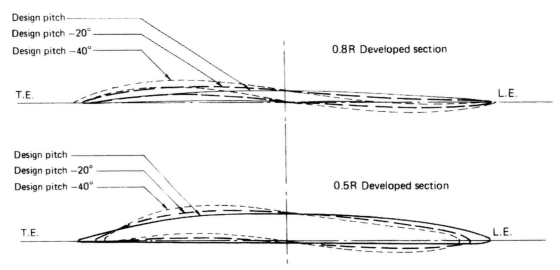

FIGURE 3.22 Section distortion due to changes of pitch angle.

in Figure 3.22, which shows the distortion found in the section definition of a North Sea ferry propeller blade at the 0.5R and 0.8R sections for pitch change angles of 20° and 40°.

Rusetskiy[8] has also addressed this problem of section distortion at off-design conditions from the point of view of mean line distortion. He developed a series of construction curves to approximate the distortion of the mean line for a given pitch change angle from design geometrical data. This technique is suitable for hand calculation purposes.

A similar problem to the one just described also exists in the definition of planar or 'straight-cut' sections through a blade. Such data are often required as input to NC machinery and other quality control operations. Klein[9] provides a treatment of this and other geometric problems.

3.11 MISCELLANEOUS CONVENTIONAL PROPELLER GEOMETRY TERMINOLOGY

In keeping with many aspects of marine engineering and naval architecture use is made in propeller technology of several terms which need further clarification.

The terms 'right-' and 'left-handed propellers' refer to the direction of rotation. In the case of a right-handed propeller, this type of propeller rotates in a clockwise direction, when viewed from astern, and thus describes a right-handed helical path. Similarly, the left-handed propeller rotates in an anticlockwise direction describing a left-handed helix.

The face and back of propellers are commonly applied terms both to the propeller in its entirety and also to the section geometry. The face of the propeller is that part of the propeller seen when viewed from astern and along the shaft axis. Hence the 'faces' of the blade sections are those located on the pressure face of the propeller when operating in its ahead design condition. Conversely, the backs of the propeller blades are those parts of the propeller seen when viewed from ahead in the same way. The backs of the helical sections, located on the backs of the propeller blades, are the same as the suction surfaces of the aerofoil in the normal design conditions.

REFERENCES AND FURTHER READING

1. *ITTC Dictionary of Ship Hydrodynamics.* Maritime Technology Monograph No. 6. RINA; 1978.
2. Holst CP. *The Geometry of the Screw Propeller*; 1924. Leiden.
3. Rosingh WHCE. Over de constructie en sterk-teberekening van hoogbelaste scheepsschroeven. Schip en Werf; 1937;103−21.
4. Abbott IH, Doenhoff AE von. *Theory of Wing Sections.* New York: Dover; 1959.
5. Brockett T. *Minimum Pressure Envelopes for Modified NACA 66 Sections With NACA a = 0.8 Camber and BU Ships Type I and Type II Sections.* DT NSRDC Report No. 1780; February 1966.
6. Oossanen P van. *Calculation of Performance and Cavitation Characteristics of Propellers, Including Effects of Non-Uniform Flow and Viscosity.* NSMB Publication 457; 1974.
7. Hawdon L, Carlton JS, Leathard FI. *The analysis of controllable pitch propeller characteristics at off-design conditions.* Trans. I. Mar. E.; 1975.
8. Rusetskiy AA. *Hydrodynamics of Controllable Pitch Propellers.* Leningrad: Shipbuilding Publishing House; 1968.
9. Klein JL. *A rational approach to propeller geometry.* Propellers '75 Symposium, Trans. SNAME; 1975.

Chapter 4

The Propeller Environment

Chapter Outline

4.1 Density of Water	47	4.6 Dissolved Gases in Sea Water	50
4.2 Salinity	48	4.7 Surface Tension	51
4.3 Water Temperature	49	4.8 Weather	51
4.4 Viscosity	49	4.9 Silt and Marine Organisms	55
4.5 Vapor Pressure	50	References and Further Reading	56

Sea water is a complex natural environment and it is the principal environment in which marine propellers operate. However, it is not the only environment since many ships and boats, some of considerable size, are designed to operate on inland lakes and waterways. Consequently, the properties of both fresh and sea water are of interest to the propulsion engineer.

This chapter considers the nature and physical properties of both fresh and sea water. The treatment, however, is brief as the subject of water properties is adequately covered by other standard texts on fluid mechanics[1-2] and oceanography.[3-4] As a consequence, the information presented here is intended to be both an *aide-memoire* to the reader of more detailed texts and a condensed source of reference material for the practicing designer and engineer.

4.1 DENSITY OF WATER

The density of sea water is a variable. Density increases with increases in salinity or pressure and with decreases in temperature. Figure 4.1 shows the relationship between density, temperature and salinity. From the figure it can be seen that temperature has a greater influence on density at a given salinity in the higher-temperature than in the lower-temperature regions. Conversely, at lower temperatures it is the salinity which has the greater effect on density since the isopleths run more nearly parallel to the temperature axis in these lower-temperature regions.

Density can normally be expected to increase with depth below the free surface. In tropical regions of the Earth a thin layer of low-density surface water is separated from the higher-density deep water by a zone of rapid density change, as seen in Figure 4.2. In the higher latitudes this change is considerably less marked. Furthermore, it

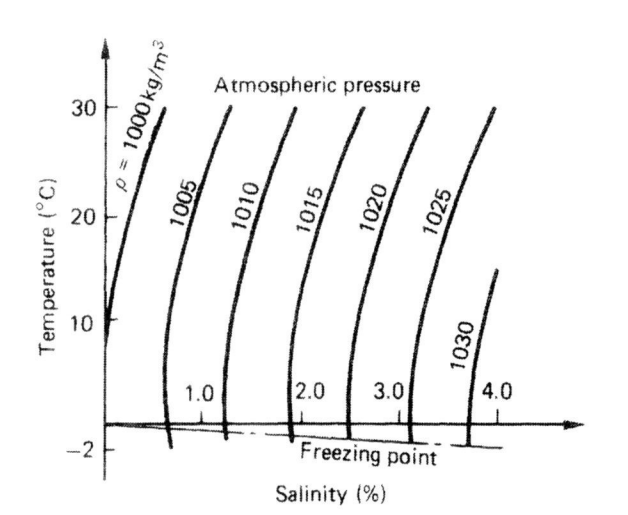

FIGURE 4.1 Variation of density with salinity and temperature at atmospheric pressure.

will be noted that the density deep in the ocean, below a depth of about 2000 m, is more or less uniform at 1027.9 kg/m^3 for all latitudes. At the surface, however, the average density varies over a range between about 1022 kg/m^3 near the equator to 1027.5 kg/m^3 in the southern latitudes, as seen in Figure 4.3. Also shown in this diagram are the average relationships of temperature and salinity for differing latitudes, from which an idea of the global variations can be deduced.

When designing propellers for ocean-going surface ships it is usual to consider a standard salinity value of 3.5 per cent. For these cases the associated density changes with temperature are given in Table 4.1.

The corresponding density versus temperature relationship for fresh water is shown in Table 4.2.

Marine Propellers and Propulsion, Third Edition.
Copyright © 2012 John Carlton. Published by Elsevier Ltd. All rights reserved.

47

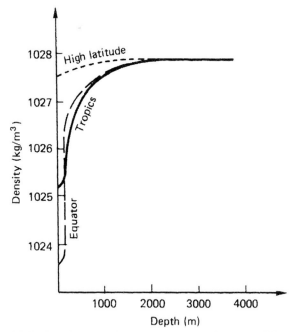

FIGURE 4.2 Typical variation of depth versus density for different global latitudes. *Reprinted with kind permission from Pergamon Press (Reference[3]).*

4.2 SALINITY

With the exception of those areas of the world where fresh water enters the sea, the salinity of the oceans generally lies between 3.4 and 3.6 per cent with an average value of 3.47 per cent by weight. Figure 4.3 indicates the average surface layer variation over the world. From this figure it can be seen that salinity is lowest near the poles, due to the influence of the polar caps, and reaches a double maximum in the region of the tropics.

It will be found that slightly higher than average values of salinity are found where evaporation rates are high, for example in the Mediterranean Sea or in the extreme case of the Dead Sea. Conversely, lower values will be found where melting ice is present or abnormally high levels of precipitation occur.

Salinity is variable with depth. In deep water the salinity is comparatively uniform and varies only between about 3.46 and 3.49 per cent.

Six principal elements account for just over 99 per cent of the dissolved solids in sea water

Chlorine	Cl^-	55.04%
Sodium	Na^+	30.61%
Sulphate	SO_4^{-2}	7.68%
Magnesium	Mg^{+2}	3.69%
Calcium	Ca^{+2}	1.16%
Potassium	K^+	1.10%
		99.28%

The relation between salinity and chlorinity was assessed in the 1960s and is taken as:

$$\text{Salinity} = 1.80655 \times \text{Chlorinity} \quad (4.1)$$

By measuring the concentration of the chlorine ion, which accounts for 55 per cent of the dissolved solids as seen above, the total salinity can be deduced from equation (4.1). The average chlorinity of the oceans is 1.92 per cent which then, from equation (4.1), gives an average salinity of 3.47 per cent.

The definition given in equation (4.1) is termed the 'absolute salinity'; however, this has been superseded by the term 'practical salinity', which is based on the electrical

FIGURE 4.3 Variation of surface temperature, salinity and density with latitude — average for all oceans. *Reprinted with kind permission from Pergamon Press (Reference[3]).*

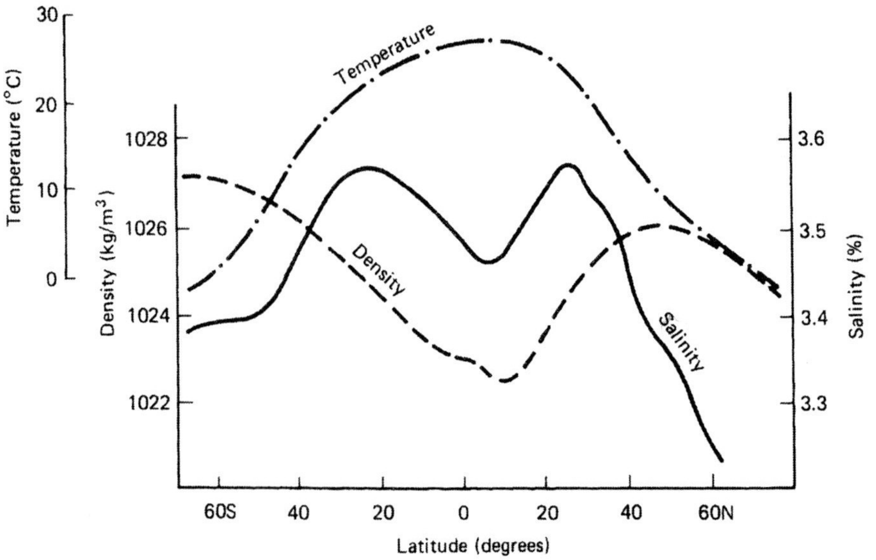

TABLE 4.1 Density Variations with Temperature (Salinity 3.5%)							
Temperature (°C)	0	5	10	15	20	25	30
Density (kg/m³)	1028.1	1027.7	1026.8	1025.9	1024.7	1023.3	1021.7

TABLE 4.2 Density Variations with Temperature (Fresh Water)							
Temperature (°C)	0	5	10	15	20	25	30
Density (kg/m³)	999.8	999.9	999.6	999.0	998.1	996.9	995.6

conductivity of sea water, since most measurements of salinity are based on this property.

4.3 WATER TEMPERATURE

The distribution of surface temperature of the ocean is zonal with lines of constant temperature running nearly parallel to the equator in the open sea. Near the coast these isotherms deflect due to the action of currents. The open sea surface temperature varies from values as high as 28°C just north of the equator down to around −2°C near the ice in the high latitudes (Figure 4.3).

The principal exchange of heat energy occurs at the air–water boundary. Surface heating is not a particularly efficient process, since convection plays little or no part in the mixing process, with the result that heating and cooling effects rarely extend below about two or three hundred meters below the surface of the sea. Consequently, below the surface the ocean can be divided broadly into three separate zones which describe its temperature distribution. First, there is an upper layer, at between 50 and 200 m below the surface, where the temperatures correspond to those at the surface. Second, there is a transition layer where the temperature drops rapidly; this layer extends down to perhaps 1000 m, and then finally there is the deep ocean region where temperature changes very slowly with depth. A typical temperature profile for low latitudes might be: 20°C at the surface; 8°C at 500 m; 5°C at 1000 m and 2°C at 4000 m.

Pickard and Emery[3] publish statistics relating to ocean water temperatures and salinities. These are reproduced here since they are useful for guidance purposes:

1. Seventy five per cent of the total volume of the oceans' water has properties within the range 0°C–6°C in temperature and 3.4 per cent to 3.5 per cent in salinity.
2. Fifty per cent of the total volume of the oceans has properties between 1.3°C and 3.8°C in temperature and salinity between 3.46 per cent and 3.47 per cent.

3. The mean temperature of the world's oceans is 3.5°C and the mean salinity is 3.47 per cent.

4.4 VISCOSITY

The resistance to the motion of one layer of fluid relative to an adjacent layer is termed the viscosity of the fluid. Consequently, relative motion between different layers in a fluid requires the presence of shear forces between the layers, which themselves are nominally parallel to the layers in the fluid.

Consider the velocity gradient shown in Figure 4.4, in which two adjacent layers in the fluid are moving with velocities u and $u + \delta u$. In this case the velocity gradient

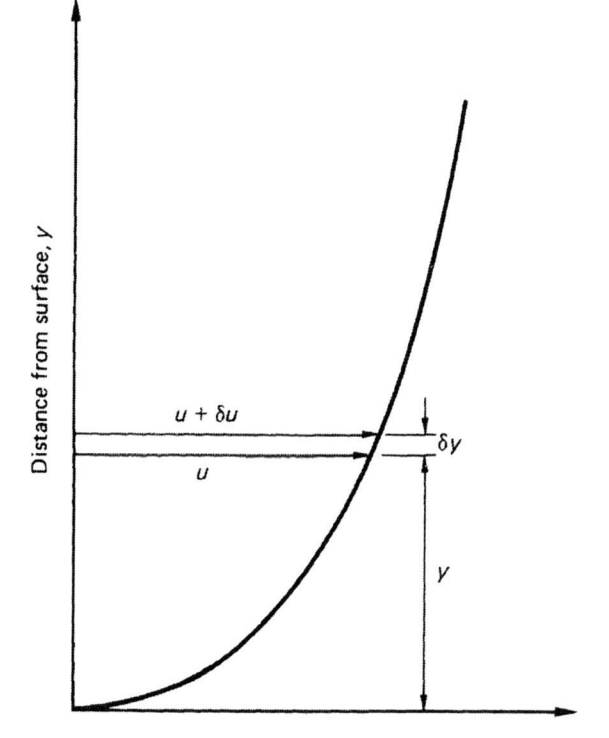

FIGURE 4.4 Typical viscous velocity gradient.

TABLE 4.3 Viscosity of Sea Water with Temperature (Salinity 3.5%)

Temperature (°C)	0	5	10	15	20	25	30
Kinematic viscosity 10^6 (m^2/s)	1.8284	1.5614	1.3538	1.1883	1.0537	0.9425	0.8493

TABLE 4.4 Viscosity of Fresh Water with Temperature

Temperature (°C)	0	5	10	15	20	25	30
Kinematic viscosity 10^6 (m^2/s)	1.7867	1.5170	1.3064	1.1390	1.0037	0.8929	0.8009

between these two layers, distance δy apart, is $\delta u/\delta y$; or $\partial u/\partial y$ in the limit. Because the layers are moving with different velocities, there will be shear forces between the layers giving rise to a shear stress τ_{yx}. Newton postulated that the tangential stress between the layers is proportional to the velocity gradient:

$$\tau_{yx} = \mu \frac{\partial u}{\partial y} \quad (4.2)$$

where μ is a constant of proportionality known as the dynamic coefficient of viscosity of the fluid. Fluids which behave with a constant coefficient of viscosity, that is independent of the velocity gradient, are termed Newtonian fluids: both fresh water and sea water behave in this way. Some drag reduction fluid additives such as long chain polymers, however, have far from constant coefficients of viscosity and are thus termed non-Newtonian fluids.

In the majority of problems concerning propeller technology we are concerned with the relationship of the fluid viscous to inertia forces as expressed by the flow Reynolds number. To assist in these studies, use is made of the term kinematic viscosity (ν) which is the ratio μ/ρ, since the viscous forces are proportional to the viscosity μ and the inertia forces to the density ρ.

For the purposes of propeller design and analysis, the values of the kinematic viscosity for sea and fresh water are given by Tables 4.3 and 4.4 respectively.

4.5 VAPOR PRESSURE

At the free surface of the water there is a movement of water molecules both in and out of the fluid. Just above the surface the returning molecules create a pressure which is known as the partial pressure of the vapor. This partial pressure, together with the partial pressures of the other gases above the liquid, makes up the total pressure just above the surface of the water. The molecules leaving the water generate the vapor pressure whose magnitude is determined by the rate at which molecules escape from the surface. When the rates of release and return of the molecules from the water are the same, the air above the water is said to be saturated and the vapor pressure equals the partial pressure of the vapor: at this condition the value of the vapor pressure is the saturation pressure. Furthermore, the vapor pressure varies with temperature since temperature influences the energy of the molecules and hence their ability to escape from the surface. If the saturation pressure increases above the total pressure acting on the fluid surface then molecules escape from the water very rapidly and the phenomenon known as boiling occurs. In this condition bubbles of vapor are formed in the liquid itself and then rise to the surface.

A similar effect to boiling occurs if the water contains dissolved gases, since when the pressure is reduced the dissolved gases are released in the form of bubbles. The reduction in pressure required for the release of bubbles is, however, less than that which will cause the liquid to boil at the ambient temperature. Within a fluid the pressure cannot generally fall below the vapor pressure at the temperature concerned since the liquid will then boil and small bubbles of vapor will form in large numbers.

Table 4.5 gives the values of the saturation vapor pressure of both sea and fresh water for a range of temperatures relevant to propeller technology.

4.6 DISSOLVED GASES IN SEA WATER

The most abundant dissolved gases which are found throughout the whole mass of the ocean are nitrogen, oxygen and argon. Additionally, there are traces of many other inert gases.

The Propeller Environment

TABLE 4.5 Saturation Vapor Pressure p_v for Fresh and Sea Water

Temperature (°C)	0.01	5	10	15	20	25	30
Fresh water p_v (Pa)	611	872	1228	1704	2377	3166	4241
Sea water p_v (Pa)	590	842	1186	1646	2296	3058	4097

The quantities of these gases which are dissolved in the ocean are a function of salinity and temperature, with the greatest amounts being found in the cooler, less saline regions. At depth all gases with the exception of oxygen tend to be retained in the saturated state by the water as it sinks from the ocean surface. In these cases it is found that the gas concentrations change little with geographic location. At the surface, the oxygen concentration is normally of the order of 0.1–0.6 per cent with values on occasion rising as high as 1 per cent. Furthermore, at the surface the water is usually very close to being saturated and consequently is sometimes found to be supersaturated in the upper 15 m or so due to photosynthesis by marine plants. Below this level oxygen tends to get consumed by living organisms and the oxidation of detritus.

When undertaking cavitation studies, particularly at model scale, it is pertinent to ask what is the correct nuclei content of the tunnel water in order to achieve realistic sea conditions. Much work has been done on this subject and Figure 4.5 shows a range of measured nuclei distributions from different sources for ocean and tunnel conditions. Weitendorf and Keller also conducted a series of nuclei distribution measurements using laser techniques on board the *Sydney Express* in 1978 as part of a much wider cavitation study. They found that the number of nuclei per unit volume having radii greater then 1 μm was broadly in agreement with the levels established by oceanographers; however, they recorded little in the way of smaller particles on these trials. In general terms, however, nuclei distribution measurements must be considered in the context of both weather and seaway and also of shallow or deep water.

In the case of full-scale cavitation studies it is sometimes quite noticeable that for nominally constant powering and sea conditions the cavitation characteristics can vary quite significantly with time. This is dependent on the spatial variations of the dissolved gas content in the water.

4.7 SURFACE TENSION

Although the subject of surface tension is normally considered to be more in the province of physicists, it does have relevance when considering the bubble dynamics and ventilations associated with cavitation.

A molecule has associated with it a 'sphere of influence' within which it attracts other molecules; this attraction is known as molecular attraction and is distinct from the gravitational attraction found between any two objects. The molecular attraction forces do not extend further than three or four times the average distance between molecules. To appreciate how surface tension forces arise consider the two molecules A and B shown in Figure 4.6. Molecule A, which is in the body of the fluid, exerts and receives a uniform attraction from all directions. However, molecule B, which is at the surface, receives its major attraction from within the fluid and so experiences a net inward force F: it is assumed here that we are considering a boundary between water and air or a vapor. This net inward force on the surface molecules increases the pressure on the main bulk of the liquid and hence needs to be balanced in order to keep the molecules in equilibrium. If the area of liquid surface increases, the number of molecules constituting that surface must also increase, and the molecules will arrive at the surface against the action of the inward force. Mechanical work is, therefore, expended in increasing the liquid surface area, which implies the existence of a tensile force in the surface.

Table 4.6 gives an indication of the values of surface tension for both fresh and sea water. However, in applying these values, it must be remembered that they can be considerably influenced by small quantities of additives, for example, detergent. In practice they can change by as much as 0.022 N/m because of contamination with oily matter.

4.8 WEATHER

The weather, or more fundamentally the air motion, caused by the dynamics of the Earth's atmosphere, influences marine propulsion technology by giving rise to additional resistance caused by both the wind and resulting disturbances to the sea surface.

The principal physical properties of air which are of concern are density and viscosity. The density at sea level for dry air is given by the relationship:

$$\rho = 0.4647 \left[\frac{p}{T}\right] \text{ kg/m}^3 \qquad (4.3)$$

where p is the barometric pressure (mmHg) and T is the local temperature (K).

For the viscosity of the air use can be made of the following relationship for dry air:

$$\mu = 170.9 \times 10^{-7} \left[\frac{393}{120 + T}\right] \left(\frac{T}{273}\right)^{3/2} \text{ Ns/m}^2 \qquad (4.4)$$

where T is the temperature (K).

When the wind blows over a surface the air in contact with the surface has no relative velocity to that surface. Consequently, a velocity gradient exists close to the solid

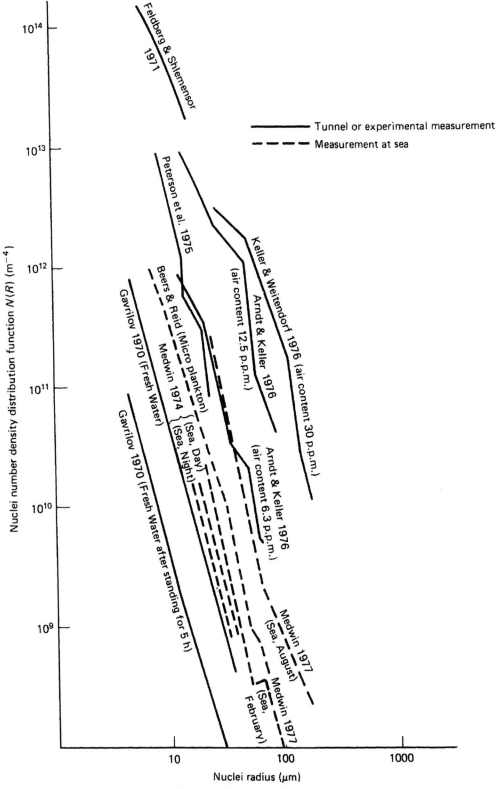

FIGURE 4.5 Nuclei density distribution.

Chapter | 4 The Propeller Environment

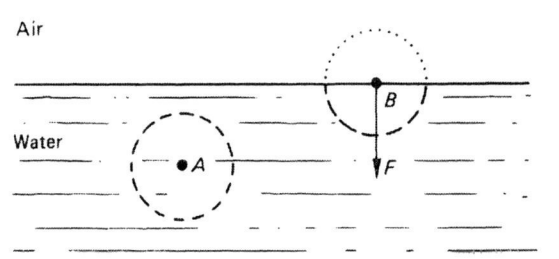

FIGURE 4.6 Molecular explanation of surface tension.

TABLE 4.6 Typical Values of Surface Tension for Sea and Fresh Water with Temperature

Temperature (°C)	0	5	10	15	20	25	30
Sea water							
(dynes/cm)	76.41	75.69	74.97	74.25	73.55	72.81	72.09
Fresh water							
(dynes/cm)	75.64	74.92	74.20	73.48	72.76	72.04	71.32

1 dyne = 10^{-5} N.

boundary in which the relative velocity of successive layers of the wind increases until the actual wind speed in the free stream is reached (Figure 4.7). Indeed the flow pattern is analogous to the boundary layer velocity distribution measured over a flat plate. To overcome problems of definition in wind speed due to surface perturbations it is normal practice to measure wind speed at a height of 10 m above the surface of either the land or the sea: this speed is often referred to as the '10 meter wind' (Figure 4.7).

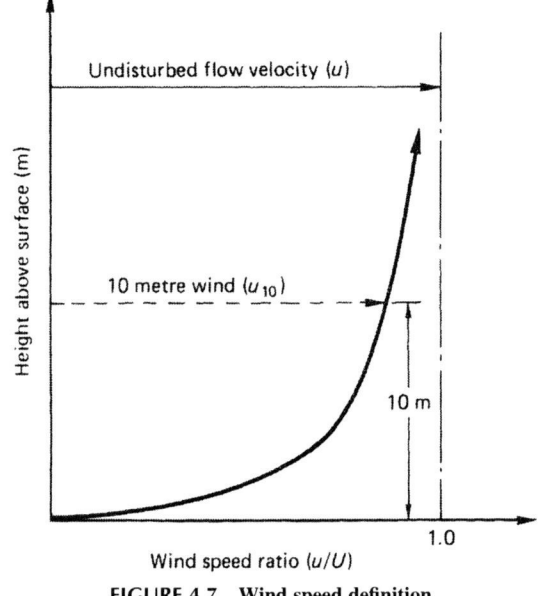

FIGURE 4.7 Wind speed definition.

As well as recording wind velocities, wind conditions are often related to the Beaufort scale, which was initially proposed by Admiral Beaufort in 1806. This scale has also been extended to give an indication of sea conditions for fully developed seas. The scale is not accurate enough for very detailed studies, since it was primarily intended as a guide to illustrate roughly what might be expected in the open sea. Nevertheless, the scale is sufficient for many purposes, both technical and descriptive; however, great care should be exercised if it is used in the reverse way, that is, for logging or reporting the state of the sea, since significant errors can be introduced into the analysis. This is particularly true in confined and restricted sea areas, such as the North Sea or English Channel, since the sea generally has two components: a surface perturbation and an underlying swell component, both of which may have differing directional bearings. Table 4.7 defines the Beaufort scale up to Force 12. Above Force 12 there are further levels defined: 13, 14, 15, 16 and 17, with associated wind-speed bands of 72—80, 81—89, 90—99, 100—108 and 109—118 knots, respectively. For these higher states descriptions generally fail except to note that conditions become progressively worse.

Until comparatively recently the only tools available to describe the sea conditions were, for example, the Beaufort scale, which as discussed relates overall sea state to observed wind, and formulae such as Stevens' formula:

$$Z = 1.5\sqrt{F} \qquad (4.5)$$

where Z is the maximum wave height in feet and F is the fetch in miles.

However, from wave records it is possible to statistically represent the sea. Using these techniques an energy spectrum indicating the relative importance of the large number of different component waves can be produced for a given sea state. Figure 4.8 shows one such example, for illustration purposes, based on the Neumann spectrum for different wind speeds and for fully developed seas. From Figure 4.8 it will be seen that as the wind speed increases, the frequency about which the maximum spectra energy is concentrated, termed the modal frequency f_0, is reduced. Many spectra have been advanced by different authorities and these will give differing results; partly because of the dependence of wave energy on the wind duration and fetch which leads to the problem of defining a fully developed sea. When the wind begins to blow short, low amplitude waves are initially formed. These develop into larger and longer waves if the wind continues to blow for a longer period of time. This leads to a time-dependent set of spectra for different wind duration, as seen in Figure 4.9. An analogous, but opposite, situation is seen when the wind dies down as the longer waves, due to their greater velocity, move out of the area, leaving only the smaller shorter waves. For continuous spectra the area

TABLE 4.7 The Beaufort Wind Scale

Number	Wind Speed at 10 m (knots)	Wind Description	Probable Mean Wave Height (m)	Noticeable Effect of Wind on Land	At Sea
0	Less than 1	Calm	None	Smoke vertical; flags still	Sea like a mirror.
1	1–3	Light air	<0.1	Smoke drifts; vanes static	Ripples with the appearance of scales are formed, but without foam crests.
2	4–6	Light breeze	0.2	Wind felt on face; leaves, flags rustle; vanes move	Small wavelets, still short but more pronounced; crests have a glassy appearance and do not break.
3	7–10	Gentle breeze	0.6	Leaves and twigs in motion; light flags extended	Large wavelets. Crests begin to break. Foam of glassy appearance perhaps scattered white horses.
4	11–16	Moderate breeze	1.0	Raises dust; moves small branches	Small waves, becoming longer; fairly frequent white horses.
5	17–21	Fresh breeze	1.9	Small trees sway	Moderate waves, taking a more pronounced long form, many white horses formed (chance of some spray).
6	22–27	Strong breeze	2.9	Large branches move; telephone wires 'sing'	Large waves begin to form; the white foam crests are more extensive everywhere (probably some spray).
7	28–33	Moderate gale	4.1	Whole trees in motion	Sea heaps up and white foam from breaking waves begins to be blown in streaks along the direction of the wind (spindrift begins to be seen).
8	34–40	Fresh gale	5.5	Twigs break off; progress impeded	Moderately high waves of greater length; edge of crests break into spindrift. The foam is blown in well-marked streaks along the direction of the wind.
9	41–47	Strong gale	7.0	Chimney pots removed	High waves. Dense streaks of foam along the direction of the wind. Sea begins to roll. Spray may affect visibility.
10	48–55	Whole gale	8.8	Trees uprooted; structural damage	Very high waves with long, overhanging crests. The resulting foam in great patches is blown in dense white streaks along the direction of the wind. On the whole, the surface of the sea takes a white appearance. The rolling of the sea becomes heavy and shock like. Visibility is affected.
11	56–64	Storm	11.0	Widespread damage	Exceptionally high waves. (Small- and medium-sized ships might, for a long time, be lost to view behind the waves.) The sea is completely covered with long white patches of foam lying along the direction of the wind. Everywhere the edges of the wave crests are blown into froth. Visibility is affected.
12	65–71	Hurricane	Over 13.0	Countryside devastated	The air is filled with foam and spray. Sea completely white with driving spray; visibility very seriously affected.

The Propeller Environment

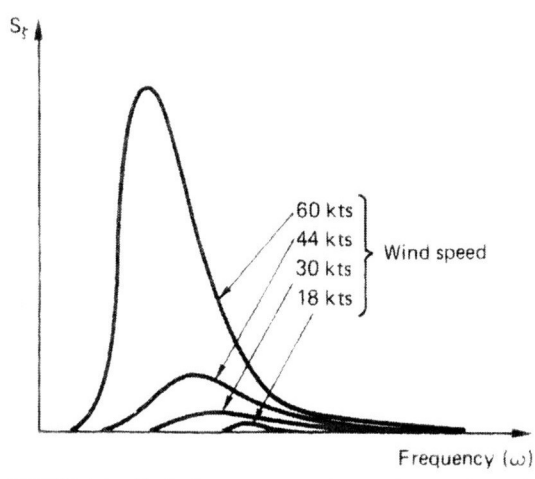

FIGURE 4.8 Typical wave spectra for varying wind speed.

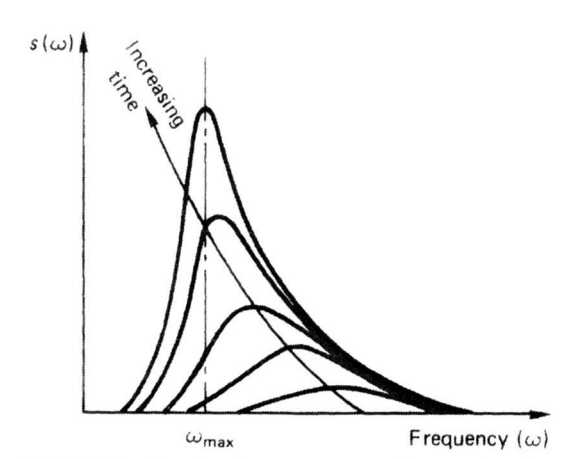

FIGURE 4.9 Growth of a wave spectra with wind duration.

under the spectrum can be shown to be equal to the mean square of the surface elevation of the water surface.

In order to study the effects of waves the energy spectrum concept provides the most convenient and rigorous approach. However, for many applications, the simpler approach of appealing directly to wave data will suffice. Typical of such data is that given by Darbyshire[5] or more recently that produced by Hogben et al.[6] which provides a wave atlas based on some 55 million visual observations from ships during the period 1854–1984. Furthermore, the World Meteorological Organization (WMO) produced a standard sea state code in 1970; this is reproduced in Table 4.8. In the context of this table, the significant wave height is the mean value of the highest third of a large number of peak–trough wave heights. It should, however, be noted that wave period does not feature in this well-established sea state definition.

4.9 SILT AND MARINE ORGANISMS

The sea, and indeed fresh water, contains a quantity of matter in suspension. This matter is of the form of small particles of sand, detritus and marine animal and vegetable life.

Particulate matter such as sand will eventually separate out and fall to the sea bottom; however, depending on its size this separation process may be measured in either hours or months. Therefore, the presence of abrasive particles must always be considered, especially in areas, such as the North Sea, which have shallow sandy bottom seas.

TABLE 4.8 World Meteorological Organization (WMO) Sea State Code

| Sea State Code | Significant wave height (m) | | Description |
	Range	Mean	
0	0	0	Calm (glassy)
1	0–0.1	0.05	Calm (rippled)
2	0.1–0.5	0.30	Smooth (wavelets)
3	0.5–1.25	0.875	Slight
4	1.25–2.5	1.875	Moderate
5	2.5–4.0	3.250	Rough
6	4.0–6.0	5.000	Very rough
7	6.0–9.0	7.500	High
8	9.0–14.0	11.500	Very high
9	Over 14.0	Over 14.00	Phenomenal

TABLE 4.9 Port Classification for Fouling[7]

Clean Ports	Fouling Ports		Cleaning Ports	
	Light	Heavy	Non-scouring	Scouring
Most UK Ports	Alexandria	Freetown	Bremen	Calcutta
Auckland	Bombay	Macassar	Brisbane	Shanghai
Cape Town	Colombo	Mauritius	Buenos Aires	Yangtze Ports
Chittagong	Madras	Rio de Janeiro	E. London	
Halifax	Mombasa	Sourabaya	Hamburg	
Melbourne	Negapatam	Lagos	Hudson Ports	
Valparaiso	Karadii		La Plata	
Wellington	Pernambuco		St Lawrence Ports	
Sydney*	Santos		Manchester	
	Singapore			
	Suez			
	Tuticorin			
	Yokohama			

*Variable conditions.

Marine animal and vegetable life covers a wide, indeed almost boundless, variety of organisms. Of particular interest to the propulsion engineer are algae, barnacles, limpets, tubeworms and weed, since these all act as fouling agents for both the hull and propeller. Christie[7] distinguishes between two principal forms of fouling: algae and animal fouling. The latter form of fouling requires the development and establishment of larvae over a period of several days, whereas algae fouling results in a slime which can take only a matter of hours to form. These growths are of course dependent on temperature, salinity and concentrations of marine bacteria in the water. Whilst no direct estimates of fouling rates are available, Evans and Svensen[8] conducted a survey which showed those areas of the world which are more prone to the fouling of hulls and propellers. Table 4.9 summarizes their findings.

REFERENCES AND FURTHER READING

1. Massey BA. *Mechanics of Fluids*. London: Van Nostrand Reinhold; 2005.
2. Anderson JD. *Fundamentals of Aerodynamics*. Berkshire, UK: McGraw-Hill; 1985.
3. Pickard GL, Emery WJ. *Descriptive Physical Oceanography, An Introduction*. Oxford: Pergamon Press; 1982.
4. Thurman V. *Introductory Oceanography*. Columbus: Charles E. Merrill; 1975.
5. Darbyshire J. *Wave statistics in the North Atlantic Ocean and on the coast of Cornwall*. The Marine Observer April 1955;**25**.
6. Hogben, Dacuna, Olliver. *Global Wave Statistics*. Teddington, UK: BMT; 1986.
7. Christie AO. *IMAEM*; 1981.
8. Evans JP, Svensen TE. *Voyage simulation model and its application to the design and operation of ships*. Trans. RINA; November 1987.

Chapter 5

The Ship Wake Field

Chapter Outline

5.1 General Wake Field Characteristics	57	**5.3 The Nominal Wake Field**	61
5.2 Wake Field Definition	59	**5.4 Estimation of Wake Field Parameters**	62
5.2.1 Velocity Ratio Method	59	**5.5 Effective Wake Field**	65
5.2.2 Taylor's Method	59	**5.6 Wake Field Scaling**	67
5.2.3 Froude Method	59	**5.7 Wake Quality Assessment**	70
5.2.4 Mean Velocity or Wake Fraction	60	**5.8 Wake Field Measurement**	72
5.2.5 Fourier Analysis of Wake Field	60	**References and Further reading**	77

A body, by virtue of its motion through the water, causes a wake field in the sense of an uneven flow velocity distribution to occur behind it; this is true whether the body is a ship, a submarine, a remotely operated vehicle or a torpedo. The wake field at the propulsor plane arises from three principal causes: the streamline flow around the body, the growth of the boundary layer over the body and the influence of any wave-making components. The latter effect naturally is dependent upon the depth of immersion of the body below the water surface. Additionally, and equally important, is the effect that the propulsor has on modifying the wake produced by the propelled body.

5.1 GENERAL WAKE FIELD CHARACTERISTICS

The wake field is strongly dependent on ship type and so each vessel may be considered to have a unique wake field. Figure 5.1 shows three wake fields for different ships. Figure 5.1(a) relates to a single-screw bulk carrier form in which a bilge vortex can be seen to be present and dominates the flow in the thwart-ship plane of the propeller disc. The flow field demonstrated by Figure 5.1(b) relates again to a single-screw vessel, but in this case to a fairly fast and fine lined vessel having a 'V'-formed after body unlike the 'U'-form of the bulk carrier shown in Figure 5.1(a). In Figure 5.1(b) it is seen, in contrast to the wake field produced by the 'U'-form hull, that a high-speed axial flow field exists for much of the propeller disc except for the sector embracing the top dead center location, where the flow is relatively slow and in some extreme cases may even reverse in direction. Definitions of 'U'- and 'V'-form hulls

are shown in Figure 5.2; however, there is no 'clear-cut' transition from one form to another, and Figure 5.1(a) and (b) represent extremes of both hull form types. Both of the flow fields discussed so far relate to single-screw hull forms and, therefore, might be expected to exhibit a reflective symmetry about the vertical center plane of the vessel. In practice, however, if asymmetric arrangements of outflows from the ship exist, for example cooling water outlets, then this reflective symmetry may be disturbed.

For a twin-screw vessel no such symmetry naturally exists, as seen by Figure 5.1(c), which shows the wake field for a twin-screw ferry. In this figure the location of the shaft supports, in this case 'A' brackets, is clearly seen, but due to the position of the shaft lines relative to the hull form, symmetry of the wake field across the 'A' bracket center line cannot be maintained. Indeed, specific attention needs to be paid to the design of the shaft supports, whether these are 'A' or 'P' brackets, bossings or gondolas, so that the flow does not become too disturbed or retarded in these locations. If this is not done vibration and noise may arise and be difficult to solve satisfactorily. This general concept is also of equal importance for single-, twin- or triple-screw ships.

It is of interest to note how the parameter ϕ, Figure 5.2, tends to influence the resulting wake field at the propeller disc of a single-screw ship. For the V-form hull (Figure 5.1(b)), one immediately notes the very high wake peak at the top dead center position of the propeller disc and the comparatively rapid transition from the 'dead-water' region to the near free stream conditions in the lower part of the disc. This is caused by the water coming from under the bottom of the ship and flowing around the curvature of the hull, so that the fluid elements which were close to the

---

Marine Propellers and Propulsion, Third Edition.
Copyright © 2012 John Carlton. Published by Elsevier Ltd. All rights reserved.

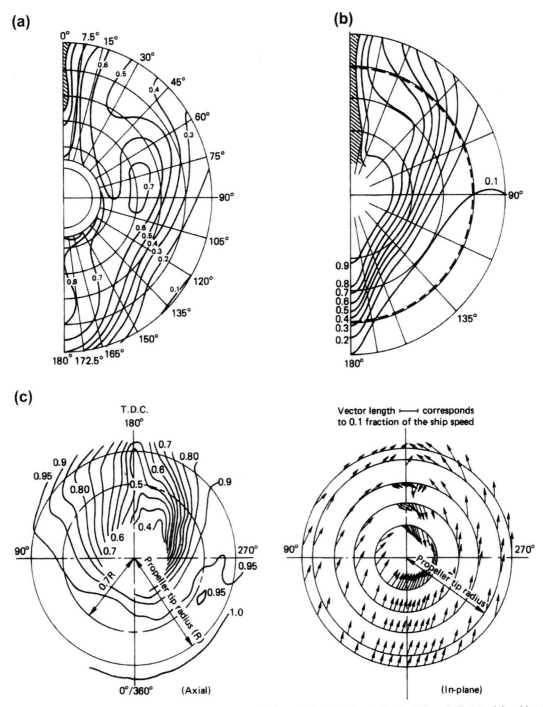

FIGURE 5.1 Typical wake field distributions: (a) axial wake field — U-form hull; (b) axial wake field — V-form hull; (c) axial and in-plane wake field — twin-screw hull. *Parts (a) and (b) reproduced with permission from[1].*

hull, and thus within its boundary layer, also remain close to the hull around the bilge and flow into the propeller close to the center plane. Consequently, a high wake peak is formed in the center plane of the propeller disc.

The alternative case of a wake field associated with an extreme U-form hull is shown in Figure 5.1(a); here the flow pattern is completely different. The water flowing from under the hull is in this case unable to follow the rapid change of curvature around the bilge and, therefore, separates from the hull surface. These fluid elements then flow upwards into the outer part of the propeller disc and the region above this separated zone is then filled with water flowing from above: this creates a downward flow close to the hull surface. The resultant downward flow close to the hull and upward flow

The Ship Wake Field

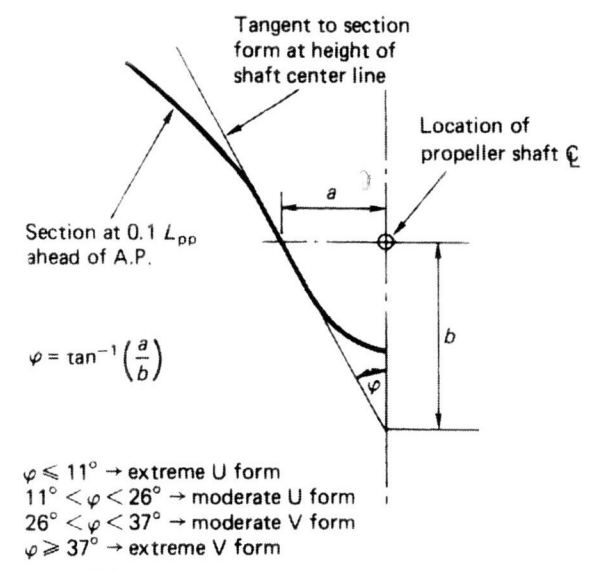

$\varphi = \tan^{-1}\left(\dfrac{a}{b}\right)$

$\varphi \leqslant 11° \rightarrow$ extreme U form
$11° < \varphi < 26° \rightarrow$ moderate U form
$26° < \varphi < 37° \rightarrow$ moderate V form
$\varphi \geqslant 37° \rightarrow$ extreme V form

FIGURE 5.2 Definition of U- and V-form hulls.

distant from the hull give rise to a rotational motion of the flow into the propeller disc which is termed the bilge vortex. The bilge vortex, therefore, is a motion which allows water particles in the boundary layer to be transported away from the hull and replaced with water from outside the boundary layer; the effect of this is to reduce the wake peak at the center plane of the propeller disc.

Over the years, in order to help designers produce acceptable wake fields for single-screw ships, several hull form criteria have been proposed, as outlined, for example, in.[1,2] Criteria of these types can be reduced to a series of guidelines, such as:

1. The angle of run of the waterlines should be kept to below 27°–30° over the entire length of run. Clearly it is useless to reduce the angle of run towards the stern post if further forward the angles increase to an extent which induces flow separation.
2. The stern post width should not exceed 3 per cent of the propeller diameter in the ranges 0.2–0.6R above the shaft center line.
3. The angle of the tangent to the hull surface in the plane of the shaft center line (see Figure 5.2) should lie within the range 11°–37°.

The detailed flow velocity fields of the type shown in Figure 5.1 and used in propeller design are almost without exception derived from model tests. Today it is still the case that some 80–85 per cent of all ships that are built do not have the benefit of a model wake field test.

The advent of computational fluid dynamics capabilities and experience in their use for ship boundary layer calculations has shown that for many hull forms it is possible to derive ship and model scale wake fields to an acceptable degree of accuracy.

5.2 WAKE FIELD DEFINITION

In order to make use of the wake field data it needs to be defined in a suitable form. There are three principal methods: the velocity ratio, Taylor and Froude methods, although today the method based on Froude's wake fraction is rarely, if ever, used. The definitions of these methods are as follows.

5.2.1 Velocity Ratio Method

Here the iso-velocity contours are expressed as a proportion of the ship speed (V_s) relative to the far-field water speed. Accordingly, water velocity at a point in the propeller disc is expressed in terms of its axial, tangential and radial components, v_a, v_t and v_r, respectively:

$$\frac{v_a}{V_s}, \quad \frac{v_t}{V_s} \quad \text{and} \quad \frac{v_r}{V_s}$$

Figure 5.1(c) is expressed using above velocity component definitions. The velocity ratio method has today become perhaps the most commonly used method of wake field representation, due first to the relative conceptual complexities the other, and older, representations have in dealing with the in-plane propeller components, and second, the velocity ratios are more convenient for data input into analytical procedures.

5.2.2 Taylor's Method

In this characterization the concept of 'wake fraction' is used. For axial velocities the Taylor wake fraction is defined as:

$$w_T = \frac{V_s - v_A}{V_s} = 1 - \left(\frac{v_A}{V_s}\right) \qquad (5.1)$$

that is, one minus the axial velocity ratio or, alternatively, it can be considered as the loss of axial velocity at the point of interest when compared to the ship speed and expressed as a proportion of the ship speed. For the other in-plane velocity components we have the following relationships:

$$w_{T_t} = 1 - \left(\frac{v_t}{V_s}\right) \quad \text{and} \quad w_{T_r} = 1 - \left(\frac{v_r}{V_s}\right)$$

However, these forms are rarely seen today, and preference is generally given to expressing the tangential and radial components in terms of their velocity ratios v_t/V_s and v_r/V_s.

Notice that in the case of the axial components the subscript 'a' is omitted from w_T.

5.2.3 Froude Method

This is similar to the Taylor characterization, but instead of using the vehicle speed as the reference velocity the Froude notation uses the local velocity at the point of interest. For example, in the axial direction we have:

$$w_F = \frac{V_S - v_a}{v_a} = \left(\frac{V_s}{v_a}\right) - 1$$

For the sake of completeness it is worth noting that the Froude and Taylor wake fractions can be transformed as follows:

$$w_F = \frac{w_T}{1 - w_T} \quad \text{and} \quad w_T = \frac{w_F}{1 + w_F}$$

5.2.4 Mean Velocity or Wake Fraction

The mean axial velocity within the propeller disc is found by integrating the wake field on a volumetric basis of the form:

$$\left.\begin{aligned}\overline{W}_T &= \frac{\int_{r_h}^{R} r \int_0^{2\pi} w_T \, d\phi \, dr}{\pi(R^2 - r_h^2)} \\ \left(\frac{\overline{v}_a}{V_s}\right) &= \frac{\int_{r_h}^{R} r \int_0^{2\pi} \left(\frac{v_a}{V_s}\right) d\phi \, dr}{\pi(R^2 - r_h^2)}\end{aligned}\right\} \quad (5.2)$$

Much debate has centered on the use of the volumetric or impulsive integral form for the determination of mean wake fraction, for example[3,4]; however, modern analysis techniques generally use the volumetric basis as a standard.

5.2.5 Fourier Analysis of Wake Field

Current propeller analysis techniques rely on being able to describe the wake field encountered by the propeller at each radial location in a reasonably precise mathematical way. Figure 5.3 shows a typical transformation of the wake field velocities at a particular radial location of a polar wake field plot, similar to those shown in Figure 5.1, into a mean and fluctuating component. Figure 5.3 then shows diagrammatically how the total fluctuating component can then be decomposed into an infinite set of sinusoidal components of various harmonic orders. This follows from Fourier's theorem, which states that any periodic function can be represented by an infinite set of sinusoidal functions. In practice, however, only a limited set of harmonic components are used, since these are sufficient to define the wake field within both the bounds of calculation and experimental accuracy: typically the first eight to ten harmonics are those which might be used, the exact number depending on the propeller blade number. A convenient way, therefore, of describing the velocity variations at a particular radius in the propeller disc is to use Fourier analysis techniques and to define the problem using the global reference frame discussed in Chapter 3. Using this basis the general approximation of the velocity distribution at a particular radius becomes:

$$\frac{v_a}{V_s} = \sum_{k=0}^{n} \left[a_k \cos\left(\frac{k\phi}{2\pi}\right) + b_k \sin\left(\frac{k\phi}{2\pi}\right) \right] \quad (5.3)$$

Equation (5.3) relates to the axial velocity ratio; similar equations can be defined for the tangential and radial components of velocity.

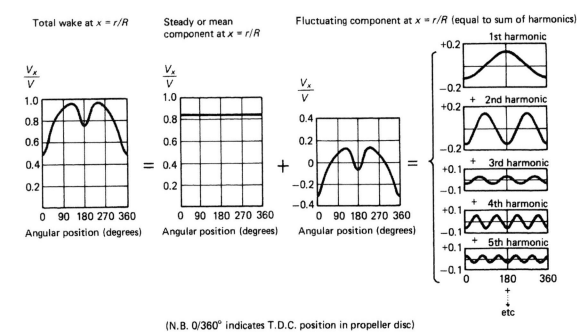

(N.B. 0/360° indicates T.D.C. position in propeller disc)

FIGURE 5.3 Decomposition of wake field into mean and fluctuating components.

5.3 THE NOMINAL WAKE FIELD

The nominal wake field is the wake field that would be measured at the propeller plane without the presence or influence of the propeller modifying the flow at the stern of the ship. The nominal wake field $\{w_n\}$ of a ship can be considered to effectively comprise three components: the potential wake, the frictional wake and the wave-induced wake, so that the total nominal wake field $\{w_n\}$ is given by

$$\{w_n\} = \{w_p\} + \{w_v\} + \{w_w\} + \{\Delta w\} \quad (5.4)$$

where the suffixes denote the above components, respectively, and the curly brackets denote the total wake field rather than values at a particular point. The component $\{\Delta w\}$ is the correlation or relative interaction component representing the non-linear part of the wake field composition.

The potential wake field $\{w_p\}$ is the wake field that would arise if the vessel were working in an ideal fluid, that is one without viscous effects. As such the potential wake field at a particular transverse plane on the body is directly calculable using analytical methods, and it matters not whether the body is moving ahead or astern. Clearly, for underwater bodies, and particularly for bodies of revolution, the calculation procedures are comparatively simpler to use than for surface ship forms. For calculations on ship forms use is made of panel methods which today form the basis of three-dimensional, inviscid, incompressible flow calculations. The general idea behind these methods is to cover the surface with three-dimensional body panels over which there is an unknown distribution of singularities; for example, point sources, doublets or vortices. The unknowns are then solved through a system of simultaneous linear algebraic equations generated by calculating the induced velocity at control points on the panels and applying the flow tangency condition. In recent years many such programs have been developed by various institutes and software houses around the world. For axisymmetric bodies in axial flow a distribution of sources and sinks along the axis will prove sufficient for the calculation of the potential wake.

In contrast to calculation methods an approximation to the potential wake at the propeller plane can be measured by making a model of the vehicle and towing it backwards in a towing tank, since in this case the viscous effects at the propeller plane are minimal.

In general, the potential wake field can be expected to be a small component of the total wake field, as shown by Harvald.[5] Furthermore, since the effects of viscosity do not have any influence on the potential wake, the shape of the forebody does not have any influence on this wake component at the stern.

The frictional wake field $\{w_v\}$ arises from the viscous nature of the water passing over the hull surface. This wake field component derives from the growth of the boundary layer over the hull surface, which, for all practical purposes, can be considered as being predominantly turbulent in nature at full-scale. To define the velocity distribution within the boundary layer it is normal, in the absence of separation, to use a power law relationship of the form:

$$\frac{v}{V} = \left(\frac{y}{\delta}\right)^n$$

where v is the local velocity at a distance y from the boundary surface, V is the free stream velocity and δ is the boundary layer thickness, which is normally defined as the distance from the surface to where the local velocity attains a value of 99 per cent of the free stream velocity. The exponent n for turbulent boundary layers normally lies within the range $1/5 - 1/9$.

A further complication within the ship boundary layer problem is the onset of separation which will occur if the correct conditions prevail in an adverse pressure gradient; that is, a pressure field in which the pressure increases in the direction of flow. Consider, for example, Figure 5.4(a), which shows the flow around some parts of the hull. At station 1 the normal viscous boundary layer has developed; further along the hull at station 2 the velocity of fluid elements close to the surface is less than at station 1, due to

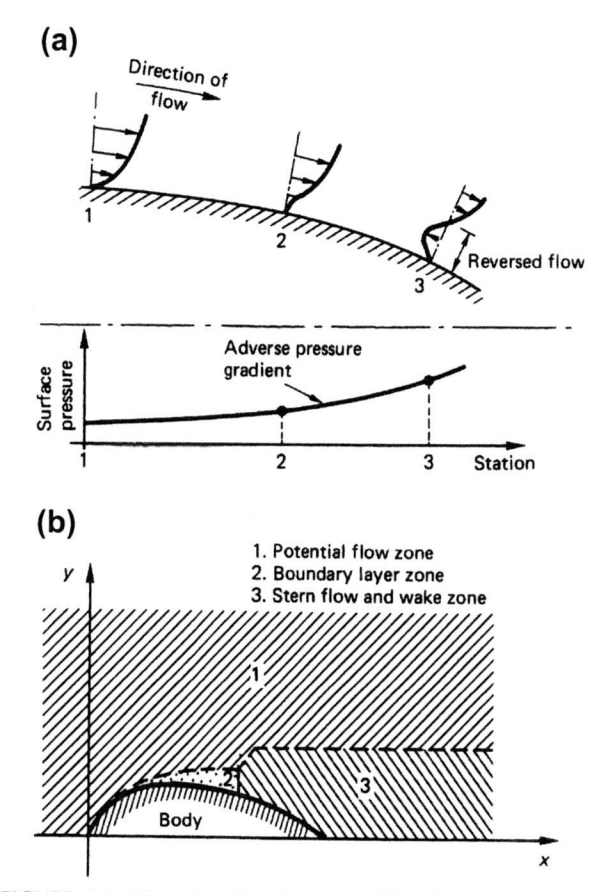

FIGURE 5.4 Flow boundary layer considerations: (a) origin of separated flow and (b) typical flow computational zones.

the steadily increasing pressure gradient. As the elements continue further downstream they may come to a stop under the action of the adverse pressure gradient, and actually reverse in direction and start moving back upstream as seen at station 3. The point of separation occurs when the velocity gradient $\partial v/\partial n = 0$ at the surface, and the consequence of this is that the flow separates from the surface leaving a region of reversed flow on the surface of the body. Re-attachment of the flow to the surface can subsequently occur if the body geometry and the pressure gradient becomes favorable.

The full prediction by analytical means of the viscous boundary layer for a ship form is a procedure which is becoming more common and successful results have in many cases been achieved using Reynolds Averaged Navier Stokes (RANS) codes. For example, Figure 5.5 illustrates an axial velocity prediction, at full-scale, of the nominal wake field for a container ship using 1.5 million cells together with k–ω SST turbulence model. In this case the 'x' positions marked on the figure show the point at which correlation with measurement positions was tested. A typical calculation procedure for a ship form may divide the hull into three primary areas for computation: the potential flow zone, the boundary layer zone and the stern flow and wake zone (Figure 5.4(b)). Today, while a significant number of wake field predictions are made in model tanks, the numbers of numerical predictions is growing at a significant rate.

The wake component due to wave action $\{w_w\}$ is due to the movement of water particles in the system of gravity waves set up by the ship on the surface of the water. Such conditions can also be induced by a vehicle operating just below the surface of the water. Consequently, the wave-induced wake field depends largely on Froude number, and is generally presumed to be of a small order for most applications. Harvald, in[6], has undertaken experiments from which it would appear that the magnitudes of $\{w_w\}$ are generally less than about 0.02 for a ship form.

5.4 ESTIMATION OF WAKE FIELD PARAMETERS

From the propeller design viewpoint the determination of the wake field in which the propeller operates is of fundamental importance. The mean wake field along with the other parameters of power, revolutions and ship speed determines the overall design dimensions of the propeller while the variability of the wake field about the mean wake influences the propeller blade section design and local pitch. A common way of determining the detailed characteristics of the wake field is from model tests; this, however, is not without problems in the areas of wake scaling and propeller interaction. In the absence, however, of either detailed wake data from model or computational origins the designer must resort to other methods of prediction; these can be in the form of regression equations, the plotting of historical analysis data derived from model or full-scale trials, or from intuition and experience, which in the case of an experienced designer must never be underestimated. In the early stages of design the methods cited above are likely to be the ones used.

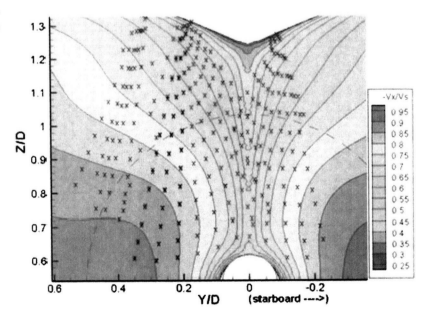

FIGURE 5.5 CFD computation of a container ship full-scale wake field. *Courtesy* Lloyd's Register.

The determination of the mean wake fraction has received much attention over the years. Harvald[6] discusses the merits of some two dozen methods developed in the period from 1896 through to the late 1940s for single-screw vessels. From this analysis he concluded that the most reliable, on the basis of calculated value versus value from model experiment, was due to Schoenherr:[7]

$$\overline{w}_a = 0.10 + 4.5 \frac{C_{pv}C_{ph}(B/L)}{(7 - C_{pv})(2.8 - 1.8C_{ph})}$$
$$+ \frac{1}{2}(E/T - D/B - k\eta)$$

where

L is the length of the ship.
B is the breadth of the ship.
T is the draught of the ship.
D is the propeller diameter.
E is the height of the propeller shaft above the keel.
C_{pv} is the vertical prismatic coefficient of the vessel.
C_{ph} is the horizontal prismatic coefficient of the vessel.
η is the angle of rake of the propeller in radians.
k is the coefficient (0.3 for normal sterns and 0.5–0.6 for sterns having the deadwood cut way).

In contrast, the more simple formula of Taylor[8] was also found to give acceptable values as a first approximation; this was

$$\overline{w}_a = 0.5C_b - 0.05$$

where C_b is the block coefficient of the vessel.

The danger with using formulae of this type and vintage today is that hull form design has progressed to a considerable extent in the intervening years. Consequently, whilst they may be adequate for some simple hull forms their use should be undertaken with great caution and is, therefore, not to be recommended as a general design tool.

Among the more modern methods that were proposed by Harvald[9] the one illustrated in Figure 5.6 is useful. This method approximates the mean axial wake fraction and thrust deduction by the following relationships:

$$\begin{aligned} \overline{w}_a &= w_1 + w_2 + w_3 \\ t &= t_1 + t_2 + t_3 \end{aligned}\Bigg\} \quad (5.5)$$

where t_1, w_1 are functions of B/L and block coefficient,

t_2, w_2 are functions of the hull forms and

t_3, w_3 are propeller diameter corrections.

Alternatively, the later work by Holtrop and Mennen[10] and developed over a series of papers resulted in the following regression formulae for single- and twin-screw vessels[10]:

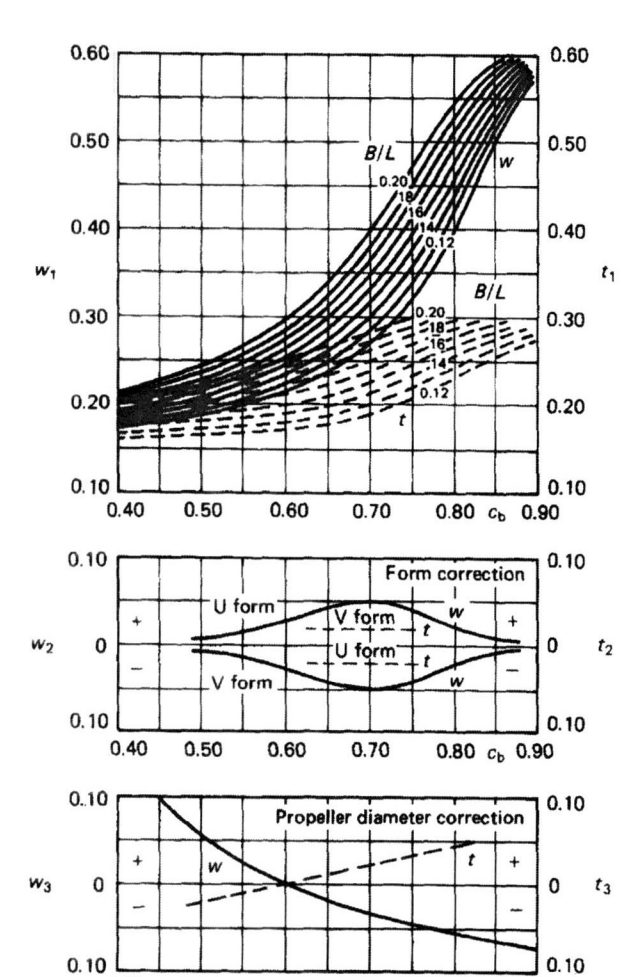

FIGURE 5.6 The wake and thrust deduction coefficient for single-screw ships. *Reproduced with permission from[9].*

Single screw:

$$\begin{aligned} \overline{w}_a &= C_9(1 + 0.015\,C_{stern})[(1 + k)C_F + C_A]\frac{L}{T_A} \\ &\quad \times (0.050776 + 0.93405\,C_{11} \\ &\quad \times \frac{[(1 + k)C_F + C_A]}{(1.315 - 1.45C_p + 0.02251cb)}\Big) \\ &\quad + 0.27915\,(1 + 0.015\,C_{stern}) \\ &\quad \times \sqrt{\frac{B}{L(1.315 - 1.45C_p + 0.02251cb)}} \\ &\quad + C_{19}(1 + 0.015C_{stern}) \end{aligned}\Bigg\} \quad (5.6)$$

Twin screw:

$$\begin{aligned} \overline{w}_a &= 0.3095C_b + 10C_b[(1 + k)C_F + C_A] \\ &\quad - 0.23\frac{D}{\sqrt{BT}} \end{aligned}$$

where:

$$C_9 = C_8 \quad (C_8 < 28)$$
$$= 32 - 16/(C_8 - 24) \quad (C_8 > 28)$$

and

$$C_8 = BS/(LDT_A) \quad (B/T_A < 5)$$
$$= S(7B/TA - 25)/(LD(B/T_A - 3)) \quad (B/T_A > 5)$$

$$C_{11} = T_A/D \quad (T_A/D < 2)$$
$$= 0.0833333\,(T_A/D)^2 + 1.33333 \quad (TA/D > 2)$$

$$C_{19} = 0.12997/(0.95 - C_B)$$
$$\quad\quad - 0.11056/(0.95 - C_p) \quad (C_p < 0.7)$$
$$= 0.18567/(1.3571 - C_M)$$
$$\quad\quad - 0.71276 + 0.38648 C_p \quad (C_p > 0.7)$$

and

Single-Screw Afterbody Form	C_{stern}
Pram with gondola	−25
V-shaped sections	−10
Normal section shape	0
U-shaped sections with Hogner stern	10

These latter formulae were developed from the results of single- and twin-screw model tests over a comparatively wide range of hull forms. The limits of applicability are referred to in the papers and should be carefully studied before using the formulae.

In the absence of model tests or computational studies the radial distribution of the mean wake field, that is the average wake value at each radial location, is difficult to assess. Traditionally, this has been approximated by the use of van Lammeren's diagrams,[11] which are reproduced in Figure 5.7. Van Lammeren's data is based on the single

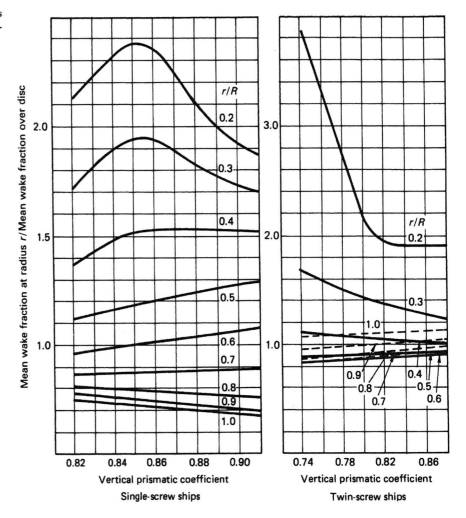

FIGURE 5.7 Van Lammeren's curves for determining the radial wake distribution. *Reproduced with permission*[6].

Chapter | 5   The Ship Wake Field

parameter of vertical prismatic coefficient, and is therefore unlikely to be truly representative for all but first approximations to the radial distribution of mean wake. Harvald[6] re-evaluated the data in which he corrected all the data to a common value of D/L of 0.004 and then arranged the data according to block coefficient and breadth-to-length ratio as shown in Figure 5.8 for single-screw models together with a correction for frame shape. In this study Harvald drew attention to the considerable scale effects that occurred between model and full-scale. He extended his work to twin-screw vessels, shown in Figure 5.9, for a diameter—length ratio of 0.03, in which certain corrections were made to the model test data partly to correct for the boundary layer of the shaft supports. The twin-screw data shown in the diagram refers to the use of bossings to support the shaft lines rather than the modern practice of 'A' and 'P' brackets.

It must be emphasized that all of these methods for the estimation of the wake field and its various parameters are at best approximations to the real situation and not a substitute for properly conducted model tests or computational analysis.

5.5 EFFECTIVE WAKE FIELD

Classical propeller theories assume the flow field to be irrotational and unbounded; however, because the propeller normally operates behind the body which is being propelled these assumptions are rarely satisfied. When the propeller is operating behind a ship the flow field in which the propeller is operating at the stern of the ship is not simply the sum of the flow field in the absence of the propeller together with the propeller-induced velocities calculated on the basis of the nominal wake. In practice a very complicated interaction takes place which gives rise to noticeable effects on propeller performance. Figure 5.10 shows the composition of the velocities that make up the total velocity at any point in the propeller disc. From the propeller design viewpoint it is the

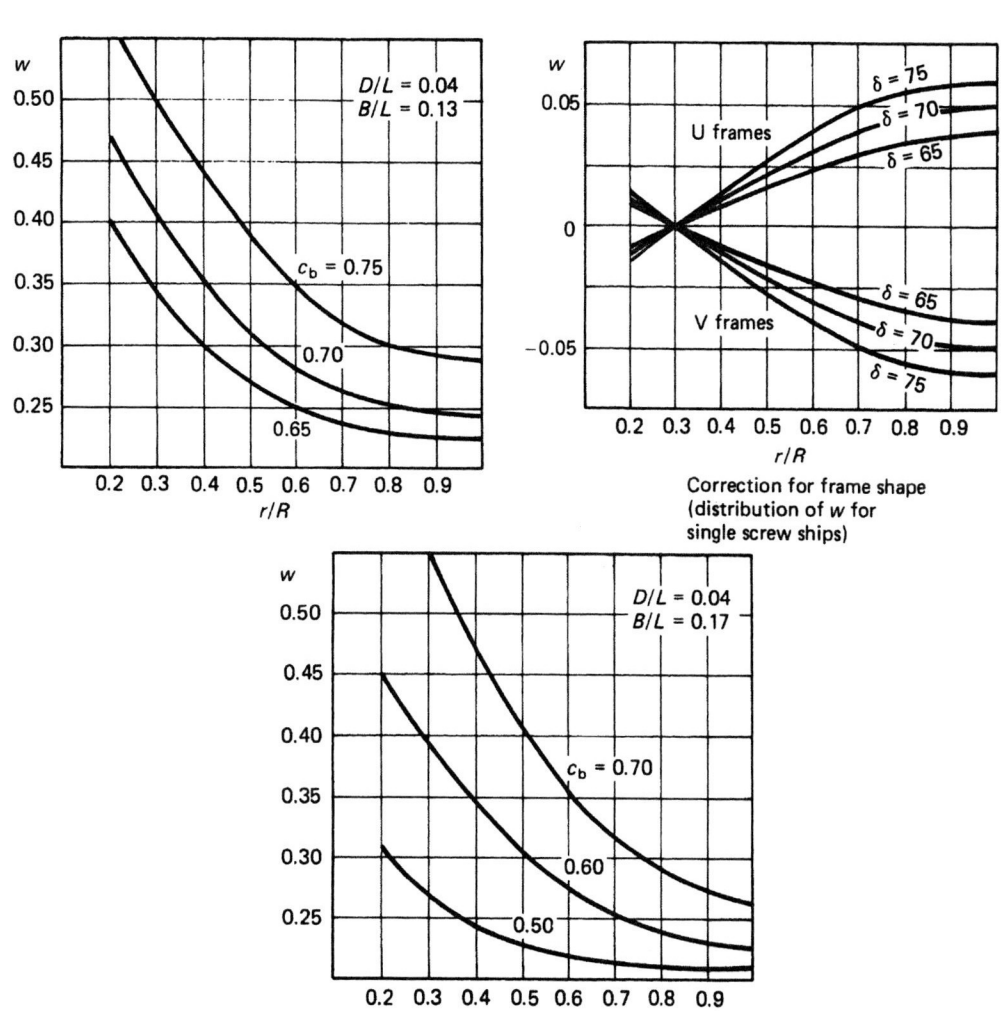

FIGURE 5.8   The radial variation of the wake coefficient of single-screw ships ($D/L = 0.04$). *Reproduced with permission[6].*

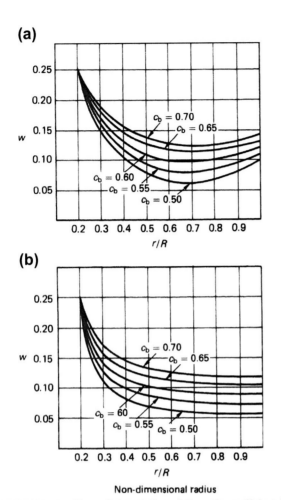

FIGURE 5.9 (a) The radial variation of the wake coefficient for models having twin screws ($D/L = 0.03$) and (b) the radial variation of the wake coefficient for twin-screw ships ($D/L = 0.03$). *Reproduced with permission*[6].

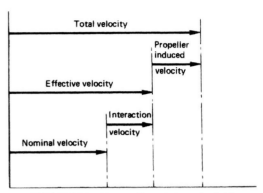

FIGURE 5.10 Composition of the wake field.

effective velocity field that is important since this is the velocity field that should be input into propeller design and analysis procedures. The effective velocity field can be seen from the figure to be defined in one of two ways:

$$\left.\begin{array}{l} \text{effective velocity} = \text{nominal velocity} \\ \qquad\qquad +\text{interaction velocity} \\ \text{or} \\ \text{effective velocity} = \text{total velocity} \\ \qquad\qquad -\text{propeller induced velocity} \end{array}\right\}$$
(5.7)

If the latter of the two relationships is used, an iterative procedure can be employed to determine the effective wake field if the total velocity field is known from measurements just ahead of the propeller. The procedure used for this estimation is shown in Figure 5.11 and has been shown to converge. However, this procedure has the disadvantage of including within it all the shortcomings of the particular propeller theory used for the calculation of the induced velocities. As a consequence this may lead to an incorrect assessment of the interaction effects arising, for example, from the differences in the theoretical treatment of the trailing vortex system of the propeller.

An alternative procedure is to use the former of the two formulations of effective velocity defined in equation (5.7). This approach makes use of the nominal wake field, for example, measured in the towing tank, this being a considerably easier measurement than that of measuring the total velocity, since for the nominal velocity measurement the propeller is absent. Several approaches to this problem have been proposed, including those known as the V-shaped segment and force-field methods. The V-shaped segment method finds its origins in the work of Huang and Groves,[12] which was based on investigations of

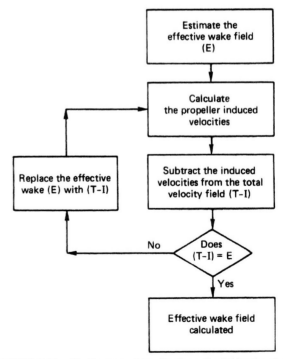

FIGURE 5.11 (T−I) approach to effective wake field estimation.

propeller—wake interaction for axisymmetric bodies. This is perhaps the simplest of all effective wake estimation procedures since it uses only the nominal wake field and principal propeller dimensions as input without undertaking detailed hydrodynamic computations. In the general case of a ship wake field, which contrasts with the axisymmetric basis upon which the method was first derived by being essentially non-uniform, the velocity field is divided into a number of V-shaped segments over which the general non-uniformity is replaced with an equivalent uniform flow. The basis of a V-shaped segment procedure is actuator disc theory, and the computations normally commence with an estimate of the average thrust loading coefficient based on a mean effective wake fraction; typically such an estimate comes from standard series open water data. From this estimate an iterative algorithm commences in which an induced velocity distribution is calculated, which then allows the associated effective velocities and their radial locations to be computed. Procedures of this type do not take into account any changes of flow structure caused by the operating propeller since they are based on the approximate interaction between a propeller and a thick stern boundary layer.

An alternative, and somewhat more complex, effective wake estimation procedure is the force-field method. Such approaches usually rely for input on the nominal wake field and the propeller thrust together with an estimate of the thrust deduction factor. These methods calculate the total velocity field by solving the Euler and continuity equations describing the flow in the vicinity of the propeller. The propeller action is modeled by an actuator disc having only an axial force component and a radial thrust distribution which is assumed constant circumferentially at each radial station. The induced velocities, which are identified within the Euler equations, can then, upon convergence, be subtracted from the total velocity estimates at each point of interest to give the effective wake distribution.

Clearly methods of effective wake field estimation such as the V-shaped segment, force-field and the (T−I) approaches are an essential part of the propeller design and analysis procedure when using classical methods. However, all of these methods lack the wider justification from being subjected to correlation, in open literature, between model and full-scale measurements. Indeed the number of vessels upon which appropriate wake field measurements have been undertaken is minimal for a variety of reasons; typically cost, availability and difficulty of measurement. The latter reason has at least been partially removed with the advent of laser-Doppler techniques which allow effective wake field measurement; nevertheless, this is still a complex and costly procedure.

The analytical treatment of effective wake prediction has gained pace during recent years. A coupled viscous and potential flow procedure was developed by Kerwin

et al.[29,30] for the design of an integrated propulsor driving an axisymmetric body. In this method the flow around the body was computed with the aid of a RANS code with the propulsor being represented by body forces whose magnitudes were estimated using a lifting surface method. As such, in this iterative procedure the RANS solver estimated the total velocity field from which the propeller-induced velocities were subtracted to derive the effective propulsor inflow. Warren et al.[31] used a similar philosophy in order to predict propulsor-induced maneuvering forces in which a RANS code was used for flow calculations over a hull, the appendages and a duct. The time averaged flow field was then input into a three-dimensional lifting surface code which estimated the time varying forces and pressures which were then re-input into the RANS solver in an iterative fashion until convergence was achieved. Hsin et al.[32] developed Kerwin's ideas to a podded propulsor system in order to predict hull—propeller interaction.

Choi and Kinnas[33,34] developed an unsteady effective wake prediction methodology by coupling an unsteady lifting surface cavitating propeller procedure with a three-dimensional Euler code. In this arrangement the propeller effect is represented by unsteady body forces in the Euler solver such that the unsteady effective wake both spatially and temporally can be estimated. Using this method it was found that the predicted total velocity distribution in front of the propeller was in good agreement with measured data. Lee et al.[35] studied rudder sheet cavitation with some success when comparing theoretical predictions with experimental observation. In this procedure a vortex lattice method was coupled to a three-dimensional Euler solver and boundary element method; the latter being used to calculate the cavitating flow around the rudder.

5.6 WAKE FIELD SCALING

Since the model of the ship, which is run in the towing tank, is tested at Froude identity, that is equal Froude numbers between the ship and model, a disparity in Reynolds number exists. This leads to a relative difference in the boundary layer thickness between the model and the full-scale ship; the model having the relatively thicker boundary layer. Consequently, for the purposes of propeller design it is necessary to scale, or contract as it is frequently termed, the wake measured on the model so that it becomes representative of that on the full-size vessel. Figure 5.12 illustrates the changes that can typically occur between the wake fields measured at model and full-scale and with and without a propeller. The results shown in Figure 5.12 relate to trials conducted on the research vessel *Meteor* in 1967 and show respectively pitot tube measurements made with a 1/14 th scale model; the full-scale vessel being towed

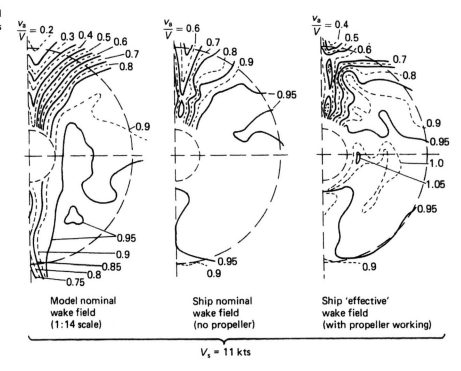

FIGURE 5.12 Comparison of model and full-scale wake fields – *Meteor* trials (1967).

without a propeller and measurements, again at full-scale, made in the presence of the working propeller.

In order to contract nominal wake fields to estimate full-scale characteristics two principal methods have been proposed in the literature and are in comparatively wide use. The first method is due to Sasajima et al.[13] and is applicable to single-screw ships. In this method it is assumed that the displacement wake is purely potential in origin and as such is independent of scale effects, and the frictional wake varies linearly with the skin friction coefficient. Consequently, the total wake at a point is considered to comprise the sum of the frictional and potential components. The total contraction of the wake field is given by

$$c = \frac{C_{\mathrm{fs}} + \Delta C_{\mathrm{fs}}}{C_{\mathrm{fm}}}$$

where C_{fs} and C_{fm} are the ship and model ITTC–1957 friction coefficients expressed by

$$C_{\mathrm{f}} = \frac{0.075}{(\log_{10} R_n - 2)^2}$$

and ΔC_{fs} is the ship correlation allowance.

The contraction in Sasajima's method is applied with respect to the center plane in the absence of any potential wake data, this being the normal case. However, for the general case the contraction procedure is shown in Figure 5.13 in which the ship frictional wake (w_{fs}) is given by

$$w_{\mathrm{fs}} = w_{\mathrm{fm}} \frac{(1 - w_{\mathrm{ps}})}{(1 - w_{\mathrm{pm}})}$$

The method was originally intended for full-form ships having block coefficients in the order of 0.8 and L/B values of around 5.7. Numerous attempts by a number of researchers have been made to generalize and improve the method. The basic idea behind Sasajima's method is to some extent based on the flat plate wake idealization; however, to account for the full range of ship forms encountered in practice, that is those with bulbous sterns, flat afterbodies above the propeller and so on, a more complete three-dimensional contraction process needs to be adopted. Hoekstra[14] developed such a procedure in the mid-1970s in which he introduced, in addition to the center plane contraction, a concentric contraction and a contraction to a horizontal plane above the propeller.

In this procedure the overall contraction factor (c) is the same as that used in the Sasajima approach. However, this total contraction is split into three component parts:

$$c = ic + jc + kc \quad (i + j + |k| = 1)$$

where i is the concentric contraction, j is the center plane contraction and k is the contraction to a horizontal surface above the propeller.

In Hoekstra's method the component contractions are determined from the harmonic content of the wake field; as such, the method makes use of the first six Fourier

Chapter | 5 The Ship Wake Field

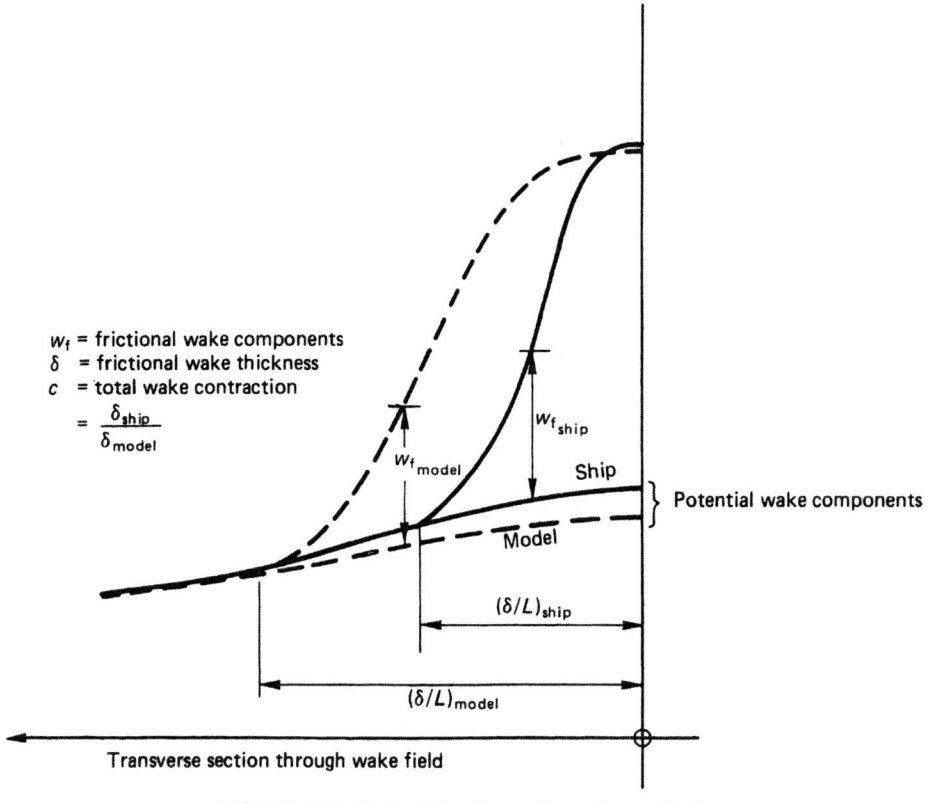

w_f = frictional wake components
δ = frictional wake thickness
c = total wake contraction
$= \dfrac{\delta_{ship}}{\delta_{model}}$

FIGURE 5.13   Basic of Sasajima wake scaling method.

coefficients of the circumferential wake field at each radius. The contraction factors are determined from the following relationships:

$$i = \frac{F_i}{|F_i| + |F_j| + |F_k|} \quad j = \frac{F_j}{|F_i| + |F_j| + |F_k|}$$

$$\text{and} \quad k = \frac{F_k}{|F_i| + |F_j| + |F_k|}$$

in which

$$F_i = \int\limits_{r_{hub}}^{2R} S_i(r)\, dr, F_j = \int\limits_{r_{hub}}^{2R} S_j(r)\, dr$$

$$\text{and } F_k = \int\limits_{r_{hub}}^{2R} S_k(r)\, dr$$

with

$$S_i = 1 - A_0$$
$$+ \begin{cases} A_2 + A_4 + A_6 - \dfrac{1}{2}S_k \text{ if } S_k \geq S_j \\ -S_j + (A_2 + A_4 + A_6) \text{ if } S_k < S_j \end{cases}$$

$$S_j = -[A_2 + A_4 + A_6$$
$$+ |\max(A_2\cos 2\phi + A_4\cos 4\phi$$
$$+ A_6\cos\phi)|] \quad (\phi \neq 0,\ \pi,\ 2\pi)$$

$$S_k = 2(A_1 + A_3 + A_5)$$

where A_n ($n = 0, 1, \ldots, 6$) are the Fourier coefficients and at the hub S_i is taken as unity with $S_j = S_k = 0$.

The method as proposed by Hoekstra also makes an estimation of the scale effect on the wake peak velocity in the center plane and for the scale effect on any bilge vortices that may be present. The method has been shown to give reasonable agreement in a limited number of cases of full-scale to model correlation. However, there have been very few sets of trial results available upon which to base any firm conclusions of this or any other wake field scaling procedure.

Figure 5.14 essentially draws the discussions of effective wake and wake scaling together. In most design or analysis situations the engineer is in possession of the model nominal wake field and wishes to derive the ship or full-scale effective wake field characteristics. There are essentially two routes to achieve this. The most common is to scale the derived nominal wake field from model to full-scale and then to derive the effective wake field at ship scale from the derived nominal full-scale wake.

The enhancement of computational fluid dynamic capabilities has permitted the estimation of ship wake fields by numerically based RANS codes. In such codes the geometric scaling can be varied so as to represent either model or full-scale and, therefore, simplifies the model to ship nominal wake field transformation seen in Figure 5.14.

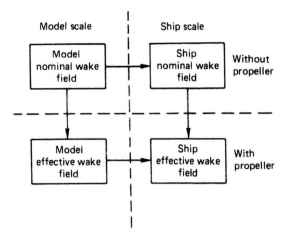

FIGURE 5.14 Relationship between model and ship wake field.

While many attempts in recent years have been made to model the propeller in the wake field estimation process, the results have ranged between disappointing and very good. The procedures used have been based on a number of approaches including actuator disc through to full modeling of the propeller geometry.

5.7 WAKE QUALITY ASSESSMENT

The assessment of wake quality is of considerable importance throughout the ship design process. Available methods generally divide themselves into two distinct categories: analytical methods and heuristic methods. Analytical methods generally use a combination of all the available wake field data (axial, tangential and radial components) to assess the flow quality, whereas heuristic methods normally confine themselves to the axial component only. Unfortunately the use of analytical methods such as those proposed by Truesdell,[15] who introduced a vorticity measure, Mockros,[16] who attempted to include the effects of turbulence into the vorticity measurement and Oswatitsch,[17] who attempted a vorticity measure for perturbed unidirectional flows, tend to be limited by commercial wake measurement practices. As a consequence heuristic assessment procedures are the ones most commonly used at the present time.

Of the many methods proposed three have tended to become reasonably widely used as an assessment basis. In 1973 van Gunsteren and Pronk[18] proposed a method based on the diagrams shown in Figure 5.15 in which the basis of the criterion is the entrance speed cavitation number and the propeller design thrust loading coefficient C_T for various values of $\Delta J/J$, that is the ratio of the fluctuation in advance coefficient to the design advance coefficient. The value ΔJ is directly related to the variation in the wake field at $0.7R$; consequently, the diagram may be used as both a propeller design and wake quality

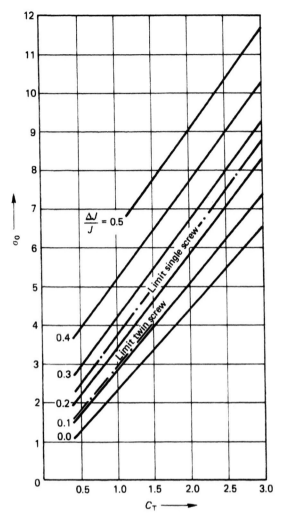

FIGURE 5.15 Van Gunsteren and Pronk assessment basis.
Reproduced with permission from Reference 18.

assessment criteria. In using this diagram it must, however, be remembered that it only takes into account the broad parameters of propeller design and the wake field characteristics and, therefore, must be used in a role commensurate with that caveat.

Huse[1] developed a set of criteria based on the characteristics of the axial velocity field. In particular his criteria address the very important area of the wake peak in the upper part of the propeller disc. His criteria are expressed as follows:

1. For large tankers and other ships with high block coefficients w_{max}, the maximum wake measured at the center plane in the range of $0.4-1.15R$ above the shaft center line should preferably be less than 0.75:

$$w_{max} < 0.75$$

2. For fine ships (block coefficients below 0.60) the w_{max} value should preferably be below 0.55:

$$w_{max} < 0.55 \quad \text{for} \quad C_b < 0.60$$

The Ship Wake Field

3. The maximum acceptable wake peak should satisfy the following relationship with respect to the mean wake at $0.7R$, $\overline{w}_{0.7}$:

$$w_{max} < 1.7\,\overline{w}_{0.7}$$

4. The width of the wake peak should also be taken into account. If the width is slightly smaller than the distance between propeller blades, pressures on the hull due to cavitation will be maximum.

From the above it is clear that Huse's criteria address the quality of wake, largely in the absence of the propeller. In practice, however, it is the propeller–wake combination that gives rise to potential propulsion and vibration problems. Odabasi and Fitzsimmons[19] have extended Huse's work in an attempt to advance wake quality assessment in this area. The criteria proposed in this latter work are as follows:

1. The maximum wake measured inside the angular interval $\theta_B = 10 + 360/Z$ degrees and in the range $0.4 - 1.15R$ around the top dead center position of the propeller disc should satisfy the following:

$$w_{max} < 0.75 \quad \text{or} \quad w_{max} < C_b$$

whichever is smaller.

2. The maximum acceptable wake peak should satisfy the following relationship with respect to the mean wake at $0.7R$:

$$w_{max} < 1.7\,\overline{w}_{0.7}$$

3. The width of the wake peak should not be less than θ_B. The definition of the wake peak for various wake distributions is shown diagrammatically in Figure 5.16.

4. The cavitation number for the propeller tip, defined as:

$$\sigma_n = \frac{9.903 - D/2 - Z_p + T_A}{0.051(\pi n D)^2}$$

and the averaged non-dimensional wake gradient at a characteristic radius, defined as

$$[\Delta w/(1 - \overline{w})]\big|_{x=1.0}$$

should lie above the dividing line of Figure 5.16. In these relationships,

D is the propeller diameter (m).

z_p is the distance between the propeller shaft axis and the base line (m).

T_A is the ship's draught at the aft-perpendicular (m).

n is the propeller rotational speed (rev/s).

Δw is the wake variation defined in Figure 5.16.

5. For the propellers susceptible to cavitation, that is near the grey area of Figure 5.17, the local wake gradient per unit axial velocity for radii inside the angular interval θ_B in the range of $0.7 - 1.15R$ should be less than unity; that is,

$$\frac{1}{(r/R)}\left|\frac{(dw/d\theta)}{(1-w)}\right| < 1.0$$

where θ is in radians.

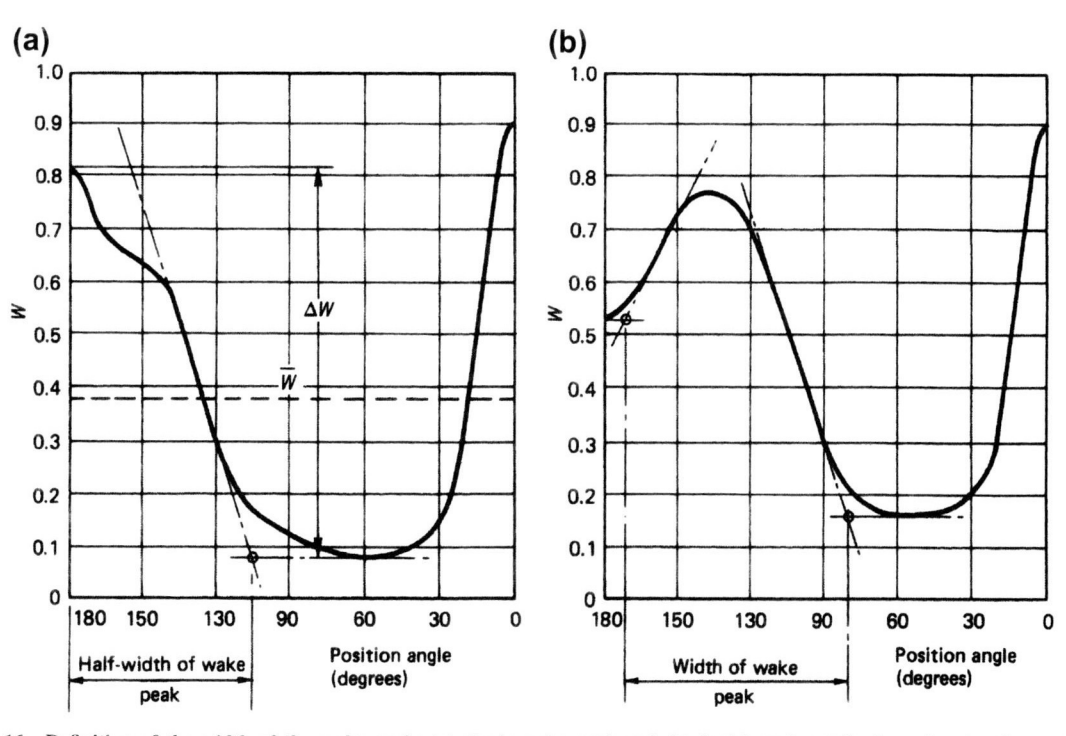

FIGURE 5.16 Definition of the width of the wake peak: (a) single-wake peak and (b) double-wake peak. *Reproduced with permission from Reference 19.*

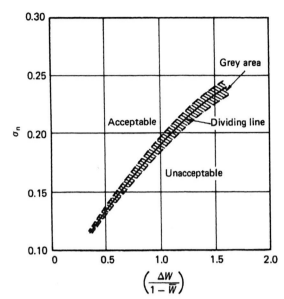

FIGURE 5.17 **Wake non-uniformity criterion.** *Reproduced with permission from Reference 19.*

The underlying reasoning behind the formulation of these criteria has been the desire to avoid high vibratory hull surface pressures, and Figure 5.17 was developed in the basis of results obtained from existing ships.

5.8 WAKE FIELD MEASUREMENT

Measurements of the wake field are required chiefly for the purposes of propeller design and for research where the various aspects of wake field scaling are being explored. Traditionally, methods of measurement have been intrusive; for example, pitot tubes, hot-wire anemometry, tufts and so on. With these methods the influence on the flow field of locating the measurement apparatus in the flow has always been the subject of much debate. In recent years, however, the use of laser-Doppler techniques has become available for both model and full-scale studies and these require only that beams of laser light are passed into the fluid.

In the case of model scale measurements detailed measurements of the wake field have largely been accomplished by using pitot tube rakes, which have in some cases been placed on the shaft in place of the model propeller, Figure 5.18(a). In these cases the rakes have been rotated to different angular positions to define the wake field characteristics. Alternatively, some experimental facilities have favored the use of a fixed pitot rake, Figure 5.18(b), in which the ends of the pitot tube rake are placed in the propeller plane. Such measurements provide quantitative data defining the nominal wake field and are based on the theory of pitot tubes which in turn is based on Bernouilli's equation. For a general point in any fluid flow the following relationship applies:

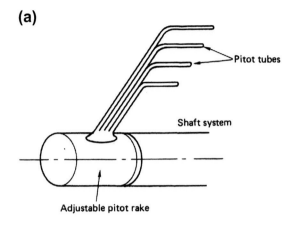

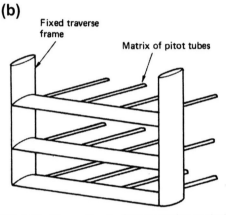

FIGURE 5.18 **Types of wake field traversing methods using pitot, total and static head tubes:** (a) rotating pitot rake located on shaft and (b) schematic fixed pitot rake.

total head(h_T) = static head(h_s) + dynamic head(h_d)

The pitot-static tube shown in Figure 5.19 essentially comprises two tubes: a total head tube and a static head tube. The opening at point A measures the total head in the direction Ox whilst the ports B, aligned in the Oy direction, measure the static head of the fluid. As a consequence from the above relationship, expressed in terms of the corresponding pressures, we have for the dynamic pressure head

$$p_d = p_T - p_s \qquad (5.8)$$

from which

$$v = \sqrt{\frac{2(p_T - p_s)}{\rho}} \qquad (5.9)$$

Depending on the type of flow problem that requires measurement, the probe is selected based on the

Chapter | 5   The Ship Wake Field

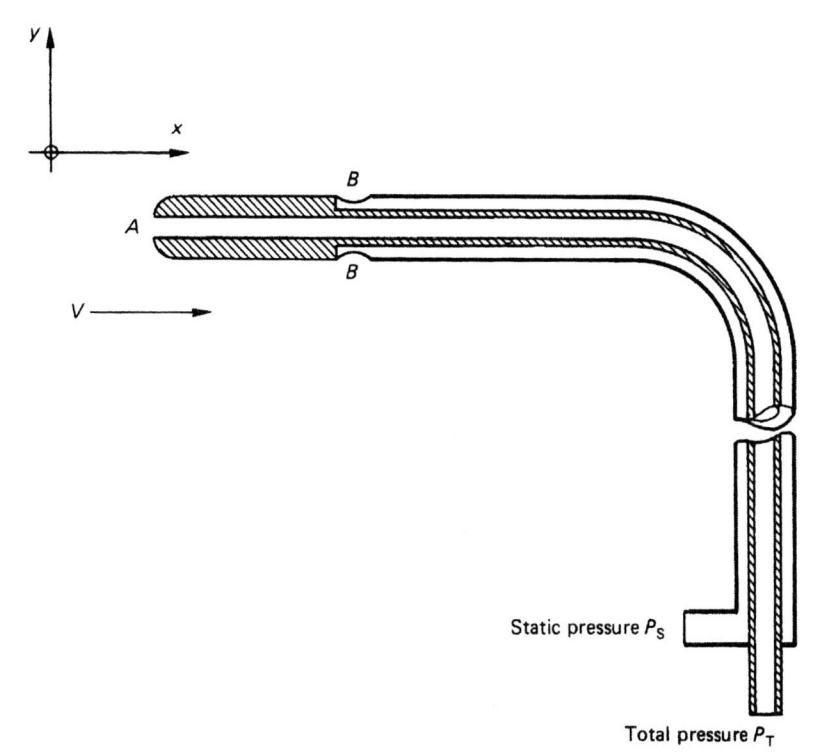

FIGURE 5.19   Pitot static probe layout.

information required and the physical space available. As such total head, static head or pitot-static tubes may be used. Clearly the former two probes only measure one pressure component, whereas the latter measures both values simultaneously. Rakes comprising combinations of total head and static head tubes are sometimes constructed to enable complete measurement to be made, or alternatively, when space is very limited, total head and static head tubes can be inserted into the flow sequentially.

When directionality of the flow is important, since the foregoing tubes are all unidirectional, special measurement tubes can be used. These normally comprise either three- or five-hole total head tubes; an example of the latter is shown in Figure 5.20. From the figure it will be seen that the outer ring of tubes are chamfered and this allows the system to become directional, since opposite pairs of tubes measure different pressures and, from previous calibration, the

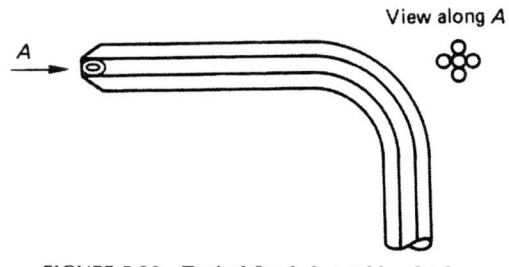

FIGURE 5.20   Typical five-hole total head tube.

differential pressures can be related to the angle of incidence of flow relative to the probe axis. References 20 and 21 should be consulted for further detailed discussion of flow measurement by total head and static head tubes, which is a specialist subject in itself.

In the case of full-scale ship wake field measurement the pitot tube principle has provided much of the data that we have at our disposal at this time. Pitot tube rakes have either been placed on the shaft in place of the propeller, see for example Canham,[22] to measure full-scale nominal wake or, alternatively, fitted to the hull just in front of the propeller to measure the inflow into the propeller;[23,24] Figure 5.21 shows this type of layout together with the five-hole tube used in this latter case. Clearly in the former case of nominal wake measurement, the ship has to be towed by another vessel, whilst in the latter case it is self-propelled. The pitot tube rakes, whether they be shaft or hull mounted, are made adjustable in the angular sense so that they can provide as comprehensive a picture as possible of the wake field.

An alternative to the measurement of flow velocity by pitot tube is to use hot-wire or hot-film anemometry techniques. Such probes rely on the cooling effect of the fluid passing over either the heated wires or hot-film to determine the flow velocities. In their most basic form the current passing through the wire is maintained constant and the flow velocity is determined by the voltage applied across the wire, since the wire resistance is dependent upon

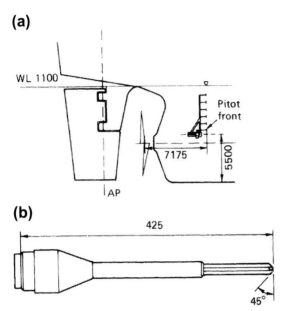

FIGURE 5.21 Full-scale wake pitot probe: (a) the mounting place of the test equipment; (b) example of one of the six, five-hole pitot tubes. *Reproduced with permission from Reference 24.*

the temperature of the wire. A more complex, but widely used, mode of operation is to employ a feed-back circuit which maintains the wire at constant resistance and as a consequence at constant temperature: the current required to do this is a function of the fluid velocity. Hot-wire anemometers, like pitot tubes, require calibration.

A typical hot-wire anemometer is shown in Figure 5.22, where two wires are arranged in an X configuration. If an 'X' wire is located such that the mean velocity is in the plane of the 'X' wire it can be used to measure both components of velocity fluctuation in that plane. The more robust hot-film anemometer comprises a heated element of a thin metallic film placed on a wedge-shaped base which is both a thermal and electric insulator. When used in water, to which it is ideally suited due to its greater robustness, the hot-film is covered with a thin layer of insulation to prevent electrical shorting problems.

In many ways the hot-film or hot-wire anemometer extends the range of fluid measurement scenarios into areas where pitot tubes tend to fail. In particular, since they are small and rapidly responding, they are ideal for measuring

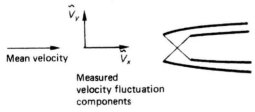

FIGURE 5.22 Hot 'X' wire anemometer.

fluctuating flows; in particular, the phenomena of transition and the structure of turbulent flows. In aerodynamic work hot-wire and film techniques have been used widely and very successfully for one-, two- and three-dimensional flow studies. Lomas[25] and Perry[26] discuss hot-wire anemometry in considerable detail whilst Scragg and Sandell[27] present an interesting comparison between hot-wire and pitot techniques. For full-scale wake field measurements no application of hot-film techniques is known to the author.

Laser-Doppler methods are advanced measurement techniques which can be applied to fluid velocity measurement problems at either model or full-scale. The laser-Doppler anemometer measures flow velocity by measuring the Doppler shift of light scattered within the moving fluid, and hence it is a non-intrusive measurement technique. The light scatter is caused by the passage of tiny particles suspended in the fluid, typically dust or fine sand grains, such that they effectively trace the streamline paths of the fluid flow. In general there are usually sufficient particles within the fluid and in many instances, at full-scale, problems of over-seeding can occur.

The operating principle of a laser-Doppler system is essentially described in Figure 5.23. In the case of a single-laser beam, Figure 5.23(a), the Doppler shift is dependent upon the velocity of the object and the relative angles between the incident and scattered light. If f_i and f_s are the frequencies of the incident and scattered beams, then the Doppler shift is given by $(f_s - f_i)$:

$$f_s - f_i = \frac{V}{\lambda} [\cos \theta_i + \cos \theta_s]$$

where λ is the wavelength of the laser.

This expression can be made independent of the position of the receiver, that is the angle θ_s, by using two laser beams of the same frequency as shown in Figure 5.23(b). This configuration leads to differential Doppler shift seen by the receiver, at some angle ϕ, as follows:

$$\text{differential Doppler shift} = \frac{2V}{\lambda} \sin\left(\frac{\theta}{2}\right) \quad (5.10)$$

The use, as in this case, of two intersection laser beams of the same frequency leads to the introduction of beam splitting optical arrangements obtaining light from a single laser.

Equation (5.10) can be considered in the context of the physically equivalent model of the interference fringes that are formed when two laser beams intersect. If the two beams, Figure 5.23(c), are of equal intensity and wavelength, the fringe pattern will appear as a series of flat elliptical discs of light separated by regions of darkness. If a particle moves through these fringes it will scatter light each time it passes through a light band at a frequency proportional to its speed. Since the separation of the fringes d is given by the expression $\lambda/(2 \sin \theta/2)$ and if the particle

The Ship Wake Field   75

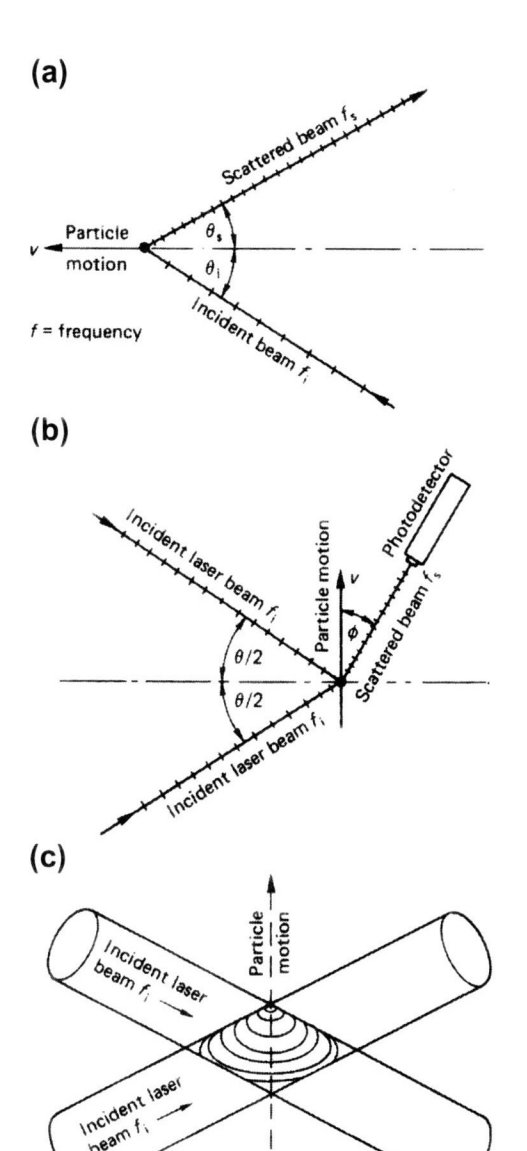

(a)

(b)

(c)

FIGURE 5.23   Laser-Doppler principle: (a) Doppler shift of a single-incident laser beam; (b) intersecting laser beam arrangement and (c) fringe pattern from two intersecting laser beams.

moves with a velocity V, it will move from one interference band to another with a frequency

$$f = \frac{2V \sin(\theta/2)}{\lambda}$$

The scattered light will, therefore, be modulated at this frequency, which is the same as the differential Doppler frequency above. Since the angle θ and the wavelength λ can be precisely defined, a measurement of the modulation frequency gives a direct measure of the velocity of the particle crossing through the intersection of the laser beams.

In terms of practical measurement capabilities several modes of operation exist. These, however, chiefly divide themselves into forward- and back-scatter techniques. Forward-scatter methods essentially place the laser and photodetector on opposite sides of the measurement point, whilst in the back-scatter mode both the laser and photodetector are on the same side. For discussion purposes four methods are of interest in order to illustrate the basic principles of the measurement procedure; these are as follows:

1. Reference beam method.
2. Differential Doppler — forward scatter.
3. Differential Doppler — backward scatter.
4. Multi-color differential Doppler.

In the case of the reference beam method, Figure 5.24(a), the photodetector is mounted coaxially with the reference laser beam in order to measure the velocity within the fluid normal to the optical axis of the instrument. In order to optimize the Doppler signal quality an adjustable neutral density filter is normally used to reduce the intensity of the reference beam.

The differential Doppler, forward-scatter mode of measurement employs two laser beams of equal intensity which are focused at a point of interest in the fluid (Figure 5.24(b)). The scattered light can then be picked up by the photodetector, which is inclined at a suitable angle α to the optical axis of the instrument: the angle α is not critical, since the detected Doppler frequency is independent of the direction of detection. This method is often employed when the intensity of the scattered light is low. Furthermore, the method has obvious advantages over the preceding one since the photodetector does not have to be located on the reference beam.

The backward-scatter differential Doppler mode (Figure 5.24(c)) permits the laser optics and the measurement optics to lie on the same side of the flow measurement point — an essential feature if full-scale ship wake measurements are contemplated. The disadvantage of this type of system is that the intensity of the back-scattered light is usually much lower than that of the forward-scattered light. This normally requires either a higher concentration of scattering particles or a higher laser power to be used to overcome this problem.

The three foregoing systems only measure velocities in one component direction. To extend this into two or more velocity components a multi-color system must be used. Figure 5.24(d) outlines a two-color back-scatter differential Doppler mode. In such a system the transmitting optics split the dual-color laser beam into converging single-color beams with a combined dual-color central beam; that is, three beams in total. The beams are then focused at the point where the measurement is required and the scattered light is returned through the receiving optics, mounted coaxially with the transmitting optics, and then diverted to photodetectors — one for each color of light. The two views

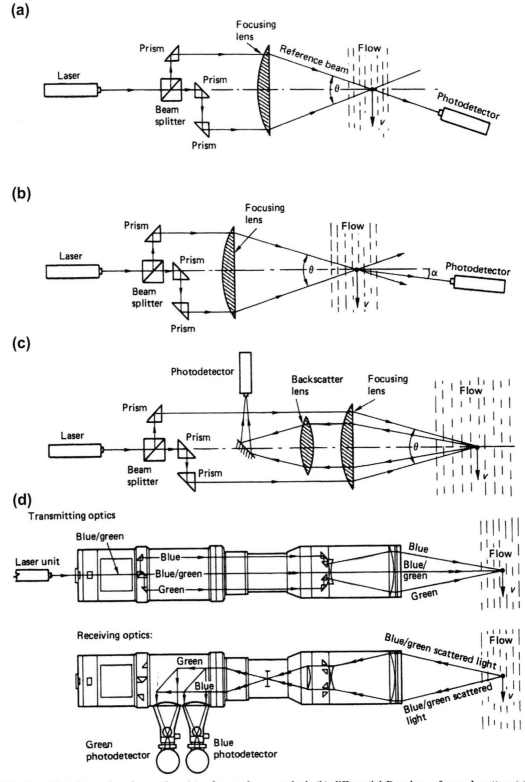

FIGURE 5.24 Laser-Doppler modes of operation: (a) reference beam method; (b) differential Doppler — forward-scatter; (c) differential Doppler — back-scatter and (d) multi-color differential Doppler mode.

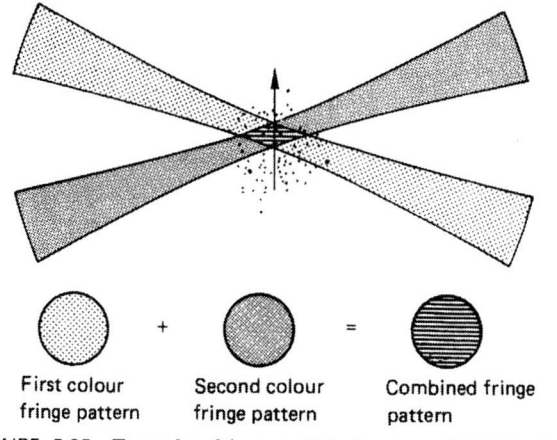

FIGURE 5.25 Two-color fringe model. *Courtesy: DANTEC Electronics Ltd.*

shown in Figure 5.24(d) are in reality a single unit containing both sets of optics. The ability of such systems to detect two velocity components can be visualized from Figure 5.25, in which the two pairs of fringe patterns are made to intersect in orthogonal planes and give a resultant fringe pattern of the type shown in the measurement volume. In this way the particles passing through this measurement volume will scatter light from both orthogonal fringe patterns.

For shipboard measurements a laser system of considerable power is required, and this requires both a carefully designed mounting system to avoid vibration problems and the provision of adequate cooling arrangements. At model scale less powerful systems are required and these can be of the forward-scatter type since the limitation of approaching the measurement from one side of the flow does not normally apply. Reference 28 provides a very good introduction to the subject of laser-Doppler anemometry.

Another non-intrusive method for wake field measurement, like laser-Doppler anemometry, is Particle Image Velocimetry (PIV). Using this method the flow is seeded with small tracer particles which are assumed to follow the streamlines within the fluid flow. The degree to which they do this is a function of Stokes Number which is defined as $Stk = \tau U_0/d_c$ where τ is the relaxation time of the particle, U_0 is the velocity of the fluid remote from the obstacle and d_c is a characteristic dimension of the obstacle around which the flow passes. In this context relaxation time (τ) is taken to be the time that the particle requires to return to equilibrium and for the particles to follow the streamlines of the flow closely, then $Stk \ll 0.1$ when tracing accuracy errors are likely to be less that 1 per cent.[36] Typically a PIV test arrangement will comprise a digital camera together with a strobe or laser having a system of optics which focuses on a limited region of the flow field. This system will be connected to a synchronizer which acts as a trigger for the control of the camera and laser and then suitable PIV software will be used to post-process the optical images.

The PIV technique relies on a planar beam of light, sometimes termed a light sheet, which illuminates the particles entrained in the flow field under consideration. The motions of the particles, from which the flow velocities are derived, are captured by pairs of images using a high-speed digital camera from which a velocity map is derived based on computations made using specialized PIV software. In cases where the flow field is small, micro PIV measurement systems are available; however, in these cases light scattering can be a problem as the seeding of the flow is achieved with fluorescent powder. To minimize this disturbance the flow is viewed through a filter, to block out the greater part of the scattered laser light, in association with high magnification lenses.

In its stereo form the PIV method can provide instantaneous three-dimensional flow vectors within a cross-section of the flow field. Additionally, the technique has the capability to achieve high spatial resolution at a particular time in contrast to LDV techniques which permit high temporal resolution at a certain instant.

REFERENCES AND FURTHER READING

1. Huse E. *Effect of Afterbody Forms and Afterbody Fins on the Wake Distribution of Single Screw Ships.* NSFI Report No. R31−74; 1974.
2. Carlton JS, Bantham I. *Full scale experience relating to the propeller and its environment.* Propellers 78 Symposium, Trans. SNAME; 1978.
3. Prohaski CW, Lammeren WPA van. *Mitstrommessung on Schiffsmodellen.* Heft 16, S.257, Schiffban; 1937.
4. Lammeren WPA van. *Analysis der voort-stuwingscomponenten in verband met het schal-effect bij scheepsmodelproven.* blz 1938;**58**.
5. Harvald SvAa. *Potential and frictional wake of ships.* Trans. RINA; 1972.
6. Harvald SvAa. *Wake of Merchant Ships.* Danish Technical Press; 1950.
7. Schoenherr KE. *Propulsion and propellers.* Principles of Naval Architecture 1939;**2**:149.
8. Taylor DW. *Speed and Power of Ships*; 1933.
9. Harvald SvAa. *Estimation of power of ships.* ISP; 1977.
10. Holtrop J. *A statistical re-analysis of resistance and propulsion data.* ISP November 1988;**31**.
11. Lammeren WPA van, Troost L, Koning JG. *Weerstand en Voortsluwing van Schepen*; 1942.
12. Huang TT, Groves NC. *Effective wake: Theory and experiment.* 13th ONR Symposium; 1980.
13. Sasajima H, Tanaka I, Suzuki T. *Wake distribution of full ships.* J Soc Nav Arch, Japan 1966;**120**.
14. Hoekstra M. *Prediction of full scale wake characteristics based on model wake survey.* Int. Shipbuilding, Prog. June 1975;(250).
15. Truesdell C. *Two measures of vorticity.* Rational Mech Anal 1953;**2**.

16. Mockros FL. The significance of vorticity, vortex motion and dissipation in turbulent fluid flows, *Dissertation*. Berkeley, CA: University of California; 1962.
17. Oswatitsch K. ÜberWirbelkennzahlen und Wirbel-musse. *Z Angew Math Phys* 1969;**20**.
18. Gunsteren LA van, Pronk C. Propeller design concepts. *Int Shipbuilding Prog* 1973;**20**(227).
19. Odabasi AY, Fitzsimmons PA. Alternative methods for wake quality assessment. *Int Shipbuilding Prog* February 1978;**25**(282).
20. Pankhurst RC, Holder DW. *Wind Tunnel Technique*. London: Pitman; 1952.
21. Piercy NA. *Aerodynamics*. London: English Universities Press; 1937.
22. Canham HJS. Resistance, propulsion and wake tests with HMS Penelope. *Trans RINA, Spring Meeting*; 1974.
23. Naminatsu, M., Muraoka, K. Wake distribution measurements: an actual tanker and its large scale model. Proc 14th ITTC.
24. Fagerjord O. Experience from stern wake measurements during sea trial. *Norwegian Maritime Research* 1978;(2).
25. Lomas CG. *Fundamentals of Hot-wire Anemometry*. Cambridge: Cambridge University Press; 1986.
26. Perry AE. *Hot Wire Anemometry*. Oxford: Clarendon Press; 1982.
27. Scragg CA, Sandell DA. A statistical evaluation of wake survey techniques. *Int Symp on Ship Viscous Resistance, SSPA, Göteborg*; 1978.
28. Durst F, Melling A, Whitelaw JH. *Principles and Practice of Laser-Doppler Anemometry*. UK: Academic Press; 1976.
29. Kerwin JE, Keenan DP, Black SD, Diggs JG. A coupled viscous/potential flow design method for wake adapted multi-stage ducted propulsors. *Trans SNAME* 1994;**102**.
30. Kerwin JE, Taylor TE, Black SD, McHugh GP. A coupled lifting surface analysis technique for marine propulsors in steady flow. *Proc Propeller/Shafting Symp, Virginia*; 1997.
31. Warren CL, Taylor TE, Kerwin JE. *Coupled viscous/potential flow method for the prediction of propulsor-induced maneuvering forces*. Virginia Beach: Proc Propeller/Shafting Symp; 2000.
32. Hsin CY, Chou SK, Chen WC. *A new propeller design method for the POD propulsion system*. Fukuoka, Japan: Proc. 24th Symp. on Naval Hydrodynam; 2002.
33. Choi JK, Kinnas SA. Prediction of non-axisymmetric effective wake by a three-dimensional Euler solver. *J Ship Res* 2001;**45**(1).
34. Choi JK, Kinnas SA. Prediction of unsteady effective wake by a Euler solver/vortex-lattice coupled method. *J Ship Res* 2003;**47**(2).
35. Lee H, Kinnas SA, Gu H, Naterajan S. *Numerical modelling of rudder sheet cavitation including propeller/rudder interaction and the effects of a tunnel*. Osaka, Japan: CAV2003; 2003.
36. Tropea, C., Yarin, A., Foss, J. *Springer Handbook of Experimental Fluid Mechanics* (ISBN 978 3 540 25141 5). Springer.

Chapter 6

Propeller Performance Characteristics

Chapter Outline

6.1 General Open Water Characteristics	**79**
6.2 The Effect of Cavitation on Open Water Characteristics	**85**
6.3 Propeller Scale Effects	**87**
6.4 Specific Propeller Open Water Characteristics	**89**
6.4.1 Fixed Pitch Propellers	89
6.4.2 Controllable Pitch Propellers	89
6.4.3 Ducted Propellers	90
6.4.4 High-Speed Propellers	91
6.5 Standard Series Data	**93**
6.5.1 Wageningen B-Screw Series	93
6.5.2 Japanese AU-Series	98
6.5.3 Gawn Series	99
6.5.4 KCA Series	99
6.5.5 Lindgren Series (Ma-Series)	102

6.5.6 Newton—Rader Series	102
6.5.7 Other Fixed Pitch Series and Data	106
6.5.8 Tests with Propellers Having Significant Shaft Incidence	107
6.5.9 Wageningen Ducted Propeller Series	110
6.5.10 Gutsche and Schroeder Controllable Pitch Propeller Series	110
6.5.11 The JD—CPP Series	110
6.5.12 Other Controllable Pitch Propeller Series Tests	112
6.6 Multi-Quadrant Series Data	**112**
6.7 Slipstream Contraction and Flow Velocities in the Wake	**119**
6.8 Behind-Hull Propeller Characteristics	**133**
6.9 Propeller Ventilation	**134**
References and Further Reading	**136**

For discussion purposes the performance characteristics of a propeller can conveniently be divided into open water and behind-hull properties. Open water characteristics relate to the description of the forces and moments acting on the propeller when operating in a uniform fluid stream which is parallel to the shaft center line; hence the open water characteristics, with the exception of inclined propeller flow problems, are steady loadings by definition. The behind-hull characteristics are those generated by the propeller when operating in a mixed wake field behind a body. Clearly these latter characteristics have both a steady and unsteady component by the very nature of the environment in which the propeller operates. In this chapter both types of characteristics will be treated separately: the discussion will initially center on the open water characteristics, since these form the basic performance parameters of the propeller, and then move on to the behind-hull characteristics that develop from the open water performance when the propeller is working behind a body in a perturbed flow.

6.1 GENERAL OPEN WATER CHARACTERISTICS

The forces and moments produced by the propeller are expressed in their most fundamental form in terms of

a series of non-dimensional characteristics for a specific geometric configuration. The non-dimensional terms used to express the general performance characteristics are as follows:

$$
\begin{aligned}
\text{thrust coefficient} \quad & K_{\mathrm{T}} = \frac{T}{\rho n^2 D^4} \\[2mm]
\text{torque coefficient} \quad & K_{\mathrm{Q}} = \frac{Q}{\rho n^2 D^5} \\[2mm]
\text{advance coefficient} \quad & J = \frac{V_{\mathrm{a}}}{nD} \\[2mm]
\text{cavitation number} \quad & \sigma = \frac{p_0 - e}{\frac{1}{2}\rho V^2}
\end{aligned}
\tag{6.1}
$$

where, in the definition of cavitation number, V is a representative velocity which can either be based on free stream advance velocity, a local velocity or propeller rotational speed. While for generalized open water studies the former basis is more likely to be encountered there are exceptions when this is not the case: notably at the bollard pull condition when $V_a = 0$ and hence $\sigma_0 \to \infty$. Consequently, care should be exercised to ascertain the velocity term being employed when using design charts or propeller characteristics for analysis purposes.

Marine Propellers and Propulsion, Third Edition.
Copyright © 2012 John Carlton. Published by Elsevier Ltd. All rights reserved.

To establish the non-dimensional groups involved in the above expressions (equations 6.1), the principle of dimensional similarity can be applied to geometrically similar propellers. The thrust of a marine propeller when working sufficiently far away from the free surface so as not to cause surface waves may be expected to depend upon the following parameters:

1. The diameter (D).
2. The speed of advance (V_a).
3. The rotational speed (n).
4. The density of the fluid (ρ).
5. The viscosity of the fluid (μ).
6. The static pressure of the fluid at the propeller station ($p_0 - e$).

where p_0 is the absolute static pressure at the shaft center line and e is the vapor pressure at ambient temperature.

Hence the thrust (T) can be assumed to be proportional to ρ, D, V_a, n, μ and ($p_0 - e$):

$$T \propto \rho^a D^b V_a^c n^d \mu^f (p_0 - e)^g$$

Since the above equation must be dimensionally correct it follows that

$$MLT^{-2} = (ML^{-3})^a L^b (LT^{-1})^c (T^{-1})^d$$
$$\times (ML^{-1}T^{-1})^f (ML^{-1}T^{-2})^g$$

and by equating indices for M, L and T we have

for mass M: $1 = a + f + g$
for length L: $1 = -3a + b + c - f - g$
for time T: $-2 = -c - d - f - 2g$

from which it can be shown that

$a = 1 - f - g$
$b = 4 - c - 2f - 2g$
$d = 2 - c - f - 2g$

Hence from the above we have

$$T \propto \rho^{(1-f-g)} D^{(4-c-2f-2g)} V_a^c n^{(2-c-f-2g)} \mu^f (p_0 - e)^g$$

from which

$$T = \rho n^2 D^4 \left(\frac{V_a}{nD}\right)^c \cdot \left(\frac{\mu}{\rho n D^2}\right)^f \cdot \left(\frac{p_0 - e}{\rho n^2 D^2}\right)^g$$

These non-dimensional groups are known by the following:

$$\text{thrust coefficient} \quad K_T = \frac{T}{\rho n^2 D^4}$$

$$\text{advance coefficient} \quad J = \frac{V_a}{nD}$$

$$\text{Reynolds number} \quad R_n = \frac{\rho n D^2}{\mu}$$

$$\text{cavitation number} \quad \sigma_0 = \frac{p_0 - e}{\frac{1}{2}\rho n^2 D^2}$$

$$\therefore K_T \propto \{J, R_n, \sigma_0\}$$

that is

$$K_T = f(J, R_n, \sigma_0) \quad (6.2)$$

The derivation for propeller torque K_Q is an analogous problem to that of the thrust coefficient just discussed. The same dependencies in this case can be considered to apply, and hence the torque (Q) of the propeller can be considered by writing it as a function of the following terms:

$$Q = \rho^a D^b V_a^c n^d \mu^f (p_0 - e)^g$$

and hence by equating indices, as before, we arrive at

$$Q = \rho n^2 D^5 \left(\frac{V_a}{nD}\right)^c \left(\frac{\mu}{\rho n D^2}\right)^f \left(\frac{p_0 - e}{\rho n^2 D^2}\right)^g$$

which reduces to

$$K_Q = g(J, R_n, \sigma_0) \quad (6.3)$$

where the torque coefficient

$$K_Q = \frac{Q}{\rho n^2 D^5}$$

With the form of the analysis chosen the cavitation number and Reynolds number have been non-dimensionalized by the rotational speed. These numbers could equally well be based on advance velocity, so that

$$\sigma_0 = \frac{p_0 - e}{\frac{1}{2}\rho V^2} \quad \text{and} \quad R_n = \frac{\rho V D}{\mu}$$

Furthermore, by selecting different groupings of indices in the dimensional analysis it would be possible to arrive at an alternative form for the thrust loading:

$$T = \rho V_a^2 D^2 \phi(J, R_n, \sigma_0)$$

which gives rise to the alternative form of thrust coefficient C_T defined as

$$C_T = \frac{T}{\frac{1}{2}\rho V_a^2 (\pi D^2/4)} = \frac{8T}{\pi \rho V_a^2 D^2} \quad (6.4)$$

$$C_T = \Phi(J, R_n, \sigma_0)$$

Similarly it can be shown that the power coefficient C_P can also be given by

$$C_P = \phi(J, R_n, \sigma_0) \quad (6.5)$$

In cases where the propeller is sufficiently close to the surface, so as to disturb the free surface or to draw air, other dimensionless groups will become important. These will

Chapter | 6   Propeller Performance Characteristics

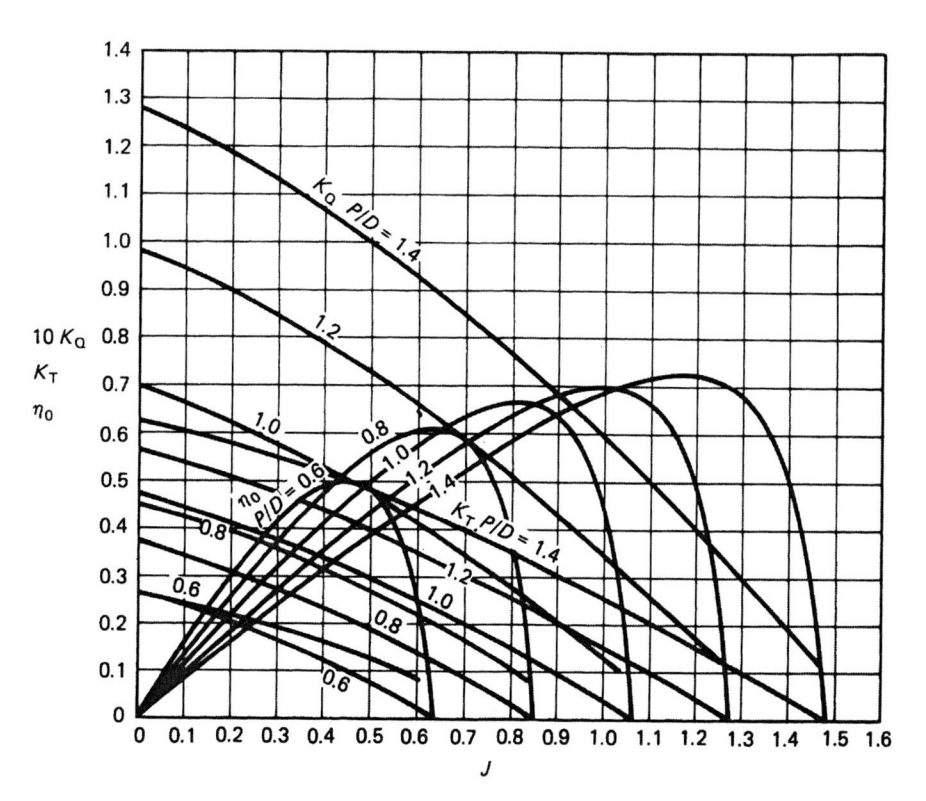

FIGURE 6.1   Open water diagram for Wageningen B5-75 screw series. *Courtesy: MARIN.*

principally be the Froude and Weber numbers and these can readily be shown to apply by introducing gravity and surface tension into the foregoing dimensional analysis equations for thrust and torque.

A typical open water diagram for a set of fixed pitch propellers working in a non-cavitating environment at forward, or positive, advance coefficient is shown in Figure 6.1. This figure defines, for the particular propeller, the complete set of operating conditions at positive advance and rotational speed since the propeller, under steady conditions, can only operate along the characteristic line defined by its pitch ratio *P/D*. The diagram is general in the sense that, subject to scale effects, it is applicable to any propeller having the same geometric form as the one for which the characteristic curves were derived. However, the subject propeller may have a different diameter or scale ratio and can work in any other fluid subject to certain Reynolds number effects. When, however, the K_T, K_Q versus *J* diagram is used for a particular propeller of a given geometric size and working in a particular fluid medium, the diagram, since the density of the fluid and the diameter then become constants, effectively reduces from the general definitions of K_T, K_Q and *J* to a particular set of relationships defining torque, thrust, revolutions and speed of advance as follows:

$$\left\{ \frac{Q}{n^2}, \frac{T}{n^2} \right\} \; versus \; \left(\frac{V_a}{n} \right)$$

The alternative form of the thrust and torque coefficient which stems from equations (6.4) and (6.5), and which is based on the advance velocity rather than the rotational speed, is defined as follows:

$$C_T = \frac{T}{\frac{1}{2}\rho A_0 V_a^2}$$

$$C_P = \frac{PD}{\frac{1}{2}\rho A_0 V_a^2}$$

(6.6)

From equations (6.6) it can be readily deduced that these thrust loading and power loading coefficients can be expressed in terms of the conventional thrust and torque coefficient as follows:

$$C_T = \frac{8}{\pi}\frac{K_T}{J^2}$$

and

$$C_P = \frac{8}{\pi}\frac{K_Q}{J^3}$$

(6.7)

The open water efficiency of a propeller (η_0) is defined as the ratio of the thrust horsepower to delivered horsepower:

$$\eta_0 = \frac{\text{THP}}{\text{DHP}}$$

Now since

$$\text{THP} = TV_a$$

and

$$\text{DHP} = 2\pi nQ$$

where T is the propeller thrust, V_a the speed of advance, n the rotational speed of the propeller and Q, the torque. Consequently, with a little mathematical manipulation we may write

$$\eta_o = \frac{TV_a}{2\pi nQ}$$

that is

$$\eta_o = \frac{K_T}{K_Q}\frac{J}{2\pi} \qquad (6.8)$$

The K_Q, K_T versus J characteristic curves contain all of the information necessary to define the propeller performance at a particular operating condition. Indeed, the curves can be used for design purposes for a particular basic geometry when the model characteristics are known for a series of pitch ratios. This, however, is a cumbersome process and to overcome these problems Admiral Taylor derived a set of design coefficients termed B_p and δ; these coefficients, unlike the K_T, K_Q and J characteristics, are dimensional parameters and so considerable care needs to be exercised in their use. The terms B_p and δ are defined as follows:

$$B_P = \frac{(\text{DHP})^{1/2}N}{V_a^{2.5}}$$

$$\delta = \frac{ND}{V_a} \qquad (6.9)$$

where

DHP is the delivered horsepower in British or metric units depending on the diagram used;
N is the propeller rpm;
V_a is the speed of advance (knots);
D is the propeller diameter (ft).

From Figure 6.2, which shows a typical propeller design diagram, it can be seen that it essentially comprises a plotting of B_p, as abscissa, against pitch ratio as ordinate with lines of constant δ and open water efficiency superimposed. This diagram is the basis of many design procedures for marine propellers, since the term B_p is usually known from the engine and ship characteristics. From the figure a line of optimum propeller open water efficiency can be seen as being the locus of the points on the diagram which have the highest efficiency for a given value of B_p. Consequently, it is possible with this diagram to select values of δ and P/D to maximize the open water efficiency η_o for a given powering condition as defined by the B_p parameter. Hence a basic propeller geometry can be derived in terms of diameter D, since $D = \delta V_a/N$, and P/D. Additionally, this diagram can be used for a variety of other propeller design purposes such as rotational speed selection; however, these aspects of the design process will be discussed later in the chapter on propeller design (Chapter 22).

It will be seen that the B_p versus δ diagram is limited to the representation of forward speeds of advance only, that is, where $V_a > 0$, since $B_p \to \infty$ when $V_a = 0$. This limitation is of particular importance when considering the design of tugs and other similar craft, which can be expected to spend an important part of their service duty at zero ship speed, termed bollard pull, while at the same time developing high powers. To overcome this problem, a different sort of design diagram was developed from the fundamental K_T, K_Q versus J characteristics, so that design and analysis problems at or close to zero speed of advance can be satisfactorily considered. This diagram is termed the $\mu - \sigma$ diagram, and a typical example is shown in Figure 6.3. In this diagram the following relationships apply:

$$\mu = n\sqrt{\frac{\rho D^5}{Q}}$$

$$\phi = V_a\sqrt{\frac{\rho D^3}{Q}} \qquad (6.10)$$

$$\sigma = \frac{TD}{2\pi Q}$$

where

D is the propeller diameter (m)
Q is the delivered torque (kgf m)
ρ is the mass density of water (kg/m^3)
T is the propeller thrust (kgf)
n is the propeller rotational speed (rev/s)
V_a is the ship speed of advance (m/s).

Diagrams of the type shown in Figure 6.3 are non-dimensional in the same sense as those of the fundamental K_T, K_Q characteristics and it will be seen that the problem of zero ship speed, that is when $V_a = 0$, has been removed, since the function $\phi \to 0$ as $V_a \to 0$. Consequently, the line on the diagram defined by $\phi = 0$ represents the bollard pull condition for the propeller. It is important, however, not to confuse propeller thrust with bollard pull, as these terms are quite distinct and mean different things. Propeller thrust and bollard pull are exactly what the terms imply; the former relates to the hydrodynamic thrust produced by the propeller, whereas the latter is the pull the vessel can exert through a towline

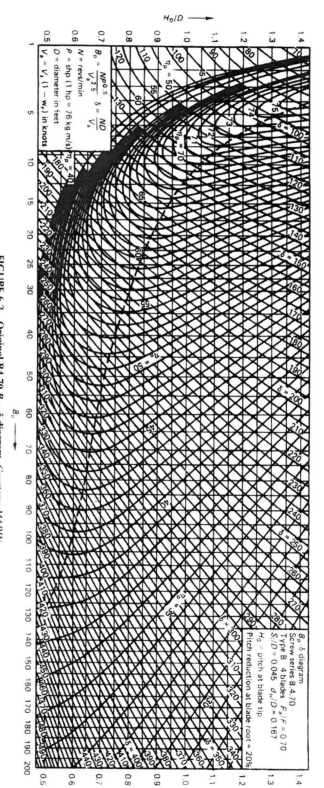

FIGURE 6.2 Original B4-70 B_p–δ diagram. *Courtesy: MARIN.*

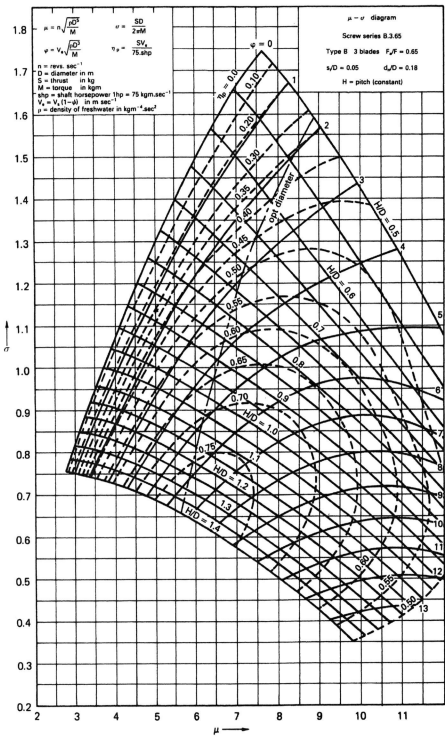

FIGURE 6.3 Original B3.65 μ–σ diagram. *Courtesy: MARIN.*

on some other stationary object. Bollard pull is always less than the propeller thrust by a complex ratio, which is dependent on the underwater hull form of the vessel, the depth of water, the distance of the vessel from other objects, and so on.

In the design process it is frequently necessary to change between coefficients, and to facilitate this process. Tables 6.1 and 6.2 have been produced in order to show some of the more common relationships between the parameters.

Chapter | 6 Propeller Performance Characteristics

TABLE 6.1 Common Functional Relationships (British Units)

$$K_Q = 9.5013 \times 10^6 \left(\frac{P_D}{N^3 D^5} \right) \text{ (salt water)}$$

$$B_P = 23.77 \sqrt{\frac{\rho K_Q}{J^5}}$$

$$J = \frac{1.1.33 \, V_a}{ND} = \frac{101.33}{\delta}$$

$$\mu = \frac{1}{\sqrt{K_Q}} = 3.2442 \times 10^{-4} \sqrt{\frac{N^3 D^5}{P_D}} \text{(salt water)}$$

$$\phi = \frac{J}{\sqrt{K_Q}} \, J\mu$$

$$\sigma = \frac{\eta_o}{J} = \frac{\eta_o \mu}{\phi} = \frac{K_T}{2\pi K_Q}$$

where:

P_D is the delivered horsepower in Imperial units
Q is the delivered torque at propeller in (lbf ft)
T is the propeller thrust (lbf)
N is the propeller rotational speed in (rpm)
N is the propeller rotational speed in (rev/s)
D is the propeller diameter in (ft)
V_a is the propeller speed of advance in (knots)
v_a is the propeller speed of advance in (ft/s)
ρ is the mass density of water (1.99 slug/ft³ sea water; 1.94 slug/ft³ for fresh water).

TABLE 6.2 Common Functional Relationships (Metric Units)

$$K_Q = 2.4669 \times 10^4 \left(\frac{P_D}{N^3 D^5} \right) \text{ (salt water)}$$

$$B_P = 23.77 \sqrt{\frac{\rho K_Q}{J^5}}$$

$$J = \frac{30.896 \, V_a}{ND} = \frac{101.33}{\delta}$$

$$\mu = \frac{1}{\sqrt{K_Q}} = 6.3668 \times 10^{-3} \sqrt{\frac{N^3 D^5}{P_D}} \text{(salt water)}$$

$$\phi = \frac{J}{\sqrt{K_Q}} \, J\mu$$

$$\sigma = \frac{\eta_o}{J} = \frac{\eta_o \mu}{\phi} = \frac{K_T}{2\pi K_Q}$$

where:

P_D is the delivered horsepower (metric units)
Q is the delivered propeller torque (kp m)
T is the propeller thrust (kp)
N is the propeller rotational speed (rpm)
n is the propeller rotational speed (rev/s)
D is the propeller diameter (m)
V_a is the propeller speed of advance (knots)
v_a is the propeller speed of advance (m/s)
ρ is the mass density of water (104.48 sea water) (101.94 fresh water)

Note the term σ in Tables 6.1 and 6.2 and in equation (6.10) should not be confused with cavitation number, which is an entirely different concept. The term σ in the above tables and equation (6.10) relates only to the $\mu - \sigma$ diagram, which is a non-cavitating diagram.

6.2 THE EFFECT OF CAVITATION ON OPEN WATER CHARACTERISTICS

Cavitation, which is a two-phase flow phenomenon, is discussed more fully in Chapter 9; however, it is pertinent here to recognize the effect that cavitation development can have on the propeller open water characteristics.

Cavitation for the purposes of generalized analysis is defined by a free stream cavitation number σ_0, which is the ratio of the static to dynamic head of the flow. For our purposes in this chapter we will consider a cavitation number based on the static pressure at the shaft center line and the dynamic head of the free stream flow ahead of the propeller:

$$\sigma_0 = \frac{\text{static head}}{\text{dynamic head}} = \frac{p_0 - e}{\frac{1}{2}\rho V_a^2}$$

Consequently, a non-cavitating flow is one where $(p_0 - e) \gg \{1/2\}\rho V_a$, that is one where σ_0 is large. As σ_0 decreases in value cavitation takes more effect as demonstrated in Figure 6.4. This figure illustrates the effect that cavitation has on the K_T and K_Q curves and, for guidance purposes only, shows a typical percentage of cavitation on the blades experienced at various cavitation numbers in uniform flow. It is immediately apparent from the figure that moderate levels of cavitation do not affect the propulsion performance of the propeller and significant cavitation activity is necessary in order to suffer thrust and torque breakdown. Furthermore, it will frequently be noted that the K_T and K_Q curves rise marginally above the non-cavitating line just prior to their rapid decline after thrust or torque breakdown.

It is not necessarily important to associate the other problems of cavitation, for example, hull-induced vibration and erosion of the blade material, with the extent of cavitation necessary to cause thrust and torque performance breakdown. Relatively small extents of cavitation, given the correct conditions, are often sufficient to give rise to these problems.

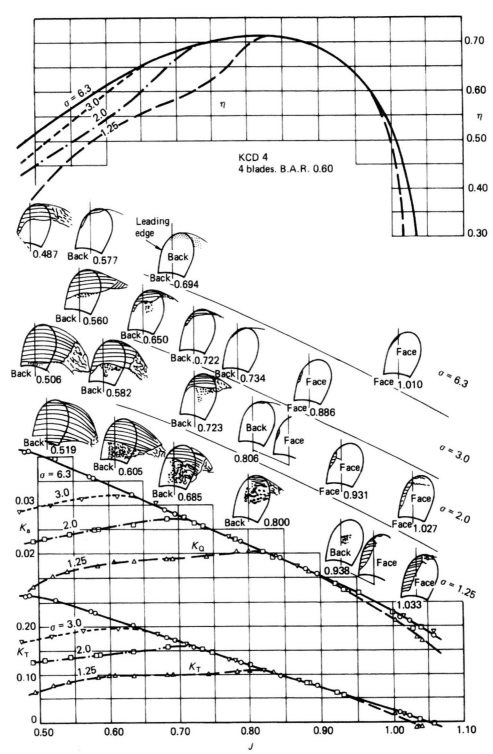

FIGURE 6.4 Curves of K_T, K_Q and η and cavitation sketches for **KCD4**. *Reproduced from Reference 15.*

6.3 PROPELLER SCALE EFFECTS

Open water characteristics are frequently determined from model experiments on propellers run at high-speed and having diameters of the order of 200–300 mm. It is, therefore, reasonable to pose the question of how the reduction in propeller speed and increase in diameter at full-scale will affect the propeller performance characteristics. Figure 6.5 shows the principal features of scale effect, from which it can be seen that while the thrust characteristic is largely unaffected the torque coefficient is somewhat reduced for a given advance coefficient.

The scale effects affecting performance characteristics are essentially viscous in nature and are primarily due to boundary layer phenomena dependent on Reynolds number. Due to the methods of testing model propellers and the consequent changes in Reynolds number between model and full-scale, or indeed a smaller model and a larger model, there can arise a different boundary layer structure to the flow over the blades. While it is generally recognized that most full-scale propellers will have a primarily turbulent flow over the blade surface this need not be the case for the model where characteristics related to laminar flow can prevail over significant parts of the blade.

In order to quantify the effect of scale on the performance characteristics of a propeller an analytical procedure is clearly required. There is, however, no common agreement as to which is the best procedure. In a survey conducted by the 1987 ITTC it was shown that from a sample of 22 organizations, 41 per cent used the ITTC−1978 procedure; 23 per cent made corrections based on correlation factors developed from experience; 13 per cent, who dealt with vessels having open shafts and struts, made no correction at all; a further 13 per cent

endeavored to scale each propulsion coefficient, whilst the final 10 per cent scaled the open water test data and then used the estimated full-scale advance coefficient.

At present the principal analytical tool available is the ITTC−1978 performance prediction method, which is based on a simplification of Lerbs' equivalent profile procedure. Lerbs showed that a propeller can be represented by the characteristics of an equivalent section at a non-dimensional radius of around $0.70R$ or $0.75R$, these being the two sections normally chosen. The method calculates the change in propeller performance characteristics as follows.

The revised thrust and torque characteristics are given by

$$\left.\begin{array}{l} K_{T_s} = K_{T_m} - \Delta K_T \\ K_{Q_s} = K_{Q_m} - \Delta K_Q \end{array}\right\} \tag{6.11}$$

where the scale corrections ΔK_T and ΔK_Q are given by

$$\Delta K_T = -0.3\Delta C_D\left(\frac{P}{D}\right)\left(\frac{cZ}{D}\right)$$

$$\Delta K_Q = 0.25\Delta C_D\left(\frac{cZ}{D}\right)$$

and in equation (6.11) the suffixes s and m denote the full-scale ship and model test values respectively. The term ΔC_D relates to the change in drag coefficient introduced by the differing flow regimes at model- and full-scale, and is formally written as

$$\Delta C_D = C_{DM} - C_{DS}$$

where

$$C_{DM} = 2\left(1 + \frac{2t}{C}\right)\left[\frac{0.044}{(R_{nx})^{1/6}} - \frac{5}{(R_{nx})^{2/3}}\right]$$

and

$$C_{DS} = 2\left(1 + \frac{2t}{c}\right)\left(1.89 + 1.62\log_{10}\left(\frac{c}{K_p}\right)\right)^{-2.5}$$

In these relationships t/c is the section thickness to chord ratio; P/D is the pitch ratio; c is the section chord length and R_{nx} is the local Reynolds number, all relating to the section located $0.75R$. The blade roughness K_p is taken as 30×10^{-6} m.

In this method it is assumed that the full-scale propeller blade surface is hydrodynamically rough and the scaling procedure considers only the effect of Reynolds number on the drag coefficient.

An alternative approach to the use of equation (6.11) has been proposed by Vasamarov[1] in which the correction

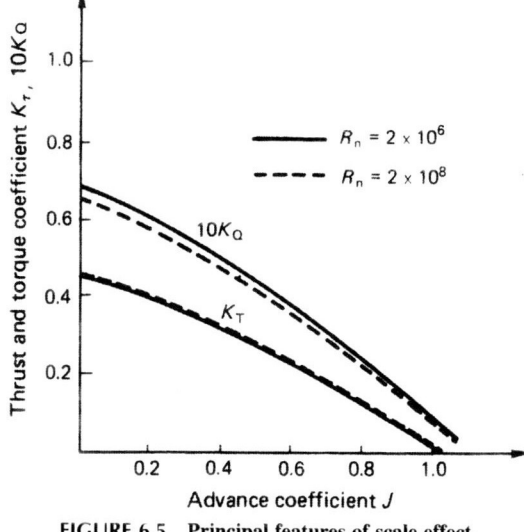

FIGURE 6.5   Principal features of scale effect.

for the Reynolds effect on propeller open water efficiency is given by

$$\eta_{os} = \eta_{om} - F(J)\left[\left(\frac{1}{R_{nm}}\right)^{0.2} - \left(\frac{1}{R_{ns}}\right)^{0.2}\right] \quad (6.12)$$

where

$$F(J) = \left(\frac{J}{J_0}\right)^{\alpha}$$

From the analysis of the function $F(J)$ from open water propeller data, it has been shown that J_0 can be taken as the zero thrust advance coefficient for the propeller. Consequently, if model tests are undertaken at two Reynolds numbers and the results analyzed according to equation (6.12); then the function $F(J)$ can be uniquely determined.

Yet another approach has been proposed (Reference 2) in which the scale effect is estimated using open water performance calculations for propellers having similar geometric characteristics to the Wageningen B Series.

The results of the analysis are presented in such a way as

$$\left.\begin{aligned} 1 - \frac{K_T}{K_{T_I}} &= f(R_n, K_T) \\ 1 - \frac{\eta_0}{\eta_{0_I}} &= g(R_n, K_T) \end{aligned}\right\} \quad (6.13)$$

where the suffix I represents the values of K_T and η_0 for an ideal fluid. Consequently, if model values of the thrust and torque at the appropriate advance coefficient are known, that is K_{T_m}, K_{Q_m}, together with the model Reynolds number, then from equation (6.13) we have

$$\frac{K_{T_m}}{K_{T_I}} = 1 - f(R_{n_m}, K_{T_m})$$

$$\Rightarrow K_{T_I} = \frac{K_{T_m}}{(1 - f(R_{n_m}, K_{T_m}))} = \left.\frac{K_{T_m}}{1 - \left(1 - \frac{K_{T_m}}{K_{T_I}}\right)}\right|_{R_{n_m}}$$

Similarly

$$\eta_{0_I} = \left.\frac{\eta_{0_m}}{1 - \left(1 - \frac{\eta_{0_m}}{\eta_{0_I}}\right)}\right|_{R_{n_m}}$$

From which the ideal values of K_{T_I} and η_{0_I} can be determined for the propeller in the ideal fluid. Since the effect of scale on the thrust coefficient is usually small and the full-scale thrust coefficient will lie between the model and ideal values the following assumption is made:

$$K_{T_S} \simeq \left(\frac{K_{T_M} + K_{T_I}}{2}\right)$$

that is the mean value, and since the full-scale Reynolds number R_{n_s} is known, the functions

$$f(R_{n_s}, K_{T_s}) \quad \text{and} \quad g(R_{n_s}, K_{T_s})$$

can be determined from which the full-scale values of K_{T_s} and η_{0_s} can be determined from equation (6.13):

$$K_{T_s} = K_{T_I}[1 - f(R_{n_s}, K_{T_s})]$$
$$\eta_{0_s} = \eta_{0_I}[1 - g(R_{n_s}, K_{T_s})]$$

from which the full-scale torque coefficient can be derived as follows:

$$K_{Q_s} = \frac{J}{2\pi}\frac{K_{T_s}}{\eta_{0_s}}$$

The essential difference between these latter two approaches is that the scale effect is assumed to be a function of both Reynolds number and propeller loading rather than just Reynolds number alone as in the case of the present ITTC procedure. It has been shown that significant differences can arise between the results of the various procedures. Scale effect correction of model propeller characteristics is not a simple procedure and attention needs to be paid to the effects of the flow structure in the boundary layer and the variations of the lift and drag characteristics within the flow regime. With regard to the general question of scaling, the above methods were primarily intended for non-ducted propellers operating on their own and the subject of scaling is still not fully understood. Although the problem is complicated by the differences in friction and lift coefficient, the scale effect is less predictable due to the quantity of both laminar flow in the boundary layer and the separation over the blade surfaces. Consequently, there is the potential for the extrapolation process from model to full-scale to become unreliable since only averaged amounts of laminar flow are taken into account in the present estimation procedures.

To try and overcome this difficulty a number of techniques have been proposed, particularly those involving leading edge roughness and the use of trip wires, but these procedures still lack rigor in their application to extrapolation. Bazilevski,[41] in a range of experimental conditions using trip wires of 0.1 mm diameter located at 10 per cent of the chord length from the leading edge, showed that the experimental scatter on the measured efficiency could be reduced from 13.6 to 1.5 per cent with the use of turbulence stimulation. It was found that trip wires placed on the suction surface of the blades were more effective than those placed on the pressure face and that the effectiveness of the trip wire was dependent upon the ratio of wire diameter to the boundary layer thickness. Boorsma, in Reference 42, considered an alternative method of turbulence tripping by the use of sand grain roughness on the leading edge based on the correlation of a sample of five propellers. In his work he showed that the rotation rate correlation factor at constant power could be reduced from 2.4 to 1.7 per cent and, furthermore, concluded that turbulence tripping was not always effective at the inner blade radii.

It is often considered convenient in model experiments to perform model tests at a higher rotational speed than would be required by strictly adhering to the Froude identity. If this is done this then tends to minimize any flow separation at the trailing edge or laminar flow on the suction side of the blade. Such a procedure is particularly important when the propeller is operating in situations where relatively low turbulence levels are encountered in the inflow and where stable laminar flow is likely to be present. Such a situation may be found in cases where tractor thrusters or podded propulsors are being investigated. Ball and Carlton, in Reference 43, show examples of this type of behavior relating to model experiments with podded propulsors.

Clearly compound propellers such as contra-rotating screws and ducted propellers will present particular problems in scaling. In the case of the ducted propeller the interactions between the propeller, the duct and the hull are of particular concern and importance. In addition there is also some evidence to suggest that vane wheels are particularly sensitive to Reynolds number effects since both the section chord lengths and the wheel rotational speed are low, which can cause difficulty in interpreting model test data.

Holtrop[44] proposed that the scaling of structures like ducts can be addressed by considering the interior of the duct as a curved plate. In this analysis an assumed axial velocity of nP_{tip} is used to determine a correction to the longitudinal towing force ΔF given by

$$\Delta F = 0.5\rho_m \Pi (nP_{\text{tip}})^2 (C_{\text{Fm}} - C_{\text{Fs}}) c_m D_m$$

where n, P_{tip}, c and D are the rotational speed, pitch at the blade tip, duct chord and diameter, respectively.

In the case of podded propulsor housings the problem is rather more complicated in that there is a dependence upon a number of factors. For example, the shape of the housing and its orientation with respect to the local flow, the interaction with the propeller wake and the scale effects of the incident flow all have an influence on the scaling problem.

6.4 SPECIFIC PROPELLER OPEN WATER CHARACTERISTICS

Before proceeding to outline the various standard series available to the propeller designer or analyst, it is helpful to briefly consider the types of characteristic associated with each of the principal propeller types, since there are important variants between, say, fixed pitch and

controllable pitch propellers or non-ducted and ducted propellers.

6.4.1 Fixed Pitch Propellers

The preceding discussions in this chapter have used as examples the characteristics relating to fixed pitch propellers since these are the simplest form of propeller characteristics. Figure 6.1 is typical of this type of propeller in that the propeller, in the absence of significant amounts of cavitation, as already discussed, is constrained to operate along a single set of characteristic thrust and torque lines relevant to the pitch ratio.

6.4.2 Controllable Pitch Propellers

With the controllable pitch propeller the additional variable of pitch angle introduces a three-dimensional nature to the propeller characteristics, since the total characteristics comprise sets of K_T and K_Q versus J curves for each pitch angle as seen in Figure 6.6. Indeed, for analysis purposes the performance characteristics can be considered as forming a surface, in contrast to the single line for the fixed pitch propeller.

When analyzing the performance of a controllable pitch propeller at off-design conditions use should not be made of fixed pitch characteristics beyond say 5° or 10° from design pitch since the effects of section distortion, discussed in Chapter 3, can considerably affect the performance characteristics.

A further set of parameters arises with controllable pitch propellers and these are the blade spindle torques, a knowledge of which is of importance when designing the blade actuating mechanism. The total spindle torque, which is the torque acting about the spindle axis of the blade and which requires either to be balanced by the hub mechanism in order to hold the blades in the required pitch setting or, alternatively, to be overcome when a pitch change is required, comprises three components as follows:

$$Q_s(J, \Delta\theta) = Q_{\text{SH}}(J, \Delta\theta) + Q_{\text{SC}}(n, \Delta\theta) + Q_{\text{SF}}(J, \Delta\theta)$$

$$(6.14)$$

where

Q_s is the total spindle torque at a given value of J and $\Delta\theta$;

Q_{SH} is the hydrodynamic component of spindle torque due to the pressure field acting on the blade surfaces;

Q_{SC} is the centrifugal component resulting from the blade mass distribution;

Q_{SF} is the frictional component of spindle torque resulting from the relative motion of the surfaces within the blade hub.

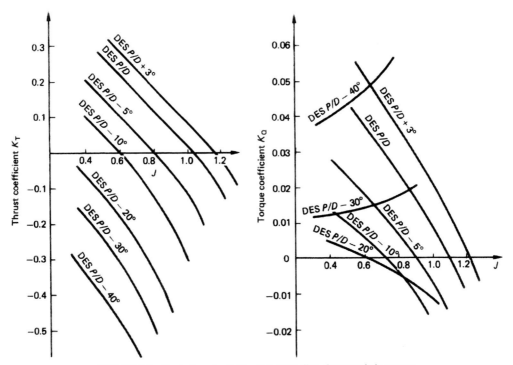

FIGURE 6.6 Typical controllable pitch propeller characteristic curves.

The latter component due to friction depends both on the geometry of the hub mechanism and the system of forces and moments generated by the blade surface pressure fields and mass distribution acting on the blade palm.

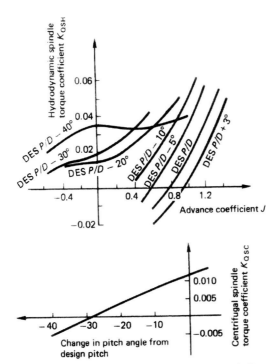

FIGURE 6.7 Typical controllable pitch propeller spindle torque characteristic curves.

Figure 6.7 shows typical hydrodynamic and centrifugal blade spindle torque characteristics for a controllable pitch propeller. In Figure 6.7 the spindle torques are expressed in the coefficient form of K_{QSH} and K_{QSC}. These coefficients are similar in form to the conventional propeller torque coefficient in so far as they relate to the respective spindle torques as follows:

$$K_{QSH} = \frac{Q_{SH}}{\rho n^2 D^5}$$
$$K_{QSC} = \frac{Q_{SC}}{\rho_m n^2 D^5}$$
(6.15)

where ρ is the mass density of water and ρ_m is the mass density of the blade material. Clearly, since the centrifugal component is a mechanical property of the blade only, it is independent of advance coefficient. Hence K_{QSC} is a function of $\Delta\theta$ only.

6.4.3 Ducted Propellers

While the general aspects of the discussion relating to non-ducted, fixed and controllable pitch propellers apply to ducted propellers, the total ducted propulsor thrust is split into two components: the algebraic sum of the propeller and duct thrusts and any second-order interaction effects. To a first approximation, therefore, the total propulsor thrust T can be written as

$$T = T_p + T_n$$

Propeller Performance Characteristics

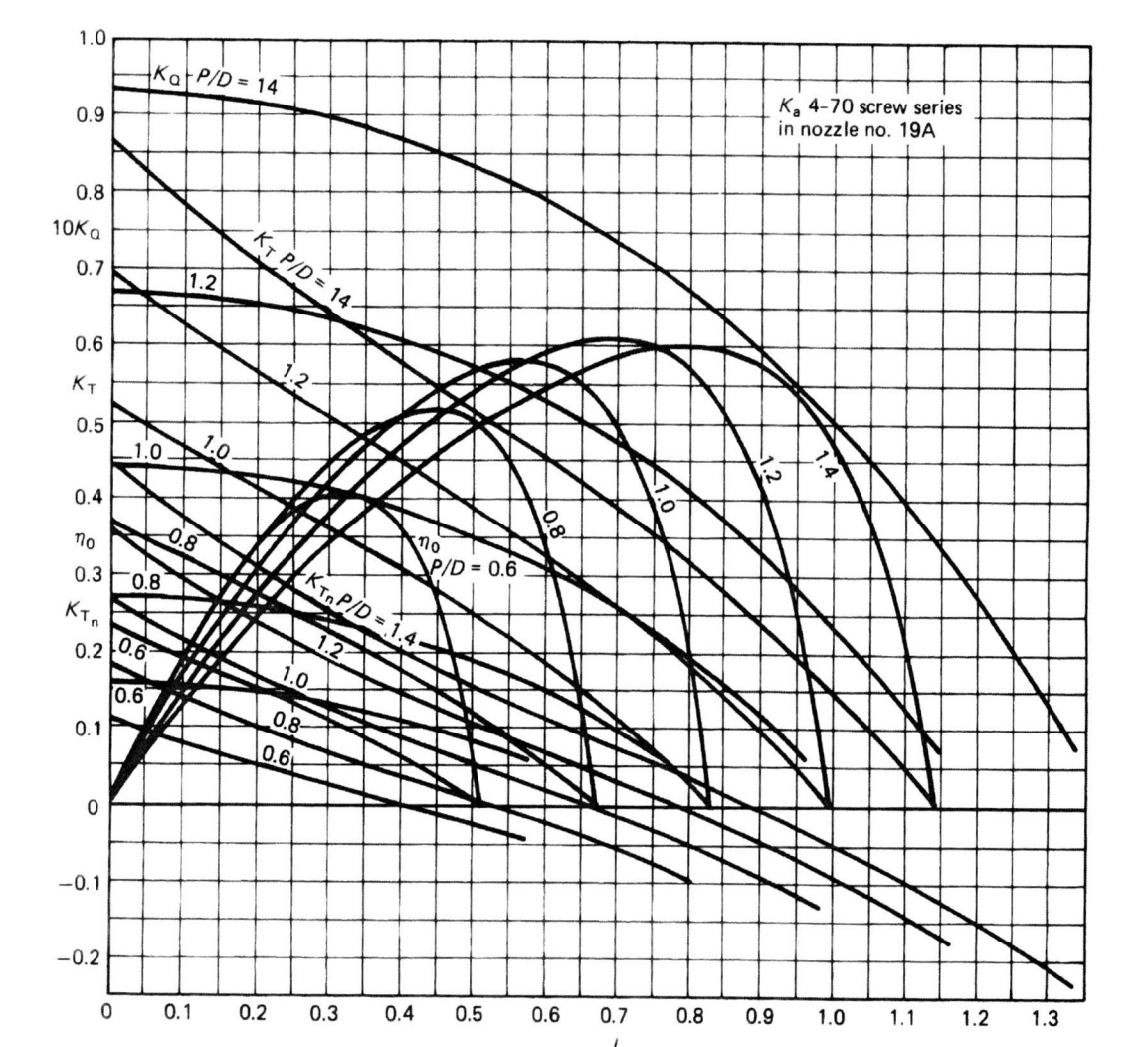

FIGURE 6.8 Open water test results of K_a 4–70 screw series with nozzle no. 19A. *Courtesy: MARIN.*

where T_p is the propeller thrust and T_n is the duct thrust.

In non-dimensional form this becomes

$$K_T = K_{TP} + K_{TN} \qquad (6.16)$$

where the non-dimensionalization factor is $\rho n^2 D^4$ as before for a force.

The results of model tests normally present values of K_T and K_{TN} plotted as a function of advance coefficient J as shown in Figure 6.8 for a fixed pitch ducted propulsor. The torque characteristic is, of course, not split into components since the propeller itself absorbs all of the torque of the engine. In general, the proportion of thrust generated by the duct to that of the total propulsor thrust is a variable over the range of advance coefficient. In merchant practice by far the greater majority of ducted propellers are designed with accelerating ducts, as discussed previously in Chapter 2. For these duct forms the ratio of K_{TN}/K_T is of the order of 0.5 at the bollard pull or zero advance coefficient condition;

however this usually falls to around 0.05 or 0.10 at the design free-running condition. Indeed, if the advance coefficient is increased to a sufficiently high level, then the duct thrust will change sign, as seen in Figure 6.8, and act as a drag; however, this situation is unlikely to arise in normal practice. When decelerating ducts are used, analogous conditions arise, but the use of these ducts is confined to certain specialist cases, normally those having a low radiated noise requirement.

6.4.4 High-Speed Propellers

With high-speed propellers much of what has been said previously will apply depending upon the application. However, the high-speed propeller will be susceptible to two other factors. The first is that cavitation is more likely to occur, and consequently the propeller type and section blade form must be carefully considered in so far as any super-cavitating blade section requirements need to be met.

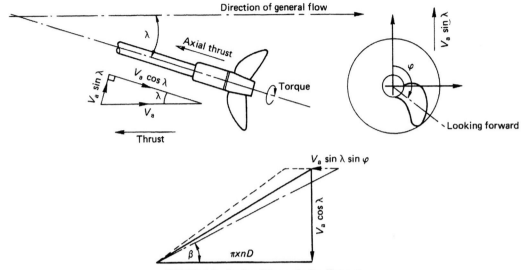

FIGURE 6.9 Inclined flow velocity diagram.

The second factor is that many high-speed propellers are fitted to shafts with considerable rake angles. This rake angle, when combined with the flow directions, gives rise to two flow components acting at the propeller plane as seen in Figure 6.9. The first of these is parallel to the shaft and has a magnitude $V_a \cos(\lambda)$ and the second is perpendicular to the shaft with a magnitude $V_a \sin(\lambda)$ where λ is the relative shaft angle as shown in the figure. It will be appreciated that the second, or perpendicular, component immediately presents an asymmetry when viewed in terms of propeller relative velocities. This is because on one side of the propeller disc the perpendicular velocity component is additive to the propeller rotational velocity whilst on the other side it is subtractive (see Figure 6.9). This then gives rise to a differential loading of the blades as they rotate around the propeller disc which causes a thrust eccentricity

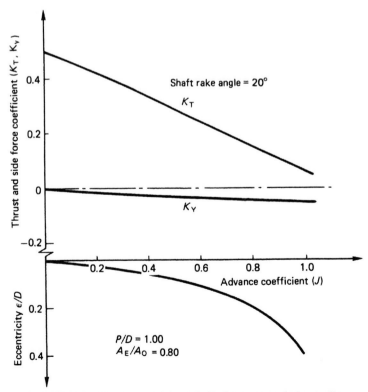

FIGURE 6.10 Thrust eccentricity and side forces on a raked propeller.

TABLE 6.3 Fixed Pitch, Non-ducted Propeller Series Summary

Series	Number of Propellers in Series	Range of Parameters			D(mm)	r_h/R	Cavitation Data Available	Notes
		Z	A_E/A_O	P/D				
Wageningen B-series	$\simeq 120$	2−7	0.3−1.05	0.6−1.4	250	0.169	No	Four-bladed propeller has non-constant pitch distribution
Au-series	34	4−7	0.4−0.758	0.5−1.2	250	0.180	No	
Gawn-series	37	3	0.2−1.1	0.4−2.0	508	0.200	No	
KCA-series	$\simeq 30$	3	0.5−1.25	0.6−2.0	406	0.200	Yes	
Ma-series	32	3 and 5	0.75−1.20	1.0−1.45	250	0.190	Yes	
Newton−Rader series	12	3	0.5−1.0	1.05−2.08	254	0.167	Yes	
KCD-series	24	3−6 (mainly 4)	0.587 principal 0.44−0.8	0.6−1.6	406	0.200	Yes	Propellers not geosyms
Meridian series	20	6	0.45−1.05	0.4−1.2	305	0.185	Yes	Propellers not geosyms

and side force components. Figure 6.10 demonstrates these features which of course will apply generally to all propellers working in non-uniform flow but are more noticeable with high-speed propellers due to the speeds and inclinations involved. The magnitude of these eccentricities can be quite large; for example, in the case of unity pitch ratio with a shaft rake of 20°, the transverse thrust eccentricity indicated by Figure 6.10 may well reach 0.40R. Naturally, due to the non-uniform tangential wake field the resulting cavitation pattern will also be anti-symmetric.

6.5 STANDARD SERIES DATA

Over the years there have been a number of standard series of propellers tested in many different establishments around the world. To discuss them all in any detail would clearly be a large undertaking requiring considerable space; consequently, those most commonly used today by propeller designers and analysts are referenced here.

The principal aim in carrying out systematic propeller tests is to provide a database to help the designer understand the factors which influence propeller performance and, in some cases, the inception and form of cavitation on the blades under various operating conditions.

A second purpose is to provide design diagrams, or charts, which will assist in selecting the most appropriate

dimensions of actual propellers to suit full-size ship applications.

The purpose here is not to provide the reader with an exhaustive catalogue of results but to introduce the various model series in terms of their nature and extent and provide suitable references from which the full details can be found. Table 6.3 summarizes the fixed pitch, non-ducted propeller series referenced here to enable rapid selection of the appropriate series for a particular set of circumstances.

6.5.1 Wageningen B-Screw Series

This is the most extensive and perhaps widely used propeller series. The series was originally presented in a set of papers presented by Troost[3−5] in the late 1940s and amongst many practitioners it is still referred to as the 'Troost series'. Over the years the model series has been added to so as to provide a comprehensive fixed pitch, non-ducted propeller series. From analysis of the early results it was appreciated that a certain unfairness between the various design diagrams existed and this was considered to result from the scale effects resulting from the different model tests. This led to a complete re-appraisal of the series in which the differences in test procedures were taken into account and the results of this work were presented by van Lammeren et al.[6]

TABLE 6.4 Extent of the Wageningen B-Screw Series (Taken From Reference 6)

Blade Number (Z)	Blade Area Ratio A_E/A_O					
2	0.30					
3	0.35	0.50	0.65	0.80		
4	0.40	0.55	0.70	0.85	1.00	
5	0.45	0.60	0.75			1.05
6	0.50	0.65	0.80			
7	0.55	0.70	0.85			

The extent of the series in terms of a blade number versus blade area ratio matrix is shown in Table 6.4 from which it may be seen that the series numbers some 20 blade area—blade number configurations. The geometry of the series is shown in Table 6.5, from which it can be seen that a reasonably consistent geometry is maintained between the members of the series with only a few anomalies; notably the non-constant nature of the face pitch near the root of the four-blade series and the blade outline of the three-bladed propellers. For completeness purposes Figure 6.11 shows the geometric outline of the B5 propeller set. Note that the propellers of this series are generally referred to by the notation $BZ \cdot y$, where B denotes the 'B'-series, Z is the blade number and y is the blade expanded area. The face pitch ratio for the series lies in the range 0.6–1.4.

The results of the fairing exercise reported by Oosterveld paved the way for detailed regression studies on the performance characteristics given by this model series. Oosterveld and van Oossanen[7] reported the findings of this work in which the open water characteristics of the series are represented at a Reynolds number 2×10^6 by an equation of the following form:

$$\left. \begin{array}{l} K_Q = \sum_{n=1}^{47} C_n(j)^{S_n}(P/D)^{t_n}(A_E/A_O)^{u_n}(z)^{v_n} \\ K_T = \sum_{n=1}^{39} C_n(j)^{S_n}(P/D)^{t_n}(A_E/A_O)^{u_n}(z)^{v_n} \end{array} \right\} \quad (6.17)$$

where the coefficients are reproduced in Table 6.6.

To extend this work further so that propeller characteristics can be predicted for other Reynolds numbers within the range 2×10^6 to 2×10^9 a set of corrections of the following form was derived:

$$\left\{ \begin{array}{l} K_T(R_n) \\ K_Q(R_n) \end{array} \right\} = \left\{ \begin{array}{l} K_T(R_n = 2 \times 10^6) \\ K_Q(R_n = 2 \times 10^6) \end{array} \right\} + \left\{ \begin{array}{l} \Delta K_T(R_n) \\ \Delta K_Q(R_n) \end{array} \right\}$$

$$(6.18)$$

where

$\Delta K_T =$
  0.000353485
  $- 0.00333758 \, (A_E/A_O) J^2$
  $- 0.00478125 \, (A_E/A_O)(P/D) J$
  $+ 0.000257792 \, (\log R_n - 0.301)^2 \cdot (A_E/A_O) J^2$
  $+ 0.0000643192 (\log R_n - 0301)(P/D)^6 J^2$
  $- 0.0000110636 (\log R_n - 0.301)^2 (P/D)^6 J^2$
  $- 0.0000276305 (\log R_n - 0.301) Z(A_E/A_O) J^2$
  $+ 0.0000954 (\log R_n - 0.301) Z(A_E/A_O)(P/D) J$
  $+ 0.0000032049 (\log R_n - 0.301) Z^2 (A_E/A_O)$
  $\times (P/D)^3 J$

$\Delta K_Q =$
  -0.000591412
  $+ 0.00696898(P/D)$
  $- 0.0000666654 Z(P/D)^6$
  $+ 0.0160818(A_E/A_O)^2$
  $- 0.000938091 (\log R_n - 0301)(P/D)$
  $- 0.00059593 (\log R_n - 0.301)(P/D)^2$
  $+ 0.0000782099 (\log R_n - 0.301)^2 (P/D)^2$
  $+ 0.0000052199 (\log R_n - 0.301) Z(A_E/A_O) J^2$
  $- 0.00000088528 (\log R_n - 0.301)^2 Z(A_E/A_O) \times (P/D) J$
  $+ 0.0000230171 (\log R_n - 0301) Z(P/D)^6$
  $- 0.00000184341 (\log R_n - 0301)^2 Z(P/D)^6$
  $- 0.00400252 (\log R_n - 0.301)(A_E/A_O)^2$
  $+ 0.000220915 (\log R_n - 0.301)^2 (A_E/A_O)^2$

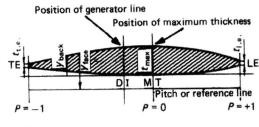

LE = leading edge
TE = trailing edge
MT = location of maximum thickness
DI = location of directrix

The Wageningen series is a general purpose, fixed pitch, non-ducted propeller series which is used extensively for

Propeller Performance Characteristics

TABLE 6.5 Geometry of the Wageningen B-Screw Series (taken from Reference 7)

Dimensions of Four-, Five-, Six- and Seven-Bladed Propellers

r/R	$\frac{c}{D} \cdot \frac{Z}{A_E/A_O}$	a/c	b/c	$t/D = A_r - B_r Z$	
				A_r	B_r
0.2	1.662	0.617	0.350	0.0526	0.0040
0.3	1.882	0.613	0.350	0.0464	0.0035
0.4	2.050	0.601	0.351	0.0402	0.0030
0.5	2.152	0.586	0.355	0.0340	0.0025
0.6	2.187	0.561	0.389	0.0278	0.0020
0.7	2.144	0.524	0.443	0.0216	0.0015
0.8	1.970	0.463	0.479	0.0154	0.0010
0.9	1.582	0.351	0.500	0.0092	0.0005
1.0	0.000	0.000	0.000	0.0030	0.0000

Dimensions for Three-Bladed Propellers

r/R	$\frac{c}{D} \cdot \frac{Z}{A_E/A_O}$	a/c	b/c	$t/D = A_r - B_r Z$	
				A_r	B_r
0.2	1.633	0.616	0.350	0.0526	0.0040
0.3	1.832	0.611	0.350	0.0464	0.0035
0.4	2.000	0.599	0.350	0.0402	0.0030
0.5	2.120	0.583	0.355	0.0340	0.0025
0.6	2.186	0.558	0.389	0.0278	0.0020
0.7	2.168	0.526	0.442	0.0216	0.0015
0.8	2.127	0.481	0.478	0.0154	0.0010
0.9	1.657	0.400	0.500	0.0092	0.0005
1.0	0.000	0.000	0.000	0.0030	0.0000

Values of V_1 for Use in the Equations

r/R P	−1.0	−0.95	−0.9	−0.8	−0.7	−0.6	−0.5	−0.4	−0.2	0
0.7−1.0	0	0	0	0	0	0	0	0	0	0
0.6	0	0	0	0	0	0	0	0	0	0
0.5	0.0522	0.0420	0.0330	0.0190	0.0100	0.0040	0.0012	0	0	0
0.4	0.1467	0.1200	0.0972	0.0630	0.0395	0.0214	0.0116	0.0044	0	0
0.3	0.2306	0.2040	0.1790	0.1333	0.0943	0.0623	0.0376	0.0202	0.0033	0
0.25	0.2598	0.2372	0.2115	0.1651	0.1246	0.0899	0.0579	0.0350	0.0084	0
0.2	0.2826	0.2630	0.2400	0.1967	0.1570	0.1207	0.0880	0.0592	0.0172	0
0.15	0.3000	0.2824	0.2650	0.2300	0.1950	0.1610	0.1280	0.0955	0.0365	0

r/R P	+1.0	+0.95	+0.9	+0.85	+0.8	+0.7	+0.6	+0.5	+0.4	+0.2	0
0.7−1.0	0	0	0	0	0	0	0	0	0	0	0
0.6	0.0382	0.0169	0.0067	0.0022	0.0006	0	0	0	0	0	0
0.5	0.1278	0.0778	0.0500	0.0328	0.0211	0.0085	0.0034	0.0008	0	0	0

(Continued)

TABLE 6.5 Geometry of the Wageningen B-Screw Series (taken from Reference 7)—cont'd

0.4	0.2181	0.1467	0.1088	0.0833	0.0637	0.0357	0.0189	0.0090	0.0033	0	0
0.3	0.2923	0.2186	0.1760	0.1445	0.1191	0.0790	0.0503	0.0300	0.0148	0.0027	0
0.25	0.3256	0.2513	0.2068	0.1747	0.1465	0.1008	0.0669	0.0417	0.0224	0.0031	0
0.2	0.3560	0.2821	0.2353	0.2000	0.1685	0.1180	0.0804	0.0520	0.0304	0.0049	0
0.15	0.3860	0.3150	0.2642	0.2230	0.1870	0.1320	0.0920	0.0615	0.0384	0.0096	0

Values of V_2 for Use in the Equations

r/R P	−1.0	−0.95	−0.9	−0.8	−0.7	−0.6	−0.5	−0.4	−0.2	0
0.9–1.0	0	0.0975	0.19	0.36	0.51	0.64	0.75	0.84	0.96	1
0.85	0	0.0975	0.19	0.36	0.51	0.64	0.75	0.84	0.96	1
0.8	0	0.0975	0.19	0.36	0.51	0.64	0.75	0.84	0.96	1
0.7	0	0.0975	0.19	0.36	0.51	0.64	0.75	0.84	0.96	1
0.6	0	0.0965	0.1885	0.3585	0.5110	0.6415	0.7530	0.8426	0.9613	1
0.5	0	0.0950	0.1865	0.3569	0.5140	0.6439	0.7580	0.8456	0.9639	1
0.4	0	0.0905	0.1810	0.3500	0.5040	0.6353	0.7525	0.8415	0.9645	1
0.3	0	0.0800	0.1670	0.3360	0.4885	0.6195	0.7335	0.8265	0.9583	1
0.25	0	0.0725	0.1567	0.3228	0.4740	0.6050	0.7184	0.8139	0.9519	1
0.2	0	0.0640	0.1455	0.3060	0.4535	0.5842	0.6995	0.7984	0.9446	1
0.15	0	0.0540	0.1325	0.2870	0.4280	0.5585	0.6770	0.7805	0.9360	1

r/R P	+1.0	+0.95	+0.9	+0.85	+0.8	+0.7	+0.6	+0.5	+0.4	+0.2	0
0.9–1.0	0	0.0975	0.1900	0.2775	0.3600	0.51	0.6400	0.75	0.8400	0.9600	1
0.85	0	0.1000	0.1950	0.2830	0.3660	0.5160	0.6455	0.7550	0.8450	0.9615	1
0.8	0	0.1050	0.2028	0.2925	0.3765	0.5265	0.6545	0.7635	0.8520	0.9635	1
0.7	0	0.1240	0.2337	0.3300	0.4140	0.5615	0.6840	0.7850	0.8660	0.9675	1
0.6	0	0.1485	0.2720	0.3775	0.4620	0.6060	0.7200	0.8090	0.8790	0.9690	1
0.5	0	0.1750	0.3056	0.4135	0.5039	0.6430	0.7478	0.8275	0.8880	0.9710	1
0.4	0	0.1935	0.3235	0.4335	0.5220	0.6590	0.7593	0.8345	0.8933	0.9725	1
0.3	0	0.1890	0.3197	0.4265	0.5130	0.6505	0.7520	0.8315	0.8020	0.9750	1
0.25	0	0.1758	0.3042	0.4108	0.4982	0.6359	0.7415	0.8259	0.8899	0.9751	1
0.2	0	0.1560	0.2840	0.3905	0.4777	0.6190	0.7277	0.8170	0.8875	0.9750	1
0.15	0	0.1300	0.2600	0.3665	0.4520	0.5995	0.7105	0.8055	0.8825	0.9760	1

A_r, B_r = constants in equation for t/D.
a = distance between leading edge and generator line at r.
b = distance between leading edge and location of maximum thickness.
c = chord length of blade section at radius r.
t = maximum blade section thickness at radius r.

design and analysis purposes. A variant of the series, designated the BB-series, was introduced since it was felt that the B-series had tip chord lengths that were not entirely representative of modern practice. Accordingly, the BB-series had a re-defined blade outline with wider tips than the parent form. However, the BB-series, of which only a few members exist, has not been widely used.

Chapter | 6 Propeller Performance Characteristics

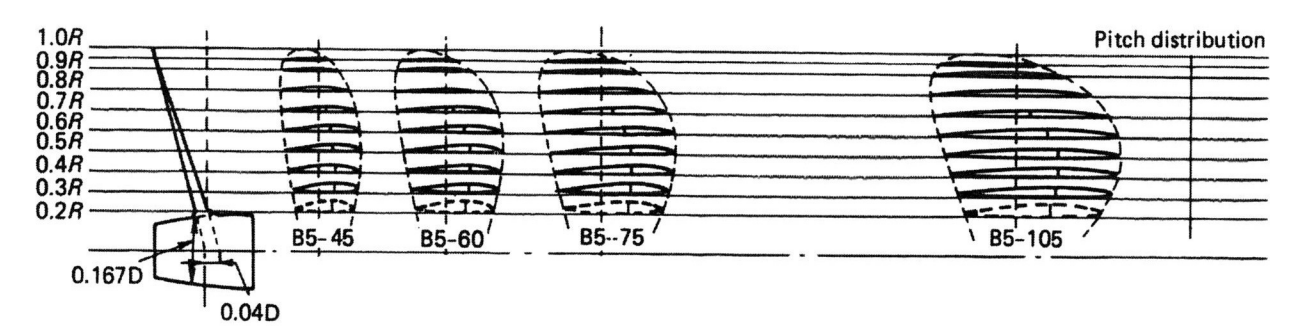

FIGURE 6.11 General plan of B5-screw series. *Reproduced with permission from Reference 6.*

TABLE 6.6 Coefficients for the K_T and K_Q Polynomials Representing the Wageningen B-Screw Series for a Reynolds Number of 2×10^6 (taken from Reference 7)

	Thrust (K_T)					Torque (K_Q)				
n	$C_{s,t,u,v}$	s (J)	t (P/D)	u (A_E/A_O)	υ(Z)	N	$C_{s,t,u,v}$	s (J)	t (P/D)	u (A_E/A_O)	υ(Z)
1	+0.00880496	0	0	0	0	1	+0.00379368	0	0	0	0
2	−0.204554	1	0	0	0	2	+0.00886523	2	0	0	0
3	+0.166351	0	1	0	0	3	−0.032241	1	1	0	0
4	+0.158114	0	2	0	0	4	+0.00344778	0	2	0	0
5	−0.147581	2	0	1	0	5	−0.0408811	0	1	1	0
6	−0.481497	1	1	1	0	6	−0.108009	1	1	1	0
7	+0.415437	0	2	1	0	7	−0.0885381	2	1	1	0
8	+0.0144043	0	0	0	1	8	+0.188561	0	2	1	0
9	−0.0530054	2	0	0	1	9	−0.00370871	1	0	0	1
10	+0.0143481	0	1	0	1	10	+0.00513696	0	1	0	1
11	+0.0606826	1	1	0	1	11	+0.0209449	1	1	0	1
12	−0.0125894	0	0	1	1	12	+0.00474319	2	1	0	1
13	+0.0109689	1	0	1	1	13	−0.00723408	2	0	1	1
14	−0.133698	0	3	0	0	14	+0.00438388	1	1	1	1
15	+0.00638407	0	6	0	0	15	−0.0269403	0	2	1	1
16	−0.00132718	2	6	0	0	16	+0.0558082	3	0	1	0
17	+0.168496	3	0	1	0	17	+0.0161886	0	3	1	0
18	−0.0507214	0	0	2	0	18	+0.00318086	1	3	1	0
19	+0.0854559	2	0	2	0	19	+0.015896	0	0	2	0
20	−0.0504475	3	0	2	0	20	+0.0471729	1	0	2	0
21	+0.010465	1	6	2	0	21	+0.0196283	3	0	2	0
22	−0.00648272	2	6	2	0	22	−0.0502782	0	1	2	0
23	−0.00841728	0	3	0	1	23	−0.030055	3	1	2	0

(Continued)

TABLE 6.6 Coefficients for the K_T and K_Q Polynomials Representing the Wageningen B-Screw Series for a Reynolds Number of 2×10^6 (taken from Reference 7)—cont'd

	Thrust (K_T)					Torque (K_Q)				
n	$C_{s,t,u,v}$	s (J)	t (P/D)	u (A_E/A_O)	v(Z)	N	$C_{s,t,u,v}$	s (J)	t (P/D)	u (A_E/A_O)	v(Z)
24	+0.0168424	1	3	0	1	24	+0.0417122	2	2	2	0
25	−0.00102296	3	3	0	1	25	−0.0397722	0	3	2	0
26	−0.0317791	0	3	1	1	26	−0.00350024	0	6	2	0
27	+0.018604	1	0	2	1	27	−0.0106854	3	0	0	1
28	−0.00410798	0	2	2	1	28	+0.00110903	3	3	0	1
29	−0.000606848	0	0	0	2	29	−0.000313912	0	6	0	1
30	−0.0049819	1	0	0	2	30	+0.0035985	3	0	1	1
31	+0.0025983	2	0	0	2	31	−0.00142121	0	6	1	1
32	−0.000560528	3	0	0	2	32	−0.00383637	1	0	2	1
33	−0.00163652	1	2	0	2	33	+0.0126803	0	2	2	1
34	−0.000328787	1	6	0	2	34	−0.00318278	2	3	2	1
35	+0.000116502	2	6	0	2	35	+0.00334268	0	6	2	1
36	+0.000690904	0	0	1	2	36	−0.00183491	1	1	0	2
37	+0.00421749	0	3	1	2	37	+0.000112451	3	2	0	2
38	+0.0000565229	3	6	1	2	38	−0.0000297228	3	6	0	2
39	−0.00146564	0	3	2	2	39	+0.000269551	1	0	1	2
						40	+0.00083265	2	0	1	2
						41	+0.00155334	0	2	1	2
						42	+0.000302683	0	6	1	2
						43	−0.0001843	0	0	2	2
						44	−0.000425399	0	3	2	2
						45	+0.0000869243	3	3	2	2
						46	−0.0004659	0	6	2	2
						47	+0.0000554194	1	6	2	2

$$\left. \begin{array}{l} Y_{face} = V_1(t_{max}\, t_{t.e.}) \\ Y_{back} = (V_1 + V_2)(t_{max}\, t_{t.e.}) + t_{t.e.} \end{array} \right\} \text{ for } P \leq 0$$

and

$$\left. \begin{array}{l} Y_{face} = V_1(t_{max} - t_{t.e.}) \\ Y_{back} = (V_1 + V_2)(t_{max} - t_{t.e.}) + t_{l.e.} \end{array} \right\} \text{ for } P \geq 0$$

Referring to the diagram, note the following:

Y_{face}, Y_{back} = vertical ordinate of a point on a blade section on the face and on the back with respect to the pitch line.

t_{max} = maximum thickness of blade section.
$t_{t.e.}$, $t_{l.e.}$ = extrapolated blade section thickness at the trailing and leading edges.
V_1, V_2 = tabulated functions dependent on r/R and P.
P = non-dimensional co-ordinate along pitch line from position of maximum thickness to leading edge (where $P = 1$), and from position of maximum thickness to trailing edge (where $P = -1$).

6.5.2 Japanese AU-Series

This propeller series is in many ways a complementary series to the Wageningen B-series; however, outside

Propeller Performance Characteristics

TABLE 6.7 Members of the AU-series (taken from Reference 8)

Number Blades	4		5		6			7	
Model propellers numbers	1305 – 1309	1310 – 1314	1128 – 1132	1133 – 1137	1189 – 1192	1193 – 1196	1197 – 1200	1144	1147
Pitch ratio	0.5, 0.6, 0.8, 1.0, 1.2		0.4, 0.6, 0.8, 1.0, 1.2		0.5, 0.7, 0.9, 1.1			0.8	
Blade section	AU		AU		AU	AU$_W$	AU$_W$	AU	
Diameter (m)	0.250		0.250		0.250			0.250	
Expanded area ratio	0.40	0.55	0.50	0.65	0.70	0.70	0.55	0.65	0.758
Boss ratio	0.180		0.180		0.180			0.180	
Blade thickness ratio	0.050		0.050		0.050			0.050	
Rake angle	10°	0°	10°	0°	10°	0°		10°	0°

Japan it has not gained the widespread popularity of the B-series. The series reported by Reference 8 comprises some propellers having a range of blade numbers from four to seven and blade area ratios in the range 0.40–0.758. Table 6.7 details the members of the series and Table 6.8, the blade geometry. The propeller series, as its name implies, has AU-type aerofoil sections and was developed from an earlier series having Unken-type sections.

6.5.3 Gawn Series

This series of propellers whose results were presented by Gawn[9] comprised a set of 37 three-bladed propellers covering a range of pitch ratios from 0.4–2.0 and blade area ratios from 0.2–1.1.

The propellers of this series each had a diameter of 503 mm (20 in.), and by this means many of the scale effects associated with smaller diameter propeller series have been attenuated. Each of the propellers has a uniform face pitch; segmental blade sections; constant blade thickness ratio, namely 0.060, and a boss diameter of 0.2D. The developed blade outline was of elliptical form with the inner and outer vertices at 0.1R and the blade tip, respectively. Figure 6.12 shows the outline of the propellers in this series. The entire series were tested in the No. 2 towing rank at A.E.W. Haslar within a range of slip from 0–100 per cent: to achieve this the propeller rotational speed was in the range 250–500 rpm. No cavitation characteristics are given for the series.

The propeller series represents a valuable dataset, despite the somewhat dated propeller geometry, for undertaking preliminary design studies for warships and other high-performance craft due to the wide range of P/D and A_E/A_O values covered. Blount and Hubble[10] in considering methods for the sizing of small craft propellers developed a set of regression coefficients of the form of equation (6.17) to represent the Gawn series. The coefficients for this series are given in Table 6.9 and it is suggested that the range of applicability of the regression study should be for pitch ratio values from 0.8–1.4, although the study was based on the wider range of 0.6–1.6. Inevitably, however, some regression formulations of model test data tend to deteriorate towards the outer limits of the data set, and this is the cause of the above restriction.

6.5.4 KCA Series

The KCA series, or as it is sometimes known, the Gawn–Burrill series[11] is in many ways a complementary series to the Gawn series introduced above. The KCA series comprise 30 three-bladed, 406 mm (16 in.) models embracing a range of pitch ratios from 0.6–2.0 and blade area ratios from 0.5–1.1. Thus the propellers can be seen to cover a similar range of parameters to the Gawn series in that they have the same upper limits for P/D and A_E/A_O, but slightly curtailed lower limits. The propellers of the KCA series all had uniform face pitch, segmental sections over the outer half of the blade, and in the inner half, the flat faces of the segmental sections were lifted at the leading and trailing edges. The blade thickness fraction of the parent screw, shown in Figure 6.13, was 0.045. The boss diameter of the series was $0.2D$.

This propeller series was tested in the cavitation tunnel at the University of Newcastle upon Tyne, England, and consequently, since the cavitation tunnel was used rather

TABLE 6.8 Blade Geometry of the AU-series (taken from Reference 8)

Dimensions of AU-4 Series Propeller

r/R		0.2	0.3	0.4	0.5	0.6	0.66	0.7	0.8	0.9	0.95	1.0	
Width of blade as % of maximum blade width	From generator line to trailing edge	27.96	33.45	38.76	43.54	47.96	49.74	51.33	52.39	48.49	42.07	17.29	Maximum blade width at 0.66 $r/R = 0.226D$ for $A_E/A = 0.40$
	From generator line to leading edge	38.58	14.25	48.32	50.80	51.15	50.26	48.31	40.53	25.13	13.35		
	Total blade width	66.54	77.70	87.08	94.34	99.11	100.00	96.64	92.92	73.62	55.62		
Blade thickness as % of D		4.06	3.59	3.12	2.65	2.18	1.90	1.71	1.24	0.77	0.54	0.30	Maximum blade thickness at proportional axis = 0.050D
Distance of the point of maximum thickness from the leading edge as % of blade width		32.0	32.0	32.0	32.5	34.9	37.9	40.2	45.4	48.9	50.0		

Offset of AU-Type Propeller

(1) Ordinates of X-Value are given as % of Blade Width

(2) Ordinates of Y-Value are Given as % of Y_{max}

r/R		0	2.00	4.00	6.00	10.00	15.00	20.00	30.00	32.00	40.00	50.00	60.00	70.00	80.00	90.00	95.00	100.00
0.20	X																	
	Y_0	35.00	51.85	59.75	66.15	76.05	85.25	92.20	89.80	100.00	97.75	89.95	78.15	63.15	45.25	25.30	15.00	4.50
	Y_U		24.25	19.05	15.00	10.00	5.40	2.35										
0.30	X	0	2.00	4.00	6.00	10.00	15.00	20.00	30.00	32.00	40.00	50.00	60.00	70.00	80.00	90.00	93.00	100.00
	Y_0	35.00	51.85	59.75	66.15	76.05	85.25	92.20	99.80	100.00	97.75	89.95	78.15	63.15	45.25	25.30	15.00	4.50
	Y_U		24.25	19.05	15.00	10.00	5.40	2.35										
0.40	X	0	2.00	4.00	6.00	10.00	15.00	20.00	30.00	32.00	40.00	50.00	60.00	70.00	80.00	90.00	95.00	100.00
	Y_0	35.00	51.85	59.75	66.15	76.05	85.25	92.20	99.80	100.00	97.75	89.95	78.15	63.15	45.25	25.30	15.00	4.50
	Y_U		24.25	19.05	15.00	10.00	5.40	2.35										
0.50	X	0	2.03	4.06	6.09	10.16	15.23	20.31	30.47	32.50	40.44	50.37	60.29	70.22	80.15	90.07	95.04	100.00
	Y_0	35.00	51.85	59.75	66.15	76.05	85.25	92.20	99.80	100.00	97.75	89.95	78.15	63.15	45.25	25.30	15.00	4.50
	Y_U		24.25	19.05	15.00	10.00	5.40	2.35										

	X	0	2.18	4.36	6.54	10.91	16.36	21.81	32.72	34.90	42.56	52.13	61.70	71.28	80.85	90.43	95.21	100.00
0.60	Y_O	34.00	49.60	58.00	64.75	75.20	84.80	91.80	99.80	100.00	97.75	89.95	78.15	63.15	45.25	25.30	15.00	4.50
	Y_U		23.60	18.10	14.25	9.45	5.00	2.25										
	X	0	2.51	5.03	7.54	12.56	18.84	25.12	37.69	40.20	47.23	56.03	64.82	73.62	82.41	91.21	95.60	100.00
0.70	Y_O	30.00	42.90	52.20	59.90	71.65	82.35	90.60	99.80	1000.00	97.75	89.95	78.15	63.15	45.25	25.30	15.00	4.50
	Y_U		20.50	15.45	11.95	7.70	4.10	1.75										
	X	0	2.84	5.68	8.51	14.19	21.28	28.38	42.56	45.40	51.82	59.85	67.88	75.91	83.94	91.97	95.99	100.00
0.80	Y_O	21.00	32.45	41.70	50.10	64.60	78.45	88.90	99.80	100.00	97.75	89.95	78.15	63.15	45.25	25.30	15.00	4.50
	Y_U		14.00	10.45	8.05	5.05	2.70	1.15										
	X	0	3.06	6.11	9.17	15.28	22.92	30.56	45.85	48.90	54.91	62.42	69.94	77.46	84.97	92.49	96.24	100.00
0.90	Y_O	8.30	21.10	31.50	40.90	57.45	74.70	87.45	99.70	100.00	98.65	92.75	83.00	69.35	51.85	30.80	19.40	6.85
	Y_U		4.00	2.70	2.05	1.20	0.70	0.30										
	X	0	3.13	6.25	9.38	15.63	23.44	31.25	46.87	50.00	55.88	63.23	70.59	77.94	85.30	92.65	96.32	100.00
0.95	Y_O	6.00	19.65	30.00	39.60	56.75	74.30	87.30	99.65	100.00	99.00	93.85	84.65	71.65	54.30	33.50	21.50	8.00
	Y_U																	

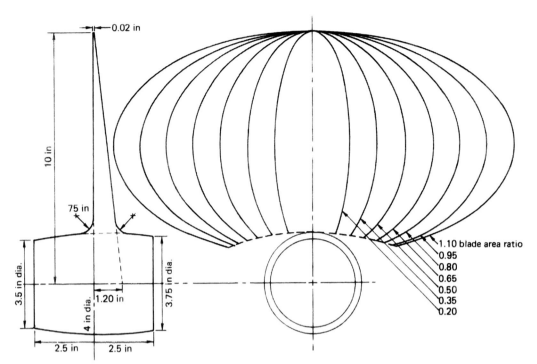

FIGURE 6.12 Blade outline of the Gawn series. *Reproduced with permission from Reference 9.*

than the towing tank, the propeller series was tested at a range of different cavitation numbers. The range used gave a series of six cavitation numbers, based on the free stream advance velocity, as follows: 5.3, 2.0, 1.5, 0.75 and 0.50. As a consequence, using this series it is possible to study the effects of the global cavitation performance of a proposed propeller design.

In order to assist in design studies using the KCA series, Emerson and Sinclair[12] have presented $B_p - \delta$ diagrams for the series both at non-cavitating and cavitating conditions, together with additional thrust and torque data for a BAR of 1.25 and *P/D* of unity.

Despite a lack of data at very low advance coefficients due to the experimental limitation of the cavitation tunnel, the KCA series of propellers, when used in conjunction with the Gawn series, provides an immensely valuable set of data upon which to base design studies of high-speed or naval craft.

6.5.5 Lindgren Series (Ma-Series)

Lindgren, working at SSPA in the 1950s, tested a series of three- and five-bladed propellers embracing a range of *P/D* ratios from 1.00−1.45 and developed area ratios from 0.75−1.20 (Reference 13). The series, designated the Ma-series, is shown in Table 6.10 from which it is seen that a total of 32 propellers were tested.

The propellers are all constant pitch with modified elliptical blade forms and sections of approximately circular back profiles. The diameter of the propellers is 250 mm, which is smaller than either of the two previous series and the boss diameter of the series is 0.19*D*. The thickness fraction of this propeller series varies between the members of the series, and is shown in Table 6.11.

The propellers of this series were tested in both a towing tank and cavitation tunnel and, consequently, provide a reasonably comprehensive set of data for preliminary study purposes. The data is presented in both K_T, K_Q versus *J* form and also in design diagram form. Although the basic design of the Ma-series propellers can be considered to be somewhat dated, it does provide a further complementary set of data to the Gawn and Gawn−Burrill results for the design of high-performance and naval craft.

6.5.6 Newton−Rader Series

The Newton−Rader series embraces a relatively limited set of twelve, three-bladed propellers intended for high-speed craft. The series (Reference 14) was designed to cover pitch ratios in the range 1.0−2.0 and blade area ratios from about 0.5−1.0.

The parent model of the series, based on a design for a particular vessel, had a diameter of 254 mm (10 in.). The principal features of the parent design were a constant face pitch ratio of 1.2 and a blade area ratio of 0.750, together with a non-linear blade thickness distribution having a blade thickness fraction of 0.06. The blade section form was based on the NACA $a = 1.0$ mean line with a quasi-

TABLE 6.9 Blount and Hubble Coefficients for Gawn Propeller Series — Equation (6.17) (taken from Reference 10)

	Thrust (K_T)						Torque (K_Q)				
n	C_n	s (J)	t (P/D)	u(EAR)	v(Z)	n	C_n	s (J)	t (P/D)	u (EAR)	v(Z)
1	−0.0558636300	0	0	0	0	1	0.0051589800	0	0	0	0
2	−0.2173010900	1	0	0	0	2	0.0160666800	2	0	0	0
3	0.2605314000	0	1	0	0	3	−0.0441153000	1	1	0	0
4	0.1581140000	0	2	0	0	4	0.0068222300	0	2	0	0
5	−0.1475810000	2	0	1	0	5	−0.0408811000	0	1	1	0
6	−0.4814970000	1	1	1	0	6	−0.0773296700	1	1	1	0
7	0.3781227800	0	2	1	0	7	−0.0885381000	2	1	1	0
8	0.0144043000	0	0	0	1	8	0.1693750200	0	2	1	0
9	−0.0530054000	2	0	0	1	9	−0.0037087100	1	0	0	1
10	0.0143481000	0	1	0	1	10	0.0051369600	0	1	0	1
11	0.0606826000	1	1	0	1	11	0.0209449000	1	1	0	1
12	−0.0125894000	0	0	1	1	12	0.0047431900	2	1	0	1
13	0.0109689000	1	0	1	1	13	−0.0072340800	2	0	1	1
14	−0.1336980000	0	3	0	0	14	0.0043838800	1	1	1	1
15	0.0024115700	0	6	0	0	15	−0.0269403000	0	2	1	1
16	−0.0005300200	2	6	0	0	16	0.0558082000	3	0	1	0
17	0.1684960000	3	0	1	0	17	0.0161886000	0	3	1	0
18	0.0263454200	0	0	2	0	18	0.0031808600	1	3	1	0
19	0.0436013600	2	0	2	0	19	0.0129043500	0	0	2	0
20	−0.0311849300	3	0	2	0	20	0.0244508400	1	0	2	0
21	0.0124921500	1	6	2	0	21	0.0070064300	3	0	2	0
22	−0.0064827200	2	6	2	0	22	−0.0271904600	0	1	2	0
23	−0.0084172800	0	3	0	1	23	−0.0166458600	3	1	2	0
24	0.0168424000	1	3	0	1	24	0.0300449000	2	2	2	0
25	−0.0010229600	3	3	0	1	25	−0.0336974900	0	3	2	0
26	−0.0317791000	0	3	1	1	26	−0.0035002400	0	6	2	0

(Continued)

TABLE 6.9 Blount and Hubble Coefficients for Gawn Propeller Series — Equation (6.17) (taken from Reference 10)—cont'd

	Thrust (K_T)						Torque (K_Q)				
n	C_n	s (J)	t (P/D)	u(EAR)	v(Z)	n	C_n	s (J)	t (P/D)	u (EAR)	v(Z)
27	0.0186040000	1	0	2	1	27	−0.0106854000	3	0	0	1
28	−0.0041079800	0	2	2	1	28	0.0011090300	3	3	0	1
29	−0.0006068480	0	0	0	2	29	−0.0003139120	0	6	0	1
30	−0.0049819000	1	0	0	2	30	0.0035895000	3	0	1	1
31	0.0025963000	2	0	0	2	31	−0.0014212100	0	6	1	1
32	−0.0005605280	3	0	0	2	32	−0.0038363700	1	0	2	1
33	−0.0016365200	1	2	0	2	33	0.0126803000	0	2	2	1
34	−0.0003287870	1	6	0	2	34	−0.0031827800	2	3	2	1
35	0.0001165020	2	6	0	2	35	0.0033426800	0	6	2	1
36	0.0006909040	0	0	1	2	36	−0.0018349100	1	1	0	2
37	0.0042174900	0	3	1	2	37	0.0001124510	3	2	0	2
38	0.0000565229	3	6	1	2	38	−0.0000297228	3	6	0	2
39	−0.0014656400	0	3	2	2	39	0.0002695510	1	0	1	2
						40	0.0008326500	2	0	1	2
						41	0.0015533400	0	2	1	2
						42	0.0003026830	0	6	1	2
						43	−0.0001843000	0	0	2	2
						44	−0.0004253990	0	3	2	2
						45	0.0000869243	3	3	2	2
						46	−0.0004659000	0	6	2	2
						47	0.000554194	1	6	2	2

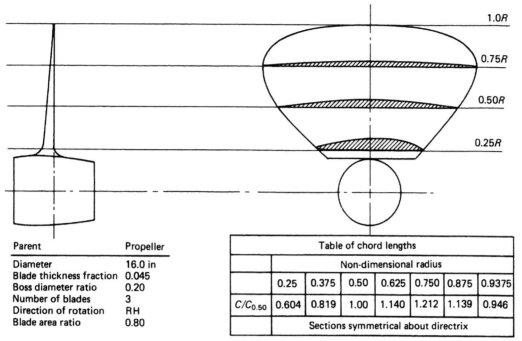

FIGURE 6.13 KCA blade outline.

elliptic thickness form superimposed. The series was designed in such a way that the propellers in the series should have the same camber ratio distribution as the parent propeller. Since previous data and experience was limited with this type of propeller it was fully expected that the section form would need to be modified during the tests. This expectation proved correct and the section form was modified twice on the parent screw to avoid the onset of face cavitation; the modification essentially involved the cutting back and 'lifting' of the leading edge. These modifications were carried over onto the other propellers of the series, which resulted in the series having the characteristics shown in Table 6.12 and the blade form shown in Figure 6.14.

Each of the propellers of the series was tested in a cavitation tunnel at nine different cavitation numbers; 0.25, 0.30, 0.40, 0.50, 0.60, 0.75, 1.00, 2.50 and 5.5. For the

TABLE 6.10 Propellers of the Ma-series

Three-Bladed Propellers

P/D	A_E/A_O	0.75	0.90	1.05	1.20
1.000		*	*	*	*
1.150		*	*	*	*
1.300		*	*	*	*
1.450		*	*	*	*

Five-Bladed Propellers

P/D	A_E/A_O	0.75	0.90	1.05	1.20
1.000		*	*	*	*
1.152		*	*	*	*
1.309		*	*	*	*
1.454		*	*	*	*

TABLE 6.11 Newton–Rader Series

	Blade Thickness Fraction	
A_D/A_O	$Z = 3$	$Z = 5$
0.75	0.063	0.054
0.90	0.058	0.050
1.05	0.053	0.046
1.20	0.053	0.042

TABLE 6.12 Extent of the Newton–Rader Series

BAR	Face Pitch Ratio			
0.48	1.05	1.26	1.67	2.08
0.71	1.05	1.25	1.66	2.06
0.95	1.04	1.24	1.65	2.04

Note: Box indicates resultant parent form.

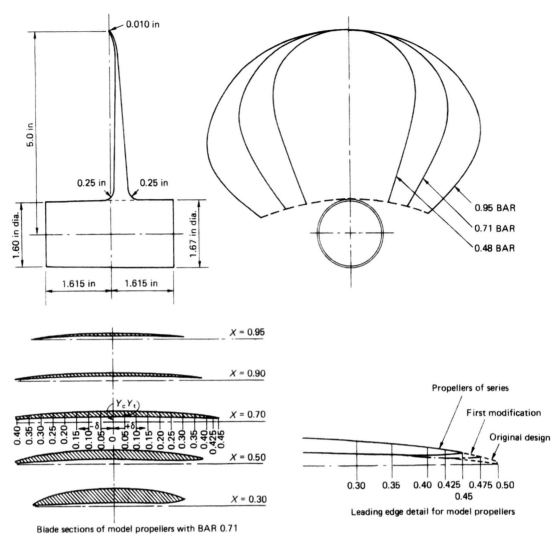

FIGURE 6.14 Newton–Rader series blade form. *Reproduced with permission from Reference 14.*

tests the Reynolds number ranged from about 7.1×10^5 for the narrow-bladed propeller through to 4.5×10^6 for the wide-bladed design at $0.7R$. The results of the series are presented largely in tabular form by the authors.

This series is of importance for the design of propellers, usually for relatively small craft, where significant cavitation is expected.

6.5.7 Other Fixed Pitch Series and Data

Apart from the major fixed pitch propeller series there have been numerous smaller studies which provide useful data, either for design purposes or for research or correlation studies. Among these other works, the KCD propeller series,[15,16] the Meridian series,[12] the contra-rotating series of MARIN and SSPA[17,18] and the DTMB research skewed propeller series[19] are worthy of specific mention.

The KCD series originally comprised a series of models for which 'interesting' full-scale results were available, and the purpose of the series was to try and correlate the observed phenomena in the tunnel and the results of particular experiences with ships. All the model propellers in this series had diameters of 406.4 mm (16 in.) and the first three members of the KCD series had a blade area ratio of 0.6, in association with three-, four- and five-blades respectively. These propellers were tested at a range of cavitation numbers in the Newcastle University tunnel in order to study the open water performance of the propellers under cavitating conditions. The results shown in Figure 6.4 relate to the KCD4 propellers of this series. After a further nine years of testing various designs[16] the series had grown to some 17 members. Of these members, six, including the parent KCD4R, had a common blade area ratio and blade number of 0.587 and four had a range of pitch ratios from 0.6–1.6.

These propellers were used to define a set of K_T, K_Q versus J diagrams and B_p versus δ charts for a series of cavitation numbers of 8.0, 6.0, 4.0 and 2.0. The remaining propellers of the series were used to explore the effects of geometric changes such as moderate amounts of skew, radial pitch variations and blade outline changes on cavitation performance. Hence the series presents an interesting collection of cavitation data for merchant ship propeller designs.

The Meridian series,[12] so called since it was derived from the proprietary design of Stone Manganese Marine Ltd, comprised four parent models having BARs of 0.45, 0.65, 0.85 and 1.05. For each parent model five mean pitch ratios 0.4, 0.6, 0.8, 1.0 and 1.2 were tested so as to cover the useful range of pitch ratios for each blade area ratio. All the propellers had a diameter of 304.8 mm (12 in.) and six blades with a boss diameter of $0.185R$. The parent propellers are not geosims of each other and consequently interpolation between propellers of different blade area ratios for general use becomes rather more complicated than for a completely geometrically similar series. As with the KCD series, this series was tested at a range of cavitation numbers resulting in the presentation of open water data in the form of K_T, K_Q diagrams and $B_p-\delta$ charts under cavitating conditions.

Over the years interest has fluctuated in contra-rotating propellers as a means of ship propulsion. This has led to model tests being undertaken at a variety of establishments around the world. Two examples of this are the MARIN series[17] and the SSPA series.[18] The MARIN series comprised five sets of propulsors with a four-bladed forward propeller and a five-bladed aft propeller. The after propeller was designed with a smaller diameter than the forward screw, the diameter reduction being consistent with the expected slipstream contraction at the design condition. The range of pitch ratios of the forward propeller at $0.7R$ radius spanned the range 0.627–1.235 with a constant expanded area ratio of 0.432 and the after propeller dimensions varied with respect to the flow conditions leaving the forward screw. Non-cavitating open water characteristics were presented for the series.

The SSPA series[18] comprised a family of contra-rotating propellers having a forward propeller of four blades with a developed area ratio of 0.40 and an aft propeller of five blades with a developed area ratio of 0.5. The forward propellers all had diameters of 250 mm and used section forms constructed from NACA 16 profiles and $a = 0.8$ mean lines. The pitch ratios of the leading propeller ranged from 0.8–1.4 and the tests were conducted in the SSPA No. 1 cavitation tunnel. Consequently, open water data is presented along with design diagrams, together with some cavitation data in homogeneous flow.

Boswell[19] presented cavitation tunnel and open water results for a series of skewed propellers. The series comprised four propellers having maximum projected skew angles of 0°, 36°, 72° and 108° at the propeller tip. The propellers each had a diameter of 304.8 mm (12 in.), five blades, an A_E/A_O of 0.725, and NACA $a = 0.8$ mean lines with 66 modified profiles, similar chordal and thickness distributions and the same design conditions; they were given the NSRDC designation of propellers 4381 (0° skew), 4382 (36° skew), 4383 (72° skew) and 4384 (108° skew). The geometry of this series of propellers, in view of their importance in propeller research, is given in Table 6.13. For each of these propellers open water K_T, K_Q versus J results are presented together with cavitation inception speed. These propellers, although giving certain useful information about the effects of skew, find their greatest use as research propellers for comparing the results of theoretical methods and studies. Indeed these propellers have found widespread application in many areas of propeller technology.

6.5.8 Tests with Propellers Having Significant Shaft Incidence

As discussed earlier in this chapter, the effects of operating a propeller at an oblique angle to the incident flow introduce significant side forces and thrust eccentricity. Several experimental studies into this effect have been undertaken, and notable are those by Gutsche,[20] Taniguchi et al.,[21] Bednarzik,[22] Meyne and Nolte[23] and Peck and Moore.[24]

Gutsche worked with a series of six three-bladed propellers: three having developed area ratios of 0.35 and the others 0.80 each in association with three pitch ratios: 0.5, 1.0 and 1.5. The propellers, all having a diameter of 200 mm, were tested at shaft angle inclinations of 0°, 20° and 30° over a range of approximately 0–100 per cent slip in the non-inclined shaft angle position.

Taniguchi et al.[21] used a series of five three-bladed propellers having a diameter of 230 mm. Three of the propellers had a pitch ratio of 1.286 whilst the remaining two had pitch ratios at $0.7R$ of 1.000 and 1.600. For the three propellers embracing the range of pitch ratios the expanded area ratio was held constant at 0.619, whilst for the three propellers having the same pitch ratio the expanded area ratio was varied as follows: 0.619, 0.514 and 0.411. All of the propellers had Tulin supercavitating sections. Results of K_Q and K_T are presented for six cavitation numbers, each at three angles of incidence: 0°, 4° and 8°. No side force or eccentricity data is given.

TABLE 6.13 Blade Geometry of DTNSRDC Propellers 4381, 4382, 4383 and 4384 (taken from Reference 43, Chapter 8)

Characteristics of DTNSRDC Propeller 4381

Number of blades, Z: 5

Hub diameter ratio: 0.2

Expanded area ratio: 0.725

Section mean line: NACA $a = 0.8$

Section thickness distribution: NACA 66 (modified)

Design advance coefficient, J_A: 0.889

r/R	c/D	P/D	θ_s (deg)	X_s/D	t_0/D	f_0/c
0.2	0.174	1.332	0	0	0.0434	0.0351
0.25	0.202	1.338	0	0	0.0396	0.0369
0.3	0.229	1.345	0	0	0.0358	0.0368
0.4	0.275	1.358	0	0	0.0294	0.0348
0.5	0.312	1.336	0	0	0.0240	0.0307
0.6	0.337	1.280	0	0	0.0191	0.0245
0.7	0.347	1.210	0	0	0.0146	0.0191
0.8	0.334	1.137	0	0	0.0105	0.0148
0.9	0.280	1.066	0	0	0.0067	0.0123
0.95	0.210	1.031	0	0	0.0048	0.0128
1.0	0	0.995	0	0	0.0029	–

Characteristics of DTNSRDC Propeller 4382

Number of blades, Z: 5

Hub diameter ratio: 0.2

Expanded area ratio: 0.725

Section mean line: NACA $a = 0.8$

Section thickness distribution: NACA 66 (modified)

Design advance coefficient, J_A: 0.889

r/R	c/D	P/D	θ_s (deg)	X_s/D	t_0/D	f_0/c
0.2	0.174	1.455	0	0	0.0434	0.0430
0.25	0.202	1.444	2.328	0.0093	0.0396	0.0395
0.3	0.229	1.433	4.655	0.0185	0.0358	0.0370
0.4	0.275	1.412	9.363	0.0367	0.0294	0.0344
0.5	0.312	1.361	13.948	0.0527	0.0240	0.0305
0.6	0.337	1.285	18.378	0.0656	0.0191	0.0247
0.7	0.347	1.200	22.747	0.0758	0.0146	0.0199
0.8	0.334	1.112	27.145	0.0838	0.0105	0.0161
0.9	0.280	1.027	31.575	0.0901	0.0067	0.0134
0.95	0.210	0.985	33.788	0.0924	0.0048	0.0140
1.0	0	0.942	36.000	0.0942	0.0029	–

TABLE 6.13 Blade Geometry of DTNSRDC Propellers 4381, 4382, 4383 and 4384 (taken from Reference 43, Chapter 8)—cont'd

Characteristics of DTNSRDC Propeller 4383

Number of blades, Z: 5

Hub diameter ratio: 0.2

Expanded area ratio: 0.725

Section mean line: NACA $a = 0.8$

Section thickness distribution: NACA 66 (modified)

Design advance coefficient, J_A: 0.889

r/R	c/D	P/D	θ_s (deg)	X_s/D	t_0/D	f_0/c
0.2	0.174	1.566	0	0	0.0434	0.0402
0.25	0.202	1.539	4.647	0.0199	0.0396	0.0408
0.3	0.229	1.512	9.293	0.0390	0.0358	0.0407
0.4	0.275	1.459	18.816	0.0763	0.0294	0.0385
0.5	0.312	1.386	27.991	0.1078	0.0240	0.0342
0.6	0.337	1.296	36.770	0.1324	0.0191	0.0281
0.7	0.347	1.198	45.453	0.1512	0.0146	0.0230
0.8	0.334	1.096	54.245	0.1651	0.0105	0.0189
0.9	0.280	0.996	63.102	0.1745	0.0067	0.0159
0.95	0.210	0.945	67.531	0.1773	0.0048	0.0168
1.0	0	0.895	72.000	0.1790	0.0029	–

Characteristics of DTNSRDC Propeller 4384

Number of blades, Z: 5

Hub diameter ratio: 0.2

Expanded area ratio: 0.725

Section mean line: NACA $a = 0.8$

Section thickness distribution: NACA 66 (modified)

Design advance coefficient, J_A: 0.889

r/R	c/D	P/D	θ_s (deg)	X_s/D	t_0/D	f_0/c
0.2	0.174	1.675	0	0	0.0434	0.0545
0.25	0.202	1.629	6.961	0.0315	0.0396	0.0506
0.3	0.229	1.584	13.921	0.0612	0.0358	0.0479
0.4	0.275	1.496	28.426	0.1181	0.0294	0.0453
0.5	0.312	1.406	42.152	0.1646	0.0240	0.0401
0.6	0.337	1.305	55.199	0.2001	0.0191	0.0334
0.7	0.347	1.199	68.098	0.2269	0.0146	0.0278
0.8	0.334	1.086	81.283	0.2453	0.0105	0.0232
0.9	0.280	0.973	94.624	0.2557	0.0067	0.0193
0.95	0.210	0.916	101.300	0.2578	0.0048	0.0201
1.0	0	0.859	108.000	0.2578	0.0029	–

Bednarzik[22] uses a similar test series arrangement to Taniguchi in that three of his propeller series have the same pitch ratio of 0.60 with varying developed area ratios of 0.35, 0.55 and 0.75. Each of the remaining two propellers has a developed area ratio of 0.55, but pitch ratios of 1.00 and 1.40 respectively. The propeller diameters are all 260 mm and each has three blades with a hub-to-tip diameter ratio of 0.3. The propellers are tested over a range of shaft inclinations of 0°, 5°, 15° and 25° with side force and eccentricity data being presented in addition to conventional K_Q and K_T coefficients.

Meyne and Nolte[23] considered a 355.46 mm diameter, four-bladed propeller having a hub ratio of $0.328R$ and an expanded blade area ratio of 0.566. The pitch ratio of the propeller was varied from 1.0 to 1.6 in a single step and tests were made with shaft inclination of 0°, 6°, 9° and 12°. Results of K_Q, K_T, side force and eccentricity are given by the authors.

Peck and Moore[24] used four 254 mm (10 in.) diameter, four-bladed propellers having nominal pitch ratios of 0.8, 1.0, 1.2 and 1.4, respectively. Measurements were made over a range of cavitation numbers at 0°, 7.5° and 15° shaft inclinations and side force data is presented in addition to the other performance data.

6.5.9 Wageningen Ducted Propeller Series

A very extensive set of ducted propeller standard series tests has been conducted at MARIN in the Netherlands over the years and these have been reported in several publications. The best source material for this series can be found in References 25 and 26.

The extent of the series can be judged from Table 6.14 which itemizes the tests conducted within this series, while Figure 6.15 shows the profiles of the various duct forms tested. In general it can be considered that ducts No. 2 through No. 24 and No. 37 represent accelerating ducts whilst those numbered 30–36 represent decelerating duct forms. In merchant practice the ducts most commonly encountered are the 19A and 37 since they are both relatively easy to fabricate and have a number of desirable hydrodynamic features. Ease of fabrication of the duct is essential; the feature which helps this significantly is the use of straight lines, wherever possible, in the profiles shown in Figure 6.15. The profile ordinates of ducts No. 19A and No. 37 are given in Table 6.15. Three principal propeller types have been used; the B-series propeller in duct Nos 2–11, the K_a-series propellers in ducts Nos 1–24 and No. 37 with limited work using the B-series propeller for duct 19A, and the K_d-series propeller for duct Nos 30–36. The details of the K_a-series propellers are reproduced in Table 6.16, and Figure 6.16 shows the general forms of the propellers for this series. These propellers have a diameter of 240 mm.

Typical of the results derived from this series are the characteristics shown in Figure 6.8. However, regression polynomials have been developed to express K_T, K_{TN} and K_Q as functions of P/D and J. The form of the polynomials is as follows:

$$\begin{aligned}
K_T = {} & A_{0,0} + A_{0,1} J + \cdots + A_{0,6} J^6 \\
& + A_{1,0}\left(\frac{P}{D}\right)^2 + A_{1,1}\left(\frac{P}{D}\right) J + \cdots + A_{1,6}\left(\frac{P}{D}\right) J^6 \\
& + A_{2,0}\left(\frac{P}{D}\right)^2 + A_{2,1}\left(\frac{P}{D}\right)^2 J + \cdots + A_{2,6}\left(\frac{P}{D}\right)^2 J^6 \\
& \cdots \\
& + A_{6,0}\left(\frac{P}{D}\right)^6 + A_{6,1}\left(\frac{P}{D}\right)^6 J + \cdots + A_{6,6}\left(\frac{P}{D}\right)^6 J^6 \\
K_{TN} = {} & B_{0,0} + B_{0,1} J + \cdots + B_{6,6}\left(\frac{P}{D}\right)^6 J^6 \\
K_Q = {} & C_{0,0} + C_{0,1} J + \cdots + C_{6,6}\left(\frac{P}{D}\right)^6 J^6
\end{aligned}$$
(6.19)

where the coefficients A, B and C are given in Table 6.17 for the 19A and 37 duct profiles with the K_a 4–70 propeller.

6.5.10 Gutsche and Schroeder Controllable Pitch Propeller Series

The Gutsche and Schroeder propeller series[27] comprises a set of five, three-bladed controllable pitch propellers. The propellers were designed according to the Gawn series[9] with certain modifications; these were that the blade thickness fraction was reduced to 0.05 and the inner blade chord lengths were restricted to allow the blades to be fully reversing. Additionally, the boss radius was increased to $0.25D$ in order to accommodate the boss actuating mechanism.

The propeller series was designed to have a diameter of 200 mm and three of the propellers were produced with a design P/D of 0.7 and having varying developed area ratios of 0.48, 0.62 and 0.77. The remaining two propellers of the series had blade area ratios of 0.62 with design pitch ratios of 0.5 and 0.9. The three propellers with a design P/D of 0.7 were tested for both positive and negative advance speed over a range of pitch ratios at $0.7R$ of 1.5, 1.25, 1.0, 0.75, 0.5, 0, −0.5, −0.75 and −1.00. The remaining two propellers of the series were tested at the limited P/D range of 1.00, 0.50, −0.50 and −1.00.

6.5.11 The JD–CPP Series

The JD–CPP series is also a three-bladed controllable pitch series comprising 15 model propellers each having a diameter of 267.9 mm. The propellers are

TABLE 6.14 Ducted Propeller Configurations Tested at MARIN Forming the Ducted Propeller Series (taken from Reference 26)

Nozzle Number	L/D	S/L	Open Water Test	Multi-Quadrant 2Q Measurement	Multi-Quadrant 4Q Measurement	Azimuth ⊙
2	0.67	0.15	B 4–55			
3	0.50	0.15	B 4–55			
4	0.83	0.15	B 4–55			
5	0.50	0.15	B 4–55			
6	0.50	0.15	B 4–55			
7	0.50	0.15	B 4–55			
7	0.50	0.15	B 4–10			
7	0.50	0.15	B 4–70			
7	0.50	0.15	B 2–30			
7	0.50	0.15	B 3–50			
7	0.50	0.15	B 5–60			
8	0.50	0.15	B 4–55			
10	0.40	0.15	B 4–55			
11	0.30	0.15	B 4–55			
18	0.50	0.15	B 4–55			
19	0.50	0.15	B 4–55			
20	0.50	0.15	B 4–55			
19A	0.50	0.15	Ka 3–50*			
19A	0.50	0.15	Ka 3–65*			
19A	0.50	0.15	Ka 4–55*			
19A	0.50	0.15	Ka 4–70*	Ka 4–70*	Ka 4–70*	Ka 4–70
19A	0.50	0.15	Ka 5–75*			
19A	0.50	0.15	B 4–70*			
21	0.70	0.15	Ka 4–70*			
22	0.80	0.15	Ka 4–70*			
23	0.90	0.15	Ka 4–70*			
24	1.00	0.15	Ka 4–70*			
37	0.50	0.15	Ka 4–70*	Ka 4–70*	Ka 4–70*	
30	0.60	0.15	Kd 5–100			
31	0.60	0.15	Kd 5–100			
32	0.60	0.15	Kd 5–100			
33	0.60	0.15	Kd 5–100			
34	0.60	0.09	Kd 5–100			
35	0.9	0.1125	Kd 5–100			
36	1.2	0.0750	Kd 5–100			

⊙ Tests at different incidence angles.
*Mathematical representation of test data available.

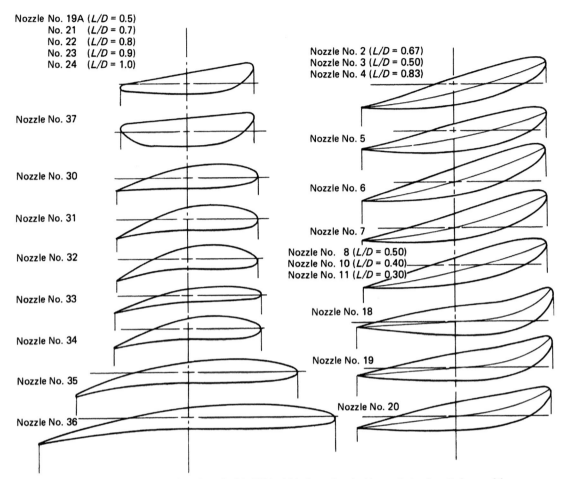

FIGURE 6.15 Duct outlines described in Table 6.14. *Reproduced with permission from Reference 26.*

split into three groups of five having expanded area ratios of 0.35, 0.50 and 0.65. The propellers all have a boss diameter of 0.28D and each of the members of the expanded area groups have design pitch ratios of 0.4, 0.5, 0.8, 1.0 and 1.2, respectively. As in the case of the Gutsche series the blade thickness fraction is 0.05. The blade design pitch distribution is constant from the tip to 0.6R but is reduced in the inner region of the blade near the root.

The propeller series, presented by Chu *et al.*[28] was tested at the Shanghai Jiao Tong University and measurements were made over a range of 50° of pitch change distributed about the design pitch setting. Results are presented for thrust, torque and hydrodynamic spindle torque coefficient for the series in non-cavitating conditions. The range of conditions tested extends to both positive and negative advance coefficients. Hence, by including spindle torque data this series is one of the most complete for controllable pitch propeller hydrodynamic study purposes and to aid studies polynomial regression coefficients have also been given by the authors.

6.5.12 Other Controllable Pitch Propeller Series Tests

In general the open water characteristics of controllable pitch propeller series have been very largely neglected in the open literature. This is particularly true of the spindle torque characteristics. Apart from the two series mentioned above which form the greatest open literature data source for controllable pitch propellers in off-design conditions, there have been other limited amounts of test data presented. Amongst these are Yazaki,[29] Hansen[30] and Miller.[31] Model tests with controllable pitch propellers in the Wageningen duct forms 19A, 22, 24, 37 and 38 are presented in Reference 40.

6.6 MULTI-QUADRANT SERIES DATA

Discussion so far has centered on the first quadrant performance of propellers. That is, for propellers working with positive rotational speed and forward or zero advance velocity. This clearly is the conventional way of operating

TABLE 6.15 Duct Ordinates for 19A and 37 Duct Form

Duct Profile No. 19A

LE

x/L	0	0.0125	0.025	0.050	0.075	0.100	0.150	0.200	0.25	0.30	0.40	0.50	0.60	0.70	0.80	0.90	0.95	**TE** 1.00
y_i/L	0.1825	0.1466	0.1280	0.1087	0.0800	0.0634	0.0387	0.0217	0.0110	0.0048	0	0	0	0.0029	0.0082	0.0145	0.0186	0.0236
y_u/L		0.2072	0.2107	0.2080						Straight line							0.0636	

Duct Profile No. 37

LE

x/L	0	0.0125	0.025	0.050	0.075	0.100	0.150	0.200	0.25	0.30	0.40	0.50	0.60	0.70	0.80	0.90	0.95	**TE** 1.00
y_i/L	0.1833	0.1500	0.1310	0.1000	0.0790	0.0611	0.0360	0.0200	0.0100	0.0040	0	0	0	0.0020	0.0110	0.0380	0.0660	0.1242
y_u/L	0.1833	0.2130	0.2170	0.2160						Straight line							0.1600	0.1242

Duct upper ordinate = propeller radius + clearance + y_u.
Duct inner ordinate = propeller radius + tip clearance + y_i.

TABLE 6.16 Details of the K_a-series Propellers (taken from Reference 26)

Dimensions of the K_a-Screw Series

r/R		0.2	0.3	0.4	0.5	0.6	0.7	0.8	0.9	1.0	
Length of the blade sections in percentages	from center line to trailing edge	30.21	36.17	41.45	45.99	49.87	52.93	55.04	56.33	56.44	Length of blade section at 0.6R $=1.969\dfrac{D}{Z}\dfrac{A_E}{A_O}$
of the maximum	from center line to leading edge	36.94	40.42	43.74	47.02	50.13	52.93	55.04	56.33	56.44	
	length of the blade section at 0.6R total length	67.15	76.59	85.19	93.01	100.00	105.86	110.08	112.66	112.88	
Maximum blade thickness in percentages of the diameter		4.00	3.52	3.00	2.45	1.90	1.38	0.92	0.61	0.50	Maximum thickness at centre of shaft $=0.049D$
Distance of maximum thickness from leading edge in percentages of the length of the sections		34.98	39.76	46.02	49.13	49.98	—	—	—	—	

Ordinates of the K_a-Screw Series

Distance of the Ordinates From the Maximum Thickness

	From Maximum Thickness to Trailing Edge					From Maximum Thickness to Leading Edge					
r/R	100%	80%	60%	40%	20%	20%	40%	60%	80%	95%	100%
Ordinates for the Back											
0.2	—	38.23	63.65	82.40	95.00	97.92	90.83	77.19	55.00	27.40	—
0.3	—	39.05	66.63	84.14	95.86	97.63	90.06	75.62	53.02	27.57	—
0.4	—	40.56	66.94	85.69	96.25	97.22	88.89	73.61	50.00	25.83	—
0.5	—	41.77	68.59	86.42	96.60	96.77	87.10	70.46	45.84	22.24	—
0.6	—	43.58	68.26	85.89	96.47	96.47	85.89	68.26	43.58	20.44	—
0.7	—	45.31	69.24	86.33	96.58	96.58	86.33	69.24	45.31	22.88	—
0.8	—	48.16	70.84	87.04	96.76	96.76	87.04	70.84	48.16	26.90	—
0.9	—	51.75	72.94	88.09	97.17	97.17	88.09	72.94	51.75	31.87	—
1.0	—	52.00	73.00	88.00	97.00	97.00	88.00	73.00	52.00	32.31	—
Ordinates for the Face											
0.2	20.21	7.29	1.77	0.1	—	0.21	1.46	4.37	10.52	20.62	33.33
0.3	13.85	4.62	1.07	—	—	0.12	0.83	2.72	6.15	10.30	21.18
0.4	9.17	2.36	0.56	—	—	—	0.42	1.39	2.92	4.44	13.47
0.5	6.62	0.68	0.17	—	—	—	0.17	0.51	1.02	1.53	7.81

Note: The percentages of the ordinates relate to the maximum thickness of the corresponding section.

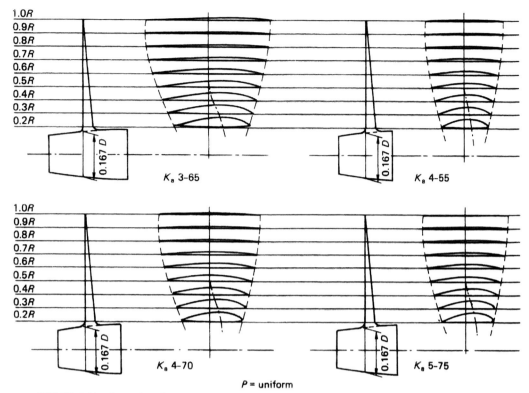

FIGURE 6.16 General outline of the K_a-series propeller. *Reproduced with permission from Reference 26.*

a propeller in the ahead operating condition, but for studying maneuvering situations or astern performance of vessels other data is required.

In the case of the fixed pitch propeller it is possible to define four quadrants based on an advance angle

$$\beta = \tan^{-1}\left(\frac{V_a}{0.7\pi nD}\right) \qquad (6.20)$$

as follows:

1st quadrant: Advance speed — ahead, rotational speed — ahead. This implies that the advance angle β varies within the range $0 \leq \beta \leq 90°$
2nd quadrant: Advance speed — ahead, rotational speed — astern. In this case β lies in the range $90° < \beta \leq 180°$
3rd quadrant: Advance speed — astern, rotational speed — astern. Here β lies in the range of $180° < \beta \leq 270°$
4th quadrant: Advance speed — astern, rotational speed — ahead. Where β is within the range $270° < \beta \leq 360°$

Provided sufficient experimental data is available it becomes possible to define a periodic function based on the advance angle β to define the thrust and torque characteristics of the propeller in each of the quadrants. In this context it should be noted that when $\beta = 0°$ or $360°$ then this defines the ahead bollard pull condition and when $\beta = 180°$ this corresponds to the astern bollard pull situation. For $\beta = 90°$ and $270°$, these positions relate to the condition when the propeller is not rotating and is being dragged ahead or astern through the water, respectively. Figure 6.17 clarifies this notation for fixed pitch propellers.

For multi-quadrant studies the advance angle notation offers a more flexible representation than the conventional advance coefficient J; since when the propeller rpm is 0, such as when $\beta = 90°$ or $270°$, then $J \to \infty$. Similarly, the thrust and torque coefficients need to be modified in order to prevent similar problems from occurring and the following are derived:

$$C_T^* = \frac{T}{\frac{1}{2}\rho V_r^2 A_O}$$

and

$$C_Q^* = \frac{Q}{\frac{1}{2}\rho V_r^2 A_O D}$$

where V_r is the relative advance velocity at the $0.7R$ blade section. Consequently, the above equations can be written explicitly as

$$\left.\begin{array}{l} C_T^* \dfrac{T}{(\pi/8)\rho[V_a^2 + (0.7\pi nD)^2]D^2} \\ \text{and} \\ C_Q^* \dfrac{T}{(\pi/8)\rho[V_a^2 + (0.7\pi nD)^2]D^3} \end{array}\right\} \qquad (6.21)$$

TABLE 6.17 Coefficients for Duct Nos 19A and 37 for Equation (6.19) — K_a Propeller 4–70 (taken from Reference 26)

	x	y	Nozzle No. 19A Axy	Nozzle No. 19A Bxy	Nozzle No. 19A Cxy	Nozzle No. 37 Axy	Nozzle No. 37 Bxy	Nozzle No. 37 Cxy
0	0	0	+0.030550	+0.076594	+0.006735	−0.0162557	−0.016806	+0.016729
1		1	−0.148687	+0.075223				
2		2		−0.061881	−0.016306			
3		3	−0.391137	−0.138094				
4		4		−0.007244	−0.077387			
5		5		−0.370620				
6		6		+0.323447			−0.099544	+0.030559
7	1	0		−0.271337		+0.598107		+0.048424
8		1	−0.432612	−0.687921		−1.009030	−0.548253	−0.011118
9		2		+0.225189	−0.024012		+0.230675	−0.056199
10		3						
11		4						
12		5						
13		6		−0.081101				
14	2	0	+0.667657	+0.666028		+0.085087	+0.460206	+0.084376
15		1				+0.425585		
16		2	+0.285076	+0.734285	+0.005193			+0.045637
17		3						−0.042003
18		4						
19		5						
20		6						
21	3	0	−0.172529	−0.202467	+0.046605		−0.215246	−0.008652
22		1						
23		2		−0.542490				
24		3						
25		4						
26		5				−0.021044		
27		6		−0.016149				
28	4	0			−0.007366		+0.042997	
29		1						
30		2						
31		3		+0.099819				
32		4						
33		5						
34		6						
35	5	0						
36		1		+0.030084		−0.038383		

Chapter | 6 Propeller Performance Characteristics

TABLE 6.17 Coefficients for Duct Nos 19A and 37 for Equation (6.19) — K_a Propeller 4–70 (taken from Reference 26)—cont'd

	x	y	Nozzle No. 19A			Nozzle No. 37		
			Axy	Bxy	Cxy	Axy	Bxy	Cxy
37		2						
38		3						
39		4						
40		5						
41		6						
42	6	0			−0.001730			
43		1	−0.017283		−0.000337			−0.001176
44		2	−0.001876	+0.000861		+0.014992		+0.002441
45		3						
46		4						
47		5						
48		6						
49	0	7				+0.036998	+0.051753	−0.012160

Note the asterisks in equation (6.21) are used to avoid confusion with the free stream velocity-based thrust and torque coefficients C_T and C_Q defined in equations (6.4)–(6.6).

Results plotted using these coefficients take the form shown in Figures 6.18–6.20 for the Wageningen B4-70 screw propeller series. These curves, as can be seen, are periodic over the range $0° \leq \beta \leq 360°$ and, therefore, lend themselves readily to a Fourier type representation. Van Lammeren et al.[6] suggest a form:

$$\left. \begin{array}{l} C_T^* = \sum\limits_{k=0}^{20} [A_k \cos(k\beta) + B_k \sin(k\beta)] \\ C_Q^* = \sum\limits_{k=0}^{20} [A_k \cos(k\beta) + B_k \sin(k\beta)] \end{array} \right\} \quad (6.22)$$

When evaluating the off-design characteristics using open water data it is important to find data from a model which is close to the design under consideration. From the Wageningen data it will be seen that blade area ratio has an important influence on the magnitude of C_T^* and C_Q^* in the two regions $40° < \beta < 140°$ and $230° < \beta < 340°$. In these regions the magnitude of C_Q^* can vary by as much as a factor of three, at model-scale, for a blade area ratio change from 0.40 to 1.00. Similarly, the effect of pitch ratio will have a considerable influence on C_Q^* over almost the entire range of β as seen in Figure 6.18. Blade number does not appear to have such pronounced effects as pitch ratio or expanded area ratio and, therefore, can be treated as a less significant variable.

Apart from the Wageningen B-Screw series there have, over the years, been other studies, undertaken on propellers operating in off-design conditions. Notable amongst these are Conn[32] and Nordstrom;[33] these latter works, however, are considerably less extensive than the Wageningen data cited above.

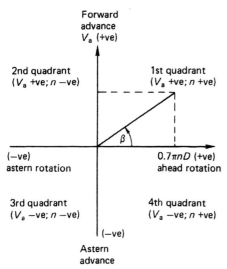

FIGURE 6.17 Four-quadrant notation.

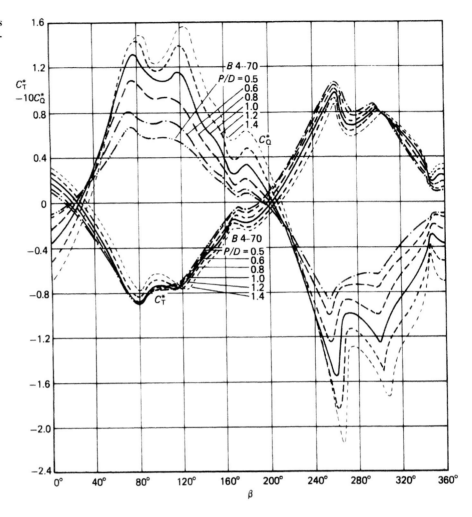

FIGURE 6.18 Open water test results with B4-70 screw series in four quadrants. *Reproduced from Reference 6.*

With regard to ducted propellers a 20-term Fourier representation has been undertaken (Reference 25) for the 19A and 37 ducted systems when using the K_a 4–70 propeller and has been shown to give a satisfactory correlation with the model test data. Consequently, as with the case of the non-ducted propellers the coefficients C_T^*, C_Q^* and C_{TN}^* are defined as follows:

$$\begin{Bmatrix} C_T^* \\ C_Q^* \\ C_{TN}^* \end{Bmatrix} = \sum_{k=0}^{20} [A_k \cos (k\beta) + B_k \sin (k\beta)] \quad (6.23)$$

The corresponding tables of coefficients are reproduced from (Reference 25) in Tables 6.18 and 6.19 for the 19A and 37 ducts, respectively.

As might be expected the propeller, in this case the $K_a = 4$–70, still shows the same level of sensitivity to P/D, and almost certainly to $A_E A_O$, with β as did the non-ducted propellers. However, the duct is comparatively insensitive to variations in P/D except in the region $-20° \leq 20°$.

Full details of the Wageningen propeller series can be found in Reference 40.

In the case of the controllable pitch propeller the number of quadrants reduces to two since this type of propeller is unidirectional in terms of rotational speed. Using the fixed pitch definition of quadrants the two of interest for controllable pitch propellers are the first and fourth, since the advance angle lies in the range $90° \leq \beta \leq -90°$. As discussed earlier, the amount of standard series data for controllable pitch propellers in the public domain is comparatively small; the work of Gutsche and Schroeder,[27] and Yazaki[29] being perhaps the most prominent. Strom-Tejsen and Porter[34] undertook an analysis of the Gutsche–Schroeder three-bladed c.p.p. series, and by applying regression methods to the data derived equations of the form:

$$\begin{Bmatrix} C_T^* \\ C_Q^* \end{Bmatrix} = \sum_{l=0}^{L} R_{l,2}(z) \sum_{m=0}^{M} P_{m,10}(y) \sum_{n=0}^{N} \{a_{l,m,n} \cos (n\beta) + b_{l,m,n} \sin (n\beta)\} \quad (6.24)$$

where

$$y = \{[P_{0.7}/D]_{\text{set}} + 1.0\}/0.25$$

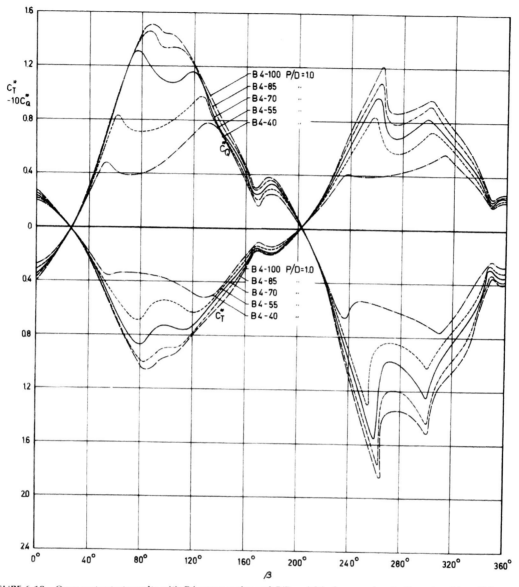

FIGURE 6.19 Open water test results with B4-screw series and $P/D = 1.0$ in four quadrants. *Reproduced from Reference 6.*

and

$$z = (A_D/A_O) - 0.50)/0.15$$

and $R_{l,2}$ (z) and $P_{m,10}$ (y) are orthogonal polynomials defined by

$$P_{m,n}(x) = \sum_{k=0}^{m} (-1)^k \binom{m}{k} \binom{m+k}{k} \frac{x^{(k)}}{n^{(k)}}$$

The coefficients a and b of equation (6.24) are defined in Table 6.20 for use in the equations; however, it has been found that it is unnecessary for most purposes to use the entire table of coefficients and that fairing based on $L = 2$, $M = 4$ and $N = 14$ provides sufficient accuracy.

6.7 SLIPSTREAM CONTRACTION AND FLOW VELOCITIES IN THE WAKE

When a propeller is operating in open water the slipstream will contract uniformly as shown in Figure 6.21(a). This contraction is due to the acceleration of the fluid by the propeller and, consequently, is dependent upon the thrust exerted by the propeller. The greater the thrust produced by the propeller for a given speed of advance, the more the slipstream will contract.

Nagamatsu and Sasajima[35] studied the effect of contraction through the propeller disc and concluded that the contraction could be represented to a first approximation by the simple momentum theory relationship:

$$\frac{D_O}{D} = [0.5(1 + (1 + C_T)^{1/2})]^{1/2} \qquad (6.25)$$

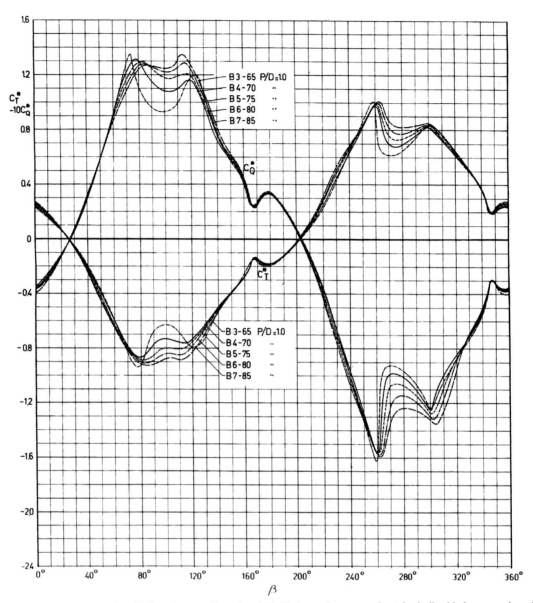

FIGURE 6.20 Open water test results with B-series propellers of variable blade number, approximately similar blade area ratio and *P/D* = 1.0. *Reproduced from Reference 6.*

where

D_O is the diameter of the slipstream far upstream
D is the diameter of the propeller disc
C_T is the propeller thrust coefficient.

Figure 6.21(b) shows the correlation found by Nagamatsu and Sasajima for both uniform and wake flow conditions. While the uniform flow results fit the curve well, as might be expected, and the wake flow results also show broad agreement, it must be remembered that our understanding of the full-interaction effects is still far from complete at this time.

The flow in the slipstream of the propeller is complex and a great deal has yet to be understood. Leathard[36] shows measurements of field point velocity studies conducted on the KCD19 model propeller which formed a member of the KCD-series discussed in Section 6.5.7 and was tested at the University of Newcastle upon Tyne. The measurements were made using an assembly of rotating pitot tubes and the results are shown in terms of the axial distribution of hydrodynamic pitch angle over a range of plus or minus seven propeller diameters as seen in Figure 6.22 for an advance coefficient of 0.80 which corresponds to the optimum efficiency condition. Studies by Keh-Sik[37] using non-intrusive laser-Doppler anemometry techniques on a series of NSRDC research propellers show similar patterns. Figure 6.23 shows the changes in slipstream radius, hydrodynamic pitch angle of the tip vortex and the

TABLE 6.18 Fourier Coefficients for K_a 4–70 Propeller in 19A Duct (Oosterveld[25])

		P/D = 0.6		P/D = 0.8		P/D = 1.0		P/D = 1.2		P/D = 1.4	
	K	A(K)	B(K)	A(K)	B(K)	A(K)	B(K)	A(K)	B(K)	A(K)	B(K)
C^*_{TN}	0	−0.14825E+0	+0.00000E+0	−0.13080E+0	+0.00000E+0	−0.10985E+0	+0.00000E+0	−0.90888E−1	+0.00000E+0	−0.73487E−1	+0.00000E+0
	1	+0.84697E−1	−0.10838E+1	+0.10985E+0	−0.10708E+1	+0.14064E+0	−0.10583E+1	+0.17959E+0	−0.11026E+1	+0.22861E+0	−0.98101E+0
	2	+0.16700E+0	−0.18023E−1	+0.15810E+0	+0.24163E−1	+0.15785E+0	+0.47284E−1	+0.14956E+0	+0.61459E−1	+0.14853E+0	+0.71510E−1
	3	+0.96610E−3	+0.11825E+0	+0.18367E−1	+0.12784E+0	+0.45544E−1	+0.13126E+0	+0.65675E−1	+0.13715E+0	+0.75328E−1	+0.14217E+0
	4	+0.14754E−1	−0.70713E−2	+0.16168E−1	−0.14064E−1	−0.51639E−2	−0.77539E−2	+0.52107E−2	−0.17280E−1	+0.34084E−2	−0.22675E−1
	5	−0.11806E−1	+0.62894E−1	−0.37402E−2	+0.76213E−1	−0.25560E−2	+0.93507E−1	−0.68232E−2	+0.96579E−1	−0.11643E−2	+0.91082E−1
	6	−0.14888E−1	+0.11519E−1	−0.11736E−1	+0.13259E−1	−0.60502E−2	+0.92520E−2	−0.62896E−2	+0.58809E−2	+0.18576E−3	−0.40283E−2
	7	+0.7331 lE−2	+0.17070E−2	+0.25483E−2	−0.42300E−1	+0.67368E−2	−0.14828E−1	+0.18178E−1	−0.22587E−1	+0.26970E−1	−0.22759E−1
	8	+0.75022E−2	+0.22990E−2	+0.12350E−2	−0.26246E−2	+0.68571E−2	−0.96554E−2	+0.60694E−2	−0.14819E−1	+0.20616E−2	−0.16727E−1
	9	−0.15128E−1	+0.13458E−1	−0.20772E−2	+0.16328E−1	+0.47245E−2	+0.96216E−2	+0.61942E−2	−0.10398E−1	+0.78666E−2	+0.86970E−2
	10	+0.33002E−2	+0.54810E−3	+0.69749E−2	−0.33979E−3	+0.23591E−2	−0.75453E−3	+0.26482E−2	−0.29324E−2	+0.46912E−2	−0.47515E−2
	11	+0.31416E−2	+0.42076E−2	+0.59838E−2	+0.23506E−2	+0.87912E−2	+0.24453E−2	+0.12137E−1	+0.40913E−2	+0.14771E−1	+0.22828E−2
	12	−0.21144E−2	−0.57232E−2	−0.14599E−2	−0.69497E−2	+0.11968E−2	−0.87981E−2	−0.35705E−2	−0.44436E−2	−0.75056E−2	−0.49383E−2
	13	+0.29438E−2	+0.74689E−2	+0.83533E−2	+0.61925E−2	+0.83808E−2	+0.18184E−2	+0.32985E−2	−0.12190E−2	+0.14983E−2	−0.25924E−2
	14	+0.33857E−3	−0.84815E−4	+0.11093E−1	+0.35046E−3	−0.82098E−3	+0.20077E−2	−0.88652E−3	−0.22551E−2	+0.24058E−2	−0.25143E−2
	15	+0.41236E−2	−0.13374E−2	+0.41885E−2	−0.11571E−2	+0.27371E−2	−0.33070E−2	+0.69807E−2	−0.32272E−2	+0.55647E−2	−0.33659E−2
	16	+0.16259E−2	−0.91934E−3	−0.12438E−3	−0.32566E−3	−0.26121E−3	−0.79201E−3	−0.17560E−3	+0.17553E−2	−0.38178E−2	+0.28153E−2
	17	+0.12759E−2	+0.27412E−2	+0.38034E−2	+0.63420E−3	+0.19133E−2	−0.36311E−3	+0.21643E−2	+0.14875E−2	+0.26704E−2	−0.22162E−3
	18	+0.20647E−2	−0.10198E−2	+0.90073E−3	−0.22749E−2	+0.32290E−3	−0.19377E−2	+0.35362E−3	+0.45353E−4	+0.15745E−2	−0.53749E−3
	19	+0.34157E−2	+0.19845E−2	+0.31147E−2	−0.36805E−3	+0.15223E−2	−0.12135E−2	+0.25772E−2	−0.88702E−3	+0.24500E−3	−0.35190E−2
	20	−0.58703E−3	−0.13980E−2	−0.106633E−3	−0.12350E−2	−0.10151E−2	−0.31678E−3	−0.18279E−2	−0.94609E−3	−0.42370E−4	−0.42846E−3
C^*_{TN}	0	−0.14276E+0	+0.00000E+0	−0.12764E+0	+0.00000E+0	−0.11257E+0	+0.00000E+0	−0.10166E+0	+0.00000E+0	−0.86955E−1	+0.00000E+0
	1	−0.55946E−2	−0.21875E+0	+0.68679E−3	−0.24100E+0	+0.93340E−2	−0.26265E+0	+0.18593E−1	−0.27769E+0	+0.30046E−1	−0.29799E+0
	2	+0.15519E+0	+0.10114E−1	+0.14639E+0	+0.18919E−1	+0.13788E+0	+0.27587E−1	+0.13408E+0	+0.35459E−1	+0.12651E+0	+0.43403E−1

(Continued)

TABLE 6.18 Fourier Coefficients for K_a 4–70 Propeller in 19A Duct (Oosterveld[25])—cont'd

K	P/D = 0.6 A(K)	P/D = 0.6 B(K)	P/D = 0.8% A(K)	P/D = 0.8% B(K)	P/D = 1.0 A(K)	P/D = 1.0 B(K)	P/D = 1.2 A(K)	P/D = 1.2 B(K)	P/D = 1.4 A(K)	P/D = 1.4 B(K)
3	+0.15915E−1	+0.47120E−1	+0.23195E−1	+0.55513E−1	+0.33223E−1	+0.65262E−1	+0.43767E−1	+0.72317E−1	+0.55034E−1	+0.83309E−1
4	+0.66633E−2	−0.58914E−1	+0.10292E−1	−0.12453E−1	+0.12672E−1	−0.40234E−2	+0.13604E−1	−0.83408E−2	+0.19376E−1	−0.14571E−1
5	+0.89343E−3	−0.12958E−2	+0.86651E−2	−0.14748E−2	+0.14250E−1	+0.10255E−2	+0.18658E−1	+0.44854E−2	+0.22082E−1	+0.43398E−2
6	−0.38876E−2	−0.20824E−2	−0.39124E−2	−0.21899E−2	−0.30407E−3	−0.32045E−2	+0.26598E−2	−0.37642E−2	+0.76282E−2	−0.39256E−2
7	+0.10976E−1	+0.52475E−2	+0.15984E−1	+0.56694E−2	+0.19888E−1	−0.21752E−2	+0.24097E−1	+0.75727E−3	+0.31821E−1	−0.23504E−2
8	+0.31959E−2	−0.15428E−2	+0.46295E−2	−0.49911E−2	+0.48334E−2	−0.59535E−2	+0.47924E−2	−0.88802E−2	+0.51835E−2	−0.13633E−1
9	+0.14201E−2	+0.23580E−2	+0.68371E−3	+0.13631E−2	+0.28427E−2	+0.90664E−3	+0.36556E−2	+0.40541E−2	−0.38898E−2	−0.14000E−2
10	+0.13507E−2	+0.23491E−3	+0.16740E−2	+0.12877E−2	+0.32326E−2	−0.10222E−2	+0.39850E−2	−0.12811E−2	+0.49300E−2	−0.28212E−2
11	+0.50526E−2	−0.26868E−2	+0.83495E−2	−0.35176E−2	+0.97693E−2	−0.48133E−2	+0.10643E−1	−0.55230E−2	−0.10731E−1	−0.77360E−2
12	−0.90855E−3	−0.45635E−2	−0.77063E−3	−0.55571E−2	−0.28378E−3	−0.56355E−2	+0.25495E−3	−0.63566E−2	+0.11388E−2	−0.68665E−2
13	+0.42758E−4	−0.44554E−4	+0.14936E−2	+0.31438E−3	+0.29395E−2	−0.18248E−2	+0.29347E−2	−0.25338E−2	+0.31378E−2	−0.42392E−2
14	+0.42084E−3	−0.18564E−4	+0.11017E−2	−0.11179E−2	+0.53177E−3	−0.20263E−2	+0.36599E−3	−0.20504E−2	−0.82607E−3	−0.33252E−2
15	+0.20269E−2	−0.80547E−3	+0.16804E−2	−0.24577E−2	+0.16229E−2	−0.30382E−2	+0.13115E−2	−0.38485E−2	−0.17537E−4	−0.45496E−2
16	−0.79748E−3	−0.10170E−2	−0.10338E−2	−0.55484E−3	−0.27265E−3	−0.11128E−2	−0.13511E−2	−0.63908E−3	−0.36227E−2	−0.12282E−2
17	+0.97452E−3	−0.46721E−4	+0.22409E−2	+0.10968E−3	+0.20276E−2	−0.15327E−2	+0.17101E−2	−0.10819E−2	−0.22400E−3	−0.15759E−2
18	+0.48897E−3	−0.17088E−3	+0.11000E−2	−0.73945E−3	+0.35477E−3	−0.12433E−2	+0.33765E−3	−0.96321E−3	−0.58416E−3	−0.77655E−6
19	+0.84347E−3	−0.60673E−3	+0.48406E−3	−0.15400E−2	+0.39082E−3	−0.20069E−2	−0.39681E−2	−0.20969E−2	−0.12806E−2	−0.17787E−2
20	−0.39298E−3	−0.36317E−3	−0.33008E−3	+0.22408E−3	−0.92513E−3	−0.48842E−3	−0.11814E−2	−0.19298E−3	−0.19870E−2	+0.49570E−3
C_Q 0	+0.17084E−1	+0.00000E+0	+0.19368E−1	+0.00000E+0	+0.35189E−1	+0.00000E+0	+0.43800E−1	+0.00000E+0	+0.73202E−1	+0.00000E+0
1	+0.10550E+0	−0.78070E+0	+0.17050E+0	−0.99912E+0	+0.24406E+0	−0.11717E+1	+0.35299E+0	−0.12949E+1	+0.47301E+0	−0.14062E+1
2	−0.27380E−1	+0.38134E−1	−0.11901E−1	+0.31924E−1	−0.73880E−2	+0.51155E−1	−0.10917E−1	+0.59030E−1	−0.33300E−1	+0.71683E−1

Chapter Propeller Performance Characteristics

3	−0.11827E−1	+0.74292E−1	−0.25601E−2	+0.81384E−1	+0.28260E−1	+0.89069E−1	+0.47062E−1	+0.93540E−1	+0.62786E−1	+0.11449E+0
4	+0.28671E−1	−0.13568E−1	+0.17763E−1	−0.35096E−3	−0.55959E−2	−0.65670E−2	−0.10779E−1	−0.61148E−2	−0.19511E−1	−0.13400E−1
5	+0.42504E−2	+0.66595E−1	+0.82085E−2	+0.10631E 0	+0.26558E−3	+0.14204E+0	−0.10193E−1	+0.16121E+0	−0.27569E−1	+0.17547E 0
6	−0.78835E−2	+0.10330E−1	−0.34336E−2	+0.15116E−1	+0.11368E−1	+0.77052E−2	−0.88824E−3	+0.14624E−1	−0.38296E−2	+0.25715E−1
7	−0.70981E−2	−0.17885E−1	−0.24534E−1	−0.27045E−1	−0.47401E−1	−0.36091E−1	−0.37893E−1	−0.53549E−1	−0.23310E−1	−0.54967E−1
8	+0.76691E−2	−0.36187E−2	−0.10289E−2	−0.59389E−2	−0.65686E−2	+0.42036E−2	−0.70346E−2	−0.31589E−2	−0.84525E−2	−0.12576E−1
9	−0.12506E−1	+0.10015E−1	−0.74938E−2	+0.11085E−1	−0.74990E−2	+0.21139E−1	−0.81030E−2	+0.14382E−1	−0.48956E−2	+0.13084E−1
10	−0.70343E−2	+0.55926E−2	+0.15444E−2	+0.83787E−2	+0.12873E−1	+0.13095E−1	+0.72622E−2	+0.99836E−2	+0.48544E−2	+0.10733E−1
11	−0.10254E−1	+0.69688E−2	−0.14173E−1	+0.15192E−1	+0.46502E−2	+0.30961E−1	−0.54390E−2	+0.38781E−1	−0.71945E−2	+0.44142E−1
12	+0.25186E−2	−0.47676E−2	−0.28034E−2	−0.51507E−2	−0.46676E−2	−0.99459E−2	−0.20060E−2	−0.46749E−2	−0.53185E−2	−0.10945E−2
13	+0.96613E−2	+0.88889E−2	+0.14246E−1	+0.14836E−1	+0.33438E−2	+0.17921E−1	+0.39281E−2	+0.14944E−1	+0.13281E−2	+0.12209E−1
14	+0.14934E−2	+0.49081E−2	+0.34663E−2	+0.25041E−2	+0.22046E−2	−0.81917E−2	−0.65256E−3	+0.15414E−1	−0.63253E−2	−0.14074E−2
15	−0.28323E−2	−0.58150E−4	+0.20764E−2	−0.78615E−3	+0.70034E−2	−0.78428E−2	+0.15414E−1	+0.22275E−2	+0.68695E−2	+0.17837E−2
16	−0.30360E−2	+0.52044E−2	−0.29424E−2	−0.24526E−2	+0.39147E−1	+0.72661E−2	+0.30356E−2	+0.71826E−2	+0.18071E−1	+0.37948E−2
17	+0.20889E−2	+0.12522E−2	+0.30149E−2	−0.32187E−3	+0.73719E−2	−0.47316E−2	+0.59073E−2	+0.10229E−2	−0.15725E−1	+0.49971E−2
18	+0.31929E−2	+0.33109E−2	+0.27714E−2	−0.48551E−3	−0.94083E−3	−0.25731E−2	+0.41433E−2	−0.59201E−2	+0.10168E−1	−0.42398E−2
19	−0.91635E−3	+0.52446E−2	+0.18423E−3	+0.35455E−2	+0.60560E−2	+0.11136E−2	+0.46102E−2	−0.14814E−2	+0.81504E−3	−0.77298E−2
20	−0.23922E−2	−0.20591E−2	−0.81634E−4	−0.34936E−3	−0.42390E−3	−0.15470E−2	−0.57423E−3	−0.43092E−2	+0.14051E−2	−0.34485E−2

TABLE 6.19 Fourier Coefficients of K_a 4–70 Propeller in No. 37 Duct (Oosterveld[25])

	K	P/D = 0.6 A(K)	P/D = 0.6 B(K)	P/D = 0.8 A(K)	P/D = 0.8 B(K)	P/D = 1.0 A(K)	P/D = 1.0 B(K)	P/D = 1.2 A(K)	P/D = 1.2 B(K)	P/D = 1.4 A(K)	P/D = 1.4 B(K)
C_T^*	0	−0.78522E−1	+0.00000E+0	−0.81169E−1	+0.00000E+0	−0.78681E−1	+0.00000E+0	−0.60256E−1	+0.00000E+0	−0.47437E−1	+0.00000E+0
	1	+0.91962E−1	−0.12241E+1	+0.12849E+0	−0.11842E+1	+0.17005E+0	−0.11152E+1	+0.22360E+0	−0.10687E+1	+0.26393E+0	−0.10004E+1
	2	+0.96733E−1	−0.10805E−1	+0.11331E+0	+0.58341E−3	+0.12604E+0	+0.20371E−1	+0.12353E+0	+0.29643E−1	+0.11478E+0	+0.46145E−1
	3	−0.14657E−2	+0.16207E+0	+0.15131E−1	+0.16441E+0	+0.24444E−1	+0.15275E+0	+0.24086E−1	+0.14275E+0	+0.47309E−1	+0.14074E+0
	4	+0.10810E−1	+0.10642E−2	+0.41567E−2	+0.62103E−2	−0.69987E−2	+0.28881E−2	−0.14518E−1	−0.14016E−1	−0.11061E−1	−0.21940E−1
	5	−0.20708E−1	+0.78648E−1	−0.16220E−1	+0.76506E−1	−0.52998E−1	+0.78299E−1	−0.62461E−2	+0.73413E−1	+0.11308E−1	+0.67294E−1
	6	−0.80316E−2	+0.14098E−1	−0.11305E−1	+0.96359E−2	−0.77500E−2	+0.18865E−1	−0.43441E−2	+0.70950E−3	−0.83647E−3	−0.44987E−2
	7	+0.11052E−1	−0.11329E−1	+0.93452E−2	−0.18036E−1	+0.67088E−2	−0.24665E−1	+0.17726E−1	−0.26735E−1	+0.24933E−1	−0.25518E−1
	8	+0.21070E−2	−0.52596E−2	+0.17779E−2	−0.12146E−1	+0.78818E−2	−0.82956E−2	+0.10820E−1	−0.10309E−1	+0.19552E−2	−0.12518E−1
	9	−0.16466E−1	+0.11815E−1	−0.62214E−2	+0.92879E−2	+0.83058E−2	+0.15085E−1	+0.89902E−2	+0.15399E−1	+0.60531E−2	+0.14151E−1
	10	+0.85238E−3	−0.23771E−2	+0.46290E−2	+0.40488E−2	+0.20833E−2	+0.15879E−2	−0.24474E−2	−0.72466E−2	−0.28748E−2	+0.12588E−2
	11	+0.39384E−2	+0.64113E−2	+0.69293E−2	+0.73891E−2	+0.72262E−2	+0.95129E−2	+0.51620E−2	+0.93292E−2	+0.64118E−2	+0.55618E−2
	12	−0.32905E−2	+0.50027E−2	−0.20445E−2	−0.40761E−2	−0.58329E−3	−0.73249E−2	−0.48962E−2	−0.39241E−2	−0.48164E−2	−0.53289E−2
	13	+0.25672E−2	+0.60467E−2	+0.59366E−2	+0.59307E−2	+0.88467E−2	+0.34931E−2	+0.80184E−2	+0.55616E−2	+0.64267E−2	+0.44079E−2
	14	+0.24770E−2	−0.28242E−2	−0.44055E−3	−0.22663E−2	−0.29559E−2	−0.63570E−2	−0.25019E−2	−0.31513E−3	−0.40358E−2	+0.48467E−3
	15	+0.62208E−2	+0.20489E−2	+0.72301E−2	−0.15148E−2	+0.11530E−1	−0.22474E−2	+0.14983E−1	−0.17566E−2	+0.16051E−1	−0.18905E−2
	16	+0.34143E−3	+0.31069E−3	−0.53198E−3	−0.13262E−3	−0.83057E−3	+0.25069E−2	−0.28220E−3	+0.18409E−2	−0.32816E−2	+0.29965E−2
	17	+0.19780E−2	+0.63925E−3	+0.26809E−2	+0.40086E−2	+0.31339E−2	−0.12990E−2	+0.23533E−2	+0.29180E−2	+0.30250E−2	−0.37761E−2
	18	+0.60762E−3	−0.22082E−2	−0.85582E−3	−0.11431E−2	−0.13290E−2	−0.41905E−3	−0.64457E−3	+0.63270E−3	−0.13567E−2	+0.32763E−2
	19	+0.34488E−2	+0.30421E−2	+0.32728E−2	−0.25883E−2	+0.30666E−2	−0.28288E−2	+0.12248E−2	−0.31003E−2	+0.31794E−2	−0.44109E−2
	20	−0.17166E−2	−0.52892E−3	−0.11347E−2	−0.93290E−4	−0.13749E−2	−0.45929E−3	−0.17391E−2	−0.31826E−3	−0.44946E−3	+0.10459E−2
C_{TN}^*	0	−0.75854E−1	+0.00000E+0	−0.85104E−1	+0.00000E+0	−0.80432E−1	+0.00000E+0	−0.72310E−1	+0.00000E+0	−0.63893E−1	+0.00000E+0
	1	+0.91152E−2	−0.34397E+0	+0.15122E−1	−0.33237E+0	+0.29904E+0	−0.32774E+0	+0.47220E−1	−0.32899E+0	+0.60260E−1	−0.33060E+0
	2	+0.85316E−1	−0.60863E−1	+0.10325E+0	+0.24803E−2	+0.10546E+0	+0.81952E−2	+0.10455E+0	+0.17983E−1	+0.10016E+0	+0.28880E−1

3	$+0.47203E-2$	$+0.82506E-1$	$+0.15649E-1$	$+0.89761E-1$	$+0.21277E-1$	$+0.93073E-1$	$+0.23912E-1$	$+0.93061E-1$	$+0.33369E-1$	$+0.99036E-1$
4	$+0.31838E-2$	$+0.51816E-2$	$+0.75680E-3$	$+0.45883E-2$	$-0.14818E-2$	$+0.38289E-2$	$-0.32961E-1$	$-0.63572E-2$	$-0.17785E-2$	$-0.10075E-1$
5	$+0.45464E-2$	$+0.99282E-2$	$+0.92408E-2$	$+0.26548E-2$	$+0.15667E-1$	$+0.21950E-3$	$+0.24863E-1$	$+0.74431E-3$	$-0.38604E-1$	$+0.71186E-1$
6	$+0.31828E-2$	$+0.59292E-3$	$-0.25160E-8$	$-0.42819E-2$	$-0.17592E-2$	$-0.63045E-2$	$+0.17700E-3$	$-0.30252E-2$	$+0.43713E-2$	$-0.48584E-1$
7	$+0.95481E-2$	$-0.19148E-2$	$+0.18041E-1$	$-0.24385E-1$	$+0.25671E-1$	$-0.23125E-2$	$+0.30933E-1$	$-0.26122E-2$	$+0.35035E-1$	$-0.18671E-1$
8	$-0.19432E-2$	$-0.17686E-2$	$-0.14737E-2$	$-0.45051E-2$	$-0.90447E-3$	$-0.36290E-2$	$+0.15955E-2$	$-0.64419E-2$	$+0.50841E-3$	$-0.10093E-1$
9	$+0.58607E-2$	$-0.78198E-3$	$+0.48237E-2$	$-0.15233E-2$	$+0.88204E-2$	$-0.90853E-3$	$+0.12406E-1$	$+0.33704E-3$	$+0.12764E-1$	$+0.24435E-4$
10	$-0.15047E-2$	$-0.30445E-2$	$-0.99338E-3$	$-0.15358E-2$	$-0.18008E-2$	$-0.18997E-2$	$-0.13918E-2$	$-0.29805E-2$	$-0.24911E-3$	$-0.22708E-2$
11	$+0.38003E-2$	$-0.27783E-2$	$+0.48625E-2$	$-0.32041E-2$	$+0.81643E-2$	$-0.40480E-2$	$+0.10216E-1$	$-0.50133E-2$	$+0.13320E-1$	$-0.77347E-2$
12	$-0.63250E-3$	$+0.17514E-2$	$-0.11657E-2$	$-0.17286E-4$	$-0.97356E-3$	$-0.34197E-3$	$+0.16203E-4$	$-0.28858E-2$	$-0.18539E-2$	$-0.40802E-2$
13	$+0.10102E-2$	$-0.60004E-3$	$+0.43615E-2$	$-0.80871E-3$	$+0.69039E-2$	$-0.13522E-2$	$+0.72150E-2$	$-0.28057E-2$	$+0.77057E-2$	$-0.20523E-2$
14	$-0.28923E-3$	$-0.32771E-3$	$-0.38004E-3$	$-0.21661E-2$	$-0.56244E-3$	$-0.22322E-2$	$-0.39249E-3$	$-0.11737E-2$	$+0.22148E-3$	$-0.21040E-2$
15	$+0.26588E-2$	$+0.16418E-3$	$+0.32422E-2$	$-0.21791E-3$	$+0.45475E-2$	$-0.14808E-2$	$+0.48746E-2$	$-0.34043E-2$	$+0.51024E-2$	$-0.47028E-2$
16	$-0.11302E-3$	$-0.14323E-2$	$-0.37155E-3$	$-0.38196E-3$	$-0.11481E-2$	$-0.21174E-2$	$-0.20546E-2$	$-0.26606E-2$	$-0.19265E-2$	
17	$+0.19966E-2$	$-0.11456E-2$	$+0.22704E-2$	$-0.13466E-2$	$+0.31160E-2$	$-0.28678E-2$	$+0.31669E-2$	$-0.31834E-2$	$+0.24555E-2$	$-0.39615E-2$
18	$-0.56644E-3$	$-0.74530E-3$	$-0.11253E-2$	$-0.23762E-3$	$-0.11878E-2$	$-0.25410E-3$	$-0.13671E-2$	$-0.11222E-2$	$-0.26363E-2$	$-0.60252E-3$
19	$+0.95964E-3$	$-0.83559E-3$	$+0.19759E-2$	$-0.10735E-2$	$+0.19158E-2$	$-0.25631E-2$	$+0.16721E-2$	$-0.26716E-2$	$+0.10883E-2$	$-0.27663E-2$
20	$-0.58527E-3$	$-0.54179E-4$	$-0.96980E-3$	$-0.57005E-5$	$-0.13129E-2$	$+0.22029E-4$	$-0.19020E-2$	$-0.49627E-3$	$-0.28978E-2$	$+0.10214E-3$
C_Q 0	$+0.14884E-1$	$+0.00000E+0$	$+0.20089E-1$	$+0.00000E+0$	$-0.30767E-1$	$+0.00000E+0$	$+0.44351E-2$	$+0.00000E+0$	$+0.64033E-1$	$+0.00000E+0$
1	$+0.10044E+0$	$-0.79096E+0$	$+0.16636E+0$	$-0.99219E+0$	$+0.24472E+0$	$-0.11315E+1$	$+0.34230E+0$	$-0.12562E+1$	$+0.45620E+0$	$-0.13383E+1$
2	$-0.25182E-1$	$+0.12206E-1$	$-0.18383E-1$	$+0.12892E-1$	$-0.11316E-1$	$+0.33712E-1$	$-0.18087E-1$	$+0.55298E-1$	$-0.26747E-1$	$+0.57075E-1$
3	$-0.10918E-1$	$+0.90718E-1$	$-0.19051E-1$	$+0.95272E-1$	$-0.81658E-2$	$+0.90343E-1$	$-0.35568E-2$	$+0.92837E-1$	$+0.16152E-1$	$+0.89051E-1$
4	$+0.27502E-1$	$+0.72669E-2$	$+0.16808E-1$	$+0.16045E+1$	$+0.16208E-2$	$+0.12113E-1$	$+0.86632E-2$	$-0.63786E-2$	$-0.49373E-2$	$-0.15846E-1$	$-0.97724E-2$
5	$-0.26072E-2$	$+0.57653E-2$	$+0.52434E-2$	$+0.94354E-2$	$+0.86632E-2$	$+0.12138E+0$	$-0.54322E-2$	$+0.94572E-3$	$-0.31662E-2$	$-0.19336E-1$	$+0.15575E+0$
6	$-0.11409E-1$	$+0.11032E-3$	$-0.11019E-1$	$-0.36169E-2$	$-0.34936E-2$	$-0.35712E-1$	$-0.46075E-1$	$-0.35793E-1$	$-0.45159E-1$	$-0.28670E-1$	$+0.25599E-2$
7	$+0.93808E+3$	$-0.99388E-2$	$-0.26942E-1$	$-0.22539E-1$	$-0.30457E-4$	$-0.66651E-2$	$+0.63862E-2$	$+0.10295E-1$	$+0.55030E-2$	$-0.42572E-1$	
8	$+0.82783E-2$	$-0.22892E-2$	$+0.94780E-2$	$-0.30457E-4$	$+0.21436E-2$	$+0.14021E-1$	$+0.63862E-2$	$+0.10295E-1$	$+0.56348E-2$	$+0.28700E-3$	$+0.29638E-2$
9	$-0.17756E-1$	$+0.53796E-2$	$-0.19689E-2$	$+0.21436E-2$	$+0.61510E-2$	$+0.14021E-1$	$+0.22438E-2$	$+0.10295E-1$	$+0.55030E-2$	$-0.49652E-1$	$+0.24850E-1$
10	$-0.42598E-2$	$+0.36876E-2$	$+0.16447E-2$	$+0.37463E-2$	$+0.51947E-2$	$+0.28095E-2$	$-0.26275E-2$	$+0.23267E-2$	$+0.23913E-2$	$+0.30607E-2$

(Continued)

TABLE 6.19 Fourier Coefficients of K_a 4–70 Propeller in No. 37 Duct (Oosterveld[25])—cont'd

K	P/D = 0.6 A(K)	P/D = 0.6 B(K)	P/D = 0.8 A(K)	P/D = 0.8 B(K)	P/D = 1.0 A(K)	P/D = 1.0 B(K)	P/D = 1.2 A(K)	P/D = 1.2 B(K)	P/D = 1.4 A(K)	P/D = 1.4 B(K)
11	−0.46664E−2	+0.96603E−2	−0.14766E−1	+0.21398E−1	−0.13702E−1	+0.32828E−1	−0.17427E−1	+0.38309E−1	−0.20976E−1	+0.41837E−1
12	+0.10278E−2	+0.41719E−3	−0.22670E−2	+0.19168E−2	+0.12766E−2	−0.19195E−2	−0.21233E−2	−0.69986E−2	+0.14416E−2	−0.32116E−2
13	+0.20667E−2	+0.72900E−2	+0.93023E−2	+0.76358E−2	+0.61680E−2	+0.88817E−2	+0.98031E−2	+0.14268E−1	+0.39953E−2	+0.17848E−1
14	+0.18501E−2	+0.96970E−3	+0.41823E−2	−0.33249E−2	+0.12713E−2	−0.71309E−2	+0.38115E−2	−0.77495E−2	−0.13605E−2	+0.30114E−2
15	+0.26112E−2	+0.87227E−3	+0.77675E−2	+0.24934E−2	+0.14570E−1	+0.30977E−2	+0.22608E−1	+0.32200E−2	+0.28437E−1	+0.75977E−2
16	−0.32505E−2	+0.21002E−3	−0.13283E−2	+0.39206E−3	+0.44360E−2	+0.64517E−3	−0.37227E−2	+0.69956E−2	+0.19337E−2	+0.27087E−2
17	−0.77389E−3	+0.28832E−2	−0.92032E−3	−0.83670E−3	+0.18140E−2	−0.22876E−2	+0.14353E−3	−0.21184E−2	+0.75472E−2	−0.13255E−2
18	+0.13220E−2	+0.70445E−3	+0.25952E−3	−0.31653E−2	−0.43127E−2	−0.98748E−4	+0.16375E−2	−0.84958E−3	+0.29007E−2	+0.30258E−3
19	+0.16856E−2	+0.39547E−2	+0.23462E−2	+0.41032E−2	+0.38794E−2	+0.10718E−2	+0.51490E−2	−0.14700E−2	+0.32369E−2	−0.37368E−2
20	−0.48127E−3	+0.17791E−2	+0.69823E−3	+0.20375E−2	+0.14728E−3	−0.23283E−2	+0.26719E−2	−0.27758E−2	+0.32247E−2	−0.23794E−2

TABLE 6.20 Coefficients for Strom-Tejsen Polynomials Defining the Gutsche–Schroeder Series — Thrust (taken from Reference 34)

	A Coefficients A(L, M, N) × 10^6						B Coefficients B(L, M, N) × 10^6					
L = 0	M = 0	1	2	3	4	5	M = 0	1	2	3	4	5
N												
0	3655	−8255	−2412	3344	−531	−468	0	0	0	0	0	0
1	21920	−325109	2702	4081	344	646	−781450	−58146	94995	−3948	−9	4189
2	3788	4004	1217	2091	−1045	−266	−10445	11179	6492	−10669	1714	1954
3	33766	55497	6639	−13267	−7191	852	65631	19985	−53191	−14561	4949	4129
4	−4176	−2115	−2876	−1533	1651	97	7884	−7946	−4414	6304	138	−1549
5	−19958	38265	−6849	−2361	758	−160	75207	−9133	−6045	11322	714	−2566
6	492	6657	3006	−2619	411	−101	−2706	1972	2150	−2104	−806	814
7	8923	−18381	5533	−1756	−2055	391	−25357	−13015	18698	−4754	3027	1404
8	5	−5883	−3671	3389	−176	−40	1551	−1903	−95	1112	135	−99
9	7832	5642	6948	−1283	−276	−75	29992	5151	2897	1913	−209	−106
10	482	4105	2507	−1989	−580	373	931	−585	−679	19	88	−322
11	626	−9723	−1720	−431	−1745	−444	3172	−3470	10337	6767	−1973	−585
12	226	−4453	−909	1458	538	−275	−1335	242	1977	−1158	158	478
13	5755	−13923	−1571	−1958	−2669	348	5390	3640	396	−2450	−510	−137
14	365	2814	−321	−593	−190	−24	1196	−64	−1544	565	122	−504
15	1559	−7313	−45	−701	−314	−284	137	3618	−3589	810	−883	−820
16	−334	−1561	47	414	522	−266	−712	−111	731	−147	113	137
17	1274	−6396	−3567	930	−132	−153	−1327	4113	−3339	−1200	−368	−149
18	916	40	−473	120	−158	−81	553	−250	−94	24	−304	76
19	−225	−1818	−1704	1930	213	10	−1357	2334	−4036	−1506	183	−329
20	−348	−207	275	145	131	36	−565	576	−254	233	6	29
21	−601	719	−1132	1511	903	−257	−1833	2294	−1620	−1176	256	−106
22	515	−262	−680	407	−74	−158	−228	233	282	−415	−54	130
23	−932	1075	−1922	1391	608	−99	−514	468	−610	−798	76	97
24	−291	267	146	−17	55	86	−143	176	132	−107	−36	103

(Continued)

TABLE 6.20 Coefficients for Strom-Tejsen Polynomials Defining the Gutsche–Schroeder Series – Thrust (taken from Reference 34)—cont'd

	A Coefficients A(L, M, N) × 10^6						B Coefficients B(L, M, N) × 10^6					
N L = 1	M = 0	1	2	3	4	5	M = 0	1	2	3	4	5
0	930	−457	2417	627	−1599	268	0	0	0	0	0	0
1	5560	36199	−2215	−5043	−1334	−1448	87660	−3013	−2277	−4736	4304	−2672
2	−278	−4286	1106	−2442	1082	786	2340	−1514	−5003	3866	336	−1044
3	−16436	−7937	4940	3197	890	2011	−39708	−14817	2086	−447	−2085	807
4	3837	−1713	1185	1923	−300	−1020	−2190	1872	5151	−3950	76	311
5	16902	−1515	1015	758	10	−159	22508	11572	2165	5732	−1117	−390
6	−1803	778	−3364	50	372	195	−82	−437	−2086	530	340	291
7	−7349	−7633	−6192	1604	246	−566	−2373	−497	−4279	−210	763	−743
8	55	688	4577	−2291	382	63	−1426	3428	207	−199	−205	−202
9	2745	6168	1898	326	−635	36	−14199	1281	−4762	−4345	1906	1198
10	−149	−1956	−417	−261	−473	365	1878	−3528	−597	371	523	−182
11	−3305	4518	−1406	397	1630	−0	8888	995	1792	1752	−1344	−657
12	56	1434	931	157	−163	114	−431	1907	−848	1352	−770	−16
13	−1063	2444	822	904	455	123	−4444	−5828	−419	−1940	284	267
14	804	−2297	−1016	907	−225	−202	−1275	1578	−1128	−512	624	−253
15	2553	3274	3444	−354	59	−2	2303	1520	1760	567	165	255
16	−490	2436	−737	259	−66	36	941	−516	−214	245	186	−63
17	−1944	−752	−441	−724	110	100	1462	−644	2032	2322	−439	−103
18	−199	−628	−170	18	167	−96	−878	700	365	−360	−11	−27
19	1837	−1653	−129	−232	−585	118	−161	−1086	−460	−595	41	177
20	201	753	−578	151	−53	−34	373	−395	160	−514	320	−29
21	−375	−433	803	27	−5	117	542	912	−3	853	−84	−134
22	−744	572	270	−444	270	−59	51	−221	484	149	−167	11
23	51	−1265	−864	−367	−178	5	−675	143	−507	327	39	251
24	564	−488	−41	−56	−49	16	233	−836	657	−414	−117	143

Chapter | 6 Propeller Performance Characteristics

	L = 2									
0	1858	-2160	1565	-716	-262	0	0	0	0	0
1	-5491	4872	-2650	641	196	4845	-4263	5372	1176	-256
2	1235	-1831	1248	323	-59	-4966	2823	-4738	852	819
3	3549	-17	1886	261	951	-355	873	701	345	967
4	-1689	2875	-1366	151	17	3842	-1954	2584	126	-777
5	-2078	-4121	1146	-527	-696	-1541	1516	-1287	-731	-993
6	119	-491	-337	659	-75	-1526	626	-539	-203	218
7	1389	2406	-2698	814	-85	1449	-2144	1010	-135	156
8	504	115	1228	-362	-305	1370	-354	582	42	-333
9	-720	938	1982	-624	-0	-1523	1073	-1157	-413	493
10	-238	-142	-409	101	132	-226	25	-55	102	113
11	66	-1373	-295	230	576	701	858	1258	151	-410
12	648	-755	264	-91	-58	-293	847	-932	102	67
13	-751	1405	-431	518	-248	260	-904	71	-255	321
14	-146	391	-707	256	-154	253	-594	591	41	-92
15	-326	-47	256	158	129	-801	-365	159	387	17
16	103	-338	569	-707	138	-91	371	-448	-120	177
17	145	-298	-287	-55	216	569	538	68	63	-156
18	174	-98	-545	-129	39	135	408	358	68	-153
19	-517	141	196	39	-235	114	-323	127	-100	-81
20	17	-102	-204	180	-102	-281	-412	-125	-71	36
21	596	-591	56	59	169	49	-37	-80	-156	-10
22	108	43	32	-144	34	-19	341	-86	-10	102
23	11	-202	107	-40	10	60	128	259	13	-61
24	-107	-11	96	16	87	-71	-266	9	-20	-42

Coefficients for Strom–Tejsen polynomials defining the Gutsche–Schrouder Series – Torque

	L = 0									
0	-5224	7097	-3358	-4473	3041	-860	0	0	0	0	0
1	-283498	105460	-391193	-7431	-400	-2300	241457	-1211300	-80321	65830	13946	-5256
2	-2361	5401	4755	-1329	339	305	1447	-3138	-308	5690	-4903	1186

(Continued)

TABLE 6.20 Coefficients for Strom-Tejsen Polynomials Defining the Gutsche–Schroeder Series – Thrust (taken from Reference 34)—cont'd

		A Coefficients A(L, M, N) × 10⁶						B Coefficients B(L, M, N) × 10⁶					
N	M = 0	1	2	3	4	5		M = 0	1	2	3	4	5
3	4626	30645	31836	2313	−764	−328		−28541	57594	4960	−28659	−9074	44
4	6544	−13318	−2369	5157	−3336	126		−304	−1966	4951	−3924	558	956
5	30438	−21864	21469	−6644	−4951	−2600		−21066	127800	14368	−4137	−415	2594
6	1011	−142	7201	−2119	346	661		−643	3704	−3358	−1047	2824	−1403
7	−20754	3236	−28269	1160	1133	1634		−2371	−45356	−23439	17969	3844	487
8	−4584	3992	−4253	−2827	1937	−840		−906	960	−2239	3190	−1866	273
9	3803	24969	1468	5987	−1828	−1517		−6720	78975	9414	2656	−2003	−1103
10	4882	−4494	3950	2282	−1158	199		1005	−1106	1313	−577	−124	65
11	−13344	60	−21041	−2939	21	2		−3240	−3368	−5230	4511	4211	−156
12	−4039	3206	−3752	−613	236	116		−962	1337	−2601	1732	−172	−96
13	−10403	14236	−11831	−4347	94	−644		1834	13753	7111	−5097	−1390	−594
14	2500	−890	1653	1003	−246	−145		400	440	197	2	−246	36
15	−5294	3369	−9874	−1982	331	−427		1856	−5692	3713	−3019	−305	−234
16	−1246	−579	−1381	474	−352	199		−525	−151	−521	−8	358	−21
17	−6537	−1127	−5471	−3833	696	578		3379	−9884	2963	−3672	−770	230
18	245	1555	75	−72	139	−57		−168	1014	422	−381	−145	112
19	−650	35	2755	−1385	1061	206		1634	−6289	1642	−2826	−1180	87
20	−216	−798	−392	601	−164	20		151	−1090	152	−264	485	−127
21	−489	4630	2405	−1581	1048	511		2020	−7368	312	−311	−706	96
22	−244	794	−171	−308	96	−12		389	−146	393	−16	−228	39
23	1430	−1749	5017	−943	638	256		60	−2317	−1202	328	−593	423
24	461	−827	700	67	−20	85		−11	−425	25	81	131	71
L = 1													
0	9424	−13356	−731	3382	−1996	116		0	0	0	0	0	0
1	55888	14526	49411	−3400	−7754	−1769		−55003	119608	7412	5292	5423	6134
2	−14831	23859	−5164	832	1620	−7		−4005	9943	−6587	−1534	4015	−1675

Propeller Performance Characteristics

3	−19531	−23447	−11081	7816	9892	1883	10736	−28008	−21466	−7133	−1292	−4745
4	1132	−2698	1536	−6920	1830	−852	−1316	−3353	8037	−3160	−2432	1771
5	−1976	34238	4648	−2054	−4018	−94	1676	29233	10863	894	−894	527
6	4286	−5416	−1003	8609	−2272	528	3025	−155	−6340	4801	284	−801
7	222	−18644	−7190	−735	455	−378	−648	−21742	7485	−3452	1759	1824
8	−8048	13227	−5046	−2936	−742	593	1337	−4706	4231	−1606	−128	−186
9	−1191	3493	4079	−3133	2881	474	4855	−11699	−6139	−2188	−1211	−1549
10	6779	−11303	3004	2760	672	−422	−952	1091	1858	−1658	−167	840
11	8417	−2428	9459	3462	−2476	348	−1140	1265	2616	2964	−292	1888
12	−3001	6360	−1543	−1351	−488	−6	735	−1188	−285	18	590	−442
13	382	−7205	−626	2438	440	306	−3191	−2228	−6197	−639	886	−560
14	−1161	−636	20	−199	602	283	697	−1836	2483	−1214	−79	305
15	2804	8763	5562	2238	230	−104	316	5018	1043	−439	65	87
16	2436	−1925	2635	−624	−481	−6	659	−1328	278	775	−650	213
17	102	−3301	−2018	830	−1060	−8	−1287	2977	1019	2547	1139	530
18	−957	−1255	−318	79	−52	353	267	−294	−130	313	−153	59
19	−2432	3712	−2846	−615	653	−57	−447	1497	−1377	−2460	6	−449
20	1417	−408	1703	−403	−104	−169	−424	894	−522	−179	357	−197
21	659	1995	672	202	−124	−292	797	−371	3233	713	192	166
22	−202	−1418	−66	−14	100	112	53	301	−919	982	−321	38
23	−1253	−1514	−1940	−1300	−25	−142	41	−1084	211	−209	43	−285
24	373	484	98	165	−211	−17	−818	1674	−1168	45	205	−80
L = 2												
0	−667	2038	637	−1161	1244	−298	0	0	0	0	0	0
1	554	−7720	−1326	−4857	453	−706	−558	−4300	−2556	3381	−2542	1485
2	−3194	2831	−1895	−1012	−677	574	−774	344	895	810	−1930	862
3	2703	−1389	6164	12655	1152	1605	−403	1131	−78	5553	1034	−576
4	4377	−5899	4002	1540	−964	260	106	−1226	1498	−1122	515	−22
5	−3174	990	−8664	−7321	−2829	−1039	1598	−313	3787	−4531	2320	−30

(Continued)

TABLE 6.20 Coefficients for Strom-Tejsen Polynomials Defining the Gutsche–Schroeder Series — Thrust (taken from Reference 34)—cont'd

	A Coefficients A(L, M, N) × 10^6						B Coefficients B(L, M, N) × 10^6					
N	M = 0	1	2	3	4	5	M = 0	1	2	3	4	5
6	−1162	−89	−842	−810	−122	119	697	819	−1737	204	918	−673
7	−1301	3368	3117	454	1490	54	−2252	630	−2404	−3766	−3161	59
8	4	1418	1358	−890	848	−340	−1226	714	−245	908	−1073	−794
9	2261	−2467	713	2376	1003	−25	2195	−3687	679	2615	1205	−33
10	705	−2529	−568	899	−748	336	1219	−1191	507	117	−335	−61
11	−87	1335	992	−2572	−1117	134	1840	2341	4119	967	−82	130
12	−553	2154	193	−288	106	−123	−1260	1486	−939	263	178	100
13	−345	−2443	−715	878	1046	61	−1409	−2660	−3194	−954	−610	−23
14	−145	−776	−551	230	82	78	618	−325	−148	277	−274	−75
15	1407	−360	−174	−499	−18	−517	1964	241	2290	936	−163	51
16	226	618	104	110	−212	−177	−368	169	132	−131	109	175
17	−292	−1007	95	−500	−171	51	−1376	−95	−1849	−961	−361	176
18	−368	−65	−404	165	119	130	58	−30	−55	38	−157	32
19	−26	−181	−775	678	191	1	−94	713	−357	361	−307	113
20	293	63	70	155	−307	23	235	−334	379	−180	138	6
21	89	1069	−54	−271	−466	−11	−62	−128	230	−324	10	12
22	−95	−189	5	70	2	87	−71	−169	−31	−55	7	−27
23	−418	−341	−235	254	133	148	−425	−41	−302	−432	−53	87
24	180	−157	112	−57	9	−25	148	128	71	−33	112	−87

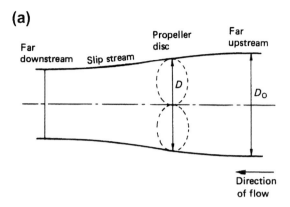

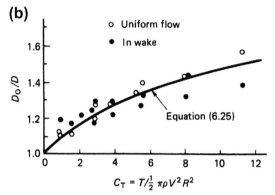

FIGURE 6.21 Slipstream contraction: (a) contraction of slipstream and (b) relation between contraction flow and propeller thrust.

field point velocities close to the trailing edge of the NSRDC 4383 propeller working at its design J of 0.889. The propeller has a skew of $72°$ which accounts for the slipstream non-dimensional radius starting at unity at a distance of some $0.35R$ behind the propeller. This propeller is one of the series, referred to in Section 6.5.7, and tested originally by Boswell.[19]

6.8 BEHIND-HULL PROPELLER CHARACTERISTICS

The behind-hull propeller characteristics, so far as powering is concerned, have been traditionally accounted for by use of the term relative rotative efficiency η_r. This term, which was introduced by Froude, accounted for the difference in power absorbed by the propeller when working in a uniform flow field at a given speed and that absorbed when working in a mixed wake field having the same mean velocity:

$$\eta_r = \frac{\text{power absorbed in open water of speed} V_a}{\text{power absorbed in mixed wake field of mean velocity } V_a} \quad (6.26)$$

Normally the correction defined by this efficiency parameter is very small and η_r is usually close to unity unless there is some particularly abnormal characteristic of the wake field. Typically, one would expect to find η_r in the range $0.96 < \eta_r < 1.04$.

As a consequence of this relationship the behind-hull efficiency (η_b), that is the efficiency of the propeller when working behind a body, is defined as

$$\eta_b = \eta_o \cdot \eta_r$$
$$\eta_b = \frac{\eta_r K_T J}{2\pi K_Q} \quad (6.27)$$

Considerations such as relative rotative efficiency are clearly at the global level of ship propulsion. At the more

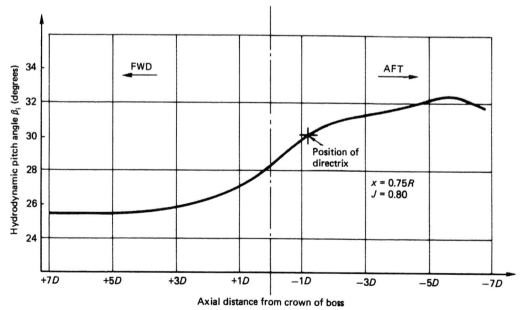

FIGURE 6.22 Axial distribution of hydrodynamic pitch for KCD19 propeller.

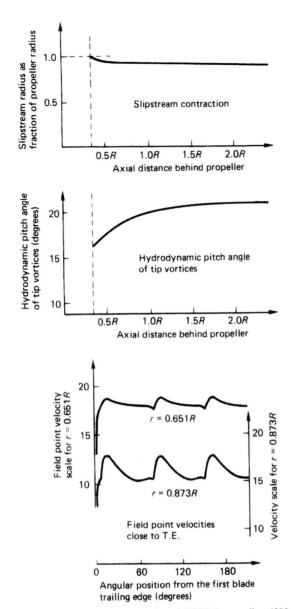

FIGURE 6.23 Slipstream properties of NSRDC propeller 4383 at design advance.

detailed level there is much still to be understood about the nature of the interactions between the propeller, its induced and interaction velocities and the wake field in which it operates.

As might be expected the effect of the mixed wake field is to induce on the propeller a series of fluctuating load components due to the changing nature of the flow incidence angles on the blade sections. Figure 11.4 shows a typical example of the variation in thrust acting on the blade of a single-screw container ship due to operation of the propeller in the wake field. The asymmetry is caused by the tangential velocity components of the wake field, which act in opposite senses in each half of the propeller disc for a single screw ship. Clearly such considerations apply to the torque forces on the blade and also the hydrodynamic spindle torque in the case of a controllable pitch propeller. Figure 6.24 shows the resulting bearing forces, that is those reacted by the bearings of the vessel, which are the sum of the individual blade components at each shaft angular position. From the figure it is seen that not only is there a thrust and torque fluctuation as derived from individual blade loads, similar to that shown in Figure 11.4, but also loads in the vertical and horizontal directions, F_Y and F_Z, and moments M_Z and M_Y. In Figure 11.5 the orbit of the thrust eccentricity relative to the shaft center line is shown for a merchant ship. These orbits define the position of the thrust vector in the propeller disc at a given instant; it should, however, be noted that the thrust vector marches around the orbit at blade rate frequency.

In addition to the blade loadings the varying incidence angles around the propeller disc introduce a fluctuating cavitation pattern over the blades. Typical of such a pattern is that shown in Figure 6.25, from which it is seen that the wake-induced asymmetry also manifests itself here in the growth and decay of the cavity volume.

6.9 PROPELLER VENTILATION

Propeller ventilation can have a significant influence on the performance characteristics of a propeller. Koushan[45] has investigated these effects in relation to a non-ducted thruster. He showed that in a ventilated condition even when the propeller is well submerged the loss of thrust can be as much as 40 per cent. In the corresponding condition of partial-submergence this loss may rise to as high as 90 per cent. Moreover, the mechanism of ventilation can take many forms; for example, it may be from the direct drawing of air from the water surface or, alternatively, it could be that the air uses some other path such as down the surface of A or P brackets or some other appendage and then passes to the propeller.

Scale effects are particularly influential in assessing the propeller characteristics and, in particular, the influence of the Weber (We), Depth Froude (Fn_p) and Ventilation (σ_v) numbers need to be considered. In this context these numbers are defined as:

$$We = nD\sqrt{(\rho D/S)}$$
$$Fnp = \pi nD/\sqrt{(gh)}$$
$$\sigma_v = 2gh/(V_R^2)$$

where

S is the air–water surface tension
h is the propeller shaft immersion
V_R is the propeller section inflow velocity uncorrected for induction effects and normally referred to the $0.7R$ radial location.

Propeller Performance Characteristics

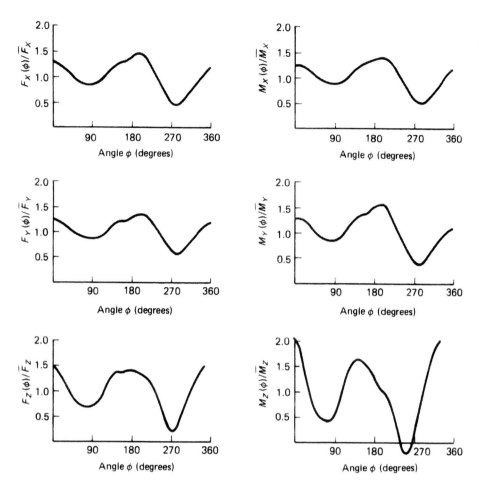

FIGURE 6.24 Typical fluctuation in bearing forces and moments for a propeller working in a wake field.

The remaining symbols ρ, g, n and D have their usual meaning.

In the case of the Weber number, because surface tension is a significant parameter in model testing, it has a significant influence on the measured results. Shiba,[46] based on a large set of model measurements, concluded that if the Weber number is greater than 180 then its effect is probably insignificant. Below that critical number, however, it was concluded that less or delayed ventilation might be observed at model-scale when compared to full-scale.

When the propeller breaks the surface or is close to the free surface and generates a system of local waves and then the Froude depth number assumes importance. In the case of the ventilation number, this is essentially a cavitation number as discussed in Chapter 9 in which the normal static vapor pressure is replaced with ambient pressure. From a little algebraic manipulation of the relationships defined above, it can be seen that if the advance coefficient of a particular test is defined and then one of either the Froude depth number or the ventilation

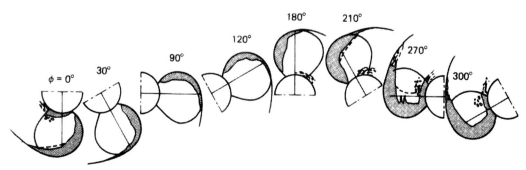

FIGURE 6.25 Cavitation pattern on the blades of a model propeller operating in a wake field. *Reproduced partly from Reference 39.*

number is satisfied then the other coefficient will also be satisfied.

REFERENCES AND FURTHER READING

1. Vasamarov KG, Minchev AD. *A Propeller Hydro-Dynamic Scale Effect*. BSHC Report No. PD-83-108; 1983.
2. Voitkounski YI, editor. *Ship Theory Handbook, 1*. Leningrad: Sudostroeme; 1985.
3. Troost L. Open water test series with modern propeller forms. *Trans. NECIES* 1938;**54**.
4. Troost L. Open water test series with modern propeller forms. II. Three bladed propellers. *Trans. NECIES*; 1940.
5. Troost L. Open water test series with modern propeller forms. III. Two bladed and five bladed propellers — Extension of the three and four bladed B-series. *Trans. NECIES* 1951;**67**.
6. Lammeren WPA van, Manen JD van, Oosterveld MWC. The Wageningen B-Screw series. *Trans. SNAME*; 1969.
7. Oosterveld MWC, Oosannen P van. Further computer-analysed data of the Wageningen B-Screw series. *ISP* July 1975;**22**.
8. Yazaki A. *Design diagrams of modern four, five, six and seven-bladed propellers developed in Japan*. 4th Naval Hydrodynamics Symp. Washington: National Academy of Sciences; 1962.
9. Gawn RWL. Effect of pitch and blade width on propeller performance. *Trans. RINA*; 1952.
10. Blount DL, Hubble EN. Sizing for segmental section commercially available propellers for small craft. *Propellers 1981 Conference, Trans. SNAME*; 1981.
11. Gawn RWL, Burrill LC. Effect of cavitation on the performance of a series of 16 in. model propellers. *Trans. RINA*; 1957.
12. Emerson A, Sinclair L. Propeller design and model experiments. *Trans. NECIES*; 1978.
13. Lingren H. *Model Tests With a Family of Three and Five Bladed Propellers*. SSPA Paper No. 47, Göteborg; 1961.
14. Newton RN, Rader HP. Performance data of propellers for high speed craft. *Trans. RINA* 1961;**103**.
15. Burrill LC, Emerson A. Propeller cavitation: Some observations from the 16 in. propeller tests in the New King's College cavitation tunnel. *Trans. NECIES* 1963;**79**.
16. Burrill LC, Emerson A. Propeller cavitation: Further tests on 16 in. propeller models in the King's College cavitation tunnel. *Trans. NECIES*; 1978.
17. Manen JE van, Oosterveld MWC. *Model Tests on Contra-Rotating Propellers*. NSMB Wageningen: Publication No. 317; 1969.
18. Bjarne E. Systematic studies of control-rotating propellers for merchant ships. *IMAS Conference, Trans. I. Mar. E*; 1973.
19. Boswell RJ. *Design, Cavitation Performance, and Open Water Performance of a Series of Research Skewed Propellers*. NSRDC Report No. 3339; March 1971.
20. Gutsche F. Untersuchung von Schiffsschrauben in Schräger Anstromung. *Schiffbanforschung* 1964;**3**.
21. Taniguchi K, Tanibayashi H, Chiba N. *Investigation into the Propeller Cavitation in Oblique Flow*. Mitsubishi Technical Bulletin No. 45; March 1969.
22. Bednarzik R. Untersuchungen über die Belaslungs-schwankungen am Einzelflügel schrägangeströmter Propeller. *Schiffbanforschung* 1969;**8**.
23. Meyne K, Nolte A. Experimentalle Untersuchungen der hydro-dynamischen Kräfte und Momente an einem Flügel eines Schiffspropellers bei schräger Anströmung. *Schiff und Hafen* 1969;**5**.
24. Peck JG, Moore DH. Inclined-shaft propeller performance characteristics. *Trans. IESS. SNAME, Spring Meeting*; 1973.
25. Oosterveld MWC. *Wake Adapted Ducted Propellers*. NSMB Wageningen Publication No. 345; June 1970.
26. Oosterveld MWC. *Ducted propeller characteristics*. London: RINA Symp. on Ducted Propellers; 1973.
27. Gutsche FA, Schroeder G. Freifahruersuche an Propellern mit festen und verstellbaren Flügeln 'voraus' und 'zuruck'. *Schiffbauforschung* 1963;**2**(4).
28. Chu C, Chan ZL, She YS, Yuan VZ. The 3-bladed JD-CPP series. *4th Lips Propeller Symp*; October 1979.
29. Yazaki A. *Model Tests on Four Bladed Controllable Pitch Propellers*. Tokyo: Ship Research Institute; March 1964. Paper 1.
30. Hansen EO. *Thrust and Blade Spindle Torque Measurements of Five Controllable Pitch Propeller Designs for MS0421*. NSRDC Report No. 2325; April 1967.
31. Miller ML. Spindle Torque Test on Four CPP Propeller Blade Designs for MS0421. *DTMB Report No.* July 1964;**1837**.
32. Conn JFC. Backing of propellers. *Trans. IESS*; 1923.
33. Nordstrom HF. *Screw Propeller Characteristics*. Göteborg: SSPA Publication No. 9; 1948.
34. Strom-Tejsen J, Porter RR. Prediction of controllable-pitch propeller performance in off-design conditions. *Third Ship Control System Symp*; 1972. Paper VII B-1, Bath, UK.
35. Nagamatsu T, Sasajima T. Effect of propeller suction on wake. *J. Soc. Nav. Arch. Japan* 1975;**137**.
36. Leathard FI. *Analysis of the performance of marine propellers, with particular reference to controllable pitch propellers*. MSc Thesis. Newcastle University; 1974.
37. Keh-Sik M. *Numerical and Experimental Methods for the Prediction of Field Point Velocities around Propeller Blades*. MIT, Department Ocean Engineering Report No. 78-12; June 1978.
38. Kerwin JE, Chang-Sup L. Prediction of Steady and UnsteadyMarine Propeller Performance by Numerical Lifting-Surface Theory. *Trans. SNAME*; 1978. Paper No. 8, Annual Meeting.
39. Oossanen P van. *Calculation of Performance and Cavitation Characteristics of Propellers Including Effects of Non-Uniform Flow and Viscosity*. MARIN Publication No. 457; 1970.
40. Kuiper G. The Wageningen propeller series. *MARIN*; May 1992.
41. Bazilevski YS. On the propeller blade turbulization in model tests. *Laventiv Lectures, St Petersburg*; 2001::201-6.
42. Boorsma A. *Improving full scale ship powering predictions by application of propeller leading edge roughness*. MSc Thesis, Delft; December 2000.
43. Ball W, Carlton JS. Podded propulsor shaft loads from free-running model experiments in calm-water and waves. *Int. J. Maritime Eng.* 2006;**148**(Part A4).
44. Holtrop J. *Extrapolation of propulsion tests for ships with appendages and complex propulsors*. Mar. Technology 2001;**38**.
45. Koushan K. *Dynamics of ventilated propeller blade loading on thrusters*. London: WTMC Conf. Trans. I. Mar. EST; 2006.
46. Shiba H. *Air Drawing of Marine Propellers*. Japan: Transportation Technical Research Institute, Report No. 9; August 1953.

Chapter 7

Theoretical Methods — Basic Concepts

Chapter Outline
- 7.1 Basic Aerofoil Section Characteristics 139
- 7.2 Vortex Filaments and Sheets 141
- 7.3 Field Point Velocities 142
- 7.4 The Kutta Condition 144
- 7.5 The Starting Vortex 145
- 7.6 Thin Aerofoil Theory 146
- 7.7 Pressure Distribution Calculations 149
- 7.8 Boundary Layer Growth Over an Aerofoil 154
- 7.9 The Finite Wing 158
- 7.10 Models of Propeller Action 160
- 7.11 Source and Vortex Panel Methods 163
- 7.12 Euler, Lagrangian and Navier–Stokes Methods 164
- References and Further Reading 166

Theoretical methods to predict the action of propellers began to develop in the latter part of the nineteenth century. Perhaps the most notable of these early works was that of Rankine, with momentum theory, which was then closely followed by the blade element theories of Froude. The modern theories of propeller action, however, had to await the more fundamental works in aerodynamics of Lanchester, Kutta, Joukowski, Munk and Prandtl in the early years of the nineteenth century before their development could begin.

Lanchester, an English automobile engineer and self-styled aerodynamicist, was the first to relate the idea of circulation with lift and he presented his ideas to the Birmingham Natural History and Philosophical Society in 1894. He subsequently wrote a paper to the Physical Society, who declined to publish these ideas. Nevertheless, he published two books, *Aerodynamics* and *Aerodonetics*, in 1907 and 1908 respectively. In these books, which were subsequently translated into German and French, we find the first mention of vortices that trail downstream of the wing tips and the proposition that these trailing vortices must be connected by a vortex that crosses the wing: the first indication of the 'horse-shoe' vortex model.

It appears that quite independently of Lanchester's work, Kutta developed the idea that lift and circulation were related; however, he did not give the quantitative relation between these two parameters. It was left to Joukowski, working in Russia in 1906, to propose the relation

$$L = \rho V \Gamma \quad (7.1)$$

This has since become known as the Kutta–Joukowski theorem. History shows that Joukowski was completely unaware of Kutta's note on the subject, but in recognition of both their contributions the theorem has generally been known by their joint names.

Prandtl, generally acclaimed as the father of modern aerodynamics, extended the work of aerodynamics into finite wing theory by developing a classical lifting line theory. This theory evolved to the concept of a lifting line comprising an infinite number of horse-shoe vortices as sketched in Figure 7.1. Munk, a colleague of Prandtl at Gottingen, first introduced the term 'induced drag' and also developed the aerofoil theory which has produced such exceptionally good results in a wide variety of subsonic applications.

From these beginnings the development of propeller theories started, slowly at first but then gathering pace through the 1950s and 1960s. These theoretical methods, whether aimed at the design or analysis problem, have all had the common aim of predicting propeller performance by means of a mathematical model which has inherent assumptions built into it. Consequently, these mathematical models of propeller action rely on the same theoretical basis as that of aerodynamic wing design, and therefore appeal to the same fundamental theorems of sub-sonic aerodynamics or hydrodynamics. Although aerodynamics is perhaps the wider ranging subject in terms of its dealing with a more extensive range of flow speeds, for example subsonic, supersonic and hypersonic flows, both non-cavitating hydrodynamics and aerodynamics can be considered to be the same subject provided the Mach number does not exceed a value of round 0.4–0.5. This is the point where the effects of compressibility in air start to become appreciable.

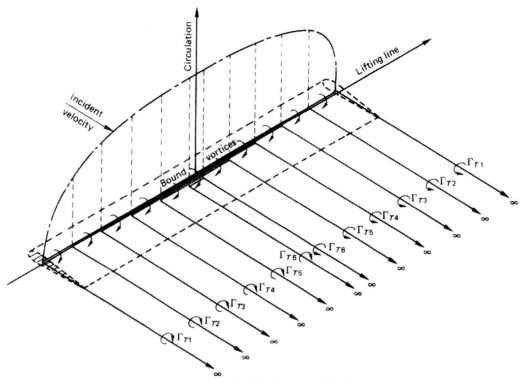

FIGURE 7.1 Prandtl's classical lifting line theory.

This book is not a treatise on general fluid mechanics and, therefore, it will not deal in detail with the more fundamental and abstract ideas of fluid dynamics. For these matters the reader is referred to references[1-4]. In both this chapter and the next we are concerned with introducing the various theoretical methods of propeller analysis in order to provide a basis for further reading or work. However, in order to do this certain prerequisite theoretical ideas are needed, some of which can be useful analytical tools in their own right. To meet these requirements the subject is structured into two parts; this chapter deals with the basic theoretical concepts necessary to evolve and understand the theories of propeller action which are then discussed in more detail in Chapter 8: Table 7.1 shows this structure.

TABLE 7.1 Outline of Chapters 7 and 8	
Chapter 7	**Chapter 8**
Basic Concepts and Theoretical Methods	*Propeller Theories*
General Introduction	Momentum Theory
Experimental Single and Cascade Aerofoil Characteristics	Blade Element Theory
Vortex Filaments and Sheets	Burrill Analysis Method
Field Point Velocities	Lerbs Method
Kutta Condition	Early Design Methods — Burrill and Eckhardt and Morgan
Kelvin's Theorem	Heavily Loaded Propellers (Glover)
Thin Aerofoil Theory	Lifting Surface Models (Morgan et al., van-Gent, Breslin)
Pressure Distribution Calculations	Advanced Lifting Line Lifting
NACA Pressure	Surface Hybrid Models
Distribution	Vortex Lattice Models (Kerwin)
Approximation	Boundary Element Methods
Boundary Layer Growth over Aerofoil	Special Propeller Types:
Finite Wing and	Controllable Pitch
Downwash	Ducted Propellers
Hydrodynamic Models of Propeller Action	Contra-rotating
Vortex and Source Panel Methods	Super-cavitating

Theoretical Methods – Basic Concepts

The review of the basic concepts will of necessity be in overview terms consistent with this being a book concerned with the application of fluid mechanics to the marine propeller problem. Furthermore, the discussion of the propeller theories, if conducted in a detailed and mathematically rigorous way, would not be consistent with the primary aim of this book and would also require many books of this size to do justice to them. Accordingly, the important methods will be discussed sufficiently for the reader to understand their essential features, uses and limitations and references will be given for further detailed study. Additionally, where several complementary methods exist within a certain class of theoretical methods, only one will be discussed with references being given to the others.

7.1 BASIC AEROFOIL SECTION CHARACTERISTICS

Before discussing the theoretical basis for propeller analysis it is perhaps worth spending time considering the experimental characteristics of wing sections, since these are in essence what the analytical methods are attempting to predict.

Figure 7.2 shows the experimental results for a two-dimensional aerofoil having National Advisory Committee for Aeronautics (NACA) 65 thickness form superimposed on an $a = 1.0$ mean line. The figure shows the lift, drag and pitching moment characteristics of the section as a function of angle of attack and for different Reynolds numbers. In this instance the moment coefficient is taken about the quarter chord point; this point is frequently chosen since it is the aerodynamic center under the assumptions of thin aerofoil theory, and in practice lies reasonably close to it. The aerodynamic center is the point where the resultant lift and drag forces are assumed to act and hence do not influence the moment, which is camber profile and magnitude related. The lift, drag and moment coefficients are given by the relationships

$$C_L = \frac{L}{\frac{1}{2}\rho A V^2}$$

$$C_D = \frac{D}{\frac{1}{2}\rho A V^2}$$

(7.2)

and

$$C_M = \frac{M}{\frac{1}{2}\rho A l V^2}$$

in which A is the wing area, l is a reference length, V is the free stream incident velocity, ρ the density of the fluid, L and D are the lift and drag forces, perpendicular and parallel respectively to the incident flow, and M is the pitching moment defined about a convenient point.

These coefficients relate to the whole wing section and as such relate to average values for a finite wing section.

For analysis purposes, however, it is important to deal with the elemental values of the aerodynamic coefficients,

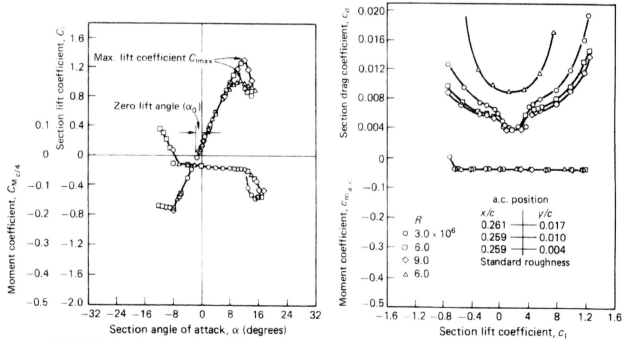

FIGURE 7.2 Experimental single aerofoil characteristics (NACA 65–209). *Reproduced with permission from Reference 11.*

and these are denoted by the lower case letters c_l, c_d, c_m, given by

$$c_l = \frac{L'}{\frac{1}{2}\rho c V^2}$$
$$c_d = \frac{D'}{\frac{1}{2}\rho c V^2} \quad (7.3)$$

and

$$c_m = \frac{M'}{\frac{1}{2}\rho c^2 V^2}$$

in which c is the section chord length and L', D' and M' are the forces and moments per unit span.

Returning now to Figure 7.2, it will be seen that while the lift slope is not influenced by Reynolds number, the maximum lift coefficient $C_{L\max}$ is dependent upon R_n. The quarter chord pitching moment is also largely unaffected by Reynolds numbers over the range of non-stalled performance and the almost constant nature of the quarter chord pitching moment over the range is typical. There is, by general agreement, a sign convention of the aerodynamic moments which states that moments which tend to increase the incidence angle are considered positive, while those which decrease the incidence angle are negative. Moments acting on the aerofoil can also be readily transferred to other points on the blade section, most commonly the leading edge or, in the case of a controllable pitch propeller, the spindle axis. With reference to the simplified case shown in Figure 7.3 it can be seen that

$$M'_{LE} = \frac{-cL'}{4} + M'_{c/4} = -x_{cp}L' \quad (7.4)$$

Clearly, in the general case of Figure 7.3 both the lift and drag would need to be resolved with respect to the angle of incidence to obtain a valid transfer of moment.

In equation (7.4) the term x_{cp} is defined as the center of pressure of the aerofoil and is the location of the point where the resultant of the distributed load over the section effectively acts. Consequently, if moments were taken about the center of pressure the integrated effect of the distributed loads would be zero. The center of pressure is an extremely variable quantity; for example, if the lift is zero, then by equation (7.4) it will be seen that $x_{cp} \to \infty$, and this tends to reduce its usefulness as a measurement parameter.

The drag of the aerofoil as might be expected from its viscous origin is strongly dependent on Reynolds number: this effect is seen in Figure 7.2. The drag coefficient c_d shown in this figure is known as the profile drag of the section and it comprises both a skin friction drag c_{df} and a pressure drag c_{dp}, both of which are due to viscous effects. However, in the case of a three-dimensional propeller blade or wing there is a third drag component, termed the induced drag, c_{di}, which arises from the free vortex system. Hence the total drag on the section is given by equation (7.5):

$$c_d = c_{df} + c_{dp} + c_{di} \quad (7.5)$$

The results shown in Figure 7.2 also show the zero lift angle for the section, which is the intersection of the lift curve with the abscissa; as such, it is the angle at which the aerofoil should be set relative to the incident flow in order to give zero lift. The propeller problem, however, rather than dealing with the single aerofoil in isolation, is concerned with the performance of aerofoils in cascades. By this is meant a series of aerofoils, the blades in the case of the propeller, working in sufficient proximity to each other so that they mutually affect each other's hydrodynamic characteristics. The effect of cascades on single aerofoil performance characteristics is shown in Figure 7.4. From

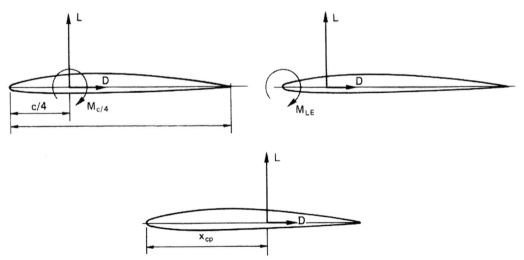

FIGURE 7.3 Moment and force definitions for aerofoils.

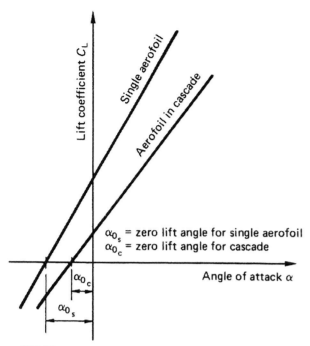

FIGURE 7.4 Effect of cascade on single aerofoil properties.

the figure it is seen that both the lift slope and the zero lift angle are altered. In the case of the lift slope this is reduced from the single aerofoil case, as is the magnitude of the zero lift angle. As might be expected, the section drag coefficient is also influenced by the proximity of the other blades; however, this results in an increase in drag.

7.2 VORTEX FILAMENTS AND SHEETS

The concept of the vortex filament and the vortex sheet is central to the understanding of many mathematical models of propeller action. The idea of a vortex flow, Figure 7.5(a), is well known and is considered in great detail by many standard fluid mechanics textbooks. It is, however, worth recalling the sign convention for these flow regimes which states that a positive circulation induces a clockwise flow. For the purposes of developing propeller models, this two-dimensional vortex flow has to be extended into the concept of a line vortex or vortex filament as shown in Figure 7.5(b).

The line vortex is a vortex of constant strength Γ acting along the entire length of the line describing its path through space; in the case of propeller technology this space will be three-dimensional. With regard to vortex filaments Helmholtz, the German mathematician, physicist and physician, established some basic principles of inviscid vortex behavior which have generally become known as Helmholtz's vortex theorems:

1. The strength of a vortex filament is constant along its length.
2. A vortex filament cannot end in a fluid. As a consequence the vortex must extend to the boundaries of the fluid which could be at $\pm \infty$ or, alternatively, the vortex filament must form a closed path within the fluid.

These theorems are particularly important since they govern the formation and structure of inviscid vortex propeller models.

The idea of the line vortex or vortex filament can be extended to that of a vortex sheet. For simplicity at this stage we will consider a vortex sheet comprising an infinite number of straight line vortex filaments side by side as shown in Figure 7.6. Although we are here considering straight line vortex filaments the concept is readily extended to curved vortex filaments such as might form a helical surface, as shown in Figure 7.7. Returning, however, to Figure 7.6, consider the sheet 'end-on' looking in the direction Oy. If the strength of the vortex sheet is defined, per unit length, over the sheet as $\gamma(s)$, where s is the distance measured along the vortex sheet in the edge view, we can

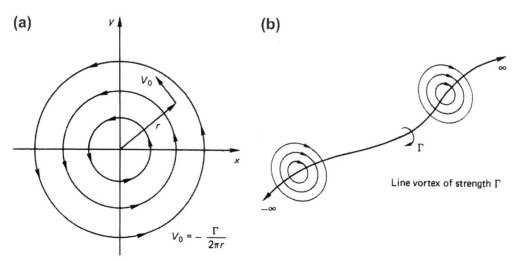

FIGURE 7.5 Vortex flows: (a) two-dimensional vortex and (b) line vortex.

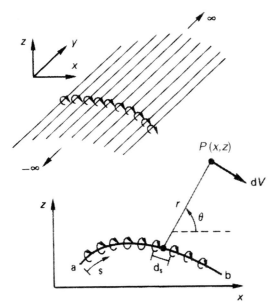

FIGURE 7.6 Vortex sheet.

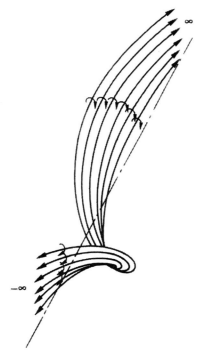

FIGURE 7.7 Helical vortex sheet.

then write for an infinitesimal portion of the sheet, ds, the strength as being equal to $\gamma \, ds$. This small portion of the sheet can then be treated as a distinct vortex strength which can be used to calculate the velocity at some point P in the neighborhood of the sheet. For the point $P(x,z)$ shown in Figure 7.6 the elemental velocity dV, perpendicular to the direction r, is given by

$$dV = \frac{-\gamma \, ds}{2\pi r} \tag{7.6}$$

Consequently, the total velocity at the point P is the summation of the elemental velocities at that point arising from all the infinitesimal sections from a to b.

The circulation Γ around the vortex sheet is equal to the sum of the strengths of all the elemental vortices located between a and b, and is given by

$$\Gamma = \int_a^b \gamma \, ds \tag{7.7}$$

In the case of a vortex sheet there is a discontinuity in the tangential component of velocity across the sheet. This change in velocity can readily be related to the local sheet strength such that if we denote upper and lower velocities immediately above and below the vortex sheet, by u_1 and u_2 respectively, then the local jump in tangential velocity across the vortex sheet is equal to the local sheet strength:

$$\gamma = u_1 - u_2$$

The concept of the vortex sheet is instrumental in analyzing the properties of aerofoil sections and finds many applications in propeller theory. For example, one such theory of aerofoil action might be to replace the aerofoil with a vortex sheet of variable strength, as shown in Figure 7.8. The problem then becomes to calculate the distribution of $\gamma(s)$ so as to make the aerofoil surface become a streamline to the flow.

These analytical philosophies were known at the time of Prandtl in the early 1920s; however, they had to await the advent of high-speed digital computers some forty years later before solutions on a general basis could be attempted.

In addition to being a convenient mathematical device for modeling aerofoil action, the idea of replacing the aerofoil surface with a vortex sheet also has a physical significance. The thin boundary layer which is formed over the aerofoil surface is a highly viscous region in which the large velocity gradients produce substantial amounts of vorticity. Consequently, there is a distribution of vorticity along the aerofoil surface due to viscosity and the philosophy of replacing the aerofoil surface with a vortex sheet can be construed as a way of modeling the viscous effects in an inviscid flow.

7.3 FIELD POINT VELOCITIES

The field point velocities are those fluid velocities that may be in either close proximity to or remote from the body of interest. In the case of a propeller the field point velocities

FIGURE 7.8 Simulation of an aerofoil section by a vortex sheet.

Theoretical Methods — Basic Concepts

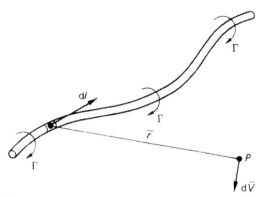

FIGURE 7.9 Application of the Biot–Savart law to a general vortex filament.

are those that surround the propeller both upstream and downstream of it.

The classical mathematical models of propeller action are today normally based on systems of vortices combined in a variety of ways in order to give the desired physical representation. As a consequence a principal tool for calculating field point velocities is the Biot–Savart law. This law is a general result of potential theory and describes both electromagnetic fields and inviscid, incompressible flows. In general terms the law can be stated, Figure 7.9, as the velocity $d\overline{V}$ induced at a point P of radius r from a segment $d\overline{s}$ of a vortex filament of strength Γ given by

$$d\overline{V} = \frac{\Gamma}{4\pi} \frac{d\overline{l} \times \overline{r}}{|\overline{r}|^3} \qquad (7.8)$$

To illustrate the application of the Biot–Savart law, two common examples of direct application to propeller theory are cited: the first is a semi-infinite line vortex and the second is a semi-infinite regular helical vortex. Both of these examples commonly represent systems of free vortices emanating from the propeller.

First, the semi-infinite line vortex. Consider the system shown in Figure 7.10, which shows a segment $d\overline{s}$ of a straight line vortex originating at O and extending to infinity in the positive x-direction. Note that in practice,

according to Helmholtz's theorem, the vortex could not end at the point O but must be joined to some other system of vortices. However, for our purposes here it is sufficient to consider this part of the system in isolation. Now the velocity induced at the point P distant r from $d\overline{s}$ is given by equation (7.8) as

$$d\overline{V} = \frac{\Gamma}{4\pi} \frac{\sin\theta \; ds}{r^2}$$

from which the velocity at P is written as

$$V_P = \frac{\Gamma}{4\pi} \int_{\theta=\alpha}^{\theta=0} \frac{\sin\theta \; ds}{r^2}$$

and since $s = h(\cot\theta - \cot\alpha)$ we have

$$V_P = \frac{\Gamma}{4\pi} \int_{\theta=\alpha}^{0} \sin\theta \; d\theta$$

that is

$$V_P = \frac{\Gamma}{4\pi h}(1 - \cos\alpha) \qquad (7.9)$$

The direction of V_P is normal to the plane of the paper, by the definition of a vector cross product.

In the second case of a regular helical vortex the analysis becomes a little more complex due to geometry considerations, although the concept is the same. Consider the case where a helical vortex filament starts at the propeller disc and extends to infinity having a constant radius and pitch angle, as shown in Figure 7.11. From equation (7.8) the velocity at the point P due to the segment $d\overline{s}$ is given by

$$d\overline{u} = \frac{\Gamma}{4\pi|a|^3}(d\overline{s} \times \overline{a})$$

and from the geometry of the problem we can derive from

$$\overline{a} = a_x \overline{i} + a_y \overline{j} + a_z \overline{k}$$

that

$$\overline{a} = -r\sin(\theta + \phi)\overline{i} - (y + y_0)\overline{j}$$
$$+ (r_0 - r\cos(\theta + \phi))\overline{k}$$

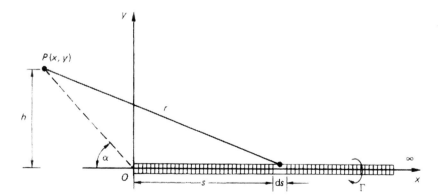

FIGURE 7.10 The Biot–Savart law applied to a semi-infinite line vortex filament.

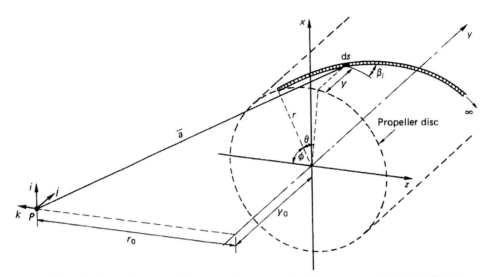

FIGURE 7.11 The application of Biot–Savart law to a semi-infinite regular helical vortex filament.

Similarly,

$$\bar{s}(\theta) = r\sin(\theta+\phi)\bar{i} + r\theta\tan\beta_i\bar{j} + r\cos(\theta+\phi)\bar{k}$$

from which we can derive

$$d\bar{u} = \frac{\Gamma}{4\pi|a|^3}$$

$$\times \begin{vmatrix} i & j & k \\ r\cos(\theta+\phi) & r\theta\tan\beta_i & -r\sin(\theta+\phi) \\ -r\sin(\theta+\phi) & -(y+y_0) & r_0 - r\cos(\theta+\phi) \end{vmatrix}$$

where the scalar a is given by

$$[(y+y_0)^2 + r^2 + r_0^2 - 2r_0 r\cos(\theta+\phi)]^{3/2}$$

Hence the component velocities u_x, u_y and u_z are given by the relations

$$u_x = \frac{r\Gamma}{4\pi}$$
$$\times \int_0^\infty \frac{\tan\beta_i(r\cos(\theta+\phi)) - (y+y_0)\sin(\theta+\phi)}{[(y+y_0)^2 + r^2 + r_0^2 - 2rr_0\cos(\theta+\phi)]^{3/2}} d\theta$$

$$u_y = \frac{r\Gamma}{4\pi}$$
$$\times \int_0^\infty \frac{r - r_0\cos(\theta+\phi)}{[(y+y_0)^2 + r^2 + r_0^2 - 2rr_0\cos(\theta+\phi)]^{3/2}} d\theta$$

$$u_z = \frac{r\Gamma}{4\pi}$$
$$\times \int_0^\infty \frac{r\tan\beta_i\sin(\theta+\phi) - (y+y_0)\cos(\theta+\phi)}{[(y+y_0)^2 + r^2 + r_0^2 - 2rr_0\cos(\theta+\phi)]^{3/2}} d\theta$$

(7.10)

These two examples are sufficient to illustrate the procedure behind the calculation of the field point velocities in inviscid flow. Clearly these principles can be extended to include horse-shoe vortex systems, irregular helical vortices, that is, ones where the pitch and radius vary, and other more complex systems as required by the modeling techniques employed.

It is, however, important to keep in mind, when applying these vortex filament techniques to calculate the velocities at various field points, that they are simply conceptual hydrodynamic tools for synthesizing more complex flows of an inviscid nature. As such they are a convenient means of solving Laplace's equation, the equation governing these types of flow, and are not by themselves of any great significance. However, when a number of vortex filaments are used in conjunction with a free stream flow function it becomes possible to synthesize a flow which has a practical propeller application.

7.4 THE KUTTA CONDITION

For potential flow over a cylinder we know that, depending on the strength of the circulation, a number of possible solutions are attainable. A similar situation applies to the theoretical solution for an aerofoil in potential flow; however, nature selects just one of these solutions.

In 1902, Kutta made the observation that the flow leaves the top and bottom surfaces of an aerofoil smoothly at the trailing edge. This, in general terms, is the Kutta condition. More specifically, however, this condition can be expressed as follows:

1. The value of the circulation Γ for a given aerofoil at a particular angle of attack is such that the flow leaves the trailing edge smoothly.

2. If the angle made by the upper and lower surfaces of the aerofoil is finite, that is non-zero, then the trailing edge is a stagnation point at which the velocity is zero.
3. If the trailing edge is 'cusped', that is the angle between the surfaces is zero, the velocities are non-zero and equal in magnitude and direction.

By returning to the concept discussed in Section 7.2, in which the aerofoil surface was replaced with a system of vortex sheets and where it was noted that the strength of the vortex sheet $\gamma(s)$ was variable along its length, then according to the Kutta condition the velocities on the upper and lower surfaces of the aerofoil are equal at the trailing edge. Then from equation (7.7) we have

$$\gamma_{(TE)} = u_1 - u_2$$

which implies, in order to satisfy the Kutta condition

$$\gamma_{(TE)} = 0 \quad (7.11)$$

7.5 THE STARTING VORTEX

Kelvin's circulation theorem states that the rate of change of circulation with time around a closed curve comprising the same fluid element is zero. In mathematical form this is expressed as

$$\frac{D\Gamma}{Dt} = 0 \quad (7.12)$$

This theorem is important since it helps explain the generation of circulation about an aerofoil. Consider an aerofoil at rest as shown by Figure 7.12(a); clearly in this case the circulation Γ about the aerofoil is zero. Now as the aerofoil begins to move the streamline pattern in this initial transient state looks similar to that shown in Figure 7.12(b). From the figure we observe that high-velocity gradients are formed at the trailing edge and these will lead to high levels of vorticity. This high vorticity is attached to a set of fluid elements which will then move downstream as they move away from the trailing edge. As they move away this thin sheet of intense vorticity is unstable and consequently tends to roll up to give a point vortex which is called the starting vortex (Figure 7.12(c)). After a short period of time the flow stabilizes around the aerofoil, the flow leaves the trailing edge smoothly and the vorticity tends to decrease and disappear as the Kutta condition establishes itself. The starting vortex has, however, been formed during the starting process, and then continues to move steadily downstream away from the aerofoil.

If we consider for a moment the same contour comprising the same fluid elements both when the aerofoil is at rest and also after some time interval when the aerofoil is in steady motion, Kelvin's theorem tells us that the circulation remains constant. In Figure 7.12(a) and (c) this implies that

$$\Gamma_1 = \Gamma_2 = 0$$

for the curves C_1 and C_2 which embrace the same fluid elements at different times, since $\Gamma_1 = 0$ when the aerofoil

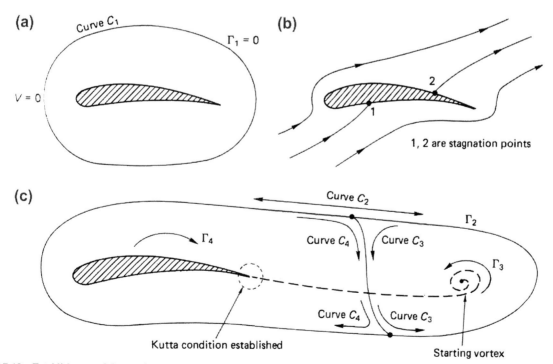

FIGURE 7.12 Establishment of the starting vortex: (a) aerofoil at rest; (b) streamlines on starting prior to Kutta condition being established and (c) conditions at some time after starting.

was at rest. Let us now consider C_2 split into two regions, C_3 enclosing the starting vortex and C_4 the aerofoil. Then the circulation around these contours Γ_3 and Γ_4 is given by

$$\Gamma_3 + \Gamma_4 = \Gamma_2$$

but since $\Gamma_2 = 0$, then

$$\Gamma_4 = -\Gamma_3 \qquad (7.13)$$

which implies that the circulation around the aerofoil is equal and opposite to that of the starting vortex.

In summary, therefore, we see that when the aerofoil is started large velocity gradients at the trailing edge are formed leading to intense vorticity in this region which rolls up downstream of the aerofoil to form the starting vortex. Since this vortex has associated with it an anti-clockwise circulation it induces a clockwise circulation around the aerofoil. This system of vortices builds up during the starting process until the vortex around the aerofoil gains the correct strength to satisfy the Kutta condition, at which point the shed vorticity ceases and steady conditions prevail around the aerofoil. The starting vortex then trails away downstream of the aerofoil.

These conditions have been verified experimentally by flow visualization studies on many occasions; the classic pictures taken by Prandtl and Tietjens[5] are typical and well worth studying.

7.6 THIN AEROFOIL THEORY

Figure 7.8 showed the simulation of an aerofoil by a vortex sheet of variable strength $\gamma(s)$. If one imagines a thin aerofoil such that both surfaces come closer together, it becomes possible, without significant error, to consider the aerofoil to be represented by its camber line with a distribution of vorticity placed along its length. When this is the case the resulting analysis is known as thin aerofoil theory, and is applicable to a wide class of aerofoils, many of which find application in different aspects of propeller technology.

Consider Figure 7.13, which shows a distribution of vorticity along the camber line of an aerofoil. For the camber line to be a streamline in the flow field the component of velocity normal to the camber line must be zero along its entire length. This implies that

$$V_n + \omega_n(s) = 0 \qquad (7.14)$$

where V_n is the component of free stream velocity normal to the camber line, see inset in Figure 7.13; and $\omega_n(s)$ is the normal velocity induced by the vortex sheet at some distance s around the camber line measured from the leading edge.

If we now consider the components of equation (7.14) separately, from Figure 7.13 it is apparent, again from the inset, that at any point Q along the camber line,

$$V_n = V \sin\left[\alpha + \tan^{-1}\left(-\frac{dz}{dx}\right)\right]$$

For small values of α and dz/dx, which are conditions of thin aerofoil theory and are almost always met in steady propeller theory, the general condition that $\sin \theta \simeq \tan \simeq \theta$ holds and, consequently, we may write for the above equation

$$V_n = V\left[\alpha - \left(\frac{dz}{dx}\right)\right] \qquad (7.15)$$

where α, the angle of incidence, is measured in radians.

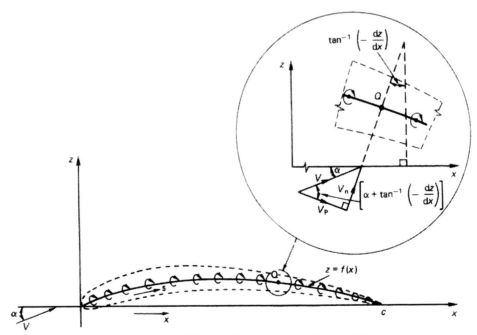

FIGURE 7.13 Thin aerofoil representation of an aerofoil.

Theoretical Methods – Basic Concepts

Now consider the second term in equation (7.14), the normal velocity induced by the vortex sheet. We have previously stated that dz/dx is small for thin aerofoil theory, hence we can assume that the camber–chord ratio will also be small. This enables us to further assume that normal velocity at the chord line will be approximately that at the corresponding point on the camber line and to consider the distribution of vorticity along the camber line to be represented by an identical distribution along the chord without incurring any significant error. Furthermore, implicit in this assumption is that the distance s around the camber line approximates the distance x along the section chord. Now, to develop an expression for $\omega_n(s)$ consider Figure 7.14, which incorporates these assumptions.

From equation (7.6) we can write the following expression for the component of velocity $d\omega_n(x)$ normal to the chord line resulting from the vorticity element $d\xi$ whose strength is $\gamma(\xi)$:

$$d\omega_n(x) = -\frac{\gamma(\xi)d\xi}{2\pi(x-\xi)}$$

Hence the total velocity $\omega_n(x)$ resulting from all the contributions of vorticity along the chord of the aerofoil is given by

$$\omega_n(x) = -\int_0^c \frac{\gamma(\xi)d\xi}{2\pi(x-\xi)}$$

Then by substituting this equation together with equation (7.15) back into equation (7.14), we derive the fundamental equation of thin aerofoil theory

$$\frac{1}{2\pi}\int_0^c \frac{\gamma(\xi)d\xi}{(x-\xi)} = V\left[\alpha - \left(\frac{dz}{dx}\right)\right] \quad (7.16)$$

This equation is an integral equation whose unknown is the distribution of vortex strength $\gamma(\xi)$ for a given incidence angle α and camber profile. In this equation ξ, as in all of the previous discussion, is simply a dummy variable along the Ox axis or chord line.

In order to find a solution to the general problem of a cambered aerofoil, and one of practical importance to the propeller analyst, it is necessary to use the substitutions

$$\xi = \frac{c}{2}(1 - \cos\theta)$$

which implies $d\xi = (c/2)\sin\theta\, d\theta$ and

$$x = \frac{c}{2}(1 - \cos\theta_0)$$

which then transforms equation (7.16) into

$$\frac{1}{2\pi}\int_0^\pi \frac{\gamma(\theta)\sin\theta\, d\theta}{\cos\theta - \cos\theta_0} = V\left[\alpha - \left(\frac{dz}{dx}\right)\right] \quad (7.17)$$

In this equation the limits of integration $\theta = \pi$ correspond to $\xi = c$ and $\theta = 0$ to $\xi = 0$ can be deduced from the above substitutions.

Now the solution of equation (7.17), which obeys the Kutta condition at the trailing edge, that is $\gamma(\pi) = 0$, and makes the camber line a streamline to the flow, is found to be

$$\gamma(\theta) = 2V\left[A_0\left(\frac{1+\cos\theta}{\sin\theta}\right) + \sum_{n=1}^\infty A_n \sin(n\theta)\right] \quad (7.18)$$

in which the Fourier coefficients A_0 and A_n can be shown, as stated below, to relate to the shape of the camber line and the angle of the incidence flow by the substitution of equation (7.18) into (7.17) followed by some algebraic manipulation:

$$A_0 = \alpha - \frac{1}{\pi}\int_0^\pi \left(\frac{dz}{dx}\right)d\theta_0$$

$$A_n = \frac{2}{\pi}\int_0^\pi \left(\frac{dz}{dx}\right)\cos(n\theta_0)d\theta_0 \quad (7.18a)$$

For the details of this manipulation the reader is referred to any standard textbook on aerodynamics.

In summary, therefore, equations (7.18) and (7.18a) define the strength of the vortex sheet distributed over a camber line of a given shape and at a particular incidence angle so as to obey the Kutta condition at the trailing edge. The restrictions to this theoretical treatment are that:

1. The aerofoils are two-dimensional and operating as isolated aerofoils.

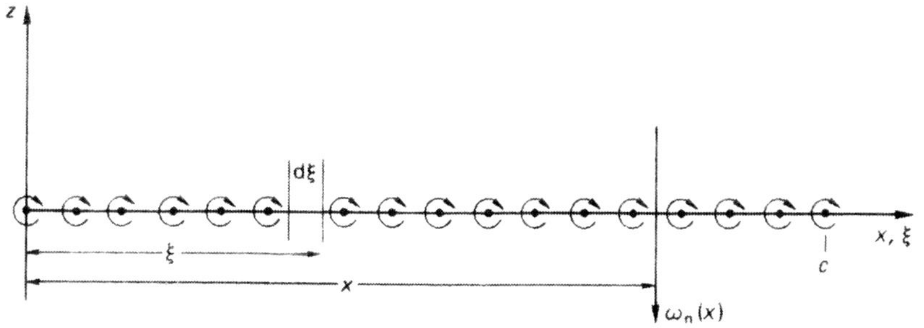

FIGURE 7.14 Calculation of induced velocity at the chord line.

2. The thickness and camber chord ratios are small.
3. The incidence angle is also small.

Conditions (2) and (3) are normally met in propeller technology, certainly in the outer blade sections. However, because the aspect ratio of a propeller blade is small and all propeller blades operate in a cascade, Condition (1) is never satisfied and corrections have to be introduced for this type of analysis, as will be seen later.

With these reservations in mind, equation (7.18) can be developed further, so as to obtain relationships for the normal aerodynamic properties of an aerofoil.

From equation (7.7) the circulation around the camber line is given by

$$\Gamma = \int_0^c \gamma(\xi) d\xi$$

which, by using the earlier substitution of $\xi = (c/2)(1 - \cos\theta)$, takes the form

$$\Gamma = \frac{c}{2} \int_0^c \gamma(\theta) \sin\theta \, d\theta \qquad (7.19)$$

from which equation (7.18) can be written as

$$\Gamma = cV \left[A_0 \int_0^\pi (1 + \cos\theta) d\theta + \sum_{n=1}^\infty A_n \int_0^\pi \sin\theta \sin(n\theta) d\theta \right]$$

which, by reference to any table of standard integrals, reduces to

$$\Gamma = cV \left[\pi A_0 + \frac{\pi}{2} A_1 \right] \qquad (7.20)$$

Now by combining equations (7.1) and (7.3), one can derive an equation for the lift coefficient per unit span as

$$c_l = \frac{2\Gamma}{Vc}$$

from which we derive from equation (7.20)

$$c_l = \pi [2A_0 + A_1] \qquad (7.21)$$

Consequently, by substituting equations (7.18a) into (7.21) we derive the general thin aerofoil relation for the lift coefficient per unit span as

$$c_l = 2\pi \left[\alpha + \frac{1}{\pi} \int_0^\pi \left(\frac{dz}{dz} \right) (\cos\theta_0 - 1) d\theta_0 \right] \qquad (7.22)$$

Equation (7.22) can be seen as a linear equation between c_l and α for a given camber geometry by splitting the terms in the following way:

$$c_l = \underbrace{2\pi\alpha}_{\text{Lift slope}} + \underbrace{2 \int_0^\pi \left(\frac{dz}{dx}\right)(\cos\theta_0 - 1) d\theta_0}_{\text{Lift at zero incidence}}$$

in which the theoretical life slope

$$\frac{dc_l}{d\alpha} = 2\pi/\text{rad} \qquad (7.23)$$

Figure 7.15 shows the thin aerofoil characteristics schematically plotted against experimental single and cascaded aerofoil results. From the figure it is seen that the actual lift slope curve is generally less than 2π.

The theoretical zero lift angle α_0 is the angle for which equation (7.22) yields a value of $c_l = 0$. As such it is seen that

$$\alpha_0 = -\frac{1}{\pi} \int_0^\pi \left(\frac{dz}{dx}\right)(\cos\theta_0 - 1) d\theta_0 \qquad (7.24)$$

Again from Figure 7.15 it is seen that the experimental results for zero lift angle for single and cascaded aerofoils are less than those predicted by thin aerofoil theory.

Thin aerofoil theory also predicts the pitching moment of the aerofoil. Consider Figure 7.16, which shows a more detailed view of the element of the vortex sheet shown in

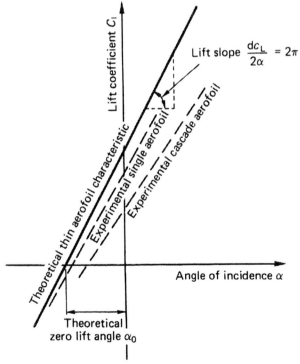

FIGURE 7.15 Thin aerofoil and experimental aerofoil characteristics.

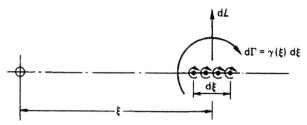

FIGURE 7.16 Calculation of moments about the leading edge.

Figure 7.14. From Figure 7.16 we see that the moment per unit span of the aerofoil is given by

$$M'_{LE} = -\int_0^c \xi(dL) = -\rho V \int_0^c \xi \gamma(\xi) d\xi$$

which, by substituting in the distribution of vorticity given by equation (7.18) and again using the transformation $\xi = (c/2)(1 - \cos\theta)$, gives

$$M'_{LE} = -\frac{\rho V^2 c^2}{2} \left[\int_0^\pi A_0 (1 - \cos^2\theta) d\theta \right.$$
$$+ \int_0^\pi \sum_{n=1}^\infty A_n \sin\theta \sin(n\theta) d\theta$$
$$\left. - \int_0^\pi \sum_{n=1}^\infty A_n \sin\theta \cos\theta \sin(n\theta) d\theta \right]$$

which, by solving in an analogous way to that for c_l and using the definition of the moment coefficient given in equation (7.3), gives an expression for the pitching moment coefficient about the leading edge of the aerofoil as

$$c_{m_{LE}} = -\frac{\pi}{2}\left[A_0 + A_1 - \frac{A_2}{2}\right]$$

or by appeal to equation (7.21)

$$c_{m_{LE}} = -\left[\frac{c_l}{4} + \frac{\pi}{4}(A_1 - A_2)\right] \qquad (7.25)$$

and since from equation (7.4)

$$c_{m_{LE}} = -\frac{c_l}{4} + c_{m_{c/4}}$$

we may deduce that

$$c_{m_{c/4}} = \frac{\pi}{4}[A_2 - A_1] \qquad (7.26)$$

Equation (7.26) demonstrates that, according to thin aerofoil theory, the aerodynamic center is at the quarter chord point, since the pitching moment at this point is dependent only on the camber profile (see equation (7.18a) for the basis of the coefficients A_1 and A_2) and independent of the lift coefficient.

Equations (7.23)–(7.26) are significant results in thin aerofoil theory and also in many branches of propeller analysis. It is therefore important to be able to calculate these parameters readily for an arbitrary aerofoil. The theoretical lift slope curve presents no problem, since it is a general result independent of aerofoil geometry. However, this is not the case with the other equations and the integrals behave badly in the region of the leading and trailing edges. To overcome these problems various numerical procedures have been developed over the years. In the case of the theoretical zero lift angle, Burrill[6] and Hawdon et al.[7] developed a tabular method based on the relationship

$$\alpha_0 = \frac{1}{c}\sum_{n=1}^{19} f_n(x) y_n(x) \text{ degrees} \qquad (7.27)$$

where the chordal spacing is given by

$$x_n = \frac{cn}{20} \quad (n = 1, 2, 3, 4, ..., 20)$$

The multipliers $f_n(x)$ are given in Table 7.2 for both sets of references. The Burrill data is sufficient for most conventional aerofoil shapes; however, it does lead to inaccuracies when dealing with 'S' shaped sections, such as might be encountered when analyzing controllable pitch propellers in off-design pitch settings. This is due to it being based on a trapezoidal rule formulation. The Hawdon relationship was designed to overcome this problem by using a second-order relationship. Systematic wind tunnel tests with camber lines ranging from a parabolic form to a symmetrical 'S' shape showed this latter relationship to agree to within 0.5 per cent of the thin aerofoil results.

With regard to the pitching moment coefficient a similar approximation method was developed by Pankhurst.[8] In this procedure the pitching moment coefficient is given by the relationship

$$c_{m_{c/4}} = \frac{1}{c}\sum_{n=1}^{14} B_n(y_b(x_n) + y_f(x_n)) \qquad (7.28)$$

where y_b and y_f are the back and face ordinates of the aerofoil at each of the x_n chordal spacings. The coefficients B_n are given in Table 7.3.

7.7 PRESSURE DISTRIBUTION CALCULATIONS

The calculation of the pressure distribution about an aerofoil section having a finite thickness has traditionally been undertaken by making use of conformal transformation methods. Theodorsen[9,10] recognized that most wing forms have a general resemblance to each other and since a transformation of the type

TABLE 7.2 Zero Lift Angle Multiplies for use with Equation (7.27)

n	x_c	$f_n(x)$ Burrill	$f_n(x)$ Hawdon et al.
1	0.15 (LE)	5.04	5.04
2	0.10	3.38	3.38
3	0.15	3.01	3.00
4	0.20	2.87	2.85
5	0.25	2.81	2.81
6	0.30	2.84	2.84
7	0.35	2.92	2.94
8	0.40	3.09	3.10
9	0.45	3.32	3.33
10	0.50	3.64	3.65
11	0.55	4.07	4.07
12	0.60	4.64	4.65
13	0.65	5.44	5.46
14	0.70	6.65	6.63
15	0.75	8.59	8.43
16	0.80	11.40	11.40
17	0.85	17.05	17.02
18	0.90	35.40	−22.82
19	0.95 (TE)	186.20	310.72

TABLE 7.3 Pitching Moment Coefficient Multipliers for Equation (7.28) (Taken From Reference 11)

n	x_n	B_n
1	0 (LE)	−0.119
2	0.025	−0.156
3	0.05	−0.104
4	0.10	−0.124
5	0.20	−0.074
6	0.30	−0.009
7	0.40	0.045
8	0.50	0.101
9	0.60	0.170
10	0.70	0.273
11	0.80	0.477
12	0.90	0.786
13	0.95	3.026
14	1.00 (TE)	−4.289

$$\zeta = z + \frac{a^2}{z}$$

transforms a circle in the z-plane (complex plane) into a curve resembling a wing section in the ζ plane (also a complex plane), most wing forms can be transformed into nearly circular forms. He derived a procedure that evaluated the flow about a nearly circular curve from that around a circular form and showed this process to be a rapidly converging procedure. The derivation of Theodorsen's relationship for the velocity distribution about an arbitrary wing form is divided into three stages as follows:

1. The establishment of relations between the flow in the plane of the wing section (ζ-plane) and that of the 'near circle' plane (z'-plane).
2. The derivation of the relationship between the flow in z'-plane and the flow in the true circle plane (z-plane).
3. The combining of the two previous stages into the final expression for the velocity distribution in the ζ-plane in terms of the ordinates of the wing section.

The derivation of the final equation for the velocity distribution, equation (7.29), can be found in Abbott and van Doenhoff[11] for the reader who is interested in the details of the derivation. For our purposes, however, we merely state the results as

$$v = \frac{V[\sin(\alpha_0 + \phi) + \sin(\alpha_0 + \varepsilon_T)][1 + (d\varepsilon/d\theta)]e^{\psi_0}}{\sqrt{\{(\sinh^2\psi + \sin^2\theta)[1 + (d\psi/d\theta)^2]\}}} \quad (7.29)$$

where v is the local velocity on any point on the surface of the wing section and V is the free stream velocity.

In order to make use of equation (7.29) to calculate the velocity at some point on the wing section it is necessary to define the co-ordinates of the wing section with respect to a line joining a point which is located midway between the nose of the section and its center of curvature to the trailing edge. The co-ordinates of these leading and trailing points are taken to be $(-2a, 0)$ and $(2a, 0)$ respectively with $a = 1$ for convenience. Next the values of θ and ψ are found from the co-ordinates (x, y) of the wing section as follows:

$$2\sin^2\theta = p + \sqrt{\left[p^2 + \left(\frac{y}{a}\right)^2\right]} \quad (7.30)$$

with

$$p = 1 - \left(\frac{x}{2a}\right)^2 - \left(\frac{y}{2a}\right)^2$$

and

$$\left.\begin{array}{l} y = 2a \sinh\psi \sin\theta \\ x = 2a \sinh\psi \cos\theta \end{array}\right\} \quad (7.31)$$

Theoretical Methods — Basic Concepts

The function $\psi_0 = (1/2\pi)\int_0^{2\pi} \psi\, d\theta$ has then to be determined from the relationship between ψ and θ. A first approximation to the parameter ε can be found by conjugating the curve of ψ against θ using the relationship

$$\varepsilon(\phi) = \frac{1}{n}\sum_{k=1}^{n}\left(\psi_{-k} - \psi_k\right)\cot\left(\frac{k\pi}{2n}\right) \quad (7.32)$$

with

$$\psi_k = \psi\left(\phi + \frac{k\pi}{n}\right)$$

where the co-ordinates in the z-plane are defined by $z = ae^{(\lambda + i\phi)}$.

For most purposes a value $n = 40$ will give sufficiently accurate results.

Finally the values of $(d\varepsilon/d\theta)$ and $(d\psi/d\theta)$ are determined from the curves of ε and ψ against θ and hence equation (7.29) can be evaluated, usually in terms of v/V.

For many purposes the first approximation to ε is sufficiently accurate; however, if this is not the case then a second approximation can be made by plotting ψ against $\theta + \varepsilon$ and re-working the calculation from the determination of the function ψ_0. Pope[12] gives a good account of the details of the calculation procedure.

This procedure is exact for computations in ideal fluids; however, the presence of viscosity in a real fluid leads to discrepancies between experiment and calculation. The growth of the boundary layer over the section effectively changes the shape of the section, and one result of this is that the theoretical rate of changes of lift with angle of incidence is not realized. Pinkerton[13] found that fair agreement with the experiment for the NACA 4412 aerofoil could be obtained by effectively distorting the shape of the section. The amount of the distortion is determined by calculating the increment $\Delta\varepsilon_T$ required to avoid infinite velocities at the trailing edge after the circulation has been adjusted to give the experimentally observed lift coefficient. This gives rise to a modified function:

$$\varepsilon_\alpha = \varepsilon + \frac{\Delta\varepsilon_T}{2}(1 - \cos\theta) \quad (7.33)$$

where ε is the original inviscid function and ε_α is the modified value of the section.

Figure 7.17 shows the agreement obtained from the NACA 4412 pressure distribution using the Theodorsen and Theodorsen with Pinkerton correction methods.

The Theodorsen method is clearly not the only method of calculating the pressure distribution around an aerofoil section. It is one of a class of inviscid methods; other methods commonly used are those by Riegels and Wittich[14] and Weber.[15] The Weber method was based originally on the earlier work of Riegels and Wittich, which in itself was closely related to the works of Goldstein, Thwaites and Watson, and provides a readily calculable

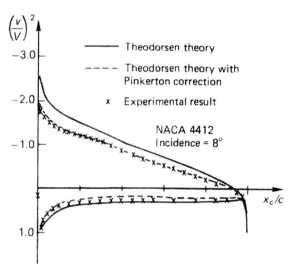

FIGURE 7.17 Comparison of theoretical and experimental pressure distributions around an aerofoil.

procedure at either 8, 16 or 32 points around the aerofoil. In this method the location of the calculation points is defined by a cosine function, so that a much greater distribution of calculation points is achieved at the leading and trailing edges of the section. Comparison of the methods with those based on the formulations of Theodorsen, Riegels–Wittich and Weber shows little variation for the range of aerofoils of interest to propeller designers, so the choice of method reduces to one of personal preference for the user. The inviscid approach was extended to the cascade problem by Wilkinson.[16] In addition to the solutions of the aerofoil pressure distribution problem discussed here, the use of numerical methods based on vortex panel methods have been shown to give useful and reliable results. These will be introduced later in the chapter.

The calculation of the viscous pressure distribution around an aerofoil is a particularly complex procedure, and rigorous methods such as those by Firmin[17] need to be employed. Indeed the complexity of these methods has generally precluded them from propeller analysis and many design programs in favor of more approximate methods. Moreover, in recent years the growing use of RANS codes has largely obviated this need.

If the section thickness distribution and camber line are of standard forms for which velocity distributions are known, such as the NACA forms, then the resulting velocity distribution can be readily approximated. The basis of the approximation is that the load distribution over a thin section may be considered to comprise two components:

1. A basic load distribution at the ideal angle of attack.
2. An additional distribution of load which is proportional to the angle of attack as measured from the ideal angle of attack.

The basic load distribution is a function only of the shape of the thin aerofoil section, and if the section is considered only to be the mean line, then it is a function only of the mean line geometry. Hence, if the parent camber line is modified by multiplying all of the ordinates by a constant factor, then the ideal design of attack α_i and the design lift coefficient c_{li} of the modified camber line are similarly derived by multiplying the parent values by the same factor.

The second distribution cited above results from the angle of attack of the section and is termed the additional load distribution; theoretically this does not contribute to any additional moment about the quarter chord point of the aerofoil. In practice there is a small effect since the aerodynamic center in viscous flow is usually just astern of the quarter chord point. This additional load distribution is dependent to an extent on the aerofoil shape and is also non-linear with incidence angle but can be calculated for a given aerofoil shape using the methods cited earlier in this chapter. The non-linearity with incidence angle, however, is small and for most marine engineering purposes can be assumed linear. As a consequence, additional load distributions are normally calculated only for a series of profile forms at a representative incidence angle and assumed to be linear for other values.

In addition to these two components of load, the actual section thickness form at zero incidence has a velocity distribution over the surface associated with it, but this does not contribute to the external load produced by the aerofoil. Accordingly, the resultant velocity distribution over the aerofoil surface can be considered to comprise three separate and, to a first approximation, independent components, which can be added to give the resultant velocity distribution at a particular incidence angle. These components are as follows:

1. A velocity distribution over the basic thickness form at zero incidence.
2. A velocity distribution over the mean line corresponding to the load distribution at its ideal angle of incidence.
3. A velocity distribution corresponding to the additional load distribution associated with the angle of incidence of the aerofoil.

Figure 7.18 demonstrates the procedure, and the velocity distributions for standard NACA aerofoil forms can be obtained from reference 11. By way of example, Table 7.4 shows the relevant data for a NACA 16—006 basic thickness form and a NACA $a = 0.8$ modified mean line. It will be seen that this data can be used principally in two ways: first, given a section form at incidence, to determine the resulting pressure distribution, and second, given the section form and lift coefficient, to determine the appropriate design incidence and associated pressure distribution.

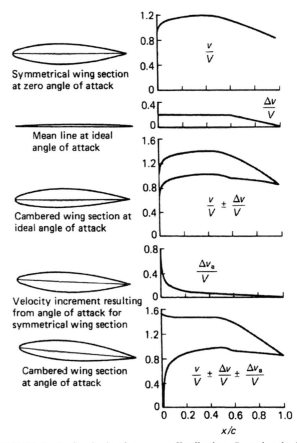

FIGURE 7.18 **Synthesis of pressure distribution.** *Reproduced with permission from Reference 11.*

In the first case for a given maximum camber of the subject aerofoil the values of c_{l_i}, α_i, $c_{m_{c/4}}$ and the $\Delta v/V$ distribution are scaled by the ratio of the maximum camber-chord ratio, taking into account any flow curvature effects as shown in Table 7.4.

In the case of the $a = 0.8$ (modified) mean line:

$$\text{camber scale factor } (S_c) \simeq \frac{y/c \text{ of actual aerofoil}}{0.06651} \quad (7.34)$$

The values of v/V relating to the basic section thickness velocity distribution at zero incidence can be used directly from the appropriate table relating to the thickness form. However, the additional load velocity distribution requires modification since that given in Table 7.4 relates to a specific lift coefficient c_l: in many cases this lift coefficient has a value of unity, but this needs to be checked reference 11 for each particular application in order to avoid serious error. Since the data given in references 11 relates to potential flow, the associated angle of incidence for the distribution can be calculated as

$$\alpha_* = \frac{C_{L_*}}{2\pi} \quad (7.35)$$

TABLE 7.4 Typical NACA Data for Propeller Type Sections

x (% c)	y (% c)	$(v/V)^2$	v/V	$\Delta v_a/V$
0	0	0	0	5.471
1.25	0.646	1.050	1.029	1.376
2.5	0.903	1.085	1.042	0.980
5.0	1.255	1.097	1.047	0.689
7.5	1.516	1.105	1.051	0.557
10	1.729	1.108	1.053	0.476
15	2.067	1.112	1.055	0.379
20	2.332	1.116	1.057	0.319
30	2.709	1.123	1.060	0.244
40	2.927	1.132	1.064	0.196
50	3.000	1.137	1.066	0.160
60	2.917	1.141	1.068	0.130
70	2.635	1.132	1.064	0.104
80	2.099	1.104	1.051	0.077
90	1.259	1.035	1.017	0.049
95	0.707	0.962	0.981	0.032
100	0.060	0	0	0

LE radius: 0.176 % c

NACA 16−006 basic thickness form

$c_{l_i} = 1.0 \quad \alpha_i = 1.40° \quad c_{m_{c/4}} = -0.219$

x (% c)	Y (% c)	dy_c/dx	P_R	$\Delta v/V = P_R/4$
0	0			
0.5	0.281	0.47539		
0.75	0.396	0.44004		
1.25	0.603	0.39531		
2.5	1.055	0.33404		
			1.092	0.273
5.0	1.803	0.27149		
7.5	2.432	0.23378		
10	2.981	0.20618		
15	3.903	0.16546		
20	4.651	0.13452		
			1.096	0.274
25	5.257	0.10873		
30	5.742	0.08595		
35	6.120	0.06498		
40	6.394	0.04507	1.100	0.275
45	6.571	0.02559		
50	6.651	0.00607		
55	6.631	−0.01404	1.104	0.276
60	6.508	−0.03537		
65	6.274	−0.05887	1.108	0.277
70	5.913	−0.08610	1.108	0.277
75	5.401	−0.12058	1.112	0.278
80	4.673	−0.18034	1.112	0.278
85	3.607	−0.23430	0.840	0.210
90	2.452	−0.24521	0.588	0.147
95	1.226	−0.24521	0.368	0.092
100	0	−0.24521	0	0

Data for NACA mean line $a = 0.8$ (modified)

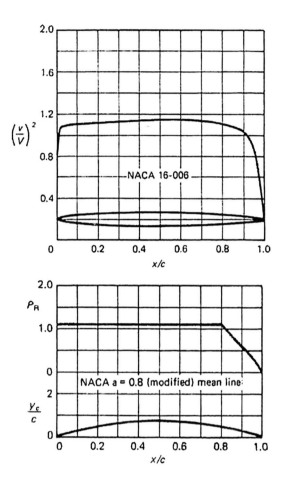

Hence the $\Delta v_a/V$ distribution has to be scaled by a factor of additional load scale factor
$$(S_A) = \left(\frac{\alpha - \alpha_i}{\alpha_*}\right) \quad (7.36)$$

The resultant velocity distribution over the surface of the aerofoil is then given by

$$(u/V)_U = \frac{v}{V} + S_c\left(\frac{\Delta v}{V}\right) + S_A\left(\frac{\Delta v_a}{V}\right)$$
$$(u/V)_L = \frac{v}{V} - S_c\left(\frac{\Delta v}{V}\right) - S_A\left(\frac{\Delta v_a}{V}\right) \quad (7.37)$$

where the suffixes U and L relate to the upper and lower aerofoil surfaces respectively.

In the second case cited above, of a given section form and desired lift coefficient, an analogous procedure is adopted in which the camber scale factor, equation (7.34), is applied to the $\Delta v/V$ distribution. However, in this case equation (7.36) is modified to take the form

$$S_A = \left(\frac{C_L - C_{L_i}}{C_{L*}}\right) \quad (7.37a)$$

The resultant surface velocity distribution is then calculated using equation (7.37).

The pressure distribution around the aerofoil is related to the velocity distribution by Bernoulli's equation:

$$p_\infty + \frac{1}{2}\rho V^2 = p_L + \frac{1}{2}\rho u^2 \qquad (7.38)$$

where p_∞ and p_L are the static pressures remote from the aerofoil and at a point on the surface where the local velocity is u respectively. Then by rearranging equation (7.38), we obtain

$$p_L - p_\infty = \frac{1}{2}\rho(V^2 - u^2)$$

and dividing by the free stream dynamic pressure $\tfrac{1}{2}\rho V^2$, where V is the free stream velocity far from the aerofoil, we have

$$\frac{p_L - p_\infty}{\frac{1}{2}\rho V^2} = \left[1 - \left(\frac{u}{V}\right)^2\right] \qquad (7.39)$$

$\left[(p_L - p_\infty)/\tfrac{1}{2}\rho V^2\right]$ is termed the pressure coefficient (C_P) for a point on the surface of the aerofoil; hence in terms of this coefficient equation (7.39) becomes

$$C_P = \left[1 - \left(\frac{u}{V}\right)^2\right] \qquad (7.40)$$

7.8 BOUNDARY LAYER GROWTH OVER AN AEROFOIL

Classical theoretical methods of the type outlined in this chapter very largely ignore the viscous nature of water by introducing the inviscid assumption early in their development. The viscous behavior of water, however, provides a generally small but, nevertheless, significant force on the propeller blade sections and, as such, needs to be taken into account in calculation methods.

Traditionally viscous effects have been taken into account in a global sense by considering the results of model tests on standard aerofoil forms and then plotting faired trends. Typical in this respect are the drag characteristics derived by Burrill[6] which were based on the NACA and other data available at that time. For many propeller sections, typically in the tip region, where the thickness to chord values are low, and also for non-conventional propulsors, it becomes necessary either to extrapolate data, develop new data or establish reliable calculation procedures. Of these options the first can clearly lead to errors, the second can be expensive, which leaves the third as an alternative course of action.

Boundary layer theory in the general sense of its application to the aerofoil problem and the complete solution of the Navier–Stokes equations is a complex matter. The power of modern computers, however, has eased this constraint considerably by permitting the introduction of RANS and Large Eddy Simulation (LES) capabilities. Nevertheless, these methods when used for more than trivial problems demand considerable computational capacity and speed.

In the context of the physical development of the subject, Schlichting[18] gives a very rigorous and thorough discussion of the boundary layer and its analysis and the reader is referred to this work, since all that can be provided within the confines of this chapter is an introduction to the subject in the context of propeller performance.

For the general case of an aerofoil, the boundary layer development commences from the leading edge, or more specifically, the forward stagnation point. In these early stages of development the flow around the section is normally laminar; however, after a period of time the flow undergoes a transition to a fully turbulent state. The time, or alternatively the distance around the section, at which the transition takes place is a variable dependent upon the flow velocities involved and the blade surface texture: in the case of a full-size propeller these times and distances are very short, but in the case of a propeller model this need not be the case. When transition takes place between the laminar and turbulent flow it takes place over a finite distance and the position of the transition is of considerable importance to the growth of the boundary layer. Figure 7.19 shows the typical growth of a boundary layer over a symmetrical aerofoil, as found from experiment. It will be seen that in this case the boundary layer thickens rapidly between $0.27c$ and $0.30c$, which is a common feature in the presence of an adverse pressure gradient, and is also associated with the transition from laminar to turbulent flow.

The work of a number of researchers has established that the boundary layer transition process can be characterized into a number of stages for a quiet boundary layer flow over a smooth surface. Moving downstream from the stable laminar flow near the leading edge an unstable two-dimensional series of Tollmien–Schlichting waves start to develop from which three-dimensional unstable waves and hairpin eddies are then generated. Vortex breakdown in high localized shear regions of the flow are then seen to occur after which cascading vortex breakdown into fully three-dimensional flow fluctuations becomes apparent. At these locally intense fluctuations turbulence spots then appear after which they then coalesce into a fully turbulent flow regime.

Separation is a phenomenon which occurs in either the laminar or turbulent flow regimes. In the case of laminar flow the curvature of the upper surface of the aerofoil may be sufficient to initiate laminar separation, and under certain conditions the separated laminar layer may undergo the transition to turbulent flow with the characteristic rapid thickening of the layer. The increase in thickness may be

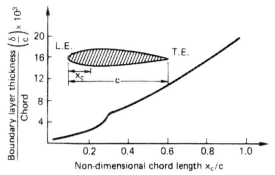

FIGURE 7.19 Typical growth of boundary layer thickness over an aerofoil section.

sufficient to make the lower edge of the shear layer contact the aerofoil surface and reattach as a turbulent boundary layer, as seen in Figure 7.20. This has the effect of forming a separation bubble which, depending on its size, will have a greater or lesser influence on the pressure distribution over the aerofoil. Owen and Klanfer[19] suggested a criterion that if the Reynolds number based on the displacement thickness ($R_{n\delta}$) of the boundary layer is greater than 550, then a short bubble, of the order of 1 per cent of the chord, forms and has a negligible effect on the pressure distribution. If $R_{n\delta} < 400$, then a long bubble, ranging from a few per cent of the chord up to almost the entire chord length, forms. In the case of the turbulent flow regime the flow will separate from the surface of the aerofoil in the presence of an adverse pressure gradient, this being one where the pressure increases in magnitude in the direction of flow. Here, as the fluid close to the surface, which is traveling at a lower speed than fluid further away from the surface due to the action of the viscous forces, travels downstream it gets slowed up to a point where it changes direction and becomes reversed flow. The point where this velocity first becomes zero, apart from the fluid layer immediately in contact with the surface whose velocity is zero by definition, is termed the stagnation point. Figure 7.21 shows three possible flow regimes about an aerofoil: the first is of a fully attached flow comprising a laminar and turbulent

part, whilst the second, Figure 7.21(b), illustrates a laminar separation condition without reattachment, and the final flow system, Figure 7.21(c), shows a similar case to Figure 7.21(a) but having turbulent separation near the trailing edge. Figure 7.22 shows in some detail the structure and definitions used in the analysis of boundary layers.

Van Oossanen[20] establishes a useful boundary layer approximation for aerofoil forms commonly met in propeller technology. The laminar part of the boundary layer is dealt with using Thwaites' approximation, which results from the analysis of a number of exact laminar flow solutions:

$$\frac{V_s \theta^2}{v} = \frac{0.45}{V_s^5} \int_0^s V_s^5 \, ds \qquad (7.41)$$

in which V_s is the velocity at the edge of the boundary layer at a point s around the profile from the stagnation point and θ is the momentum thickness.

From the momentum thickness calculated by equation (7.41) a parameter m can be evaluated as follows:

$$m = -\frac{dV_s}{ds}\left(\frac{\theta^2}{v}\right) \qquad (7.42)$$

where v is the kinematic viscosity of water.

Curl and Skan[21] defined a relationship between the form parameter H and m together with a further shear stress parameter l; these values are shown in Figure 7.23. Consequently, the boundary layer displacement thickness $\delta*$ and wall shear stress τ_w can be calculated from

$$\left.\begin{array}{r}\delta* = \theta H(m) \\ \dfrac{\tau_w}{\rho V_s^2} = \dfrac{l(m)v}{V_s \theta}\end{array}\right\} \qquad (7.43)$$

Separation of the laminar boundary layer is predicted to occur when $m = 0.09$.

To determine when the laminar–turbulent transition takes place the method developed by Michel, and extended by Smith,[22] appears to work reasonably well for profiles

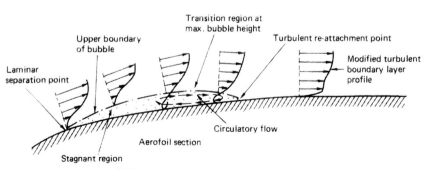

FIGURE 7.20 Laminar separation bubble.

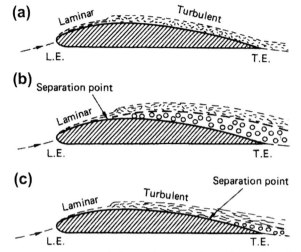

FIGURE 7.21 Schematic flow regimes over the suction surface of an aerofoil: (a) fully attached laminar followed by turbulent boundary layer flow over suction surface; (b) laminar, leading edge separation without reattachment of flow over suction surface and (c) laminar followed by turbulent boundary layer with separation near the trailing edge.

having a peaked minimum in the pressure distribution. For flat pressure distribution profiles, however, the method is less accurate. According to the correlation upon which the Michel–Smith method is based, laminar–turbulent transition is predicted to occur when the Reynolds number based on momentum thickness R_θ reaches the critical value given by

$$R_\theta = 1.174 R_s^{0.46} \tag{7.44}$$

in which

$$R_s = \frac{sv}{\nu} \quad \text{and} \quad R_\theta = \frac{\theta V_s}{\nu}$$

and V and V_s are the free stream and local velocities respectively, s is the distance of the point under consideration around the surface of the foil from the stagnation point and θ is the momentum thickness.

Van Oossanen suggests that the validity of this criterion can be considered to be in the range $10^5 \leq R_s \leq 10^8$.

For the turbulent part of the boundary layer, which is principally confined to the region of increasing pressure for most aerofoils, the method proposed by Nash and Macdonald[23] provides a useful assessment procedure. In this method the turbulent boundary layer is characterized by a constant value pressure gradient parameter Π and a corresponding constant value shape factor G along the body. These parameters are defined by

$$\left. \begin{array}{l} \Pi = \dfrac{\delta^*}{\tau_w}\left(\dfrac{dp}{ds}\right) \\ \text{and} \\ G = \sqrt{\left(\dfrac{\rho V_s^2}{\tau_w}\right)\left[1 - \dfrac{1}{H}\right]} \end{array} \right\} \tag{7.45}$$

where dp/ds is the pressure gradient at the edge of the boundary layer and $H = \delta^*/\theta$.

Nash showed that a good empirical fit to experimental data gave rise to a unique function $G(\Pi)$ defined as

$$G = 6.1\sqrt{(\Pi - 1.81)} - 1.7 \tag{7.46}$$

To establish the growth of the turbulent boundary layer over the aerofoil surface in two dimensions it is necessary to integrate the momentum-integral equation

$$\frac{d}{ds}(\rho V_s^2 \theta) = \tau_w(1 + \Pi) \tag{7.47}$$

This equation, which can be written as

$$\frac{d\theta}{dx_s} = -(H+2)\frac{\theta}{V_s}\left(\frac{dV_s}{ds}\right) + \frac{\tau_w}{\rho V_s^2} \tag{7.47a}$$

if used in association with Nash's skin friction law for incompressible flow,

$$\frac{\tau_w}{\rho V_s^2} = \left[2.4711 \ln\left(\frac{V_s\theta}{\nu}\right) + 475 + 1.5G + \frac{1727}{(G^2+200)} - 16.87\right]^{-2} \tag{7.48}$$

can be used to calculate the growth of the turbulent boundary layer from the point of transition. At the

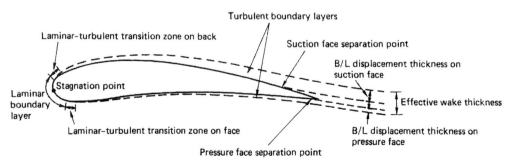

FIGURE 7.22 Boundary layer structure.

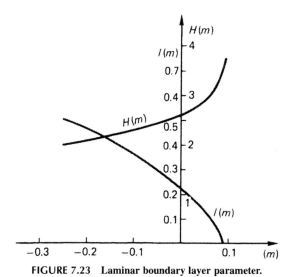

FIGURE 7.23 Laminar boundary layer parameter.

where

α_2 is the two-dimensional angle of attack,
α_{02} is the two-dimensional zero lift angle and
α_{02p} is the two-dimensional zero lift angle from thin aerofoil theory.

Equation (7.50) is in contrast to the simpler formulations used by Burrill[6], which are based on the geometric thickness to chord ratio of the section. Therefore, these earlier relationships should be used with some caution, since the lift slope and zero lift angle correction factors are governed by the growth of the boundary layer over the aerofoil to a significant degree.

With the increasing use of computational fluid dynamics in propeller and ship flow analysis problems a number of turbulence models are encountered. For example, these might include:

- The $k-\varepsilon$ model in either the standard or Chen and Kim extended model.
- The $k-\omega$ model in the standard, Wilcox modified or Menter's baseline model.
- Menter's one equation model.
- The RNG $k-\varepsilon$ turbulence model.
- Reynolds stress models.
- Menter's SST $k-\omega$ turbulence model.
- The Splalart–Allmaras turbulence model.

In the case of the $k-\varepsilon$ model it was found that if the generic turbulent kinetic energy equation was coupled to either a turbulence dissipation or turbulence length scale modeling equation, then it gave improved performance. The energy and dissipation equations as formulated by Jones and Launder[24] rely on five empirical constants, one of which controls the eddy viscosity and two others, which are effectively Prandtl numbers, which relate the eddy diffusion to the momentum eddy viscosity. Sadly these constants are not universal constants for all flow regimes but when combined with the continuity and momentum equations form the basis of the $k-\varepsilon$ model for the analysis of turbulent shear flows. The $k-\omega$ model is not dissimilar in its formulation to the $k-\varepsilon$ approach, but instead of being based on a two equation approach its formulation is centered on four equations. The Reynolds stress models, frequently called second-order closure, form a rather higher level approach than either the $k-\varepsilon$ or $k-\omega$ approaches in that they model the Reynolds stresses in the flow field. In these models the eddy viscosity and velocity gradient approaches are discarded and the Reynolds stresses are computed directly by either an algebraic stress model or a differential equation for the rate of change of stress. Such approaches are computationally intensive but, in general, the best Reynolds stress models yield superior results for complex flows, particularly where separation and reattachment are involved. Moreover, even for attached

transition point, given by equation (7.44), the continuity of momentum thickness is assumed to give a Reynolds number based on momentum thickness greater than 320: if this is not the case, then the momentum thickness is increased so as to give a value of 320. In order to start the calculation procedure at the transition point, which is an iteration involving θ, Π, G, τ_w and H in equations (7.45) and (7.46), an initial value of $G = 6.5$ can be assumed.

Turbulent separation is predicted to occur when

$$\frac{\tau_w}{\rho V_s} < 0.0001 \qquad (7.49)$$

Van Oossanen[20] has shown that the resulting magnitude of the effective wake thickness (Figure 7.22) of the aerofoil has a significant effect on the lift slope curve and the zero lift angle correlation factor. As such, a formulation of lift slope and zero lift angle correlation factors based on the effective boundary layer thickness was derived using the above analytical basis and represented the results of wind tunnel tests well. These relationships are

$$\frac{dc_l}{d\alpha_2} = 7.462 - \sqrt{\left[135.2\left(\frac{t_s}{c}\right) - 2.899\right]}$$

$$\frac{\alpha_{02}}{\alpha_{02p}} = 6.0 - 5.0\left[\frac{y_{ss} + \delta_{ss}^*}{y_{sp} + \delta_{sp}^*}\right] \qquad (7.50)$$

for $(y_{ss} + \delta_{ss}^*) < (y_{sp} + \delta_{sp}^*)$

and

$$\frac{\alpha_{02}}{\alpha_{02p}} = 1.2 - 0.2\left[\frac{y_{ss} + \delta_{ss}^*}{y_{sp} + \delta_{sp}^*}\right]$$

for $(y_{ss} + \delta_{ss}^*) > (y_{sp} + \delta_{sp}^*)$

boundary layers the Reynolds stress models surpass the k–ε model results.

The boundary layer contributes two distinct components to the aerofoil drag. These are the pressure drag (D_p) and the skin friction drag (D_f). The pressure drag, sometimes referred to as the form drag, is the component of force, measured in the drag direction, due to the integral of the pressure distribution over the aerofoil. If the aerofoil was working in an inviscid fluid, then this integral would be zero – this is d'Alembert's well-known paradox. However, in the case of a real fluid the pressure distribution decreases from the inviscid prediction in the regions of separated flow and consequently gives rise to non-zero values of the integral. The skin friction drag, in contrast, is the component of the integral of the shear stresses τ_w over the aerofoil surfaces, again measured in the drag direction. Hence the viscous drag of a two-dimensional aerofoil is given by:

2D viscous drag = skin friction drag
+ pressure drag that is,

$$D_v = D_f + D_p \qquad (7.51)$$

7.9 THE FINITE WING

Discussion so far has largely been based on two-dimensional, infinite aspect ratio, aerofoils. Aspect ratio is taken in the sense defined in the classical aerodynamic way:

$$AR = \frac{b^2}{A} \qquad (7.52)$$

where b is the span of the wing and A is the plan form area. Marine propellers and all wing forms clearly do not possess the infinite aspect ratio attribute; indeed marine propellers generally have quite low aspect ratios. The consequence of this is that for finite aspect ratio wings and blades the flow is not two-dimensional but has a spanwise component. This can be appreciated by studying Figure 7.24 and by considering the mechanism by which lift is produced. On the pressure surface of the blade the pressure is higher than for the suction surface. This clearly leads to a tendency for the flow on the pressure surface to 'spill' around onto the suction surface at the blade tips. Therefore, there is a tendency for the streamlines on the pressure surface of the blade to deflect outwards and inwards on the suction surface (Figure 7.24(a) and (b)). Hence the flow moves from a regime which is two-dimensional in the case of the infinite aspect ratio wing case to become a three-dimensional problem in the finite blade. The tendency for the flow to 'spill' around the tip establishes a circulatory motion at the tips as seen in Figure 7.24(b), and this creates the trailing vortex which is seen at each wing or

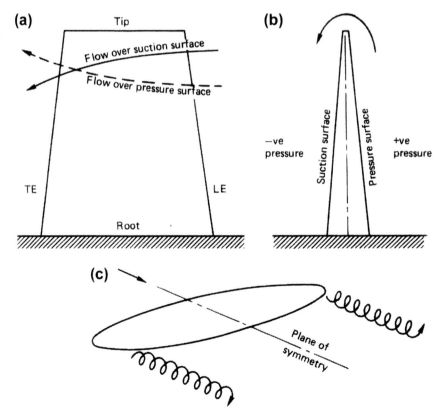

FIGURE 7.24 Flow over a finite aspect ratio wing: (a) plan view of blade; (b) flow at blade tip and (c) schematic view of wing tip vortices.

Theoretical Methods – Basic Concepts

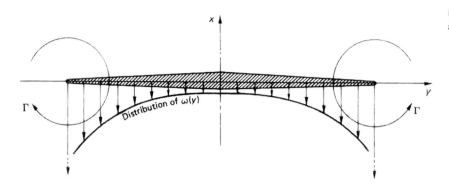

FIGURE 7.25 Downwash distribution for a pair of tip vortices on a finite wing.

blade tip, and is sketched in Figure 7.24(c). These tip vortices trail away downstream and their strength is clearly dependent upon the pressure differential, or load distribution, over the blade.

One consequence of the generation of trailing vortices is to produce an additional component of velocity at the blade section called downwash. For the case of the two wing tip vortices shown in Figure 7.24(c), the distribution of downwash $\omega(y)$ along the chord is shown in Figure 7.25. This distribution derives from the relationship:

Downwash at any point $y =$

Contribution from the left-hand vortex

$+$ Contribution from the right-hand vortex

that is,

$$\omega(y) = -\frac{\Gamma}{4\pi}\left[\frac{1}{(b/2+y)} + \frac{1}{(b/2-y)}\right]$$

$$\omega(y) = \frac{\Gamma}{4\pi}\left[\frac{b}{(b/2)^2 - y^2}\right] \quad (7.53)$$

where the span of the aerofoil is b. The downwash velocity $\omega(y)$ combines with the incident free stream velocity V to produce a local velocity which is inclined to the free stream velocity at the blade section, as shown in Figure 7.26, by an angle α_i.

Consequently, although the aerofoil is inclined at a geometric angle of attack α to the free stream flow, the section is experiencing a smaller angle of attack α_{eff} such that

$$\alpha_{\text{eff}} = \alpha - \alpha_i \quad (7.54)$$

Since the local lift force is by definition perpendicular to the incident flow, it is inclined at an angle α_i to the direction of the incident flow. Therefore, there is a component of this lift force D_i which acts parallel to the free stream's flow, and this is termed the 'induced drag' of the section. This component is directly related to the lift force and not to the viscous behavior of the fluid.

As a consequence, we can note that the total drag on the section of an aerofoil of finite span comprises three distinct components, as opposed to the two components of

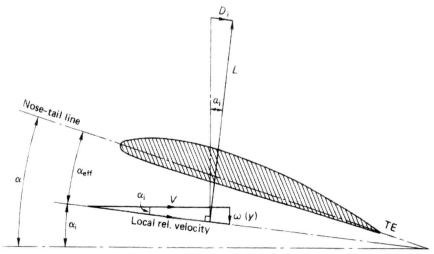

FIGURE 7.26 Derivation of induced drag.

FIGURE 7.27 Formation of trailing vortices.

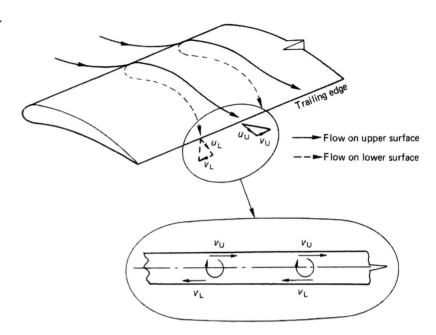

equation (7.51) for the two-dimensional aerofoil, as follows:

$$\text{total 3D drag} = \text{skin friction drag} + \text{pressure drag} + \text{induced drag} \quad (7.55)$$

that is,

$$D = D_v + D_f + D_i$$

where the skin friction drag D_f and the pressure drag and D_p are viscous contributions to the drag force.

In Figure 7.24 it was seen that a divergence of the flow on pressure and suction surfaces took place. At the trailing edge, where these streams combine, the difference in

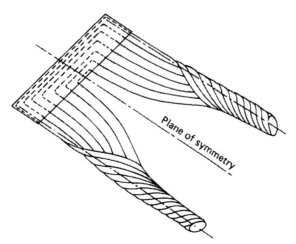

FIGURE 7.28 Schematic roll-up of trailing vortices.

spanwise velocity will cause the fluid at this point to roll up into a number of small streamwise vortices which are distributed along the entire span of the wing, as indicated by Figure 7.27. From this figure it is seen that the velocities on the upper and lower surfaces can be resolved into a spanwise component (v) and an axial component (u). It is the difference in spanwise components on the upper and lower surfaces, v_U and v_L respectively, that give rise to the shed vorticity as sketched in the inset to the diagram. Although vorticity is shed along the entire length of the blade these small vortices roll up into two large vortices just inboard of the wing or blade tips and at some distance from the trailing edge as shown in Figure 7.28. The strength of these two vortices will of course, by Helmholtz's theorem, be equal to the strength of the vortex replacing the wing itself and will trail downstream to join the starting vortex, Figure 7.12; again, in order to satisfy Helmholtz's theorems. Furthermore, it is of interest to compare the trailing vortex pattern of Figure 7.28 with that of Prandtl's classical model, shown in Figure 7.1, where roll-up of the free vortices was not considered. Although Figure 7.28 relates to a wing section, it is clear that the same hydrodynamic mechanism applies to a propeller blade form.

7.10 MODELS OF PROPELLER ACTION

Many models directed towards describing the action of the marine propeller tend to solve the potential flow problem, subject to viscous constraints, defined by the propeller operating at a particular advance and rotational speed. In one case, the design problem, the aim is to define the required blade geometry for a given set of operating conditions. In the other case, the analysis problem, which is the inverse of the design problem, the geometry is defined and the resulting load and flow condition is calculated.

The propeller analysis problem, for example, is formulated by considering the propeller to function in an unbounded incompressible fluid and to have an inflow which is defined as the effective flow field. This effective flow field, which was discussed in Chapter 5, is defined in terms of a fixed Cartesian reference frame, and the propeller, whose shaft axis is coincident with one of the axes of the effective flow field reference frame, is defined with respect to a rotating reference frame (Chapter 3). For the purposes of analysis the propeller is considered to comprise a number of identical blades symmetrically arranged around a boss, which is assumed to rotate with constant speed. The boss is either idealized as an axisymmetric body or, alternatively, ignored: this latter option was normally the case with many of the earlier theoretical models.

The definition of the blade geometry within the analytical model is normally based on the locus of the mid-chord line of each of the sections. This locus is defined with respect to one axis of the rotating reference frame by its radial distribution of rake and skew. Having defined this locus, the leading and trailing edges of the blade can then be defined by laying off the appropriate half chord lengths along each of the helix lines at the defined radii. These helix lines are defined by the radial distribution of the section nose—tail pitch. This process effectively defines the section nose—tail lines in space, whereupon the blade section geometry can be defined in terms of the radial and chordwise distribution of camber and section thickness.

The solution of the hydrodynamic problem of defining the velocity potential at a point on the surface of the blades can be expressed, in the same way as any other incompressible flow problem around a lifting body, as a surface integral over the blade surfaces and the wake by employing Green's formula. For analysis purposes this generalized formulation of the propeller analysis problem can be defined as a distribution of vorticity and sources over the blades together with a distribution of free vorticity in the wake of the propeller defining the vortex sheets emanating from the blades. The distribution of sources, and by implication sinks, is to represent in hydrodynamic terms the flow boundaries defined by the blade surface geometry. In some propeller analysis formulations the distributions of vorticity are replaced by an equivalent distribution of normal dipoles in such a way that the vortex strength is defined by the derivative of the strength of the dipoles. The completion of the definition of the analysis problem is then made by the imposition of the Kutta condition at the trailing edge, thereby effectively defining the location in space of the vortex sheets, and the introduction of a boundary layer approximation.

In order to solve the analysis problem it is frequently the case that the longitudinal and time-dependent properties of the effective inflow field are ignored. When this assumption is made the flow can be considered to be cyclic, and as a consequence the effective inflow defined in terms of the normal Fourier analysis techniques.

The foregoing description of the analysis problem, the design problem being essentially the inverse of the analysis situation but conducted mostly under a mean inflow condition, defines a complex mathematical formulation, the solution of which has been generally attempted only in comparatively recent times, as dictated by introduction of enhanced computational capabilities. Having said this, there were early examples of solutions based on these types of approach, notably those by Strescheletzky, who achieved solutions using an 'army' of technical assistants armed with hand calculators. As previously discussed, the solution of the propeller problem is essentially similar to any other incompressible flow problem about a three-dimensional body and, in particular, there is a close connection between subsonic airplane wing and marine propeller technologies. Indeed the latter, although perhaps older, relied for much of its development on aerodynamic theory. Notwithstanding the similarities, there are significant differences, the principal ones being the helical nature of marine propeller flow and the significantly lower aspect ratio of the blades.

For discussion purposes, however, the models of propeller action normally fall into one of five categories as follows:

1. Momentum or blade element models.
2. Lifting line models.
3. Lifting surface models.
4. Boundary element models.
5. Computational fluid dynamic models.

With regard to these five models, the momentum and blade element approaches will be introduced and discussed in the next chapter. The next two models will be briefly introduced here in the context of basic principles prior to discussion of particular approaches or combinations of approaches in Chapter 8. The discussion of computational fluid dynamic models will also be left until the next chapter.

The lifting line model is perhaps the simplest mathematical model of propeller action in that it assumes the aerofoil blade sections to be replaced by a single line vortex whose strength varies from section to section. The line, which is a continuous line in the radial direction about which vortices act, is termed the lifting line, and is normally considered to pass through the aerodynamic centers of the sections; this, however, is not always the case, especially

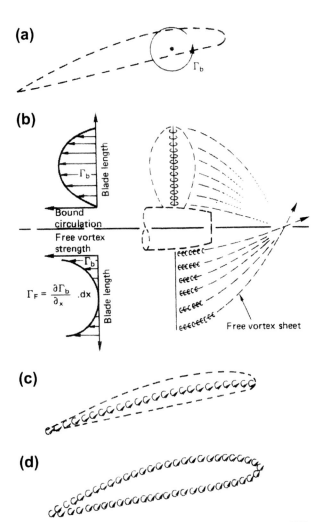

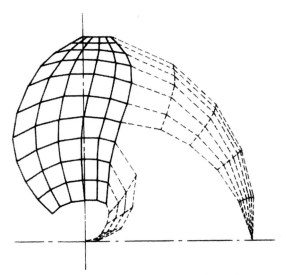

FIGURE 7.30 Vortex lattice model of propeller.

FIGURE 7.29 Hydrodynamic models of propeller action: (a) lifting line; (b) lifting line model of propeller action; (c) lifting surface and (d) surface vorticity.

with the earlier theories, where the directrix was frequently used as the location for the lifting line. Figure 7.29(a) demonstrates for a particular section the lifting line concept in which the aerofoil and its associated geometry is replaced by a single point. The lifting line concept is ideal for airplane propellers on account of their high aspect ratio, but for the marine propeller, with low aspect ratios and consequent strong three-dimensional effects over the wide blades, this approach, although simple, does have significant shortcomings. Since the strength of the bound vortices vary in the radial direction, then to satisfy Helmholtz's theorem, free vortices are shed from bound vortices along the lifting line whose strengths are given by

$$\Gamma_F = \left(\frac{d\Gamma_b}{dr}\right) \Delta r \qquad (7.56)$$

where Γ_b is the bound vortex strength and r is the radial position on the propeller. Figure 7.29(b) also outlines this concept and the similarity with Prandtl's classical lifting line theory for wings, shown in Figure 7.1, can be appreciated.

A higher level of blade representation is given by the lifting surface model. Rather than replacing the aerofoil section with a single bound vortex, as in the lifting line case, here the aerofoil is represented by an infinitely thin, bound vortex sheet. This bound vortex sheet is used to represent the lifting properties of the blade, in a manner analogous to thin aerofoil theory, and in the later theories the section thickness geometry is represented by source–sink distribution in order to estimate more correctly the section surface pressure distributions for cavitation prediction purposes. Such a model, normally referred to as a lifting surface model, is shown schematically in Figure 7.29(c). Models of this type present an order of numerical complexity greater than those for the lifting line concept; however, they too provide an attractive compromise between the simpler models and a full surface vorticity model. In this latter class of models the distribution of vorticity is placed around the section as seen in Figure 7.29(d), and thereby takes into account the effects of section thickness more fully than otherwise would be the case. Clearly, in both the lifting surface and surface vorticity models the strengths of the component vortices must be such that they generate the required circulation at each radial station.

Vortex lattice models, Figure 7.30, represent one of the more recent developments in propeller theoretical models. In this figure the solid lines represent the blade model and the hatched line the model of the propeller wake in terms of the roll-up of the vortices from the tip and root sections. Vortex lattice models make use of the concept of straight line segments of vortices joined together to cover the propeller blade with a system of vortex panels. The

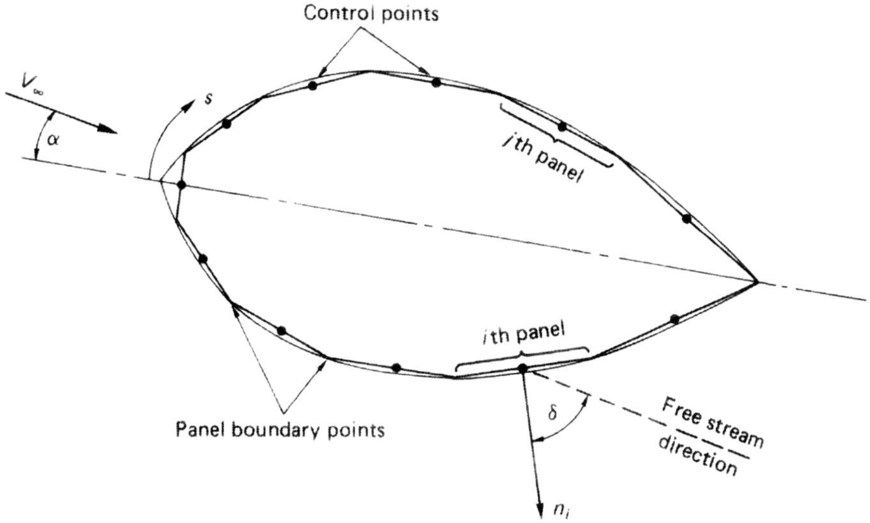

Theoretical Methods — Basic Concepts

FIGURE 7.31 Source panel solution method.

velocities at the control points, defined in each panel, over the blade are expressed in terms of the unknown strengths of the vortices. Then by applying a flow tangency condition at each control point the vortex strengths over the blade can be calculated and the flow problem solved. Hence vortex lattice models are a subclass of lifting surface models and consider the flow problem using a discrete rather than continuous system of singularities over the blade. This makes the computations somewhat less onerous.

In order to move towards the full surface vorticity concept idealized in Figure 7.29(d), much interest has been generated in the use of panel methods to provide a solution to the propeller design and analysis problems. The next section in this chapter considers the underlying principles of panel methods.

7.11 SOURCE AND VORTEX PANEL METHODS

Classical hydrodynamic theory shows that flow about a body can be generated by using the appropriate distributions of sources, sinks, vortices and dipoles distributed both within and about itself. Increased computational power has led to the development of panel methods, and these have now become commonplace for the solution of potential flow problems about arbitrary bodies.

In the case where the body generates no lift the flow field can be computed by replacing the surface of the body with an appropriate distribution of source panels (Figure 7.31). These source panels effectively form a source sheet whose strength varies over the body surface in such a way that the velocity normal to the body surface just balances the normal component of the free stream velocity. This condition ensures that no flow passes through the body and its surface becomes a streamline of the flow field. For practical computation purposes, the source strength λ_j is assumed to be constant over the length of the jth panel but allowed to vary from one panel to another. The mid-point of the panel is taken as the control point at which the resultant flow is required to be a tangent to the panel surfaces, thereby satisfying the flow normality condition defined above. The end points of each panel, termed the boundary points, are coincident with

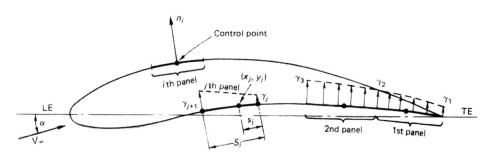

FIGURE 7.32 Vortex panel solution method.

those of the neighboring panels and consequently form a continuous surface.

From an analysis of this configuration of n panels as shown in Figure 7.31, the total velocity at the surface at the ith control point, located at the mid-point of the ith panel, is given by the sum of the contributions of free stream velocity V and those of the source panels:

$$V_i = V_\infty \sin \delta + \sum_{j=1}^{n} \frac{\lambda_j}{2\pi} \int_j \frac{\partial}{\partial s}(\ln r_{ij}) \mathrm{d}s_j \quad (7.57)$$

from which the associated pressure coefficient is given by

$$C_{Pi} = 1 - \left(\frac{V_i}{V_\infty}\right)^2 \quad (7.57a)$$

In equation (7.57) r is the distance from any point on the jth panel to the mid-point on the ith panel and s is the distance around the source sheet.

When the body for analysis is classified as a lifting body, the alternative concept of a vortex panel must be used since the source panel does not possess the circulation property which is essential to the concept of lift generation. The modeling procedure for the vortex panel is analogous to that for the source panel in that the body is replaced by a finite number of vortex panels as seen in Figure 7.32. On each of the panels the circulation density is varied from one boundary point to the other and is continuous over the boundary point. In these techniques the Kutta condition is easily introduced and is generally stable unless large numbers of panels are chosen on an aerofoil with a cusped trailing edge.

As in the case of the source panels the boundary points and control points are located on the surface of the body; again with control points at the mid-panel position. At these control points the velocity normal to the body is specified so as to prevent flow through the aerofoil. Using this approach the velocity potential at the ith control point (x_i, y_i) is given by

$$\phi(x_i, y_i) = V_\infty (x_i \cos \alpha + y_i \sin \alpha)$$
$$- \sum_{j=1}^{m} \int \frac{\gamma(s_j)}{2\pi} \tan^{-1}\left(\frac{y_i - y_j}{x_i - x_j}\right) \mathrm{d}s_j \quad (7.58)$$

for a system of m vortex panels, with

$$\gamma(s_j) = \gamma_j + (\gamma_{j+1} - \gamma_j)\frac{s_j}{S_j} \quad (7.58a)$$

Methods of this type — and the outline discussed here is but one example — can be used in place of the classical methods, for example Theodorsen or Weber, to calculate the flow around aerofoils. Typically for such a calculation one might use fifty or so panels to obtain the required accuracy, which presents a fairly extensive numerical task as compared to the classical approach. However, the absolute number of panels used for a particular application is dependent upon the section thickness—chord ratio in order to preserve the stability of the numerical solution. Nevertheless, methods of this type do have considerable advantages when considering cascades or aerofoils with flaps, for which exact methods are not available.

In a similar manner to the classical two-dimensional methods, source and vortex panel methods can be extended to three-dimensional problems. However, for the panel methods the computations become rather more complex but no new concepts are involved.

7.12 EULER, LAGRANGIAN AND NAVIER–STOKES METHODS

The term computational fluid dynamics is used to mean a number of computational procedures: in some cases from the most elementary to the more complex. In the sense of significant fluid mechanics computations the term is usually used to describe the use of either Eulerian or Navier–Stokes models: in the former case relating to inviscid computations, while in the later, taking account of the influence of viscosity and thermal conduction.

For the Euler model it is considered that the flow within a three-dimensional space obeys the following two principles:

1. The conservation of mass, which assumes that mass can neither be created nor destroyed. As such, this proposition gives rise to the *continuity equation* of fluid mechanics.
2. Adherence to Newton's Second Law of motion. This states that the force acting on a moving object is equal to the product of its mass and acceleration and this leads to the *momentum equation* of fluid mechanics.

While the continuity and momentum equations as formulated by Euler are sufficient for the analysis of inviscid and incompressible flows, if thermodynamic considerations are necessary in the solution of the problem at hand then the energy equation, formulated in the nineteenth century, is required:

3. That energy is conserved. As one form of energy decreases then other forms increase so as to maintain a balance. From this the principle of *conservation of energy* is derived for fluid systems.

These three principles when applied to an unsteady three-dimensional inviscid fluid system give rise to the Euler equations as follows:

Continuity equation

$$\frac{\partial \rho}{\partial t} + \nabla \cdot (\rho V) = 0$$

Momentum equations

$$\rho \frac{Du}{Dt} = -\frac{\partial p}{\partial x}$$

$$\rho \frac{Dv}{Dt} = -\frac{\partial p}{\partial y}$$

$$\rho \frac{Dw}{Dt} = -\frac{\partial p}{\partial z}$$

Energy equation

$$\rho \frac{D(e + V^2/2)}{Dt} = \rho \dot{q} - \nabla \cdot (pV)$$

where

$$\frac{D}{Dt} = \frac{\partial}{\partial t} + (V \cdot \nabla)$$

that is, the local derivative plus the convective derivative. In the above equations u, v and w are the velocities in the Cartesian directions x, y and z respectively; ρ is the fluid density; p is the pressure; V is the velocity of the fluid element; e is the internal kinetic energy per unit volume; $\dot{q}$ is the volumetric rate of heat addition and t is time.

These equations form a set of coupled non-linear partial differential equations and are collectively known as the Euler equations and for which there is no general analytical solution. Lagrange, along with Laplace to some extent, endeavored to simplify the Euler equations in order to make them approximations which could be solved. Lagrange considered the governing equations and wrote them in such a way that they pertained to the moving elements such that the pressure and velocity of the element could be described as a function of time. In effect, each element could be considered separately in terms of its instantaneous location in space and this approach is known as the Lagrangian Method. Lagrange's second contribution was to introduce the velocity potential (φ) and the stream function (Ψ) from which the flow velocity can be calculated. When introduced into the Euler equations this resulted in being able to make a number of simplifications to the solution for certain types of flow. In this context the velocity potential is defined along a streamline at any point where the velocity is V, then if δs is an incremental distance along the streamline then the velocity potential between two points A and B on the streamline is given by:

The velocity potential of B relative to A $= \varphi = \int_A^B V ds$

In contrast, to define the stream function consider a point A located on a streamline in the fluid and joined by an arbitrarily defined line to fixed point O. The stream function (ψ) of A is equal to the flow or flux across the line OA given by:

$$\varphi = \int_O^A V_0 \delta l$$

where V_0 is the normal velocity across the elemental length (δl) of the arbitrary line OA.

In the development of solutions for partial differential equations, Laplace, although not overly concerned with fluid mechanics, defined an important equation of mathematical physics as:

$$\frac{\partial^2 G}{\partial x^2} + \frac{\partial^2 G}{\partial y^2} + \frac{\partial^2 G}{\partial z^2} = 0$$

By using Lagrange's stream function ψ for an inviscid, incompressible flow this then satisfied Laplace's equation as follows:

$$\frac{\partial^2 \varphi}{\partial x^2} + \frac{\partial^2 \varphi}{\partial y^2} = 0$$

Furthermore, if the flow is irrotational the velocity potential also satisfies Laplace's equation as being:

$$\frac{\partial^2 \phi}{\partial x^2} + \frac{\partial^2 \phi}{\partial y^2} + \frac{\partial^2 \phi}{\partial z^2} = 0$$

If the flow instead of being inviscid now has a finite value for viscosity the physical foundation of the Eulerian model still applies; however, further terms to account for the friction and thermal conduction influences have to be introduced in all but the continuity equations. The resulting equations, termed the Navier–Stokes equations, are as follows:

Continuity equation:

$$\frac{\partial \rho}{\partial t} + \nabla \cdot (\rho V) = 0$$

Momentum equations:
In the x-direction

$$\rho \frac{Du}{Dt} = -\frac{\partial p}{\partial x} + \frac{\partial \tau_{xx}}{\partial x} + \frac{\partial \tau_{yx}}{\partial y} + \frac{\partial \tau_{zx}}{\partial z}$$

In the y-direction

$$\rho\frac{Dv}{Dt} = -\frac{\partial p}{\partial y} + \frac{\partial \tau_{xy}}{\partial x} + \frac{\partial \tau_{yy}}{\partial y} + \frac{\partial \tau_{zy}}{\partial z}$$

In the z-direction

$$\rho\frac{Dw}{Dt} = -\frac{\partial p}{\partial z} + \frac{\partial \tau_{xz}}{\partial x} + \frac{\partial \tau_{yz}}{\partial y} + \frac{\partial \tau_{zz}}{\partial z}$$

Energy equation:

$$\rho\frac{D(e + V^2/2)}{Dt} = \rho q + \frac{\partial}{\partial x}\left(k\frac{\partial T}{\partial x}\right) + \frac{\partial}{\partial y}\left(k\frac{\partial T}{\partial y}\right)$$
$$+ \frac{\partial}{\partial z}\left(k\frac{\partial T}{\partial z}\right) - \nabla \cdot pV + \frac{\partial(u\tau_{xx})}{\partial x}$$
$$+ \frac{\partial(u\tau_{yx})}{\partial y} + \frac{\partial(u\tau_{zx})}{\partial z} + \frac{\partial(v\tau_{xy})}{\partial x}$$
$$+ \frac{\partial(v\tau_{yy})}{\partial y} + \frac{\partial(v\tau_{zy})}{\partial z} + \frac{\partial(w\tau_{xz})}{\partial x}$$
$$+ \frac{\partial(w\tau_{yz})}{\partial y} + \frac{\partial(w\tau_{zz})}{\partial z}$$

where

$$\tau_{xy} = \tau_{yx} = \mu\left(\frac{\partial v}{\partial x} + \frac{\partial u}{\partial y}\right)$$

$$\tau_{yz} = \tau_{zy} = \mu\left(\frac{\partial w}{\partial y} + \frac{\partial v}{\partial z}\right)$$

$$\tau_{zx} = \tau_{xz} = \mu\left(\frac{\partial u}{\partial z} + \frac{\partial w}{\partial x}\right)$$

$$\tau_{xx} = \lambda(\nabla \cdot V) + 2\mu\frac{\partial u}{\partial x}$$

$$\tau_{yy} = \lambda(\nabla \cdot V) + 2\mu\frac{\partial v}{\partial y}$$

$$\tau_{zz} = \lambda(\nabla \cdot V) + 2\mu\frac{\partial w}{\partial z}$$

In this set of equations T is the temperature; k is the thermal conductivity; μ is the coefficient of viscosity; τ_{ij} are the flow shear stresses; λ is the bulk viscosity coefficient and the remaining symbols have the same meaning as for the Euler equations.

All of these methods find considerable application in the solution of fluid mechanics problems associated with marine propeller and ship propulsion problems. This applies to the determination of the flow around ships, the calculation of loadings of appendages and to increasing the understanding of cavitation dynamics. In some of these contexts, particularly the latter, the relatively new approach of Large Eddy Simulation (LES) is beginning to be explored. Although a very powerful technique to aid the understanding of vortical structures in turbulence because it permits the capture deterministically of the formation and development of coherent vortices and structures, from a mathematical perspective the LES problem is not well posed. Nevertheless, it is particularly helpful in attempting an understanding of the cavitation collapse dynamics discussed later in Chapter 9.

REFERENCES AND FURTHER READING

1. Milne-Thomson LM. *Theoretical Aerodynamics*. New York: Dover; 1973.
2. Milne-Thomson LM. *Theoretical Hydrodynamics*. Macmillan; 1968.
3. Lighthill J. *An Informal Introduction to Theoretical Fluid Mechanics*. Oxford; 1986.
4. Anderson JD. *Fundamentals of Aerodynamics*. McGraw-Hill; 1985.
5. Prandtl L, Tietjens OG. *Applied Hydro- and Aeromechanics*. United Engineering Trustees; 1934, also Dover, 1957.
6. Burrill LC. Calculation of marine propeller performance characteristics. Trans NECIES 1944;**60**.
7. Hawdon L, Carlton JS, Leathard FI. The analysis of controllable pitch propellers characteristics at off-design conditions. Trans I Mar E 1976;**88**.
8. Pankhurst RC. *A Method for the Rapid Evaluation of Glauert's Expressions for the Angle of Zero Lift and the Moment at Zero Lift*. R. & M. No. 1914; March 1944.
9. Theodorsen T. *Theory of Wing Sections of Arbitrary Shape*. NACA Report No. 1931;**411**.
10. Theodorsen T, Garrick IE. General Potential Theory of Arbitrary Wing Sections. *NACA Report No.* 1933;**452**.
11. Abbott H, Doenhoff AE van. *Theory of Wing Sections*; 1959. Dover.
12. Pope A. *Basic Wing and Airfoil Theory*. McGraw-Hill; 1951.
13. Pinkerton RM. *Calculated and Measured Pressure Distributions over the Mid-Span Sections of the NACA 4412 Airfoil*. NACA Report No. 563; 1936.
14. Riegels F, Wittich H. *Zur Bereschnung der Druck-verteilung von Profiles*. Jahrbuch der Deutschen Luftfahrtforschung; 1942.
15. Weber J. *The Calculation of the Pressure Distribution on the Surface of Thick Cambered Wings and the Design of Wings with Given Pressure Distributions*. R. & M. No. 3026. HMSO; June 1955.
16. Wilkinson DH. *A Numerical Solution of the Analysis and Design Problems for the Flow Past of one or more Aerofoils in Cascades*. R. & M. No. 3545. HMSO; 1968.
17. Firmin MCP. *Calculation of the Pressure Distribution, Lift and Drag on Aerofoils at Subcritical Conditions*. RAE Technical Report No. 72235; 1972.
18. Schlichting H. *Boundary Layer Theory*. McGraw-Hill; 1979.
19. Owen RR, Klanfer L. *RAE Report Aero 2508*; 1953.
20. Oossanen P van. *Calculation of Performance and Cavitation Characteristics of Propellers Including Effects of Non-Uniform Flow and Viscosity*. NSMB Publication 457; 1970.

21. Curle N, Skau SW. *Approximation methods for predictory separation properties of laminar boundary layers*. Aeronaut. Quart 1957;**3**.
22. Smith AMO. *Transition, pressure gradient, and stability theory*. Proc 9th Int Congress of App Mech Brussels 1956;**4**.
23. Nash JF, Macdonald AGJ. *The Calculation of Momentum Thickness in a Turbulent Boundary Layer at Mach Numbers up to Unity*. ARC Paper C.P. No. 1967;**963**.
24. Jones WP, Launder BE. *The prediction of laminarisation with a two-equation model of turbulence*. Int J Heat Mass Trans 1972;**15**.

Chapter 8

Theoretical and Analytical Methods Relating to Propeller Action

Chapter Outline

- 8.1 Momentum Theory — Rankine (1865); R.E. Froude (1887) — 169
- 8.2 Blade Element Theory — W. Froude (1878) — 171
- 8.3 Propeller Theoretical Development (1900–1930) — 172
- 8.4 Burrill's Analysis Procedure (1944) — 174
- 8.5 Lerbs Analysis Method (1952) — 180
- 8.6 Eckhardt and Morgan's Design Method (1955) — 183
- 8.7 Lifting Surface Correction Factors — Morgan *et al.* — 188
- 8.8 Lifting Surface Models — 193
- 8.9 Lifting Line–Lifting Surface Hybrid Models — 194
- 8.10 Vortex Lattice Methods — 194
- 8.11 Boundary Element Methods — 198
- 8.12 Methods for Specialist Propulsors — 200
- 8.13 Computational Fluid Dynamics Analysis — 202
- References and Further Reading — 204

The preceding chapter introduced and discussed the basic building blocks upon which the theory of propeller action has been based. This chapter, as outlined in Table 7.1, will now build upon that foundation and outline the important theories. However, as only a cursory glance at the propeller literature will reveal, this is a vast subject, and therefore in this chapter we will discuss classes of propeller theory and use one method from the class as being representative of that class for discussion purposes. The choice of method for this purpose will perhaps say more about the author's own usage and preferences rather than represent any general consensus about the superiority of the method. In each case, however, references will be given to other methods in the class under discussion.

The theoretical methods have, as far as possible, been introduced in their chronological order so as to underline the thread of development of the subject through time. Therefore, it is logical to start with the earliest attempt by Rankine[1] at producing a basis for propeller action.

8.1 MOMENTUM THEORY — RANKINE (1865); R.E. FROUDE (1887)

Rankine proposed a simple theory of propeller action based on the axial motion of water passing through the propeller disc. The theory did not concern itself with the geometry of the propeller which was producing the thrust and, consequently, his work is not very useful for blade design purposes. It does, however, lead to some general conclusions about propeller action which have subsequently been validated by more recent propeller theoretical methods and experiment.

The assumptions upon which Rankine based his original theory are as follows:

1. The propeller works in an ideal fluid and, therefore, does not experience energy losses due to frictional drag.
2. The propeller can be replaced by an actuator disc, and this is equivalent to saying that the propeller has an infinite number of blades.
3. The propeller can produce thrust without causing rotation in the slipstream.

The actuator disc concept is very common in earlier propeller theories and can be considered to be a notional disc having the same diameter as the propeller but of infinitesimal thickness. This disc, which is located at the propeller plane, is considered to absorb all of the power of the engine and dissipate this power by causing a pressure jump across the two faces of the disc and hence an increase of total head of the fluid.

Rankine's original theory, which is based on the above three assumptions, is generally known as the 'axial momentum theory'. R.E. Froude in his subsequent work[2] removed the third assumption and allowed the propeller to impart a rotational velocity to the slipstream, and thereby to become a more realistic model of propeller action. The subsequent theory is known either as the Rankine–Froude momentum theory or the general momentum theory of propellers. Here we shall follow the more general case of momentum theory, which is based on the first two assumptions only.

Figure 8.1 shows the general case of a propeller which has been replaced by an actuator disc and is working inside a streamtube; in the figure the flow is proceeding from left to right. Stations A and C are assumed to be far upstream and downstream respectively of the propeller, and the actuator disc is located at station B. The static pressure in the slipstream at stations A and C will be the local static pressure p_0, and the increase in pressure immediately behind the actuator disc is Δp, as also shown in Figure 8.1. Now the power P_D absorbed by the propeller and the thrust T generated are equal to the increase in kinetic energy of the slipstream per unit time and the increase in axial momentum of the slipstream respectively:

$$\left. \begin{array}{l} P_D = (\dot{m}/2)[V_C^2 - V_A^2] \\ T = \dot{m}[V_C - V_A] \end{array} \right\} \quad (8.1)$$

where $\dot{m}$ is the mass flow per unit time through the disc, from which it can be shown that

$$P_D = \frac{1}{2}T[V_C + V_A] \quad (8.2)$$

However, the power P_D is also equal to the work done by the thrust force of the propeller:

$$P_D = TV_B \quad (8.3)$$

Then, by equating equations (8.2) and (8.3), we find that the velocity at the propeller disc is equal to the mean of the velocities far upstream and downstream of the propeller:

$$V_B = \frac{1}{2}T[V_C + V_A] \quad (8.4)$$

If V_B and V_C are expressed in terms of the velocity V_A far upstream as follows:

$$\left. \begin{array}{l} V_B = V_A + aV_A \\ V_C = V_A + a_1V_A \end{array} \right\} \quad (8.5)$$

where a and a_1 are known as the axial inflow factors at the propeller disc and in the ultimate wake (far upstream), then by combining equation (8.4) with equation (8.5) we derive that

$$a_1 = 2a \quad (8.6)$$

Equations (8.4)–(8.6) lead to the important result that half the acceleration of the flow takes place before the propeller disc and the remaining half after the propeller disc. As a consequence of this result it follows that the slipstream must contract between the conditions existing far upstream and those existing downstream in order to satisfy the continuity equation of fluid mechanics.

From equation (8.1) the thrust is given by

$$T = \dot{m}[V_C - V_A]$$

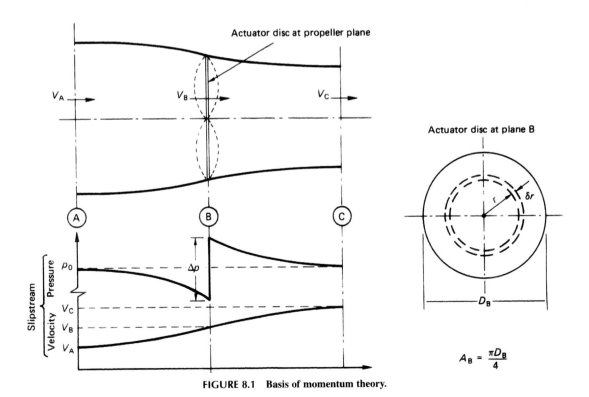

FIGURE 8.1 Basis of momentum theory.

and by combining this equation with equation (8.4) and the continuity equation

$$\rho V_B A_B = \rho V_A V_A \qquad (8.7)$$

we can derive that

$$C_T = \frac{T}{\rho A_B V_A^2} = 2\left(\frac{D_A}{D_B}\right)^2\left[\left(\frac{D_A}{D_B}\right)^2 - 1\right]$$

from which it can be shown that the contraction at the propeller plane D_B/D_A is given in terms of the propeller thrust coefficient C_T as

$$\left(\frac{D_B}{D_A}\right) = \frac{1}{\sqrt{\left[\frac{1}{2}\left(1 + \sqrt{(1 + 2C_T)}\right)\right]}} \qquad (8.8)$$

A similar, although slightly more complex, result can be derived for the contraction of the slipstream after the propeller. Figure 6.21 showed some experimental correlation by Nagamatsu and Sasajima with this formula which, as can be seen, is derived purely from axial momentum consideration.

If it is conjectured that the increase in pressure Δp is due to the presence of an angular velocity ω in the slipstream immediately behind the propeller disc, then the angular velocity of the water relative to the propeller blades, immediately ahead and astern of the propeller, is respectively Ω and $\Omega - \omega$. Then from Bernoulli's theorem the increase in pressure Δp at a particular radius r is given by

$$\Delta p = \rho\left(\Omega - \frac{1}{2}\omega\right)\omega r^2$$

Also the elemental thrust dT acting at some radius r is

$$dT = 2\pi r \, \Delta p \, dr$$

which, by writing $\omega = 2a'\Omega$, where a' is a rotational inflow factor, reduced to

$$dT = 4\pi \rho r^3 \Omega^2 (1 - a') a' \qquad (8.9)$$

Now the elemental torque dQ at the same radius r is equal to the angular momentum imparted to the slipstream per unit time within the annulus of thickness dr, namely

$$dQ = \dot{m}\omega r^2 = 2\pi \rho r^3 \omega \, dr = 4\pi \rho r^3 V_A \Omega (1 + a) a' \, dr \qquad (8.10)$$

The ideal efficiency of the blade element (η_i) is given by

$$\eta_i = \frac{\text{thrust horsepower}}{\text{delivered horsepower}} = \frac{V_A}{\Omega}\frac{dT}{dQ} \qquad (8.11)$$

hence, by substituting equations (8.9) and (8.10) into equation (8.11), we have

$$\eta_i = \frac{(1 - a')}{(1 + a)} \qquad (8.12)$$

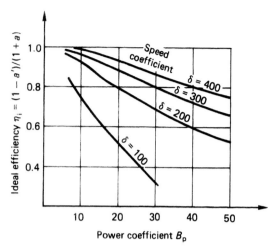

FIGURE 8.2 Ideal propeller efficiency from general momentum theory.

It can further be shown from this theory that for maximum efficiency the value of η_i should be the same for all radii. Equation (8.12) leads to the second important result of momentum theory, which is that there is an upper bound on the efficiency of an ideal, friction-less propeller. The ideal efficiency is a measure of the losses incurred by the propeller because the changes in momentum necessary to generate the required forces are accompanied by changes in kinetic energy. Figure 8.2 shows the relationships obtained from general momentum theory between η_i and the normal propulsion coefficients of B_p and δ for comparison purposes with actual experimental propeller results, remembering of course that the curves of η_i represent an upper bound for the efficiency value.

If the rotational assumption had not been removed in the derivation for η_i, then the ideal efficiency would have been found to be

$$\eta_i = \frac{1}{(1 + a)} \qquad (8.13)$$

8.2 BLADE ELEMENT THEORY — W. FROUDE (1878)

In contrast to the work of Rankine, W. Froude[3] developed a quite different model of propeller action, which took account of the geometry of the propeller blade. In its original form the theory did not take account of the acceleration of the inflowing water from its far upstream value relative to the propeller disc. This is somewhat surprising, since this could have been deduced from the earlier work of Rankine; nevertheless, this omission was rectified in subsequent developments of the work.

Blade element theory is based on dividing the blade up into a large number of elementary strips, as seen in

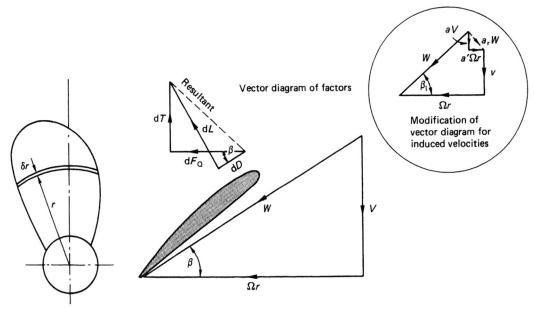

FIGURE 8.3 Blade element theory.

Figure 8.3. Each of these elementary strips can then be regarded as an aerofoil subject to a resultant incident velocity W.

The resultant incident velocity was considered to comprise an axial velocity V together with a rotational velocity Ωr, which clearly varies linearly up the blade. In the normal working condition the advance angle β is less than the blade pitch angle θ at the section, and hence gives rise to the section having an angle of incidence α. The section will, therefore, experience lift and drag forces from the combination of this incidence angle and the section zero lift angle, from which one can deduce that, for a given section geometry, the elemental thrust and torques are given by

$$\left. \begin{array}{l} dT = \dfrac{1}{2}\rho Z c W^2 (c_l \cos\beta - c_d \sin\beta)\, dr \\[6pt] dQ = \dfrac{1}{2}\rho Z c W^2 (c_l \sin\beta + c_d \cos\beta)\, r\, dr \end{array} \right\} \quad (8.14)$$

where Z and c are the number of blades and the chord length of the section respectively.

Now since the efficiency of the section η is given by

$$\eta = \frac{V}{\Omega}\frac{dT}{dQ}$$

then by writing $c_d/c_l = \gamma$ and substituting equation (8.14) into this expression for efficiency, we derive that

$$\eta = \frac{\tan\beta}{\tan(\beta + \gamma)} \quad (8.15)$$

Consequently, this propeller-theoretical model allows the thrust and torque to be calculated provided that the appropriate values of lift and drag are known. This, however, presented another problem since the values of c_l and c_d could not be readily calculated for the section due to the difficulty in establishing the effective aspect ratio for the section.

The basic Froude model can, as mentioned previously, be modified to incorporate the induced velocities at the propeller plane. To do this the advance angle β is modified to the hydrodynamic pitch angle β_i, and the velocity diagram shown in Figure 8.3 is amended accordingly to incorporate the two induced velocities, as shown in the inset to the figure.

Although Froude's work of 1878 failed in some respects to predict propeller performance accurately, it was in reality a great advance, since it contained the basic idea upon which all modern theory is founded. It was, however, to be just over half a century later before all of the major problems in applying these early methods had been overcome.

8.3 PROPELLER THEORETICAL DEVELOPMENT (1900–1930)

Immediately prior to this period propeller theory was seen to have been developing along two quite separate paths, namely the momentum and blade element theories. This led to inconsistent results; for example, blade element theory suggests that propeller efficiency will tend towards 100 per cent if the viscous forces are reduced to zero, whereas momentum theory, which is an inviscid approach, defines a specific limit on efficiency which is dependent upon speed of advance and thrust coefficient (Figure 8.2). This and other discrepancies led to a combination of the two theories by some engineers in which the induced velocities are determined by momentum theory and the analysis then conducted by the blade element theory. Although this was successful in many respects, none of these combined theories were entirely satisfactory.

In Chapter 7 it was seen how Lanchester and Prandtl put forward the concept that the lift of a wing was due to the development of circulation around the section and that a system of trailing vortices emanated from the blade and its tips. Nevertheless, it was not until 1919 that Prandtl had the necessary experimental confirmation of this new vortex theory. The application of this method to the propeller problem led to the conclusion that free vortices must spring from the tips of the propeller when operating. However, if due allowance is made for the induced velocities caused by these free vortices, which have a helical form, the forces at the blade elements will be the same as in two-dimensional flow. Consequently, the lift and drag coefficients for the section could be obtained from two-dimensional wind tunnel test data provided the results were corrected for aspect ratio according to Prandtl's formula. As this is standard wind tunnel technique, this then opened up a fund of aeronautical data for propeller design purposes.

The total energy loss experienced by the propeller comprises the losses caused by the creation of kinetic energy in the slipstream due to the effects of induced drag and the losses resulting from the motion of sections in a viscous fluid; that is, their profile drag. This latter component can be minimized by proper attention to the design of the blade sections; however, this is not the case with the losses resulting from the induced drag. The induced drag is a function of the design conditions, and in order to maximize the efficiency of a propeller it is necessary to introduce a further parameter which will ensure that the induced drag is minimized. Betz[4,5] established the basic minimum energy loss condition by analyzing the vortex system in the slipstream of a lightly loaded propeller having an infinite number of blades and working in a uniform stream. He established that the induced drag is minimized when the vortex sheets far behind the propeller are of constant pitch radially; in formal terms this leads to the condition for each radius:

$$x\pi D \tan \beta_e = \text{constant} \quad (8.16)$$

where β_e is the pitch angle of the vortex sheet at each radius far behind the propeller in the ultimate wake.

It follows from the Betz condition that for a uniform stream propeller the vortex sheets will leave the lifting line at constant pitch and undergo a deformation downstream of the propeller until they finally assume a larger constant pitch in the ultimate wake. Furthermore, for the propeller working in the uniform stream condition the direction of the resultant induced velocity $a_r W$, in the inset of Figure 8.3, is normal to the direction of the inflow velocity W at the lifting line. This result is known as the 'condition of normality' for a propeller working in an ideal fluid. In an appendix to Betz's paper Prandtl established a simple method for correcting the results for propellers having a small number of blades. This was based on the results of a model which replaced a system of vortex sheets by a series of parallel lines with a regular gap between them.

The total circulation at any radius on a propeller blade derived from the Betz condition relates to the infinitely bladed propeller. For such a propeller it can be shown[6] that the induced velocities at the propeller disc are the same as those derived from simple axial momentum theory. In the case of the 'infinite' blade number the helical vortex sheets emanating from each blade are 'very close' together; however, for the real case of a propeller with a small number of blades, the trailing vortices are separated from each other. Hence, the mean induced velocity in the latter case, when considered about a circumferential line at some radius in the propeller disc, is less than the local induced velocity at the blades. Prandtl's earlier relationship for the mean velocity at a particular radius, when compared to the velocity of the lifting line, was

$$K = \frac{2}{\pi}\cos^{-1}\left[\exp\left\{-\frac{Z}{2\tan\beta_i}(1-x)\sqrt{(1+\tan^2\beta_i)}\right\}\right] \quad (8.17)$$

This important effect was subsequently studied by Goldstein[7] who considered the flow past a series of true helicoidal surfaces of infinite length. He obtained an expression for the ratio between the mean circulation taken around an annulus at a particular radius and the circulation at the helicoidal surfaces. These values take the form of a correction factor K and an example is shown in Figure 8.4 for a four-bladed propeller. From this figure it is readily seen that the corrections have more effect the greater the non-dimensional radius of the propeller. This is to be expected, since the distance between the blade sections is greater for the outer radii, and since the induced velocity at the lifting line would normally be a maximum it follows that the mean induced velocity through an annulus at a particular radius will be less than the value at the lifting line, the ratio being given by the Goldstein factor. At zero pitch angle the value of the correction factor is clearly unity.

Goldstein's work was based on the model of a propeller which has zero boss radius. In practice this is clearly not the case and Tachmindji and Milam[8] subsequently made a detailed set of calculations for propellers with blade numbers ranging from 3–6 and having a finite hub radius of $0.167R$. Table 8.1 defines these values, from which it is seen that values at the outer radii are broadly comparable,

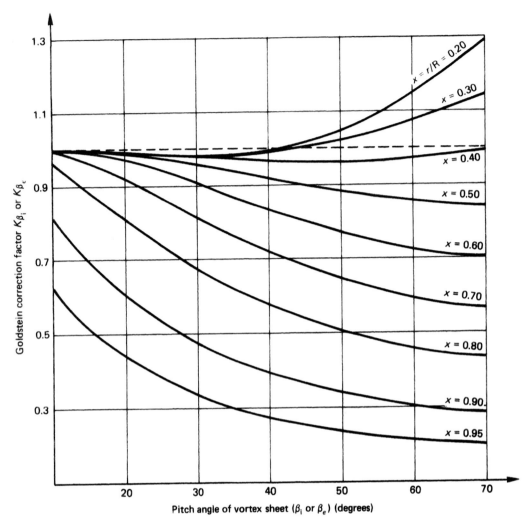

FIGURE 8.4 Goldstein correction factors for a four-bladed propeller.

as might be expected, whereas those at the inner radii change considerably.

With the work of Goldstein, Betz, Prandtl and Lanchester, the basic building blocks for a rational propeller theory were in place by about 1930. Although Perring[9] and Lockwood Taylor[10] established theories of propeller action, it was left to Burrill[11] to establish an analytical method which gained comparatively wide acceptance within the propeller community.

8.4 BURRILL'S ANALYSIS PROCEDURE (1944)

Burrill's procedure is essentially a strip theory method of analysis which combines the basic principles of the momentum and blade element theories with aspects of the vortex analysis method. As such the method works quite well for moderately loaded propellers working at or near their design condition; however, for heavily and lightly loaded propellers the correlation with experimental results is not so good, although over the years several attempts have been made to improve its performance in these areas; for example, Sontvedt.[12]

In developing his theory, Burrill considered the flow through an annulus of the propeller, as shown in Figure 8.5. From considerations of continuity through the three stations identified (far upstream, propeller disc and far downstream) one can write, for the flow through the annulus in one revolution of the propeller,

$$2\pi \rho r_0 \frac{V_a}{n} \delta r_0 = 2\pi \rho r_1 \frac{V_a(1 + K_{\beta_i} a)}{n} \delta r_1$$
$$= 2\pi \rho r_2 \frac{V_a(1 + 2K_\varepsilon a)}{n} \delta r_2$$

where $K_{\beta i}$ and K_ε are the Goldstein factors at the propeller disc and in the ultimate wake respectively; V_a is the speed of

TABLE 8.1 Goldstein–Tachminji Correction Factors for $x_n = 0.167$

Blade Number = 3

	r/R							
	0.950	0.900	0.800	0.700	0.600	0.500	0.400	0.300
0	1.000	1.000	1.000	1.000	1.000	1.000	1.000	1.000
5	0.729	0.902	0.990	0.999	1.000	1.000	1.000	1.000
10	0.543	0.732	0.916	0.979	0.995	0.997	0.996	0.988
15	0.443	0.612	0.818	0.925	0.971	0.985	0.982	0.963
20	0.372	0.523	0.725	0.854	0.928	0.961	0.963	0.934
25	0.320	0.454	0.645	0.782	0.876	0.929	0.943	0.911
30	0.280	0.401	0.578	0.716	0.822	0.893	0.923	0.896
35	0.249	0.358	0.523	0.658	0.771	0.857	0.904	0.888
40	0.225	0.325	0.479	0.609	0.726	0.823	0.888	0.886
45	0.206	0.298	0.442	0.569	0.686	0.793	0.874	0.891
50	0.191	0.277	0.413	0.536	0.654	0.767	0.863	0.902
55	0.179	0.260	0.390	0.509	0.627	0.745	0.855	0.916
60	0.170	0.247	0.372	0.488	0.605	0.728	0.850	0.935
65	0.162	0.236	0.357	0.471	0.589	0.715	0.849	0.957
70	0.157	0.228	0.347	0.459	0.577	0.707	0.850	0.982

Blade Number = 4

	r/R							
	0.950	0.900	0.800	0.700	0.600	0.500	0.400	0.300
0	1.000	1.000	1.000	1.000	1.000	1.000	1.000	1.000
5	0.804	0.949	0.997	0.999	1.000	1.000	1.000	1.001
10	0.620	0.810	0.959	0.993	0.998	0.999	0.999	0.997
15	0.514	0.696	0.890	0.966	0.989	0.994	0.992	0.983
20	0.440	0.609	0.813	0.921	0.969	0.983	0.982	0.964
25	0.385	0.539	0.742	0.868	0.938	0.967	0.970	0.946
30	0.341	0.483	0.679	0.814	0.902	0.948	0.959	0.933
35	0.307	0.437	0.624	0.763	0.864	0.927	0.950	0.926
40	0.279	0.400	0.578	0.717	0.828	0.906	0.944	0.927
45	0.257	0.369	0.539	0.678	0.795	0.886	0.941	0.935
50	0.240	0.345	0.507	0.644	0.766	0.869	0.941	0.951
55	0.225	0.325	0.481	0.617	0.741	0.854	0.944	0.973
60	0.214	0.309	0.460	0.594	0.721	0.843	0.949	1.000
65	0.205	0.297	0.440	0.576	0.705	0.834	0.956	1.033
70	0.198	0.288	0.431	0.562	0.694	0.829	0.965	1.068

(Continued)

TABLE 8.1 Goldstein–Tachminji Correction Factors for $x_n = 0.167$—cont'd

Blade Number = 5

	r/R							
	0.950	0.900	0.800	0.700	0.600	0.500	0.400	0.300
0	1.000	1.000	1.000	1.000	1.000	1.000	1.000	1.000
5	0.858	0.973	0.992	1.000	1.000	1.000	1.001	1.002
10	0.681	0.864	0.980	0.998	0.999	0.999	0.999	1.000
15	0.572	0.759	0.932	0.984	0.995	0.997	0.996	0.992
20	0.496	0.675	0.872	0.957	0.985	0.991	0.990	0.980
25	0.438	0.606	0.811	0.919	0.968	0.982	0.982	0.966
30	0.393	0.549	0.753	0.876	0.944	0.971	0.975	0.956
35	0.356	0.502	0.701	0.834	0.918	0.960	0.970	0.949
40	0.326	0.462	0.656	0.794	0.891	0.948	0.969	0.949
45	0.302	0.430	0.617	0.758	0.865	0.937	0.971	0.957
50	0.282	0.403	0.584	0.727	0.842	0.928	0.977	0.972
55	0.266	0.382	0.557	0.700	0.822	0.920	0.986	0.994
60	0.258	0.364	0.535	0.678	0.805	0.914	0.997	1.023
65	0.243	0.351	0.571	0.660	0.791	0.911	1.011	1.058
70	0.235	0.331	0.503	0.646	0.781	0.909	1.025	1.095

Blade Number = 6

	r/R							
	0.950	0.900	0.800	0.700	0.600	0.500	0.400	0.300
0	1.000	1.000	1.000	1.000	1.000	1.000	1.000	1.000
5	0.897	0.986	1.000	1.000	1.000	1.000	1.001	1.002
10	0.730	0.902	0.990	0.999	1.000	1.000	1.000	1.001
15	0.620	0.808	0.958	0.992	0.998	0.998	0998	0.997
20	0.543	0.728	0.912	0.976	0.992	0.994	0.993	0.989
25	0.484	0.661	0.859	0.949	0.982	0.989	0.988	0.979
30	0.437	0.604	0.808	0.917	0.967	0.983	0.983	0.970
35	0.398	0.556	0.761	0.883	0.949	0.976	0.980	0.964
40	0.367	0.516	0.718	0.849	0.930	0.970	0.980	0.964
45	0.341	0.483	0.680	0.818	0.911	0.965	0.985	0.970
50	0.320	0.455	0.647	0.789	0.893	0.961	0.993	0.982
55	0.303	0.432	0.620	0.765	0.878	0.959	1.004	1.002
60	0.289	0.413	0.598	0.744	0.864	0.958	1.019	1.029
65	0.278	0.398	0.579	0.727	0.854	0.959	1.036	1.061
70	0.269	0.386	0.565	0.714	0.845	0.961	1.053	1.098

Compiled from Reference 8.

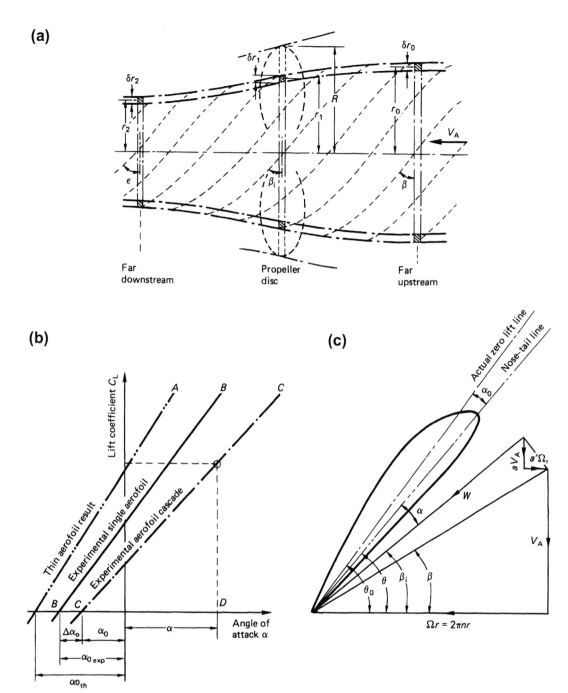

FIGURE 8.5 The Burrill analysis procedure: (a) slipstream contraction model; (b) lift evaluation and (c) flow vectors and angles.

advance of the uniform stream and a is the axial induction factor.

From this relation, by assuming that the inflow is constant at each radius, that is $a = $ constant, it can be shown that the relationship between the radii of the slipstream is given by

$$r_0 = r_1(1 + K_{\beta_i}a)^{1/2} = r_2(1 + K_\varepsilon a)^{1/2} \qquad (8.18)$$

Furthermore, it is also possible, by considering the momentum of the fluid in relation to the quantity flowing through the annular region, to define a relation for the thrust acting on the fluid

$$dT = V\, dQ$$

for which the thrust at the propeller disc can be shown to be

$$dT = 4\pi r_1 K_\varepsilon a V_A^2 [1 + K_{\beta_i}a]\rho\, dr_1 \qquad (8.19)$$

However, by appealing to the blade element concept, an alternative expression for the elemental thrust at a particular radius r_1 can be derived as follows:

$$dT = \frac{\rho}{2} Z c (1+a)^2 \times \frac{V_A^2}{\sin^2 \beta_i} [c_1 \cos \beta_i - c_d \sin \beta_i] \, dr_1 \quad (8.20)$$

and then by equating equations (8.19) and (8.20) and in a similar manner deriving two further expressions for the elemental torque acting at the particular radius, one can derive the following pair of expressions for the axial for tangential inflow factors a and a' respectively.

$$\left. \begin{array}{l} \dfrac{a}{1+a} \left(\dfrac{1 + K_{\beta_i} a}{1+a} \right) = \left(\dfrac{c_1 \sigma_s}{2 K_\varepsilon} \right) \dfrac{\cos(\beta_i + \gamma)}{2 \sin^2 \beta_i \cos \gamma} \\[2mm] \dfrac{a'}{1-a'} \left(\dfrac{1 - K_{\beta_i} a}{1+a} \right) = \left(\dfrac{c_1 \sigma_s}{2 K_\varepsilon} \right) \dfrac{\sin(\beta_i + \gamma)}{\sin 2\beta_i \cos \gamma} \end{array} \right\} \quad (8.21)$$

in which σ_s is the cascade solidity factor defined by $Zc/(2\pi r)$ and γ is the ratio of drag to lift coefficient c_d/c_1.

In equation (8.21) the lift coefficient c_1 is estimated from the empirical relationships derived from wind tunnel tests on aerofoil sections and applied to the results of thin aerofoil theory as discussed in the preceding chapter. Thus the lift coefficient is given by

$$c_1 = 2\pi k_s \cdot k_{gs} (\alpha + \alpha_0) \quad (8.22)$$

where k_s and k_{gs} are the thin aerofoil to single aerofoil and single aerofoil to cascade correction factors for lift slope derived from wind tunnel test results and α_0 is the experimental zero lift angle shown in Figure 8.5(b). The term α is the geometric angle of attack relative to the nose–tail line of the section as seen in Figure 8.5(c). From equation (8.22) it can be seen that the lift slope curve reduces from 2π in the theoretical thin aerofoil case to $2\pi k_s k_{gs}$ in the experimental cascade situation; that is, the line CC in Figure 8.5(b). Burrill chooses to express k_s as a simple function of thickness to chord ratio, and k_{gs} as a function of hydrodynamic pitch angle β_i and cascade solidity σ_s. Subsequently, work by van Oossanen, discussed in Chapter 7, has shown that these parameters, in particular k_s and k_{gs}, are more reliably expressed as functions of the section boundary layer thicknesses at the trailing edge of the sections.

With regard to the effective angle of attack of the section, this is the angle represented by the line CD measured along the abscissa of Figure 8.5(b) and the angle $(\alpha + \alpha_0)$ on Figure 8.5(c). This angle α_e is again calculated by appeal to empirically derived coefficients as follows:

$$\alpha_e = \alpha + \alpha_0 = \alpha + \alpha_{0_{TH}} (K_{\alpha_0} - K_{g_{\alpha_0}}) \quad (8.23)$$

where α_{0TH} is the two-dimensional theoretical zero lift angle derived from equation (7.27) and K_{α_0} and $K_{g\alpha_0}$ are the single aerofoil to theoretical zero lift angle correlation factor and cascade allowance respectively. The former, as expressed by Burrill, is a function of thickness to chord ratio and position of maximum camber while the latter is a function of hydrodynamic pitch angle β_i and solidity of the cascade σ_s.

By combining equations (8.21) and (8.22) and noting that the term $(1 + K_{\beta_i} a)/(1+a)$ can be expressed as

$$1 - [(1 - K_{\beta_i}) \tan(\beta_i - \beta)/\tan \beta_i]$$

it can be shown that the effective angle of attack $(\alpha + \alpha_0)$ is given by

$$\alpha + \alpha_0 = \frac{2}{K_s K_{gs} \pi \sigma_s} K_\varepsilon \sin \beta_i \tan(\beta_i - \beta) \times \left[1 - \frac{\tan(\beta_i - \beta)}{\tan \beta_i} (1 - K_{\beta_i}) \right] \quad (8.24)$$

This equation enables, by assuming an initial value, the value of the effective angle of attack to be calculated for any given value of $(\beta_i - \beta)$, by means of an iterative process. Once convergence has been achieved the lift can be calculated from equation (8.22). The drag coefficient again is estimated from empirical data based on wind tunnel test results and this permits the calculation of the elemental thrust and torque loading coefficients:

$$\left. \begin{array}{l} dK_Q = \dfrac{\pi^3 x^4 \sigma_s}{8} (1-a')^2 (1 - \tan^2 \beta_i) c_i \dfrac{\sin(\beta_i + \gamma)}{\cos \gamma} \, dx \\[2mm] dK_T = \left(\dfrac{dK_Q}{dx} \right) \dfrac{2}{x \tan(\beta_i + \gamma)} \, dx \end{array} \right\} \quad (8.25)$$

where the rotational induced velocity coefficient a' can most conveniently be calculated from

$$a' = \frac{(\tan \beta_i - \tan \beta) \tan(\beta_i + \gamma)}{1 + \tan \beta_i \tan(\beta_i + \gamma)} \quad (8.26)$$

Figure 8.6 shows the algorithm adopted by Burrill to calculate the radial distribution of loading on the propeller blade, together with certain modifications, such as the incorporation of the drag coefficients from his later paper on propeller design.[13]

Burrill's analysis procedure represents the first coherent step in establishing a propeller calculation procedure. It works quite well for the moderately loaded propeller working at or near its design condition; however, at either low or high advance the procedure does not behave as well. In the low advance ratios the constant radial axial inflow factor, consistent with a lightly loaded propeller, must contribute to the underprediction of thrust and torque coefficient for these advance ratios. Furthermore, the Goldstein factors rely on the

Theoretical and Analytical Methods Relating to Propeller Action

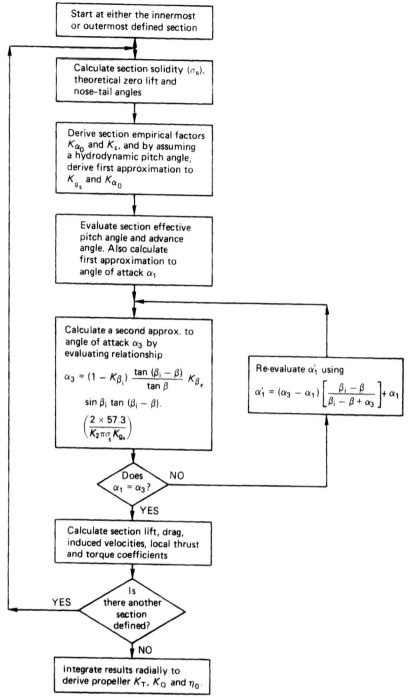

FIGURE 8.6 Burrill calculation algorithm.

conditions of constant hydrodynamic pitch and consequently any significant slipstream distortion must affect the validity of applying these factors. Alternatively, in the lightly loaded case it is known that the theory breaks down when the propeller conditions tend toward the production of ring vortices.

Clearly, since the Goldstein factors are based on the concept of zero hub radius, the theory will benefit from the use of the Tachmindji factors which incorporate a boss of radius $0.167R$. Additionally, the use of the particular cascade corrections used in the method were criticized at the time of the method's publication; however, no real alternative for use with a method of this type has presented itself.

The Burrill method represents the final stage in the development of a combined momentum—blade element approach to propeller theory. Methods published subsequent to this generally made greater use of the lifting line, and subsequently lifting surface concepts of aerodynamic theory. The first of these was perhaps due to Hill[14] and was followed by Strscheletsky;[15] however, the next significant development was that due to Lerbs,[16] who laid the basis for moderately loaded lifting line theory. Strscheletsky's work, although not generally accepted at the time of its introduction, due to the numerical complexity of his solution, has subsequently formed a basis for lifting line heavily loaded propeller analysis.

8.5 LERBS ANALYSIS METHOD (1952)

Lerbs followed the sequence of lifting line development by proposing a method of analysis for the moderately loaded propeller working in an inviscid fluid. The moderately loaded assumption requires that the influence of the induced velocities are taken into account, and as such the vortex sheets emanating from each blade differ slightly from the true helical form, this latter form only being true for the lightly loaded propeller.

The development of the model, which has to some extent become regarded as a classic representation of lifting line models, followed the work of Kawada. Kawada considered the problem of a propeller whose blades were represented by a line of constant bound vorticity from the root to the tip with a system of free helical tip vortices, one from each blade, and an axial hub vortex whose strength is equal to the sum of the tip vortex strengths. Lerbs considered the more advanced case of the blades being represented by a line of radially varying bound vorticity $\Gamma(x)$. In this case, in order to satisfy Stokes' theorem, this must give rise to the production of a vortex sheet whose strength is a variable depending upon the radius. The strength of a particular element of the vortex sheet is given by

$$\Gamma_F(x) = \left(\frac{\partial \Gamma}{\partial r}\right) dr \qquad (8.27)$$

Figure 8.7 outlines in schematic form the basis of Lerbs' model. Within the model centrifugal and slipstream contraction effects are ignored, and so the sheets comprise cylindrical vortex lines of constant diameter and pitch in the axial direction.

Unlike some previous work, within the model, prior assumptions with respect to the pitch of the vortex sheets are avoided, and consequently it is necessary to evaluate both the axial and induced velocity components, since no unique relation between them exists when the sheet form differs from the truly helical. In his approach Lerbs also considered the presence of a propeller hub in the calculation procedure, but assumed the circulation at the hub to be zero. This latter assumption clearly does not represent actual conditions on a propeller but is a computational convenience.

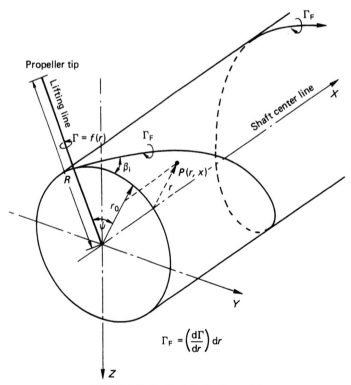

FIGURE 8.7 Basis of the Lerbs model.

Lerbs showed, by appealing to the Biot–Savart law, that Kawada's expressions for induced velocity, based on infinite vortices extending from $-\infty$ in one direction to $+\infty$ in the other, are valid for the calculation of the induced velocities at the propeller disc ($X=0$) provided that the resultant induced axial and tangential velocities are divided by two. For points in the slipstream that are not in the plane of the propeller disc, the relations governing the induced velocity are less simple. Lerbs gives the following set of equations for the axial and tangential induced velocities at a radius r in the propeller disc induced by a single free helical vortex line emanating from a radius r_0 in the propeller disc (Figure 8.7):

Axial induced velocities:
Internal points ($r < r_0$)

$$\bar{w}_{ai} = \frac{Z\Gamma_F}{4\pi k_0}\left\{1 + 2Z\frac{r_0}{k_0}\sum_{n=1}^{\infty} nI_{nz}\left(\frac{nZ}{k_0}r\right)K'_{nz}\left(\frac{nZ}{k_0}r_0\right)\right\}$$

External points ($r > r_0$)

$$\bar{w}_{ae} = \frac{Z^2\Gamma_F r_0}{2\pi k_0^2}\sum_{n=1}^{\infty} nK_{nz}\left(\frac{nZ}{k_0}r\right)I'_{nz}\left(\frac{nZ}{k_0}r_0\right)$$

Tangential induced velocities:
Internal points ($r < r_0$)

$$\bar{w}_{ti} = \frac{Z^2\Gamma_F r_0}{2\pi k_0 r}\sum_{n=1}^{\infty} nI_{nz}\left(\frac{nZ}{k_0}r\right)K'_{nz}\left(\frac{nZ}{k_0}r_0\right)$$

External points ($r > r_0$)

$$\bar{w}_{te} = \frac{Z\Gamma_F}{4\pi r}\left\{1 + 2Z\frac{r_0}{k_0}\sum_{n=1}^{\infty} nK_{nz}\left(\frac{nZ}{k_0}r\right)I'_{nz}\left(\frac{nZ}{k_0}r_0\right)\right\}$$

where $k_0 = r_0 \tan\beta_{i0}$ and I_{nz} and K_{nz} are the modified Bessel functions of the first and second kind respectively. In order to evaluate these expressions, use is made of Nicholson's asymptotic formulae to replace the Bessel functions, from which it is then possible to derive the following set of expressions after a little manipulation of the algebra involved:

$$(r < r_0)\bar{w}_{ai} = \frac{Z\Gamma_F}{2\pi k_0}(1+B_2); \quad \bar{w}_{ti} = -\frac{Z\Gamma_F}{4\pi r}B_2$$

$$(r > r_0)\bar{w}_{ae} = \frac{-Z\Gamma_F}{2\pi k_0}B_1; \quad \bar{w}_{te} = -\frac{Z\Gamma_F}{4\pi r}(1+B_1)$$

where

$$B_{1,2} = \left(\frac{1+y_0^2}{1+y^2}\right)^{0.25}\left[\frac{1}{e^{zA_{1,2}}-1} \mp \frac{1}{2Z}\frac{y_0^2}{(1+y_0^2)^{1.5}}\right.$$
$$\left. \times \log_e\left(1 + \frac{1}{e^{zA_{1,2}}-1}\right)\right]$$

and

$$A_{1,2} = \pm\left(\sqrt{(1+y^2)} - \sqrt{(1+y_0^2)}\right)$$

$$\mp \frac{1}{2}\log_e\frac{\left(\sqrt{(1+y_0^2)}-1\right)\left(\sqrt{(1+y^2)}+1\right)}{\left(\sqrt{(1+y_0^2)}+1\right)\left(\sqrt{(1+y^2)}-1\right)}$$

in which

$$y_0 = \frac{1}{\tan\beta_{i0}} \quad \text{and} \quad y = \frac{x}{x_0 \tan\beta_{i0}} \qquad (8.28)$$

The distinction between the two conditions of ($r < r_0$) and ($r > r_0$) is made since when the point of interest coincides with the radius at which free vortex emanates, that is when $r = r_0$, the velocity components tend to an infinite magnitude. To avoid this problem Lerbs introduces the concept of an induction factor, which is a non-dimensional parameter and is the ratio of the velocity induced by a helical vortex line to that produced by a semi-infinite straight line vortex parallel to the shaft axis at radius r_0. A semi-infinite vortex is, in this context, one that ranges from $z = 0$ to $+\infty$, and the velocity induced by such a straight line vortex at a radius r in the propeller disc is given by $\Gamma_F/[4\pi(r - r_0)]$. The induction factors in the axial and tangential directions are formally defined as

$$\bar{w}_a = \frac{\Gamma_F}{4\pi(r-r_0)}i_a; \quad \bar{w}_t = \frac{\Gamma_F}{4\pi(r-r_0)}i_t \qquad (8.29)$$

It will readily be seen from equation (8.29) that the velocity induced by a straight line vortex also tends to an infinite magnitude when $r = r_0$. However, when $r \to r_0$ both the velocities induced by the straight line and helical vortices are of the same order, consequently, the ratio of the velocities and hence the induction factors remain finite. When interpreted in the context of the expressions for the axial and tangential induced velocities given in equation (8.28), we have

$$i_{ai} = \frac{Zx}{x_0 \tan\beta_{i0}}\left(\frac{x_0}{x} - 1\right)(1+B_2)$$

$$i_{ae} = \frac{Zx}{x_0 \tan\beta_{i0}}\left(\frac{x_0}{x} - 1\right)B_1$$

$$i_{ti} = Z\left(\frac{x_0}{x} - 1\right)B_2 \qquad (8.30)$$

$$i_{te} = -Z\left(\frac{x_0}{x} - 1\right)(1+B_1)$$

in which the suffices i and e refer to internal and external radii relative to the nominal value x_0.

From these equations it is apparent that the induction factors do not depend upon the circulation but are simply a function of the geometry of the flow. The induction factors defined by equation (8.30) describe the induction of Z free helical vortices of non-dimensional radius x_0 at a point in the propeller plane at a non-dimensional radius x. There are, however, other induced velocities acting in this plane. These come from the bound vortices on the lifting lines, but in the case of uniform flow these cancel out provided the blades are symmetrically arranged.

In introducing the concept of a propeller hub into the analysis procedure, Lerbs used as a representation an infinitely long cylinder of radius r_h. This leads to two effects, first the definition of the circulation at the root of the blades, and second, the effect on the induced flow. With respect to the problem of the circulation at the root it is argued that for any two adjacent blades, the pressure on the face of one will tend to equalize with the suction on the back of the other. Consequently, it follows that for the purposes of this theory the circulation at the root of the blade can be written to zero. For the induced flow effect this clearly leads to the condition that the radial component of flow at the hub must be zero, since no flow can pass through the hub. Lerbs had some difficulty in incorporating this last effect; however, he derived a tentative solution by appealing to Kawada's equation for the radial component of flow and treated the problem as though the boss were located in the ultimate wake.

Equation (8.30) relates to the effects of a single vortex emanating from each blade of a propeller at a given radius. In order to generalize these relations so as to incorporate the effect of all the free vortices emanating from the propeller, it is necessary to add the contributions of all of the free vortices at the point of interest. For example, in the case of the tangential component,

$$w_t(r) = \int_{r_h}^{R} \overline{w}_t(\overline{r}_0) dr_0$$

which, when expanded in association with equations (8.27) and (8.29), gives

$$\frac{w_t}{V} = \frac{1}{2} \int_{x_h}^{1.0} \left(\frac{dG}{dx_0}\right) \frac{i_t}{(x - x_0)} dx_0 \qquad (8.31)$$

where G is the non-dimensional circulation coefficient given by $\Gamma/(\pi DV)$. The improper integrals representing the values of w_t and in the analogous expression for w_a are similar to those encountered in aerofoil theory, the difference being that in the propeller case the induction factors allow for the curvature of the vortex sheets. To establish a solution for the propeller problem Lerbs extended the work of Glauert and introduced a variable (ϕ) defined by equation (8.32), which allows a circular representation of the radial location on the lifting line:

$$x = 0.5[(1 + x_h) - (1 - x_h)\cos \phi] \qquad (8.32)$$

The distribution of circulation $G(x)$ is continuous for all but very few propellers, which must be treated separately, and the boundary values at the tip and root are known. As a consequence of this the circulation distribution can be represented by an odd Fourier series:

$$G = \sum_{m=1}^{\infty} G_m \sin(m, \phi)$$

in addition, the induction factors depend upon ϕ and ϕ_0, equation (8.32), and can be represented by an even Fourier series:

$$i(\phi, \phi_0) = \sum_{n=1}^{\infty} I_n(\phi)\cos(n\phi_0)$$

Now by combining these expressions with equation (8.31), it is possible to write expressions for the tangential and, analogously, the axial induced velocities at a radial location ϕ as follows:

$$\left. \begin{aligned} \frac{w_a}{V} &= \frac{1}{1 - x_h} \sum_{m=1}^{\infty} m G_m h_m^a(\phi) \\ \frac{w_t}{V} &= \frac{1}{1 - x_h} \sum_{m=1}^{\infty} m G_m h_m^t(\phi) \end{aligned} \right\} \qquad (8.33)$$

where

$$h_m^a(\phi) = \frac{\pi}{\sin \phi} \left[\sin(m\phi) \sum_{n=0}^{m} I_n^t(\phi)\cos(n\phi) + \cos(m\phi) \right.$$
$$\left. \times \sum_{n=m+1}^{\infty} I_n^t(\phi)\sin(n\phi) \right]$$

and

$$h_m^t(\phi) = \frac{\pi}{\sin \phi} \left[\sin(m\phi) \sum_{n=0}^{m} I_n^t(\phi)\cos(n\phi) + \cos(m\phi) \right.$$
$$\left. \times \sum_{n=m+1}^{\infty} I_n^t(\phi)\sin(n\phi) \right]$$

It should be noted that at the hub and root the functions become indefinite, that is when $\phi = 0°$ or $180°$. From l'Hôpital's rule the limits for the function become

$$h_m^{a,t}(0) = \pi \left[m \sum_{n=0}^{m} I_n^{a,t}(0) + \sum_{n=m+1}^{\infty} n I_n^{a,t}(0) \right]$$

$$h^{a,t}(180°) = -\pi \cos(m\pi)$$
$$\times \left[m \sum_{n=0}^{m} I_n^{a,t} \cos(n\pi) + \sum_{n=m+1}^{\infty} n I_n^{a,t} \cos(n\pi) \right]$$

These equations enable the induced velocity components to be related to the circulation distribution and also to the induction factors.

More recently Morgan and Wrench[99] made a significant contribution to the calculation of Lerbs' induction factors, and their method is used by many modern lifting line procedures.

Lerbs, having proposed the foregoing general theoretical model, then proceeded to apply it to two cases (Reference 16). The first was the moderately loaded, free-running propeller, whilst the second application was for wake adapted propellers.

In the first case, it is deduced from a consideration of the energy balance over the propeller and its slipstream that the pitch of the vortex sheets coincides with the hydrodynamic pitch angle of the section. Furthermore, it is shown that the Betz condition holds at the lifting line as well as in the ultimate wake, which implies the regularity of the vortex sheets in terms of their helical shape, so that the condition of normality holds. That is, that the induced velocities are normal to the incident flow on to the section. Lerbs then applied this basis to the solution of optimum and non-optimum propellers, and in this respect a strong agreement was shown to exist between this work and the earlier studies of Goldstein.

For the case of the wake adapted propeller, the propeller is considered to operate in a wake field which varies radially, but is constant circumferentially: that is, the normal design condition. For this case the optimum and non-optimum loading cases were considered which led, in the former case, to the condition

$$\frac{\tan \beta}{\tan \beta_i} = c\sqrt{\frac{1 - w_T(x)}{1 - t(x)}} \quad (8.34)$$

where $w_T(x)$ and $t(x)$ are the Taylor wake fraction and thrust deduction respectively and c is a constant which requires being determined in each case. An approximation to c can be calculated according to

$$c = \eta_i \sqrt{\frac{1 - t}{1 - w_0}} \quad (8.35)$$

where η_i is the ideal propeller efficiency and w_0 and t are the effective wake and thrust deduction factors respectively. In the case of the non-optimum propeller, in which the problem is that of determining the induced velocity components when the powering conditions, wake field and the character of the circulation distribution are known, the induced velocity components become

$$\frac{w_{a,t}}{V} = \frac{k}{1 - x_h} \sum_{m=1}^{\infty} mF_m h_m^{a,t} \quad (8.36)$$

in which the constant k, defined by equation (8.37), is determined from the given powering conditions characterized by the power coefficient C_p:

$$k^2 + k(1 - x_h)\frac{\int_{x_h}^{1} F_x[1 - w(x)]dx}{\int_{x_h}^{1} F_x\left(\sum_{m=1}^{\infty} mF_m h_m^a\right)dx} - C_{pi}\left(\frac{1 - x_h}{4z}\right)\frac{J_S}{\int_{x_h}^{1} F_x\left(\sum_{m=1}^{\infty} mF_m h_m^a\right)dx} = 0$$

and in order to determine the functions h_m^a the hydrodynamic pitch angle β_i is derived from

$$\tan \beta_i = \frac{[1 - w(x)] + \frac{k}{(1 - x_h)}\sum_{m=1}^{\infty} mF_m h_m^a}{\frac{x}{J_s} - \frac{k}{(1 - x_h)}\sum_{m=1}^{\infty} mF_m h_m^t}$$

The values of k and h_m are in this procedure determined from an iterative procedure. In the above equation the circulation characterizing function F is used since the exact distribution is unknown. The characterizing function is related to the circulation distribution by a constant term k such that

$$G(x) = kF(x) \quad (8.37)$$

Clearly, in the case of the open water propeller the terms $t(x)$ and $w(x)$ are zero for all values of the non-dimensional radius x. Hence equation (8.34) reduces to the simpler expression $\tan \beta/\tan \beta_i =$ constant, and equation (8.36) also reduces accordingly.

8.6 ECKHARDT AND MORGAN'S DESIGN METHOD (1955)

The mid-1950s saw the introduction of several design methods for propellers. This activity was largely driven by the increases in propulsive power being transmitted at that time coupled with the new capabilities in terms of the mathematical analysis of propeller action. Among these new methods were those by Burrill,[13] which to a large extent was an extension to his earlier analysis procedure, van Manen and van Lammeren[17] and Eckhart and Morgan.[18] This later procedure found considerable favor at the time of its introduction, as did the other methods in their countries of origin.

Eckhardt and Morgan's method is an approximate design method which relies on two basic assumptions of propeller action. The first is that the slipstream does not contract under the action of the propeller, whilst the second is that the condition of normality of the induced velocity applies. Both of these assumptions, therefore, confine the method to the design of light and moderately loaded propellers.

The design algorithm is outlined in Figure 8.8. In their paper the authors put forward two basic procedures for propeller design: one for open water propellers and the other for a wake adapted propeller. In either case the design commences by choosing the most appropriate diameter, obtained perhaps from standard open water series data such as the Troost series, and the blade number which will have been selected, amongst other considerations on the basis of the ship and machinery natural vibration frequencies.

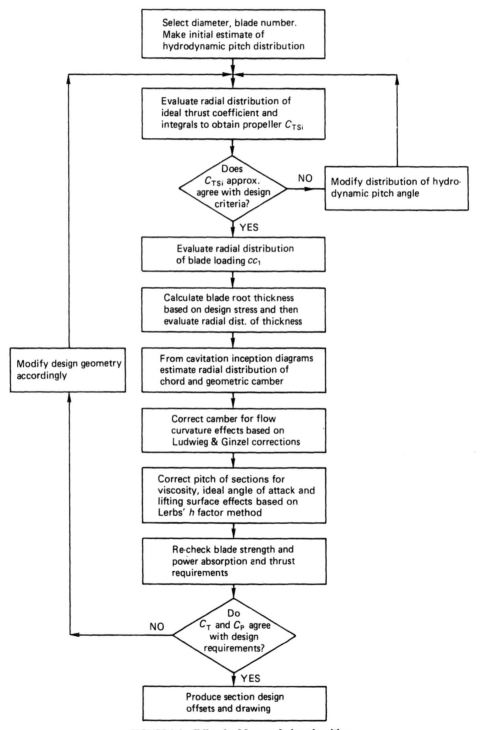

FIGURE 8.8 Edhardt–Morgan design algorithm.

Additionally, at this preliminary stage in the design the hub or boss diameter will have been chosen, as will the initial ideas on the radial distributions of rake and skew. At the time that this work was first proposed the full hydrodynamic implications of rake, but more particularly skew, were very much less well understood than they are today.

In outlining the Eckhardt and Morgan procedure, the wake adapted version of the procedure will be followed, since the open water case is a particular solution of the more general wake adapted problem. For wake adapted design purposes the circumferential average wake fraction $\overline{w}_x$ at each radius is used as an input to the design procedure. Hence the radial distribution of advance angle β can be readily calculated using equation (6.20), as can the propeller thrust loading coefficient C_{TS}, equation (6.6), but based on ship speed V_s rather than advance speed V_a. Lerbs derived a relationship for the non-viscous flow case which can be deduced from equations (8.34) and (8.35) for the hydrodynamic pitch angle based on the ideal efficiency:

$$\tan \beta_i \simeq \frac{\tan \beta}{\eta_i} \sqrt{\frac{1 - \overline{w}}{1 - \overline{w}_x}} \quad (8.38)$$

The ideal efficiency η_i is estimated from Kramer's diagrams, assuming a non-viscous thrust coefficient C_{TSi}, which is some 2–6 per cent greater than that based on the ship speed and calculated above. Kramer's diagrams are based on an extension of Goldstein's approach, and whilst some reservations have been expressed concerning their accuracy, especially at high blade numbers, they are suitable for a first approximation purpose. Following on from this initial estimate of the radial hydrodynamic pitch distribution, the elemental ideal thrust loading coefficients for each section are computed and the ideal propeller thrust coefficient C_{TSi} evaluated from

$$C_{TSi} = 8 \int_{x_h}^{1.0} Kx \frac{u_t}{2V_s} \left(\frac{x\pi}{J_s} - \frac{u_t}{2V_s} \right) dx \quad (8.39)$$

where

$$\frac{u_t}{2V_s} = \frac{(1 - w_x)\sin \beta_i \, \sin(\beta_i - \beta)}{\sin \beta}$$

and K is the Goldstein function. The latter equation follows from the implied condition of normality.

The value of C_{TSi} can then be compared to the original design assumption of being within the range of 2–6 per cent greater than C_{TS}, and if there is a significant difference a new value of the hydrodynamic pitch angle β_i at $0.7R$ can be estimated from the relationship

$$(\tan \beta_i)_{j+1} \simeq (\tan \beta_i)_j \left[1 - \frac{(C_{Ti})_{desired} - (C_{Ti})_{calc}}{5(C_{Ti})_{desired}} \right] \quad (8.40)$$

Hence, upon convergence of the values of the ideal ship thrust coefficient, the product cc_l, which effectively represents the load of the section under a given inflow condition, see equation (7.3), can be evaluated from

$$cc_l = \frac{4\pi D}{Z} \left[\frac{Kx(u_t/2V_s)}{x\pi/J_s - u_t/2V_s} \right] \cos \beta_i \quad (8.41)$$

Now in order to derive the propeller blade chord lengths from equation (8.41) it is necessary to ensure compatibility with the cavitation criteria relating to the design. Eckhardt and Morgan did this by making use of incipient cavitation charts which had been derived from theoretical pressure distribution calculations for a series of standard sections forms. The forms presented in their paper relate to the National Advisory Committee for Aeronautics (NACA) 16 section with either an $a = 0.8$ or $a = 1.0$ mean line and the NACA 66 section with an $a = 0.8$ mean line. Figure 8.9 shows the essential features of the diagram in which the minimum cavitation number at which cavitation will occur, σ_{min}, is plotted as abscissa and the function cc_l/t forms the ordinate. The thickness to chord and camber chord ratios act as parameters in the manner shown in Figure 8.9. As far as the propeller blade section cavitation criteria are concerned these are calculated at the top dead center position in the propeller disc. Consequently, the blade section cavitation numbers, based on the velocities at the lifting line, are calculated for this location in the propeller disc since these will represent a combination of the worst static head position coupled with a mean dynamic head.

With regard to section thickness, Eckhardt and Morgan base this on the calculation of the root thickness using Taylor's approach, from which they derive the radial distribution of thickness. Consequently, since the section thicknesses and their cavitation numbers are known the values of (t/c) and (y/c) can be deduced from inception diagrams of the type shown in Figure 8.9, and hence the radial distribution of chord length can be determined by a process of direct calculation followed by fairing results.

The values of camber chord ratio derived from the inception diagram are in fact two-dimensional values operating in rectilinear flow and, therefore, need to be corrected for flow curvature effects. This is done by introducing a relationship of the form

$$\frac{y}{c} = k_1 k_2 \left(\frac{y}{c}\right)' \quad (8.42)$$

in which (y/c) is the actual section camber chord ratio, $(y/c)'$ is the two-dimensional value and k_1 and k_2 are correction factors derived from the work of Ludwieg and Ginzel.[19] The factor $k_1 = f(J_i, A_E/A_O)$ whilst $k_2 = g(x, A_E/A_O)$. Ludwieg and Ginzel addressed themselves to the relative effectiveness of a camber line in curved and straight flows: in a curved flow the camber is less effective than in a straight or rectilinear flow. Their analysis was based on the effectiveness of circular arc camber lines operating at their shock-free entry conditions by evaluating the

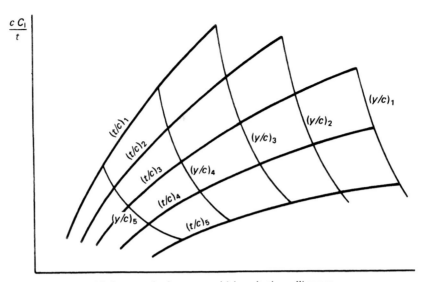

FIGURE 8.9 Cavitation inception diagram.

streamline curvature, or the change in downwash in the direction of the flow.

In addition to the camber correction factor, the pitch of the sections needs corrections for viscosity, for the ideal angle of attack of the camber line, and for the change in curvature over the chord, for which the Ludwieg and Ginzel corrections are insufficient. The first two corrections are dealt with by the authors with a single correction factor to determine a pitch correction:

$$\alpha_1 = k_3 c_1 \qquad (8.43)$$

where k_3 depends upon the shape of the mean line. With regard to the latter effect, Lerbs, using Wessinger's simplified lifting surface theory, defined the further pitch correction angle α_2. This correction is necessary since the Ludwieg and Ginzel correction was based upon considerations of the flow curvature at the mid-point of the section and experiments with propellers showed that they were under-pitched with this correction. Lerbs used Weissinger's theory in reverse since the bound circulation is known from lifting line theory. Lerbs satisfied the boundary condition at the $0.75c$ position on the chord by an additional angle of attack, assuming that the bound vortex was sited at the $0.25c$ point. This correction is made in two parts: one due to the effects of the bound vortices and the other due to the free vortex system, such that

$$\alpha_2 = \alpha_b = \alpha_f - (\alpha_i - \alpha_o) \qquad (8.44)$$

The bound vortex contribution α_b is defined by the equation

$$\alpha_b = \frac{\sin \beta_i}{2} \sum_1^z \left[\left(\frac{c}{D} \sin \theta_z - 0.7 \cos \beta_i \cos \theta_z \right) \right.$$
$$\left. \times \int_{xh}^{1.0} \frac{c \, dx}{(P/R)^3} \right] \qquad (8.45)$$

in which θ_z is the angular position of the blade and

$$\left(\frac{P}{R}\right)^3 = \left[x^2 + \left(\frac{c}{D}\right)^2 + 0.49 - 2 \right.$$
$$\left. \times \left(\frac{c}{D} \cos \theta_z \cos \beta_i + 0.7 \sin \theta_z\right) x \right]^{3/2}$$

In evaluating equation (8.45) the calculation is made for the blade in the 90°, or athwart ship position, and the effects of the other blades on the bound vortex of this blade are determined.

For the free vortex contribution α_f this is determined from the approximation

$$\alpha_f \simeq (\beta_i - \beta) \frac{2}{1 + \cos^2 \beta_i \left(\frac{2}{h} - 1\right)} \text{ rad} \qquad (8.46)$$

in which the parameter h is a function of θ_β, and θ_β is defined as

$$\theta_\beta = \tan^{-1}\left(\frac{0.7}{\sin \beta_i} \frac{D}{c}\right)$$

Given these two correction factors α_b and α_f, equation (8.44) can be evaluated to give a total pitch correction factor $\Delta P/D$. This factor is applied at $0.7R$, and the same percentage correction applied to the other blade radii. The correction is defined as

$$1 + \frac{\Delta P/D}{P/D} = \frac{\tan(\beta_i + \alpha_2)_{0.7}}{(\tan \beta_i)_{0.7}} \qquad (8.47)$$

Using this and incorporating the correction α_1 for viscosity and ideal angle of attack the final pitch can be computed as

$$P/D = \pi x \tan(\beta_i + \alpha_1)\left(1 + \frac{\Delta P/D}{P/D}\right) \qquad (8.48)$$

Theoretical and Analytical Methods Relating to Propeller Action

TABLE 8.2 Minimum Energy Loss Assumptions

	Variable Wake	Uniform Wake
Burrill	$x\pi \tan \varepsilon = $ constant	$x\pi \tan \varepsilon = $ constant
Eckhardt and Morgan	$\dfrac{\tan \beta}{\tan \beta_i} = \eta_i \sqrt{\dfrac{1 - \overline{w}_x}{1 - w}}$	$\dfrac{\tan \beta}{\tan \beta_i} = \eta_i = $ constant
Hill	$\dfrac{\tan \beta}{\tan \beta_i} = \eta_i = $ constant	$\dfrac{\tan \beta}{\tan \beta_i} = \eta_i = $ constant
van Manen et al.	$\dfrac{\tan \beta}{\tan \beta_i} = \eta_i = $ constant	$\dfrac{\tan \beta}{\tan \beta_i} = \eta_i = $ constant

With equation (8.48) the design is essentially complete; however, it is indeed essential to ensure that the final values of thrust and power coefficients agree with the initial design parameters. If this is the case, then the blade section geometry can be calculated; if not, then another iteration of the design is required.

A design method of this type produces an answer to a particular design problem; it does not, however, produce a unique solution relative to other methods. To illustrate this problem McCarthy[20] compared four contemporary design methods and the solutions they produced for two particular design problems — a single-screw tanker and a twin-screw liner. The methods compared were those of Burrill,[13] Eckhardt and Morgan, van Manen and van Lammeren[17] and the earlier work of Hill.[14] Each of the methods used contemporary calculation procedures; however, they differed in their initial minimum energy loss assumptions and also in the correction factors applied to the camber line and the section pitch angle. In this later respect Burrill used the Gutsche data, Eckhardt and Morgan as has been seen used both the Ludwieg and Ginzel camber correction and the Lerbs lifting surface correction, van Manen used Ludwieg and Ginzel and Hill used empirically derived factors. With regard to the minimum energy loss assumptions these are shown in Table 8.2.

The Burrill condition is essentially the Betz condition, whilst the methods of Eckhardt and Morgan and van Manen use the Betz condition for uniform wake. Hill initially used the Betz condition; however, the thrust distribution is then altered so as to reduce tip loading, which then moves away from the initial assumption. For the variable wake case Burrill and also Eckhardt and Morgan assume the local thrust deduction factor to be constant over the disc, whereas, in contrast to this, the van Manen method assumes a distribution

$$\frac{1 - t_x}{1 - \overline{t}} = \left(\frac{1 - w_x}{1 - \overline{w}}\right)^{1/4} \tag{8.49}$$

The changes caused by these various assumptions and corrections can be seen, for example, in Figure 8.10, which shows the radial distribution of $cc_{L/D}$ for the linear example of McCarthy. In this example all the propellers had the same blade number and diameter, and all were designed for the same powering condition assuming a van Lammeren radial distribution of mean wake fraction. While Figure 8.10 is instructive in showing the non-convergence of the methods to a single solution the relative trends will tend to change from one design example to another; consequently, when using design methods of this type the experience of the designer is an important input to the procedure, as is the analysis of the design, either by mathematical models in the various positions around the propeller disc or by model test.

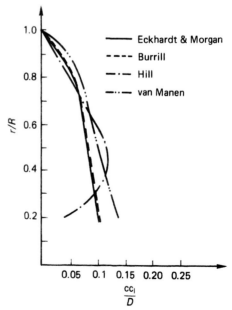

FIGURE 8.10 Comparison of propeller design methods. *Reference 20.*

8.7 LIFTING SURFACE CORRECTION FACTORS – MORGAN ET AL.

Figure 8.10 showed a divergence between design methods in the calculation of the radial loading parameter cc_l/D. If this is carried a stage further the divergence becomes increasingly larger in the determination of camber and angle of attack and hence section pitch. This is in part due to the variety of correction procedures adopted for the lifting line model: for example, those by Gutsche, Ludwieg and Ginzel and also by Lerbs.

To help in rationalizing these various methods of correcting lifting line results Morgan et al.[21] derived a set of correction factors, based on the results of lifting surface theory, for camber, ideal angle due to loading and ideal angle due to thickness.

In general terms the lifting surface approach to propeller analysis can be seen from Figure 8.11, which shows the salient features of the more recent models. Earlier models adopted a fan lattice approach and distributed singularities along the camber line, which clearly was not as satisfactory as the more recent theoretical formulations.

The mathematical model that Morgan et al. used was based on the work of Cheng[22] and Kerwin and Leopold[23] and comprised two distinct components. The blade loading model assumed a distribution of bound vortices to cover the blades and a system of free vortices to be shed from these bound vortices downstream along helical paths. The second part of the model, relating to the effects of blade thickness, assumed a network of sources and sinks to be distributed over the blades. In addition to the normal inviscid and incompressible assumptions of most propeller theories, the free stream inflow velocity was assumed to be axisymmetric and steady; that is, the propeller is assumed to be proceeding at a uniform velocity. Furthermore, in the model each of the blades is replaced by a distribution of bound vortices such that the circulation distribution varied both radially and chordally over the blades. The restriction on these bound vortices is, therefore, that the integral of the local circulation at a particular radial location between the leading and trailing edges is equal to the value $G(r)$ required by the lifting line design requirement:

$$\int_{TE}^{LE} G_b(r, c)\, dc = G(r) \qquad (8.50)$$

This is analogous to the thin aerofoil theory requirement discussed in the previous chapter. In their work the authors used the NACA $a = 0.8$ mean line distribution, since this line has the benefit of developing approximately all of its theoretical lift in viscous flow whereas the $a = 1.0$ mean line, for example, develops only about 74 per cent of its theoretical lift. This would, therefore, introduce viscous

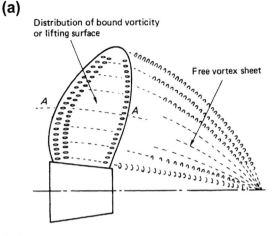

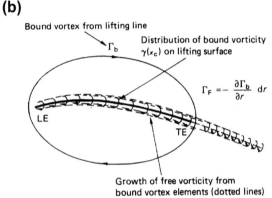

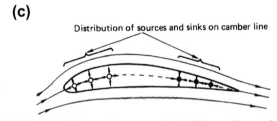

FIGURE 8.11 Lifting surface concept: (a) lifting surface model for a propeller blade; (b) lifting surface concept at section $A-A$ to simulate blade loading and (c) source–sink distribution to simulate section thickness.

flow considerations, leading to considerably more complicated numerical computations. Since the bound vorticity varies at each point on the blade surface, by Helmholtz's vortex theorems a free vortex must be shed at each point on the blade. Hence, at any given radius the strength of the free vorticity builds up at the radius until it reaches the trailing edge, where its strength is equal to the rate of change of bound circulation at that radius

$$G_f(r) = -\frac{dG(r)}{dr}\, dr \qquad (8.51)$$

which is analogous to equation (8.27).

With regard to the free vortex system at each radius they are considered to lie on a helical path defined by a constant diameter and pitch; the pitch, however, is allowed to vary in the radial direction. Therefore, since the slipstream contraction and the centrifugal effects on the shape of the free vortex system are ignored, their analysis is consistent with moderately loaded propeller theory. In their model, as with many others of this type, the boundary conditions on the blade are linearized and this implies that only small deviations exist between the lifting surface representing the blade and the hydrodynamic pitch angle. This is an analogous situation to that discussed for this aerofoil theory in Chapter 7, where the conditions are satisfied on the profile chord and not on the profile. Finally, the hub is assumed to be small in this approach and is, consequently, ignored; also the propeller rake is not considered. With respect to blade pitch angles it is assumed that the pitch of the blades and of the free trailing vortex sheets is the hydrodynamic pitch angle obtained from lifting line considerations.

The method of analysis essentially used the basis of equations (8.50) and (8.51) to define the circulation acting on the system from which the velocities at points on the lifting surface, representing the blade, can be calculated by applications of the Biot–Savart law to each of the vortex systems emanating from each blade. In this analysis the radial components of velocity are neglected. The analysis, in a not dissimilar way to that outlined previously for the thin aerofoil theory, derives from lifting surface theory the relevant flow velocities and compares them to the resulting induced velocity derived from lifting line theory. From this comparison it develops two geometric correction factors, one for the maximum camber ordinate and the other for the ideal angle of attack for use when these have been derived from purely lifting line studies. This is done using the boundary condition at each section:

$$\alpha_i(r) + \frac{\partial y_p(r, x_c)}{\partial x_c} = \frac{u_n}{V_r}(r, x_c) - \frac{u}{V_r}(r) \quad (8.52)$$

where α_i is the ideal angle of attack, y_p is the chordwise camber ordinate at a position x_c along the chord. V_r is the resultant inflow velocity to the blade section and u_n and u are resultant induced velocities normal to the section chord and induced velocities from lifting line theory. Figure 8.12 demonstrates these parameters for the sake of clarity. It is of interest to compare equation (8.52) with equation (7.15) so as to appreciate the similarities in the two computational procedures.

The camber correction $k_c(r)$ derived from this procedure is defined by

$$k_c(r) = \frac{\text{maximum camber ordinate}}{\text{maximum 2D camber ordinate}} \quad (8.53)$$

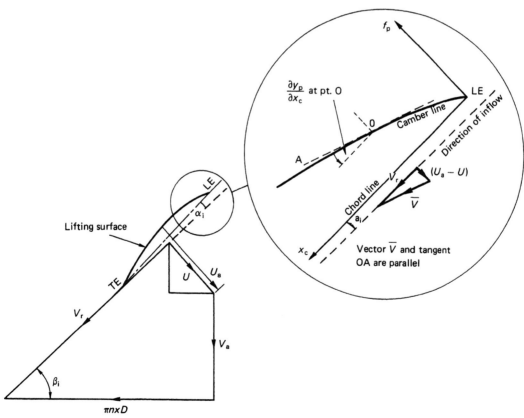

FIGURE 8.12 Boundary condition for determination of camber line.

in which the maximum two-dimensional camber ordinate is that derived from a consideration of the section lift coefficient in relation to the aerofoil data given in, say, Reference 24. At other positions along the chord the camber ordinates are scaled on a pro rata basis.

The second correction for the ideal angle of attack $k_\alpha(r)$, to give shock-free entry, is defined as follows:

$$k_\alpha(r) = \frac{\text{section ideal angle of attack}}{\text{section 2D ideal angle of attack for } c_1 = 1.0} \quad (8.54)$$

In keeping with the mathematical model the denominator of equation (8.54) relates to the NACA $a = 0.8$ mean line at an ideal angle of attack for $c_1 = 1.0$.

For the final correction of the three developed by Morgan et al., that relating to blade thickness effects, this was determined by introducing a source–sink system distributed after the manner developed by Kewin and Leopold[23] and demonstrated by Figure 8.11(c). In this case the induced velocities were again calculated at a point on the lifting surface since the source strength distribution is known from the normal linearized aerofoil theory approximation. As with the previous two corrections the radial component of velocity is ignored. The effects of blade thickness over the thin aerofoil case can then be studied by defining a further linear boundary condition, which is analogous to equation (8.52):

$$\alpha_t(r) + \frac{\partial y_{p_t}}{\partial x_c}(r, x_c) = \frac{u_{nt}}{V_r}(r, x_c) \quad (8.55)$$

where α_t is the ideal angle induced by thickness, y_{pt} is the change in camber along the chord x_c due to thickness effects and u_{nt} is the induced velocity normal to the section chord.

Calculations show that introducing a finite blade section thickness causes an increase in the inflow angle to maintain the same loading together with a change in the camber: this latter effect is, however, small for all cases except for small values of pitch ratio. The thickness correction factor $k_t(r)$ is made independent of the magnitude of the thickness by dividing it by blade thickness fraction (t_F):

$$k_t(r) = \frac{1}{t_F}\int_0^1 \frac{u_{nt}}{V_r}(r, x_c)\,dx_c \quad (8.56)$$

From equation (8.56) the required correction to the ideal angle of attack, which is added to the sum of the hydrodynamic pitch and ideal angles, is calculated by

$$\alpha_{it} = k_t(r)t_F \quad (8.57)$$

It will be seen that $k_t(r)$ is a function of propeller loading, since V_r is also a function of loading; however, this is small and can be ignored.

To provide data for design purposes Morgan et al. applied the Cheng and Kerwin and Leopold procedures to a series of open water propellers of constant hydrodynamic pitch having a non-dimensional hub diameter of $0.2R$. The lifting line calculations were based on the induction factor method of Lerbs and then the lifting surface calculations were made on the basis of the loading and pitch distributions derived from Lerbs' method.

The lifting surface corrections were calculated for propellers having four, five and six blades, expanded area ratios from 0.35–1.15, hydrodynamic pitch ratios of 0.4–2.0, and for symmetrical and skewed blades, the latter having skew angles of 7°, 14° and 21°. The design of the propellers was based on the NACA $a = 0.8$ mean line together with the NACA 66 (modified) thickness distribution. The radial thickness distribution was taken to be linear and the blade outlines were chosen to be slightly wider toward the tip than the Wageningen B series, as can be seen by comparing Table 8.3 with the appropriate data in Table 6.5.

The radial distribution of skew was chosen for this series such that the blade section mid-chord line followed a circular arc in the expanded plane. Figure 8.13 shows a typical example of the correction factors for a five-bladed propeller at $0.7R$ and having a propeller skew angle of 21°.

In general terms the three-dimensional camber and ideal angles are usually larger than the two-dimensional values when developing the same lift coefficient. The correction factors tend to increase with expanded area ratio and k_c and k_α decrease with increasing blade number. Blade thickness, in general, tends to induce a positive angle to the flow. This addition to the ideal angle is largest near the blade root and decreases to negligible values towards the blade tip: the correction increases with increasing blade

TABLE 8.3 Blade Chord Coefficient for Morgan et al. Series

r/R	C(r)
1.0	0
0.95	1.5362
0.90	1.8931
0.80	2.1719
0.70	2.2320
0.60	2.1926
0.50	2.0967
0.40	1.9648
0.30	1.8082
0.20	1.6338

Theoretical and Analytical Methods Relating to Propeller Action

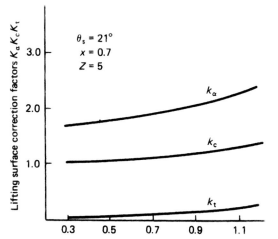

FIGURE 8.13 Lifting surface correction factors derived by Morgan et al.

Van Oossanen found that although no data was given for a blade number of seven, because of the regularity of the curves extrapolation to a blade number of seven was possible. Hence, the limits for equation (8.58) are suggested by van Oossanen as being

$$3 \leq Z \leq 7$$
$$0.35 \leq A_E/A_O \leq 1.15$$
$$0.4 \leq \pi\lambda_i \leq 2.0$$
$$0 \leq \tan\theta_{sx} \leq 1.0256 - [1.0518 - (x - 0.2)^2]$$

where θ_{sx} is the blade section skew angle and λ_i is the hydrodynamic advance coefficient $x \tan \beta_i$.

These polynomials are limited to moderate skew propellers. Cummings et al.[27] extended the range of these corrections into the highly skewed propeller range. Unfortunately however, the range of applicability is not as great as in the previous works: blade numbers 4–6, the parameter $\pi\lambda_i = 0.8$ and 1.2 and a single expanded area ratio of 0.75. van Oossanen, using similar techniques, developed polynomials for each of the cases $\pi\lambda_i = 0.8$ and 1.2, of the form

$$k_c, k_\alpha, k_t = \sum_{i=1} c_i x_i^{a_i} z^{b_i} (\theta_{sx})^{d_i} \quad (8.59)$$

number. Skew induces an inflow angle which necessitates a pitch change which is positive towards the blade root and negative towards the tip.

A polynomial representation of these correction factors offers many advantages for design purposes. To answer this need van Oossanen[25] derived by means of multiple regression analysis a polynomial representation of the correction factors calculated both by Morgan et al.[21] and Minsaas and Slattelid,[26] resulting in an expression of the form

$$k_c, k_\alpha, k_t = \sum_{i=1} C_i Z^{S_i} (\tan\theta_{sx})^{t_i} (A_E/A_O)^{u_i} (\lambda_i)^{v_i} \quad (8.58)$$

The original factors published by Cummings, Morgan and Boswell are given in Table 8.4.

TABLE 8.4 k_c, k_α and k_t Factors Derived by Cummings for Highly Skewed Propellers

Skew θ_s (%)		Correction Factors for Highly Skewed Propellers, $\pi\lambda_i = 0.8$, EAR $= 0.75$								
		Z = 4			Z = 5			Z = 6		
	r	k_c	k_α	k_t	k_c	k_α	k_t	k_c	k_α	k_t
0	0.3	1.540	1.961	0.521	1.490	1.795	0.783	1.469	1.699	1.080
	0.4	1.237	1.711	0.370	1.169	1.549	0.540	1.129	1.452	0.732
	0.5	1.259	1.591	0.253	1.143	1.430	0.357	1.076	1.311	0.474
	0.6	1.338	1.589	0.166	1.185	1.396	0.225	1.090	1.285	0.291
	0.7	1.498	1.682	0.105	1.307	1.468	0.137	1.190	1.318	0.172
	0.8	1.778	1.844	0.067	1.534	1.623	0.084	1.376	1.449	0.103
	0.9	2.389	2.356	0.046	2.062	2.057	0.058	1.842	1.832	0.070
50	0.3	1.588	7.227	0.491	1.517	5.985	0.719	1.476	5.148	0.971
	0.4	1.310	4.737	0.350	1.219	3.909	0.499	1.165	3.383	0.660
	0.5	1.310	4.182	0.237	1.186	3.504	0.328	1.110	3.050	0.425
	0.6	1.422	3.257	0.151	1.249	2.789	0.204	1.145	2.432	0.257

(Continued)

TABLE 8.4 k_c, k_α and k_t Factors Derived by Cummings for Highly Skewed Propellers—cont'd

Skew θ_s (%)		Correction Factors for Highly Skewed Propellers, $\pi\lambda_i = 0.8$, EAR = 0.75								
		Z = 4			Z = 5			Z = 6		
	r	k_c	k_α	k_t	k_c	k_α	k_t	k_c	k_α	k_t
	0.7	1.602	1.948	0.091	1.389	1.778	0.119	1.240	1.636	0.147
	0.8	1.881	0.613	0.057	1.619	0.808	0.071	1.435	0.884	0.086
	0.9	2.557	−5.100	0.045	2.188	−3.607	0.056	1.935	−2.718	0.064
100	0.3	1.838	11.914	0.477	1.683	9.804	0.683	1.585	8.396	0.912
	0.4	1.562	7.580	0.354	1.396	6.166	0.499	1.286	5.235	0.657
	0.5	1.608	6.625	0.247	1.398	5.481	0.345	1.266	4.700	0.448
	0.6	1.747	4.969	0.161	1.493	4.179	0.221	1.333	3.697	0.284
	0.7	1.949	2.427	0.097	1.648	2.258	0.432	1.154	2.039	0.167
	0.8	2.219	−0.105	0.059	1.867	0.281	0.079	1.640	0.547	0.097
	0.9	2.944	−11.674	0.050	2.446	−8.766	0.064	2.164	−6.923	0.076

Skew θ_s (%)		Correction Factors for Highly Skewed Propellers, $\pi\lambda_i = 1.2$, EAR = 0.75								
		Z = 4			Z = 5			Z = 6		
	r	k_c	k_α	k_t	k_c	k_α	k_t	k_c	k_α	k_t
0	0.3	1.640	1.815	0.382	1.565	1.687	0.580	1.532	1.607	0.827
	0.4	1.330	1.691	0.308	1.257	1.556	0.459	1.223	1.469	0.631
	0.5	1.354	1.636	0.237	1.242	1.483	0.341	1.176	1.394	0.458
	0.6	1.431	1.674	0.174	1.285	1.493	0.241	1.196	1.390	0.315
	0.7	1.573	1.749	0.122	1.392	1.556	0.161	1.277	1.416	0.205
	0.8	1.803	1.886	0.083	1.580	1.661	0.106	1.430	1.530	0.131
	0.9	2.360	2.320	0.063	2.030	2.029	0.080	1.818	1.817	0.098
50	0.3	1.714	7.726	0.345	1.605	6.187	0.517	1.552	5.610	0.733
	0.4	1.427	5.124	0.277	1.318	4.131	0.404	1.270	3.402	0.572
	0.5	1.434	4.439	0.211	1.319	3.960	0.298	1.217	3.133	0.419
	0.6	1.538	3.433	0.150	1.365	2.884	0.206	1.253	3.104	0.285
	0.7	1.703	2.078	0.101	1.485	1.889	0.134	1.361	1.596	0.179
	0.8	1.930	0.486	0.067	1.684	0.778	0.087	1.509	0.802	0.111
	0.9	2.560	−5.527	0.053	2.169	−4.112	0.069	1.918	−2.549	0.090
100	0.3	2.028	12.649	0.340	1.826	10.337	0.510	1.737	8.784	0.701
	0.4	1.733	8.241	0.274	1.548	6.645	0.401	1.347	5.612	0.540
	0.5	1.779	7.034	0.203	1.551	5.826	0.293	1.350	5.036	0.389
	0.6	1.913	5.206	0.136	1.646	4.496	0.195	1.433	3.933	0.256
	0.7	2.090	2.555	0.082	1.781	2.343	0.117	1.554	2.241	0.155
	0.8	2.337	−0.212	0.047	1.981	0.241	0.068	1.719	0.641	0.090
	0.9	2.998	−12.356	0.040	2.488	−5.222	0.057	2.006	−7.293	0.073

Taken from Reference 27.

8.8 LIFTING SURFACE MODELS

Figure 8.11 showed in a conceptual way the basis of the lifting surface model. Essentially the blade is replaced by an infinitely thin surface which takes the form of the blade camber line and upon which a distribution of vorticity is placed in both the spanwise and chordal directions. Early models of this type used this basis for their formulations and the solution of the flow problem was in many ways analogous to the thin aerofoil approach discussed in Chapter 7. Later lifting surface models then introduced a distribution of sources and sinks in the chordal directions so that, in conjunction with the incident flow field, the section thickness distribution could be simulated and hence the associated blade surface pressure field approximated. The use of lifting surface models, as indeed for other models of propeller action, is for both the solution of the design and analysis problems. In the design problem the geometry of the blade is only partially known in so far as the radial distributions of chord, rake skew and section thickness distributions are either fully or approximately known. The radial distribution of pitch and the chordwise and radial distribution of camber remain to be determined. In order to solve the design problem the source and vortex distributions representing the blades and their wake need to be placed on suitable reference surfaces to enable the induced velocity field to be calculated. Linear theories assume that the perturbation velocities due to the propeller are small compared with the inflow velocities. In this way the blades and their wake can simply be projected onto stream surfaces formed by the undisturbed flow. However, in the majority of practical design cases the resulting blade geometry deviates substantially from this assumption, and as a consequence the linear theory is generally not sufficiently accurate.

The alternative problem differs from the design solution in that the propeller geometry is completely known and we are required to determine the flow field generated under known conditions of advance and rotational speed. The analysis exercise divides into two comparatively well-defined types: the steady flow and the unsteady flow solutions. In the former case the governing equations are the same as in the design problem, with the exception that the unknowns are reversed. The circulation distribution over the blades is now the unknown. As a consequence, the singular integral which gives the velocity induced by a known distribution of circulation in the design problem becomes an integral equation in the analysis problem, which is solved numerically by replacing it with a system of linear algebraic equations. In the case of unsteady propeller flows their solution is complicated by the presence of shed vorticity in the blade wake that depends on the past history of the circulation around the blades. Accordingly, in unsteady theory the propeller blades are assumed to generate lift in gusts, for which an extensive literature exists for the general problem: for example, McCroskey,[28] Crighton[29] and the widely used Sear's function. The unsteadiness of the incident flow is characterized by the non-dimensional parameter k, termed the reduced frequency parameter. This parameter is defined as the product of the local semi-chord, and the frequency of encounter divided by the relative inflow speed. For the purposes of unsteady flow calculations the wake or inflow velocity field is characterized at each radial station by the harmonic components of the circumferential velocity distribution (Figure 5.3), and with the assumption that the propeller responds linearly to changes in inflow, the unsteady flow problem reduces to one of estimating the response of the propeller to each harmonic. In the case of a typical marine propeller the reduced frequency k corresponding to the first harmonic is of the order of 0.5, whilst the value corresponding to the blade rate harmonic will be around two or three. From classical two-dimensional theory of an aerofoil encountering sinusoidal gusts, it is known that the effects of flow unsteadiness become significant for values of k greater than 0.1. As a consequence the response of a propeller to all circumferential harmonics of the wake field is unsteady in the sense that the lift generated from the sections is considerably smaller than that predicted from the equivalent quasi-steady value and is shifted in phase relative to the inflow.

In the early 1960s many lifting surface procedures made their appearance due mainly to the various computational capabilities that became available generally at that time. Prior to this, the work of Strscheletsky,[15] Guilloton[30] and Sparenberg[31] laid the foundations for the development of the method. Pien[32] is generally credited with producing the first of the lifting surface theories subsequent to 1960. The basis of this method is that the bound circulation can be assumed to be distributed over the chord of the mean line, the direction of the chord being given by the hydrodynamic pitch angle derived from a separate lifting line calculation. This lifting line calculation was also used to establish the radial distribution of bound circulation. In Pien's method the free vortices are considered to start at the leading edge of the surface and are then continued into the slipstream in the form of helical vortex sheets. Using this theoretical model the required distortion of the chord into the required mean line can be determined by solving the system of integral equations defining the velocities along the chord induced by the system of bound and free vortices. The theory is linearized in the sense that a second approximation is not made using the vortex distribution along the induced mean line.

Pien's work was followed by that of Kerwin,[33] van Manen and Bakker,[34] Yamazaki,[35] English,[36] Cheng,[22] Murray,[37] Hanaoka,[38] van Gent[39,40] and a succession of papers by Breslin, Tsakonas and Jacobs spanning something over thirty years continuous development of the method. Indeed, much of this latter development is captured in the book by Breslin and Anderson.[100] Typical of modern lifting surface theories is that by Brockett.[41] In this method the solid boundary effects of the hub are ignored; this is consistent with the generally small magnitude of the forces being produced by the inner regions of the blade. Furthermore, in Brockett's approach it is assumed that the blades are thin, which then permits the singularities which are distributed on both sides of the blades to collapse into a single sheet. The source strengths, located on this single sheet, are directly proportional to the derivative of the thickness function in the direction of flow; conversely the vortex strengths are defined. In the method a helicoidal blade reference surface is defined together with an arbitrary specified radial distribution of pitch. The trailing vortex sheet comprises a set of constant radius helical lines whose pitch is to be chosen to correspond either to that of the undisturbed inflow or to the pitch of the blade reference surface. Brockett uses a direct numerical integration procedure for evaluating the induced velocities. However, due to the non-singular nature of the integrals over the other blades and the trailing vortex sheets the integrands are approximated over prescribed sets of chordwise and radial intervals by trigonometric polynomials. The integrations necessary for both the induced velocities and the camber line form are undertaken using predetermined weight functions. Unfortunately, the integral for the induced velocity at a point on the reference blade contains a Cauchy principal value singularity. This is solved by initially carrying out the integration in the radial direction and then factoring the singularity out in the chordwise integrand. A cosine series is then fitted to the real part of the integrand, the Cauchy principal value of which was derived by Glauert in 1948.

8.9 LIFTING LINE–LIFTING SURFACE HYBRID MODELS

The use of lifting surface procedures for propeller design purposes clearly requires the use of computers having a reasonably large capacity. Such capabilities are not always available to designers and as a consequence there has developed a generation of hybrid models essentially based on lifting line procedures and incorporating lifting surface corrections together with various cavitation prediction methods.

It could be argued that the very early methods of analysis fell, to some extent, into this category by relying on empirical section and cascade data to correct basic high aspect ratio calculations. However, the real evolution of these methods can be considered to have commenced subsequent to the development of the correction factors by Morgan *et al.*[21] The model of propeller action proposed by van Oossanen[25] typifies an advanced method of this type by providing a very practical approach to the problem of propeller analysis. The method is based on the Lerbs induction factor approach,[16] but because this was a design procedure the Lerbs method was used in the inverse sense, which is notoriously unstable. To overcome this instability in order to determine the induced velocities and circulation distribution for a given propeller geometry, van Oossanen introduced an additional iteration for the hydrodynamic pitch angle. In order to account for the effects of non-uniform flow, the average of the undisturbed inflow velocities over the blade sections is used to determine the advance angle at each blade position in the propeller disc. The effect of the variation of the undisturbed inflow velocities is accounted for by effectively distorting the geometric camber distribution. The effect of the bound vortices is also included because of their non-zero contribution to the induced velocity in a non-uniform flow. The calculation of the pressure distribution over the blades at each position in the propeller disc is conducted using the Theodorsen approach after first distorting the blade section camber and by defining an effective angle of attack such that a three-dimensional approximation is derived by use of a two-dimensional method.

So as to predict propeller performance correctly, particularly in off-design conditions, van Oossanen calculates the effect of viscosity on the lift and drag properties of the blade sections. The viscous effects on lift are accounted for by boundary layer theory, in which the lift curve slope is expressed in terms of the boundary layer separation and the zero lift angle is calculated as a function of the relative wake thickness on the suction and pressure sides. In contrast, the section drag coefficient is based on an equivalent profile analysis of the experimental characteristics of the Wageningen B series propellers.

The cavitation assessment is calculated from a boundary layer analysis and is based on the observation that cavitation inception occurs in the laminar–turbulent transition region of the boundary layer. The extent of the cavitation is derived by calculating the value of Knapp's dynamic similarity parameter for spherical cavities for growth and decline, based on the results of cavitation measurements on profiles.

This method has proved a particularly effective analysis tool for general design purposes and Figure 8.14 underlines the value of the method in estimating the extent of cavitation and its comparison with observations.

8.10 VORTEX LATTICE METHODS

The vortex lattice method of analysis is in effect a subclass of the lifting surface method. In the case of propeller design and analysis it owes its origins largely to Kerwin, working

Theoretical and Analytical Methods Relating to Propeller Action

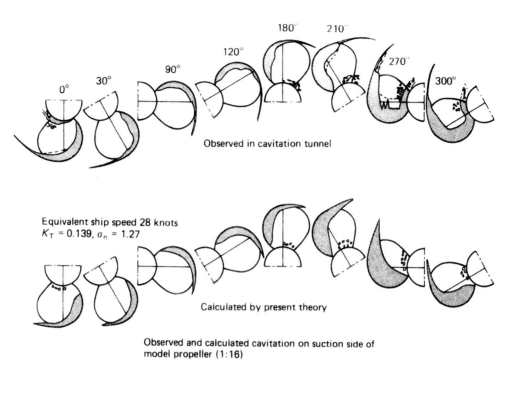

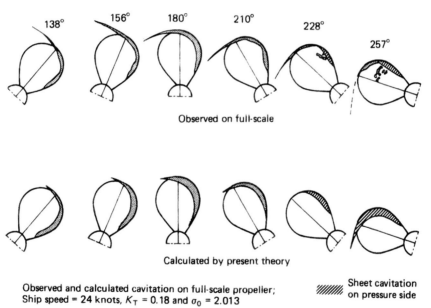

FIGURE 8.14 Comparison of observed and predicted cavitation by van Oossanen's hybrid method of propeller analysis. *Courtesy: MARIN.*

at the Massachusetts Institute of Technology, although in recent years others have taken up the development of the method: for example, Szantyr.[42]

In the vortex lattice approach the continuous distributions of vortices and sources are replaced by a finite set of straight line elements of constant strength whose end points lie on the blade camber surface (Figure 7.31). From this system of line vortices the velocities are computed at a number of suitably located control points between the elements. In the analysis problem the vortex distributions (Figure 8.15) are unknown functions of time and space, and as a consequence have to be determined from the boundary conditions of the flow regime being analyzed. The source distributions, however, can be considered to be independent

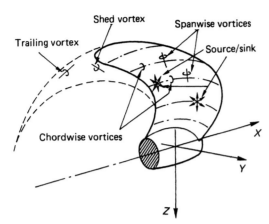

FIGURE 8.15 Basic components of lifting surface models.

of time, and their distribution over the blade is established using a stripwise application of thin aerofoil theory at each of the radial positions. As such, the source distribution is effectively known, leaving the vortex distribution as the principal unknown. Kerwin and Lee[43] consider the vortex strength at any point as a vector lying in the blade or vortex sheet which can be resolved into spanwise and chordwise components on the blades, with the corresponding components termed shed and trailing vorticity in the vortex sheets emanating from the blades (Figure 8.15). Based on this approach the various components of the vortex system can be defined with respect to time and position by applying Kelvin's theorem in association with the pressure continuity condition over the vortex wake. Hence the distributed spanwise vorticity can be determined from the boundary conditions of the problem.

In essence there are four principal characteristics of the vortex lattice model which need consideration in order to define a valid model. These are as follows:

1. The orientation of the elements.
2. The spanwise distribution of elements and control points.
3. Chordwise distribution of elements and control points.
4. The definition of the Kutta condition in numerical terms.

With regard to element distribution Greeley and Kerwin[44] proposed for steady flow analysis that the radial interval from the hub r_h to the tip R be divided into M equal intervals with the extremities of the lattice inset one-quarter interval from the ends of the blade. The end points of the discrete vortices are located at radii r_m given by

$$r_m = \frac{(R - r_h)(4m - 3)}{4M + 2} + r_h \qquad (8.60)$$
$$(m = 1, 2, 3, \ldots, M + 1)$$

In the case of the chordwise distribution of singularities they chose a cosine distribution in which the vortices and control points are located at equal intervals of $\tilde{s}$, where the chordwise variable s is given by:

$$S = 0.5(1 - \cos \tilde{s}) \qquad (0 \leq \tilde{s} \leq \pi)$$

If there are N vortices over the chord, the positions of the vortices, $S_v(n)$, and the control points, S, across the two faces of the disc (i), are given by

$$\left. \begin{array}{l} S_v(n) = 0.5\left\{1 - \cos\left[\dfrac{\left(n - \dfrac{1}{2}\right)\pi}{N}\right]\right\} \quad n = 1, 2, \ldots, N \\[2ex] \text{and} \\[1ex] S_c(i) = 0.5\left\{1 - \cos\left[\dfrac{i\pi}{N}\right]\right\} \quad i = 1, 2, \ldots, N \end{array} \right\}$$
(8.61)

With this arrangement the last control point is at the trailing edge and two-dimensional calculations show that this forces the distribution of vorticity over the chord to have the proper behavior near the trailing edge; that is, conformity with the Kutta condition. In the earlier work Kerwin and Lee[43] showed that for the solution of both steady and unsteady problems the best compromise was to use a uniform chordwise distribution of singularities together with an explicit Kutta condition:

$$S_n(n) = \frac{n - 0.75}{N} \qquad (n = 1, 2, \ldots, N) \qquad (8.62)$$

Lan[45] showed that chordwise spacing of singularity and control points proposed by equation (8.61) gave exact results for the total lift of a flat plate or parabolic camber line and was more accurate than the constant spacing arrangement, equation (8.62), in determining the local pressure near the leading edge. This choice, as defined by equation (8.61), commonly referred to as cosine spacing, can be seen as being related to the conformal transformation of a circle into a flat or parabolically cambered plate by a Joukowski transformation.

The geometry of the trailing vortex system has an important influence on the accuracy of the calculation of induced velocities on the blade. The normal approach in lifting surface theories is to represent the vortex sheet emanating from each blade as a pure helical surface with a prescribed pitch angle. Cummings,[46] Loukakis[47] and Kerwin[48] developed conceptually more advanced wake models in which the roll-up of the vortex sheet and the contraction of the slipstream were taken into account. Current practice with these methods is to consider the slipstream to comprise two distinct portions: a transition zone and an ultimate zone as shown in Figure 8.16. The transition

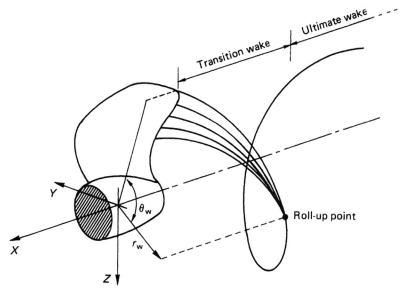

FIGURE 8.16 Deformation of wake model.

zone of the slipstream is the one where the roll-up of the trailing vortex sheet and the contraction of the slipstream are considered to occur and the ultimate zone comprises a set of Z helical tip vortices together with either a single rolled-up hub vortex or Z helical hub vortices. Hence the slipstream model is defined by some five parameters as follows (see Figure 8.16):

1. Radius of the rolled-up tip vortices (r_w).
2. Angle between the trailing edge of the blade tip and the roll-up point (θ_w).
3. Pitch angle of the outer extremity of the transition slipstream (β_T).
4. Pitch angle of the ultimate zone tip vortex helix (β_w).
5. Radius of the rolled-up hub vortices (r_{wh}) in the ultimate zone if this is not considered to be zero.

In using vortex lattice approaches it had been found that whilst a carefully designed lattice arrangement should be employed for the particular blade which is being analyzed, the other $Z-1$ blades can be represented by significantly coarser lattice without causing any important changes in the computed results. This, therefore, permits economies of computing time to be made without loss of accuracy. Kerwin[48] shows a comparison of the radial distributions of pitch and camber obtained by the vortex lattice approach and by traditional lifting surface methods (Reference 41; Figure 8.17). Although the results are very similar, some small differences are seen to occur particularly with respect to the camber at the inner radii.

The problem of vortex sheet separation and the theoretical prediction of its effects at off-design conditions have occupied the attention of many hydrodynamicists around the world. At these conditions the vortex sheet tends to

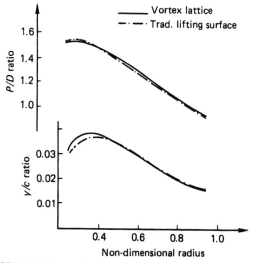

FIGURE 8.17 Comparison of results obtained between traditional lifting surface and vortex lattice methods. *Kerwin[48]*.

form from the leading edge at some radius inboard from the tip rather than at the tip. Kerwin and Lee[43] developed a somewhat simplified representation of the problem which led to a substantial improvement in the correlation of theoretical predictions with experimental results. In essence their approach is shown in Figure 8.18, in which for a conventional vortex lattice arrangement the actual blade tip is replaced by a vortex lattice having a finite tip chord. The modification is to extend the spanwise vortex lines in the tip panel as free vortex lines which join at a 'collection point', this then becomes the origin of the outermost element of the discretized vortex sheet. The position of the collection point is established by setting the pitch angle of the leading-edge free vortex equal to the mean of the

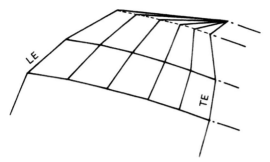

FIGURE 8.18 Simplified leading-edge vortex separation model. *Kerwin and Lee[43]*.

undisturbed inflow angle and the pitch angle of the tip vortex as it leaves the collection point. Greeley and Kerwin[44] developed the approach further by establishing a semi-empirical method for predicting the point of leading-edge separation. The basis of this method was the collapsing of data for swept wings in a non-dimensional plotting of critical leading-edge suction force, as determined from inviscid theory as a function of a local leading-edge Reynolds number, as shown in Figure 8.19. This then allowed the development of an approximate model in which the free vortex sheet was placed at a height equal to 16-blade boundary-layer thickness and the resulting change in the calculated chord wise pressure distribution found.

Lee and Kerwin *et al.* developed the vortex lattice code PUF-3 in its original form. However, the code was extended to include a number of further features amongst which were wake alignment and shaft inclination, mid-chord cavitation, thickness-load coupling, the influence of the propeller boss,

duct effects and right- and left-handed rotational options. This extended form of the code is known as MPUF—3A which is also coupled to a boundary element method to solve for the diffraction potential on the ship's hull. This latter process is done once the propeller problem has been solved and then determines the unsteady pressure fluctuations acting on the hull due to the action of the propeller.

More recently a three dimensional Euler equation solver based on a finite volume method has been developed at the University of Texas. This capability, assuming that the inflow velocity field is known sufficiently far upstream of the propeller from model tests or computations, estimates the effective wake field at the propeller. Currently, the propeller is represented by time-averaged body forces over the volume that the propeller forces are covering while they are rotating. This procedure is coupled to the MPUF—3A code in an iterative manner such that the Euler solver defines the global flow and effective wake field while the MPUF—3A code solves for the flow around the blades, including cavitation, and provides the propeller body forces to be used in the Euler solver.

8.11 BOUNDARY ELEMENT METHODS

Boundary element methods for propeller analysis have been developed in recent years in an attempt to overcome two difficulties of lifting surface analyses. The first is the occurrence of local errors near the leading edge and the second is the more widespread errors which occur near the hub where the blades are closely spaced and relatively thick. Although the first problem can to some extent be

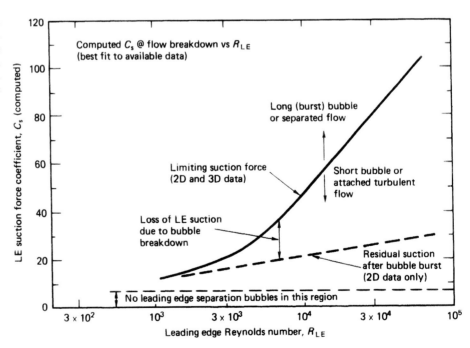

FIGURE 8.19 Empirical relationship between the value of the leading-edge suction force coefficient at the point of flow breakdown as a function of leading-edge Reynolds number. *Reproduced with permission from Reference 44.*

overcome by introducing a local correction derived by Lighthill,[49] in which the flow around the leading edge of a two-dimensional, parabolic half-body is matched to the three-dimensional flow near the leading edge derived from lifting surface theory, the second problem remains.

Boundary element methods are essentially panel methods, which were introduced in Chapter 7, and their application to propeller technology began in the 1980s. Prior to this the methods were pioneered in the aircraft industry, notably by Hess and Smith, Maskew and Belotserkovski. Hess and Valarezo[50] introduced a method of analysis based on the earlier work of Hess and Smith[51] in 1985. Subsequently, Hoshino[52] has produced a surface panel method for the hydrodynamic analysis of propellers operating in steady flow. In this method the surfaces of the propeller blades and hub are approximated by a number of small hyperboloidal quadrilateral panels having constant source and doublet distributions. The trailing vortex sheet is also represented by similar quadrilateral panels having constant doublet distributions. Figure 8.20, taken from Reference 52, shows a typical representation of the propeller and vortex sheet combination using this approach. The strengths of the source and doublet distributions are determined by solving the boundary value problems at each of the control points which are located on each panel. Within this solution the Kutta condition is obviously obeyed at the trailing edge.

Using methods of this type good agreement between theoretical and experimental results for blade pressure distributions and open water characteristics has been achieved. Indeed a better agreement of the surface pressure distributions near the blade-hub interface has been found to exist than was the case with conventional lifting surface methods.

Kinnas and his colleagues at the University of Texas, Austin, have in recent years done a considerable amount of development on boundary element codes. The initial development of the PROPCAV code in 1992 developed the boundary element method to solve for an unsteady cavitating flow around propellers which were subject to non-axisymmetric inflow conditions.[101] Subsequently, this approach has been extended to include the effects of non-cylindrical propeller bosses, mid-chord cavitation on the back and face of the propeller (Reference 102), the modeling of unsteady developed tip vortex cavitation (Reference 103) and the influence of fully unsteady trailing wake alignment (Reference 104). Good correlation has been shown to exist between the results of this computational method and the measured performance of the DTMB 4383, 72° skew propeller at model-scale for both non-cavitating and cavitating flows. Currently the effects of viscosity are estimated by using uniform values of friction coefficient applied to the wetted parts of the propeller blades; however, the code is being coupled to an integral boundary layer solver in order to better account for the effects of viscosity. This solver will both determine the friction acting on the propeller blades and estimate the influence of the viscous effects on the blade pressure distributions. Such a capability may also permit the influence of viscosity on the location of the cavity detachment in the case of mid-chord cavitation as well as on the location of the leading vortex.

A method proposed by Greco *et al.*[105] aimed at enhancing the slipstream flow prediction when using a boundary element method showed that the estimated position of the tip vortex was in good agreement with experimental data. In essence the propeller induced trailing wake was determined as part of the flow field solution using an iterative method in which the wake surface is aligned to the local flow. The numerical predictions from this method were then correlated with the vorticity field derived from laser Doppler velocity measurements made in a cavitation tunnel.

Within the framework of the MARIN-based Cooperative Research Ships organization Vaz and Bosschers have been developing a three-dimensional sheet cavitation model using a boundary element model of the marine propeller.[106] This developing approach has been tested against the results from two model propellers under steady flow conditions: the propellers being the MARINS and the INSEAN E779A propellers. In the case of the former propeller, which was designed to exhibit only sheet cavitation, two conditions were examined. At low loading the cavity extent was underpredicted but at moderate loadings the correlation was acceptable. In the second case, the INSEAN propeller had a higher tip loading than the S propeller with the cavitation having partial and super-cavitation in the tip region together with a cavitating tip vortex. For this propeller the cavity extents were predicted reasonably well. This method is currently still in its development phase and it is planned to extend the validation to behind conditions and also cavity volume variations, the latter being done through the low-frequency hull pressure pulses which are, in addition to the contribution from the non-cavitating propeller, mainly influenced by the cavity volume accelerations.

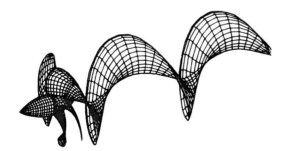

FIGURE 8.20 **Panel arrangement on propeller and trailing vortex wake for boundary element representation.** *Reproduced with permission from Reference 52.*

8.12 METHODS FOR SPECIALIST PROPULSORS

The discussion in this chapter has so far concentrated on methods of design and analysis for conventional propellers. It is also pertinent to comment on the application of these methods to specialist propulsor types: particularly controllable pitch propellers, ducted propellers, contra-rotating propellers and super-cavitating propellers.

The controllable pitch propeller, in its design pitch conditions, is in most respects identical to the conventional fixed pitch propeller. It is in its off-design conditions that special analysis procedures are required to determine the blade loads, particularly the blade spindle torque and hence the magnitude of the actuating forces required. Klaassen and Arnoldus[53] made an early attempt at describing the character of these forces and the methods of translating these into actuating forces. This work was followed by that of Gutsche[54] in which he considered the philosophical aspects of loading assumptions for controllable pitch propellers. Rusetskiy,[55] however, developed hydrodynamic models based on lifting line principles to calculate the forces acting on the blades during the braking, ring vortex and contraflow stages of controlled pitch propeller off-design performance. This procedure, whilst taking into account section distortion by means of the effect on the mean line, is a straightforward procedure which lends itself to hand calculation. The fundamental problem with the calculation of a controllable pitch propeller at off-design conditions is not that of resolving the loadings acting on the blades into their respective actuating force components, but of calculating the blade loadings on surface pressure distributions under various, and in some cases extreme, flow regimes and with the effects of blade section distortion as previously discussed in relation to propeller geometry. The basic principles of Rusetskiy's method were considered and various features enhanced by Hawdon et al.,[56] particularly in terms of section deformation and the flow over these deformed sections. Lifting line-based procedures continued to be the main method of approaching the calculation of the hydrodynamic loading components until the 1980s: the centrifugal spindle torque is a matter of propeller geometry and the frictional spindle torque which is dependent on mechanics and the magnitude of the resultant hydrodynamic and centrifugal components. Pronk[57] considered the calculation of the hydrodynamic loading by the use of a vortex lattice approach based on the general principles of Kerwin's work. In this way the computation of the blade hydrodynamics lost many of the restrictions that the earlier methods required to be placed on the calculation procedure.

As early as 1879 Parsons fitted and tested a screw propeller having a complete fixed shrouding and guide vanes. However, the theoretical development of the ducted propeller essentially started with the work of Kort.[58] In its early form the duct took the form of a long channel through the hull of the ship but soon gave way to the forerunners of the ducted propellers we know today: comprising an annular stationary aerofoil placed around the outside of a fixed or controllable pitch propeller.

Following Kort's original work, Steiss[59] produced a one-dimensional actuator disc theory for ducted propeller action; however, development of ducted propeller theory did not really start until the 1950s. Horn and Amtsberg[60] developed an earlier approach, in which the duct was replaced by a distribution of vortex rings of varying circulation along the length of the duct. Then in 1955 Dickmann and Weissinger[61] considered the duct and propeller to be a single unit replaced by a vortex system. In this system the propeller is assumed to have an infinite number of blades and constant bound vortex along the span of the blade. The slipstream is assumed to be a cylinder of constant radius and no tangential induced velocities are present in the slipstream. Despite the theoretical work the early design methods, several of which are still used today, were essentially pseudo-empirical methods. Typical of these are those presented by van Manen and co-workers[62–64] and they represent a continuous development of the subject which was based on theoretical ideas supported by the results of model tests, chiefly the K_a ducted propeller series. Theoretical development, however, continued with contributions by Morgan,[65] Dyne[66] and Oosterveld[67] for ducted propellers working in uniform and wake adapted flow conditions.

Chaplin developed a non-linear approach to the ducted propeller problem and subsequently Ryan and Glover[68] presented a theoretical design method which avoided the use of a linearized theory for the duct by using surface vorticity distribution techniques to represent both the duct and the propeller boss. The representation of the propeller was achieved by means of an extension of the Strscheletsky approach and developed by Glover in earlier studies on heavily loaded propellers with slipstream contraction.[69] The treatment of the induced velocities, however, was modified in order to take proper account of the induced velocities of the duct to achieve good correlation with experimental results. In this way the local hydrodynamic pitch angle at the lifting line was defined as

$$\beta_i = \tan^{-1}\left[\frac{V_a + u_{ap} + u_{ad}}{\omega r - u_{tp}}\right]$$

where

u_{ap} is the axial induced velocity of the propeller,
u_{ad} is the axial induced velocity of the duct and
u_{tp} is the tangential induced velocity of the propeller.

Subsequently, Caracostas[70] extended the Ryan and Glover work to off-design operation conditions using some of the much earlier Burrill[11] philosophies of propeller analysis.

Tsakonas and Jacobs[71] extended the theoretical approach to ducted propeller operation by utilizing unsteady lifting surface theory to examine the interaction of the propeller and duct when operating in a non-uniform wake field. In this work they modeled the duct and propeller geometry in the context of their camber and thickness distributions. In addition to the problem of the interactions between the duct and propeller there is also the problem of the interaction between the ducted propulsor and the body which is being propelled. Falcao de Campos[72] studied this problem in the context of axisymmetric flows. The basic approach pursued assumes the interaction flow between the ducted propulsor and the hull, which ultimately determines the performance of the duct and propeller, is inviscid in nature and can, therefore, be treated using Euler's equations of motions. Whilst this approach is valid for the global aspects of the flow, viscous effects in the boundary layers on the various components of the ducted propulsor system can be of primary importance in determining the overall forces acting on the system. As a consequence Falcao de Campos considered these aspects in some detail and then moved on to study the operation of a ducted propeller in axisymmetric shear flow. The results of his studies led to the conclusion that inviscid flow models can give satisfactory predictions of the flow field and duct performance over a wide range of propeller loadings, provided that the circulation around the duct profile can be accurately determined and a detailed account of the viscous effects on the duct can be made in the establishment of the criteria for the determination of the duct circulation. Additionally, Kerwin et al.,[107] in their extension of the MPUF–3A code to ducted propellers, developed an orifice equation model in order to take account of the viscous flow through the propeller tip–duct gap.

The main thrust of ducted propeller research has been in the context of the conventional accelerating or decelerating duct forms, including azimuthing systems, although this latter aspect has been treated largely empirically. The pumpjet is a closely related member of the ducted propeller family and has received close attention for naval applications as it exerts a considerably greater degree of control over the flow than does a conventional ducted propeller. As a result of its application to submarine propulsion much of the research is classified but certain aspects of the work are published in open literature. Two treatments of the subject are to be found in References 73 and 74 while a more recent exposition of the subject is given by Wald.[75] In this latter work equations have been derived to describe the operation of a pumpjet which is closely integrated into the hull design and ingests a portion of the hull boundary layer. From that study it was shown that maximum advantage of this system is only attained if full advantage is taken of the separation inhibiting effect of the propulsor on the boundary layer of the afterbody: a fact not to be underestimated in other propulsor configurations.

Another closely related member of the ducted propeller family is the ring propeller. This comprises a fixed pitch propeller with an integrally mounted or cast annular aerofoil at the blade tips and which has a low ring length to diameter ratio. In addition to the tip-mounted annular aerofoil designs, some of which can be found in small tugs or coasters, there have been designs proposed where the ring has been sited at some intermediate radial location on the propeller: one such example was the English Channel packet steamer Cote d'Azur, built in 1950. In this latter example the ring may not have been incorporated for purely hydrodynamic reasons given the duty and speed of the ship. Work on ring propellers has mainly been confined to model test studies and reported by van Gunsteren[76] and Keller.[77] From these studies the ring propeller is shown to have advantages when operating in off-design conditions, with restricted diameter, or by giving added protection to the blades in ice. However it has the disadvantage of giving a relatively low efficiency.

Contra-rotating propellers, as discussed in Chapter 2, have been the subject of interest which has waxed and waned periodically over the years. The establishment of theoretical methods to support contra-rotating propeller development has a long history starting with the work of Greenhill[78] who produced an analysis method for the Whitehead torpedo; however, the first major advances in the study were made by Rota[79] who carried out comparative tests with single and contra-rotating propellers on a steam launch. In a subsequent paper (Reference 80) he further developed this work by comparing and contrasting the results of the work contained in his earlier paper with some propulsion experiments conducted by Luke.[81] Little more appears to have been published on the subject until Lerbs introduced a theoretical treatment of the problem in 1955,[82] and a year later van Manen and Sentic[83] produced a method based on vortex theory supported by empirical factors derived from open water experiments. Morgan[84] subsequently produced a step-by-step design method based on Lerbs' theory. He showed that the optimum diameter can be obtained in the usual way for a single-screw propeller but assuming the absorption of half the required thrust or power and that the effect on efficiency of axial spacing between the propellers was negligible. Whilst Lerbs' work was based on lifting line principles, Murray[37] considered the application of lifting surfaces to the theory of contra-rotating propellers.

Van Gunsteren[85] developed a method for the design of contra-rotating propellers in which the interaction effects of the two propellers are largely determined with the aid of momentum theory. This approach allows the slipstream contraction effects and an allowance for the mutually induced pressures in the cavitation calculation to be taken into account in a relatively simple manner. The radial distributions of the mutually induced velocities are calculated by lifting line theory; however, the mutually induced effects are separated from self-induced effects in such a way that each propeller of the pair can be designed using a procedure for simple propellers. Agreement between this method and experimental results indicated a reasonable level of correlation.

Tsakonas *et al.*[86] have extended the development of lifting surface theory to the contra-rotating propeller problem by applying linearized unsteady lifting surface theory to propeller systems operating in uniform or non-uniform wake fields. In this latter approach the propeller blades lie on helicoidal surfaces of varying pitch, and have finite thickness distributions, together with arbitrary definitions of blade outline, camber and skew. Furthermore, the inflow field of the after propeller is modified by accounting for the influence of the forward propeller so that the potential and viscous effects of the forward propeller are incorporated in the flow field of the after propeller. It has been shown that these latter effects play an important role in determining the unsteady loading on the after propeller and as a consequence cannot be ignored. Subsequently, work at the Massachusetts Institute of Technology has extended panel methods to rotor–stator combinations.

High-speed and more particularly super-cavitating propellers have been the subject of considerable research effort. Two problems present themselves: the first is the propeller inflow and the second is the blade design problem. In the first case of the oblique flow characteristics these have to some extent been dealt with empirically, as discussed in Chapter 6. In the case of calculating the performance characteristics, the oblique flow characteristics manifest themselves as an in-plane flow over the propeller disc, whose effect needs to be taken into account. Theoretical work on what was eventually to become a design method started in the 1950s with the work of Tulin on steady two-dimensional flows over slender symmetrical bodies,[87] although super-cavitating propellers had been introduced by Posdunine as early as 1943. This work was followed by other studies on the linearized theory for super-cavitating flow past lifting foils and struts at zero cavitation number (References 88 and 89), in which Tulin used the two-term Fourier series for the basic section vorticity distribution. Subsequently Johnson[90] in developing a theoretical analysis for low drag super-cavitating sections used three- and five-term expressions. Tachmindji and Morgan[91] developed a practical design method based on a good deal of preceding research work which was extended with additional design information.[92] The general outline of the method essentially followed a similar form to the earlier design procedure set down by Eckhardt and Morgan in Reference 18.

A series of theoretical design charts for two-, three- and four-bladed super-cavitating propellers was developed by Caster.[93,94] This work was based on the two-term blade sections and was aimed at providing a method for the determination of optimum diameter and revolutions. Anderson[95] developed a lifting line theory which made use of induction factors and was applicable to normal super-cavitating geometry and for non-zero cavitation numbers. However, it was stressed that there was a need to develop correction factors in order to get satisfactory agreement between the lifting line theory and experimental results.

Super-cavitating propeller design generally requires an appeal to theoretical and experimental results − not unlike many other branches of propeller technology. In the case of theoretical methods Kinnas *et al.*[108,109] have extended their boundary element code to the modeling of super-cavitating and surface piercing propellers analysis. With regard to the experimental data to support the design of super-cavitating propellers the designer can make appeal to the works of Newton and Radar,[96] van den Voorde and Esveldt[97] and Taniguchi and Tanibayashi.[98]

8.13 COMPUTATIONAL FLUID DYNAMICS ANALYSIS

During the last ten years considerable advances have been made in the application of computational fluid dynamics to the analysis and design of marine propellers. This has now reached a point where in the analysis case useful insights into the viscous and cavitating behavior of propellers can be obtained from these methods. While progress has been made with the codes in the design case, these have not yet reached a level where these methods have gained wide acceptance but, no doubt, this will happen in the coming years.

A number of approaches for modeling the flow physics have been developed. Typically for the analysis of the flow around cavitating and non-cavitating propellers these approaches are the Reynolds Averaged Navier–Stokes (RANS) method, Large Eddy Simulation (LES) techniques, Detached Eddy Simulations (DES) and Direct Numerical Simulations (DNS). However, in terms of practical propeller computations, as distinct from research exercises, the application of many of these methods is limited by the amount of computational effort required to derive a solution. As such, the RANS codes appear to have found most favor because the computational times are rather lower than for the other methods. Most of the approaches have a number of common basic features in that they employ multi-grid acceleration and finite volume approximations. There are, nevertheless, a number of differences to be found between various practitioners in that a variety of approaches are used for the grid topology, cavitating flow modeling and turbulence modeling. In this latter context there is a range of turbulence models in use, for example $k-\varepsilon$, $k-\omega$, and Reynolds stress models are frequently seen

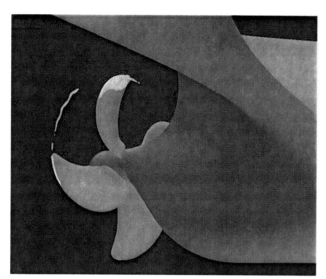

FIGURE 8.21 Propeller cavitation extent at a given position in the propeller disc using CFD methods. Courtesy *Lloyd's Register.*

being deployed, with results from the latter two methods yielding good correlations. Notwithstanding the diversity of approaches Figure 8.21 demonstrates the usefulness of the CFD approach in terms of the cavitation development over a propeller blade at a particular position in the propeller aperture.

Computational grid formation has proved a difficult area in marine propeller analysis, particularly in terms of achieving a smooth distribution of grid cells. Moreover, important structures in the flow field such as shaft lines and A-bracket structures require careful modeling with localized grid refinements. These considerations also apply to flow structures such as propeller blade tip vortices. Notwithstanding these issues, when considering propulsion test simulations these are characterized by widely different spatial and time scales for the hull and propeller.

If structured curvilinear grids are used in the modeling process this may result in a large number of cells which, in turn, may produce a complicated and time-consuming grid generation process. This has led to unstructured grids being favored since these can easily handle complex geometries and the clustering of grid cells in regions of the flow where large parameter gradients occur. Rhee and Joshi[110] analyzed a five-bladed c.p.p. propeller in open water conditions using hybrid unstructured meshes in which they used prismatic cells in the boundary layer with a system of tetrahedral cells filling in the remainder of the computational domain far from the solid boundaries. This approach allowed them to have a detailed model of the boundary layer flows while retaining many of the advantages of an unstructured mesh. In this formulation of the problem they used a $k-\omega$ turbulence model. When correlating their computed results the K_T and K_Q values were 8 per cent and 11 per cent different from the measured model test values, and while good agreement was found between the circumferential averaged axial and tangential velocities the predicted radial velocities were less accurate. Additionally, the turbulent velocity fluctuations in the wake region were also found to be underpredicted. An alternative grid generation approach for complex geometries, the Chimera technique, is becoming relatively popular.[111] In this approach simple structured grids, called sub-grids, are used for limited parts of the fluid domain and these sub-grids may overlap each other. All of these sub-grids are then embedded into a parent grid that extends across the whole fluid domain. This method has been used to address tip vortex and propeller flows as described in References 112 and 113.

Notwithstanding the present issues in the application of computational fluid dynamics methods to propeller hydrodynamics the method is gaining maturity. One of the underlying values of these types of study is in giving insights into phenomenological behavior where classical extrapolation techniques are not applicable. For example, Abdel-Maksoud and his colleagues have examined the scale effects on ducted propellers and also the influence of the hub cap shape on propeller performance.[114,115] Similarly, Wang et al.[116] have examined the three-dimensional viscous flow field around an axisymmetric body with an integrated ducted propulsor and other work has been done on podded propulsors which will be discussed in a later chapter.

In developing the method further in order to reach its full potential research is required in a number of areas. In particular, it is necessary to achieve a robust and reliable modeling of the boundary layer and similarly with wakes and two-phase flow behavior. In addition, as discussed by Kim and Rhee,[117] who analyzed the interaction between turbulence modeling and local mesh refinements, it is apparent that an adequate grid resolution of the flow field regions where vertical flow dominates is particularly important.

Kawamura et al.[118] explored the influence that turbulence modeling had on a non-cavitating and cavitating propeller performance. In their study they used a number of turbulence models in association with a commercial RANS code: in particular the turbulence models considered were the two layer RNG $k-\varepsilon$, the standard $k-\omega$ and the SST $k-\omega$. It was found, when comparing to measurements, that the calculated torque coefficients were affected by the turbulence model with standard $k-\omega$ model giving the best correlation. Li et al.[119] examined the influence of turbulence modeling on the prediction of model- and full-scale conventional and highly skewed propeller open water characteristics. In this case the models used were two equation models: SST $k-\omega$, RNG $k-\varepsilon$ and the Realizable $k-\varepsilon$ alternatives. At model-scale the prediction error was within 2 per cent and 12 per cent for the thrust and torque coefficients respectively. At full-scale in the case of the

TABLE 8.5 Comparison of the Mean and RMS Force and Torque Coefficients at J = −0.7 (Compiled From Reference 122)

	K_T	K_Q	K_{fy}	K_{fz}
Mean LES	−0.38	−0.072	0.004	−0.002
Measured mean	−0.33	−0.065	0.019	−0.006
LES root mean square value	0.067	0.012	0.061	0.057
Measured root mean square value	0.060	0.011	0.064	0.068

conventional propeller the predicted performance using the SST k–ω model differed only slightly from the two k–ε models. However, the full-scale prediction for the highly skewed propeller showed significant differences between the models. The SST k–ω indicated that the K_T value was increased by about 5 per cent with little change in K_Q whereas the two k–ε turbulence models predicted a small decrease in the thrust coefficient of the order of 0.8 per cent and in the case of the torque coefficient a decrease of around 5–6 per cent. Such a conclusion therefore indicates that the choice of turbulence model, or more particularly the underlying formulation of the turbulence model, has a dependence on the blade geometry.

More recently Sánchez-Caja et al.[120] considered the simulation of viscous flow around a ducted propeller and rudder configuration using a number of RANS-based approaches. Their conclusion was that unsteady flows can in certain circumstances be simulated by simplified RANS approaches which reduce the problem to one of steady state. The level of success of this type of approach is naturally dependent upon what the designer is trying to achieve. In their particular case of the ducted propeller–rudder configuration the more simplified approach, when compared to the time-accurate one, yielded good predictions of average thrust developed by the components of the unit. Nevertheless, to achieve these levels of correlation it was found necessary to maintain weak interactions between the stationary and rotation parts of the unit: in this way numerical flow blockage could be achieved.

With the increasing ability of computational methods to analyze propeller–hull or propeller–appendage configurations the level of interaction analysis that can be achieved is steadily increasing. This enables the designer to consider in much greater depth the influence of one component on the other.

Increasing exploration with Large Eddy Simulation (LES)[121] has resulted in approaches to the solution of complex flows taking place. For example, the near wake flow characteristics of a propeller were estimated by Bensow et al.[122] using a rotating grid and the results compared to PIV and LDV flow measurements. It was found that the LES methods can yield good qualitative insights into, for example, the development and interaction of the tip and hub vortices. Additionally, Vysohlid and Mahesh[123] computed the highly unsteady and separated flows generated in a propeller emergency stop scenario. Under such conditions the propeller while proceeding with ahead advance velocity has astern rotational speed and, therefore, the trailing edge of the propeller becomes an effective leading edge giving rise to significant separation over the blade surfaces. Moreover, because the propeller is pushing against the incident flow in these operating regimes an unstable vortex ring is produced in the vicinity of the propeller blade tips. To attempt this type of numerical modeling an unstructured Large Eddy Simulation in a rotating reference frame was used. The correlation of the computed values with model tests for the thrust, torque and side forces is shown by Table 8.5 from which it is seen that good agreement is generally achieved.

REFERENCES AND FURTHER READING

1. Rankine WJ. *On the mechanical principles of the action of propellers*. Trans RINA 1865;**6**.
2. Froude RE. *On the part played in the operation of propulsion differences in fluid pressure*. Trans RINA 1889;**30**.
3. Froude W. *On the elementary relation between pitch, slip and propulsive efficiency*. Trans. RINA 1878;**19**.
4. Betz A. *Ergebn aerodyn Vers Anst* Gottingen; 1919.
5. Betz A. *Eine Erweiterung der SchraubenTheorie*. Zreits, Flugtech; 1920.
6. Burrill LC. *On propeller theory*. Trans IESS; 1947.
7. Goldstein S. *On the vortex theory of screw propellers*. Proc Royal Soc, Ser A 1929;**123**.
8. Tachmindji AI, Milam AB. *The Calculation of the Circulation for Propellers with Finite Hub Having 3, 4, 5 and 6 Blades*. DTMB Report No. 1141; 1957.
9. Perring WGA. *The vortex theory of propellers, and its application to the wake conditions existing behind a ship*. Trans RINA 1920.
10. Taylor Lockwood J. *Screw propeller theory*. Trans NECIES; 1942.
11. Burrill LC. *Calculation of marine propeller performance characteristics*. Trans NECIES 1944;**60**.

12. Sontvedt T. *Open water performance of propellers.* Shipping World and Shipbuilder; January 1973.
13. Burrill LC. *The optimum diameter of marine propellers: a new design approach.* Trans. NECIES 1956;**72**.
14. Hill JG. *The design of propellers.* Trans. SNAME 1949;**57**.
15. Strscheletsky M. *Hydrodynamische Grundlagen zur Berechnung der Schiffschranben.* G. Braun, Karlsruhe; 1950.
16. Lerbs HW. *Moderately loaded propellers with a finite number of blades and an arbitrary distribution of circulation.* Trans SNAME 1952;**60**.
17. Manen JD van, Lammeren WPA. *The design of wake-adapted screws and their behaviour behind the ship.* Trans IESS 1955;**98**.
18. Eckhardt MK, Morgan WB. *A propeller design method.* Trans SNAME 1955;**63**.
19. Ludwieg H, Ginzel J. *Zur Theorie der Breitblattschranbe.* Aerodyn Vers, Göttingen; 1944.
20. McCarthy JH. *Comparing four methods of marine propeller design.* ASNE J.; August 1960.
21. Morgan WB, Silovic V, Denny SB. *Propeller lifting-surface corrections.* Trans. SNAME 1968;**76**.
22. Cheng HM. *Hydrodynamic Aspect of Propeller Design Based on Lifting Surface Theory Parts I and II.* DTMB Report No. 1802, 1964, 1803, 1965.
23. Kerwin JE, Leopold R. *A design theory for subcavitating propellers.* Trans SNAME 1964;**72**.
24. Abbott IH, Doenhoff AE von. *Theory of Wing Sections.* Dover; 1958.
25. Oossanen P van. *Calculation of Performance and Cavitation Characteristics of Propellers Including Effects of Non-Uniform Flow and Viscosity.* NSMB Publication No. 1974;**457**.
26. Minsaas K, Slattelid OH. *Lifting surface correction for 3 bladed optimum propellers.* ISP 1971;**18**.
27. Cummings RA, Morgan WB, Boswell RJ. *Highly skewed propellers.* Trans SNAME 1972;**20**.
28. McCroskey WJ. *Unsteady aerofoils.* Ann. Rev. Fluid Mech. 1982;**14**.
29. Crighton DG. *The Kutta conditions in unsteady flows.* Ann Rev Fluid Mech 1985;**17**.
30. Guilloton R. *Calcul des vitesses induites en vue du trace des helices.* Schiffstechnik 1957;**4**.
31. Sparenberg JA. *Application of lifting surface theory to ship screws.* Proc K Ned Akad Wet Ser B 1959;**62**(S).
32. Pien PC. *The calculation of marine propellers based on lifting surface theory.* J Ship Res 1961;**5**(2).
33. Kerwin JE. *The solution of Propeller Lifting Surface Problems by Vortex Lattice Methods.* Report Department of Ocean Engineering MIT; 1961.
34. Manen JD van, Bakker AR. *Numerical results of Sparenberg's lifting surface theory for ship screws.* Washington: Proc 4th Symp Nav Hydro; 1962.
35. Yamazaki R. *On the theory of screw propellers.* Washington: Proc 4th Symp Nav Hydro; 1962.
36. English JW. *The Application of a Simplified Lifting Surface Technique to the Design of Marine Propellers.* Ships Div Report No. SH R30/62 NPL; 1962.
37. Murray MT. *Propeller Design and Analysis by Lifting Surface Theory.* ARL; 1967.
38. Hanaoka T. *Numerical lifting surface theory of a screw propeller in non-uniform flow (Part 1 – Fundamental theory).* Rep Ship Res Inst, Tokyo 1969;**6**(5).
39. Gent W van. *Unsteady lifting surface theory of ship screws: derivation and numerical treatment of integral equation.* J Ship Res 1975;**19**(4).
40. Gent W. van. *On the Use of Lifting Surface Theory for Moderately and Heavily Loaded Ship Propellers.* NSMB Report No. 1977;**536**.
41. Brockett TE. *Lifting surface hydrodynamics for design of rotating blades.* Propellers '81 Symp SNAME; 1981.
42. Szantyr JA. *A new method for the analysis of unsteady propeller cavitation and hull surface pressures.* Trans RINA; 1984. Spring Meeting.
43. Kerwin JE. *Chang-Sup Lee. Prediction of Steady and Unsteady Marine Propeller Performance by Numercial Lifting-Surface Theory.* Trans. SNAME, Paper No. 8, Annual Meeting; 1978.
44. Greeley DA, Kerwin JE. *Numerical methods for propeller design and analysis in steady flow.* Trans. SNAME 1982;**90**.
45. Lan CE. *A quasi-vortex lattice method in thin wing theory.* J Airer 1974;**11**.
46. Cummings DE. *Vortex Interaction in a Propeller Wake.* Department of Naval Architecture Report No. 68–12. MIT; 1968.
47. Loukakis TA. *A New Theory for the Wake of Marine Propellers.* Department of Naval Architecture Report No. 71–7. MIT; 1971.
48. Kerwin JEA. *Deformed Wake Model for Marine Propellers.* Department of Ocean Engineering Rep. 76–6. MIT; 1976.
49. Lighthill MJ. *A new approach to thin aerofoil theory.* Aeronaut Quart 1951;**3**.
50. Hess JL, Valarezo WO. *Calculation of Steady Flow about Propellers by Means of a Surface Panel Method.* AIAA, Paper No. 85; 1985.
51. Hess JL, Smith AMO. *Calculation of potential flow about arbitrary bodies.* Prog Aeronaut Sci 1967;**8**.
52. Hoshino T. *Hydrodynamic analysis of propellers in steady flow using a surface panel method.* Trans Soc Nav Arch Jap;1989.
53. Klaassen H, Arnoldus W. *Actuating forces in controllable pitch propellers.* Trans I Mar E 1964.
54. Gutsche F. *Loading assumptions for controllable pitch propellers.* Schiffbanforschung 1965;**4**.
55. Rusetskiy AA. *Hydrodynamics of Controllable Pitch Propellers.* Leningrad: Idatelstvo Sudostroyeniye; 1968.
56. Hawdon L, Carlton JS, Leathard FI. *The analysis of controllable pitch propeller characteristics at off design conditions.* Trans I Mar E 1975.
57. Pronk C. *Blade Spindle Torque and Off-Design Behaviour of Controllable Pitch Propellers.* Delft University; 1980.
58. Kort L. *Der neue Dusen schraubenartrieb.* Werft Reederei Hafen 1934;**15**.
59. Steiss W. *Erweiterte Stahltheorie für Dusensehraubon mit und ne Leitapparat.* Werft Reediri und Hafen 1936;**17**.
60. Horn F, Amtsberg H. *Entwurf von Schiffdusen-systemen (Kortdusen).* Jahrbuch der Schiffbantech-nishen, Gesellschaft 1950;**44**:1961.
61. Dickmann HE, Weissinger J. *Beitrag zur Theories Optimalar Dusenscharauben (Kortdusen).* Jahrbuch der Schiffbautechnischon Gessellschaft 1955;**49**.
62. Manen JD van, Superina A. The design of screw propellers in nozzles. *ISP* March 1959;**6**.
63. Manen JD. van. *Effect of radial load distribution on the performance of shrouded propellers.* Int Shipbuilding Prog 1962;**9**(93).

64. Manen JD van, Oosterveld MWC. Analysis of ducted-propeller design. *Trans SNAME* 1966;**74**.
65. Morgan WB. *Theory of Ducted Propeller with Finite Number of Blades*. University of California, Institute of Engineering Research; May, 1961.
66. Dyne GA. *Method for the Design of Ducted Propellers in a Uniform Flow*. SSPA Report No. 1967;**62**.
67. Oosterveld MWC. *Wake Adapted Ducted Propellers*. NSMB Report No. 345; 1970.
68. Ryan PG, Glover EJ. *A ducted propeller design method: a new approach using surface vorticity distribution techniques and lifting line theory*. Trans RINA 1972;**114**.
69. Glover EJ. *Slipstream deformation and its influence on marine propeller design*. PhD Thesis, Newcastle upon Tyne; 1970.
70. Caracostas N. *Off-design performance analysis of ducted propellers*. Propellers '78 Symp Trans SNAME; 1978.
71. Tsakonas S, Jacobs WR. *Propeller-duct interaction due to loading and thickness effects*. Propellers '78 Symp Trans SNAME; 1978.
72. Falcao de Campos JAC. *On the Calculation of Ducted Propeller Performance in Axisymmetric Flows*. NSMB Report No. 696; 1983.
73. McCormick BW, Eisenhuth JJ. *The design and performance of propellers and pumpjets for underwater propulsion*. American Rocket Soc.; 1962. 17th Annual Meeting.
74. Henderson RE, McMahon JF, Wislicenus GF. *A Method for the Design of Pumpjets*. Ordnance Research Laboratory, Pennsylvania State University; 1964.
75. Wald QE. *Analysis of the integral pumpjet*. J Ship Res December 1970.
76. Gunsteren LA van. *Ringpropellers*. Canada: SNAME & I. Mar. E. Meeting; 1969.
77. Keller WH auf'm. *Comparative Tests with B-Series Screws and Ringpropellers*. NSMB Report No. 66–047. DWT; 1966.
78. Greenhill AG. *A theory of the screw propeller*. Trans RINA; 1988.
79. Rota G. *The propulsion of ships by means of contrary turning screws on a common axis*. Trans. RINA; 1909.
80. Rota G. *Further experiments on contrary turning co-axial screw propellers*. Trans RINA; 1912.
81. Luke WJ. *Further experiments upon wake and thrust deduction*. Trans. RINA; 1914.
82. Lerbs HW. *Contra-Rotating Optimum Propellers Operating in a Radially Non-Uniform Wake*. DTMB Report No. 941; 1955.
83. Manen JD van, Sentic A. Contra-rotating propellers. *ISP* 1956;**3**.
84. Morgan WB. *The design of contra-rotating propellers using Lerbs' theory*. Trans. SNAME; 1960.
85. Gunsteren LA van. *Application of momentum theory in counter-rotating propeller design*. ISP; 1970.
86. Tsakonas S, Jacobs WR, Liao P. *Prediction of steady and unsteady loads and hydrodynamic forces on counter-rotating propellers*. J Ship Res 1983;**27**.
87. Tulin MP. *Steady Two Dimensional Cavity Flow about Slender Bodies*. DTMB Report No. 834; 1953.
88. Tulin MP, Burkart MF. *Linearized Theory for Flows about Lifting Foils at Zero Cavitation Number*. DTMB Report No. 638; 1955.
89. Tulin MP. *Supercavitating flow past foils and struts*. Proc Symp on Cavitation in Hydrodynamics, NPL; 1956.
90. Johnson VE. *Theoretical Determination of Low-Drag Supercavitating Hydrofoils and Their Two-Dimensional Characteristics at Zero Cavitation Number*. NACA Report No; 1957.
91. Tachmindji AJ, Morgan WB. *The Design and Performance of Supercavitating Propellers*. DTMB Report No. 1957;**807**.
92. Tachmindji AJ, Morgan WB. *The design and estimated performance of a series of supercavitating propellers*. Washington: Proc. 2nd Symp. on Nov. Hydro., ONR; 1969.
93. Caster EB. *TMB 3-Bladed Supercavitating Propeller Series*. DTMB Report No. 1245; 1959.
94. Caster EB. *TMB 2, 3 and 4 Bladed Supercavitating Propeller Series*. DTMB Report No; 1637.
95. Anderson P. *Lifting-line theory and calculations for supercavitating propellers*. ISP; 1974.
96. Newton RN, Rader HP. *Performance data for propellers for high speed craft*. Trans RINA 1961;**103**.
97. Voorde CB, van den, Esveldt J. *Tunnel tests on supercavitating propellers*. ISP 1962;**9**.
98. Taniguchi K, Tanibayashi H. *Cavitation tests on a series of supercavitating propellers*. Japan: IAHR Symp. on Cav. Hydraulic Mach; 1962.
99. Morgan WB, Wrench JW. *Some computational aspects of propeller design*. Meth Comp Phys 1965;**4**.
100. Breslin JP, Anderson P. *Hydrodynamics of Ship Propellers*. Cambridge; 1992.
101. Kinnas SA, Fine NE. *A nonlinear boundary element method for the analysis of unsteady propeller sheet cavitation*. Seoul: Proc. 19th Symp. on Naval Hydrodyn.; 1992.
102. Young YL, Kinnas SA. *A BEM for the prediction of unsteady midchord face and/or back propeller cavitation*. J Fluid Eng 2001;**123**(2):311–9.
103. Lee HS, Kinnas SA. *Application of boundary element method in the prediction of unsteady blade sheet and developed tip vortex cavitation on marine propellers*. J Ship Res 2004;**48**(1): 15–30.
104. Lee HS, Kinnas SA. *Unsteady wake alignment for propellers in nonaxisymmetric flows*. J Ship Res 2005;**49**(3).
105. Greco L, Salvatore F, Di Felice F. *Validation of a quasi-potential flow model for the analysis of marine propellers wake*. St John's, Newfoundland: Proc. 25th Symp. on Naval Hydrodyn.; 2004.
106. Vaz G, Bosschers J. *Modelling three dimensional sheet cavitation on marine propellers using a boundary element method*. Wageningen: CAV2006 Conf.; 2006.
107. Kerwin JE, Kinnas SA, Lee J-T, Shih WZ. *A surface panel method for the hydrodynamic analysis of ducted propellers*. Trans SNAME 1987;**95**.
108. Young YL, Kinnas SA. *Numerical modelling of supercavitating propeller flows*. J Ship Res 2003;**47**(1):48–62.
109. Young YL, Kinnas SA. *Performance prediction of surface-piercing propellers*. J Ship Res 2004;**48**(4):288–304.
110. Rhee SH, Joshi S. *CFD validation for a marine propeller using an unstructured mesh based RANS method*. Honolulu: Proc. FEDSM'03; 2003.
111. Muscari R, Di Mascio A. *Simulation of the flow around complex hull geometries by an overlapping grid approach*. Japan: Proc 5th Osaka Colloquium; 2005.
112. Hsiao C-T, Chahine GL. *Numerical simulation of bubble dynamics in a vortex flow using Navier–Stokes computations and moving Chimera grid scheme*. Pasadena: CAV2001; 2001.
113. Kim J, Paterson E, Stern F. *Verification and validation and subvisual cavitation and acoustic modelling for ducted marine*

113. *propulsor*. Seoul: Proc. 8th Symp. on Numerical Ship Hydrodyn.; 2003.
114. Abdel-Maksoud M, Heinke HJ. *Scale effects on ducted propellers*. Fukuoka, Japan: Proc. 24th Symp. on Naval Hydrodyn.; 2002.
115. Abdel-Maksoud M, Hellwig K, Blaurock J. *Numerical and experimental investigation of the hub vortex flow of a marine propeller*. St John's, Newfoundland: Proc. 25th Symp. on Naval Hydrodyn.; 2004.
116. Wang T, Zhou LD, Zhang X. *Numerical simulation of 3-D viscous flow field around axisymmetric body with integrated ducted propulsion*. J Ship Mech 2003;**7**(2).
117. Kim SE, Rhee SH. *Toward high-fidelity prediction of tip-vortex around lifting surfaces*. St John's, Newfoundland: Proc. 25th Symp. on Naval Hydrodyn.; 2004.
118. Kawamura T, Watanabe T, Takekoshi Y, Maeda M, Yamaguchi H. *Numerical simulation of cavitating flow around a propeller*. J Soc Of Naval Arch of Japan 2004;**195**.
119. Li DQ, Berchiche N, Janson CE. *Influence of turbulence models on the prediction of full scale propeller open water characteristics with RANS methods*. Rome, Italy: Proc. 26th Symp. on Naval Hydrodynamics; 2006.
120. Sánchez-Caja A, Sipila TP, Pylkkänen JV. *Simulation of viscous flow around a ducted propeller with rudder using different RANS-based approaches*. Trondheim, Norway: 1st International Symp. On Marine Propulsors, Symp'09; June 2009.
121. Lesieur M, Mátais O, Comte P. *Large-Eddy Simulations of Turbulence*. ISBN 978-0-521-78124-. Cambridge University Press; 2005.
122. Bensow RE, Liefvendahl M, Wikström N. *Propeller near wake analysis using LES with a rotating mesh*. Rome, Italy: 26th Symp. on Naval Hydrodynamics; 2006.
123. Vysohlid M, Mahesh K. *Large Eddy Simulation of crashback in marine propellers*. Rome, Italy: Symp. on Naval Hydrodynamics; 2006.

Chapter 9

Cavitation

Chapter Outline

- 9.1 The Basic Physics of Cavitation — 210
- 9.2 Types of Cavitation Experienced by Propellers — 214
- 9.3 Cavitation Considerations in Design — 221
- 9.4 Cavitation Inception — 229
- 9.5 Cavitation-Induced Damage — 234
- 9.6 Cavitation Testing of Propellers — 239
- 9.7 Analysis of Measured Pressure Data from a Cavitating Propeller — 243
- 9.8 The CFD Prediction of Cavitation — 245
- References and Further Reading — 248

Cavitation is a general fluid mechanics phenomenon that can occur whenever a liquid is used in a machine which induces pressure and velocity fluctuations in the fluid. Consequently, pumps, turbines, propellers, bearings and even the human body, in for example the heart and knee joints, are all examples of machines where the destructive consequences of cavitation may occur. While cavitation sometimes has undesirable consequences, this however need not always be the case in situations such as drug delivery, the cutting of rocks or steel plates.

The history of cavitation has been traced back to the middle of the eighteenth century, when some attention was paid to the subject by the Swiss mathematician Euler in a paper read to the Berlin Academy of Science and Arts in 1754.[1] In that paper Euler discussed the possibility that a phenomenon that we would today call cavitation occurs on a particular design of water wheel and the influence this might have on its performance.

However, little reference to cavitation pertaining directly to the marine industry has been found until the mid-nineteenth century, when Reynolds wrote a series of papers[2] concerned with the causes of engine racing in screw propelled steamers. These papers introduced the subject of cavitation as we know it today by discussing the effect it had on the performance of the propeller: when extreme cases of cavitation occur, the shaft rotational speed is found to increase considerably from that expected from the normal power absorption relationships.

The trial reports of *HMS Daring* in 1894 noted this over-speeding characteristic, as did Sir Charles Parsons shortly afterwards, during the trials of his experimental steam turbine ship *Turbinia*. The results of the various full-scale experiments carried out in these early investigations showed that an improvement in propeller performance could be brought about by the increase in blade surface area. In the case of the *Turbinia*, which originally had a single propeller on each shaft and initially only achieved just under twenty knots on trials, Parsons found that to absorb the full power required on each shaft it was necessary to adopt a triple propeller arrangement to increase the surface area to the required proportions. Consequently, he used three propellers mounted in tandem on each shaft, thereby deploying a total of nine propellers distributed over the three propeller shafts. This arrangement not only allowed the vessel to absorb the full power at the correct shaft speeds, but also permitted the quite remarkable trial speed of 32.75 knots to be attained.

In an attempt to appreciate the reasons for the success of these decisions, Parsons embarked on a series of model experiments designed to investigate the nature of cavitation. To accomplish this task, Parsons constructed in 1895 an enclosed circulating channel. This apparatus allowed the testing of 2 in. diameter propellers and was a forerunner of cavitation tunnels as we know them today. However, recognizing the limitations of this tunnel, Parsons constructed a much larger tunnel fifteen years later in which he could test 12 in. diameter propeller models. Subsequently, other larger tunnels were constructed in Europe and America during the 1920s and 1930s, each incorporating the lessons learned from its predecessors. More recently a series of very large cavitation facilities have been constructed in various locations around the world. Typical of these are the depressurized towing tank at MARIN in Ede; the large cavitation tunnel at SSPA in Gothenburg, the HYCAT at HSVA in Hamburg; the Grande Tunnel Hydrodynamique at Val de Reuil in France and the Large Cavitation Channel (LCC) in Memphis, Tennessee.

Marine Propellers and Propulsion, Third Edition.
Copyright © 2012 John Carlton. Published by Elsevier Ltd. All rights reserved.

9.1 THE BASIC PHYSICS OF CAVITATION

The underlying physical process which governs the action of cavitation can, at a generalized level, be considered as an extension of the well-known situation in which a kettle of water will boil at a lower temperature when taken to the top of a high mountain. In the case of cavitation development the pressure is allowed to fall to a low level while the ambient temperature is kept constant, which in the case of a propeller is that of the surrounding sea water. Parsons had an early appreciation of this concept and he, therefore, allowed the pressure above the water level in his tunnels to be reduced by means of a vacuum pump. This enabled cavitation to appear at much lower shaft speeds, making its observation easier.

If cavitation inception were to occur when the local pressure reached the vapor pressure of the fluid then the inception cavitation number σ_i would equal the minimum pressure coefficient Cp_{min}. However, a number of other influencing factors prevent this simple relationship from being valid. For example, the ability of the fluid to withstand tensions; nuclei requiring a finite residence time in which to grow to an observable size and measurement and calculation procedures normally produce time averaged values of pressure coefficients. Consequently, the explanation of cavitation as being simply a water boiling phenomenon, although partially true, is an oversimplification of the actual physics that occur. To initially appreciate this, consider first the phase diagram for water shown in Figure 9.1. If it is assumed that the temperature is sufficiently high for the water not to enter its solid phase, then at either point B or C it could be expected that the water is in its liquid state and has an enthalpy equivalent to that state. For example, in the case of fresh water at standard pressure and at a temperature of 10°C this would be of the order of 42 kJ/kg. However, at point A, which lies in the vapor phase, the fluid would be expected to have an enthalpy equivalent to a superheated vapor, which in the example quoted above, when the pressure was dropped to say 1.52 kPa, would be in excess of 2510 kJ/kg. The differences in these figures is primarily because the fluid gains a latent enthalpy change as the liquid–vapor line is traversed so that at points B and C the enthalpies are

$$h_{B,C} = h_{fluid}(p, t)$$

and at the point A the fluid enthalpy becomes

$$h_A = h_{fluid} + h_{latent} + h_{superheat}$$

Typically for fresh water the liquid–vapor line is defined by Table 9.1.

Second, it is important to distinguish between two types of vaporization. The first is the well-known process of vaporization across a flat surface separating the liquid and its vapor. The corresponding variation in vapor pressure varies with temperature as shown in Table 9.1, and along this curve the vapor can coexist with its liquid in equilibrium. The second way in which vaporization can occur is by cavitation, which requires the creation of cavities within the liquid itself. In this case the process of creating a cavity within the liquid requires work to be done in order to form the new interface. Consequently, the liquid can be subjected to pressures below the normal vapor pressure, as defined by the liquid–vapor line in Figure 9.1, or Table 9.1, without vaporization taking place. As such, it is possible to start at a point such as C, shown in Figure 9.1, which is in the liquid phase, and reduce the pressure slowly to a value well below the vapor pressure, to reach the point A with the fluid still in the liquid phase. Indeed, in cases of very pure water, this can be extended further, so that the pressure becomes negative; when a liquid is in these over-expanded states it is said to be in a metastable phase. Alternatively it is possible to bring about the same effect at constant pressure by starting at a point B and gradually heating the fluid to a metastable phase at point A. If either of these paths, constant pressure or temperature, or indeed some intermediate path, is followed, then eventually the liquid reaches a limiting condition at some point below the

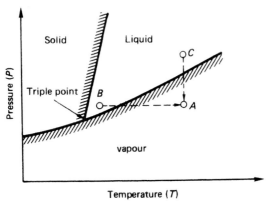

FIGURE 9.1 Phase diagram for water.

TABLE 9.1 Saturation Temperature of Fresh Water

Pressure (kPa)	0.689	6.894	13.79	27.58	55.15	101.3	110.3
Saturation temperature (C)	1.6	38.72	52.58	67.22	83.83	100.0	102.4

liquid–vapor line in Figure 9.1, and either cavitates or vaporizes.

The extent to which a liquid can be induced metastably to a lower pressure than the vapor pressure depends on the purity of the water. If water contains a significant amount of dissolved air, then as the pressure decreases the air comes out of the solution and forms cavities in which the pressure will be greater than the vapor pressure. This effect applies also when there are no visible bubbles; sub-microscopic gas bubbles can provide suitable nuclei for cavitation purposes. Hence cavitation can either be vaporous or gaseous or, perhaps, a combination of both. Consequently, the point at which cavitation occurs can be either above or below the vapor pressure corresponding to the ambient temperatures.

In the absence of nuclei a liquid can withstand considerable negative tensions without undergoing cavitation. For example, in the case of a fluid, such as water, which obeys van der Waals' equation:

$$\left(p + \frac{a}{V^2}\right)(V - b) = RT \quad (9.1)$$

a typical isotherm is shown in Figure 9.2, together with the phase boundary for the particular temperature. In addition, the definition of the tensile strength of the liquid is also shown on this figure. The resulting limiting values of the tensions that can be withstood form a wideband; for example, at room temperature, by using suitable values for a and b in equation (9.1), the tensile strength can be shown to be about 500 bars. However, some researchers have suggested that the tensile strength of the liquid is the same as the intrinsic pressure a/V^2 in equation (9.1); this yields a value of around 10 000 bars. In practice, water subjected to rigorous filtration and pre-pressurization seems to rupture at tensions of the order of 300 bars. However, when solid, non-wetted nuclei having a diameter of about 10^6 cm are present in the water it will withstand tensions of only the order of tens of bars. Even when local pressure conditions are known accurately it is far from easy to predict when cavitation will occur because of the necessity to estimate the size and distribution of the nuclei present.

Despite the extensive literature on the subject, both the understanding and predictability of bubble nucleation is a major problem for cavitation studies. There are in general two principal models of nucleation; these are the stationary crevice model and the entrained nuclei models. Nuclei in this sense refers to clusters of gas or vapor molecules of sufficient size to allow subsequent growth in the presence of reduced pressure. The stationary nuclei are normally assumed to be harbored in small crevices of adjacent walls while, in contrast, the traveling nuclei are assumed to be entrained within the mainstream of the fluid. Consequently, entrained nuclei are considered the primary source of cavitation, although of course cavitation can also be initiated from stationary nuclei located in the blade surface at the minimum pressure region. Of the nucleation models proposed those of Harvey et al.[3–6] and subsequently by others (References 7–10) are probably the most important. These models propose that entrained micro-particles in the liquid, containing in themselves unwetted acute angled micro-crevices, are a source of nucleation. This suggests that if a pocket of gas is trapped in a crevice then, if the conditions are correct, it can exist in stable equilibrium rather than dissolve into the fluid. Consider first a small spherical gas bubble of radius R in water. For equilibrium, the pressure difference between the inside and outside of the bubble must balance the surface tension force:

$$p_v - p_l = \frac{2S}{R} \quad (9.2)$$

where

p_v is the vapor and/or gas pressure (internal pressure)
p_l is the pressure of the liquid (external pressure)
S is the surface tension.

Now the smaller the bubble becomes, according to equation (9.2), the greater must be the pressure difference across the bubble. Since, according to Henry's law, the solubility of a gas in a liquid is proportional to gas pressure, it is reasonable to assume that in a small bubble the gas should dissolve quickly into the liquid. Harvey et al., however, showed that within a crevice, provided the surface is hydrophobic, or imperfectly wetted, then a gas pocket can continue to exist. Figure 9.3 shows in schematic form the various stages in the nucleation process on a micro-particle. In this figure the pressure reduces from left to right, from which it is seen that the liquid–gas interface changes from a convex to concave form and eventually the bubble in the crevice of the micro-particle grows to a sufficient size so that a part breaks away to form a bubble entrained in the body of the fluid.

Other models of nucleation have been proposed, for example those of Fox and Herzfeld[11] and Plesset,[14] and no

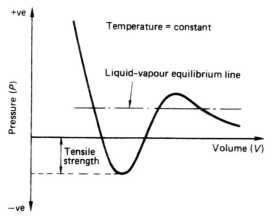

FIGURE 9.2 Van der Waals' isotherm and definition of tensile strength of liquid.

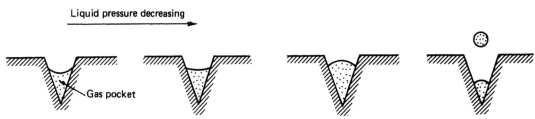

FIGURE 9.3 Nucleation model for a crevice in an entrained micro-particle. *Harvey* et al.[3-6]

doubt these also play a part in the overall nucleation process, which is still far from well understood. Fox and Herzfeld suggested that a skin of organic impurity, for example fatty acids, accumulates on the surface of a spherical gas bubble in order to inhibit the dissolving of the gas into the fluid as the bubble decreases in size; this reduction in size causes the pressure differential to increase, as seen by equation (9.2). In this way it is postulated that the nuclei can stabilize against the time when the bubble passes through a low-pressure region, at which point the skin would be torn apart and a cavity initiated. The 'skin' model has in later years been refined and improved by Yount.[12,13] Plesset's unwetted mote model suggested that such motes can provide bubble nucleation without the presence of gases other than the inevitably present vapor of the liquid. The motes, it is suggested, would provide weak spots in the fluid about which tensile failure of the liquid would occur at pressures much less than the theoretical strength of the pure liquid.

An additional complicating factor arises from the flow over propeller blades being turbulent in nature, consequently any nuclei in the center of the turbulent eddies may experience localized pressures which are rather lower than the mean or time averaged pressure that has been either calculated or measured. As a result the local pressure within the eddy formations may fall below the vapor pressure of the fluid while the average pressure remains above that level.

Cavitation gives rise to a series of other physical effects which, although of minor importance to ship propulsion, are interesting from the physical viewpoint and deserve passing mention especially with regard to material erosion. The first is sonoluminescence, which is a weak emission of light from the cavitation bubble in the final stage of its collapse. This is generally ascribed to the very high temperatures resulting from the essentially adiabatic compression of the permanent gas trapped within the collapsing cavitation bubbles. Schlieren and interferometric pictures have succeeded in showing the strong density gradients or shock waves in the liquid around collapsing bubbles. When bubbles collapse surrounding fluid temperatures as high as 100 000 K have been suggested and Wheeler[15] has concluded that temperature rises of the order of 500–800°C can occur in the material adjacent to the collapsing bubble. The collapse of the bubbles is completed in a very short space of time (milli- or even microseconds) and it has been shown that the resulting shock waves radiated through the liquid adjacent to the bubble may have a pressure difference as high as 4000 atm.

The earliest attempt to analyze the growth and collapse of a vapor or gas bubble in a continuous liquid medium from a theoretical viewpoint appears to have been made by Besant.[16] This work was to some extent ahead of its time, since bubble dynamics was not an important engineering problem in the mid-1800s and it was not until 1917 that Lord Rayleigh laid the foundations for much of the analytical work that continues to the present time.[17] His model considered the problem of a vapor-filled cavity collapsing under the influence of a steady external pressure in the liquid, and although based on an oversimplified set of assumptions, Rayleigh's work provides a good model of bubble collapse and despite the existence of more modern and advanced theories is worthy of discussion in outline form.

In the Rayleigh model the pressure p_v within the cavity and the pressure at infinity p_0 are both considered to be constant. The bubble is defined using a spherical co-ordinate system whose origin is at the center of the bubble whose initial steady state radius is R_0 at time $t = 0$. At some later time t, under the influence of the external pressure p_0 which is introduced at time $t = 0$, the motion of the bubble wall is given by

$$\frac{d^2R}{dt^2} + \frac{3}{2R}\left(\frac{dR}{dt}\right)^2 = \frac{1}{\rho R}(p_v - p_0) \qquad (9.3)$$

where ρ is the density of the fluid. By direct integration of equation (9.3), assuming that both p_v and p_0 are constant, Rayleigh described the collapse of the cavity in terms of its radius R at a time t as being

$$\left(\frac{dR}{dt}\right)^2 = \frac{2}{3}\frac{(p_0 - p_v)}{\rho}\left(\left(\frac{R_0}{R}\right)^3 - 1\right) \qquad (9.4)$$

By integrating equation (9.4) numerically it is found that the time to collapse of the cavity t_0, known as the 'Rayleigh collapse time', is

$$t_0 = 0.91468 R_0 \left(\frac{\rho}{p_0 - p_v}\right)^{1/2} \qquad (9.5)$$

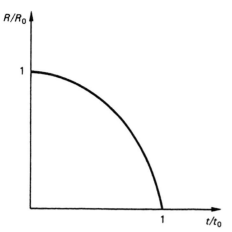

FIGURE 9.4 Collapse of a Rayleigh cavity.

because the spherical co-ordinates required the numerical analysis to be terminated as the microjet approaches the initial bubble center. Figure 9.5 shows the results of a computation of an initially spherical bubble collapsing close to a solid boundary, together with the formation of the microjet directed towards the wall.

Subsequently, Chahine has studied cloud cavity dynamics by modeling the interaction between bubbles. In his model he was able to predict the occurrence of high pressures during collapse principally by considering the coupling between bubbles in an idealized way through symmetric distributions of identically sized bubbles.

Bark in his researches at Chalmers University has demonstrated the effects of cavity rebound following the initial collapse of a cavity. Figure 9.6 shows this effect over a sequence of four-blade passages in terms of a propeller radiated hull pressure signature. From the figure it is immediately obvious that the hull pressure signature is very variable, particularly in terms of amplitude, from blade passage to blade passage. Moreover, the influence of the cavity rebound in comparison to the cavity growth and initial decay parts of the signature is significant and, if the physical conditions are correct, one or more rebound events can take place. This variability in the signature which comprises spatial, temporal and phenomenological attributes underlines the importance of analyzing these signatures correctly in order that information is not inadvertently lost when analyzing hull surface pressure signatures.

This time t presupposes that at the time $t = 0$ the bubble is in static equilibrium with a radius R_0. The relationship between bubble radius and time in non-dimensional terms is derived from the above as being

$$\frac{t}{t_0} = 1.34 \int_{R/R_0}^{1} \frac{dx}{(1/x^2 - 1)^{1/2}} \quad (9.6)$$

and the results of this equation, shown in Figure 9.4, have been shown to correspond to experimental observations of a collapsing cavity.

The Rayleigh model of bubble collapse leads to a series of significant results from the viewpoint of cavitation damage; however, because of simplifications involved it cannot address the detailed mechanism of cavitation erosion. The model shows that infinite velocities and pressures occur at the point when the bubble vanishes and in this way points towards the basis of the erosion mechanism. The search for the detail of this mechanism has led to considerable effort on the part of many researchers in recent years. Such work has introduced not only the effects of surface tension, internal gas properties and viscosity effects, but also those of bubble asymmetries which predominate during the collapse process. Typical of these advanced studies is the work of Mitchell and Hammitt[18] who also included the effects of pressure gradient and relative velocity as well as wall proximity. An alternative approach by Plesset and Chapman[19] used potential flow assumptions, thereby precluding the effects of viscosity which in the case of water is unlikely to be of major importance. Plesset and Chapman focused on the bubble collapse mechanism under the influence of wall proximity, which is of major significance in the study of cavitation damage. Their approach was based on the use of cylindrical co-ordinates as distinct from Mitchell's spherical co-ordinate approach, and this allowed them to study the microjet formation during collapse to a much deeper level

In extending the study of the physics of cavitation and, in particular, its aggressiveness towards material erosion, Fortes-Patella and her colleagues have been developing a model of cavitation action.[55–65] In essence, the method is based on the study of the pressure wave characteristics

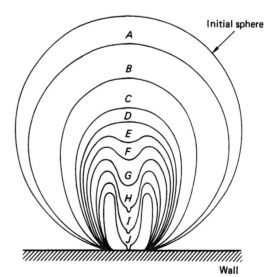

FIGURE 9.5 Computed bubble collapse (Plesset–Chapman).

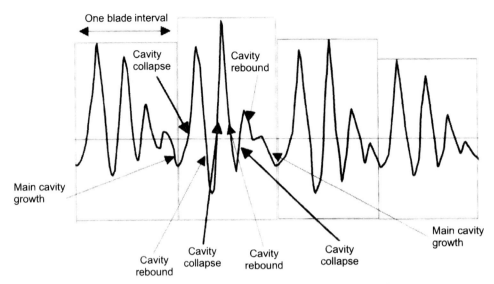

FIGURE 9.6 Influence of cavity rebound on a hull radiated pressure signature.

emitted during bubble collapse: in particular, focusing on the relationship between the initial and collapsing states. Within this work a better agreement was found between experiment and calculation for pressure wave models of erosion. They also showed that there was no influence of material on the flow pressure pulse histogram and that the number of pits normalized by surface area and time was found by experiment to be proportional to $\lambda^{2.7}$, where λ is the geometric scaling factor. This result is close to the cubic law which was noted by Lecoffre[66] and it was found that volume damage rate does not appear to significantly change with scale. It was shown that the flow speed (V) does, however, have a significant influence on the erosion in that the number of pits per unit area and time is proportional to V^5 and the pit volume, normalized on the same basis, is proportional to V^7. In this context pit depth did not seem to significantly vary with the flow speed.

The computational model developed is based on a series of energy transformations within an overall energy balance scenario as outlined in Figure 9.7. Within this model the terms P_{pot}, P_{pot}^{mat}, P_{waves}^{mat} represent the potential power of the vapor structure, the flow aggressiveness potential and the pressure wave power, respectively, while the η^* and η^{**} represent transmission efficiencies and β is the transmission factor for fluid–material interaction.

9.2 TYPES OF CAVITATION EXPERIENCED BY PROPELLERS

Cavitating flows are by definition multi-phase flow regions. The two phases that are most important are the water and its own vapor; however, in almost all cases there is a quantity of gas, such as air, which has significant effects in both bubble collapse and inception — usually most importantly in the inception mechanism. As a consequence cavitation is generally considered to be a two-phase, three-component flow regime. Knapp et al.[20] classified cavitation into fixed, traveling or vibratory forms, the first two being of greatest interest in the context of propeller technology.

A fixed cavity is one in which the flow detaches from the solid boundary of the immersed object to form a cavity or envelope which is fixed relative to the object upon which it forms and, in general, such cavities have a smooth glassy appearance. In contrast, as their name implies, traveling cavities move with the fluid flowing past the body of interest. Traveling cavities originate either by breaking away from the surface of a fixed cavity, from which they can then enter the flow stream, or from nuclei entrained within the fluid medium. Figure 9.8 differentiates between these two basic types of cavitation.

The flow conditions at the trailing edge of a cavity are not dissimilar, but rather more complicated, to those of water passing over a weir. The cavity shedding mechanism is initiated by the re-entrant jet in that it forms between the

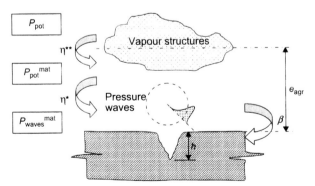

FIGURE 9.7 Basis of the Fortes-Patella et al. Model.

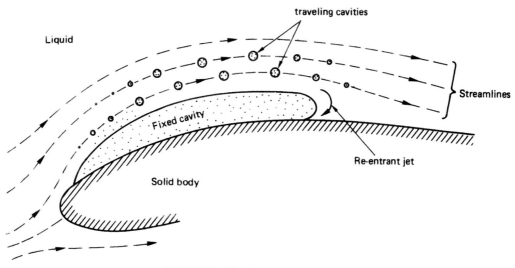

FIGURE 9.8 Fixed and traveling cavities.

underside of the cavity and the propeller blade surface. The behavior of the re-entrant jet, therefore, is of importance in the phenomenological behavior of the cavitation on the blades. In studies on twisted aerofoils undertaken by Foeth and van Terwisga[67,68] they concluded that the cavity topology principally determines the direction of the re-entrant flow and that convex cavity shapes appear to be intrinsically unstable. Moreover, the condition when the re-entrant jet reaches the leading edge of the aerofoil is not the only determinant in shedding because the side entrant jets of convex cavities have both a chordal and spanwise motion. These motions focus in the closure region of the sheet cavity where they tend to disturb the flow which then initiates a break-off from the main sheet cavity structure. Interestingly, during the collapse of the sheet cavity structures it was noted that they degenerated into vortical structures which leads to the conclusion that a mixing layer exists with its characteristic spanwise and streamwise vortices.

Uhlman[80] has studied the fully non-linear axisymmetric potential flow past a body of revolution using the boundary integral method and devised a model for the exact formulation of the re-entrant jet cavity closure condition. The results of this modeling approach were shown to be in good agreement with experimental results and were consistent with momentum flux requirements. This re-entrant jet model represents an enhancement over the earlier Riabouchinsky-type cavity closure model (Figure 9.25), since the boundary conditions entail the physical conditions of constant pressure and no flux.

The cavitation patterns which occur on marine propellers are usually referred to as comprising one or more of the following types for model propellers: sheet, bubble, cloud, tip vortex or hub vortex cavitation. Whether all of these cavity types translate across to full-scale is a matter of some conjecture: many certainly do.

Sheet cavitation initially becomes apparent at the leading edges of the propeller blades on the backs or suction surfaces if the sections are working at positive incidence angles. Conversely, if the sections are operating at negative incidence this type of cavitation may initially appear on the face of the blades. Sheet cavitation appears because when the sections are working at non-shock-free angles of incidence, large suction pressures build up near the leading edge of the blades of the 'flat plate' type of distribution shown in Figure 7.18. If the angles of incidence increase in magnitude, or the cavitation number decreases, then the extent of the cavitation over the blade will grow both chordally and radially. As a consequence the cavitation forms a sheet over the blade surface whose extent depends upon the design and ambient conditions. Figure 9.9(a) shows an example of sheet cavitation on a model propeller, albeit with tip vortex cavitation also visible. Sheet cavitation is generally stable in character, although there are cases in which a measure of instability can be observed. In these cases the reason for the instability should be sought, and if it is considered that the instability will translate to full-scale, then a cure should be attempted, as this may lead to blade erosion or unwanted pressure fluctuations.

Bubble cavitation (Figure 9.9(b)), is primarily influenced by those components of the pressure distribution which cause high suction pressures in the mid-chord region of the blade sections. Thus the combination of camber line and section thickness pressure distributions identified in Figure 7.18 have a considerable influence on the susceptibility of a propeller towards bubble cavitation. Since bubble cavitation normally occurs first in the mid-chord region of

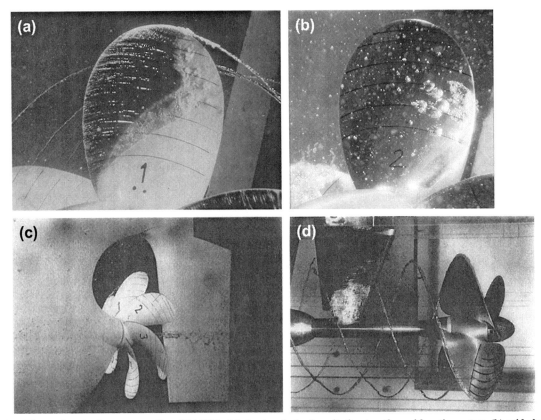

FIGURE 9.9 Types of cavitation on propellers (MARIN): (a) sheet and cloud cavitation together with a tip vortex; (b) mid-chord bubble cavitation together with a tip vortex and some leading edge streak cavitation; (c) hub vortex cavitation with traces of LE and tip vortex in top of propeller disc (*Courtesy: MARIN*) and (d) tip vortex cavitation.

the blade, it tends to occur in non-separated flows. This type of cavitation, as its name implies, appears as individual bubbles growing, sometimes quite large in character, and contracting rapidly over the blade surface.

Cloud cavitation is frequently to be found behind strongly developed stable sheet cavities and generally in moderately separated flow in which small vortices form the origins for small cavities. This type of cavitation (Figure 9.9(a) with traces on Figure 9.9(b)) appears as a mist or 'cloud' of very small bubbles and its presence should always be taken seriously.

The vortex types of cavitation, with few exceptions, occur at the blade tips, the leading edge and hub of the propeller and they are generated from the low-pressure core of the shed vortices. The hub vortex is formed by the combination of the individual vortices shed from each blade root, and although individually these vortices are unlikely to cavitate, under the influence of a converging propeller cone the combination of the blade root vortices has a high susceptibility to cavitate. When this occurs the resulting cavitation is normally very stable and appears to the observer as a rope with strands corresponding to the number of blades of the propeller. Tip vortex cavitation is normally first observed some distance behind the tips of the propeller blades. At this time the tip vortex is said to be 'unattached', but as the vortex becomes stronger, either through higher blade loading or decreasing cavitation number, it moves towards the blade tip and ultimately becomes attached. Figures 9.9(c) and (d) show typical examples of the hub and tip vortices, respectively.

In addition to the principal classes of cavitation, there is also a type of cavitation that is sometimes referred to in model test reports as 'streak' cavitation. This type of cavitation, again as its name implies, forms relatively thin streaks extending from the leading edge region of the blade chordally across the blades.

Propulsor–Hull Vortex (PHV) cavitation was reported by Huse[21] in the early 1970s. This type of cavitation may loosely be described as the 'arcing' of a cavitating vortex between a propeller tip and the ship's hull. Experimental work with flat, horizontal plates above the propeller in a cavitation tunnel shows that PHV cavitation is most pronounced for small tip clearances. In addition, it has been observed that the propeller advance coefficient also has a significant influence on its occurrence; the lower the advance coefficient the more likely PHV cavitation is to occur. Figure 9.10 shows a probable mechanism for PHV cavitation formation. In the figure it is postulated that at

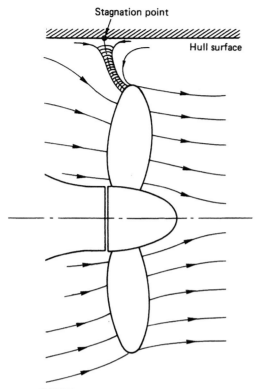

FIGURE 9.10 Basis for PHV cavitation.

high loading the propeller becomes starved of water due to the presence of the hull surface above and possibly the hull in the upper part of the aperture ahead of the propeller. To overcome this water starvation the propeller endeavors to draw water from astern, which leads to the formation of a stagnation streamline from the hull to the propeller disc, as shown. The PHV vortex is considered to form due to turbulence and other flow disturbances close to the hull, causing a rotation about the stagnation point, which is accentuated away from the hull by the small radius of the control volume forming the vortex. This theory of PHV action is known as the 'pirouette effect' and is considered to be the most likely of all the theories proposed. Thus the factors leading to the likelihood of the formation of PHV cavitation are generally considered to be:

1. low advance coefficients,
2. low tip clearance,
3. flat hull surfaces above the propeller.

Van der Kooij and co-workers[53,54] studied the problem of propeller—hull vortex cavitation for the ducted propeller case and concluded that the occurrence of PHV cavitation depended strongly on hull—duct clearance and propeller blade position.

Methods of overcoming the effects of PHV cavitation are discussed in Chapter 23.

The foregoing observations relate principally to indications gained from undertaking model tests. In recent years, however, considerably more full-scale observations have been made using both the conventional hull window penetrations and more recently using the borescope technique. This has increased the understanding of the full-scale behavior of cavitation and its correlation to model-scale testing. Figure 9.11, taken from (Reference 69), shows a consecutive sequence of borescope images taken under natural daylight conditions of the tip vortex development emanating from the propeller blades of an 8500 teu container ship. This continuous sequence, comprising eight images, was taken at a time interval of 1/25 seconds. In the figure the rising propeller blade can be clearly seen on the right-hand side of the images and the behavior of the vortices emanating from the two blades immediately preceding the rising blade can be observed on the left. The observation was made from the hull above the propeller over a period of 0.28 seconds with the ship proceeding on a steady course at constant speed. At the arbitrary time $t = 0$ the vortex from the blade immediately leading the rising blade exhibits a well-formed structure having some circumferential surface texture and small variations in radius with slight tendency towards expansion near the top of the picture. By 0.04 seconds later the cavitating structural expansion has started to grow with the expansion showing a distinct asymmetric behavior towards the propeller station. In the subsequent frames this asymmetric expansion progressively increases and exhibits a tendency for the principal area of asymmetry to become increasingly distinct from the main vortex structure. By the time $t = 0.16$ seconds a new vortex is clearly following the tendency of its predecessor as indeed the described vortex followed the behavior of its own predecessor shown at time $t = 0$.

The complexity of the tip vortex mechanisms was discussed by Carlton and Fitzsimmons[70] in relation to observations made on a number of ships. In that paper a mechanism derived from full-scale observation of LNG ship propeller cavitation was described to explain an expansive mechanism for the tip vortex structure. This was in effect an interaction between the tip vortex and the super-cavitating parts of the sheet cavity at the blade tip region where the super-cavitating part of the blade sheet cavity was rapidly expanded under the action of the tip vortex. It is, therefore, interesting to note that Lücke[71] has identified from model tests two mechanisms for tip vortex bursting: one following the conventional aerodynamic treatment of vortex bursting and the other very similar to that described at full-scale above, thereby-suggesting a possible model-full-scale similarity. While the earlier descriptions of this phenomenon centered on steady course ship operation at constant speed, the complexity of the tip vortex development was found to increase significantly when the ship began to undertake turning maneuvers in open water. An example of this behavior is shown in Figure 9.12 in which the expansive cloud seen in Figure 9.12(a) and

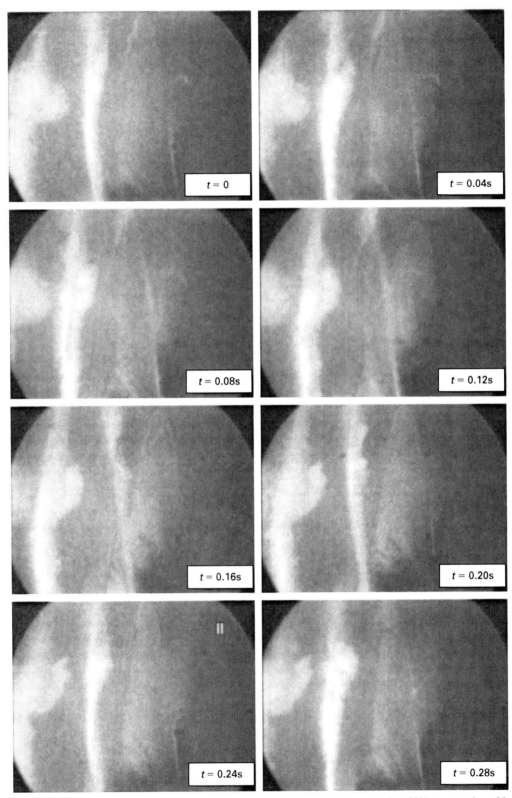

FIGURE 9.11 A sequence of images of the tip vortex emanating from the propeller of an 8500 teu container ship.

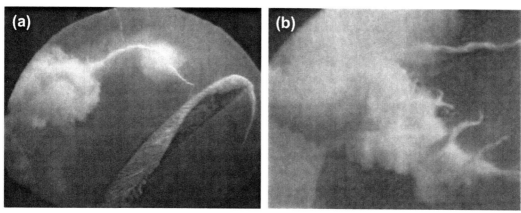

FIGURE 9.12 Full-scale cavitating sheet and vortex cavitation on an LNG ship: (a) cavitation on a straight course and (b) tip vortex behavior during turning.

developed during the cavity collapse phase under uniform straight course conditions has, in the turning maneuver, extended its trailing volume region and developed a system of ring-like vortex structures circumscribing this trailing part of the cavitating volume (Figure 9.12(b)). However, in interpreting these structures it must be recalled that only the cavitating part of the vortex structure is visible in these images and the complete vortex structure, including the cavitating and non-cavitating parts, is considerably larger.

Maneuvers have been found to generate extremely complex interactions between cavitation structures on a propeller blade and also between the propeller and the hull as well as between propellers in multi-screw ships. In the case of a high-speed, twin-screw passenger ship when undertaking berthing maneuvering in port, strong cavitation interaction was observed between the propellers. This interaction took the form of vorticity shed from one propeller blade and directing itself transversely across the ship's afterbody to interact with the cavity structures on the adjacent propeller blades. Figure 9.13 captured this interaction taking place by means of a digital camera viewing through a conventional hull window arrangement. The complexity of the cavity structure and the locus of its travel are immediately apparent.

In the case of propeller–hull interaction Figure 9.14 shows a cavitating propeller–hull vortex captured by a borescope observation in a steep buttock flow field which then entered the propeller disc of a podded propulsor. In this image the relatively strong tip vortices can be seen emanating from the propeller blades while the tip vortex rises vertically towards the hull.

Vortex interaction, particularly at off-design conditions may cause troublesome excitation of the ship structure by generating a combined harmonic and broadband signature. Figure 9.15 shows a series of images demonstrating the

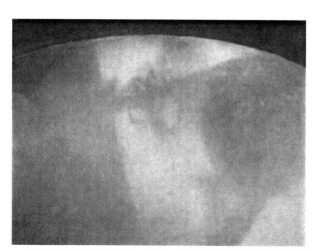

FIGURE 9.13 Cavitation interaction between propellers.

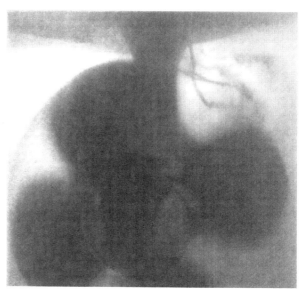

FIGURE 9.14 Example of a propeller–hull vortex emanating from a podded propulsion unit.

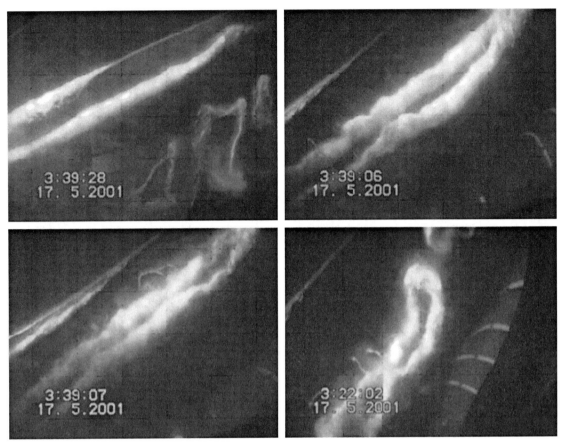

FIGURE 9.15 Vortex interaction mechanisms from a cruise ship propeller.

interaction of vortex cavitation emanating from one of the propellers of a twin-screw ship when operating at full shaft speed and reduced blade pitch at 8 knots: this is discussed further in terms of its effects and consequences in Chapter 22. From the images it can be seen that the propeller is emitting both a cavitating tip and a leading edge vortex from the blades. At first these vortices travel back in the flow field largely independently, certainly as far as perturbation to the cavitating part of the vortex structures are concerned (3:39:28): however, even at this early stage some small influence on the tip vortex can be seen.

As the vortices co-exist the mutual interference builds up (3:39:06) with the cavitating part of the vortices thickening and becoming less directionally stable as well as inducing some ring vortices which encircle both of the vortex structures. A short time later (3:39:07) ring vortex structures are being developed to a far greater degree with the vortices thickening and being surrounded by much greater coaxial cloudiness; although some cloudiness, as can be seen, was present one second earlier.

Finally, due to the interaction of the two vortices they eventually destroy their basic continuous helical form and break up into intermittent ring formations following each other along a helical track (3:22:02). However, even at this late stage some of the earlier encircling ring structures are still present together with coaxial cloudy regions around the main core of the vortex. As these new ring structures pass downstream (3:39:28) their cavitating cores lose thickness and gain a strong cloudy appearance.

In the case of a patrol boat which was powered by a triple-screw, fixed pitch propeller arrangement the underlying problem was a poor design basis for the propeller: principally a large slow-running propeller with a high P/D ratio. The propeller sections experienced high angles of attack due to the variations in the tangential component of the velocity field induced by the shaft angle, 10 degrees, and this gave rise to a set of face and root cavitation erosion issues which could not be reconciled without recourse to artificial means. The blades were designed and manufactured to ISO 484 Class S ; however, the cavitation problem was exacerbated by a lack of consistent definition of the blade root section geometry, which is outside the ISO standard, and which permitted arbitrary section forms to result in the root regions. The blade roots originally were also very close to the leading edge of the boss which caused problems in blending the blade leading edge onto the hub. This initial hub was designed with a small leading edge radius.

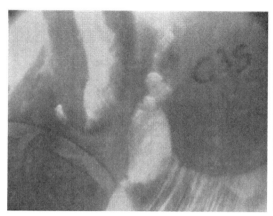 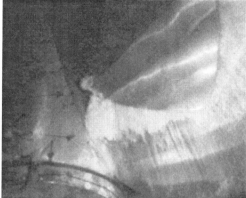

FIGURE 9.16 Root cavitation on a high speed propeller in the sprint condition.

Figure 9.16 shows a typical cavitation pattern observed on the propeller blades when near the top dead center position with the propeller operating at close to its sprint condition. From the figure a large, but relatively benign, back sheet cavity can be seen which spreads over a significant portion of the blade chord length. In the root region this sheet cavity transforms itself into a complex thick structure having a much more cloudy nature together with embedded vortex structures. Indeed, whenever the ring type structures as seen in the center of the picture have been observed, at either model- or full-scale, these have often indicated a strong erosion potential. Root cavitation structures of this type are extremely aggressive in terms of cavitation erosion and being shed from the downstream end of a root cavity may indicate a mechanism which, in association with ship speed, can produce two or more isolated erosion sites along the root chord.

Kennedy et al.[81] studied the cavitation performance of propeller blade root fillets. They found that the critical region of the fillet as far as cavitation inception was concerned was from its leading edge to around 20 per cent of the chord length. Moreover, fillet forms, for example those with small radii, which give rise to vortex structures should be avoided. However, the relatively low Reynolds number at which typical propeller model tests are conducted may not permit the observation of vortex cavitation structures in the blade roots and this presents a further difficulty for the designer.

9.3 CAVITATION CONSIDERATIONS IN DESIGN

The basic cavitation parameter used in propeller design is the cavitation number which was introduced in Chapter 6. In its most fundamental form the cavitation number is defined as

$$\text{cavitation number} = \frac{\text{static pressure head}}{\text{dynamic pressure head}} \quad (9.7)$$

The relationship has, however, many forms in which the static head may relate to the shaft center line immersion to give a mean value over the propeller disc, or may relate to a local section immersion either at the top dead center position or some other instantaneous position in the disc. Alternatively, the dynamic head may be based upon either single velocity components such as the undisturbed free stream advance velocity and the propeller rotational speed or the vectorial combination of these velocities in either the mean or local sense. Table 9.2 defines some of the more common cavitation number formulations used in propeller technology: the precise one chosen depends upon the information known or the intended purpose of the data.

The cavitating environment in which a propeller operates has a very large influence not only on the detail of the propeller design but also upon the type of propeller that is contemplated for use. For example, is it better to use a conventional, super-cavitating or surface piercing propeller design for a given application? A useful initial guide to determining the type of propeller most suited to a particular application is afforded by the diagram shown Figure 9.17, which was derived from the work of Tachmindji and Morgan. The diagram is essentially concerned with the influence of inflow velocities, propeller geometric size and static head and attempts from these parameters, grouped into advance coefficient and cavitation number, to give guidance on the best regions in which to adopt conventional and super-cavitating propellers. Clearly the 'grey' area in the middle of the diagram is dependent amongst other variables on both the wake field fluctuations and also shaft inclination angle. Should neither the conventional nor super-cavitating propeller option give a reasonable answer to the particular design problem, then the further options of waterjet or surface piercing propulsors need to be explored, since these extend the range of propulsion alternatives.

From the early works of Parsons, Barnaby and Thornycroft on models and at full-scale it was correctly concluded that extreme back or suction side cavitation of

TABLE 9.2 Common Formulations of Cavitation Numbers

Definition	Symbol	Formulation
Free stream-based cavitation no.	σ_0	$\dfrac{p_0 - p_v}{\frac{1}{2}\rho v_A^2}$
Rotational speed-based cavitation no.	σ_n	$\dfrac{p_0 - p_v}{\frac{1}{2}\rho(\pi x n D)^2}$
Mean cavitation no.	σ	$\dfrac{p_0 - p_v}{\frac{1}{2}\rho[v_A^2 + (\pi x n D)^2]}$
Local cavitation no.	σ_L	$\dfrac{p_0 - p_v + xRg\cos\theta}{\frac{1}{2}\rho[[v_A(x,\theta) + u_A(x,\theta)]^2 + [\pi x n D - v_T(x,\theta) - u_T(x,\theta)]^2]}$
		u_A and u_T are the propeller-induced velocities
		v_A and v_T are the axial and tangential wake velocities

Sometimes the local cavitation number is calculated without the influence of u_A and u_T, and also u_T when this is not known.

the type causing thrust breakdown could be avoided by increasing the blade surface area. Criteria were subsequently developed by relating the mean thrust to the required blade surface area in the form of a limiting thrust loading coefficient. The first such criterion of 77.57 kPa (11.25 lbf/in^2) was derived in the latter part of the 19th century. Much development work was undertaken in the first half of 20th century in deriving refined forms of these thrust loading criteria for design purposes; two of the best known are those derived by Burrill[22] and Keller.[23]

Burrill's method, which was proposed for fixed pitch, conventional propellers, centers around the use of the diagram shown in Figure 9.18. The mean cavitation number is calculated based on the static head relative to the shaft center line, and the dynamic head is referred to the *0.7R* blade section. Using this cavitation number $\sigma_{0.7R}$, the thrust loading coefficient τ_c is read off from Figure 9.18 corresponding to the permissible level of back cavitation desired. It should, however, be remembered that the percentage back cavitation allowances shown in the figure

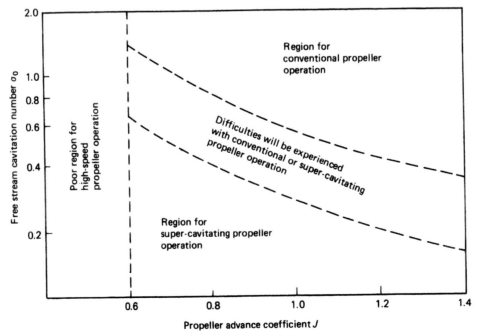

FIGURE 9.17 Zones of operation for propellers.

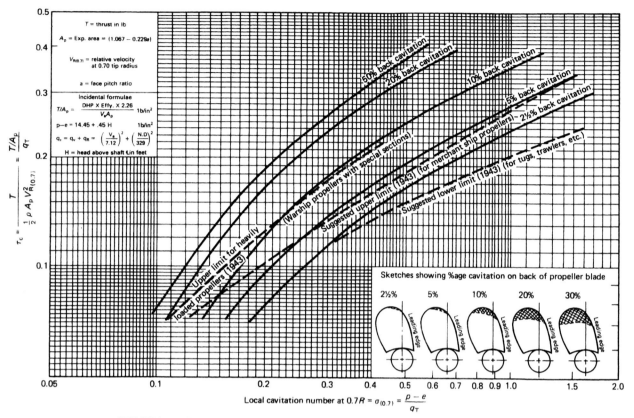

FIGURE 9.18 Burrill cavitation diagram for uniform flow. *Reproduced from Reference 22.*

are based on cavitation tunnel estimates in uniform axial flow. From the value of τ_c read off from the diagram the projected area for the propeller can be calculated from the following:

$$A_P = \frac{T}{\frac{1}{2}\tau_c \rho [V_A^2 + (0.7\pi nD)^2]} \qquad (9.8)$$

To derive the expanded area from the projected area, Burrill derived the empirical relationship which is valid for conventional propeller forms only:

$$A_E = \frac{A_P}{(1.067 - 0.229 P/D)} \qquad (9.9)$$

The alternative blade area estimate is due to Keller and is based on the relationship for the expanded area ratio:

$$\frac{A_E}{A_O} = \frac{(1.3 + 0.3Z)T}{(p_0 - p_v)D^2} + K \qquad (9.10)$$

where

p_0 is the static pressure at the shaft center line (kgf/m^2)
p_v is the vapour pressure (kgf/m^2)
T is the propeller thrust (kgf)
Z is the blade number and
D is the propeller diameter (m).

The value of K in equation (9.10) varies with the number of propellers and ship type as follows: for single-screw ships $K = 0.20$, but for twin-screw ships it varies within the range $K = 0$ for fast vessels through to $K = 0.1$ for the slower twin-screw ships.

Both the Burrill and Keller methods have been used with considerable success by propeller designers as a means of estimating the basic blade area ratio associated with a propeller design. In many cases, particularly for small ships and boats, these methods, and even more approximate ones perhaps, form the major part of the cavitation analysis; however, for larger vessels and those for which measured model wake field data is available, the cavitation analysis should proceed considerably further to the evaluation of the pressure distributions around the sections and their tendency towards cavitation inception and extent.

In Chapter 7 various methods were discussed for the calculation of the pressure distribution around an aerofoil section. The nature of the pressure distribution around an aerofoil is highly dependent on the angle of attack of the section. Figure 9.19 shows idealized typical velocity distributions for an aerofoil in a non-cavitating flow at positive, ideal and negative angles of incidence. This figure clearly shows how the areas of suction on the blade surface change to promote back, mid-chord or face cavitation in the positive, ideal or negative incidence conditions,

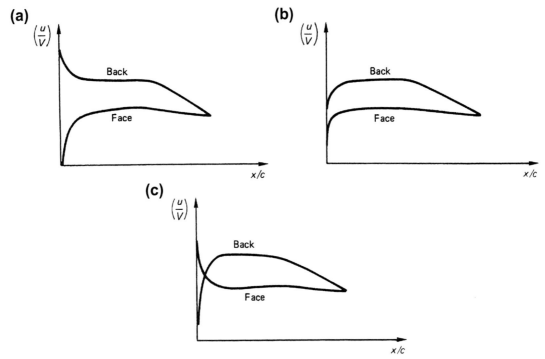

FIGURE 9.19 Typical section velocity distributions: (a) positive incidence; (b) ideal incidence and (c) negative incidence.

respectively. When cavitation occurs on the blade section the non-cavitating pressure distribution is modified with increasing significance as the cavitation number decreases. Balhan[24] showed, by means of a set of two-dimensional aerofoil experiments in a cavitation tunnel, how the pressure distribution changes. Figure 9.20 shows a typical set of results at an incidence of 5 for a Karman–Trefftz profile with thickness and camber chord ratios of 0.0294 and 0.0220, respectively. From the figure the change in form of the pressure distribution for cavitation numbers ranging from 4.0 down to 0.3 can be compared with the results from potential theory; the Reynolds number for these tests was within the range 3×10^6 to 4×10^6. The influence that these pressure distribution changes have on the lift coefficient can be deduced from Figure 9.21, which is also taken from Balhan and shows how the lift coefficient varies with cavitation number and incidence angle of the aerofoil. From this figure it is seen that at moderate to low incidence the effects are limited to the extreme low cavitation numbers, but as incidence increases to high values, 5° in propeller terms, this influence spreads across the cavitation number range significantly.

For propeller blade section design purposes the use of 'cavitation bucket diagrams' is valuable, since they capture in a two-dimensional sense the cavitation behavior of a blade section. Figure 9.22(a) outlines the basic features of a cavitation bucket diagram. This diagram is plotted as a function of the section angle of attack against the section cavitation number; however, several versions of the diagrams have been produced: typically, angle of attack may be replaced by lift coefficient and cavitation number by minimum pressure coefficient. From the diagram, no matter what its basis, four primary areas are identified: the cavitation-free area and the areas where back sheet, bubble and face cavitation can be expected. Such diagrams are produced from systemic calculations on a parent section form and several cases are supported by experimental measurement[25]. The width of the bucket defined by the parameter α_d is a measure of the tolerance of the section to cavitation-free operation. Figure 9.22(b) shows an example of a cavitation bucket diagram based on experimental results using flat-faced sections. This work, conducted by Walchner and published in 1947, clearly shows the effect of the leading edge form on the section cavitation inception characteristics. Furthermore, the correlation with the theoretical limiting line can be seen for shockless entry conditions.

While useful for design purposes the bucket diagram is based on two-dimensional flow characteristics, and can therefore give misleading results in areas of strong three-dimensional flow; for example, near the blade tip and root or in other locations where the loading changes rapidly.

Propeller design is based on the mean inflow conditions that have either been measured at model-scale or estimated empirically using procedures as discussed in Chapter 5. When the actual wake field is known, the cavitation

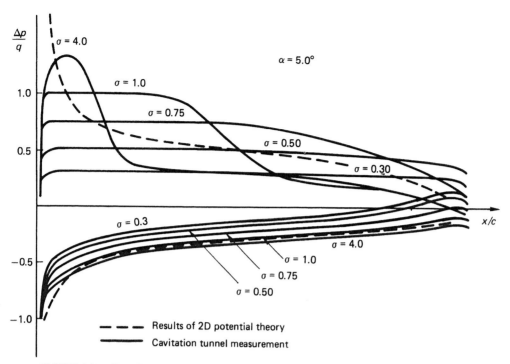

FIGURE 9.20 The effect of cavitation on an aerofoil section pressure distribution. *Reference 24.*

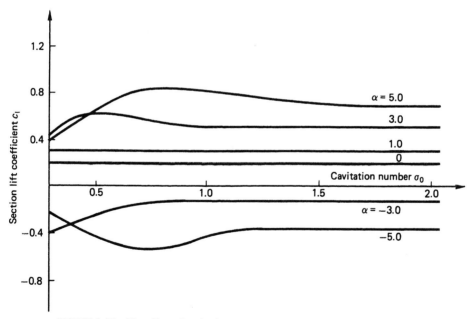

FIGURE 9.21 The effect of cavitation on the section lift coefficient. *Reference 24.*

analysis needs to be considered as the propeller passes around the propeller disc. This can be done either in a quasi-steady sense using procedures based on lifting line methods with lifting surface corrections, or by means of unsteady lifting surface and boundary element methods. The choice of method depends in essence on the facilities available to the analyst and both approaches are commonly used. Figure 8.14 shows the results of a typical analysis carried out for a twin-screw vessel.[26] Figure 8.14 also gives an appreciation of the variability that exists in cavitation extent and type on a typical propeller when operating in its design condition.

The calculation of the cavitation characteristics can be done either using the pseudo-two-dimensional aerofoil

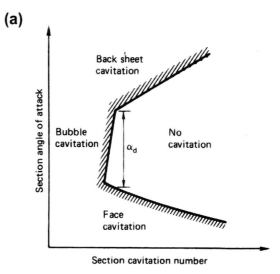
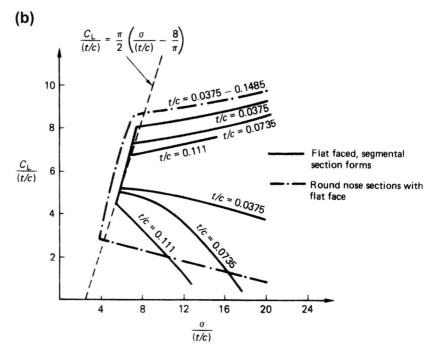

FIGURE 9.22 Cavitation 'bucket' diagrams: (a) basic features of a cavitation bucket diagram and (b) Walchner's foil experiments with flat-faced sections.

pressure distribution approach in association with cavitation criteria or using a cavitation modeling technique; the latter method is particularly important in translating propeller cavitation growth and decay into hull-induced pressures. The use of the section pressure distributions calculated from either a Theodoressen or Weber basis to determine the cavitation inception and extent has been traditionally carried out by equating the cavitation number to the section suction pressure contour as seen in Figure 9.23(a). Such analysis, however, does not take account of the time taken for a nucleus to grow from its size in the free stream to a visible cavity and also for its subsequent decline as well as the other factors discussed in Section 9.1. Although these parameters of growth and decline are far from fully understood, attempts have been made to derive engineering approximations for calculation purposes. Typical of these is that by van Oossanen[26] in which the growth and decay is based on Knapp's similarity parameter.[27] In van Oossanen's approach (Figure 9.23(b)), at a given value of cavitation number σ the nuclei are expected to grow at a position x_{c1}/C on the aerofoil and reach a maximum size at x_{c2}/C, whence the cavity starts to decline in size until it vanishes at a position x_{c3}/C. Knapp's similarity parameter, which is based on Rayleigh's equation

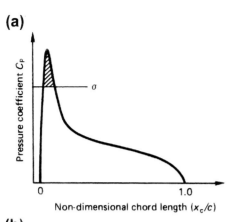

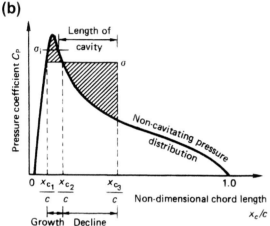

FIGURE 9.23 Determination of cavitation extent: (a) traditional approach to cavitation inception and (b) van Oossanen's approach to cavitation inception.

for bubble growth and collapse, defines a ratio K_n as follows:

$$K_n = \frac{t_D \sqrt{(\Delta p)_D}}{t_G \sqrt{(\Delta p)_G}} \quad (9.11)$$

where t and Δp are the total change times and effective liquid tension producing a change in size, respectively, and the suffixes D and G refer to decline and growth. Van Oossanen undertook a correlation exercise on the coefficient K_n for the National Advisory Committee for Aeronautics (NACA) 4412 profile, which resulted in a multiple regression-based formula for K_n as follows:

$$\log_{10} K_n = 9.407 - 84.88(\sigma/\sigma_i)^2 + 75.99(\sigma/\sigma_i)^3$$
$$- \frac{0.6507}{(\sigma/\sigma_i)} + \log_{10}(\theta_{inc}/c)$$
$$\times x \left[1.671 + 4.565(\sigma/\sigma_i) - 32(\sigma/\sigma_i)^2 \right.$$
$$\left. + 25.87(\sigma/\sigma_i)^3 - \frac{0.1384}{(\sigma/\sigma_i)} \right]$$
$$(9.12)$$

in which θ_{inc} is the momentum thickness of the laminar boundary layer at the cavitation inception location. For calculation purposes it is suggested that if the ratio (θ_{inc}/c) is greater than 0.0003 bubble cavitation occurs and for smaller values sheet cavitation results. As a consequence of equation (9.12) it becomes possible to solve equation (9.11) iteratively in order to determine the value of x_{c3} since equation (9.11) can be rewritten as

$$K_n = \frac{\int_{x_{c2}/c}^{x_{c3}/c} \frac{d(x_c/c)}{V_{x_c}(x_c/c)} \sqrt{\int_{x_{c2}/c}^{x_{c3}/c} [\sigma + C_P(x_c/c)] d(x_c/c)}}{\int_{x_{c1}/c}^{x_{c2}/c} \frac{d(x_c/c)}{V_{x_c}(x_c/c)} \sqrt{\int_{x_{c1}/c}^{x_{c2}/c} -[\sigma + C_P(x_c/c)] d(x_c/c)}}$$

where V_{xc} is the local velocity at x_c.

The starting point x_{c1} for the cavity can, for high Reynolds numbers in the range $1 \times 10^5 < R_{x_{tr}} < 6 \times 10^7$, be determined from the relationship derived by Cebeci et al.[28] as follows:

$$R_{\theta_{tr}} = 1.174 \left[1 + \frac{22\,400}{R_{x_{tr}}} \right] R_{x_{tr}}^{0.46} \quad (9.13)$$

where $R_{\theta_{tr}}$ is the Reynolds number based on momentum thickness and local velocity at the position of transition, and $R_{x_{tr}}$ is the Reynolds number based on free stream velocity and the distance of the point of transition from the leading edge. For values of $R_{\theta_{tr}}$ below this range, the relationship

$$R_{\theta_{ci}} = 4.048 \, R_{x_{ci}}^{0.368} \quad (9.14)$$

holds in the region $1 \times 10^4 < R_{x_{ci}} < 7 \times 10^5$. In this case, $R_{\theta_{ci}}$ is the Reynolds number based on local velocity and momentum thickness at the point of cavitation inception, and $R_{x_{ci}}$ is the Reynolds number based on distance along the surface from the leading edge and free stream velocity at the position of cavitation inception. Having determined the length of the cavity, van Oossanen extended this approach to try and approximate the form of the pressure distribution on a cavitating section, and for these purposes assumed that the cavity length is less than half the chord length of the section. From work on the pressure distribution over cavitating sections it is known that the flat part of the pressure distribution, Figure 9.20 for example, corresponds to the location of the actual cavity. Outside this region, together with a suitable transition zone, the pressure returns approximately to that of a non-cavitating flow over the aerofoil. Van Oossanen conjectured that the length of the transition zone is approximately equal to the length of the cavity and the resulting pressure distribution approximation is shown in Figure 9.24.

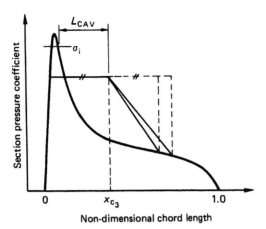

FIGURE 9.24 Van Oossanen's approximate construction of a cavitation pressure distribution on an aerofoil section.

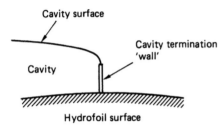

FIGURE 9.25 Riabouchinsky-type cavity termination 'wall'.

Ligtelijn and Kuiper[29] conducted a study to investigate the importance of the higher harmonics in the wake distribution on the type and extent of cavitation, and as a consequence give guidance on how accurately the wake should be modeled. Their study compared the results of lifting surface calculations with the results of model tests in a cavitation tunnel where the main feature of the wake field was a sharp wake peak. It was concluded that the lower harmonics of the wake field principally influence the cavity length prediction and that the difference between two separate calculations based on four and ten harmonics was negligible.

Considerable work has been done in attempting to model cavitation mathematically. The problem is essentially a free streamline problem, since there is a flow boundary whose location requires determination as an integral part of the solution. Helmholtz and Kirchoff in the latter part of the nineteenth century attempted a solution of the flow past a super-cavitating flat plate at zero cavitation number using complex variable theory. Subsequently, Levi-Civita extended this work to include the flow past curved bodies. The zero cavitation number essentially implied an infinite cavity, and the next step in the solution process was to introduce finite cavitation numbers which, as a consequence, introduce finite sized cavities. The finite cavity, however, requires the cavity to be terminated in an acceptable mathematical and physical manner. Several models have been proposed, amongst which the Riabouchinsky cavity termination model, which employs a 'wall' to provide closure of the cavity (Figure 9.25), and the more physically realistic re-entrant jet model (Figure 9.8), are examples. These models, most of which were developed in the late 1940s, are non-linear models which satisfy the precise kinematic and dynamic boundary conditions over the cavity surface. As a consequence considerable analytical complexity is met in their use. Tulin[30] developed a linearized theory for zero cavitation number and this was extensively applied and extended such that Geurst[31] and Geurst and Verbrugh[32] introduced the linearized theory for partially cavitating hydrofoils operating at finite cavitation numbers, and extended this work with a corresponding theory for super-cavitating hydrofoils[33].

Three-dimensional aspects of the problem were considered by Leehey[34] who proposed a theory for super-cavitating hydrofoils of finite span. This procedure was analogous to the earlier work of Geurst on two-dimensional cavitation problems in that it uses the method of matched asymptotic expansions from which a comparison can be made with the earlier work. Uhlman,[35] using a similar procedure, developed a method of analysis for partially cavitating hydrofoils of finite span. With the advent of large computational facilities significantly more complex solutions could be attempted. Typical of these is the work of Jiang[36] who examined the three-dimensional problem using an unsteady numerical lifting surface theory for super-cavitating hydrofoils of finite span using a vortex source lattice technique.

Much of the recent work is based on analytical models which incorporate some form of linearizing assumptions. However, techniques now exist, such as boundary integral or surface singularity methods, which permit the solution of a Neumann, Dirichlet or mixed boundary conditions to be expressed as an integral of the appropriate singularities distributed over the boundary of the flow field. Uhlman,[37] taking advantage of these facilities, has presented an exact non-linear numerical model for the partially cavitating flow about a two-dimensional hydrofoil (Figure 9.26). His approach uses a surface vorticity technique in conjunction with an iterative procedure to generate the cavity shape and a modified Riabouchinsky cavity termination wall to close the cavity. Comparison with Tulin and Hsu's earlier thin cavity theory[38] shows some significant deviations between the calculated results of the non-linear and linear approaches to the problem.

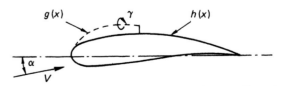

FIGURE 9.26 Uhlman's non-linear model of a two-dimensional partially cavitating flow.

Stern and Vorus[39] developed a non-linear method for predicting unsteady sheet cavitation on propeller blades by using a method which separates the velocity potential boundary value problem into a static and dynamic part. A sequential solution technique was adopted in which the static potential problem relates to the cavity, fixed instantaneously relative to the blade, while the dynamic potential solution addresses the instantaneous reaction of the cavity to the static potential and predicts the cavity deformation and motion relative to the blades. In this approach, because the non-linear character of the unsteady cavitation is preserved, the predictions from the method contain many of the observed characteristics of both steady and unsteady cavitation behavior. Based on this work two modes of cavity collapse were identified, one being a high-frequency mode where the cavity collapsed towards the trailing edge whilst the second was a low-frequency mode where the collapse was towards the leading edge.

Isay,[40] in association with earlier work by Chao, produced a simplified bubble grid model in order to account for the compressibility of the fluid, surrounding a single bubble. From this work the Rayleigh–Plesset equation (9.3) was corrected to take account of the compressibility effects of the ambient fluid as follows:

$$\frac{d^2R}{dt^2} + \frac{3}{2R}\left(\frac{dR}{dt}\right)^2 = \frac{1}{\rho R}\left(p_G - \frac{2S}{R} - p_\infty e^{-\alpha/\alpha_{**}} + p_v e^{-\alpha/\alpha_{**}}\right) \quad (9.15)$$

where p_v and p_∞ are vapor pressure and local pressure in the absence of bubbles, α is the local gas volume ratio during bubble growth, α_{**} is an empirical parameter and S is the surface tension. Furthermore, Isay showed that bubbles growing in an unstable regime reach the same diameter in a time-dependent pressure field after a short distance. This allows an expression to be derived for the bubble radius just prior to collapse. Mills[41] extended the above theory, which was based on homogeneous flow, to inhomogeneous flow conditions met within propeller technology and where local pressure is a function of time and position on the blade.

Following this theoretical approach equation (9.3) then becomes

$$\frac{3\omega^2}{2}\left(\frac{\partial R_{\phi 0}}{\partial \chi}\right)^2 + R_{\phi 0}\omega^2\left(\frac{\partial^2 R_{\phi 0}}{\partial \chi^2}\right)$$
$$= \frac{1}{\rho}[p(R_{\phi 0}) - p(\chi, \phi_0 - \chi)] \quad (9.16)$$

from which computation for each class of bubble radii can be undertaken.

In equation (9.16) χ is the chordwise co-ordinate, ω is the rate of revolution and ϕ_0 is the instantaneous blade position.

For computation purposes the gas volume $\alpha_{\phi 0}$ at a position on the section channel can be derived from

$$\frac{\alpha_{\phi 0}(\chi)}{1 + \alpha_{\phi 0}(\chi)} = \frac{4\pi}{3}\sum_{j=1}^{J}\xi_{0j}.R_{0j}^3(\chi, \phi_0 - \chi) \quad (9.17)$$

in which ξ_{0j} is the bubble density for each class and R_0 is the initial bubble size. Using equations (9.16) and (9.17) in association with a blade undisturbed pressure distribution calculation procedure (Chapter 8), the cavity extent can be estimated over the blade surface.

Vaz and Bosschers[72] adapted a partially non-linear model in which the kinematic and dynamic boundary conditions are applied in the partially cavitating flow case on the surface of the blade below the cavity surface. In contrast for the super-cavitating case the conditions are applied at the cavity surface. While the method can, in the case of the two-dimensional partial cavitation, be improved by a Taylor expansion of the velocities on the cavity surface based on the cavity thickness. Their analysis and cavitation modeling procedure when applied to the prediction of sheet cavitation in steady flow for the MARIN S propeller, designed such that sheet cavitation is only present, has shown good correlation with the experimental observations although this correlation appeared to be load dependent: showing underprediction at lower loadings. In an alternative correlation exercise with the INSEAN E779A propeller, which has a higher tip loading leading to both partial and super-cavitation in the outer regions of the blade in addition to a cavitating tip vortex, the blade cavitation extent was reasonably well predicted.

In an alternative approach to propeller sheet cavitation prediction, Sun and Kinnas[73] have used a viscous–inviscid interactive method of analysis. In their approach the inviscid wetted and cavitating flows are analyzed using a low-order potential boundary element analysis based on a thin cavity modeling approach. Then by making the assumption of a two-dimensional boundary layer acting in strips along the blade, the effects of viscosity on the wetted and cavitating flows are taken into account by coupling the inviscid model with a two-dimensional integral boundary layer analysis procedure. Comparison of the results from this procedure with the first iteration of a fully non-linear cavity approach,[74,75] which itself had been validated from a FLUENT Computational Fluid Dynamics (CFD) modeling, has shown good correlation with the differences being negligible when the cavities are thin.

9.4 CAVITATION INCEPTION

Cavitation inception is defined as taking place when nuclei, due to being subjected to reduced pressure, reach a critical size and grow explosively.

The mechanisms underlying cavitation inception are important for a number of reasons; for example, in predicting the onset of cavitation from calculations and interpreting the results of model experiments to make full-scale predictions.

Cavitation inception is a complex subject which is far from completely understood at the present time. It is dependent on a range of characteristics embracing the nuclei content of the water (see Chapter 4), the growth of the boundary layer over the propeller sections and the type of cavitation experienced by the propeller. Thus it is not only related to the environment in which the propeller is working but also to intimate details of the propeller geometry, flow velocities and the wake field.

The nuclei content of the water has been shown to be important in determining the cavitation extent over the blades of a propeller in a cavitation tunnel: in particular tip vortex cavitation is very sensitive to the nuclei content of the water. The free air content as a proportion of the nuclei content, rather than oxygen or total air content, should therefore be measured during cavitation experiments. Figure 23.4 demonstrates this somewhat indirectly in terms of the cavitation erosion rate and its variability with air content of the water. This figure implies that the structure and perhaps extent of the cavity changes with air or gas content. Kuiper[42] explored the effect of artificially introducing nuclei into the water by electrolysis techniques. When electrolysis is used the water is decomposed into its hydrogen and oxygen components, the amount of gas produced being dependent only on the current applied. The governing equations for this are

$$4H_2O + 4e \rightarrow 2H_2\uparrow + 4OH^- \text{ (at the cathode)}$$

$$4OH^- + 4e \rightarrow 2H_2O + O_2\uparrow \text{ (at the anode)}$$

Since the electrolysis method produces twice the amount of hydrogen when compared to the amount of oxygen, the cathode is generally used for the production of the bubbles. This method is also known under the name of the 'hydrogen bubble technique' for flow visualization. Kuiper showed that this technique, when introduced into the flow, can have a significant effect on the observed cavitation over the blades; the extent clearly depends on the amount of nuclei present in the water initially. Figure 9.27 demonstrates this effect for sheet cavitation observed on a propeller; similar effects can be observed with bubble cavitation — to the extent of its not being present with low nuclei concentrations and returns with enhanced nuclei content. Care, however, needs to be taken in model tests not to over-seed the flow with nuclei such that the true cavitation pattern is masked.

In the case of a full-scale propeller the boundary layer is considered to be fully turbulent except for a very small region close to the leading edge of the blade. This is not

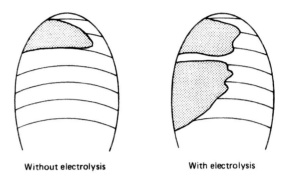

Without electrolysis   With electrolysis

FIGURE 9.27 Effect of electrolysis (nuclei content) on cavitation inception.

always the case on a model propeller, also shown by Kuiper[42] using paint pattern techniques on models. The character of the boundary layer on the suction side of a propeller blade is shown in Figure 9.28. In the region where the loading is generally highest, in the outer radii of the blade, a short laminar separation bubble AB can exist near the leading edge, causing the boundary layer over the remainder of the blade at the tip to be turbulent. A separation radius BC, whose position is dependent on the propeller loading, may also be found, as shown in the figure, below which the flow over the blade is laminar. The region CD is then a transition region whose chordwise location is dependent on Reynolds number but is generally located at some distance from the leading edge. The region aft of the line DE is a region of laminar separation at midchord due to the very low sectional Reynolds numbers at those radii in combination with thick propeller sections. The locations of the points B, C and D in specific cases are strongly dependent on the propeller geometric form, the propeller loading and flow Reynolds number. The boundary layer on the pressure face of the blades is generally considerably less complex: under normal operating

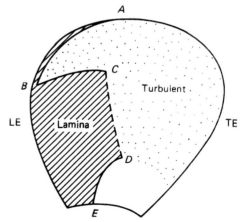

FIGURE 9.28 Schematic representation of the boundary layer on the suction side of a model propeller in open water.

conditions no laminar separation occurs and a significant laminar region may exist near the leading edge. Transition frequently occurs more gradually than on the suction surface due to the more favorable pressure gradient.

Because the boundary layer can be laminar over a considerable region of the blade and an increase in Reynolds number does not generally move the transition region to the leading edge of the blade, some testing establishments have undertaken experiments using artificial stimulation of the boundary layer to induce turbulent flow close to the leading edge. Such stimulation is normally implemented by gluing a small band of carborundum grains of the order of 60 μm at the leading edge of the blades. Figure 9.29 shows the effect of stimulating a fully turbulent boundary layer over the blades for the same propeller conditions as shown in Figure 9.27; in this case the introduction of electrolysis in addition to the leading edge roughening had little further significant effect. The use of leading edge roughening is, however, not a universally accepted technique of cavitation testing among institutes. Consequently, the interpretive experience of the institute in relation to its testing procedure is an important factor in estimating full-scale cavitation behavior.

It is frequently difficult to separate out the effects due to nuclei changes and Reynolds effects since the parameters involved change simultaneously if the tunnel velocity is altered. Furthermore, the cavitation patterns expected at full-scale are normally estimated from model test results, and consequently it is necessary to interpret the model test results. The International Towing Tank Conference (ITTC) Proceedings[43] give a distillation of the knowledge on this subject, detailed below, and in so doing consider the cases of a peaked pressure distribution, a shock-free entrance condition and a 'flat' pressure distribution. Figure 9.30, taken from[43], shows these three cases, and for each case considers the following: (i) a typical boundary layer distribution over the suction surface; (ii) a typical cavitation pattern with few nuclei at moderate Reynolds numbers, of the order of 2×10^5; (iii) the effect of increasing nuclei; (iv) the effect of increasing Reynolds number and (v) the expected full-scale extrapolation.

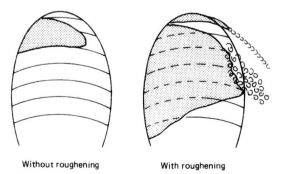

Without roughening   With roughening

FIGURE 9.29 **Effect of LE roughening (turbulence stimulation) on cavitation inception.**

For a peaked pressure distribution, Figure 9.30(a), if the flow is separated a smooth glassy sheet will be observed, whereas if the flow is attached no cavitation inception may occur, although the minimum pressure may be below the vapor pressure. In the latter case the flow is sensitive to surface irregularities and these can cause some streaks of cavitation as seen in the figure. With the water speed held constant the effect of increasing the nuclei content on the sheet cavitation in a region with a laminar separation bubble is negligible. In this case only a few nuclei enter the cavity, and therefore the increase of the partial pressure of the gas is small and cavitation inception is hardly affected. In the case of an attached laminar boundary layer in association with a peaked pressure distribution, the effect of increasing nuclei content is also small, although the number of streaks may increase. Furthermore, if the pressure peak is not too narrow some bubble cavitation may also be noticed. The effect of Reynolds number on sheet cavitation in a separated region is small; however, the appearance of the cavity becomes rather more 'foamy' at higher Reynolds numbers. In the alternative case of a region of attached laminar flow the effect of Reynolds number is indirect, as the boundary layer becomes thinner and, as a consequence, the surface irregularities become more pronounced. This has the effect of increasing the number of streaks, which at very high Reynolds numbers or speeds will tend to merge into a 'foamy' sheet. In these cases the character of the cavity at the leading edge remains streaky, with perhaps open spaces between the streaks. When extrapolating the observations of cavitation resulting from a peaked pressure distribution in the case of a smooth sheet cavity the boundary layer at model scale normally has a laminar separation bubble. As a consequence scale effects on inception and developed cavitation are likely to be small in most cases. When regions of attached laminar flow occur scale effects tend to be large. In such cases the application of leading edge roughness may be necessary or, alternatively, the tests should be conducted at higher Reynolds numbers. The cavitation streaks found in attached laminar flow regions indicate the presence of a sheet cavity at full scale, as seen in Figure 9.30(a).

In the case of a shock-free entry pressure distribution, that shown in Figure 9.30(b), the boundary layer over the model propeller for Reynolds numbers of the order of 2×10^5 at $0.7R$ changes from that seen from the peaked pressure distribution in Figure 9.30(a). For this type of pressure distribution bubble cavitation can be expected, and its extent is strongly dependent on the nuclei content of the water, as seen in Figure 9.30(b). In contrast the effect of Reynolds number is small for this kind of pressure distribution. Nevertheless, it must be remembered that the nuclei content may change with speed, as does the critical pressure of the nuclei, and this can result in an increase in bubble cavitation. Also due to the thinner boundary layer at the

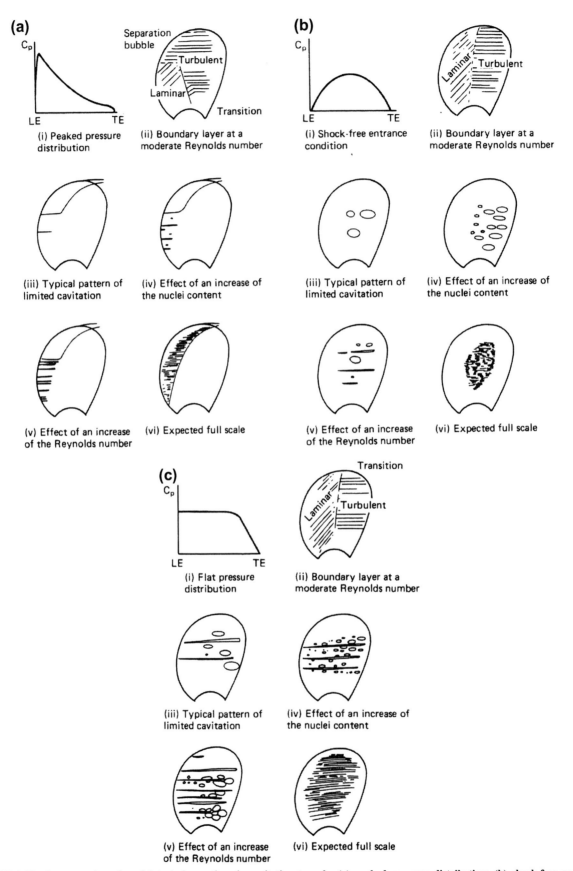

FIGURE 9.30 Interpretation of model test observations in cavitation tunnels: (a) peaked pressure distribution; (b) shock-free pressure distribution and (c) flat pressure distribution. *Reproduced from Reference 43.*

higher Reynolds number, surface irregularities may generate nuclei more readily, which can result in streak-like rows of bubble or spot-like cavitation. The scale effects for this type of pressure distribution often occur at both inception and with developed cavitation. Clearly the nuclei content at model scale should be as high as possible, as should the Reynolds number. Furthermore, the application of leading edge roughness can assist in reducing scale effect. When bubble cavitation occurs at model-scale, the full-scale cavity is expected to take the form of a 'frothy' cloud which can have consequences for the erosion performance of the blades, as will be seen in Section 9.5.

With a flat pressure distribution (Figure 9.30(c)), bubble cavitation can also be expected to occur. The bubbles reach their maximum size at or beyond the constant pressure region and long streaks of cavitation, which originate at the leading edge, may also occur. These streaks may give the appearance of merging bubble rows, so that they have a cloudy appearance, and the cavities are found to be very unstable. The effect of increasing the nuclei content has a similar effect to that for bubble cavitation in that the bubbles become smaller and more extensive. In addition, the number of streaks may increase and the cavities remain unstable. If roughness is applied, a sheet cavity is formed, and this has a somewhat cloudy appearance at its trailing edge. The influence of Reynolds number on a flat pressure distribution is particularly pronounced: the number of streaks increase, which frequently results in the formation of a sheet cavity instead of bubble cavitation. The extrapolation to full-scale results is a cloudy sheet cavitation, as seen in Figure 9.30(c).

In the foregoing discussion of cavitation inception, no mention has been made of tip vortex cavitation. Cavitation of the vortices which emanate from propeller blade tips is a rather poorly understood phenomenon; this is partly due to a general lack of understanding of the complex flow regime which exists at the propeller tip. Tip vortex cavitation is very often one of the first forms of cavitation to be observed in model tests and the prediction of the onset of this type of cavitation is particularly important in the design of 'silent' propellers, as a cavitating vortex represents a significant source of noise. The cavitating tip vortex is subject to Reynolds scaling effects and McCormick[77] proposed a scaling procedure to predict the full-scale behavior. The relationship he derived is given by:

$$\sigma_{fs}/\sigma_{ms} = (Re_{fs}/Re_{ms})^{0.35}$$

where the Reynolds number is conventionally defined as $Re = nD^2/\nu$ and the suffixes fs and ms refer to full- and model-scale, respectively. Resulting from the use of this relationship at various institutes, a number of variants are seen depending upon their experience in that indices ranging from about 0.25–0.4 are in use. In current practice values closer to the higher end of the range tend to be favored. Tip vortex cavitation occurs in the low-pressure core of the tip vortex; in a recent experimental study Arakeri et al. found a strong coupling effect between velocity and the dissolved gas content in a cavitation tunnel on the tip vortex cavitation inception. Observations in a cavitation tunnel show that the radius of the cavitating core of a tip vortex near inception is relatively constant with the distance from the blade tip. However, the strength of the tip vortex increases with distance behind the tip of the propeller due to the roll-up of the vortex sheets: this increase in strength occurs rapidly with distance in the initial stages. This explains why cavitation of the tip vortex is sometimes noticed to commence some distance from the propeller tip; however, this depends on the nature of the boundary layer over the blade in the tip region. If the boundary layer separates near the tip, then an attached tip vortex results, whereas if the boundary layer is laminar near the tip then the tip vortex is detached. Kuiper[42] suggests that the radius of the cavitating core of a tip vortex is independent of Reynolds number and nuclei content, and consequently this can be used as a basis for the determination of cavitation inception, both on the model and on the full-scale propeller. Based on data from Chandrashekhara[44] and also tests on a propeller especially designed to study tip vortex phenomena, Kuiper suggests the following relationship to give an approximation to the inception index for a tip vortex:

$$\sigma_{ni} = 0.12(P/D - J)_{0.9R}^{1.4} \cdot R_n^{0.35} \qquad (9.18)$$

It is also suggested that this relationship gives a good initial estimate for both conventional and strongly tip unloaded propellers at model-scale.

Within the general field of fluid mechanics and aerodynamics the phenomenon of vortex bursting has been extensively researched. This effect manifests itself as a sudden enlargement of the vortex, which then gives rise to a particularly confused flow regime. English[45] discusses this phenomenon in relation to the cavitating tip vortices of a series of container ship propellers.

Face cavitation has long been an anathema to propeller designers, principally because of its potential link with erosion coupled with the face of the propeller having a tensile stress field distributed over its surface. Consequently, the margin in design against face cavitation has normally been reasonably robust, perhaps of the order of $0.25K_T$, but opinion in recent years has been rather less conservative: in part due to a greater phenomenological understanding of cavitation, experience from modern propeller designs in finding that face cavitation is not as erosive as it was originally thought and also in order to give a greater flexibility to deal with the problems of back cavitation. Face sheet cavitation can have some rather different properties from those encountered on the suction side, or back, of the propeller blade. First, it should be noted that the surface pressure distribution over the section giving rise to face cavitation will normally be rather different from that causing suction side cavitation and this tends to result in the production of a rather less stable cavity formation due to it being relatively thicker. A further difference is to be found in the behavior of the re-entrant jet because face cavitation tends towards a more two-dimensional character than back sheet cavitation. Consequently, the shedding mechanism is likely to be different in the case of face cavitation, which is usually found to occur in the inner radii of the propeller as distinct from back cavitation, leaving to one-side blade root situations, which is normally found in the outer regions of the blade. Considerations of this type have led Bark[78] to develop a working hypothesis relating to the number n_s of shed cavities from the leading edge stated as follows:

A sheet cavity, particularly on the pressure side, behaving locally 2-D has a low risk for generation of erosion if for the number n_s of shed cavities per global cavitation cycle it holds that $n_s > a$, where a is an empirically determined acceptable number of shed cavities per global cavitation cycle T_{cav} as defined by the periodic inflow to the propeller blade. In the limiting case of one shedding per global cavitation cycle, a=1, the erosion can be severe, and for a low risk of erosion significantly higher values of a are required, possibly a > 10 or more.

There is an upper limit of a above which the shed cavities can be synchronized according to the behaviour discussed in Section 3.1.5, p. 53[78] and the cavitation instead can become erosive.

This criterion has been considered by Moulijn et al.[79] in in relation to a series of full- and model-scale observations and computational studies in relation to a Ro/Ro container ship and they concluded that this is a very promising tool for the prediction of the erosiveness of face cavitation but that it will require some further development.

9.5 CAVITATION-INDUCED DAMAGE

There have been few propellers designed for surface ships which do not induce cavitation at some point in the propeller disc: not all propellers, however, exhibit cavitation erosion. Therefore, it is wrong to simply equate the presence of cavitation on the blades at some point within a propeller revolution to the certainty of erosion of the propeller material.

Early cavitation damage models were largely based on an observer's experience when looking at the type of cavitation exhibited on a model propeller in a cavitation tunnel together with heuristic conclusions about its stability or, alternatively, relying on empirically derived margins. These margins were usually conservative in their prediction and as a consequence tended to promote propeller blade areas and section designs with significant margins of safety; these margins being mostly but not always justified. As such, these largely heuristic approaches in most cases took the general form:

$$\text{Damage} = f(\sigma_0, \tau_c, \text{model test cavitation type and stability}) \quad (9.19)$$

where σ_0 related to a representative cavitation number and τ_c was a thrust loading coefficient.

Cavitation erosion in the majority of cases is thought to derive from the action of traveling bubbles which either pass around the aerofoil, pass around a fixed cavity or break off from a fixed cavity. Figure 9.31(a) shows a schematic representation of the collapse of a bubble which in this case has passed around the outside of a fixed part of a cavity located on the surface of an aerofoil section. As the pressure recovers the bubble reaches the collapse point in the stagnation region behind the downstream end of the fixed cavity. The collapse mechanism generates a set of shock pressure contours, the magnitude of the pressure on each idealized contour being inversely proportional to the radius from the point of collapse. In addition to the pressures generated by bubble collapse, if the collapse mechanism takes place close to a solid boundary surface a microjet is formed which is directed towards the surface; Figure 9.31(b). The formation of a microjet in the proximity of a wall can be explained, albeit in somewhat simplified terms, by considering a spherical bubble close to a rigid wall starting to collapse. If the spherical form of the bubble were to be maintained during collapse the radial motion of the water would need to be uniform at all points around the bubble during collapse. However, the presence of the wall restricts the water flow to the collapsing bubble in the regions of the bubble adjacent to the wall.

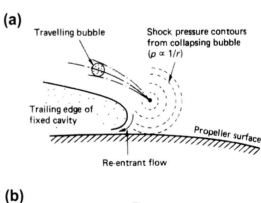

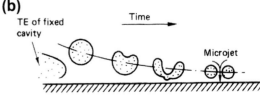

FIGURE 9.31 Erosive mechanisms formed during bubble collapse: **(a) pressure waves from bubble collapse and (b) microjet formation close to surface.**

As a consequence the upper part of the bubble, remote from the wall, tends to collapse faster, leading to a progressive asymmetry of the bubble as shown in Figures 9.5 and 9.31(b), which induces a movement of the bubble centroid towards the wall and creates a linear momentum of the bubble centroid towards the wall. This leads to an acceleration of the virtual mass of the bubble towards the wall as the collapse progresses, resulting eventually in the formation of a high-velocity microjet: with velocities thought to be up to the order of 1000 m/s.[46]

In addition to these mechanisms, the potential to encounter cavity rebound activity is also of importance. Rebound as discussed earlier is the re-growth of the vapor phase of the cavity and this mechanism is thought to provide an important contribution to the damaging process originating from the microjet and pressure wave attack. This is particularly so as the first and sometimes the second rebound pressures can in some instances significantly exceed that of the initial collapse.

The occurrence of material erosion will be dependent on a complex function which among other factors will include:

1. The energy involved in the cavity collapse mechanism.
2. The exposure time of the material surface to the cavity collapse energy.
3. The distance of the cavity collapse from the material surface.

In a practical shipboard case the energy transfer rate of the cavity collapse is not a constant with time since the cavity structures have temporal and spatial variations associated with them. The distance between the cavity collapse process and the material surface is also important, particularly for erosion initiation, and this is a variable once erosion has commenced. This is because it is likely that the eroded surface moves further from the cavity collapse location due to the erosion. The time during which the material surface is exposed to the cavity collapse energy is a significant parameter because this defines the amount of work-hardening that the material will suffer in its surface layers.

Frequently, the first stages of erosion, known as orange peeling and seen in Figure 9.32(a), will commence after a period in service and then may progress little further during the life of the propeller. In other cases the orange peeling will progress to light erosion and then again cease to progress further: Figure 9.32(b) shows relatively light erosion developing close to the propeller leading edge. In more serious instances the erosion will progress beyond light erosion into the material to a depth of some millimeters before ceasing to progress further. In one case, known to the author, of a controllable pitch propelled large passenger ship the erosion progressed to an average depth of 38.6 mm in the blade root sections before arresting itself. In severe cases of cavitation attack on the propeller blades where the collapse energies are large then the erosion will progress until full penetration of the propeller blade has been achieved; Figure 9.32(c). In some cases this process will take months or even years, while in others just a few hours.

It is has been noted that the surface morphology of cavitation erosion alters significantly depending upon whether the attack is on virgin or weld repaired material. In the latter case the resulting surface tends to be considerably smoother as can be seen by comparing Figures 9.33(a) and (b). Differences such as these imply that further dimensions may manifest themselves in the erosion process. This may be due to differences in material hardness and also in the relative electrical and hence corrosion properties of the material.

At the present time the detailed mechanisms that apply to cavitation erosion are far from well understood. There

FIGURE 9.32 Typical blade surface damage morphology (a) orange peel; (b) progression to light erosion and (c) severe erosion with some full penetration.

FIGURE 9.33 Comparison of the erosion morphologies of (a) cast and (b) welded material.

are a number of potential candidates for contributing to the mechanism and these are:

- Bubble collapse and rebound.
- Microjet formation from a collapsing bubble.
- Clouds of micro-bubbles collapsing.
- Cavitating vortices.

In the case of spherical bubble collapse it has been shown that high instantaneous internal pressures and temperatures can be generated and these may lead to strong pressure waves being generated.[82] Phillipp[83] suggested that in order to move towards the solution of the erosion problem it is necessary to consider a number of factors: the shock wave deriving from the cavity collapse, the action of the microjet and the toroidal vortex structure which forms after the microjet has impinged on the surface from the collapsing bubble. Figure 9.34 shows the measured pressures during a single bubble collapse from which it is seen that in this case the instantaneous peak pressure reached a value close to 10 MPa.[84] Furthermore, from the figure it can be seen that the peak pressure occurred in the latter stages of bubble collapse when the microjet is well formed. Interestingly, it can also be ascertained that the duration of the peak pressure event lasts only a few microseconds and this clearly has implications for any attempts at computational modeling of the processes.

In the case of clouds of micro-bubbles collapsing this phenomenon is characterized by cascades of implosions.[85]

It is thought that the pressure wave created by the collapse process of a particular bubble in the fluid then influences the collapse velocities of other bubbles in the neighborhood. This process then increases the amplitudes of the pressure waves generated from these subsequent collapse processes. The collapse mechanism includes both bubble collapse and rebound which if it occurs may progress though several cycles due to the non-equilibrium thermodynamic characteristics of the process. Reisman et al.[86] showed that depending on the geometry and nature of the cloud the shock wave that propagates into the cloud can significantly strengthen near the center due to geometric focusing and, thereby, increase the erosive potential of the total collapse.

Kawanami et al.[98] studied the process from the break-off of a cloud cavity to its eventual decay in the vicinity of a hydrofoil. In the early stage of the break-up the cloud cavity comprised a cavitating vortex core with both ends of the vortex limbs on the aerofoil. Around the vortex core were many small bubbles surrounding it and circulation could be clearly seen. Then when moving downstream the vortex core and its surrounding bubbles were observed to split into two or more segments, mostly two, and the central portion of the vortex bubble disappeared together with most of its associated bubbles. The remnants of the core, together with the congregation of bubbles, momentarily remained in an approximately normal position relative to the aerofoil. The final stage of the collapse was a movement and focusing along the axis of the remaining vapor structure towards the aerofoil after which pitting could be observed on the aerofoil.

In the case of cavitating vortices these appear in shear flows that might be found, for example, in the wakes of bluff bodies and can be the cause of severe erosion in machinery. In such cases there are two aspects which seem to suggest themselves as possible origins for their high erosive potential:

- Foamy cloud formations at the end of collapse mechanisms in which cascade processes can occur.
- The rather long duration of the loading time: typically some tens of microseconds.

In terms of the aggressiveness of the various cavitating structures many studies have been undertaken by various investigators. These tend to indicate that the pressure waves derived from the collapse of either micro-bubbles or clouds of micro-bubbles have amplitudes of around 100 MPa with durations of about 1 μs. In contrast, cavitating vortices can develop impacting jet pressures in excess of 100 MPa with durations of the order of 10 μs and above. Empirical evidence gained at full- and model-scale has shown that whenever cavitating vortical loop structures are seen in the proximity of the end of a collapsing cavitation structure the attendant erosion can be particularly severe, Figure 9.35.

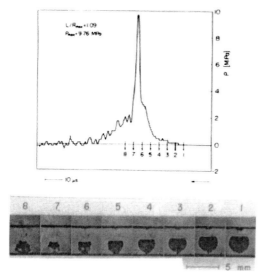

FIGURE 9.34 Pressure signature generated from a single bubble collapse. *Reference 86.*

This is an interesting observation in the context to Kawanami's studies on the break-up of cloud structures.[98]

Single or clouds of bubbles when they collapse in the proximity of a propeller blade surface inflict a time dependent distribution of loading on to the outer layers of

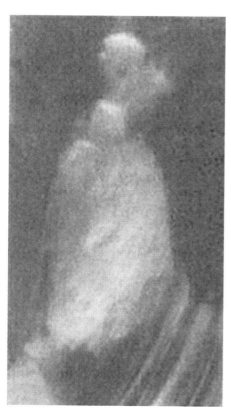

FIGURE 9.35 Loop vortex at the end of a collapsing cavitation structure (note the blade leading edge is at the bottom of the picture).

the material. Such an action induces a work hardening mechanism into the material and the general characteristics of the through thickness hardening profile can be seen in Figure 9.36. The micro-hardness measurements shown in the figure relate to a specimen from a nickel—aluminum bronze propeller material where the suction surface has been the subject of cavitation attack while the pressure surface showed no signs of erosion manifesting itself. It is seen in the figure that there are three traverses recorded:

1. In a region of relatively severe cavitation erosion on the suction surface.
2. In a region of light cavitation erosion.
3. On the pressure surface.

From Figure 9.36 it can be seen that when under heavy cavitation attack and consequent material loss, Traverse No.1, the material work hardens considerably in the surface layers. Indeed, the degree of hardening induced at the surface is such that it almost doubles over the outer 3 mm from the core material values. This implies a significant level of embrittlement of the material has taken place. When under lesser levels of cavitation attack and with less material loss the degree of work hardening becomes less, as seen for Traverse No.2. Both of these traverses were conducted on material from the propeller blade suction surface. However, when moving to the pressure surface, Traverse No.3, it is seen that the hardness near the material surface is essentially the value recorded throughout the body of the material. In this case the only hardness perturbation is that due to the mechanical finishing operations during the manufacture of the outer surface layers of the propeller blade. Analogous results have been measured in Duplex stainless steels that have experienced cavitation erosion.

When examining propeller blades that have been subjected to cavitation erosion, in some cases it has been observed that color tinting marks have been present on the surrounding material. In the relatively few instances where this has been positively identified, this is a surprising observation given that for this to have occurred extremely high local temperatures must have been experienced in the collapse process to offset the heat sink provided by the rest of the material and also the surrounding sea water. Notwithstanding these occasional full-scale observations, Figure 9.37 shows a mild steel specimen which was subjected to cavitation attack using a high-frequency vibration exciter. From the figure it is seen that oxide-induced temper coloring is visible on the material, indicating that the material in the vicinity of the erosion has experienced local temperatures of at least 300°C when immersed during the test in a simulated sea water bath maintained at 20°C. As with the full-scale observations, to achieve such a temperature rise it is likely that considerably higher local temperatures were encountered at the erosion site. Indeed,

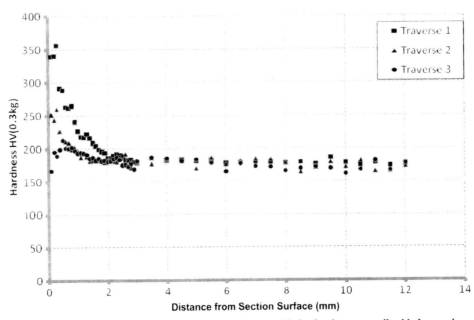

FIGURE 9.36 Micro-hardness measurements made on a nickel–aluminum propeller blade sample.

such observations tend to support the high temperature predictions of Philipp and Lauterbom.[82]

Cavitation erosion damage occurs in many forms and at different rates and first occurs in the vicinity of the cavity collapse regions and not generally at the inception point of the cavity. Bubble, vortex and cloud cavitation structures, rather than stable sheet cavitation, are thought to be most responsible for material erosion attack. At the Technical University of Delft studies on the break-up of sheet cavitation have been made using a centrally twisted aerofoil section in oscillating flows. These experiments have shown that the behavior of the re-entrant jet at the trailing edge of the sheet cavity is such as to promote line vorticity which then rapidly forms into closed vortex structures in association with the foil structure about which they were formed.

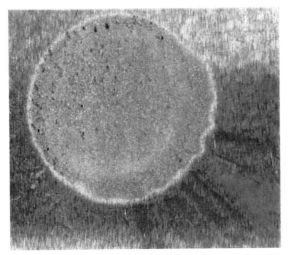

FIGURE 9.37 Oxide induced temper coloring on a mild steel specimen subjected to cavitation erosion attack.

The speed with which erosion can take place is also a variable: in some extreme cases significant material damage can occur as rapidly as in a few hours, whereas in other instances the erosion develops slowly over a period of months or years. Furthermore, in some other cases the erosion starts and then the rate of erosion falls off, so as to stabilize with no further erosion occurring. This stabilization takes place when a certain critical depth is reached and the profile of the cavity is such as to cause favorable flow conditions with the material boundary at depths beyond the destructive mechanisms of the cavity collapse. In other instances the formation of a primary erosion cavity will cause a flow disturbance sufficient to re-introduce cavitation further downstream, and this may give rise to the secondary erosion upon the collapse of this additional cavitation.

Referring back to equation 9.19 it can be seen that as understanding improves of the factors that cause a material to erode then that equation should perhaps be rewritten and include additional terms as follows:

Damage = g(Energy in the cavitation collapse mechanism,
  the pressure field distribution on surface,
  the frequency of bubble collapse,
  the level of embrittlement of the material,
  the transient temperature of collapse,
  the relative electro-potential of the erosion site,
  the material mechanical properties,
    fracture modes, ...)

(9.20)

Cavitation collapse near the trailing edge can lead to the phenomenon of 'trailing edge curl'. This type of cavitation damage, shown in Figure 9.38, is, as its name implies, a physical bending of the trailing edge of the blade. This bending of the blade is caused by the 'peening' action of cavitation collapse in the vicinity of the thinner sections of blade in the region of the trailing edge which then results in their bending. Van Manen[47] discusses this effect in some detail.

9.6 CAVITATION TESTING OF PROPELLERS

In order to study cavitation and its effects using propeller models it is necessary to ensure both geometric and flow similarity as any deviation from these requirements causes scale effects to occur. Geometric similarity requires that the model is a geosim of its full-scale counterpart and that considerable care has been taken in the model manufacture to ensure that the tolerances on design dimensions are satisfactory for model testing purposes. If the tolerances are not satisfactory, then false cavitation patterns and inception behavior will result from the model tests.

Flow or dynamic similarity is fully obtained when the effects of gravitation, viscosity, surface tension, vaporization characteristics, static pressure, velocity, fluid density, gas diffusion and so on are properly taken into account. Unfortunately, in a real flow situation using a model to represent a full-scale propeller, it is impossible to satisfy all of these parameters simultaneously. In Chapter 6 the main propeller non-dimensional groupings were derived from dimensional analysis and for the purposes of cavitation testing the primary groupings are:

$$\text{Froude number } F_n \frac{V_a}{(gD)}$$

$$\text{Reynolds number } R_n \frac{\rho V_a D}{\eta}$$

$$\text{Weber number } W_n \frac{\rho V_a^2 D}{S}$$

$$\text{Advance ratio } J \frac{V_a}{nD}$$

$$\text{Cavitation number } \sigma_0 \frac{pp_v}{\rho V_a^2}$$

FIGURE 9.38 Blade trailing edge curl.

By making the assumption that the properties of sea water and the water in the cavitation facility are identical, this being a false assumption but a close enough approximation for the present discussion purposes, it can be seen that simultaneous identity can be obtained only for the following non-dimensional groups:

1. F_n, σ_0 and J when the pressure and propeller rotational speed can be freely chosen.
2. R_n, σ_0, J and ψ (where ψ is the gas content number $d/(VD)$) where again the pressure and rotational speed can be freely chosen and the high flow speeds required for R_n present no problem.
3. W_n, σ_0, J and ϕ (where ϕ is the gas content number $cD/(\rho V_a^2)$) where once again the pressure and rotational speed can be chosen freely.

Cavitation tunnel model testing with marine propellers is normally undertaken using a K_T identity basis. This essentially requires that the cavitation number and advance coefficient are set. As the simultaneous satisfaction of the Reynolds and Froude identity is not possible, water speeds are normally chosen as high as possible to minimize the differences between model- and full-scale Reynolds number. However, running at the correct full-scale Reynolds number is generally not possible in most laboratories. Nevertheless, cavitation testing frequently attempts to follow the second group of non-dimensional coefficients identified above.

The implication of ignoring the Froude identity is that for a given radial position on the blade and angular position in the disc the local cavitation number will not be the same for model and ship. Indeed, the cavitation number identity for model and ship under these conditions is obtained at only one point, normally taken as the shaft center line. Although this secures a mean cavitation number, it does not model the cavitation conditions correctly since the conditions for cavitation inception are not the same as those required once cavitation has been formed. Newton[48] discusses the influence of the effect of Froude number on the onset of tip vortex cavitation, from which it is seen that this is significant. As a consequence, when undertaking cavitation inception studies for propellers, the correct Froude number should be modeled. In order to improve the simulation of the pressure field over the propeller disc Newton suggests using a nominal cavitation number based on the $0.7R$ position in the top dead center position.

The walls of a cavitation tunnel have an effect on the flow conditions in the test section. If the propeller is considered to be an actuator disc, that is having an infinite number of blades, then the corrections for the effect of the tunnel walls can be calculated for a non-cavitating propeller using the Wood and Harris method.[49] Van Manen has shown that, if a finite number of blades are considered, this influence is negligible for the normal ratios of propeller disc area to tunnel cross-section. Equivalent and validated corrections for cavitating propellers have, however, yet to be derived. The cavitation experienced by a propeller at the various positions in the propeller disc is fundamentally influenced by the inflow velocities and hence, by implication, the simulation of the wake field. Many methods of simulating the wake field of the vessel are used. The simplest of these is through the use of a wire gauge arrangement, termed a wake screen, located upstream of the propeller. The design of the wake screen is done on a trial and error basis in attempting to simulate the required wake fields. A more favored approach is to use a dummy model comprising a forebody and afterbody with a shortened parallel mid-body section. This produces the general character of the wake field and the 'fine tuning' is accomplished with a simplified wake screen attached to the dummy hull body. In several institutes a full model hull form can now be used in the cavitation facility.

However, by representing the measured nominal wake field of the model only part of the inflow problem is solved, since there are both scale effects on ship wake and propeller induction effects to be considered, as discussed in Chapter 5. As a consequence these other effects need to be accounted for if a proper representation of the inflow velocity field is to be correctly simulated: the methods for doing this, however, can still only be regarded as tentative.

The nuclei content of the water is an important aspect, as already discussed. Although a traditional way of measuring the total gas content in a cavitations tunnel is by the van Slyke method, various means exist by which the nuclei content can be measured; these can be divided into two main types. The first is where a sample of the water is taken from the main flow and forced to cavitate, thereby providing information on the susceptibility of the liquid to cavitate. The second comprises those employing holographic and light-scattering methods and giving information on the nuclei distribution itself. An example of the first type of method is where the tunnel water is passed through a glass venturi tube whose pressure has been adjusted in such a way that a limited number of bubbles explode in the throat of the venturi, the bubble explosions being limited to the order of twenty per second. The detection of the bubbles passing through the venturi tube is by optical means. With regard to the second group of methods, the holographic method discriminates between particulates and bubbles, and can therefore be regarded as an absolute method. It is, therefore, extremely useful for calibration purposes. However, the analysis of holograms is tedious, and therefore makes the method less useful for routine work. In the case of light-scattering techniques, these have improved considerably since the mid-1920s and their reliability for routine measurements is now adequate for practical purposes. Mées[88,89] undertook a comparison of three nuclei measurement techniques and presented a comparison of the digital on-line holographic technique, the interferometric laser imaging method and the Venturi technique. It was found that while the latter method gave directly the information on the critical pressure of the water, the interferometric might provide a useful alternative that could be adapted to very small bubble sizes, less than 100 μm.

The science, or art, of cavitation testing of model propellers was initiated by Sir Charles Parsons in his attempts to solve the cavitation problems of his steam turbine prototype vessel *Turbinia*. He constructed the first cavitation tunnel, Figure 9.39, from a copper rectangular conduit of uniform section. This conduit was formed into an 'oval' so as to form a closed circuit having a major axis of the order of 1 m. The screw shaft was inserted horizontally through a gland in the upper limb and driven externally, initially by a small vertical steam engine and later by an electric motor. Within the tunnel Parsons installed windows on either side of the tunnel and a plane mirror was fixed to an extension of the shaft, which reflected light from an arc lamp in order to illuminate the model propeller for a fixed period each revolution. The propeller diameter was 2 in. and cavitation commenced at about 1200 rpm. In constructing the tunnel Parsons recognized the importance of static pressure and made provision for the reduction of the atmospheric pressure by an air pump in order to allow cavitation to be observed at lower rotational speeds. This forerunner of the modern cavitation tunnel, constructed in 1895, is today preserved in working order in the Department of Marine Technology at the University of Newcastle upon Tyne. It is frequently cited alongside the current facility of the university, and in this way provides an interesting contrast in the developments that have taken place over the intervening years. Recognizing the limitations of his first tunnel, in 1910 Parsons constructed a larger facility at Wallsend, England,

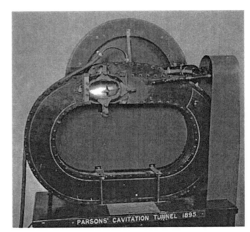

FIGURE 9.39 The first cavitation tunnel constructed by Sir Charles Parsons.

FIGURE 9.40 The second tunnel constructed by Sir Charles Parsons at Wallsend.

David Taylor Model Basin. This was followed, for example, by the building of facilities in Hamburg, Wageningen, Massachusetts Institute of Technology, Haslar (UK).

Today there are what might be termed traditional cavitation tunnels and the new breed of large tunnels that are being constructed around the world. Typical of the traditional tunnels, traditional in the sense of their size, is that shown in Figure 9.41. These tunnels are usually mounted in the vertical plane and are formed from a closed re-circulating conduit having a variable speed and pressure capability. Typical of the speed and pressure ranges of this kind of tunnel are speeds of up to 10—11m/s and pressure ranges of 10—180 kPa, giving cavitation number capabilities in the range 0.2—6.0.

in which he was able to test propeller models of up to 12 in. in diameter; Figure 9.40. The tunnel, which was a closed conduit, had a working cross-section having dimensions of 0.7m × 0.76 m and the flow rate in the test section was controlled by a circulating pump of variable speed. The propeller model was mounted on a dynamometer which was capable of measuring thrust, torque and rotational speed. Unlike its predecessor, this tunnel has not survived to the present day.

In the years that followed, several cavitation facilities were constructed in Europe and the USA. In 1929 a tunnel capable of testing 12 in. diameter propellers was built at the

Some modern cavitation facilities also have variable test sections, thereby allowing one of the appropriate dimensions to be installed into the tunnel body so as to meet the particular requirements of a measurement assignment. One such facility is that owned by SSPA, in which the test section can be varied from 2.5—9.6 m, thereby allowing hull models to be inserted into the facility. Clearly, in such cavitation tunnels the maximum velocity attainable in the working section is dependent on the test section body deployed for the measurement.

To meet the increasingly stringent demands of naval hydrodynamic research and modern merchant ship design requirements a new breed of large cavitation tunnel is making its appearance; facilities have been constructed in the USA, Germany and France. Figure 9.42 shows the

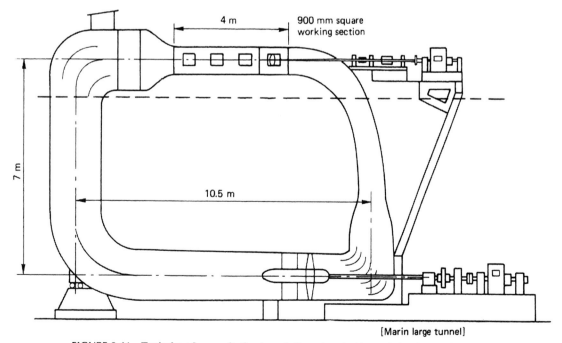

FIGURE 9.41 Typical modern cavitation tunnel. *Reproduced with permission from Reference 42.*

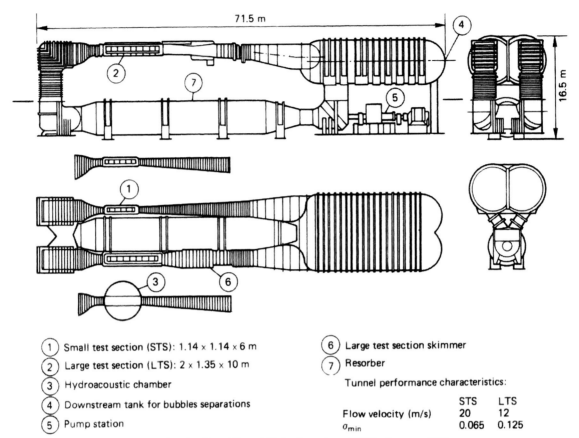

1. Small test section (STS): 1.14 × 1.14 × 6 m
2. Large test section (LTS): 2 × 1.35 × 10 m
3. Hydroacoustic chamber
4. Downstream tank for bubbles separations
5. Pump station
6. Large test section skimmer
7. Resorber

Tunnel performance characteristics:

	STS	LTS
Flow velocity (m/s)	20	12
σ_{min}	0.065	0.125

FIGURE 9.42 Grand Tunnel Hydrodynamique (GTH). *Courtesy: DCN.*

Grand Tunnel Hydrodynamique located at Le Val de Reuil, France and owned by Bassin d'Essais des Carenes de Paris. This tunnel has two parallel test sections: the larger of the two has a cross section of 2.0 m × 1.35 m and is 10 m long, whilst the smaller section has a 1.14 m square section and is 6 m long. The larger and smaller sections can give maximum flow velocities of 12 and 20 m/s, respectively, and the larger limb can be used as either a free surface or fully immersed test section. In Figure 9.42 the large downstream tank is used to remove the air produced in, or injected into, the test section. This tank has a total volume of 1600 m³ and can remove the air from dispersions with void fractions of up to 10 per cent. No bubbles larger than 100 μm can pass through the tank at its maximum flow rate. In this facility cavitation nuclei concentrations are automatically controlled by nuclei generators and measurement systems. In addition to the larger downstream tank a resorber, 5 m in diameter, ensures that no nuclei return to the test section after one revolution, and in order to reduce flow noise, the water velocities are kept below 2.5 m/s except in the test section. A second large European facility built in Germany at HSVA and called the HYKAT has been commissioned and has working section dimensions of 2.8 m × 1.6 m × 11.0 m with a maximum flow velocity of 12.6 m/s. Apart from being able to insert complete hull models of towing tank size into the tunnel, one of the major benefits of these larger tunnel facilities is their quiet operation, thereby allowing greater opportunities for noise measurement and research. Figure 9.43, from data supplied by Wietendorf and Friesel for different water gas contents,

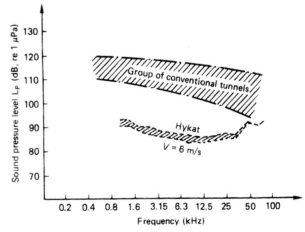

FIGURE 9.43 Comparison of background noise levels of the HYKAT facility with other tunnels.

shows the measured background noise levels of the HYKAT in relation to conventional tunnels.

The recently built facility in the USA, known as the Large Cavitation Channel (LCC) and operated by DTRC, is currently the largest facility in the world. It has a working section with a cross-section of the order of 3 m square.

Apart from cavitation tunnels, there are also the depressurized towing tank facilities which, in essence, are a conventional towing tank contained within a concrete pressure vessel which can be evacuated in order to reduce the internal air pressure. This depressurization capability has a series of air locks in order to allow personnel to travel with the carriage to make observations. Facilities such as these are owned by CSSRC in China and MARIN in the Netherlands: this latter capability has a tank dimension of 240 m × 18 m × 8 m and was designed to be evacuated to a pressure of 0.04 atm in around eight hours. A depressurized towing tank allows testing at the correct Froude number, cavitation number and advance coefficient. In addition, as is the case with the large and variable test section cavitation facilities, the flow around the complete model hull helps considerably in modeling the inflow into the propeller, although the scale effects on wake are still present. In the depressurized facility the free surface effects are readily modeled and the tank boundaries are comparatively remote from the model.

In cavitation facilities around the world, of which the above are a few examples for illustration purposes, several measurement and visualization capabilities exist for a variety of cavitation related measurements. The basic method of viewing cavitation is by the use of stroboscopic lighting. The stroboscopic lighting circuitry is triggered from the model propeller shaft rotational speed together with a multiplier and phase adjustment to account for differences in blade number and position around the disc. The traditional method of recording cavitation is to use the cavitation sketch, Figure 9.44, which is the experimenter's interpretation of the cavitation type and extent observed at various positions of the blade around the propeller disc. In many cases this method has been replaced or supplemented with the use of photographs taken under stroboscopic lighting or by video cameras, the latter being particularly useful. Figure 9.9 shows typical still images from cavitation tunnel experiments. The developments in computer technology have now made the superimposition of cavitation images from video recording and measured model hull surface pressure information possible. Savio *et al.*[87] introduced the development of a stereo imaging laser-based technique to examine cavitation structures. The computer-based technique is configured to develop three-dimensional representations of the cavity structures.

Over the years several attempts have been made to predict cavitation erosion qualitatively using 'soft surface' techniques which are applied to the blade surface: typical of this work is that of Kadoi and Sasajima,[50] Emerson[51] and Lindgren and Bjarne.[52] The techniques used have been based on the application of marine paint, soft aluminum and stencil ink. The ITTC have proposed the use of the latter. Care has, however, to be exercised in interpreting the results, in terms of both the surface used and the cavitation formation at model-scale due to the various scale effects. The method currently has relatively good correlation experience for propellers but less so for rudders.

At the present time research is being undertaken on the use of sonoluminescence in the use of cavitation studies. Sonoluminescence is generally ascribed to the high internal temperatures resulting from the essentially adiabatic compression of the permanent gas and vapor which is trapped within a collapsing cavitation bubble.

In recent years several full-scale observations of cavitation have been made. These require either the placing of observation windows in the hull, usually in several locations or the use of borescopes as discussed in Chapter 17.

9.7 ANALYSIS OF MEASURED PRESSURE DATA FROM A CAVITATING PROPELLER

The basis of the development of the radiated hull pressure signature is the acceleration of the cavity volume with respect to time modified by the self-induced component of pressure generation arising from the vibration of the ship structure or model test equipment at the point of measurement. The phenomenological processes giving rise to the hydrodynamic-based radiated parts of the signature are closely linked to the type of cavity collapse and rebound events. As such, the hydrodynamic excitation process is a time domain event and can be understood best through the pressure time series and its manipulations.

In experimental studies the pressure time signature is most commonly analyzed using a Fourier-based technique, due largely to the need to relate excitation sources to ship hull and structural response characteristics. Fourier techniques, originally developed as a curve fitting process, have as their underlying tenet the requirement of piecewise continuity of the function that is being analyzed; whether this is over a long or short-time frame. Given that this condition is satisfied, then assuming a sufficient number of terms are taken and the numerical stability of the algorithm is sufficient then the method will satisfactorily curve fit the function as a sum of transcendental functions.

To gain a phenomenological understanding of cavitation rather more than a Fourier-based curve fitting algorithm is necessary. This is for two reasons: first, a set of coefficients of transcendental component functions tell little about the structure of the underlying cavitation causing the signature, and second, and perhaps more

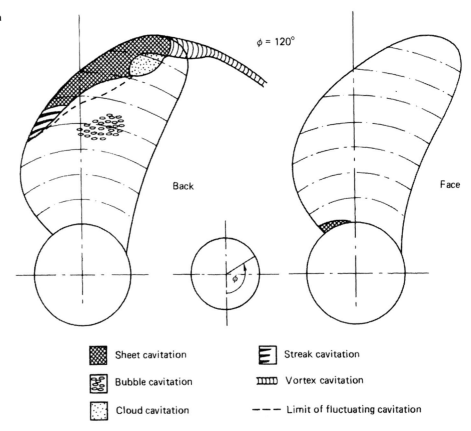

FIGURE 9.44 Typical cavitation sketch.

importantly, cavitation-based signatures are rarely uniform with respect to time. There are blade surface pressure changes which vary from blade to blade in a single revolution and also changes from one revolution to the next. These changes are random in nature and result from the interaction of the temporal changes in the flow homogeneity; the flow field, these being the sum of the steady nominal inflow field and the seaway-induced velocities, and the blade-to-blade geometric variations due to the manufacturing tolerances of the propeller blades. These changes influence both the general form of the cavity volume variation and the higher frequencies and noise generated from the random perturbations of the topological form of the underlying cavity structure. The Fourier analysis method tends, by its formulation, to average these variations out and thereby valuable physical information is lost.

If a phenomenological approach is to be adopted for the analysis of radiated hull pressure signatures, this being the most appropriate for engineering purposes, then other analytical approaches are required based on the fundamental dataset of the time series. Moreover, samples of the measured time series data should always be shown in any report: all too often such data is omitted from measurement reports whether relating to model or full-scale studies. A number of candidate analysis approaches offer themselves and among these are short-form Fourier transforms, joint time-frequency analysis, wavelet techniques and a double integral analysis of the underlying pressure signature. Experience has shown that each of these methods has shortcomings due, for the most part, to the rapid collapse of the cavity volumes in adverse wake gradients. Nevertheless, wavelet methods and the double integral technique have been shown to have some advantages when considering different aspects of the problem. While in the case of the wavelets most work has focused on standard applications of Daubechies formulations which have allowed some progress to be made, further discrimination is believed to be possible if purpose-designed wavelet forms are used to describe different cavity physical phenomena.

Notwithstanding the wavelet class of methods, the double integral approach has been shown to be the most successful at phenomenological discrimination. The pressure integration approach is essentially a time domain process, which together with visual observations of cavitation can link the dynamics of visual events with the dynamics of pressure pulses. It is clear from both ship and model-scale analysis of such data that the more severe excitation events are generated by cavitation which grows, collapses and rebounds in a small cylindrical sector of the propeller disc and slipstream which spans the wake peak. It is the passage of the propeller blades through this slow-speed region which causes the flare-up and collapse of cavity volumes on the blade and in the tip vortex shed by the advancing blade.

9.8 THE CFD PREDICTION OF CAVITATION

With the growth in computer-based capabilities, considerable effort is being expended in the development of methods to predict the extent and characteristics of the cavitation development over propeller blades.

Many of the computational fluid dynamics models that are used for cavitation studies use a barotropic equation of state using the relationship $\rho = \rho(p)$ which assumes that the mixture density is a function of the local pressure. This implies that all of the effects caused by bubble content are disregarded except for the compressibility and that the bubbly mixture can be regarded as a single-phase compressible fluid. Normally the methods employing a barotropic state law assume a continuous variation of density between liquid and vapor values in a range of pressures centered on the vapor pressure. Such approaches are attractive because of their simplicity but it has to be recognized that they assume equilibrium thermodynamics in their solution.

In practice, the transient dynamics of cavitation are important. Cavitation inception is associated with the growth of nuclei with diameters in the range 10^{-5} to 10^{-3} cm and contains mixtures of vapor and non-condensable gases. These nuclei as they travel into regions where the ambient pressure drops below the vapor pressure grow extremely rapidly and then, after forming congregations of bubbles, collapse when they encounter higher pressures. The collapse dynamics depend on a variety of factors which embrace considerations of surface tension, viscosity and non-condensable content. However, there are no cavitation models which attempt to account for all the variables present in the collapse phases of cavitation. Some models do, nevertheless, attempt to account for the non-equilibrium effects but are mostly based on the Rayleigh–Plesset equation. Typically these models are those by Schnerr and Sauer, Gerber, Zwart and Senocak and Shyy. For example, the Schneer–Sauer cavitation model solves for the vapor volume fraction using a transport equation whose source terms are derived from the Rayleigh–Plesset equation and account for the mass transfer between the vapor and liquid phases in the cavitation. The bubble radius can then be determined by assuming a bubble density number. Zwart's model also finds its origin in the Rayleigh–Plesset formulation and deploys an inter-phase mass transfer procedure.

Bulten and Oprea[76] considered the use of CFD methods for propeller tip cavitation inception. They found that, provided local mesh refinement is utilized in the tip region, which enables a detailed analysis of the flow in the tip vortex, then at model-scale cavitation inception can be predicted reasonably in the case of the DTRC 4119 propeller. Furthermore, they have extended their studies to the consideration of McCormick's scaling law for cavitation inception but suggest that further work is necessary before definitive conclusions can be drawn on the correlation. In the corresponding case of rudder cavitation prediction the multi-phase capabilities of the more advanced commercial packages have been shown to give good agreement with observation given that the inflow from the propeller is modeled with some accuracy.

A major correlation exercise was undertaken within the VIRTUE propeller modeling exercise and reported in 2008 in Rome. Within this exercise some seven computational models based on RANS and LES solvers were deployed in a benchmarking exercise based on the INSEAN E779A propeller model in both uniform and non-uniform inflow conditions. This model propeller is a four-bladed, constant pitch, conventional low-skew fixed pitch propeller having a diameter 227.2 mm and a nominal forward rake of $4°$: indeed, its design basis is to be found in a modification of the Wageningen series propellers. From the numerical computations, the predictions of thrust and torque at the defined loading condition fell within the range of -10 per cent to $+8$ per cent of the measured values at model-scale. The comparison of the predicted unsteady cavity extents with experimental observation showed a qualitative similarity but a number of quantitative differences were observed. These were considered to be of an order that could lead to differences in the prediction of hull surface pressures and erosion given, in the latter case, that suitable metallurgical failure models existed. The principal conclusions from this benchmarking exercise were that considerable further study is required in the following areas:

- Turbulence and cavitation models.
- Grid resolution.
- Numerical dissipation.
- Modeling of non-uniform flow.

In partial contrast, a study on the modeling of sheet cavitation, again for the INSEAN E779A propeller, was undertaken by Liu[90] using FLUENT 6.2 and a full cavitation model with the Singhal characterization[91] used to estimate the vapor generation and condensation rates. Within this work the influence that the effect of the non-condensable mass fraction, f_g, had on the sheet cavitation geometry was the main focus of the study. Their conclusion was that a value of $f_g = 1.5 \times 10^{-7}$ gave a good prediction when compared to experimental values. By extending their study further and using a vapor fraction equal to 0.1 the simulated cavity extents were well predicted at advance coefficients of 0.71 and 0.83.

Kanematu and Ando[92] endeavored to predict steady and unsteady cavitation using a modeling approach for the camber surfaces of vortex surfaces with sources distributed over the wetted surfaces. In this model the kinematic boundary conditions were satisfied on the camber and wetted surfaces to derive the appropriate source and vortex strengths while the cavity was represented by doublets. An iterative solution was then introduced to determine the doublet strengths and cavity thicknesses. The method was validated against two six-bladed propellers designed for the ship *Seiun Maru*: one being of conventional skew and the other highly skewed. In the case of the conventional skew the steady predictions of cavity extent correlated well with those measured at model-scale for an advance coefficient of 0.7, Figure 9.45, as did the predictions of thrust and torque with varying loading. With regard to the unsteady analyses the cavity extents and thicknesses correlated well as seen in Figure 9.46.

Sato *et al.*[94] explored the use of a commercial code using the SST k-ω turbulence model in association with unstructured grids in a small cylindrical computational domain surrounding the propeller. This model comprised some 1.8×10^6 cells and the time step used was comparatively large since for this study low-frequency phenomena were of primary interest. The propellers used encompassed a range of blade numbers, blade area and pitch ratios, and skew angles. The findings of this study suggested that the fundamental characteristics of the sheet cavitation were reasonably well predicted as was the first blade order component of the pressure fluctuation. However, some cavity extent issues were noted at the top-dead-center position and the tip vortex was not apparent in the calculations.

The efficacy of different turbulence models was considered by Hasuike *et al.*[95] Their work was based on an

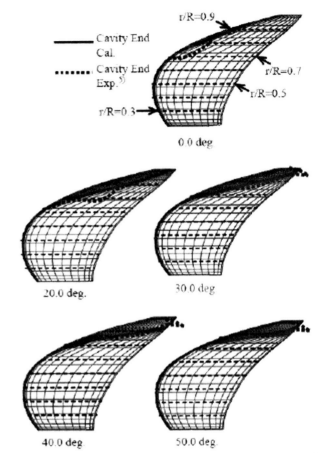

FIGURE 9.46 **Correlation of the unsteady observed and computed cavity extents.** *Reference 93.*

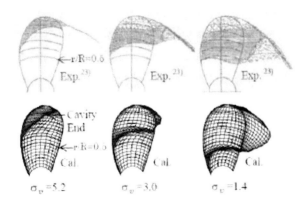

FIGURE 9.45 **Correlation of the steady observed and computed cavity extents.** *Reference 92.*

adaptive mesh refinement approach and was used to model the DTMB 4119 propeller. They concluded that while the SST k-ω turbulence model was able to predict the open water performance satisfactorily it underestimated the boundary layer thickness. In contrast the k-ε model predicted the boundary layer thickness well but underestimated the propeller open water efficiency. They then moved forward to consider unsteady cavitation generated by the *Seiun Maru* highly skewed propeller using both the barotropic and full cavitation model. The predictions agreed well with the measured data when contour cavity void fractions of 0.2 and 0.3 were taken for the barotropic and full cavitation models respectively. Ji *et al.*[96] examined through a computational fluid dynamic simulation the unsteady cavitation developed by the *Seiun Maru* highly skewed propeller at full-scale. The modeling was based on the SST k-ω turbulence simulation and iso-surfaces having a vapor volume fraction of 0.1. The estimated cavity formations at various blade angular positions were considered against those recorded from experiment and while the cavity extents were fairly represented, albeit with some underprediction, the tip vortex was not predicted.

The essential characteristics of the steady and unsteady cavitating flows around two and three dimensional hydrofoils were examined by Li et al.[97] Using the modified SST k-ω turbulence model, features such as the development of re-entrant jets, pinch-off and the shedding of vortex and cloud cavities for a two-dimensional NACA0015 aerofoil at unsteady and shedding conditions could be represented. At a higher cavitation number the model predicted a high-frequency sheet cavity together with some minor shedding at the trailing edge of the sheet cavity. Moreover this modified SST k-ω turbulence model provided enhanced correlation with the standard fluid dynamic lift, drag and shedding frequency predictions when compared with the standard SST k-ω turbulence formulation. This enhanced correlation, however, did not extend to the unsteady case. Of further interest in this study was that vortex group cavitation, secondary cavities, appeared to be observable in the simulation; Figure 9.47.

Typical of some modern design and analysis studies into cavitating flow structures associated with propellers is that shown in Figure 9.48.

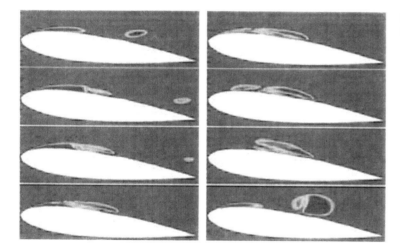

FIGURE 9.47 **The time history of cavity shape when at a vapor void fraction of 1.0.**

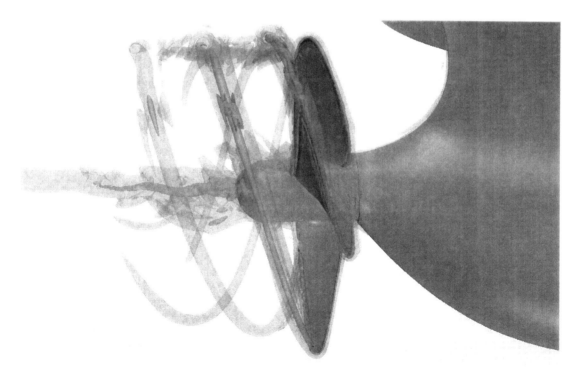

FIGURE 9.48 **Cavitation CFD analysis of a propeller.** *Courtesy Lloyd's Register.*

REFERENCES AND FURTHER READING

1. Euler M. *Théorie plus complète des machines qui sont mises en movement par la réaction de l'eau.* Berlin: L'Académie Royale des Sciences et Belles Lettres; 1756.
2. Reynolds O. *The causes of the racing of the engines of screw steamers investigated theoretically and by experiment.* Trans. INA; 1873.
3. Harvey EN, et al. *Bubble formulation in animals.* I. Physical factors. J. Cell. Comp. Physiol. 1944;**24**(1).
4. Harvey EN, et al. *Bubble formation in animals. II. Gas nuclei and their distribution in blood and tissues.* J. Cell. Comp. Physiol. 1944;**24**(1).
5. Harvey EN, et al. *Removal of gas nuclei from liquids and surfaces.* J. Amer. Chem. Soc. 1945;**67**.
6. Harvey EN, et al. *On cavity formation in water.* J. Appl. Phys. 1947;**18**(2).
7. Strasberg MJ. Acoust. Soc. Am. 1959;**31**.
8. Flynn, H.G. Physical Acoustics, Vol. 1. Academic Press, New York.
9. Winterton RHS. J. Phys. D: Appl. Phys. 1977;**10**.
10. Cram LA. *Cavitation and Inhomogeneities in Underwater Acoustics.* Berlin: Springer; 1980.
11. Fox FE, Herzfeld KF. *Gas bubbles with organic skin as cavitation nuclei.* J. Acoustic. Soc. Amer. 1954;**26**.
12. Yount DE. J. Acoust. Soc. Am. 1979;**65**.
13. Yount DE. J. Acoust. Soc. Am. 1982;**71**.
14. Plesset MS. *Bubble Dynamics.* CIT Rep. 5.23. Calif. Inst. of Tech.; February 1963
15. Wheeler WH. *Indentation of metals by cavitation.* ASME Trans., Series D 1960;**82**.
16. Besant WH. *Hydrostatics and Hydrodynamics.* Cambridge: Art ISE, Cambridge University Press; 1859.
17. Rayleigh L. *On the pressure developed in a liquid during the collapse of a spherical cavity.* Phil. Mag. 1917;**34**.
18. Mitchell TM, Hammitt FG. *Asymmetric cavitation bubble collapse.* Trans. ASME J., Fluids Eng. 1973;**95**(1).
19. Plesset MS, Chapman RB. *Collapse of an initially spherical vapour cavity in neighbourhood of a solid boundary.* J. Fluid Mech. 1971;**47**(2).
20. Knapp RT, Daily JW, Hammitt FG. *Cavitation.* New York: McGraw-Hill; 1970.
21. Huse E. *Propeller–Hull Vortex Cavitation.* Norw. Ship Model Exp. Tank Publ. No. 106; May 1971.
22. Burrill LC, Emerson A. *Propeller cavitation: Further tests on 16 in. propeller models in the King's College Cavitation Tunnel.* Trans. NECIES 1978;**195**.
23. Keller J. Auf'm. *Enige Aspecten bij het Ontwerpen van Scheepsschroeven.* Schip enWerf 1966;(24).
24. Balhan J. *Metingen aan Enige bij Scheepss-chroeven Gebruikelijke Profielen in Vlakke Stro-ming met en Zonder Cavitatie.* NSMB Publ. No. 99; 1951.
25. Shan YT. *Wing section for hydrofoils. 3. Experimental verifications.* J. Ship Res. 1985;**29**.
26. Oossanen P van. *Calculation of Performance and Cavitation Characteristics of Propellers Including Effects of Non-Uniform Flow and Viscosity.* NSMB Publ. No. 457; 1974.
27. Knapp RT. *Cavitation Mechanics and its Relation to the Design of Hydraulic Equipment.* Proc. I. Mech. E. Pt, No. 166; 1952.
28. Cebeci T, Mosinskis GJ, Smith AMO. *Calculation of Viscous Drag on Two-Dimensional and Axi-Symmetric Bodies in Incompressible Flows.* American Institute of Aeronautics and Astronautics; 1972. Paper No. 72–1.
29. Ligtelijn J Th, Kuiper G. *Intentional cavitation as a design parameter.* 2nd PRADS Symp.; 1983
30. Tulin MP. *Steady Two Dimensional Cavity Flows about Slender Bodies.* DTMB Report 834; May 1953.
31. Geurst JA. *Linearised theory for partially cavitated hydrofoils.* ISP 1959;**6**.
32. Geurst JA, Verbrugh PJ. *A note on cambered effects of a partially cavitated hydrofoil.* ISP 1959;**6**.
33. Geurst JA. *Linearised theory for fully cavitated hydrofoils.* ISP 1960;**7**.
34. Leehey P. *Supercavitating hydrofoil of finite span.* Leningrad: Proc. IUTAM Symp. on Non-Steady Flow of Water at High Speeds; 1971.
35. Uhlman JS. *A partially cavitated hydrofoil of finite span.* J.F.E. 1978;**100**.
36. Jiang CW. *Experimental and theoretical investigations of unsteady supercavitating hydrofoils of finite span.* PhD Thesis. MIT; 1977.
37. Uhlman JS. *The surface singularity method applied to partially cavitating hydrofoils.* 5th Lips Symp.; May 1983
38. Tulin MP, Hsu CC. *The theory of leading-edge cavitation on lifting surfaces with thickness.* Symp. on Hydro. of Ship and Offshore Prop. Syst.; March 1977.
39. Stern F, Vorus WG. *A non-linear method for predicting unsteady sheet cavitation on marine propellers.* J. Ship. Res. 1983;**17**.
40. Isay WH. *Kavitation.* Hamburg: SchiffahrtsVerlag 'Hansa'; 1981.
41. Mills L. *Die Anwendung der Blasendy namik auf die theoretische mit experimentellen Daten.* PhD Thesis, Hamburg; 1991.
42. Kuiper G. *Cavitation Inception on Ship Propeller Models.* Publ. No. 655, NSMB; 1981.
43. *Report of Cavitation Committee.* Proc. of 18th ITTC, Kobe; 1987.
44. Chandrashekhara N. *Analysis of the tip vortex cavitation inception at hydrofoils and propellers.* Schiffstechnik, **112**.
45. English J. *Cavitation induced hull surface pressures – Measurements in a water tunnel.* London: RINA Symp. on Propeller Induced Vibration; 1979.
46. Brunton JH. *Cavitation damage.* Proc. Third Int. Cong. on Rain Erosion; August 1970.
47. Manen JD van. *Bent trailing edges of propeller blades of high powered single screw ships.* ISP 1963;**10**.
48. Newton RN. *Influence of cavitation on propeller performance.* Int. Ship. Prog. August 1961;**8**.
49. Wood R McK, Harris RG. *Some Notes on the Theory of an Airscrew Working in a Wind Channel.* A.R.C., R&M No. 662; 1920.
50. Kadoi H, Sasajima T. *Cavitation erosion production using a 'soft surface'.* ISP June 1978;**25**.
51. Emerson A. *Cavitation Patterns and Erosion –; Model – Full Scale Comparison.* Appl. 6, Rep. of Cav. Comm. 14th ITTC; 1975.
52. Lindgren HB, Bjarne E. *Studies of propeller cavitation erosion.* Edinburgh: Conf. on Cavitation, I. Mech. E.; 1976.
53. Gent, W. van, Kooij, J. *Influence of Hull Inclination and Hull–Duct Clearance on Performance, Cavitation and Hull Excitation of a Ducted Propeller: Part 1.* NSMB; 1980.
54. Kooij J. van der, Berg W. van den. *Influence of Hull Inclination and Hull–Duct Clearance on Performance, Cavitation and Hull Excitation of a Ducted Propeller: Part II.* NSMB; 1983.
55. Fortes-Patella R, Reboud JL. *A new approach to evaluate the cavitation erosion power.* 3rd Int. Symp. on Cavitation (CAV1998); 1998.

56. Fortes-Patella R, Reboud JL. *Energetical approach and impact efficiency in cavitation erosion*. 3rd Int. Symp. on Cavitation (CAV1998); 1998.
57. Fortes-Patella R, Challier G, Reboud JL. *Study of pressure wave emitted during spherical bubble collapse*. San Francisco: Proc. 3rd ASME/JSME Joint Fluids Eng. Conf.; July 1999.
58. Fortes-Patella R, Reboud JL, Archer A. *Cavitation damage measurement by 3D laser profilometry*. Wear 2000;**246**:59–67.
59. Challier G, Fortes-Patella R, Reboud JL. *Interaction between pressure waves and spherical bubbles: discussion about cavitation erosion mechanism*. Boston: 2000 ASME Fluids Eng. Summer Conf.; June 2000.
60. Coutier-Delgosha O, Fortes-Patella R, Reboud JL. *Evaluation of the turbulence model influence on the numerical simulations of unsteady cavitation*. Trans. ASME January 2003;**125**.
61. Lohrberg H, Stoffel B, Fortes-Patella R, Reboud JL. *Numerical and experimental investigations on the cavitating flow in a cascade of hydrofoils*. 4th Int. Symp. on Cavitation (CAV2001); 2001.
62. Lohrberg H, Stoffel B, Coutier-Delgosha O, Fortes-Patella R, Reboud JL. *Experimental and numerical studies on a centrifugal pump with curved blades in cavitating condition*. 4th Int. Symp. on Cavitation (CAV 2001); 2001.
63. Fortes-Patella R, Challier G, Reboud JL. *Cavitation erosion mechanism: numerical simulation of the interaction between pressure waves and solid boundaries*. 4th Int. Symp. on Cavitation (CAV 2001); 2001.
64. Coutier-Delgosha O, Reboud JL, Fortes-Patella R. *Numerical study of the effect of the leading edge shape on cavitation around inducer blade sections*. 4th Int. Symp. on Cavitation (CAV 2001); 2001.
65. Coutier-Delgosha O, Perin J, Fortes-Patella R, Reboud JL. *A numerical model to predict unsteady cavitating flow behaviour in inducer blade cascades*. Osaka: 5th Int. Symp. on Cavitation (CAV 2003); November 2003.
66. Lecoffre Y. *Cavitation erosion, hydrodynamic scaling laws, practical methods of long term damage prediction*. CAV1995 Conf., Deauville; 1995.
67. Foeth E-J, Terwisga T van. *The structure of unsteady cavitation. Part I. Observations of an attached cavity on a three dimensional hydrofoil*. Wageningen: 6th Int. Symp. on Cavitation CAV2006; September 2006.
68. Foeth E-J, Terwisga T van. *The structure of unsteady cavitation. Part II: Applying time-resolve PIV to attached cavitation*. Wageningen: 6th Int. Symp. on Cavitation CAV2006; September 2006.
69. Carlton JS, Fitzsimmons PA. *Full scale observations relating to propellers*. Wageningen: 6th Int. Symp. on Cavitation CAV2006; September 2006.
70. Carlton JS, Fitzsimmons PA. *Cavitation: Some full scale experience of complex structures and methods of analysis and observation*. Proc. 27th ATTC, St John's Newfoundland; August 2004.
71. Lücke T. *Investigations of propeller tip vortex bursting*. Genoa: Proc. Nav 2006 Conf.; June 2006.
72. Vaz G, Bosschers J. *Modelling three dimensional sheet cavitation on marine propellers using a boundary element method*. Wageningen: 6th Int. Symp. on Cavitation CAV2006; September 2006.
73. Sun H, Kinnas SA. *Simulation of sheet cavitation on propulsor blades using a viscous/inviscid interactive method*. Wageningen: 6th Int. Symp. on Cavitation CAV2006; September 2006.
74. Kinnas SA, Mishima S, Brewer WH. *Nonlinear analysis of viscous flow around cavitating hydrofoils*. 20th Symp. on Naval Hydrodynamics, University of California; 1994.
75. Brewer WH, Kinnas SA. *Experimental and viscous flow analysis on a partially cavitating hydrofoil*. J. Ship Res. 1997;**41**:161–71.
76. Bulten N, Oprea AL. *Evaluation of McCormick's rule propeller tip cavitation inception based on CFD results*. Wageningen: 6th Int. Symp. on Cavitation CAV2006; September 2006.
77. McCormick BW. *On cavitation produced by a vortex trailing from a lifting surface*. ASME J. Basic Eng.; 1962:369–70. and *A study of the minimum pressure in a trailing vortex system*. PhD Thesis, State University, Pennsylvania. 1962.
78. Bark G, Berchiche N, Grekula M. *Application of Principles for Observation and Analysis of Eroding Cavitation – The EROCAV Observation Handbook*. 3.1 ed.). Department of Naval Architecture and Ocean Engineering, Chalmers University of Technology; 2004.
79. Moulijn JC, Friesch J, Genderen M van, Junglewitz A, Kuiper G. *A criterion for the erosive-ness of face cavitation*. Wageningen: 6th Int. Symp. on Cavitation CAV2006; September 2006.
80. Uhlman JS. *A note on the development of a nonlinear axisymmetric re-entrant jet cavitation model*. J. Ship Res. September 2006;**50**(3).
81. Kennedy JL, Walker DL, Doucet JM, Randell T. *Cavitation performance of propeller blade root fillets*. Trans. SNAME; 1993.
82. Philipp A, Lauterbom W. *Cavitation erosion by single laser-produced bubbles*. J. of Fluid Mechanics 1998;**361**:75–116.
83. Phillipp A. *Single bubble erosion on a solid surface*. Deauville: CAV'95 Int. Symp. on Cavitation; 1995.
84. Shima A, Takayama K, Tomita Y, Ohsawa N. *Mechanism of impact pressure generation from spark generated bubble collapse near a wall*. AIAA Journal 1983;**21**:55–65.
85. Franc JP, Michel J-M. *Fundamentals of Cavitation*. Dordrecht: Kluwer Academic Press; 2004.
86. Reisman GE, Wang Y-C, Brennen CE. *Observations of shock waves in cloud cavitation*. J. of Fluid Mechanics 1998;**335**:255–85.
87. Savio L, Vivam M, Ferrando M, Carrera G. *Propeller cavitation 3D reconstruction through stereo-vision algorithms*. France: 1st Int. Conf. on Advanced Model Measurement Technology for the EU Maritime Industry; 2009.
88. Mées L, Lebrun D, Allano D, Françoise Walle F, Lecoffre Y, Boucheron R, Fréchou D. *Development of interferometric techniques for nuclei size measurement in cavitation tunnel*. Pasadena, California: 28th Symp. on Naval Hydrodynamics; 2010.
89. Mées L, Lebrun D, Fréchou D, Boucheron R. *Interferometric laser imaging technique applied to nuclei size measurements in cavitation tunnel. Advanced Model Measurement Technology for the Maritime Industry AMT11*. University of Newcastle upon Tyne; 2011.
90. Liu D, Hong F, Zhao F, Zhang Z. *The CFD analysis of propeller sheet cavitation*. Nantes, France: 8th Int. Conf. on Hydrodynamics (ICHD'08); 2008.
91. Singhal AK, Athavale MM, Li H, Jiang Y. *Mathematical basis and validation of the full cavitation model*. Journal of Fluids Engineering 2002;**124**(3):617–24.
92. Kanemaru T, Ando J. *Calculation of steady cavitation on a marine propeller using a simple surface panel method*. Journal of the Japan Society of Naval Architects and Ocean Engineers 2008;**7**:151–60.

93. Kanemaru T, Ando J. *Calculation of unsteady cavitation on a marine propeller using a simple surface panel method.* Journal of the Japan Society of Naval Architects and Ocean Engineers 2009;**9**:45–53.
94. Sato K, Oshima A, Egashira H, Takano S. *Numerical prediction of cavitation and pressure fluctuation around marine propeller.* Michigan, USA: Proc. 7th Int. Symp. on Cavitation (CAV 2009) Ann Arbor; 2009.
95. Hasuike N, Yamasaki S, Ando J. *Numerical study on cavitation erosion risk of marine propellers operating in wake flow.* Michigan, USA: Proc. 7th Int. Symp. on Cavitation (CAV 2009) Ann Arbor; 2009.
96. Ji B, Luo X-W, Wu Y-L, Liu S-H, Xu H-Y, Oshima A. *Numerical investigation of unsteady cavitating turbulent flow around a full scale marine propeller.* Shanghai, China: 9th Int. Conf. on Hydrodynamics (ICHD'10); 2010.
97. Li D-Q, Grekula M, Lindell P. *A modified SST k-ω turbulence model to predict the steady and unsteady sheet cavitation on 2D and 3D hydrofoils.* Michigan, USA: Proc. 7th Int. Symp. on Cavitation (CAV 2009) Ann Arbor; 2009.
98. Kawanami Y, Kato H, Yamaguchi H, Maeda M, Nakasumi S. *Inner structure of cloud cavity on a foil section.* 4th Int. Symp. on Cavitation (CAV2001); 2001.

Chapter 10

Propeller Noise

Chapter Outline

10.1 Physics of Underwater Sound	251	10.5 Transverse Propulsion Unit Noise	263
10.2 Nature of Propeller Noise	255	10.6 Measurement of Radiated Noise	263
10.2.1 Non-Cavitating Propeller Noise	257	10.7 Noise in Relation to Marine Mammals	264
10.2.2 Cavitation Noise	258	10.7.1 Marine Mammals	265
10.3 Noise Scaling Relationships	261	10.7.2 Marine Mammal Phonation	266
10.4 Noise Prediction and Control	262	References and Further Reading	268

The noise produced by a propeller, in terms of both its intensity and its spectral content, has been of considerable importance to warship designers and military strategists for many years. However, in recent years the subject has assumed a growing importance in the merchant shipping sector. An early compendium of knowledge pertaining to ship noise is given in Reference 1.

Prior to considering the noise characteristics generated by marine propellers it is useful to briefly consider the basic nature and physics of underwater sound and its propagation.

10.1 PHYSICS OF UNDERWATER SOUND

The speed of sound in water is some 4.3–4.4 times greater than that in air at locations close to sea level and Table 10.1 shows some typical values. The speeds shown in this table are approximate since some variations occur with ambient conditions. For a more precise determination of the speed of sound in sea water use can be made of an equation based on the work of Lovett,[2] which relates the speed of sound to the temperature, salinity, latitude and the depth at which the speed is required. This relationship has the following form:

$$C(z, S, T, \phi) = 1449.05 + 4.57T - 0.0521T^2$$
$$+ 0.00023T^2$$
$$+ (1.333 - 0.0126T + 0.00009T^2)$$
$$\times (S - 35)$$
$$+ (1.333 - 0.025T)(1 - 0.0026 \cos \phi)z$$
$$+ (0.213 - 0.01T)(1 - 0.0026 \cos \phi)^2 z^2$$
$$\times (0.1 - 0.00026 \cos \phi)Tz$$

(10.1)

where

T is the ambient temperature (°C)
S is the salinity in parts per thousand
z is the depth (km)
ϕ is the latitude (deg)

This regression equation is essentially valid for all oceanic waters down to a depth of around 4 km and has a standard deviation of 0.02 m/s.

As a direct consequence of the speed increase in water compared to that in air, the acoustic wavelengths in water will be greater than in air by the same factor, since:

$$\text{Wavelenght}(\lambda) = \frac{\text{speed of sound}}{\text{frequency}} \quad (10.2)$$

The transmissibility of sound in water is considerably affected by the frequency of the noise source. In general, high frequencies in water are strongly attenuated with increasing distance from the source, whilst the lower

TABLE 10.1 Speed of Sound in Air and Water Close to Sea Level

Medium	Speed (m/s)
Air at 21°C	344
Fresh water	1480
Salt water at 21°C and 3.5% salinity	1520

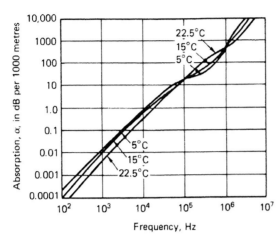

FIGURE 10.1 Sound absorption in sea water. *Reproduced with permission from Reference 23.*

frequencies tend to travel further and, therefore, are rather more serious from the ship radiated noise viewpoint. This is demonstrated in Figure 10.1, which shows the variation in absorption factor, measured in dB per 1000 m, over the range of frequencies 10^2-10^7 Hz.

Noise levels are measured using the decibel scale. While the original definition of the decibel was based on power ratios:

$$\text{dB} = 10\log_{10}(W/W_0)\text{dB} \quad (10.3)$$

where W_0 is a reference power, the use of the scale has widened from its original transmission line theory application to become a basis for many measurements of different quantities having a dynamic range of more than one or two decades. In the context of noise assessment, the sound pressure level (L_p) is the fundamental measure of sound pressure, and it is defined in terms of a pressure ratio as follows:

$$L_P = 20\log_{10}(P/P_0)\text{dB} \quad (10.4)$$

where P is the pressure measured at a point of interest and P_0 is a reference pressure set normally to 20 µPa in air and 1µPa in other media. Table 10.2 shows a conversion of the decibel scale into pressure ratios. These data can also be used when two sound pressure levels L_{p1} and L_{p2} are given and their difference (ΔL_p) can be expressed independently of the reference pressure P_0 as follows:

$$\Delta L_P = L_{p2} - L_{p1} = 20[\log_{10}(P_2/P_0) - \log_{10}(P_1/P_0)]$$

that is

$$\Delta L_P = 20[\log_{10}(P_1/P_2)] \quad (10.5)$$

From Table 10.2 it is seen that a change of, say, 6 dB causes either a doubling or halving of the sound pressures experienced. Similarly, a change of 12 dB effectively either quadruples or quarters the pressure levels.

TABLE 10.2 Decibel to Power Ratio Conversion

Pressure Ratio	−dB+	Pressure Ratio
1.000	0.0	1.000
0.989	0.1	1.012
0.977	0.2	1.023
0.966	0.3	1.035
0.955	0.4	1.047
0.933	0.6	1.072
0.912	0.8	1.096
0.891	1.0	1.122
0.841	1.5	1.189
0.794	2.0	1.259
0.708	3.0	1.413
0.631	4.0	1.585
0.562	5.0	1.778
0.501	6.0	1.995
0.447	7.0	2.239
0.398	8.0	2.512
0.355	9.0	2.818
0.316	10.0	3.162
0.251	12.0	3.981
0.200	14.0	5.012
0.158	16.0	6.310
0.126	18.0	7.943
0.100	20.0	10.000
0.0316	30.0	31.62
0.0100	40.0	100.0
0.0032	50.0	316.2
10^{-3}	60.0	10
10^{-4}	80.0	10
10^{-5}	100.0	10

Because the human ear does not respond equally to all frequencies within the audible noise range a weighting scale was devised to correct the actual physical pressure levels to those interpreted by the ear. This weighted scale is generally known as the A-weighting scale and its effect is shown by Figure 10.2 for the audible sound range of about 20 Hz–20 kHz. From this figure a marked fall-off in response can be seen for frequencies less than about

Chapter | 10 Propeller Noise

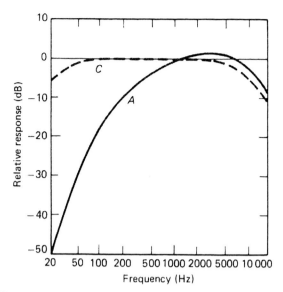

FIGURE 10.2 Filter characteristics for *A*- and *C*-weighted sound levels.

1000 Hz. Although many other weighting scales have been proposed few have gained widespread acceptance with perhaps the exception of the C-weighting scale which, as may be seen in Figure 10.2, is nearly flat between about 90 Hz through to 3000 Hz with a relative weighting of 0 dB in this interval. Nevertheless this latter scale is seldom used in marine technology.

In order to give some notion of how noise expressed in the dBA scale relates to common experience, Table 10.3 cites a few common examples. With regard to marine mammals a considerable variation in noise levels can be encountered. For example, in the case of the song of the bowhead whale this is typically in the range of 158–189 dB, re 1 μPa at 1 m, within a frequency range of 20–500 Hz. In contrast, a ringed seal might phonate at levels of 95–116 dB, re 1 μPa at 1 m, in the much higher frequency range of 0.4–16 kHz and harbor porpoises phonate at frequencies as high as 150 kHz.

In general, acoustic measurements make use of third-octave and octave filters in order to define noise spectra. A third-octave filter is one in which the ratio of the upper to the lower passband limits, that is the range of frequencies the filter will allow to pass through it, is $2^{1/3}$ (i.e. 1.2599). In the case of the octave filter, this ratio between the upper and lower passband limits is two and it is normally centered, as are the third-octave filters, on one of the preferred center frequencies in ISO R266. These center frequencies are calculated from $10^{n/10}$, where n is the band number: in practice nominal values are used to identify the center frequencies and Table 10.4 lists the set of third-octave and octave passbands relating to the audible range for convenience of reference.

While the subject of propeller noise is important to both the merchant and naval worlds the reasons for this importance derive from different origins. The exception to this statement is in the case of oceanographic and research vessels, which have similar noise requirements to naval vessels, in that they use instrumentation with ranges of the order of up to 10 kHz. In the merchant service the increasing awareness of the health hazards caused by the long-term exposure to high noise levels has led to the formulation of recommended levels of noise in different areas of a merchant vessel by the International Maritime Organization (IMO). The 1981 IMO Code on noise levels[3] defines maximum levels of noise for crew spaces as shown in Table 10.5.

In addition to defined levels of noise of this type there are further considerations of passenger comfort and annoyance in, for example, cruise liners and ferries. In order to appreciate the magnitude of the propeller noise problem it is instructive to compare the results of full-scale measurements on a variety of ships, recorded inside the hull but close to the propeller with the levels quoted in Table 10.4. The measurements reported by Flising[4] are shown in Figure 10.3 for a variety of ship types ranging from larger tankers to Rhine push boats, and it can be seen that levels of the order of 100–110 dBA were frequently noted in the lower-frequency bands. These measurements have been recorded in locations close to the propeller, such as in the aft peak tank and near the aft peak bulkhead. By considering this figure, which shows noise levels of the order of 100 dBA at the aft peak region of the vessel, it can be seen by reference to Table 10.4 that these sound pressure levels have to be considerably reduced by the time they reach a hospital or cabin location in the vessel according to the IMO code.

The origins of the naval interest in the subject of noise stem from a set of rather different constraints. These are largely twofold: first, there is interference from the noise generated by the vessel on its own sensors and weapons systems, and second, there is the radiated noise, which is transmitted from the ship to the far field, and by which the ship can be detected by an enemy. In this latter context a ship noise signature of a few tens of watts could be sufficient to give an enemy valuable information at

TABLE 10.3 Some Common Examples of Noise Levels	
Open country far from city or other interference	20–30 dBA
Quiet residential areas at night	30–50 dBA
Quiet residential areas during the day	40–50 dBA
Light traffic	50–60 dBA
Petrol driven lawn mower	90–100 dBA
Discotheque	110–120 dBA

TABLE 10.4 Third-Octave and Octave Passbands

Band Number	Nominal Center Frequency	Third-octave Passband	Octave Passband
1	1.25 Hz	1.12–1.41 Hz	
2	1.6	1.41–1.78	
3	2	1.78–2.24	1.41–2.82 Hz
4	2.5	2.24–2.82	
5	3.15	2.82–3.55	
6	4	3.55–4.47	2.82–5.62
7	5	4.47–5.62	
8	6.3	5.62–7.08	
9	8	7.08–8.91	5.62–11.2
10	10	8.91–11.2	
11	12.5	11.2–14.1	
12	16	14.1–17.8	11.2–22.4
13	20	17.8–22.4	
14	25	22.4–28.2	
15	31.5	28.2–35.5	22.4–44.7
16	40	35.5–44.7	
17	50	44.7–56.2	
18	63	56.2–70.8	44.7–89.1
19	80	70.8–89.1	
20	100	89.1–112	
21	125	112–141	89.1–178
22	160	141–178	
23	200	178–224	
24	250	224–282	178–355
25	315	282–355	
26	400	355–447	
27	500	447–562	355–708
28	630	562–708	
29	800	708–891	
30	1000	891–1120	708–1410
31	1250	1120–1410	
32	1600	1410–1780	
33	2000	1780–2240	1410–2820
34	2500	2240–2820	
35	3150	2820–3550	
36	4000	3550–4470	2820–5620
37	5000	4470–5620	

Chapter | 10 Propeller Noise

TABLE 10.4 Third-Octave and Octave Passbands—cont'd

Band Number	Nominal Center Frequency	Third-octave Passband	Octave Passband
38	6300	5620–7080	
39	8000	7080–8910	5620–11200
40	10 kHz	8910–11200	
41	12.5 kHz	11.2–14.1 kHz	
42	16 kHz	14.1–17.8 kHz	11.2–22.4 kHz
43	20 kHz	17.8–22.4 kHz	

a considerable range. Indeed, by undertaking noise signature analysis at remote locations it is possible not only to determine which class of vessel has been located, but if sufficient information is known about the character of each signature, it is also possible to identify the particular ship. Clearly, the ultimate goal of a warship designer must be to make the ship's signature vanish into the background noise of the sea, which comprises contributions from the weather, marine life and also other shipping from a wide geographical area.

This leads to an important distinction in the types of noise that are generated by the various components of a ship. These are termed self- and radiated noise and it is convenient to define these as follows:

Self-noise The noise, from all shipboard sources, generated by the subject vessel and considered in terms of the effect it has on the vessel's own personnel and equipment.

Radiated noise The noise generated by the ship and experienced at some point distant from the ship by which its detection, recognition or influence on the environment could be initiated.

TABLE 10.5 Maximum Noise Levels Permitted on a Ship According to the 1981 IMO Code

Location	Level (dBA)
Engine room	110
Workshops	85
Machinery control room	75
Navigating bridge	65
Mess room	65
Recreation room	65
Cabins and hospital	60

When considering the noise generated by ships it is useful to place it in the context of the ambient noise level in deep water. Wenz[5] and Perrone[6] considered the ambient noise levels in deep water and the results of their work are shown in Figure 10.4 as measured from omnidirectional receivers. From this figure it is seen that below about 20 Hz ocean turbulence and seismic noise predominate, whereas in the range 20 Hz to around 200 Hz the major contributions are from distant shipping and biological noise. Above 500 Hz to around 20 kHz the agitation of the local sea surface is the strongest source of ambient noise and beyond 50 kHz thermal agitation of the water molecules becomes an important noise source, where the noise spectrum level increases at around 6 dB/octave. In the case of shallow water the noise levels can be considerably higher due to possible heavier concentrations of shipping, nearby surf and waves breaking, higher biological noise, shore-based noises, and so on.

10.2 NATURE OF PROPELLER NOISE

There are five principal mechanisms by which a propeller can generate pressure waves in water and hence give rise to a noise signature. These are:

1. The displacement of the water by the propeller blade profile.
2. The pressure difference between the suction and pressure surfaces of the propeller blade when they are rotating.
3. The flow over the surfaces of the propeller blades.
4. The periodic fluctuation of the cavity volumes caused by operation of the blades in the variable wake field behind the vessel.
5. The sudden collapse process associated with the life of a cavitation bubble or vortex.

Clearly, the first three causes are associated with the propeller in either its cavitating or non-cavitating state; however, they are non-cavitating effects. The latter two

FIGURE 10.3 Sound pressure levels in hull close to propeller. *Reproduced with permission from Reference 4.*

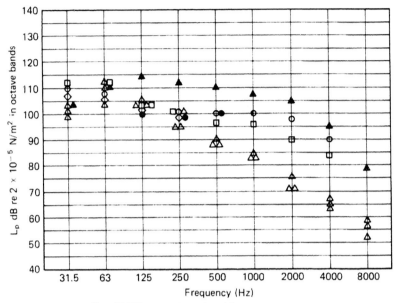

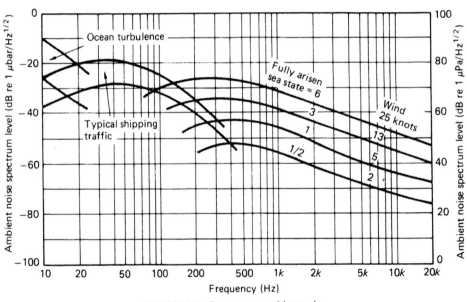

FIGURE 10.4 Deep-water ambient noise.

causes are cavitation-dependent phenomena, and therefore occur only when the propeller is experiencing cavitation.

Propeller noise can, therefore, be considered as comprising two principal constituents: a non-cavitating and a cavitating component. In terms of the noise signature of a vessel, prior to cavitation inception, all components of noise arising from the machinery, hull and propeller are important. Subsequent to cavitation inception, whilst the hull and machinery sources need careful consideration, the propeller noise usually becomes the dominant factor. Figure 10.5 typifies this latter condition, in which the self-noise generated at the sonar dome of a warship is seen. This

Chapter | 10 Propeller Noise

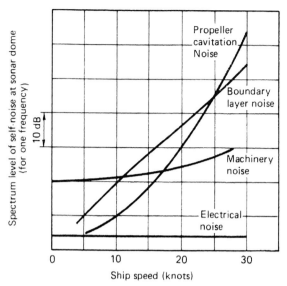

FIGURE 10.5 Example of the variations in self-noise as a function of ship speed due to propeller, boundary layer machinery. *Reproduced with permission from Reference 24.*

figure shows the comparative noise levels at this location of the hull boundary layer, the machinery, electrical noise and the propeller. When studying this figure it should be remembered that the propeller, in this case, is at the opposite end of the vessel to the sonar dome, and hence the importance of the propeller as a noise source can be fully appreciated and is seen to dominate at speeds above 25 knots.

Propeller noise comprises a series of periodic components, or tones, at blade rate and its multiples, together with a spectrum of high-frequency noise due to cavitation and blade boundary layer effects. Figure 10.6 shows a radiated cavitating propeller noise spectrum based on a third-octave band analysis; the sound pressure levels are referred to 1 µPa level in keeping with the normal practice. Within this noise spectrum the blade rate noise is commonly below the audible threshold, although not below sensor detection limits: typically in the case of a four-bladed propeller operating at say 250 rpm this gives a blade rate frequency of 16.7 Hz, which is just below the normally recognized human audible range of about 20–20000 Hz.

To consider noise generation further, it is convenient to address separately the issues of non-cavitating and cavitating noise. The former, although not considered for most merchant ships, is of considerable interest in the case of research ships and naval vessels which rely on being able to operate quietly in order to undertake their work or detect potential threats. For these latter cases, a designer endeavors to extend the non-cavitating range of operation of the vessel as far as possible.

Blake[21] offers a detailed treatment of the analysis of both non-cavitating and cavitating noise for marine applications.

10.2.1 Non-Cavitating Propeller Noise

The marine propeller in its non-cavitating state and in keeping with other forms of turbo-machinery produces a noise signature of the type sketched in Figure 10.7. It is seen from this figure that there are distinct tones associated with the blade frequencies together with a broadband noise

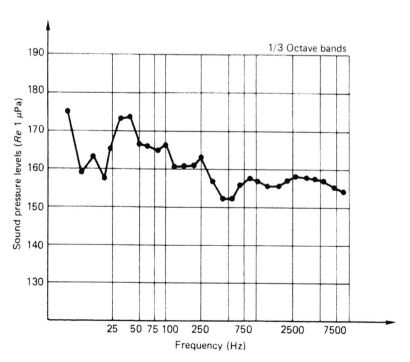

FIGURE 10.6 Radiated cavitation noise spectra measured outside a hull at full-scale.

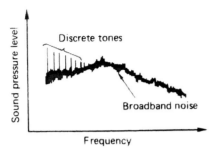

FIGURE 10.7 Idealized non-cavitating noise spectrum.

at higher frequencies. The broadband spectrum comprises components derived from inflow turbulence into the propeller and various boundary layer and edge effects such as vortex shedding and trailing edge noise.

For analysis purposes there are some similarities between the marine propeller as a noise source and both air propellers and helicopter rotors. A marine propeller can, for the purposes of noise prediction, be considered as a compact noise source since the product of the wave number times the radius is much less than unity. The wave number is defined as the frequency divided by the speed of sound. This considerably simplifies the analytical assessment of the noise characteristics from that of, say, a helicopter rotor. However, this simplification is perhaps balanced by the greater density of water because this increases the entrained mass of the blades relative to their mass in air and, therefore, their flexibility becomes a significant consideration in terms of the radiation efficiency.

With regard to the blade rate noise, the propeller is normally operating behind a vessel or underwater vehicle and works in a circumferentially varying wake field. This causes a fluctuating angle of incidence to occur on the blade sections which can be represented as a gust normal to the blade when considered relative to the propeller blade. From this gust model an expression can be generated for the far-field radiated source pressure.

The analysis of the broadband components is different. In the blade rate problem the unsteadiness is caused by the circumferential variation in the wake field; however, in the inlet turbulence case we need to consider the level of turbulence in the incident flow. This implies that the wake harmonics associated with this feature become a function of time and not necessarily just the analysis position in the propeller disc. To accommodate this feature, the turbulence velocity spectrum has to be incorporated into the analysis procedure to describe the flow and derive an expression for the radiated pressure due to this component.

Trailing edge noise is perhaps the least well understood of the broadband noise mechanisms, since it involves a detailed knowledge of the flow around the trailing edge of the section. The role of viscosity within the boundary layer

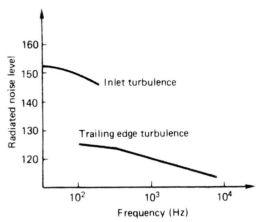

FIGURE 10.8 Typical radiated noise levels from a rigid hydrofoil moving in disturbed water. *Reference 20.*

is a crucial parameter in estimating the levels of radiated noise produced and is at present the subject of research. Blake, however, in his study of the subject gives an appreciation of the relative levels of trailing edge and inlet turbulence noise: Figure 10.8 is taken from his work for illustration purposes.

The problem of optimizing marine propellers for noise in sub-cavitating conditions by theory is still in its infancy since the complete solution requires both a detailed viscous flow calculation over the propeller blades together with an inlet turbulence spectrum, in addition to the normal wake field data. Jenkins[7] discusses this problem further in the context of the non-cavitating marine propeller.

In addition to the foregoing effects, there are also hydroelastic and fouling effects which need consideration in non-cavitating noise analysis.

10.2.2 Cavitation Noise

Just prior to visible cavitation inception it has been observed that in a fairly narrow frequency range the measured noise levels increase.

The collapse of cavitation bubbles creates shock waves and hence generates noise. This is essentially 'white noise' covering a frequency band up to around 1 MHz. From the theoretical viewpoint, the problem of noise radiation by cavitation was approached until recently from the behavior of a single cavitation bubble such that the bubble dynamics were considered in a variable pressure field (Reference 8); for example, along the surface of a propeller blade section. Under these conditions the bubble will undergo volume fluctuations and as a consequence radiate acoustic energy. Using this approach the spectral power density of a set of bubbles becomes the product of the number of bubbles per unit time and the spectral energy density due to the growth and collapse of a single bubble, assuming that the bubbles occur as random events. Such models, however, only

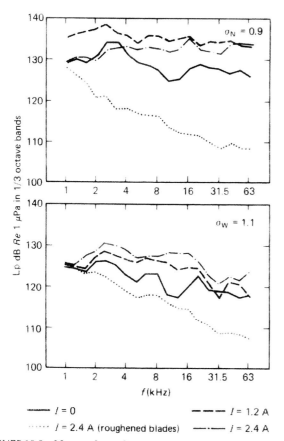

FIGURE 10.9 Measured sound pressure spectra, results with smooth blades, comparison with results with roughened blades (I = electrolysis current). Reproduced with modification from Reference 9, with permission from ASME.

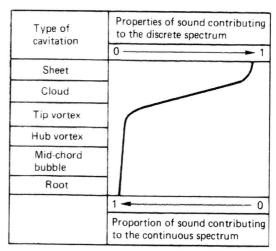

FIGURE 10.10 Relative contribution of different cavitation types to the sound power spectrum. Reproduced with modification from Reference 22, with permission.

partially predict the real behavior of cavitating propeller blades and tend to fail in their prediction capability at very high bubble densities. Work by van der Kooij[9] and Arakeri and Shanmuganathan[10] support this conclusion. Van der Kooij shows by means of model tests on smooth and roughened blades that the noise generated by bubble cavitation initially increases with an increasing number of bubbles and then falls off very markedly when a large number of bubbles are present. Figure 10.9, taken from Reference 8, shows this effect. In the case of the smooth blade different bubble densities were induced by a varying electrolysis current ranging from 0–2.4 A, while in the case of the roughened blade a large number of bubbles were generated from the application of artificial leading edge roughness in association with an electrolysis current of 2.4 A.

Figure 10.10, based on Reference 21, shows in a schematic way the relative contribution of different cavitation types to the sound power spectrum. Hence an appreciation can be gained of the influence that a particular cavitation type has on either the continuous or discrete noise spectra.

The prediction of noise from cavitation by theoretical means is more complex than for the non-cavitating propeller and, as a consequence, most prediction is done using model propellers operating in a cavitation tunnel. At present the inability of theoretical methods to take account of the detailed boundary layer and cavitation dynamics tends to limit their value.

Matusiak[25] has developed a theoretical procedure to evaluate the noise induced by a cavitating propeller having mostly sheet cavitation. The noise propagation model used in the method is a linear acoustic approximation giving spherical spreading for an unbounded homogeneous medium. This approach produces a broadband propeller-induced pressure spectrum which has been shown to generally correlate with measured results for cavitating propellers. The high-frequency noise emissions of a non-cavitating propeller are not considered and because the method is based on potential flow analysis it does not take into account viscous flow effects and, therefore, does not predict the noise measured at low Reynolds numbers. Similarly, the lack of viscous flow computation may also contribute to prediction problems with leading edge cavitation effects of highly skewed propellers and the influence of tip vortices are disregarded. Choi et al.[26] have examined the noise emissions from vortex cavitation bubbles. This has involved the growth, splitting and collapse of vortex cavitation bubbles for a single vortex; a vortex that experiences a pressure drop and recovery and a vortex that is interacting with another stronger vortex. While the qualitative dynamics were similar the inception, dynamics and noise production of the resulting bubbles suggested that scaling could not be achieved using the normal flow parameters. They found that the properties of the surrounding pressure field and the line vortices had a strong influence on the bubble phenomenology even when the changes were modest. Salvatore et al.[27] have proposed a method to model unsteady cavitation and noise. Their model is based on a potential flow formulation and a sheet cavitation model is superimposed in order to estimate the transient cavity patterns over the propeller blades. With regard to the noise emissions from the propeller these estimates are based on the Ffowcs-Williams and Hawkings[28] equation which describes the acoustic pressure field generated by lifting bodies under arbitrary motion and the subsequent work of di Francescantonio.[29]

The noise emitted by a cavitating propeller depends on the type of cavitation present at the particular operating condition. For example, back, face, hub and tip vortex cavitation types all have different noise signatures, as seen in Figure 10.11; taken from Sunnersjö.[11] From this figure the wide range of noise spectra derived from the same propeller when at four particular load conditions can be noted.

Noise measurements are now a regular feature of many cavitation tunnel test programs. The purpose of these tests can be to compare the noise spectra derived from different load conditions for the same propeller; comparisons between different propellers or the full-scale prediction of the noise spectra under different characteristic load conditions for a particular design. However, when a noise study is undertaken in a cavitation tunnel the presence of the tunnel walls influences the results to an extent that the results are not, without correction, representative of the free field

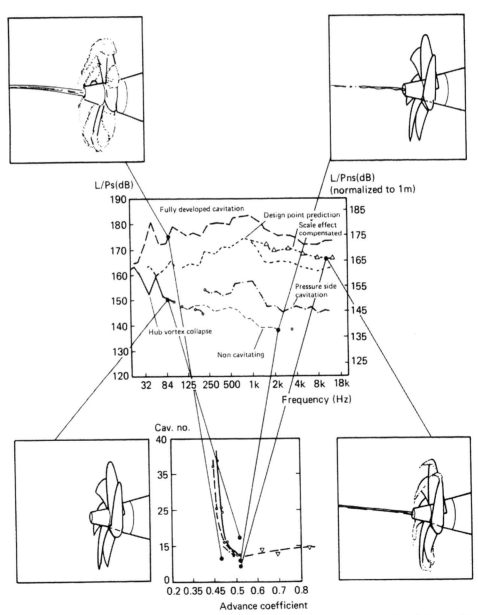

FIGURE 10.11 Effect of cavitation type of noise spectra. *Reproduced with permission from Reference 11.*

conditions. As a consequence, a correction factor has to be developed by substituting a calibrated noise source in place of the propeller so that a comparison can be made as to what the noise level would have been in the free field without the tunnel walls. This leads to the definition of a transfer function for the particular configuration of the form

$$P_{\text{ff}} = \phi P_t \tag{10.6}$$

where P_{ff} is the required free field noise spectrum, P_t is the measured noise spectrum in the tunnel and ϕ is the transfer function between the tunnel and free field.

10.3 NOISE SCALING RELATIONSHIPS

The basic requirement for deriving the full-scale noise prediction from model measurement is that the cavitation dynamics between model- and full-scale are identical. Scaling laws and their relation to bubble dynamics are discussed by many researchers and a full treatment of these can be obtained from (References 12 and 13). The scaling laws are based on the production of the pressure waves produced by a pulsating spherical bubble and immersed in an infinite volume of water. From this type of model the pressure at some point remote from the cavity or bubble can be expressed as follows:

$$P(r,t) = \frac{\rho}{3r}\left(\frac{d^2 R^3}{dt^2}\right) \tag{10.7}$$

in which R is the cavity radius, t is the time, r is distance from the center of the cavity and ρ is the water density.

The relevant scaling laws are then derived from the transformation of the variables in equation (10.7) and a relationship can be derived which is consistent with the approximation of a first-order model for the scaling of the continuous part of the power spectrum:

$$\frac{G_s(f_s)}{G_m(f_m)} = \left(\frac{r_m D_s}{r_s D_m}\right)^2 \left(\frac{\rho_s}{\rho_m}\right)^{1/2} \left(\frac{\Delta P_s}{\Delta P_m}\right)^{3/2} \lambda \tag{10.8}$$

where the suffixes m and s refer to model and ship scale respectively.

If this equation is applied to the measured sound pressure p in a frequency band Δf about a center frequency f and, furthermore, if the analysis bandwidth Δf is a constant percentage of the center frequency f (i.e. $\Delta f = af$, where a = constant as in Table 10.3) then equation (10.8) reduces to

$$\left[\frac{p_s(f_s, af_s)}{p_m(f_m, af_m)}\right]^2 = \left(\frac{r_m D_s}{r_s D_m}\right)^2 \left(\frac{\Delta P_s}{\Delta P_m}\right)^2 \tag{10.9}$$

Equation (10.9) can be shown to be valid both for spectral lines and for the continuous part of the spectrum. This reinforces the need to use a constant percentage bandwidth for the analysis of propeller noise. Equation (10.9) can be transformed from its dependence on the pressure difference Δp, which drives cavity collapse, to a dependence on propeller shaft speed n by assuming that the cavitation extents are identical between model- and full-scale at equal cavitation numbers:

$$\left[\frac{p_s(f_s, af_s)}{p_m(f_m, af_m)}\right]^2 = \left(\frac{r_m D_s}{r_s D_m}\right)^2 \left(\frac{\rho_s}{\rho_m}\right)^2 \left(\frac{n_s D_s}{n_m D_m}\right)^4 \tag{10.10}$$

With regard to frequency scaling, in relation to the first two sources of noise mentioned earlier, that is the displacement of the water by the blade profile and the pressure difference across the blade, it is clear that these are linked by the blade rate frequency. As a consequence of this, inverse shaft speed provides a suitable reference. However, in the case of a cavitating blade the collapse process cannot be directly linked to blade frequency. To overcome this problem a suitable time reference can be derived from the Rayleigh formula for the collapse time of a vapor-filled cavity:

$$T_c = 0.915 R_{\max} \left(\frac{\rho}{\Delta p}\right)^{1/2} \tag{10.11}$$

where $R_{\max}$ is the cavity radius, Δp is the pressure difference and ρ is the density. If this equation is then non-dimensionalized, the frequency scaling law becomes

$$\frac{f_s}{f_m} = \frac{n_s}{n_m}\left(\frac{\sigma_s}{\sigma_m}\right)^{1/2} \tag{10.12}$$

This is in contrast to the frequency scaling law for the non-cavitating or indeed the slowly fluctuating cavitating process:

$$\frac{f_s}{f_m} = \frac{n_s}{n_m} \tag{10.13}$$

However, it can be seen that provided σ_s and σ_m are the same, which is a prerequisite for cavitation similarity, together with the implied assumption concerning the cavity extents and dynamics, then equations (10.12) and (10.13) become identical and equation (10.13) suffices, and then becomes the counterpart of equation (10.10) for propeller noise scaling.

The assumptions concerning cavity extents at equal cavitation numbers are reasonable under well-developed cavitation conditions. However, close to incipient cavitation this is not always the case since many model tests suffer from scale effects which require that $\sigma_s > \sigma_m$ for equivalent cavitation extents. Under such conditions equations (10.10) and (10.13) do not apply and, therefore, appeal should be made to References 13 and 14.

The scaling relationships (10.10) and (10.13) are entirely applicable for fully developed cavitation states at multiples of blade frequency below about one-fifth blade rate. However, above this value, due to simplifications made in their deviation, they should be considered as a first approximation only.

10.4 NOISE PREDICTION AND CONTROL

If a definitive prediction of the noise spectra emission from a particular propeller—ship combination is required, then model test studies are, at the present time, the only realistic means of achieving this. Bark[15] discusses the correlation achievable and the reasons likely for any disparity of correlation between model and full scale. Figure 10.12, taken from Reference 15, demonstrates a good correlation of the non-dimensional noise in third-octave bands using mean rms levels. The diagram shows results for several speed conditions and a single gas content $\alpha/\alpha_s = 0.4$. In the figure the non-dimensional noise level $L(K_p)$ is given by

$$L(K_p) = 20 \log 10^6 K_p = 20 \log \left[\frac{p_{rms} \times 10^6}{\rho n^2 D^2}\right] \quad (10.14)$$

In his study, Bark suggests that the best correlation was found with the highest water velocities and that the influence of gas content in the range $0.4 < \alpha/\alpha_s < 0.7$ was not particularly large. Clearly, however, if this were to be extended to too high a value, then the high-frequency noise would be damped by the gas bubbles. This type of effect was demonstrated by van der Kooij[9]. From Figure 10.12 it is seen that the spectrum shape is similar in both model- and full-scale, although certain deviations will be noted in the frequency scaling, which can be attributed either to wave reflection at the hull or to differences in the cavitation scaling assumptions.

If model tests cannot be undertaken or contemplated, for whatever reason, it is still possible to make estimates of the propeller noise based on previous measurements. This type of prediction is, however, not as accurate as that based on model tests. Consequently, it needs to be used with care as the values derived are based on historical data, sometimes quite old, and therefore may not be strictly applicable to new project studies. Typical of this type of method is data on surface ship radiated noise spectra made during the Second World War and reported in a compendium issued by the US Office of Scientific Research and Development in 1945. These measurements were based on results from American, British and Canadian ranges, and the results converted into source levels at 1 m relative to 1 μPa using a process with an error bound of the order of 3 dB. This resulted in an expression for the noise level L_s at a given frequency being defined by the relationship

$$L_s = L'_s + 20(1 - \log f) \quad 100\,\text{Hz} < f < 10\,\text{kHz} \quad (10.15)$$

where L'_s is the overall level measured in the band from 100 Hz—10 kHz and f is the frequency in Hz.

Subsequent measurements made after the war on a variety of cargo vessels and tankers showed that deviations in the overall level L'_s occurred throughout much of the spectrum of the order of 1—3 dB. Re-analysis of the original measurements by Ross,[16] together with further more modern data, showed that the term L'_s could be expressed as a function of the propeller tip speed and blade number, and the use of the following expression for ships over 100 m in length was proposed:

$$L'_s \simeq 175 + 60 \log \frac{U_T}{25} + \log \frac{Z}{4} \quad (10.16)$$

where U_T is the tip speed in the range 15—50 m/s and Z is the number of cavitating blades. As a consequence equations (10.15) and (10.16) can be used to derive an approximation for the noise spectrum, referred to a source level at 1 m. In developing this expression it must be recalled that at the time ships were mostly fitted with conventional propeller designs and, as a consequence,

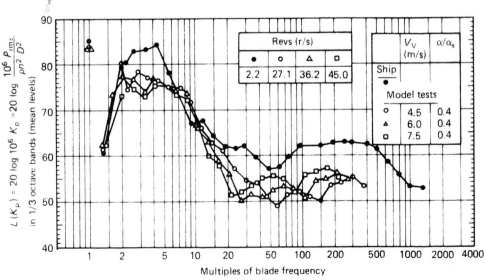

FIGURE 10.12 Non-dimensional noise presented as $L(K_p)$ in third-octave bands (mean rms levels); Comparison of full-scale data (filled symbols) with model data at three water velocities (open symbols). *Reproduced with permission from Reference 15.*

equation (10.16) would not be applicable to propellers with advanced blade loading designs.

The control of the noise emitted from a propeller can be done either by attempting a measure of control by re-designing the propeller blade surfaces or, assuming no further design improvement is possible, by attempting acoustic suppression through the vessel. Either of the methods is applicable to the merchant service because they are largely concerned with self-noise in the majority of cases. In the naval applications the concern with radiated noise frequently dictates that the source of the noise is suppressed.

The suppression of the noise within the vessel is a matter of calculating the noise paths through the vessel and designing an appropriate suppression system: as such this aspect lies outside the scope of this text and reference should be made to documents such as References 17 and 18. Where it is required that the noise be suppressed at source, this can be achieved by the consideration, for example, of any one or a combination of the following alternatives:

1. Re-design of the hull form to improve the wake field.
2. Change in radial distribution of skew.
3. Change in radial pitch distribution.
4. Adjustment of the general section profiles.
5. Changes to the leading edge and/or trailing edge geometry.
6. Changes to the section chord lengths.

10.5 TRANSVERSE PROPULSION UNIT NOISE

Transverse propulsion units are recognized as a major source of noise when they are in use for docking maneuvers. Units of this type, as sources of noise, are integrated into the hull structure rather than having a fluid medium between them and the hull surface as in the case of the propeller. In Chapter 14 the design of transverse propulsion units is discussed in some detail; however, the prediction of the likely levels of noise from these units is considered here.

A noise prediction method for controllable pitch units developed by the Institute of Applied Physics, Delft[19] is based on a large number of measurements on-board different types of ships. In essence the noise emitted by the transverse propulsion unit is defined as

$$L_P(P_B, \Delta\theta, L/D, \mathscr{L}) = L_{P0}(P_B, \mathscr{L}_0) - L_{P\Delta}(\Delta\theta, L/D, \Delta\mathscr{L}) \quad (10.17)$$

where

L_p = level of noise predicted at the point of interest (dBA).
L_{p0} = base level of noise at full power (P_B) in a standard cabin located near the thruster on the tanktop ($\mathscr{L}_0$).
$L_{p\Delta}$ = the change of noise level due to part load or pitch ($\Delta\theta$), tunnel length/diameter ratio (L/D) and position in ship ($\Delta\mathscr{L}$) as defined by deck (D_K) or frame number (F_r).

From a regression analysis based on the results of Reference 19 it was found that the base level of noise in equation (10.17) can be defined by the following relationship:

$$L_{p0}(P_B, \mathscr{L}_0) = 108.013 - 7.074K + 10.029K^2 \\ + (24.058 - 4.689K + 0.615K^2) \\ \times \log_{10}(P_B)$$

(10.17a)

where

K = tunnel center line immersion ratio I/D with K in the range $1 < K < 3$.
P_B = maximum continuous rating of the unit.

With regard to the change in noise level, $L_{p\Delta}$ can thus be defined as

$$L_{P\Delta}(\Delta\theta, L/D, \Delta\mathscr{L}) = L_P(\Delta\theta) + L_P(L/D) + L_P(\Delta\mathscr{L})$$

(10.17b)

in which

$$L_P(\Delta\theta) = 26.775 - 13.387 \log_{10}(\Delta P_B)$$
$$L_P(L/D) = 4.393 + 14.593 \log_{10}(L/D)$$

and

$$L_P(\Delta\mathscr{L}) = 10 \exp[0.904827 + 0.968977 \log_{10}(D_k) \\ - 0.348142(\log_{10}(D_k))^2] \\ + 10 \exp[-0.222330 - 0.126009(D_k) \\ + 0.007657(D_k)^2] F_r$$

ΔP_B = percentage MCR at which the unit is working
L/D is in the range $1 < L/D < 10$
D_κ is in the range $1 < D_\kappa < 5$ ($D_\kappa = 0$ at the tanktop)
Fr is in the range $1 < Fr < 50$.

Clearly, in using a formula of the type described by equation (10.17), care needs to be exercised, particularly in respect of the absolute accuracy; however, the relationship does provide guidance as to the noise levels that can be expected.

10.6 MEASUREMENT OF RADIATED NOISE

The measurement of the radiated noise is an important aspect of the trials of surface warships and submarines in the context of sonar detection and torpedo navigation systems. Furthermore, such trials form an important stage in the development of future designs of vessels. In addition to warships, certain specialist vessels such as research ships also have a radiated noise control requirement and so benefit from noise emission trials.

Radiated noise measurements are conducted on noise ranges especially constructed for the purpose. For the purposes of the measurement at least two hydrophones should be used: one directly under the track of the vessel and another a distance to one side of the track – not less than 100 m from the track. Water depth is important and the hydrophone located on the vessel's track should not be at a depth of less than 20 m, and if the water depth lies between 20 and 60 m then it should be planted on the bottom. For regions where the depth is greater than 60 m, the hydrophone should be located at a depth from the surface of one-third of the water depth. Furthermore, the trial noise ranges must be selected so that the acoustic background levels are well below the likely levels of the quietest machinery to be evaluated and the level of bottom reflection is insignificant.

The purpose of the two hydrophones is different. The one residing on the track of the vessel is primarily intended for the detailed study of the noise spectra over the frequency range 10–1200 Hz, whereas the beam hydrophone looks at the wider frequency range of 10–80 000 Hz in connection with sonar detection and torpedo acquisition risk. The performance characteristics of the hydrophones are particularly important and reference should be made to agreed codes of practice[20]: details the standardization agreement of NATO for these purposes and also for the conduct of the trials.

For the purposes of these trials, as in the case of normal power absorption trials, Chapter 16, the vessel needs to be maintained at steady conditions with the minimum use of helm. Records of nominal speed, actual speed, main engine and propeller shaft speeds, propeller pitch, vibration characteristics and so on need to be maintained during and prior to the trials to establish cavitation inception speeds of the propulsors. In addition, weather records need to be kept since weather can have a considerable effect on the measured noise spectrum: particularly if it is raining. Accurate shaft rotational speeds and vibration spectra of the more important main and auxiliary items of machinery should be recorded. Where large changes in operating draught occur the vessel should also be ranged in at least the extreme operating conditions. Additionally, it is also useful to measure, by means of over-side measurements, the noise spectra from the vessel when the ship is moored between buoys.

When undertaking noise range measurements it is often desirable to make measurements on reciprocal ship headings. When this is done, and to ensure that the ship conditions are as near the same as possible for both runs on the range, the turning of the ship at the ends of the range should be done as gradually as possible in the manner discussed in Chapter 17. Maneuvers such as Williamson Turns, which were designed for life-saving purposes, should be avoided when turning the ship at the end of a particular run on a noise range because they significantly disturb the dynamic equilibrium of the ship.

As well as the relatively simple measurement configurations for noise measurement, more advanced capabilities can also be deployed for more detailed signature analysis purposes. Typical of these latter capabilities are those located in Loch Goil and Loch Fyne in Scotland.

In addition to the types of fixed location measurement procedures already referenced, there is also the portable noise measurement buoy method which can be particularly useful for making rapid measurements at sea in deep water. The procedure is essentially to drop a sonar buoy in the sea during relatively calm weather conditions and then, knowing the position of the sonar buoy, travel back past the buoy at the desired speed conditions and at a known distance-off. Then, knowing the co-ordinates and the ship's line of travel it becomes a relatively simple matter to derive a noise spectrum for the ship. Moreover, some buoy-based methods have been calibrated against the definitive range measurements, such as Loch Goil, and the results found to compare favorably.

10.7 NOISE IN RELATION TO MARINE MAMMALS

At the 58th Session of the Marine Environmental Protection Committee of the International Maritime Organisation, a proposal for a work program was tabled by the United States of America. The program was directed towards promoting action to minimize the incidental introduction of noise from commercial shipping operations into the marine environment to reduce potential adverse impacts on marine life.

The proposal claimed that a significant proportion of the anthropogenic noise input into the oceans is attributable to the increasing number and size of commercial ships operating over wide-ranging geographical areas of the world. Moreover, it claimed that the noise generated by these sources has the potential to disturb the behavior and critical life functions of marine animals. Within the deep ocean environment there is evidence to suggest that the noise levels have increased in recent years. Hildebrand[30] drew the conclusion that from measurements made at the San Nicolas SOSUS Array in the Pacific Ocean this increase was of the order of 3 dB over the decade to 2004.

The noise signatures emitted by ships are variable. They vary, amongst other factors, with the type and age of the ship, the ship's speed and the type of propulsor deployed. Figure 10.13 shows the upper and lower bounds of typical emitted noise spectra derived from a variety of source

Chapter | 10 Propeller Noise

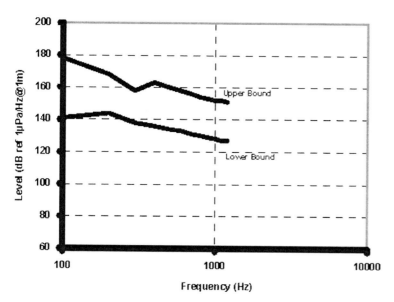

FIGURE 10.13 Bounds of ship noise spectra (15 ships).

measurements for a range of ship types, sizes and ages over the last two decades.

The actual signature, however, produced by a ship varies considerably with the power required to propel the ship and the propeller type. This in turn influences the flow and cavitation development over the propeller blades. In the case, for example, of a cruise ship the overall sound level of the ship may increase by between 6 per cent and 12 per cent when the ship increases its speed from 10 to 20 knots and the resulting signature emitted is a combination of the emissions from the diesel generators, HVAC, electric propulsion and the propulsors.

Within the literature, propeller singing is often cited as a source of noise emission from ships. In the relatively few instances when singing is encountered the phenomenon is caused by the shedding of vortices from the propeller tip and outer trailing edge region, which may then excite high-frequency blade resonances. Singing, however, is normally so annoying to the ship's crew that at the next dry-docking, or indeed at a special docking if the noise is too intrusive, minor modifications are made to the trailing edges of the propeller blades in order to cure the source of the noise emission.

10.7.1 Marine Mammals

Marine mammals evolved from land animals, making the transition from land to water approximately 55 million years ago and adapted to exploit the deep oceans over a period of 15 million years. The evolutionary process brought about adaptations to their respiratory and auditory systems, limb structure and additional specialisms to other body parts. There are three orders of marine mammals: *Cetacea*, *Sirenia* and *Carnivora*. The first two are exclusively marine mammals and the third includes some groups that live on land and in water, such as polar bears, and others that live only on land, such as dogs.

The living *Cetaceans*, which are all carnivorous, are extremely diverse. However, there are two main groups and these are Toothed whales (*Odontocetes*) and Baleen whales (*Mysticetes*). Toothed whales are the most numerous members of the *Cetacea*. They comprise all dolphins and porpoises, and the sperm, killer, pilot, beluga, narwhal and beaked whales. They are different from Baleen whales in that they have teeth, have only a single blowhole and are mostly smaller, with the sperm whale being the largest at around 18 m in length. Toothed whales actively hunt prey, often using echolocation for this purpose. In contrast Baleen whales are less diverse than the Toothed whales, encompassing bowhead, right, blue, fin, sei, Bryde's, minke, humpback and grey whales. Unlike the Toothed whales they have two blow holes, baleen plates and are generally much larger in size. These baleen plates or 'whalebone' are made of keratin and arranged like densely packed combs creating a filter to sift zooplankton and small fish from the seawater.

The Order *Sirenia* includes the only herbivores of the sea: the Manatee and Dugong. As their colloquial name 'sea cow' would suggest, they are bulky, rotund and rather slow-moving. They live in shallow, warm or tropical, coastal or river waters.

Pinnipeds are part of the order *Carnivores*. They are from a less ancient lineage than the *Cetacea* and *Sirenians*, and are hence less specialized for deep water living. The main three families within this are: True or Earless seals (*Phocids*), Fur or Eared seals and Sea lions (*Otariids*) and the Walrus (*Odobenids*). True seals are more specialized for water than Fur seals and Sea lions: they have no external

FIGURE 10.14 Example of the spectrum of a right whale call. *Reference 31.*

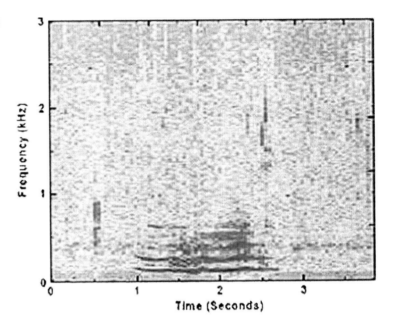

ear flaps, are more streamlined and tend to be larger than *Otariids*. However, they are much clumsier on land. A defining characteristic of the Fur seals and Sea Lions, *Otariids*, is the presence of small external ears. Additionally, they generally have thicker fur, less blubber and longer necks than the *Phocids*. They are much better suited to life on land, being able to use their rear flippers to walk and lift their bodies clear of the ground. Consequently, they tend to spend a greater proportion of their time on land than the *Phocids*, especially when they are with pups. The Walrus family shares some characteristics with both *Phocids* and *Otariids*; nevertheless they are immediately identifiable from their huge tusks, whiskers and bulk. They have no external ears, but walk on their hind flippers and spend significant periods of time on ice.

10.7.2 Marine Mammal Phonation

The term *phonation* is used to describe the sounds produced by marine mammals since some marine mammal sounds are not produced in the larynx, in the way that humans produce speech.

For most marine mammals, hearing and sound is important for a number of reasons:

- Avoidance of predators.
- Communication and social behavior.
- Echolocation.
- Foraging for food.
- Navigation.
- Overall awareness of their environment.
- Parental care.
- Reproduction.

Marine mammals have basically the same ear structure as terrestrial mammals, although this has been adapted to suit an aquatic environment. In addition, the brains of marine mammals are thought to process the sound in the same way as humans.

The ears are not the only potentially delicate structures in marine mammals that could be impacted by sound. Fundamentally any airspace in the body may be susceptible to pressure and acoustic effects. While the ear is an obvious candidate, other airspaces such as the lungs, airways and sinuses also need to be considered. For example, acoustically induced resonance in the lungs of human divers can be potentially harmful and this has been demonstrated at frequencies of around 70 Hz. Additionally, in *Cetaceans* it is likely that there is more than one pathway to the inner ear. There is substantial evidence to suggest that high frequency sounds are channeled through sound conducting tissues located in the lower jaw of *Odontocetes*, especially for echolocation tasks. The lower jaw bones typically contain mandibular fats, which are good conductors of sound and well impedance matched for the task underwater. However, lower frequency sounds probably follow the more conventional route of transmission into the middle and inner ear. Figures 10.14 and 10.15[31] illustrate the principal types of phonation from marine mammals: the first being for communication while the latter is for echolocation purposes. In both spectra it is seen that there is a considerable quantity of data transmitted and received at differing frequencies.

Toothed whales, dolphins and porpoises are mostly sociable; traveling and feeding in groups, often co-operating to maximize feeding potential. Dolphin phonations have been the most comprehensively studied. They produce

Chapter 10 Propeller Noise

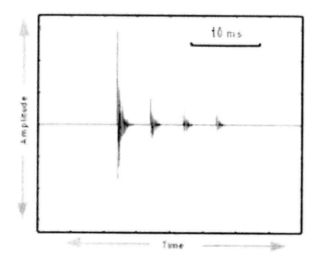

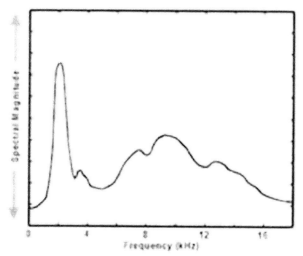

FIGURE 10.15 Idealization of sperm whale click and spectrum. *Reference 31.*

complex 'whistles' around 10 kHz for communication purposes and 'clicks' at 100 kHz for echolocation, by which they navigate and detect prey. Whistles are also used as signature calls so that the mammals can identify each other: a function that is especially important between mothers and calves. This is the general pattern of phonation for *Odontocetes*, with a few exceptions:

- Porpoises produce even higher ultrasounds, around 130 kHz and some species have not been proven to whistle. Some porpoises produce low frequency clicks (~2 kHz) for communication instead.
- Sperm whales only produce clicks, with most sound energy around 3 kHz and 10.5 kHz.
- Killer whale phonations are of a lower frequency: their whistles are around 10.5 kHz and their pulsed calls are mostly around 4 kHz, but can vary from 500 Hz–25 kHz.

The larger Baleen whales typically produce relatively low frequency (~20 Hz–1000 Hz) continuous tone or sweeping 'moan' phonations, with some species producing infrasonics: frequencies that are too low for humans to hear. Many sounds are quite simple, apart from when humpback and bowhead whales sing. The songs are produced by solitary males as reproductive displays and travel long distances underwater Reference 32. While recognizing that these sounds are low frequency and therefore travel long distances through the ocean, this may suggest that the hearing mechanism of these, and perhaps other, mammals is particularly sensitive.

There has been no conclusive evidence that Baleen whales echolocate, although there is speculation that the reverberation of their low frequency calls from the sea bed provides them with a picture of their environment, such as ice or islands, thus helping them to navigate (References 32 and 33).

It is unknown how important sound is to the lives of Manatees and Dugong; they are usually very quiet except for calls between mother and calf and to signal danger. Recorded calls have been in the general range 1 kHz–10 kHz.

Pinnipeds produce phonations above and below water. Pinnipeds that mate on land tend to be more vocal on land and limited to clicks and barks in the frequency range 100 Hz–4 kHz underwater. In contrast, the True seals that mate underwater produce a greater range and variety of phonations underwater from grunts below 100 Hz, to clicks, which may be for echolocation, up to 150 kHz. It is known that all Pinnipeds use sound to establish and maintain the bond between mother and pup and acoustic contact is especially important when the two become separated (Reference 32).

It is relatively easy, with the aid of hydrophones, to measure the apparent range of communication between marine mammals of the same species. What is uncertain is the actual breadth of the hearing of each of the species and whether within that broader range distress can be caused by being subjected to certain frequencies and noise amplitudes. It is known that in certain circumstances some marine mammals are reluctant to approach certain large ships too closely, but whether this is because of distress caused by noise or a natural reluctance to come too close to another unrecognized large moving maritime object is again unclear. In contrast, other of the smaller marine mammals (typically, dolphins, porpoises and killer whales) seem to enjoy the presence of a ship as they are frequently seen in the company of ships for quite long distances. As such, in these later cases it would seem that the noise emissions from the ship are not too distressing for them.

While little is known about the auditory thresholds and audiograms for the larger mammals, a body of information is available for the smaller marine mammals since these can be handled and subjected to experimentation more easily. Figure 10.16 is typical of the data available in the form of an audiogram and for those species already studied is typically non-linear with respect to frequency. For marine mammals the audiogram is typically u-shaped; implying

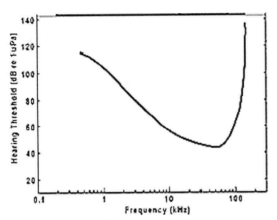

FIGURE 10.16 Example of a dolphin audiogram.

that at the upper and lower frequency ends of the hearing range hearing is less sensitive than in the middle of the frequency range. Indeed, such a hearing sensitivity is similar to that of humans, characterized by the A-weighted hearing scale. Indeed, such a finding is not surprising due to the similarity of the ear structure for both humans and marine mammals. In the case of the marine mammals studied to date this general pattern seems to hold true, although frequency and sensitivity ranges vary between the various marine mammal frequencies.

When compiling the known data on the Toothed and Baleen whales, Pinnipeds and *Sirenians* (Reference 34) it is clear that there is an abundance of communication and navigational signals produced by these mammals within the frequency range 100 Hz–100 kHz with, in the case of the Baleen whales, frequencies down to 10Hz. It must also be noted that the mechanisms for emitting and receiving the various types of signal are likely to be different and, moreover, there is likely to be an interval within the spectrum between the two types of signal. While within each of the ranges associated with the various species, some estimate can be made for regions of particular sensitivity from the available audiograms; such a broad range clearly encompasses the frequency ranges emitted by ships. As was noted in Figure 10.13, where full-scale noise data bounds in the frequency range 100 Hz to around 1250 Hz were shown, the true extent of the noise spectrum from a propeller is considerably greater and extends up to 100 kHz and beyond, although with reduced noise levels. Additionally, propulsors emit low frequency noise down to the first blade rate frequency and lower. While these frequencies are normally below the threshold of human hearing they can border on the lower end of the frequency ranges used by Baleen whales.

In terms of the contribution of cavitation to these noise emission spectra, a significant amount of work has been undertaken in recent years, both in naval and merchant ship communities. Each type of cavitation has a characteristic noise signature associated with it, depending upon the cavity dynamics that are involved. For example, in the case of sheet cavities the decay of the primary structures into vortex structures and then, subsequently, into systems of micro-bubbles produces characteristic signatures: probably through synchronized collapse of all or part of the system of bubbles — the actual mechanism being far from fully understood at the present time. While there is much that can be learnt from naval practice in the design of quieter merchant ship propellers, it must be recognized that there is a fundamental difference in the design philosophy of both types of ship. In the former case, the concept of cavitation inception speed features prominently where the propeller is designed to operate in a sub-cavitating mode up to a certain ship speed: typically, 10 or 15 knots depending on the naval requirement. To achieve this several iterations of the design process, including cavitation tunnel testing, are normally required. In the case of the merchant propeller, consideration of propulsion efficiency is dominant and the presence of cavitation is accepted provided that it does not promote either significant ship internal vibration or erosion of the propeller blades or rudder. Nevertheless, the cavitation structural dynamics on the propeller blades and in the slipstream, which will influence the radiated hull surface pressures and the erosive potential of the cavitation, will also have a significant effect on the far field noise emissions. As such, there is likely to be some synergy in these merchant ship cavitation influences.

Considerable further work is required at full-scale on the observation of the cavitating structures on ship's propellers since, due to scale effects, these structures generally vary from model-scale predictions. Such studies then need to be related to the noise emissions measured using sonar buoys or sea-bed arrays and thence to the effect that they might have on marine mammals; again about which there is much to learn, particularly for the larger creatures.

REFERENCES AND FURTHER READING

1. *Physics of Sound in the Sea*. Department of the Navy; 1965. NAVMAT P-9675.
2. Lovett. *J Acoust Soc Am* 1978;63.
3. *Code on Noise Levels on Board Ships*. International Maritime Organization; 1981.
4. Flising A. Noise reduction in ships. *Trans I Mar E*; 1978.
5. Wenz JJ. *Acoust Soc Am* 1962;34.
6. Perrone I. *J Acoust Soc Am* 1969;46.
7. Jenkins CJ. *Optimising propellers for noise*. Advances in Underwater Technology, Ocean Science and Offshore Engineering,. Tech. Common to Aero and Marine Engineering;1988;15.
8. Fitzpatrick HM, Strasberg M. *Hydrodynamic sources of sound*. Washington, DC: Proc. 1st Symp. on Naval Hydrodyn.; September 1956.
9. Kooij J van der. *Sound generation by bubble cavitation on ship model propellers; the effects of leading edge roughness*. Proc. 2nd

Int. Symp. on Cavitation and Multiphase Flow Noise, ASME; December 1986.
10. Arakeri VH, Shanmuganthan V. *On the evidence for the effect of bubble interference on cavitation noise.* J Fluid Mech 1985;**159**.
11. Sunnersjö S. *Propeller noise prediction.* SSPA Highlights 1986;**2**.
12. Baiter HJ. *Aspects of cavitation noise.* Wageningen: High Powered Propulsion of Large Ships Symp., NSMB; 1974.
13. Strasberg, M. *Propeller cavitation noise after 35 years of study.* Symp. of Noise and Fluids Eng., Proc. ASME, Atlanta.
14. Blake WK, Sevik MM. *Recent developments in cavitation noise research.* Int. Symp. on Cavitation Noise, ASME; November 1982.
15. Bark G. *Prediction of Propeller Cavitation Noise from Model Tests and Its Comparison with Full Scale Data.* SSPA Report No. 103; 1986.
16. Ross D. *Mechanics of Underwater Noise.* Oxford: Pergamon Press; 1976.
17. *Ship Design Manual – Noise.* BSRA; 1982.
18. Morrow RT. *Noise Design Methods for Ships, Trans RINA.* Paper No 7, Spring Meeting; 1988.
19. Dekker N. *Lips Newslett.*; March 1987.
20. *Standards for Use When Measuring and Reporting Radiated Noise Characteristics of Surface Ships, Submarines, Helicopters, Etc. in Relation to Sonar Detection and Torpedo Acquisition Risk.* NATO; November 1984. STANAG No. 1136.
21. Blake WK. *Aero-Hydroacoustics for Ships, Vol. I and II.* DTNSRDC. Report No. 84/010; June 1984.
22. Baiter HJ. *Advanced views of cavitation noise. Int.* Symp: on Propulsors and Cavitation; Hamburg; 1992.
23. Brüel, Kjaer. *Noise Pocket Handbook.*
24. Oossanen P van. *Calculation of Performance and Cavitation Characteristics of Propellers Including Effects of Non-uniform Flow and Viscosity.* Publication No. 457; 1974.
25. Matusiak J. *Pressure and noise induced by a cavitating marine screw propeller.* PhD Thesis. Espoo: VTT; 1992.
26. Choi J, Chang N, Yakushiji R, Dowling DR, Ceccio SL. *Dynamics and noise emission of vortex cavitation bubbles in single and multiple vortex flow.* Wageningen: 6th Int. Symp. on Cavitation (CAV 2006); September 2006.
27. Salvatore F, Testa C, Ianniello S, Pereira F. *Theoretical modelling of unsteady cavitation and induced noise.* Wageningen: 6th Int. Symp. on Cavitation (CAV 2006); September 2006.
28. Ffowcs-Williams JE, Hawkins DL. *Sound generated by turbulence and surfaces in arbitrary motion. Phil Trans Royal Soc* 1969;**A264**:321–42.
29. Francescantonio P. di, *The prediction of sound scattered by moving bodies.* Munich: 1st AIAA/CEAS Aero-acoustics Conf., AIAA Paper 95-112; May 1995.
30. Hildebrand J. *Large vessels as sound sources 1: Radiated sound and ambient noise in nearshore/continental shelf environments.* Washington DC: NOAA Vessel-Quieting Symp.; May 2007.
31. Gould JC. *Underwater acoustic sensitivity of marine mammals: a guide for the consideration of acoustic impacts.* Private Communication; February 2006.
32. Richardson WJ, Greene CR, Malme CI, Thomson DH. *Marine Mammals and Noise.* Academic Press; 1995.
33. Nachtigall, P.E. *Marine Mammals, Hearing and Sound.* Advisory Committee on Acoustic Impacts on Marine Mammals.
34. Carlton JS, Dabbs E. *The Influence of Ship Underwater Noise Emissions on Marine Mammals.* Lloyd's Register Technology Days, Paper No.9., Lloyd's Register; 2009.

Chapter 11

Propeller, Ship and Rudder Interaction

Chapter Outline

- 11.1 **Bearing Forces and Moments** — 271
  - 11.1.1 Propeller Weight — 272
  - 11.1.2 Dry Propeller Inertia — 274
  - 11.1.3 Added Mass, Inertia and Damping — 275
  - 11.1.4 Propeller Forces and Moments — 279
  - 11.1.5 Propeller Forces and Moments Induced by Turning Maneuvers — 287
  - 11.1.6 Out-of-Balance Forces and Moments — 288
- 11.2 **Hydrodynamic Interaction** — 288
  - 11.2.1 Non-Cavitating Blade Contribution — 289
  - 11.2.2 Cavitating Blade Contribution — 289
  - 11.2.3 Influence of the Hull Surface — 290
  - 11.2.4 Methods for Predicting Hull Surface Pressures — 291
- 11.3 **Propeller–Rudder Interaction** — 293
  - 11.3.1 The Single-Phase Approach — 294
  - 11.3.2 The Two-Phase Approaches — 294
  - 11.3.3 Model Testing — 296
  - 11.3.4 Some Full-Scale Remedial Measures — 297
- **References and Further Reading** — 297

The propeller interacts with the ship and rudder in a variety of ways. In the case of the propeller–ship interaction this is effected either through the coupling between the shafting system and the vessel or via pressure pulses transmitted through the water from the propeller to the hull surface. For this type of interaction the forces and moments can be considered to comprise both a steady and a fluctuating component. The steady or constant components of the interaction originate from attributes such as propeller weight, inertia and the mean wake field, while the fluctuating interactions derive principally from the variations in the wake field generated by the ship and within which the propeller has to operate. Notwithstanding the wake-induced variations there are also loading perturbations induced by the ship's sea motions and maneuvering activities. There is a further source of variation to the induced loadings and that is where significant propeller out-of-balance forces and moments are present.

In the alternative case of propeller–rudder interaction there is both the effect of the rudder on the pressure field surrounding the propeller and the efflux of the propeller on the rudder including the effects of cavitation. Additionally, the hull profile above the rudder station also influences the flow field over the rudder and, as a consequence, influences both the rudder and the propeller–rudder interaction.

For general convenience of discussion propeller–ship interaction can be considered in two separate categories. The first comprises the forces and moments transmitted through the shafting system, frequently termed somewhat loosely, 'bearing forces', while the second includes the forces experienced by the ship that are transmitted through the water in the form of pressure waves; these being termed 'hydrodynamic forces'. These two classes of interaction will, therefore, be considered separately in Sections 11.1 and 11.2 respectively and the propeller–rudder interaction discussion will then follow in Section 11.3.

11.1 BEARING FORCES AND MOMENTS

The loadings experienced by the vessel which come under the heading of bearing forces are listed under generalized headings in Table 11.1. It will be seen that they form a series of mechanical and hydrodynamically based forces and moments, all of which are either reacted at the bearings of the shafting system or change the vibratory properties of the shafting system in some way: for example, by altering the inertia or mass of the system. In the case of bearing reactions, the propeller generated forces and moments are

TABLE 11.1 Propeller Bearing Forces

Propeller weight and center of gravity
Dry propeller inertia
Added mass, inertia and damping
Propeller forces and moments
Out-of-balance forces and moments

supported by the lubrication film in the bearings which, in turn, is supported by the mechanical structure of the bearings and their seatings. As a consequence, in the analysis of marine shafting systems it is important also to recognize that, in addition to the influence of the propeller, the stiffness and damping of the lubrication film in the bearings can have important effects in terms of shafting response.[1]

11.1.1 Propeller Weight

The weight of a propeller needs to be calculated for each marine installation and is usually presented by manufacturers in terms of its dry weight. The dry weight, as its name implies, is the weight of the propeller in air, while the weight reacted by the shafting when the vessel is afloat is somewhat less. This is due to the Archimedean upthrust resulting from the displacement of the water by the volume of the propeller material. Hence, the effective weight of the propeller experienced by the ship's tail shaft is

$$W_E = W_D - U \quad (11.1)$$

where

W_E is the effective propeller weight
W_D is the weight of the propeller in air
U is the Archemedean upthrust.

The dry weight of the propeller W_D, which represents a constant downward force by its nature unless any out-of-balance occurs, is calculated from the propeller detailed geometry. This calculation is carried out in two parts; first the blade weight including an allowance for fillets, and in the case of a controllable pitch propeller the blade palm also, and secondly the calculation of the weight of the propeller boss or hub.

The blade weight calculation is essentially performed by means of a double integration over the blade form. The first integration evaluates the area of each helicoidal section by integration of the section thickness distribution over the chord length. Hence for each helical section the area of the section is given by

$$A_x = c \int_0^1 t(x_c) \, dx_c \quad (11.2)$$

where

x_c is the non-dimensional chordal length
$t(x_c)$ is the section thickness at each chordal location
c is the section chord length
A_x is the section area at the radial position $x = r/R$.

This integration can most conveniently be accomplished in practice by a Simpson's numerical integration over about eleven ordinates, as shown in Figure 11.1. Whilst this procedure will give an adequate estimate of the section areas for most detailed purposes, it is often useful to be able to bypass this stage of the calculation for quick estimates. This can be done by defining an area coefficient C_A which derives from equation (11.2) as follows:

$$C_A = \frac{A_x}{c t_{\max}} \quad (11.3)$$

where $t_{\max}$ is the section maximum thickness.

The area coefficient is the ratio of the section area to the rectangle defined by the chord length and section maximum thickness. Table 11.2 gives a typical set of C_A values for fixed and controllable pitch propellers which may be used for estimates of the section area A_x using equation (11.3).

The second integration to be performed is the radial quadrature of the section areas A_x between the boss, or hub radius, and the propeller tip. This integration gives the blade volume for conventional blade forms as follows:

$$V' = \int_{rh}^{R} A_x \, dr \quad (11.4)$$

As before this integration can readily be performed numerically using a Simpson's procedure as shown in Figure 11.1. Additionally, the radial location of the center of gravity of the blade can also be estimated for conventional blades by the incorporation of a series of moment arms, as shown in Figure 11.1. However, for non-conventional blade forms it is advisable to evaluate these parameters by means of higher-order geometric definition and interpolation coupled with numerical integration procedures.

The blade volume V calculated from equation (11.4) needs to be corrected for the additional volume of the blade fillets and a factor of the order of 2–5 per cent would be reasonable for most cases. The weight of the blades can then be determined from

$$W_b = \rho_m Z V \quad (11.5)$$

where

Z is the number of blades
ρ_m is the material density
V is the volume of one blade corrected for the fillets.

For the controllable pitch propeller the blade weight W_b is further corrected for the weight of the blade palm, the evaluation of which is dependent upon the specific geometry of the palm. The dry propeller weight W_D is then the sum of the blade weights and the boss or hub weight. In the case of the fixed pitch propeller the boss weight can normally be calculated from approximating the boss form by a series of concentric annular cylinders or by the first theorem of Guldemus: that is, volume = area × the distance traveled by its centroid. For the controllable pitch

Chapter | 11 Propeller, Ship and Rudder Interaction

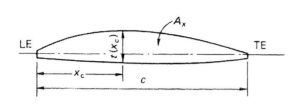

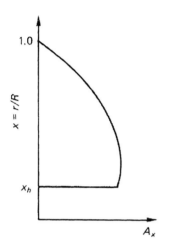

x_c	$t(x_c)$	Simp. mult. (SM)	SM $\times t(x_c)$
0.0	t_{LE}	1/2	$t_{LE}/2$
0.1	$t_{0.1c}$	2	$2t_{0.1c}$
0.2	$t_{0.2c}$	1	$t_{0.2c}$
0.3	$t_{0.3c}$	2	$2t_{0.3c}$
.	.	.	.
.	.	.	.
0.9	$t_{0.9c}$	2	$2t_{0.9c}$
1.0	t_{TE}	1/2	$t_{TE}/2$
			Σ_0

$$A_x = \frac{0.2c\,\Sigma_0}{3}$$

x	A_x	SM	SM $\times A_x$	1st moment arm	SM $\times A_x \times$ 1st. Mt. A
1.0	$A_{1.0}$	1/2	$A_{1.0}/2$	0	0
x_2	A_{x_2}	2	$2A_{x_2}$	1	$2A_{x_2}$
x_3	A_{x_3}	1	A_{x_3}	2	$2A_{x_3}$
x_4	A_{x_4}	2	$2A_{x_4}$	3	$6A_{x_4}$
.	.	.	.	.	.
.	.	.	.	.	.
x_{10}	$A_{x_{10}}$	2	$2A_{x_{10}}$	9	$18A_{x_{10}}$
x_h	A_{x_h}	1/2	$A_{x_h}/2$	10	$5A_{x_h}$
			Σ_1		Σ_2

Blade volume $V' = \dfrac{(R - r_h)\,\Sigma_1}{15}$

Position of centroid from tip $= \dfrac{\Sigma_2 (R - r_h)}{10\Sigma_1} = x_{g_1}$

Position of centroid from shaft $C_L = R - x_{g_1} = x_{g_0}$

FIGURE 11.1 Calculation of blade volume and centroid.

propeller the calculation of the hub weight is a far more complex matter since it involves the computation of the weights of the various internal and external components of the hub and the oil present in the hub: thus each hub has to be treated on its own particular merits.

The resulting dry weight of the propeller W_D is then given by

$$W_D = W_b + W_H \tag{11.6}$$

where W_H is the boss or hub weight.

The Archimedean upthrust U is readily calculated for the blades and the boss of a fixed pitch propeller as

$$U = \frac{\rho}{\rho_m} W_D$$

where ρ is the density of the water and it is assumed that the boss is a homogeneous solid mass, such as might be experienced with an oil injection fitted propeller, in contrast to a conventional key fitted boss with a lightning chamber.

Hence from equation (11.1) the effective weight W_E of the propeller is given by

$$W_E = W_D \left[1 - \frac{\rho}{\rho_m}\right] \tag{11.7}$$

Since for sea water and nickel–aluminum bronze the ratio ρ/ρ_m is about 0.137, it can be seen that the upthrust is about one-seventh of the propeller dry weight.

In the case of the controllable pitch propeller equation (11.7) does not apply since the hub weight W_H in equation (11.6) is the sum of the internal weights and not derived from a homogeneous mass; similarly for the fixed pitch propeller with the non-homogeneous boss. As a consequence the upthrust derives from the following rewrite of the upthrust equation:

$$U = \frac{\rho}{\rho_m} W_B + \rho V_H = \rho\left(\frac{W_B}{\rho_m} + V_H\right)$$

where V_H is the external volume of the propeller hub.

TABLE 11.2 Approximate Section Area Coefficient Values

Non-Dimensional Radius x = r/R	Area Coefficient C_A	
	Fixed Pitch	Controllable Pitch
0.95	0.78	0.78
0.90	0.74	0.74
0.80	0.72	0.72
0.70	0.71	0.71
0.60	0.71	0.71
0.50	0.71	0.71
0.40	0.70	Fair to 0.8 at hub radius
0.30	0.70	
0.20	0.69	

11.1.2 Dry Propeller Inertia

The evaluation of the dry propeller inertia is in effect an extension of the calculation procedure outlined in Figure 11.1. At its most fundamental the mechanical inertia is the sum of all of the elemental masses in the propeller multiplied by the square of their radii of gyration. This, however, is not a particularly helpful definition for practically calculating the inertia of a propeller.

For many practical purposes it is sufficient for conventional propellers to extend the table shown by Figure 11.1 to that shown by Table 11.3 to include the summations Σ_1, Σ_2 and Σ_3. From this table the moment of inertia of a blade can be estimated about the blade tip from the following equation:

$$I_{\text{Tip}} = \frac{2}{3}\sum_3 \left(\frac{R - r_h}{10}\right)^3$$

and by using the parallel axes theorem of applied mechanics the moment of inertia of the blade can be deduced about the shaft center line as

$$I_{\text{OX}} = \rho_m k \left[I_{\text{Tip}} - \frac{2}{3}\sum_1 (R - r_h)\{l_{gt}^2 - l_{g0}^2\}\right] \quad (11.8)$$

where

ρ_m is the density of the material
l_{gt} is the distance of the centroid from the blade tip
l_{g0} is the distance of the centroid from the shaft center line
k is the allowance for the fillets.

For a fixed pitch propeller, the estimation of the boss inertia is relatively straightforward, since it can be approximated by a series of concentric cylinders to derive the inertia I_H about the shaft center line. In the case of the controllable pitch propeller hub, however, the contribution of each component of I_H has to be estimated separately. When this has been done the dry moment of inertia of the propeller can be found as follows:

$$I_O = I_H + ZI_{\text{OX}} \quad (11.9)$$

TABLE 11.3 Calculation of Moment of Inertia of a Blade

x	A_x	SM	$A_x \times$ SM	1st Moment arm	$A_x \times$ SM $\times$ 1st MA.	2nd M.A.	$A_x \times$ SM $\times$ 2nd M.A.
1.0	$A_{1.0}$	1/2	$A_{1.0/0.5}$	0	0	0	0
x_2	A_{x_2}	2	$2A_{x_2}$	1	$2A_{x_2}$	1	$2A_{x_2}$
x_3	A_{x_3}	1	A_{x_3}	2	$2A_{x_3}$	2	$4A_{x_3}$
x_4	A_{x_4}	2	$2A_{x_4}$	3	$6A_{x_4}$	3	$18A_{x_4}$
x_5	A_{x_5}	1	A_{x_5}	4	$4A_{x_5}$	4	$16A_{x_5}$
x_6	A_{x_6}	2	$2A_{x_6}$	5	$10A_{x_6}$	5	$50A_{x_6}$
x_7	A_{x_7}	1	A_{x_7}	6	$6A_{x_7}$	6	$36A_{x_7}$
x_8	A_{x_8}	2	$2A_{x_8}$	7	$14A_{x_8}$	7	$98A_{x_8}$
x_9	A_{x_9}	1	A_{x_9}	8	$8A_{x_9}$	8	$64A_{x_9}$
x_{10}	$A_{x_{10}}$	2	$2A_{x_{10}}$	9	$18A_{x_{10}}$	9	$162A_{x_{10}}$
x_h	A_{x_h}	1/2	$A_{x_h/0.5}$	10	$5A_{x_h}$	10	$50A_{x_h}$
			$\Sigma_1 =$ _____		$\Sigma_2 =$ _____		$\Sigma_3 =$ _____

11.1.3 Added Mass, Inertia and Damping

When a propeller is immersed in water the effective mass and inertia characteristics of the propeller change when vibrating as part of a shafting system due to the presence of water around the blades. Additionally, there is also a damping term to consider deriving from the propeller's vibration in water. This mode of vibration considers the global properties of the propeller as a component of the line shafting and, therefore, it is a vibratory behavior distinct from the individual vibration of the blades which is discussed separately in the chapter on Propeller Design (Chapter 22).

The global vibration characteristics of a propeller are governed by two hydrodynamic effects. The first of these is that the propeller is excited by variations of hydrodynamic loading due to its operation in a non-uniform wake field. The second is a reaction loading caused by the vibration behavior of the propeller which introduces a variation in the section angle of attack and, in turn, produces variations to the hydrodynamic reaction load. In the case of a metal propeller, provided the variations in angle of attack are small the vibratory loading can be considered to vary linearly and the principle of superposition applied. This implies that independent evaluations of the excitation load and reaction load are possible; however, this is unlikely to be the case for a composite propeller. As a consequence, to derive the excitation load caused by a metallic propeller working in a wake field in the absence of any vibration motion, only the steady state rotation of the propeller is considered. To derive the reaction loading the propeller is considered to be a rigid body vibrating in a homogeneous steady flow. Since the forces and moments generated by a vibrating propeller are assumed to vary linearly with the magnitude of the vibratory motion, the forces and moments can be determined per unit motion, termed propeller coefficients, whose magnitude can be determined either by calculation from lifting surface or vortex lattice methods or by experiment.

In general terms a propeller vibrates in the six rigid body modes defined in Figure 11.2, where δ_i and ϕ_i refer to the displacements and rotations respectively. Assuming that the propeller vibrates as a rigid body and operates in a non-homogeneous wake field the consequent vibratory component of lift gives rise to forces F_i and moments Q_i about the Cartesian reference frame. Now as the propeller is vibrating in water, it experiences the additional hydrodynamic force and moment loadings f_i and q_i due to its oscillating motion: these additional terms give rise to the added mass and damping coefficients. Because of the linearizing assumption, these forces and moments can be considered as deriving from the propeller's vibratory motion in a uniform wake field, and the equation of motion for the vibrating propeller can be written as

$$M \ddot{\mathbf{x}} = \mathbf{f_e} + \mathbf{f_H} + \mathbf{f_S} \quad (11.10)$$

where $\mathbf{x}$, $\mathbf{f_e}$, $\mathbf{f_H}$ and $\mathbf{f_S}$ are the displacement, excitation, additional hydrodynamic force and the external excitation (e.g. shaft forces and moments from the engine and transmission) vectors respectively given by

$$\mathbf{x} = \begin{bmatrix} \delta_x, \delta_y, \delta_z, \phi_x, \phi_y, \phi_z \end{bmatrix}^T$$

$$\mathbf{f_e} = \begin{bmatrix} F_x, F_y, F_z, Q_x, Q_y, Q_z \end{bmatrix}^T$$

$$\mathbf{f_H} = \begin{bmatrix} f_x, f_y, f_z, q_x, q_y, q_z \end{bmatrix}^T$$

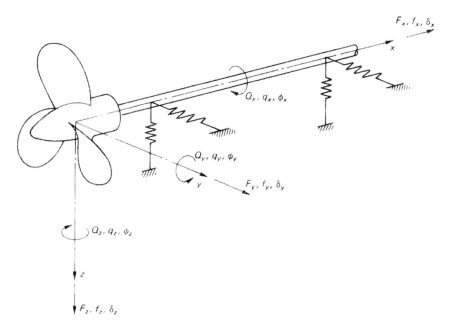

FIGURE 11.2 Propulsion shafting vibration parameters.

and the propeller mass matrix M is the diagonal matrix

$$M = \begin{bmatrix} m & 0 & 0 & 0 & 0 & 0 \\ 0 & m & 0 & 0 & 0 & 0 \\ 0 & 0 & m & 0 & 0 & 0 \\ 0 & 0 & 0 & I_{xx} & 0 & 0 \\ 0 & 0 & 0 & 0 & I_{yy} & 0 \\ 0 & 0 & 0 & 0 & 0 & I_{zz} \end{bmatrix}$$

in which $I_{yy} = I_{zz}$ since they are the diametral mass moments of inertia, I_{xx} is the polar moment about the shaft axis and m the mass of the propeller.

Now the additional hydrodynamic force vector $\mathbf{f_H}$ depends upon the displacements, velocities and accelerations of the propeller and so can be represented by the classical vibration theory relationship as:

$$\mathbf{f_H} = -M_a \ddot{\mathbf{x}} - C_p \dot{\mathbf{x}} - K_p \mathbf{x} \qquad (11.11)$$

in which the matrices M_a and C_p are the added mass and damping matrices respectively and K_p is a stiffness matrix depending upon the immersion of the propeller. If the propeller is fully immersed, then the matrix $K_p = 0$ and need not be considered further: this would not be the case, for example, with a surface piercing propeller. As a consequence, for a deeply immersed propeller equation (11.11) can be simplified to

$$\mathbf{f_H} = -M_a \ddot{\mathbf{x}} - C_p \ddot{\mathbf{x}} \qquad (11.11(a))$$

Then by combining equations (11.11(a)) and (11.10) the resulting equation of motion for the propeller is derived:

$$[M + M_a]\ddot{\mathbf{x}} + C_p \dot{\mathbf{x}} - \mathbf{f_S} = \mathbf{f_e} \qquad (11.12)$$

The forms of the matrices M_a and C_p are identical, each having a full leading diagonal of linear and rotational terms with a set of non-diagonal coupling terms. The added mass matrix has the form

$$M_a = \begin{bmatrix} m_{11} & 0 & 0 & m_{41} & 0 & 0 \\ 0 & m_{22} & -m_{32} & 0 & m_{52} & -m_{62} \\ 0 & m_{32} & m_{22} & 0 & m_{62} & m_{52} \\ m_{41} & 0 & 0 & m_{44} & 0 & 0 \\ 0 & m_{52} & -m_{62} & 0 & m_{55} & -m_{65} \\ 0 & m_{62} & m_{52} & 0 & m_{65} & m_{55} \end{bmatrix}$$

(11.13)

From this matrix it can be seen that several of the terms, for example m_{22} and m_{33}, have identical values, and hence this represents the simplest form of interactions between orthogonal motions. The matrix, as can be seen, is symmetrical, with the exception of four sign changes which result from the 'handedness' of the propeller. An alternative form of the matrix in equation (11.13), which demonstrates the physical meaning of the terms in relation to Figure 11.2 can be seen in equation (11.13(a)). In addition, the physical correspondence of the terms is also revealed by this comparison:

$$M_a = \begin{bmatrix} F_x/\ddot{\delta}_x & 0 & 0 & F_x/\ddot{\phi}_x & 0 & 0 \\ 0 & F_y/\ddot{\delta}_y & -F_y/\ddot{\delta}_z & 0 & F_y/\ddot{\phi}_y & -F_y/\ddot{\phi}_z \\ 0 & F_z/\ddot{\delta}_y & F_z/\ddot{\delta}_z & 0 & F_z/\ddot{\phi}_y & F_z/\ddot{\phi}_z \\ M_x/\ddot{\delta}_x & 0 & 0 & M_x/\ddot{\phi}_x & 0 & 0 \\ 0 & M_y/\ddot{\delta}_y & -M_y/\ddot{\delta}_z & 0 & M_y/\ddot{\phi}_y & -M_y/\ddot{\phi}_z \\ 0 & M_z/\ddot{\delta}_y & M_z/\ddot{\delta}_z & 0 & M_z/\ddot{\phi}_y & M_z/\ddot{\phi}_z \end{bmatrix}$$

(11.13(a))

Similarly for the damping matrix C_p the same comparison can be made as follows:

$$C_p = \begin{bmatrix} c_{11} & 0 & 0 & c_{41} & 0 & 0 \\ 0 & c_{22} & -c_{32} & 0 & c_{52} & -c_{62} \\ 0 & c_{32} & c_{22} & 0 & c_{62} & c_{52} \\ c_{41} & 0 & 0 & c_{44} & 0 & 0 \\ 0 & c_{52} & -c_{62} & 0 & c_{55} & -c_{65} \\ 0 & c_{62} & c_{52} & 0 & c_{65} & c_{55} \end{bmatrix}$$

$$= \begin{bmatrix} F_x/\dot{\delta}_x & 0 & 0 & F_x/\dot{\phi}_x & 0 & 0 \\ 0 & F_y/\dot{\delta}_y & -F_y/\dot{\delta}_z & 0 & F_y/\dot{\phi}_y & -F_y/\dot{\phi}_z \\ 0 & F_z/\dot{\delta}_y & F_z/\dot{\delta}_z & 0 & F_z/\dot{\phi}_y & F_z/\dot{\phi}_z \\ M_x/\dot{\delta}_x & 0 & 0 & M_x/\dot{\phi}_x & 0 & 0 \\ 0 & M_y/\dot{\delta}_y & M_y/\dot{\delta}_z & 0 & M_y/\dot{\phi}_y & M_y/\dot{\phi}_z \\ 0 & M_y/\dot{\delta}_y & M_z/\dot{\delta}_z & 0 & M_z/\dot{\phi}_y & M_z/\dot{\phi}_z \end{bmatrix}$$

(11.14)

When considering the vibratory characteristics of the propulsion shafting the coefficients in equations (11.13) and (11.14) need careful evaluation and it is insufficient to use arbitrary values, for example 0.25 times the polar dry inertia for m_{44} in equation (11.13), as many of these parameters vary considerably with differences in propeller design. Six added mass and damping terms are needed for coupled torsional axial motion, and these are $\{m_{11}, m_{44}, m_{41}, c_{11}, c_{44} \text{ and } c_{41}\}$.

Alternatively,

$$\left\{ \frac{F_x}{\ddot{\delta}_x}, \frac{M_x}{\ddot{\phi}_x}, \left(\frac{F_x}{\ddot{\phi}_x} = \frac{M_x}{\ddot{\delta}_x}\right), \frac{F_x}{\dot{\delta}_x}, \frac{M_x}{\dot{\phi}_x}, \left(\frac{F_x}{\dot{\phi}_x} = \frac{M_x}{\dot{\delta}_x}\right) \right\}$$

For lateral motion of the shafting system twelve terms are needed, which can be separated into two groups. The first group is where the forces and moments in the lateral directions are in the same direction as the motion: $\{m_{22}, m_{55}, m_{52}, c_{22}, c_{55} \text{ and } c_{52}\}$

that is,

$$\left\{ \left(\frac{F_y}{\ddot{\delta}_y} = \frac{F_y}{\ddot{\delta}_z}\right), \left(\frac{M_y}{\ddot{\phi}_y} = \frac{M_z}{\ddot{\phi}_z}\right), \left(\frac{F_y}{\ddot{\phi}_y} = \frac{F_z}{\ddot{\phi}_z} = \frac{M_y}{\ddot{\delta}_y} = \frac{M_z}{\ddot{\delta}_z}\right), \right.$$
$$\left(\frac{F_y}{\dot{\delta}_y} = \frac{F_z}{\dot{\delta}_z}\right), \left(\frac{M_y}{\dot{\phi}_y} = \frac{M_z}{\dot{\phi}_z}\right) \text{ and}$$
$$\left. \left(\frac{F_y}{\dot{\phi}_y} = \frac{F_z}{\dot{\phi}_z} = \frac{M_y}{\dot{\delta}_y} = \frac{M_z}{\dot{\delta}_z}\right) \right\}$$

and the second group has forces and moments in the lateral directions which are normal to the direction of motion: $\{m_{32}, m_{65}, m_{62}, c_{32}, c_{65} \text{ and } c_{62}\}$

that is,

$$\left\{ \left(\frac{F_y}{\ddot{\delta}_z} = \frac{F_z}{\ddot{\delta}_y}\right), \left(\frac{M_y}{\ddot{\phi}_z} = \frac{M_z}{\ddot{\phi}_y}\right), \left(\frac{F_y}{\ddot{\phi}_z} = \frac{F_z}{\ddot{\phi}_y} = \frac{M_y}{\ddot{\delta}_z} = \frac{M_z}{\ddot{\delta}_y}\right), \right.$$

$$\left(\frac{F_y}{\dot{\delta}_z} = \frac{F_z}{\dot{\delta}_y}\right), \left(\frac{M_y}{\dot{\phi}_z} = \frac{M_z}{\dot{\phi}_y}\right) \text{ and }$$

$$\left. \left(\frac{F_y}{\dot{\phi}_z} = \frac{F_z}{\dot{\phi}_y} = \frac{M_y}{\dot{\delta}_z} = \frac{M_z}{\dot{\delta}_y}\right) \right\}$$

In the past considerable research efforts have been devoted to determining some or all of these coefficients. In the early years, the added axial mass, polar entrained inertia and the corresponding damping coefficients were the prime candidates for study (i.e. m_{11}, m_{44}, c_{11} and c_{44}). However, latterly all of the coefficients have been studied more easily by taking advantage of modern analytical and computational capabilities. While knowledge of these components is a prerequisite for shafting system analysis, it must be emphasized that they should be applied in conjunction with the corresponding coefficients for the lubrication films in the bearings, particularly the stern tube bearing, as these are known to have a significant effect on the shaft vibration characteristics in certain circumstances.

Archer,[2] in attempting to solve marine shafting vibration problems, considered the question of torsional vibration damping coefficient, c_{44}. Archer derived an approximation based on the open water characteristics of the Wageningen B-Screw series as they were presented at that time. He argued that when torsional vibration is present the changes in the rotational speed of the shaft are so rapid and the inertia of the ship so great that they can be regarded as taking place at constant advance speed. This implies that the propeller follows a law in which the torque Q and rotational speed n are connected by a law of the form $Q \propto n^r$, where $r > 2$. If such a relation is assumed to hold over the range of speed variation resulting from the torsional vibration of the propeller, then by differentiating at a constant speed of advance V_a

$$\left.\frac{\partial Q}{\partial \omega}\right|_{V_a=\text{const.}} = \frac{1}{2\pi}\left.\frac{\partial Q}{\partial n}\right|_{V_a=\text{const.}} = K$$

where K, the propeller damping coefficient, is a constant. Hence

$$K = \frac{1}{2\pi}\frac{\partial}{\partial n}(bn^r)$$

where b is constant, and

$$K = a\frac{Q}{N}$$

where a is constant and equal to $9.55r$.

Now by taking $K_Q = f(J)$, Archer derived an expression for the index r as:

$$r = \left[2 - \frac{J(dK_Q/dJ)}{K_Q}\right]$$

which can be solved by appeal to the appropriate open water torque characteristic of the propeller under consideration. To this end, Archer gives a series of some nine diagrams to aid solution.

Lewis and Auslander[3] considered the longitudinal and torsional motions of a propeller and, as a result of conducting a series of experiments, supported by theory, derived a set of empirically based formulae for the entrained polar moment of inertia, the entrained axial mass both with and without rotational constraint and a coupling inertia factor.

Burrill and Robson[4] some two years later again considered this problem and produced a method of analysis — again based on empirical relations, albeit supported by a background theory — which has found favor for many years in some areas of the propeller manufacturing and consultancy industry. The basis of the Burrill and Robson approach was the derivation of experimental coefficients for a series of some forty-nine 16 in. propellers which were subject to torsional and axial excitation. To apply this approach to an arbitrary propeller design Burrill generalized the procedure as shown in Table 11.4.

From which the entrained inertia I_e about the shaft axis, equivalent to the m_{44} term in equation (11.13), can be estimated as

$$I_e = \frac{\pi\rho}{48}ZK_I R^3 \sum_I \quad (11.15)$$

Similarly, for the axial entrained mass m_a, equivalent to m_{11} in equation (11.13),

$$m_a = \frac{\pi\rho}{48}ZK_A R \sum_A \quad (11.16)$$

From Table 11.4 it can be seen that the blade form parameters $\sum_I$ and $\sum_A$ are the result of two integrations radially along the blade. The empirical factors K_A and K_I are given in Table 11.5.

An analysis of the hydrodynamic coefficients based on unsteady propeller theory was undertaken by Schwanecke.[5] This work resulted in the production of a set of calculation factors based on principal propeller dimensions such as blade area ratio, pitch ratio and blade number. The equations derived by Schwanecke are as follows:

(A) Added mass coefficients

$$m_{11} = 0.2812\frac{\pi\rho D^3}{Z}\left(\frac{A_e}{A_o}\right)^2 \text{ kg s}^2/\text{m}$$

TABLE 11.4 Derivation of Burrill blade Form Parameters for Entrained Inertia and Mass

$x = r/R$	p	$\theta = \tan^{-1}(p/2\pi r)$	$(cx \sin \theta)^2$	SM	$SM \times (cx \sin \theta)^2$	$(c \cos \theta)^2$	SM	$SM \times (c \cos \theta)^2$
0.250	p_1	θ_1		1/2			1/2	
0.375	p_2	θ_2		2			2	
0.500	p_3	θ_3		1			1	
0.625	p_4	θ_4		2			2	
0.750	p_5	θ_5		1			1	
0.875	p_6	θ_6		2			2	
1.000	p_7	θ_7		1/2			1/2	
				$\sum_I =$	=		$\sum_A =$	

TABLE 11.5 The Burrill and Robson K_A and K_I Factors

BAR/Z	K_I	K_A
0.11	0.893	0.969
0.12	0.845	0.920
0.13	0.805	0.877
0.14	0.772	0.841
0.15	0.741	0.808
0.16	0.714	0.773
0.17	0.691	0.750
0.18	0.670	0.725
0.19	0.650	0.702
0.20	0.631	0.691
0.21	0.611	0.660
0.22	0.592	0.639
0.23	0.573	0.619
0.24	0.555	0.600
0.25	0.538	0.582
0.26	0.522	0.564
0.27	0.506	0.546
0.28	0.490	0.529
0.29	0.476	0.512
0.30	0.462	0.498
0.31	0.448	0.484
0.32	0.435	0.470
0.33	0.423	0.457
0.34	0.412	0.444
0.35	0.400	0.432
0.36	0.398	0.420

$$m_{22} = m_{33} = 0.6363 \frac{\rho D^3}{\pi Z} \left(\frac{P}{D}\right)^2 \left(\frac{A_e}{A_o}\right)^2 \text{ kg s}^2/\text{m}$$

$$m_{44} = 0.0703 \frac{\rho D^5}{\pi Z} \left(\frac{P}{D}\right)^2 \left(\frac{A_e}{A_o}\right)^2 \text{ kp s}^2 \text{ m}$$

$$m_{55} = m_{66} = 0.0123 \frac{\pi \rho D^5}{Z} \left(\frac{A_e}{A_o}\right)^2 \text{ kp s}^2 \text{ m}$$

$$m_{41} = -0.1406 \frac{\rho D^4}{Z} \left(\frac{P}{D}\right) \left(\frac{A_e}{A_o}\right)^2 \text{ kp s}^2$$

$$m_{52} = 0.0703 \frac{\rho D^4}{Z} \left(\frac{P}{D}\right) \left(\frac{A_e}{A_o}\right)^2 \text{ kp s}^2$$

$$m_{62} = 0.0408 \frac{\rho D^4}{Z^2} \left(\frac{P}{D}\right) \left(\frac{A_e}{A_o}\right)^3 \text{ kp s}^2$$

$$m_{65} = 0.0030 \frac{\pi \rho D^5}{Z^2} \left(\frac{A_e}{A_o}\right)^2 \text{ kp s}^2 \text{ m}$$

(B) Damping coefficients

$$c_{11} = 0.0925 \rho \omega D^3 \left(\frac{A_e}{A_o}\right) \text{ kp s/m}$$

$$c_{22} = c_{33} = 0.1536 \frac{\rho \omega D^2}{\pi} \left(\frac{P}{D}\right)^2 \left(\frac{A_e}{A_o}\right) \text{ kp s/m}$$

$$c_{44} = 0.0231 \frac{\rho \omega D^5}{\pi} \left(\frac{P}{D}\right)^2 \left(\frac{A_e}{A_o}\right) \text{ kp s m}$$

$$c_{55} = c_{66} = 0.0053 \pi \rho \omega D^5 \left(\frac{A_e}{A_o}\right) \text{ kp s m}$$

$$c_{41} = -0.0463 \rho \omega D^4 \left(\frac{P}{D}\right) \left(\frac{A_e}{A_o}\right) \text{ kp s}$$

$$c_{52} = 0.0231 \rho \omega D^4 \left(\frac{P}{D}\right) \left(\frac{A_e}{A_o}\right) \text{ kp s}$$

$$c_{62} = 0.0981 \frac{\rho \omega D^4}{Z} \left(\frac{P}{D}\right) \left(\frac{A_e}{A_o}\right)^2 \text{ kp s}$$

$$c_{65} = 0.0183 \frac{\pi \rho \omega D^5}{Z} \left(\frac{A_e}{A_o}\right)^2 \text{ kp s m}$$

$$c_{35} = 0.1128 \frac{\rho \omega D^4}{Z} \left(\frac{P}{D}\right) \left(\frac{A_e}{A_o}\right)^2 \text{ kp s}$$

With regard to the damping coefficients, Schwanecke draws a distinction between the elements c_{26} and c_{35} which is in contrast to some other contemporary works.

The foregoing discussion relates specifically to fixed pitch propellers and the dependency on pitch is clearly evident in all of the formulations. Clearly, therefore, a controllable pitch propeller working at a reduced off-design pitch setting has a lower entrained inertia than when at design pitch. Van Gunsteren and Pronk[6] suggested a reduction, of the order of that shown in Table 11.6, based on results from a series of controllable pitch propellers.

The provision of a reliable database of either theoretical or experimental results has always been a problem in attempting to correlate the calculations of the elements of the matrices defined in equations (11.13) and (11.14) with experimental data. Experimental data, either at full- or model-scale, is difficult to obtain, and only limited data is available — notably the work of Burrill and Robson.[4] Hylarides and van Gent,[7] however, attempted to rectify this problem to some extent from the theoretical viewpoint by considering calculations based on a number of propellers from the Wageningen B-Screw series. The calculations were based on unsteady lifting surface theory for four-bladed propellers of the series in order to derive the coefficients shown in Table 11.7. Examination of this table clearly shows the fallacy, noted earlier, of using fixed percentages in vibration calculations for rigorous shafting behavior analysis purposes.

Parsons and Vorus,[8] using the work of Hylarides and van Gent as a basis, investigated the correlation that could be achieved by calculating the added mass and damping estimates from lifting surface and lifting line procedures. In addition they also examined the implications of changing the blade skew from the standard B-Screw series design, as well as that of changing the vibration frequency. Their work resulted in a series of regression based formula based on the Wageningen B-Screw series geometry and suitable for initial design purposes. These regression equations, which are based on a lifting line formulation, have the form

$$\begin{Bmatrix} m_{ij} \\ c_{ij} \end{Bmatrix} = C_1 + C_2(A_E/A_O) + C_3(P/D) + C_4(A_E/A_O)^2 + C_5(P/D)^2 + C_6(A_E/A_O)(P/D) \quad (11.17)$$

The coefficients C_1 to C_6 are given in Tables 11.8 to 11.11 for the four-, five-, six-, and seven-bladed Wageningen B series propellers respectively. The range of application of equation (11.17) is for expanded area ratios in the range 0.5–1.0 and pitch ratios in the range 0.6–1.2. Parsons and Vorus also developed a set of lifting surface corrections which can be applied to equation (11.17) to improve the accuracy of the estimate and these are given by Table 11.12. In Table 11.12 the blade aspect ratio AR is given by

$$AR = \frac{0.22087 Z}{A_E/A_O}$$

where Z is the blade number. The correction factors given by Table 11.12 are introduced into equation (11.17) as follows:

$$\begin{Bmatrix} m_{ij} \\ c_{ij} \end{Bmatrix}_{\text{Lifting surface}} = \begin{Bmatrix} m_{ij} \\ c_{ij} \end{Bmatrix}_{\text{eqn (11.7)}} \times LSC \quad (11.18)$$

11.1.4 Propeller Forces and Moments

The mean and fluctuating forces and moments produced by a propeller working in the ship's wake field have to be reacted at the bearings and, therefore, form a substantial contribution to the bearing forces. In the early stages of design the main components of the force (F_x) and moment (M_x) (Figure 11.3) are calculated from open water propeller data assuming a mean wake fraction for the vessel. However, as the design progresses and more of the detailed propeller geometry and structure of the wake field emerge more refined estimates must be made.

The effective thrust force of a propeller is seldom, if ever, directed along the shaft axis. This is due to the effects of the wake field and possibly any shaft inclination relative to the flow (see Chapter 6). In general, the line of action of

TABLE 11.6 Typical Reduction in Entrained Inertia at Off-Design Pitch Settings

Percentage of design pitch setting	0	20	40	60	80	100
Percentage of entrained polar moment (m_{44}) at design pitch	4	5	15	36	66	100

TABLE 11.7 Dimensionless Values of the Propeller Coefficients

Propeller Type	B4-40-50	B4-40-80	B4-40-120	B4-70-50	B4-70-80	B4-70-120	B4-100-50	B4-100-80	B4-100-120
A_E/A_O	0.40	0.40	0.40	0.70	0.70	0.70	1.00	1.00	1.00
P/D	0.50	0.80	1.20	0.50	0.80	1.20	0.50	0.80	1.20
Axial Vibrations									
$F_x/\dot{\delta}_x$	$-6.24\ 10^{-1}$	$-5.95\ 10^{-1}$	$-5.42\ 10^{-1}$	$-8.37\ 10^{-1}$	$-7.27\ 10^{-1}$	$-6.30\ 10^{-1}$	$-8.53\ 10^{-1}$	$-6.71\ 10^{-1}$	$-6.12\ 10^{-1}$
$F_x/\ddot{\delta}_x$	$-2.91\ 10^{-2}$	$-2.74\ 10^{-2}$	$-2.31\ 10^{-2}$	$-8.37\ 10^{-2}$	$-7.34\ 10^{-2}$	$-6.01\ 10^{-2}$	$-1.34\ 10^{-1}$	$-1.18\ 10^{-1}$	$-9.68\ 10^{-2}$
$F_x/\dot{\phi}_x$	$4.96\ 10^{-2}$	$7.58\ 10^{-2}$	$1.04\ 10^{-1}$	$6.66\ 10^{-2}$	$9.26\ 10^{-2}$	$1.20\ 10^{-1}$	$5.82\ 10^{-2}$	$8.54\ 10^{-2}$	$1.17\ 10^{-1}$
$F_x/\ddot{\phi}_x$	$2.31\ 10^{-3}$	$3.48\ 10^{-3}$	$4.42\ 10^{-3}$	$6.66\ 10^{-3}$	$9.34\ 10^{-3}$	$1.15\ 10^{-2}$	$1.07\ 10^{-2}$	$1.50\ 10^{-2}$	$1.85\ 10^{-2}$
$M_x/\dot{\delta}_x$	$4.96\ 10^{-2}$	$7.58\ 10^{-2}$	$1.04\ 10^{-1}$	$6.66\ 10^{-2}$	$9.26\ 10^{-2}$	$1.20\ 10^{-1}$	$5.82\ 10^{-2}$	$8.54\ 10^{-2}$	$1.17\ 10^{-1}$
$M_x/\ddot{\delta}_x$	$2.31\ 10^{-3}$	$3.48\ 10^{-3}$	$4.42\ 10^{-3}$	$6.66\ 10^{-3}$	$9.34\ 10^{-3}$	$1.15\ 10^{-2}$	$1.07\ 10^{-2}$	$1.50\ 10^{-2}$	$1.85\ 10^{-2}$
$M_x/\dot{\phi}_x$	$-3.95\ 10^{-3}$	$-9.65\ 10^{-3}$	$-1.98\ 10^{-2}$	$-5.30\ 10^{-3}$	$-1.18\ 10^{-2}$	$-2.30\ 10^{-2}$	$-4.63\ 10^{-3}$	$-1.09\ 10^{-2}$	$-2.23\ 10^{-2}$
$M_x/\ddot{\phi}_x$	$-1.84\ 10^{-4}$	$-4.43\ 10^{-4}$	$-8.44\ 10^{-4}$	$-5.30\ 10^{-4}$	$-1.19\ 10^{-3}$	$-2.19\ 10^{-3}$	$-8.48\ 10^{-4}$	$-1.91\ 10^{-3}$	$-3.53\ 10^{-3}$
Transverse Vibrations, Loads and Motions Parallel									
$F_y/\dot{\delta}_y$	$-2.69\ 10^{-2}$	$-5.56\ 10^{-2}$	$-1.12\ 10^{-1}$	$-3.67\ 10^{-2}$	$-7.01\ 10^{-2}$	$-1.42\ 10^{-1}$	$-5.11\ 10^{-2}$	$-8.15\ 10^{-2}$	$-1.62\ 10^{-1}$
$F_y/\ddot{\delta}_y$	$-1.72\ 10^{-3}$	$-3.56\ 10^{-3}$	$-5.97\ 10^{-3}$	$-3.94\ 10^{-3}$	$-8.66\ 10^{-3}$	$-1.51\ 10^{-2}$	$-4.02\ 10^{-3}$	$-1.46\ 10^{-2}$	$-2.57\ 10^{-2}$
$F_y/\dot{\phi}_y$	$-2.38\ 10^{-2}$	$-3.55\ 10^{-2}$	$-5.02\ 10^{-2}$	$-2.92\ 10^{-2}$	$-4.41\ 10^{-2}$	$-6.33\ 10^{-2}$	$-3.90\ 10^{-2}$	$-5.28\ 10^{-2}$	$-7.31\ 10^{-2}$
$F_y/\ddot{\phi}_y$	$-1.35\ 10^{-3}$	$-1.93\ 10^{-3}$	$-2.33\ 10^{-3}$	$-3.04\ 10^{-3}$	$-4.67\ 10^{-3}$	$-5.89\ 10^{-3}$	$-4.95\ 10^{-3}$	$-7.89\ 10^{-3}$	$-1.01\ 10^{-2}$
$M_y/\dot{\delta}_y$	$-2.60\ 10^{-2}$	$-3.64\ 10^{-2}$	$-5.06\ 10^{-2}$	$-3.18\ 10^{-2}$	$-4.40\ 10^{-2}$	$-6.28\ 10^{-2}$	$-4.36\ 10^{-2}$	$-5.23\ 10^{-2}$	$-7.17\ 10^{-2}$
$M_y/\ddot{\delta}_y$	$-1.31\ 10^{-3}$	$-1.88\ 10^{-3}$	$-2.28\ 10^{-3}$	$-3.12\ 10^{-3}$	$-4.64\ 10^{-3}$	$-5.82\ 10^{-3}$	$-5.06\ 10^{-3}$	$-7.86\ 10^{-3}$	$-9.97\ 10^{-3}$
$M_y/\dot{\phi}_y$	$-3.51\ 10^{-2}$	$-3.26\ 10^{-2}$	$-3.09\ 10^{-2}$	$-3.88\ 10^{-2}$	$-3.79\ 10^{-2}$	$-3.71\ 10^{-2}$	$-4.90\ 10^{-2}$	$-4.44\ 10^{-2}$	$-4.20\ 10^{-2}$
$M_y/\ddot{\phi}_y$	$-1.66\ 10^{-3}$	$-1.51\ 10^{-3}$	$-1.27\ 10^{-3}$	$-3.74\ 10^{-3}$	$-3.58\ 10^{-3}$	$-3.12\ 10^{-3}$	$-5.96\ 10^{-3}$	$-5.90\ 10^{-3}$	$-5.21\ 10^{-3}$
Transverse Vibrations, Loads and Motions Naturally Perpendicular									
$F_z/\dot{\delta}_y$	$2.65\ 10^{-3}$	$2.23\ 10^{-3}$	$2.97\ 10^{-3}$	$4.77\ 10^{-3}$	$9.99\ 10^{-4}$	$9.27\ 10^{-4}$	$1.77\ 10^{-2}$	$5.26\ 10^{-3}$	$2.30\ 10^{-3}$
$F_z/\ddot{\delta}_y$	$6.18\ 10^{-5}$	$-6.45\ 10^{-5}$	$-8.26\ 10^{-5}$	$2.96\ 10^{-5}$	$1.42\ 10^{-5}$	$1.81\ 10^{-6}$	$5.74\ 10^{-4}$	$5.50\ 10^{-5}$	$-2.14\ 10^{-5}$
$F_z/\dot{\phi}_y$	$1.23\ 10^{-2}$	$1.24\ 10^{-2}$	$1.63\ 10^{-2}$	$1.62\ 10^{-2}$	$1.70\ 10^{-2}$	$2.76\ 10^{-2}$	$2.70\ 10^{-2}$	$2.52\ 10^{-2}$	$4.20\ 10^{-2}$
$F_z/\ddot{\phi}_y$	$4.19\ 10^{-4}$	$2.75\ 10^{-4}$	$7.45\ 10^{-5}$	$1.13\ 10^{-3}$	$8.79\ 10^{-4}$	$5.56\ 10^{-4}$	$1.80\ 10^{-3}$	$1.49\ 10^{-3}$	$1.05\ 10^{-3}$
$M_z/\dot{\delta}_y$	$-6.22\ 10^{-3}$	$-6.07\ 10^{-3}$	$-4.21\ 10^{-3}$	$-5.33\ 10^{-3}$	$-6.10\ 10^{-3}$	$-1.27\ 10^{-3}$	$-1.40\ 10^{-3}$	$-4.88\ 10^{-3}$	$3.71\ 10$
$M_z/\ddot{\delta}_y$	$-4.79\ 10^{-4}$	$-5.27\ 10^{-4}$	$-4.91\ 10^{-4}$	$-9.11\ 10^{-4}$	$-1.06\ 10^{-3}$	$-1.01\ 10^{-3}$	$-1.41\ 10^{-3}$	$-1.67\ 10^{-3}$	$-1.57\ 10^{-3}$
$M_z/\dot{\phi}_y$	$3.51\ 10^{-3}$	$2.64\ 10^{-3}$	$3.83\ 10^{-3}$	$6.09\ 10^{-3}$	$5.18\ 10^{-3}$	$9.44\ 10^{-3}$	$1.13\ 10^{-2}$	$8.91\ 10^{-3}$	$1.64\ 10^{-2}$
$M_z/\ddot{\phi}_y$	$-1.01\ 10^{-4}$	$-1.53\ 10^{-4}$	$-1.89\ 10^{-4}$	$6.19\ 10^{-6}$	$-1.53\ 10^{-4}$	$-2.52\ 10^{-4}$	$-1.74\ 10^{-5}$	$-1.99\ 10^{-4}$	$-3.37\ 10^{-4}$

TABLE 11.8 Regression Equation Coefficients for B4 Propellers ($E \pm N = \times 10^{\pm N}$)[8]

Parameter Component	C_1	C_2	C_3	C_4	C_5	C_6
Torsional/Axial						
m_{44}	0.30315E−2	−0.80782E−2	−0.40731E−2	0.34170E−2	0.43437E−3	0.99715E−2
m_{41}	0.12195E−2	0.17664E−1	−0.85938E−2	−0.23615E−1	0.94301E−2	−0.26146E−1
m_{11}	−0.62948E−1	0.17980	0.58719E−1	0.17684	−0.21439E−2	−0.15395
c_{44}	−0.35124E−1	0.81977E−1	0.32644E−1	−0.41863E−1	0.60813E−2	−0.37170E−1
c_{41}	0.13925	−0.48179	−0.14175	0.27711	−0.94311E−2	0.17407
c_{11}	0.32017	0.29375E+1	−0.90814	−0.19719E+1	0.53868	−0.65404
Lateral: Parallel						
m_{55}	−0.26636E−2	0.61911E−2	0.26565E−2	0.77133E−2	−0.66326E−3	−0.40324E−2
m_{52}	−0.19644E−2	−0.47339E−2	0.45533E−2	0.89144E−2	−0.44606E−2	0.11823E−1
m_{22}	0.17699E−1	−0.59698E−1	−0.18823E−1	0.29066E−1	−0.33316E−2	0.73554E−1
c_{55}	−0.63518E−2	0.22851	−0.31365E−1	−0.14332	0.25084E−1	−0.49546E−1
c_{52}	−0.11690	0.36582	0.10076	−0.21326	0.18676E−3	−0.12515
c_{22}	−0.35968	0.87537	0.29734	−0.47961	0.14001E−1	−0.33732
Lateral: Perpendicular						
m_{65}	0.12333E−3	0.35676E−2	−0.35561E−3	−0.36381E−2	0.65794E−3	−0.17943E−2
m_{62}	−0.17250E−2	0.64561E−2	0.19195E−2	−0.40546E−2	0.40439E−3	−0.47506E−2
m_{32}	−0.99403E−2	0.23315E−1	0.10895E−1	−0.11360E−1	−0.71528E−3	−0.15718E−1
c_{65}	0.59756E−1	−0.18982	−0.17653E−1	0.82400E−1	0.61804E−2	−0.80790E−2
c_{62}	0.78572E−1	−0.18627	−0.37105E−1	0.11053	0.17847E−1	−0.55900E−1
c_{32}	0.14397	−0.32322	−0.15348E−1	0.24992	0.14289E−1	−0.21254

TABLE 11.9 Regression Equation Coefficients for B5 Propellers ($E \pm N = \times 10^{\pm N})^8$

Parameter Component	C_1	C_2	C_3	C_4	C_5	C_6
Torsional/Axial						
m_{44}	0.27835E−2	−0.71650E−2	−0.37301E−2	0.30526E−2	0.46275E−3	0.85327E−2
m_{41}	−0.26829E−3	0.17208E−1	−0.55064E−2	−0.21012E−1	0.72960E−2	−0.22840E−1
m_{11}	−0.47372E−1	0.13499	0.43428E−1	0.15666	0.41444E−1	−0.12404
c_{44}	−0.30935E−1	0.69382E−1	0.27392E−1	−0.37293E−1	0.63542E−2	−0.21635E−1
c_{41}	0.14558	−0.44319	−0.17025	0.24558	0.14798E−1	0.12226
c_{11}	0.16202	0.30392E+1	−0.59068	−0.17372E+1	0.37998	−0.71363
Lateral: Parallel						
m_{55}	−0.18541E−2	0.40694E−2	0.20342E−2	0.72761E−2	−0.47031E−3	−0.33269E−2
m_{52}	−0.20455E−3	−0.73445E−2	0.26857E−2	0.95299E−2	−0.34485E−2	0.10863E−1
m_{22}	0.17180E−1	−0.54519E−1	−0.17894E−1	0.27151E−1	−0.19451E−2	0.62180E−1
c_{55}	−0.25532E−2	0.20018	−0.22067E−1	−0.10971	0.18255E−1	−0.43517E−1
c_{52}	−0.98481E−1	0.28632	0.10154	−0.15975	−0.10484E−1	−0.79238E−1
c_{22}	−0.27180	0.61549	0.24132	−0.33370	0.10475E−1	−0.19101
Lateral: Perpendicular						
m_{65}	−0.51073E−3	0.36044E−2	0.12804E−3	−0.30064E−2	0.25624E−3	−0.11174E−2
m_{62}	−0.18142E−2	0.56442E−2	0.15906E−2	−0.35420E−2	0.72381E−4	−0.27848E−2
m_{32}	−0.68895E−2	0.16524E−1	0.62244E−2	−0.87754E−2	−0.38255E−3	−0.82429E−2
c_{65}	0.33407E−1	−0.99682E−1	−0.68219E−2	0.22669E−1	0.56635E−2	−0.22639E−1
c_{62}	0.33507E−1	−0.66800E−1	−0.73992E−2	0.33112E−1	0.13061E−1	−0.80862E−1
c_{32}	0.15158E−1	−0.10109E−1	0.63232E−1	0.50161E−1	0.35720E−2	−0.27201

TABLE 11.10 Regression Equation Coefficients for B6 Propellers ($E \pm N = \times 10^{\pm N})^8$

Parameter Component	C_1	C_2	C_3	C_4	C_5	C_6
Torsional/Axial						
m_{44}	0.23732E−2	−0.62877E−2	−0.30606E−2	0.27478E−2	0.29060E−3	0.73650E−2
m_{41}	−0.17748E−2	0.14993E−1	−0.51316E−2	−0.18451E−1	0.64733E−2	−0.19096E−1
m_{11}	−0.39132E−1	0.10862	0.37308E−1	0.13359	−0.33222E−3	−0.10387
c_{44}	−0.27873E−1	0.61760E−1	0.23242E−1	−0.35004E−1	0.70046E−2	−0.11641E−1
c_{41}	0.14228	−0.41189	−0.1770	0.22644	0.26626E−1	0.83269E−1
c_{11}	0.11113	0.29831E+1	−0.44133	−0.15696E+1	0.28560	−0.66976
Lateral: Parallel						
m_{55}	−0.16341E−2	0.33153E−2	0.19742E−2	0.64129E−2	−0.52004E−3	−0.29555E−2
m_{52}	−0.40692E−4	−0.68309E−2	0.24412E−2	0.86298E−2	−0.30852E−2	0.92581E−2
m_{22}	0.13668E−1	−0.46198E−1	−0.12970E−1	0.24376E−1	−0.29068E−2	0.52775E−1
c_{55}	0.63116E−3	0.18370	−0.17663E−1	−0.92593E−1	0.13964E−1	−0.37740E−1
c_{52}	−0.85805E−1	0.24006	0.10020	−0.13176	−0.16091E−1	−0.52217E−1
c_{22}	−0.24147	0.52269	0.23271	−0.28526	0.68691E−2	0.13848
Lateral: Perpendicular						
m_{65}	−0.46261E−3	0.27610E−2	0.11516E−3	−0.21853E−2	0.13978E−3	−0.66927E−3
m_{62}	−0.13553E−2	0.40432E−2	0.11434E−2	−0.25422E−2	−0.10052E−4	−0.17182E−2
m_{32}	−0.45682E−2	0.10797E−1	0.41254E−2	−0.57931E−2	−0.33138E−3	−0.50724E−2
c_{65}	0.21438E−1	−0.54389E−1	−0.73262E−2	−0.54776E−2	0.76065E−2	−0.24165E−1
c_{62}	0.14903E−1	−0.10434E−1	−0.40106E−2	−0.31096E−2	0.14738E−1	−0.83101E−1
c_{32}	−0.15579E−1	0.83912E−1	0.52866E−1	−0.24903E−1	0.93828E−2	−0.24420

TABLE 11.11 Regression Equation Coefficients for B7 Propellers ($E \pm N = \times 10^{\pm N}$)^8

Parameter Component	C_1	C_2	C_3	C_4	C_5	C_6
Torsional/Axial						
m_{44}	0.21372E−2	−0.56155E−2	−0.27388E−2	0.24553E−2	0.26675E−3	0.64805E−2
m_{41}	−0.50233E−3	0.13927E−1	−0.41583E−2	−0.16454E−1	0.56027E−2	−0.17030E−1
m_{11}	−0.32908E−1	0.88748E−1	0.32596E−1	0.11886	−0.96860E−3	−0.87831E−1
c_{44}	−0.24043E−1	0.51680E−1	0.18585E−1	−0.31175E−1	0.75424E−2	−0.10541E−2
c_{41}	0.14003	−0.37358	−0.18904	0.20133	0.40056E−1	0.45135E−1
c_{11}	0.34070E−1	0.29353E+1	−0.24280	−0.13929E+1	0.17571	−0.65123
Lateral: Parallel						
m_{55}	−0.14132E−2	0.26715E−2	0.18052E−2	0.56906E−2	−0.52314E−3	−0.24846E−2
m_{52}	0.17646E−3	−0.65252E−2	0.19867E−2	0.78350E−2	−0.26907E−2	0.82990E−2
m_{22}	0.12144E−1	−0.40599E−1	−0.11616E−1	0.21395E−1	−0.24429E−2	0.46140E−1
c_{55}	−0.26383E−2	0.17179	−0.29958E−2	−0.79085E−1	0.50612E−2	−0.33233E−1
c_{52}	−0.78069E−1	0.20492	0.10121	−0.10980	−0.21608E−1	−0.28950E−1
c_{22}	−0.20348	0.42553	0.19690	−0.23664	0.11822E−1	−0.69910E−1
Lateral: Perpendicular						
m_{65}	−0.38383E−3	0.19693E−2	0.17327E−3	−0.15326E−2	0.26748E−4	−0.36439E−3
m_{62}	−0.96395E−3	0.28059E−2	0.80259E−3	−0.17783E−2	−0.37139E−4	−0.10340E−2
m_{32}	−0.29000E−2	0.69281E−2	0.24977E−2	−0.37440E−2	−0.12487E−3	−0.30407E−2
c_{65}	0.10617E−1	−0.24040E−1	−0.19931E−2	−0.20636E−1	0.60272E−2	−0.26183E−1
c_{62}	0.15152E−2	−0.20965E−1	−0.36583E−2	−0.22580E−1	0.11790E−1	−0.80079E−1
c_{32}	−0.36770E−1	0.13433	0.56417E−1	−0.62648E−1	0.74649E−2	−0.22397

TABLE 11.12 Lifting Surface Corrections for Parsons and Vorus Added Mass and Damping Equations[8]

LSC(m_{44}) =	$0.61046 + 0.34674(P/D)$ $+ 0.60294(AR)^{-1} - 0.56159(AR)^{-2}$ $- 0.80696(P/D)(AR)^{-1}$ $+ 0.45806(P/D)(AR)^{-2}$	LSC(m_{55}) =	$-0.1394 + 0.89760(AR)$ $+ 0.34086(P/D) - 0.15307(AR)^2$ $- 0.36619(P/D)(AR) + 0.70192(P/D)(AR)^2$
LSC(m_{41}) =	$0.65348 + 0.28788(P/D)$ $+ 0.39805(AR)^{-1} - 0.42582(AR)^{-2}$ $- 0.61189(P/D)(AR)^{-1}$ $+ 0.33373(P/D)(AR)^{-2}$	LSC(m_{52}) =	$0.0010398 + 0.66020(AR)$ $+ 0.39850(P/D) - 0.10261(AR)^2$ $- 0.34101(P/D)(AR) + 0.060368(P/D)(AR)^2$
LSC (m_{11}) =	$0.61791 + 0.23741(PD)$ $+ 0.42253(AR)^{-1} - 0.43911(AR)^{-2}$ $- 0.46697(P/D)(AR)^{-1}$ $+ 0.25124(P/D)(AR)^{-2}$	LSC(m_{22}) =	$0.78170 + 0.36153(AR-2)$ $- 0.19256(P/D)(AR-2)$ $+ 0.17908(P/D)(AR-2)^2$ $- 0.16110(AR-2)^2$ $- 0.061038(P/D)^2(AR-2)$
LSC(c_{44}) =	$0.82761 - 0.41165(AR)^{-1}$ $+ 1.2196(P/D)(AR)^{-1} + 6.3993(AR)^{-3}$ $- 13.804(P/D)(AR)^{-3} - 6.9091(AR)^{-4}$ $+ 15.594(P/D)(AR)^{-4}$	LSC(c_{55}) =	$0.78255 + 0.061046(AR)$ $- 2.5056(AR)^{-3} + 1.6426(AR)^{-4}$ $+ 1.8440(P/D)(AR)^{-4}$
LSC(c_{41}) =	$0.80988 - 0.63077(AR)^{-2}$ $+ 1.3909(P/D)(AR)^{-1} + 7.5424(AR)^{-3}$ $- 15.689(P/D)(AR)^{-3} - 8.0097(AR)^{-4}$ $+ 17.665(P/D)(AR)^{-4}$	LSC(c_{52}) =	$1.0121 + 0.73647(AR)^{-2}$ $- 3.8691(AR)^{-3}$ $- 1.5129(P/D)(AR)^{-3} + 3.0614(AR)^{-4}$ $+ 3.0984(P/D)(AR)^{-4}$
LSC(c_{11}) =	$0.82004 - 0.67190(AR)^{-2}$ $+ 1.3913(P/D)(AR)^{-1} + 7.7476(AR)^{-1}$ $- 16.807(P/D)(AR)^{-3} - 8.2798(AR)^{-4}$ $+ 19.121(P/D)(AR)^{-4}$	LSC (c_{22}) =	$0.84266 + 6.7849(AR)^{-2}$ $+ 0.12809(P/D)(AR)^{-1}$ $- 21.030(AR)^{-3} - 3.3471(P/D)(AR)^{-3}$ $+ 15.842(AR)^{-4} + 5.1905(P/D)(AR)^{-4}$

the effective thrust force will be raised above the shaft axis as a direct result of the slower water velocities in the upper part of the propeller disc. Furthermore, due to the effects of the tangential velocity components, the effective thrust force is unlikely to lie on the plane of symmetry of the axial wake field. The thrust eccentricity $e_T(t)$ is the distance from the shaft center line to the point through which the effective thrust force acts. Thus it has two components, one in the thwart ship direction $e_{Ty}(t)$ and the other in the vertical direction $e_{Tz}(t)$ such that,

$$e_T^2(t) = e_{Ty}^2(t) + e_{Tz}^2(t) \tag{11.19}$$

Equation (11.19) also underlines the functionality of the thrust eccentricity with time (t) since as each blade rotates around the propeller disc it is continuously encountering a different inflow field and hence the moments and forces are undergoing a cyclic change in their magnitude.

In Chapter 5 it was seen that the wake field could be expressed as the sum of a set of Fourier components. Since the blade sections operate wholly within this wake field the cyclic lift force generated by the sections and the components of this force resolved in either the thrust or torsional directions can also be expressed as the sum of Fourier series expansions:

$$\left. \begin{array}{l} F(t) = F_{(0)} + \sum_{k=1}^{n} F_{(k)} \cos(\omega t + \phi_k) \\ \text{and} \\ M(t) = M_{(0)} + \sum_{k=1}^{n} M_{(k)} \cos(\omega t + \phi_k) \end{array} \right\} \tag{11.20}$$

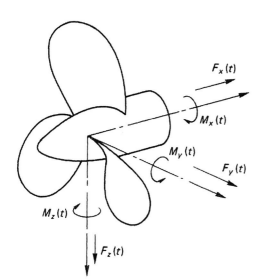

FIGURE 11.3 Hydrodynamic forces and moment activity on a propeller.

Hence the resulting forces and moments can all be expressed as a mean component plus the sum of a set of harmonic components. In the case of high shaft inclinations the movement of the effective thrust force can be considerable in the thwart direction (see Chapter 6). In such cases the term thrust eccentricity can also apply to that direction. Whilst the discussion, for illustration purposes, so far has centered on the thrust force, a similar set of arguments also applies to the shaft torque and the other orthogonal forces and moments.

In the absence of high shaft inclination the magnitude of the bearing forces depend on the characteristics of the wake field, the geometric form of the propeller (in particular the skew and blade number), the ship speed, rate of ship turning and the rotational speed of the propeller. Indeed, for a given application, the forces and moments generated by the fixed pitch propeller are, in general, proportional to the square of the revolutions, since for a considerable part of the upper operational speed range the vessel will work at a nominally constant advance coefficient. Some years ago an investigation by theoretical means of the dynamic forces at blade and twice blade frequency was carried out on twenty ships.[9] The results of these calculations are useful for making preliminary estimates of the dynamic forces at the early stages of design and Table 11.13 shows the results of these calculations in terms of the mean values and their ranges. From the table it will be seen that each of the six loading components are expressed in terms of the mean thrust T_0 or the mean torque Q_0.

The majority of the theoretical methods discussed in Chapter 8 can be applied to calculate the bearing forces. In essence, the calculation is essentially a classical propeller analysis procedure conducted at incremental steps around the propeller disc. However, probably the greatest bar to absolute accuracy is the imprecision with which the wake field is known due to the scale effects between model-scale, at which the measurement is carried out, and full-scale. Sasajima[10] specifically developed a simplified quasi-steady method of establishing the propeller forces and moments which relied on both the definition of the propeller open water characteristics and a weighted average wake distribution. A wake field prediction is clearly required for this level of analysis, and a radial weighting function is then applied to the wake field at each angular position, the weighting being applied in proportion to the anticipated thrust loading distribution. Improved results in this analysis procedure are also found to exist by averaging the wake over the chord using a weight function similar to the vortex distribution over the flat plate. The expressions derived by

TABLE 11.13 Typical First and Second Order Dynamic Forces for Preliminary Estimation Purposes

				Blade Number		
				4	5	6
	Thrust	$F_{X(1)}$	Mean	$0.084T_0$	$0.020T_0$	$0.036T_0$
			Range	$\pm 0.031T_0$	$\pm 0.006T_0$	$\pm 0.0024T_0$
	Vertical force	$F_{Z(1)}$	Mean Range	$0.008T_0 \pm 0.004T_0$	$0.011T_0 \pm 0.009T_0$	$0.003T_0 \pm 0.002T_0$
Blade rate frequency component	Horizontal force	$F_{Y(1)}$	Mean Range	$0.012T_0 \pm 0.011T_0$	$0.021T_0 \pm 0.016T_0$	$0.009T_0 \pm 0.004T_0$
	Torque	$M_{X(1)}$	Mean	$0.062Q_0$	$0.0011Q_0$	$0.030Q_0$
			Range	$\pm 0.025Q_0$	$\pm 0.0008Q_0$	$\pm 0.020Q_0$
	Vertical moment	$M_{Z(1)}$	Mean Range	$0.075Q_0 \pm 0.050Q_0$	$0.039Q_0 \pm 0.026Q_0$	$0.040Q_0 \pm 0.015Q_0$
	Horizontal moment	$M_{Y(1)}$	Mean Range	$0.138Q_0 \pm 0.090Q_0$	$0.125Q_0 \pm 0.085Q_0$	$0.073Q_0 \pm 0.062Q_0$
	Thrust	$F_{X(2)}$	Mean	$0.022T_0$	$0.017T_0$	$0.015T_0$
			Range	$\pm 0.004T_0$	$\pm 0.003T_0$	$\pm 0.002T_0$
	Vertical force	$F_{Z(2)}$	Mean Range	$0.008T_0 \pm 0.004T_0$	$0.002T_0 \pm 0.002T_0$	$0.001T_0 \pm 0.001T_0$
Twice blade rate frequency component	Horizontal force	$F_{Y(2)}$	Mean Range	$0.00T_0 \pm 0.001T_0$	$0.006T_0, \pm 0.003T_0$	$0.003T_0 \pm 0.001T_0$
	Torque	$M_{X(2)}$	Mean	$0.016Q_0$	$0.014Q_0$	$0.010Q_0$
			Range	$\pm 0.010Q_0$	$\pm 0.008Q_0$	$\pm 0.002Q_0$
	Vertical moment	$M_{Z(2)}$	Mean Range	$0.019Q_0 \pm 0.013Q_0$	$0.012Q_0 \pm 0.011Q_0$	$0.007Q_0 \pm 0.002Q_0$
	Horizontal moment	$M_{Y(2)}$	Mean Range	$0.040Q_0 \pm 0.036Q_0$	$0.080Q_0 \pm 0.040Q_0$	$0.015Q_0 \pm 0.002Q_0$

Sasajima for the fluctuating forces around the propeller disc are shown below:

$$\tilde{K}_T(\theta) = \frac{1}{Z}\sum_{i=1}^{z} K_T(J(\theta + \theta_i))$$

$$\tilde{K}_Q(\theta) = \frac{1}{Z}\sum_{i=1}^{z} K_Q(J(\theta + \theta_i))$$

$$\tilde{F}_y(\theta) = -\frac{2}{Z\xi_f}\sum_{i=1}^{z} K_Q(J(\theta + \theta_i))\cdot\cos(\theta + \theta_i)$$

$$\tilde{F}_z(\theta) = \frac{2}{Z\xi_f}\sum_{i=1}^{z} K_Q(J(\theta + \theta_i))\cdot\sin(\theta + \theta_i)$$

$$\tilde{M}_y(\theta) = -\frac{\xi_f}{2Z}\sum_{i=1}^{z} K_T(J(\theta + \theta_i))\cdot\cos(\theta + \theta_i)$$

$$\tilde{M}_v(\theta) = -\frac{\xi_f}{2Z}\sum_{i=1}^{z} K_T(J(\theta + \theta_i))\cdot\sin(\theta + \theta_i)$$

where

$$\{\tilde{F}_y, \tilde{F}_z\} = \frac{\{F_y, F_z\}}{\rho n^2 D^4}, \{\tilde{M}_y, \tilde{M}_z\} = \frac{\{M_y, M_z\}}{\rho n^2 D^2}$$

$\xi_f = \dfrac{r_f}{R}$; non-dimensional radius of the loading point.

$\theta_i = \dfrac{2\pi(i-1)}{z}$

Z = number of blades.
$J(\theta + \theta_i)$ = advance coefficient at each angular position of each blade.
$K_T(J)$, $K_Q(J)$ = open water characteristics of the propeller.

Figure 11.4 shows a typical thrust fluctuation for a propeller blade of a single-screw ship; the asymmetry noted is due to the tangential components of the wake field acting in conjunction with axial components. Furthermore, Figure 11.5 illustrates a typical locus of the thrust eccentricity $e_T(t)$; the period of travel around this locus is of

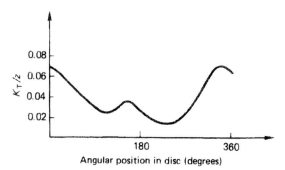

FIGURE 11.4 Typical propeller thrust fluctuation.

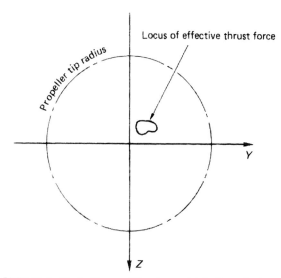

FIGURE 11.5 Typical locus of thrust eccentricity for a single screw vessel.

course dependent upon the blade number, because of symmetry, and is therefore $1/nz$. The computation of this locus generally requires the computation of the effective centers of thrust of each blade followed by a combination of these centers together with their thrusts to form an equivalent moment arm for the total propeller thrust.

11.1.5 Propeller Forces and Moments Induced by Turning Maneuvers

When a ship undertakes a turning maneuver the wake field into the propeller suffers a distortion due the asymmetry of the flow field around the ship. Measurements undertaken in the latter years of the last century by Lloyd's Register on a series of cruise ships using underwater telemetry methods showed that the bending propeller-generated forces and moments can be considerable, particularly for twin-screw ships. In a typical case for an inward turning, twin-screw cruise ship when undertaking a turn to port, the port propeller bearing experienced increase in the resultant force magnitude of two to three times the normal free running, straight course loading. Moreover, the direction of the force vector moved from approximately a five o'clock position, when viewed from aft, to somewhere between the three and four o'clock positions. In the case of the starboard propeller, on the outside of the turn, the relative magnitude increased by a factor of four to five and the direction changed from a generally seven o'clock position to the six o'clock position. It was also noted that in some cases there were differing temporal variations in the loading signature as the maneuvers progressed.

More recently Vartdal et al.[25] have shown similar trends in a study on a number of ships in steady state and transient operating conditions. In particular they note the importance of these loadings in the development of satisfactory shaft alignment conditions for ships and in particular the effect on, and life expectancy of, the after stern tube bearings.

Clearly measurements such as those cited above in which the forces and moments significantly change during ship maneuvers will also have a strong influence on the positions of the effective center of action of the forces which have to be reacted to by the ship's system of bearings. When these changes in loadings alter the effective centers of reaction in the bearing significantly then changes to the lateral vibration resonant frequencies have been known to occur in long flexible shaft lines.

While the calculation of these off-design conditions is potentially possible the wake field in which the propeller is operating is difficult to predict either from model measurements or from CFD methods. In the former case measurement would be required in a large seakeeping basin in order to correctly simulate the motion of the ship; however, there is the transient nature of the scale effects to contend with in this approach. The alternative case of CFD prediction offers promise in order to capture the full-scale situation, but there is still some way to progress in steady state full ship propeller rudder simulations before the transient turning maneuver can be undertaken.

11.1.6 Out-of-Balance Forces and Moments

A marine shafting system will experience a set of significant out-of-balance forces and couples if either the propeller becomes damaged so as to alter the distribution of mass or the propeller has not been balanced prior to installation.

For large propellers ISO 484/1[11] defines a requirement for static balancing to be conducted such that the maximum permissible balancing mass m_b at the tip of the propeller is governed by the equation

$$m_b \leq C \frac{M}{RN^2} \text{ kg or } KM \text{ kg, whichever is the smaller}$$

where

M is the mass of the propeller (kg)
R is the tip radius (m)
N is the designed shaft speed (rpm).

The coefficients C and K are defined in Table 11.14 according to the manufacturing class.

TABLE 11.14 ISO Balance Constants

ISO Class	S	I	II	III
C	15	25	40	75
K	0.0005	0.001	0.001	0.001

For the larger-diameter propellers a static balance procedure is normally quite sufficient and will lead to a satisfactory level of out-of-balance force which can both be accommodated by the bearings and also will not cause undue vibration to be transmitted to the vessel.

In the case of smaller propellers, ISO 484/2[12] applies for diameters between 0.80 and 2.50 m, the ISO standard also calls for static balance without further definition. For many of these smaller propellers this is a perfectly satisfactory procedure; however, there exists a small subset of high rotational speed, high blade area ratio propellers where dynamic balance is advisable. With these propellers, because of their relatively long axial length, considerable out-of-balance couples can be exerted on the shafting if this precaution is not taken.

When a propeller suffers the loss of a significant part or the whole of a blade through either blade mechanical impact or material fatigue and a spare propeller or repair capability is not available, then in order to minimize the vibration that will result from the out-of-balance forces the opposite blade of an even number bladed propeller should be similarly reduced. In the case of an odd bladed propeller, a corresponding portion should be removed from the two opposite blades so as to conserve balance. The amount to be removed can conveniently be calculated from the tables of Figure 11.1 used in association with a vectorial combination of the resulting centrifugal forces. The vibration resulting from the out-of-balance forces of a propeller will be first shaft order in frequency.

Additional out-of-balance forces can also be generated by variations in the blade-to-blade manufacturing tolerances of built-up and controllable pitch propellers. These are generally first order and of small magnitude; however, they can on occasions have noticeable effects. This is discussed further in Chapter 26.

11.2 HYDRODYNAMIC INTERACTION

The hydrodynamic interaction between the propeller and the hull originates from the passage of the blades beneath or in the vicinity of the hull and includes the cavitation dynamics on the surfaces of the blades. The pressure differences caused by these two types of action are then transmitted through the water to produce a fluctuating pressure over the hull surface which, because it acts over a finite area of the hull, produces an excitation force on the vessel. As a consequence the analysis of the hydrodynamic interaction can most conveniently be considered in three parts which eventually combine with the others, provided their phase angles are respected, to form the total pressure signal on the hull surface. These component parts are detailed in Table 11.15.

For the purposes of discussion here the pressures originating from the rotating propeller will form the main focus

Chapter | 11 Propeller, Ship and Rudder Interaction

TABLE 11.15 Hydrodynamic Interaction Components

Pressure from the passage of the non-cavitating blade $p_0(t)$
Pressure from the cavity volume variations on each blade $p_c(t)$

The effect of the hull surface on the free space pressure signal — termed the solid boundary factor (SBF)

of the discussion as distinct from the resultant forces on the hull. This latter force field is essentially the integration of the pressure field over the hull surface, taking into account the curvature and form of the hull in the region of the propeller.

11.2.1 Non-Cavitating Blade Contribution

The contribution to the total pressure signal on the hull from the passage of the non-cavitating blade is in the form of a continuous time series $p_0(t)$ and is generally considerably smaller than the cavitating component developed by the propeller. Therefore, for many ships the non-cavitating component will be overwhelmed by the cavitating component once the cavitation inception point is passed. This point is generally well below the design point for the vessel; however, some ship's propellers are designed to have high cavitation inception points to satisfy a design constraint such as might be found in some naval and research ship applications.

In the case of the non-cavitating propeller the pressure signature derives from the thickness of the propeller blades and the hydrodynamic loading over the surfaces of the blades. Huse[13] proposed a method in which the thickness effect is accounted for by an equivalent symmetrical profile, which is defined by a distribution of sources and sinks located along the chord line of the equivalent profile. The pressure signal derived in this way varies linearly with the equivalent profile thickness and the method is shown to give a good agreement with experiment.

The contribution from the blade loading may be considered in two portions: a contribution from the mean hydrodynamic loading and one from the fluctuating load component. In the case of the mean loading a continuous layer of dipoles distributed along the section mean line to simulate shockless entry may be used while the fluctuating loading can be simulated by dipoles clustered at the theoretical thin aerofoil aerodynamic center of the section; that is, at $0.25c$ from the leading edge.

The most important of the propeller parameters in determining the non-cavitating pressure signal are considered to be the blade number and the blade thickness. The pressure $p_0(t)$ can be expressed as:

$$p_0(t) = -\rho \frac{\partial \phi(t)}{\partial t}$$

in which

ρ is the density of the water
$\phi(t)$ is the velocity potential
t is time.

Several theoretical solutions, in addition to the work of Huse, have appeared in the literature. By way of contrast, the alternative approach by Breslin and Tsakonas[14] provides a good example of these other approaches.

11.2.2 Cavitating Blade Contribution

In Chapter 9 it was seen that a propeller may be subjected to many forms of cavitation that depend on the propeller operating point, the characteristics of the wake field and the detailed propeller geometry. Typically the propeller may experience on its blades any of the following: suction side sheet cavitation; tip vortex cavitation, which may collapse off the blade, as indeed can suction side sheet cavitation, and both cavitation types may interact to produce additional complexities to the excitation signature; cloud cavitation; pressure face cavitation or propeller—hull vortex cavitation. Clearly it is difficult to develop a unifying analytical treatment which will embrace all of these cavitation types, although the underlying physics of the pressure transmission processes are largely similar.

The need for such an all embracing treatment can to some extent be reduced by consideration of the cavitation types and their known effects on vibration. For example, face cavitation does not normally contribute significantly to the overall hull excitation, although it may contribute to high-frequency noise emissions from the propeller. Similarly, the hub vortex, unless it is particularly strong, does not normally significantly contribute to the hull pressure fluctuations, although it may cause excitation of the rudder which in turn can lead to hull excitation. The tip vortex may contribute to excitation at multiples of the blade rate frequency in addition to broadband characteristics and this can be of particular concern for vessels having hulls which extend well aft of the propeller station. Additionally English[15] has drawn attention to certain cases in which instabilities can arise in the tip vortex that cause the apparent expansive behavior of the tip vortex. This behavior is now thought to be due to super-cavitating sheet—tip vortex interaction and in these cases excitation and noise frequencies above blade rate are experienced. The problem of calculating the effects of the contribution of tip vortex cavitation stems largely from the work necessary to define reliable tip vortex inception behavior models: indeed much research is currently progressing in this field.

The cavitating blade contribution to the hull pressure field is considered to derive principally from the pulsation of the suction side sheet and tip vortex cavities. These types of cavitation may collapse either on or off the blade; in the former case they are generally responsible for blade rate and the first three or four harmonic frequencies of the excitation spectrum; whereas in the latter case the collapse extends the spectrum to the higher harmonics of blade rate frequency and broadband. In addition clouds of cavities may accompany the sheet cavitation, particularly during its collapse phases, and these have complex dynamics which contribute principally to the broadband part of the excitation spectrum.

In order to calculate the effects of the cavity volume variations on the cavitating pressures it is necessary to model the volume of the cavity on the blade. This is often done by constructing a system of sources, the strengths of which vary with blade angular position, so as to model the changing cavitation volume as the propeller rotates in the wake field. From a model of this type it is possible to derive an expression for the velocity potential of the sources with time from which an expression for the blade rate harmonics of the pressures can be developed. An expression developed by Breslin[16] has the following form:

$$p_{c_{qz}} = -\frac{\rho Z^3}{2\pi} \cdot \frac{(q\omega)^2}{R_p} \mathrm{Re}[V_{qz} e^{iqz\phi}] \qquad (11.21)$$

where

q is the harmonic order (blade rate, $q = 1$)
ρ is fluid mass density
Z is the blade number
ω is the angular velocity
R_p is the distance of the field point $\sqrt{(r^2 + x^2)}$
V_{qz} is the qzth harmonic component of the complex amplitude of the cavity volume.
ϕ is the blade position angle.

Consequently by expressing V_{qz} in its complex form as

$$V_{qz} = a_{qz} + ib_{qz}$$

and extracting the real part it can be shown that:

$$p_{c_{qz}} = \frac{\rho Z^3}{2\pi} \frac{(q\omega)^2}{R_p} \sqrt{(a_{qz}^2 + b_{qz}^2)} \cdot \cos(qz\phi + \varepsilon) \qquad (11.21\mathrm{a})$$

where $\varepsilon = \tan^{-1}(b_{qz}/a_{qz})$.

From this expression it can be seen that the asymptotic pressure due to the cavity volume variation at a particular blade rate harmonic frequency depends upon the blade rate harmonic of the cavity volume. Furthermore, the dependence of the field point pressure on the inverse of the distance R_p:

$$p_{cqz} \propto \frac{1}{R_p}$$

becomes apparent and clearly demonstrates how the field point pressures from the cavitating propeller decay with increasing distance from the propeller: hence the advantage of providing adequate clearances around the propeller become apparent. This proportionality for the cavitating propeller of $p_c \propto R_p^{-1}$ is in contrast to that for a non-cavitating propeller since this is more closely expressed by $p_0 \propto R_p^{-2.5}$. As a consequence this component decays more rapidly than the cavitating component with distance.

Expressions of the type shown in equation (11.21) or (11.21(a)) are a simplification of the actual conditions and as such are only valid at distances that are large compared to the propeller radius. This can lead to some difficulty when calculating the field point pressures in cases where the tip clearances are small.

Skaar and Raestad[17] developed an expression based on similar assumptions to that of equation (11.21) which provides further insight into the behavior of the cavitating pressure signature. Their relationship is

$$p_c \simeq \frac{\rho}{4\pi} \frac{1}{R_p} \frac{\partial^2 V}{\partial t^2} \qquad (11.22)$$

in which V is the total cavity volume.

From this expression it can further be seen that the cavitating pressure signature is proportional to the second derivative of the total cavitation volume variation with time. This demonstrates why the pressures encountered upon the collapse of the cavity, which is usually more violent, are greater than those experienced when the cavity is growing. In this context, with a single-screw vessel having a right-handed propeller the pressures measured above the propeller plane are generally greater on the starboard side than on the port side. Skaar and Raestad show the equivalence of equation (11.22) to that of (11.21) in the discussion to Reference 17.

The dependence of the pressure on the $\partial^2 V/\partial t^2$ term shows the implied dependence of the pressure on the quality of the wake field and consequently any steps taken to improve the wake field at the design stage are very likely to have a beneficial effect on the pressure signature. The alternative is to approach the problem from the blade design viewpoint. In this case attention to the radial and chordal distribution of loading, the skew distribution and the blade area are all known parameters that have a significant influence on the cavitating pressure impulses. Blade number, unlike the non-cavitating pressure impulses, in general has a very limited influence on the cavitating pressure characteristics. Nevertheless, blade modifications are frequently attenuating the symptoms rather than the more fundamental problem of a poor wake field.

11.2.3 Influence of the Hull Surface

The discussion so far has concerned itself with the free field pressures from cavitating and non-cavitating propellers. When a solid boundary is introduced into the vicinity of the propeller then the pressures acting on that boundary are altered significantly. For example, if a rigid flat plate is introduced at a distance above the propeller, then the fluctuating pressure acting on the plate surface will be twice that of the free field pressure. This leads to the concept of a solid boundary factor (SBF) which is defined as follows:

$$\text{SBF} = \frac{\text{pressure acting on the boundary surface}}{\text{free field pressure in the absence of the solid boundary}} \quad (11.23)$$

In the case of a flat plate, which is of infinite stiffness, the solid boundary factor is equal to 2. However, in the case of a ship form a lesser value would normally be expected due to the real hull stiffness being different from a rigid flat plate and the influence of pressure release at the sea surface. Garguet and Lepeix[18] discussed the problems associated with putting the SBF equal to 2 in giving misleading results to calculations. Subsequently, Ye and van Gent,[19] using a potential flow calculation with panel methods, suggested a value of 1.8 as being more appropriate for ship calculations.

In practice the solid boundary factor can be considered as a composite factor having one component S_b which takes into account the hull form and another S_f which accounts for the proximity of the free surface. Hence equation (11.23) can be written as the product of two components:

$$\text{SBF} = S_b \cdot S_f \quad (11.24)$$

Wang[20] and Huse[21] have both shown that the dominant factor is the proximity of the free surface S_f. In their work the variability of the solid boundary factor is discussed and it falls from a value of just under 2.0 at the shaft center line to a value of zero at the free surface—hull interface. Huse gives relationships for S_b and S_f for an equivalent clearance ratio of 0.439 which are as follows:

$$\left.\begin{array}{l} S_b = 2.0 + 0.0019\alpha - 0.00024\alpha^2 \\ S_f = 9.341\delta - 30.143\delta^2 + 33.19\delta^3, 0 < \delta \leq 0.35 \\ S_f = 1.0 \text{ for } \delta > 0.35 \end{array}\right\} \quad (11.25)$$

where

α is the inclination of the section with respect to the horizontal measured in the thwart direction.

and

δ is the ratio of the field point immersion depth to the shaft immersion depth.

11.2.4 Methods for Predicting Hull Surface Pressures

In essence there are three methods for predicting hull surface pressures; these are by means of empirical methods, by calculations using theoretical or numerical methods and by experimental measurements.

With regard to the empirical class of methods the most well known and adaptable is that by Holden et al.[22] This method is based on the analysis of some 72 ships for which full-scale measurements were made prior to 1980 and, as such, some caution needs to be exercised in using this method for certain modern ship hull forms. The method was intended for a first estimate of the likely hull surface pressures using a conventional propeller form. Holden proposed the following regression-based formula for the estimation of the non-cavitating and cavitating pressures respectively:

and

$$\left.\begin{array}{l} p_0 = \dfrac{(ND)^2}{70} \dfrac{1}{Z^{1.5}} \cdot \left(\dfrac{K_0}{d/R}\right) \text{N/m}^2 \\[2mm] p_c = \dfrac{(ND)^2}{160} \cdot \dfrac{V_s(w_{T\max} - w_e)}{\sqrt{(h_a + 10.4)}} \cdot \left(\dfrac{K_c}{d/R}\right) \text{N/m}^2 \end{array}\right\} \quad (11.26)$$

N is the propeller rpm.
D is the diameter (m).
V_s is the ship speed (m/s).
Z is the blade number.
d is the distance from $r/R = 0.9$ to a position on the submerged hull when the blade is at the top dead center position (m).
R is the propeller radius (m).
$w_{T\max}$ is the maximum value of the Taylor wake fraction in the propeller disc.
w_e is the mean effective full scale Taylor wake fraction
h_a is the depth to the shaft centre line.
and K_0 and K_c are given respectively by the relationships

$$K_0 = 1.8 + 0.4(d/R) \text{ for } d/R \leq 2$$
$$\text{and} \quad K_c = 1.7 - 0.7(d/R) \text{ for } d/R < 1$$
$$K_c = 1.0 \text{ for } d/R > 1$$

The total pressure impulse which combines both the cavitating and non-cavitating components of equation (11.26) acting on a local part of the submerged hull is then found from

$$p_z = \sqrt{(p_0^2 + p_c^2)} \quad (11.27)$$

Empirical methods of this type are particularly useful as a guide to the expected pressures. They should not, however, be regarded as a definitive prediction, because differences, sometimes quite substantial, will occur when correlated with full-scale measurements. For example, equation (11.26) gives results having a standard deviation of the order of 30 per cent when compared to the base measurement set from which it was derived.

The tip vortex if either excessive or unstable can be the origin of high radiated unsteady pressures. These pressures are often at frequencies higher than blade rate and are not necessarily at integer multiples of blade rate. Raestad,[29] based on experience with the noise and vibration characteristics of some 15 cruise ships and ferries developed a Tip Vortex Index (TVI) method of evaluating the proposed propeller–ship interaction. The method considers the results from full-scale measurements in relation to the propeller design parameters; particularly in relation to the blade tip loading.

In the case of theoretical or numerical calculations more detail can be taken into account and this is conducive to achieving a higher level of accuracy. The theoretical models which would be used in association with this form of analysis are those which can be broadly grouped into the lifting surface or vortex lattice categories. In particular, unsteady lifting surface theory is a basis for many advanced theoretical approaches in this field. Notwithstanding the ability of analytical methods to provide an answer, care must be exercised in the interpretation of the results, since these are particularly influenced by factors such as wake scaling procedures; the description of the propeller model and the hull surface; the distribution of solid boundary factors and the harmonic order of the pressures considered in the analysis. Furthermore, propeller calculation procedures normally assume a rigid body condition for the hull and as a consequence do not account for the self-induced pressures resulting from hull vibration: these have to be taken into account by other means, typically finite element models of the hull structure. As a consequence of all of these factors considerable care must be exercised in interpreting the results and the method used should clearly be subjected to a validation process.

In a series of papers van Wijngaarden,[23,24] has considered the forward and inverse scattering problem based on an acoustic boundary element solution of the Hemholtz equation. In the former, once the propeller source strength is known then the computational procedure is directed towards establishing the hull surface pressure distribution, while in the latter approach the reverse is the case. Such methods are particularly useful at various stages in design and analysis. The forward scattering solution is of value during the initial design processes since knowing the theoretical propeller cavitation characteristics an initial estimate of the hull surface pressure distribution can be made. However, as the design progresses and model tests have perhaps been carried out to determine the hull surface pressures then the inverse solution can be helpful.

Model measurement methods of predicting hull surface pressures can be conducted in either cavitation tunnels or other specialized facilities such as depressurized towing tanks. Originally the arrangement in a cavitation tunnel comprised a simple modeling of the hull surface by a flat or angled plate above a scale model of the propeller. Although this technique is still sometimes employed in some establishments, a more enlightened practice with moderately sized cavitation tunnels is to use a dummy model having a shortened center body, as shown in Figure 11.6. However, in the largest facilities the towing tank model is used. The advantage of using a model of the actual hull form is twofold: first it assists in modeling the flow of water around the hull surface and only requires wake screens, which are essentially arrangements of wire mesh, for fine tuning purposes of the wake field and, secondly, it makes the interpretation of the measured hull surface pressures easier

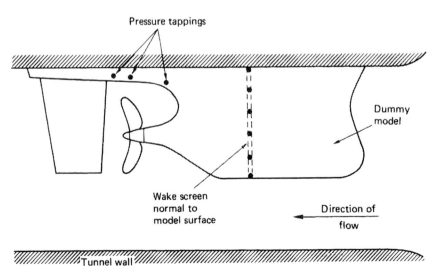

FIGURE 11.6 Dummy model and propeller in a cavitation tunnel.

since the real hull form is simulated. In order to measure the hull surface pressures arrays of pressure tappings, normally comprising between 8 and 32 sensors depending upon the establishment and the test requirements, are inserted into the model hull surface above and around the propeller station. Some commercial model experiments have a limited array of pressure tapping points. When this is the case accurate estimates of the total resultant force are difficult and as a consequence the largest matrix of pressure measurement points possible should be employed in model experiments. Additionally, it should be recalled that the hull-induced pressure distribution from the propeller, excluding the self-induced effect, comprises two parts: the cavitating and the non-cavitating part. With regard to the cavitating part the pressure field is approximately in phase over the ship's afterbody; however, the non-cavitating contribution has a strongly varying phase distribution across the hull surface, particularly in the athwart ship direction. As such, when properly deployed the larger sensor arrays help in discriminating between these components.

To interpret model test results appeal can be made to dimensional analysis. From this method it can be shown that the pressure at a point on the hull surfaces above a propeller has a dependence on the following set of dimensional parameters:

$$p = \rho n^2 D^2 \Phi\{J, K_T, \sigma, R_n, F_n, (z/D)\} \qquad (11.28)$$

where

J is the advance coefficient.
K_T is the propeller thrust coefficient.
σ is the cavitation number.
R_n is the Reynolds number.
F_n is the Froude number.

and

z is the distance from the propeller to the point on the hull surface.

In equation (11.28), the quantities ρ, n and D have their normal meaning.

As a consequence of this relationship a pressure coefficient K_p can be defined as

$$K_p = \frac{p}{\rho n^2 D^2} \qquad (11.29)$$

which has the functional dependence defined in equation (11.28).

Equation (11.28) defines the hull surface pressure as a function of propeller loading, cavitation number, geometric scaling and Reynolds and Froude identity. Assuming that the geometric scaling, cavitation and thrust identity have all been satisfied, the hull surface pressure at ship scale can be derived from equation (11.29) as follows, using the suffixes m and s for model and ship respectively.

By assuming the identity of K_p between model- and full-scale we may write:

$$\frac{p_s}{p_m} = \frac{\rho_s}{\rho_m}\left(\frac{n_s}{n_m}\right)^2\left(\frac{D_s}{D_m}\right)^2$$

and for Froude identity ($F_{ns} = F_{nm}$)

$$\frac{p_s}{p_m} = \frac{\rho_s}{\rho_m}\left(\frac{D_m}{D_s}\right)^2 \cdot \frac{\rho_s}{\rho_m}\lambda \qquad (11.30)$$

where λ is the model-scale of the propeller.

Equation (11.30) implies that the Reynolds condition over the blades has also been satisfied, which is clearly not the case if the Froude identity is satisfied. Hence, it is important to ensure that the correct flow conditions exist over the propeller blades in order to ensure a representative cavitation pattern over the blades and pressure coefficient on the model as discussed in Chapter 9.

Recognizing that the wake field may potentially present a significant source of error when endeavoring to predict hull excitation forces from model tests van Wijngaarden,[30] explored ways of reducing this error. From this study it was proposed that the ship model, because of Reynolds Number dissimilarities, should not necessarily replicate the full-scale ship's lines but should be designed based on an inverse principle using RANS codes. This would enable a model geometric hull form to be produced which creates a far closer approximation to the ship scale wake field. This procedure was applied to a container ship with encouraging results.

11.3 PROPELLER–RUDDER INTERACTION

Propeller–rudder interaction essentially takes two forms. The first is due to the interference to the pressure field in which the propeller is operating by the presence of the rudder in the flow field. In general this influence is relatively small unless the propeller–rudder clearances are particularly small and, therefore, would not normally be taken into account when undertaking propeller calculations. The exception to this is if a full computational fluid dynamics study was being undertaken of the afterbody and propulsion system. Nevertheless, it is a real effect which can be demonstrated in model tests. The second influence is the effect that the propeller flow field has on the rudder. This sometimes manifests itself in cavitation erosion of the rudder structure, particularly in the case of container, LNG, and other fast ships and craft. Alternatively, vibration of the rudder assembly may be a result of either a forced or resonant character by virtue of the rudder working in the flow field generated by the propeller. The design of rudders and control surfaces, while outside the scope of this work, is addressed in Reference 26, which contains much useful experimental data.

The problem of rudder erosion and its avoidance, together with a number of other related matters, is

FIGURE 11.7 Example of rudder erosion.

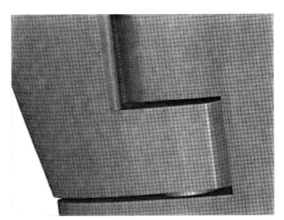

FIGURE 11.8 Detail of the mesh at the rudder–horn connection.

discussed in Reference 27. Figure 11.7 shows a typical example of cavitation erosion on a rudder. While such problems can be addressed by classical hydrodynamic approaches, CFD methods assist in permitting a greater amount of detail to be developed within the solution. There are two principal CFD-based approaches available: a single-phase and a two-phase approach, and both have shown good correlation with model- and full-scale performance.

11.3.1 The Single-Phase Approach

In this approach a numerical model of the rudder is set up based on the assumption of a steady, single-phase flow condition with the vapor phase of the fluid not explicitly modeled. When generating the computational domain it has been found that a *trimmed-cell* technology, to optimize the quality and distribution of cells in the flow volume, provides a useful basis for the analysis. Within this approach the mesh comprises two parts, the extrusion layer adjacent to specified surfaces and the core volume. The mesh structure contains regular hexahedral cells, of zero skewness, in the core volume and trimmed cells and hexahedra with corners and edges removed, adjacent to the outer edge of the extrusion layer. Figure 11.8 shows a typical detail of the trimmed cells at the surfaces of the horn–rudder connection and, moreover, the fine mesh needed to resolve curved edges is visible through the gap. The extrusion layer, which can be imagined as a thick wrap around the surface, comprises regular hexahedral cells which are orthogonal to the surface of the rudder and thereby ensure high-quality cells to resolve the boundary layer flow.

With regard to the detail of the mesh it has been found that an unstructured grid with a finer mesh applied throughout the main region of the flow development, in particular near the rudder surface, and with a coarser grid applied towards the edges of the domain, proves satisfactory for these types of computation. Typically, the required size for the final computational mesh would deploy the use of around 1.4×10^6 cells.

The boundary conditions for this type of computational model are derived from the velocity field generated by the propeller and hull boundary layer. In most cases it has been found satisfactory to use the nominal wake components measured in a towing tank and then convert these to an effective wake field for input into a suitable propeller lifting surface capability in order to define the inflow into the rudder. Scaling of the regions outside the propeller slipstream is also required.

The rudder surfaces are modeled using a combined hybrid approach which switches between low Re and high Re modeling options depending on the size of y^+ in the near-wall cell. This approach has been found to produce better results in the near-wall region as it is less influenced by the quality of the grid distribution. A two-equation k-ω or a k-ωSST turbulence modeling approach is normally the basis for modeling turbulence. Typical results in the form of contour plots of the cavitation number, based on the local total static pressure and the free stream velocity over the surface of the rudder, are presented in Figure 11.9.

For correlation purposes the computational results shown in Figure 11.9 are compared against the model test observations for 4° of inboard rudder rotation. Results of this type suggest that a single-phase modeling approach can be used to identify problem areas on the rudder and indicate the possibility of cavitation inception. Furthermore, since this type of computation is not too computationally demanding it can be routinely used as a design tool.

11.3.2 The Two-Phase Approaches

While the cavitation number or some derivative of it, as discussed in Section 11.3.1, can be used as an indicator of

Propeller, Ship and Rudder Interaction

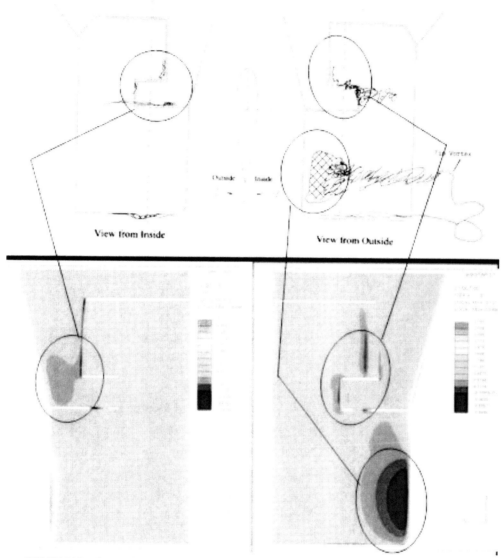

FIGURE 11.9 Comparison between model tests and computations for a four inboard rudder rotation.

the likelihood of cavitation occurring, it is preferable, although computationally more expensive, to model the propeller–rudder interaction problem without the single-phase restriction. This is because the existence of cavitation can change both the local and global flow behavior and this is not taken into account by single-phase flow computations.

The type of computational model described previously can be further enhanced by utilizing a cavitation model based on the Rayleigh bubble model, equation (9.3), combined with a free surface interface capturing method. In this approach the liquid–vapor mixture is treated as a continuum with varying material properties; for example, density and viscosity. A scalar variable is defined in this procedure to represent the cavitation strength, its value denoting the volume ratio of vapor over the fluid mixture and cavitation is assumed to take place when the pressure in the liquid falls below a critical pressure, p_{crit}: expressed as:

$$p_{\text{crit}} = p_v - \frac{4\sigma}{3R*}$$

where p_v is the saturation vapor pressure at the local temperature, σ is the liquid surface tension coefficient and $R*$ is the characteristic size of the cavitation nuclei.

Figure 11.10 shows computed results for the same 4° inboard rudder rotation that was shown in Figure 11.9, but in this latter case at ship scale. In Figure 11.9 the cavitation zones appearing on the rudder surface are clearly identifiable. This is because the inside part of the rudder exhibits

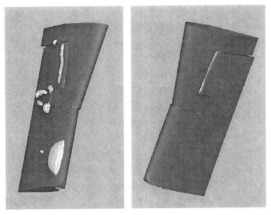

FIGURE 11.10 Full-scale cavitation predictions for the semi-spade rudder of Figure 11.9.

cavitation mainly along the gap edges and at the bottom surface of the rudder, again in good agreement with observations.

This type of capability to predict cavitation occurrence can be exploited at the rudder design stage in order to improve the design and achieve a significant reduction in likelihood of cavitation erosion. Moreover, it is possible to use predictive modeling techniques of this type to explore the presence of cavitation during normal ship operation. This can be accomplished through the construction of an operation diagram which shows the probability of cavitation occurring at a number of critical locations on the rudder throughout the typical auto-pilot range of turning angles. Such analyses have been shown to be most valuable in preventing erosion from taking place.

Notwithstanding the forgoing discussion which has been based on a propeller–rudder combination with an incident wake field, the power of many modern computational facilities permits an attempt at the solution of complete hull–propeller–rudder model studies. Such an ability assists in reducing the dependence in scaling incumbent on the earlier partial solution dictated by less capable computers. However, when using these integrated solutions it must, at this stage in CFD development, be remembered that the full description of the propeller using these techniques does still present some problems. By way of example Figure 11.11 demonstrates the value of CFD modeling in endeavoring to gain an appreciation of the flow configurations developed by a propeller–rudder–hull combination.

11.3.3 Model Testing

In many cases of relatively low-powered, slow-speed ships a separate model test program to determine the cavitation characteristics of the rudder may not be necessary. However, there are some ship types where this is not the case.

To achieve an acceptable solution for high-powered ships a careful design strategy comprising elements of computation and model testing is helpful in minimizing the risk of cavitation erosion problems at full scale. Such a strategy might involve the measurement of the rudder incident flow field generated by the propeller and the ship's boundary layer, recognizing that this latter flow field component will require some modification from model-scale values due to scale effects that will be present. Having defined this inflow field a first iteration for the rudder geometry design can then be produced which may suggest the desirability of a contoured leading edge, in contrast to the normal straight line leading edge, so as to permit the

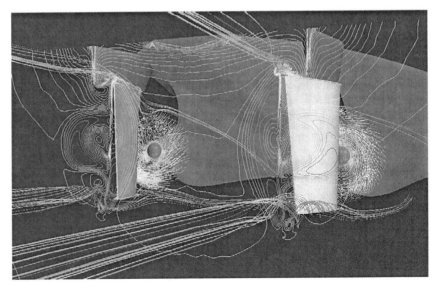

FIGURE 11.11 CFD flow field computation in the afterbody region of a ship. *Courtesy Lloyd's Register.*

greatest cavitation-free incidence ranges to be obtained for the components of the rudder.

Upon completion of this design definition, large-scale model cavitation tests should then be carried out to estimate the full-scale characteristics of the design, both in overall terms as well as in the detailed behavior of the design around the pintle housings and the interfaces between the rudder and horn. This latter aspect may also benefit from an even larger-scale model test in cases where particularly onerous conditions are encountered involving only the center pintle and gap regions of the rudder assembly. Within the testing phase of the design it is important to carefully evaluate the influence that the normal range of auto-pilot rudder angles has on the cavitation dynamics since these angular variations may strongly influence the erosion potential of the design. Furthermore, to assess this erosion potential a paint erosion technique might form an integral part of the testing program; however, the reliability of this technique for rudder erosion prediction is not yet as good as similar procedures for propeller blades.

Following the model testing phase then a second iteration of the design can be made in which detailed geometric changes can be introduced in the knowledge of the individual cavitation bucket assessments, in the case of semi-spade rudders, of the horn, pintle and blade regions of the rudder. Depending on the extent of the changes required then a further model test may also be desirable.

11.3.4 Some Full-Scale Remedial Measures

When cavitation erosion has been experienced after a ship has entered service there are a number of options available to attenuate the effects of the cavitation which do not involve complete redesign.

Stainless steel cladding: This technique has been tried both as wide sheets of stainless steel and also as a sequence of adjacent narrow strips. Experience favors the use of narrow strips since the wider strips have been found to detach in service. General experience, however, with this method is mixed and is very dependent upon the severity of the attack in terms of the energy transfer within the collapse process, the quality of welding and the general flow conditions prevailing.

Twisted rudders: The US Navy developed a design methodology for continuously twisted full-spade rudders and has proven them in service on the Arleigh Burke (DDG51) class of Frigates (Reference 28). The rudder designs were evaluated in the LCC test facility and provided a 7° increase in the cavitation-free envelope at 31 knots. Stepped and continuously twisted rudders have also been introduced to merchant ships and appear to be beneficial in reducing erosion problems induced by the propeller slipstream and course-keeping operations.

Scissor plates: These flat plates are placed in the horizontal gap between the rudder horn and the blade of a semi-balanced rudder. They have been very successful in reducing erosion in these regions of the rudder.

Flow spoilers: Such devices have been advocated for combating erosion on pintle housings and the forward facing edges of the rudder blade, immediately behind the rudder horn. There are few reports on their effectiveness and experience with these systems has been inconclusive.

Profiled leading edge transitions between the rudder leading edge and the base of the rudder: Fast vessels should avoid a 90° angle between the rudder leading edge and a flat base plate, since sheet and vortex cavitation have been observed in these regions and resulted in erosion and corrosion of the base plate within about 25 per cent of the rudder chord from the leading edge. Fairings in this region need to be carefully designed to cater for the full range of auto-pilot course-keeping angles. There is also some evidence that the provision of a radius on the bottom plate of the rudder can be helpful in both reducing cavitation erosion and energy losses due to vortex generation.

The annular gap: This gap between the aft surface of the horn and the moveable blade may be reduced in size by fitting vertical strips which block the passage of any flow within this gap. This approach seeks to reduce the cross-flow angle of attack onto the forward facing edges of the rudder blade.

REFERENCES AND FURTHER READING

1. Jakeman RW. *Analysis of dynamically loaded hydrodynamic journal bearings with particular reference to the mis-aligned marine stern tube bearing*. PhD Thesis, Cranfield; 1988.
2. Archer S. *Torsional vibration damping coefficients for marine propellers*. Engineering; May 1955.
3. Lewis FM, Auslander J. *Virtual inertia of propellers*. J. Ship Res.; March 1960.
4. Burrill LC, Robson W. *Virtual mass and moment of inertia of propellers*. Trans. NECIES 1962;**78**.
5. Schwanecke H. *Gedanken wur Frage de Hydrodynamischen Erregungen des Propellers und der Wellenleitung*. STG 1963;**57**.
6. Gunsteren LA van, Pronk C. *Determining propeller speed and moment of inertia*. Diesel and Gas Turbine Progress; November/December 1973.
7. Hylarides S, Gent W van. *Propeller hydrodynamics and shaft dynamics*. Symp. on High Powered Propulsion of Large Ships, NSMB; 1974.
8. Parsons MG, Vorus WS. *Added mass and damping estimate for vibrating propellers*. Propellers '81 Symp. Trans. SNAME; 1981.
9. *Vibration Control in Ships*. Oslo: Veritec; 1985.
10. Sasajima T. *Usefulness of quasi-steady approach for estimation of propeller bearing forces*. Propellers '78 Symp. Trans. SNAME; 1978.
11. ISO 484/1. *Shipbuilding — Ship Screw Propellers — Manufacturing Tolerances — Part 1: Propellers of Diameter Greater than 2.50 m*; 1981.

12. ISO 484/2. *Shipbuilding — Ship Screw Propellers — Manufacturing Tolerances — Part 2: Propellers of Diameter Between 0.80 and 2.50m Inclusive*; 1981.
13. Huse E. *The magnitude and distribution of propeller induced surface forces on a single screw ship model*. Ship Model Tank Publication No. 100; 1968. Trondheim.
14. Breslin JP, Tsakonas S. *Marine propeller pressure field due to loading and thickness effects*. Trans. SNAME 1959;**67**.
15. English J. *Cavitation induced hull surface pressures — Measurements in water tunnels*. Symp. on Propeller Induced Ship Vibration, Trans. RINA; 1979.
16. Breslin, J.P. *Discussion on Reference 17*.
17. Skaar KT, Raestad AE. *The relative importance of ship vibration excitation forces*. Symp. on Propeller Induced Ship Vibration, Trans. RINA; 1979.
18. Garguet M, Lepeix R. *The problem of influence of solid boundaries on propeller-induced hydro-dynamic forces*. Symp. on High Powered Propulsion of Large Ships, Wageningen; 1974.
19. Ye Y, Gent W van. *The Solid Boundary Factor in Propeller Induced Hull Excitation — Characteristics and Calculations*. MARIN Report No. 50812-2-RF; May, 1988.
20. Wang G. *The influence of solid boundaries and free surface on propeller-induced pressure fluctuations*. Norwegian Mar. Res. 1981;**9**.
21. Huse E. *Cavitation induced excitation forces on the hull*. Trans. SNAME 1982;**90**.
22. Holden KO, Fagerjord O, Frostad R. *Early design-stage approach to reducing hull surface forces due to propeller cavitation*. Trans. SNAME; November 1980.
23. Wijngaarden E van. *Recent developments in predicting propeller-induced hull pressure pulses*. London: Proc. 1st Int. Ship Noise and Vibration Conf.; June 2005.
24. Wijngaarden E van. *Determination of propeller source strength from hull-pressure measurements*. London: Proc. 2nd Int. Ship Noise and Vibration Conf.; June 2006.
25. Vartdal BJ, Gjestland T, Arvidsen Tl. *Lateral propeller forces and their effects on shaft bearings*. 1st Int. Symp. on Marine Propellers Smp'09, Trondheim; June 2009.
26. Molland AF, Turnock SR. *Marine Rudders and Control Surfaces: Principles, Data, Design and Applications (ISBN 978-0-75-066944-3)*. Oxford Butterworth-Heinemann; 2007.
27. Carlton JS, Fitzsimmons PA, Radosavljevic D, Boorsma A. *Factors influencing rudder erosion: Experience of computational methods, model tests and full scale observations*. Genoa: NAV 2006 Conf.; June 2006.
28. Shen YT, et al. *A twisted rudder for reduced cavitation*. J. Ship Res. December 1997;**41**(4).
29. Raestad AE. Tip vortex index — *An engineering approach to propeller noise prediction*. Naval Architect; July/August 1996::11–6.
30. Wijngaarden E, van. *Prediction of propeller-induced hull-pressure fluctuations*. PhD Thesis, TU Delft; 2011.

Chapter 12

Ship Resistance and Propulsion

Chapter Outline

12.1 Froude's Analysis Procedure	300
12.2 Components of Calm Water Resistance	301
12.2.1 Wave-Making Resistance R_W	302
12.2.2 The Contribution of the Bulbous Bow	305
12.2.3 Transom Immersion Resistance	307
12.2.4 Viscous form Resistance	307
12.2.5 Naked Hull Skin Friction Resistance	308
12.2.6 Appendage Skin Friction	309
12.2.7 Viscous Resistance	309
12.3 Methods of Resistance Evaluation	311
12.3.1 Traditional and Standard Series Analysis Methods	311
Taylor's Method (1910–1943)	312
Ayre's Method (1942)	312
Auf'm Keller Method	312
Harvald Method	313
Standard Series Data	314
12.3.2 Regression-Based Methods	314
12.3.3 Direct Model Test	316
General Procedure for Model Tests	316
Resistance Tests	317
Open Water Tests	318
Propulsion Tests	319
Flow Visualization Tests	319
Model Test Facilities	319
Two-Dimensional Extrapolation Method	319
Three-Dimensional Extrapolation Method	321
12.3.4 Computational Fluid Dynamics	322
12.4 Propulsive Coefficients	323
12.4.1 Relative Rotative Efficiency	324
12.4.2 Thrust Deduction Factor	324
12.4.3 Hull Efficiency	325
12.4.4 Quasi-Propulsive Coefficient	325
12.5 The Influence of Rough Water	325
12.6 Restricted Water Effects	327
12.7 High-Speed Hull form Resistance	328
12.7.1 Standard Series Data	328
12.7.2 Model Test Data	329
Multi-Hull Resistance	329
12.8 Air Resistance	330
References and Further Reading	330

Prior to the mid-nineteenth century comparatively little was known about the laws governing the resistance of ships and the power required to develop a particular speed. Brown[1] gives an account of the problems of that time and depicts the role of William Froude, who can justly be considered to be the father of ship resistance studies. An extract from Brown's account reads as follows:

... In the late 1860s Froude was a member of a committee of the British Association set up to study the problems of estimating the power required for steamships. They concluded that model tests were unreliable and often misleading and that a long series of trials would be needed in which actual ships were towed and the drag force measured. Froude wrote a minority report pointing out the cost of such a series of trials and the fact that there could never be enough carried out to study all possible forms. He believed that he could make sense from the results of model tests and carried out a series of experiments in the River Dart to prove his point. By testing models of two different shapes and three different sizes he was able to show that there were two components of resistance, one due to friction and the other to wave-making and that these components obeyed different scaling laws. Froude was now sufficiently confident to write to Sir Edward Reed (Chief Constructor of the Navy) on 24 April 1868, proposing that an experiment tank be built and a two year programme of work be carried out. After due deliberation, in February 1870, Their Lordships approved the expenditure of £2000 to build the world's first ship model experiment tank at Torquay and to run it for two years. The first experiment was run in March 1872 with a model of HMS Greyhound. Everything was new. The carriage was pulled along the tank at constant speed by a steam engine controlled by a governor of Froude design. For this first tank he had to design his own resistance dynamometer and followed this in 1873 by his masterpiece, a propeller dynamometer to measure thrust, torque and rotational speed of model propellers. This dynamometer was made of wood, with brass wheels and driving bands made of leather boot laces. It continued to give invaluable service until 1939 when its active life came to an end with tests of propellers for the fast minelayers...

... William Froude died in 1879, having established and developed a sound approach to hull form design, made a major contribution to the practical design of ships, developed new experiment techniques and trained men who were to spread the Froude tradition throughout the world. William was succeeded as Superintendent AEW by Edmund Froude, his son, whose first main task was to plan a new establishment since the Torquay site was too small and the temporary building was nearing the end of its life. Various sites were considered but the choice fell on Haslar, Gosport, next to the Gunboat Yard, where AEW, then known as the Admiralty Marine Technology Establishment (Haslar) or AMTE(H),* remains to this day. A new ship tank, 400 ft long, was opened in 1877 ...

... Edmund was worried about the consistency of results being affected by the change to Haslar. He was a great believer in consistency, as witness a remark to Stanley Goodall, many years later, 'In engineering, uniformity of error may be more desirable than absolute accuracy'. As Goodall said 'That sounds a heresy, but think it over'. Froude took two measures to ensure consistent results; the first, a sentimental one, was to christen the Haslar tank with water from Torquay, a practice repeated in many other tanks throughout the world. The flask of Torquay water is not yet empty – though when Hoyt analysed it in 1978 it was full of minute animal life! The more practical precaution was to run a full series of tests on a model of HMS Iris at Torquay just before the closure and repeat them at Haslar. This led to the wise and periodical routine of testing a standard model, and the current model, built of brass in 1895, is still known as Iris, though very different in form from the ship of that name. Departures of the Iris model resistance from the standard value are applied to other models in the form of the Iris Correction. With modern water treatment the correction is very small but in the past departures of up to 14.5 per cent have been recorded, probably due to the formation of long chain molecules in the water reducing turbulence in the boundary layer. Another Froude tradition, followed until 1960, was to maintain water purity by keeping eels in the tanks. This was a satisfactory procedure, shown by the certification of the tank water as emergency drinking water in both World Wars, and was recognised by an official meat ration, six pence worth per week, for the eels in the Second World War! ...

So much then for the birth of the subject as we know it today and the start of the tradition of 'christening' a new towing tank from the water of the first tank, sited at Froude's home at Chelston Cross at Crockington near Torquay.

12.1 FROUDE'S ANALYSIS PROCEDURE

William Froude[2] recognized that ship models of geometrically similar form would create similar wave systems, albeit at different speeds. Furthermore, he showed that the smaller models had to be run at slower speeds than the larger models in order to obtain the same wave pattern. His work showed that for a similarity of wave pattern between two geometrically similar models of different size the ratio of the speeds of the models was governed by the relationship

$$\frac{V_1}{V_2} = \sqrt{\frac{L_1}{L_2}} \qquad (12.1)$$

By studying the comparison of the specific resistance curves of models and ships Froude noted that they exhibited a similarity of form although the model curve was always greater than that for the ship (Figure 12.1). This led Froude to the conclusion that two components of resistance were influencing the performance of the vessel and that one of these, the wave-making component R_w, scaled with $V/\sqrt{L}$ and the other did not. This second component, which is due to viscous effects, derives principally from the flow of the water around the hull but also is influenced by the air flow and weather acting on the above-water surfaces. This second component was termed the frictional resistance R_F.

Froude's major contribution to the ship resistance problem, which has remained useful to the present day, was his conclusion that the two sources of resistance might be separated and treated independently. In this approach, Froude suggested that the viscous resistance could be calculated from frictional data whilst that wave-making resistance R_w could be deduced from the measured total resistance R_T and the calculated frictional resistance R_F as follows:

$$R_W = R_T - R_F \qquad (12.2)$$

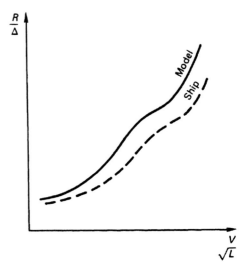

FIGURE 12.1 Comparison of a ship and its model's specific resistance curves.

---

* As from April 1991 AMTE(H) became part of the Defence Research Establishment Agency (DREA) and is now part of Qinetiq.

In order to provide data for calculating the value of the frictional component Froude performed his famous experiments at the Admiralty-owned model tank at Torquay. These experiments entailed towing a series of planks ranging from 10–50 ft in length, having a series of surface finishes of shellac varnish, paraffin wax, tin foil, graduation of sand roughness and other textures. Each of the planks was 19 in. deep and $\frac{3}{16}$ in. thick and was ballasted to float on its edge. Although the results of these experiments suffered from errors due to temperature differences, slight bending of the longer planks and laminar flow on some of the shorter planks, Froude was able to derive an empirical formula which would act as a basis for the calculation of the frictional resistance component R_F in equation (12.2). The relationship Froude derived took the form

$$R_F = fSV^n \quad (12.3)$$

in which the index n had the constant value of 1.825 for normal ship surfaces of the time and the coefficient f varied with both length and roughness, decreasing with length but increasing with roughness. In equation (12.3), S is the wetted surface area.

As a consequence of this work Froude's basic procedure for calculating the resistance of a ship is as follows:

1. Measure the total resistance of the geometrically similar model R_{TM} in the towing tank at a series of speeds embracing the design $V/\sqrt{L}$ of the full-size vessel.
2. From this measured total resistance subtract the calculated frictional resistance values for the model R_{FM} in order to derive the model wave-making resistance.
3. Calculate the full-size frictional resistance R_{FS} and add this to the full-size wave-making resistance R_{WS}, scaled from the model value, to obtain the total full-size resistance R_{TS}

$$R_{TS} = R_{WM}\left(\frac{\Delta_S}{\Delta_M}\right) + R_{FS} \quad (12.4)$$

In equation (12.4) the suffixes M and S denote model- and full-scale, respectively and Δ is the displacement.

The scaling law of the ratio of displacements derives from Froude's observations that when models of various sizes, or a ship and its model, were run at corresponding speeds dictated by equation (12.1), their resistances would be proportional to the cubes of their linear dimensions or, alternatively, their displacements. This was, however, an extension of a law of comparison which was known at that time.

Froude's law, equation (12.1), states that the wave-making resistance coefficients of two geometrically similar hulls of different lengths are the same when moving at the same $V/\sqrt{L}$ value, V being the ship or model speed and L being the waterline length. The ratio $V/\sqrt{L}$ is termed the speed-length ratio and is of course dimensional; however, the dimensionless Froude number can be derived from it to give

$$F_n = \frac{V}{\sqrt{(gL)}} \quad (12.5)$$

in which g is the acceleration due to gravity (9.81 m/s^2). Care needs to be exercised in converting between the speed length ratio and the Froude number:

$$F_n = 0.3193 \frac{V}{\sqrt{L}} \quad \text{where } V \text{ is in m/s; } L \text{ is in meters}$$

$$F_n = 0.1643 \frac{V}{\sqrt{L}} \quad \text{where } V \text{ is in knots; } L \text{ is in meters}$$

Froude's work with his plank experiments was carried out prior to the formulation of the Reynolds number criteria and this undoubtedly led to errors in his results: for example, the laminar flow on the shorter planks. Using dimensional analysis, after the manner shown in Chapter 6, it can readily be shown today that the resistance of a body moving on the surface, or at an interface of a medium, can be given by

$$\frac{R}{\rho V^2 L^2} = \phi\left\{\frac{VL\rho}{\mu}, \frac{V}{\sqrt{(gL)}}, \frac{V}{a}, \frac{\sigma}{g\rho L^2}, \frac{p_0 - p_v}{\rho V^2}\right\} \quad (12.6)$$

In this equation the left-hand side term is the resistance coefficient C_R while on the right-hand side of the equation:

The 1st term is the Reynolds number R_n.
The 2nd term is the Froude number F_n (equation (12.5)).
The 3rd term is the Mach number M_a.
The 4th term is the Weber number W_e.
The 5th term is the Cavitation number.

For the purposes of ship propulsion the 3rd and 4th terms are not generally significant and can, therefore, be neglected. Hence equation (12.6) reduces to the following for all practical ship purposes:

$$C_R = \phi\{R_n, F_n, \sigma_0\} \quad (12.7)$$

In which,

ρ is the density of the water
μ is the dynamic viscosity of the water
p_0 is the free stream undisturbed pressure
p_v is the water vapor pressure.

12.2 COMPONENTS OF CALM WATER RESISTANCE

In the case of a vessel which is undergoing steady motion at slow speeds, that is where the ship's weight balances the

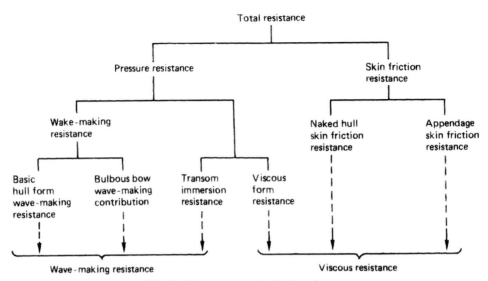

FIGURE 12.2 Components of ship resistance.

displacement upthrust without the significant contribution of hydrodynamic lift forces, the components of calm water resistance can be broken down into the contributions shown in Figure 12.2. From this figure it is seen that the total resistance can be decomposed into two primary components, pressure and skin friction resistance, and these can then be broken down further into more discrete components. In addition to these components there is of course the added resistance due to rough weather and air resistance: these are, however, dealt with separately in Sections 12.5 and 12.8, respectively.

Each of the components shown in Figure 12.2 can be studied separately, provided that it is remembered that each will have an interaction on the others and, therefore, as far as the ship is concerned, need to be considered in an integrated way.

12.2.1 Wave-Making Resistance R_W

Lord Kelvin[3-5] in 1904 studied the problem of the wave pattern caused by a moving pressure point. He showed that the resulting wave system comprises a divergent set of waves together with a transverse system which are approximately normal to the direction of motion of the moving point. Figure 12.3 shows the system of waves so

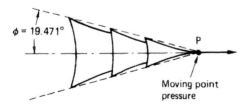

FIGURE 12.3 Wave pattern induced by a moving-point pressure in calm water.

formed. The pattern of waves is bounded by two straight lines which in deep water are at an angle ϕ to the direction of motion of the point, where ϕ is given by

$$\phi = \sin^{-1}\left(\frac{1}{3}\right) = 19.471°$$

The interference between the divergent and transverse systems gives the observed wave their characteristic shape and, since both systems move at the same speed, the speed of the vessel, the wavelength λ between successive crests is

$$\lambda = \frac{2\pi}{g}V^2 \qquad (12.8)$$

The height of the wave systems formed decreases fairly rapidly as they spread out laterally because the energy contained in the wave is constant and it has to be spread out over an increasingly greater length. More energy is absorbed by the transverse system than by the divergent system, and this disparity increases with increasing speed.

A real ship form, however, cannot be represented adequately by a single moving pressure point as analyzed by Kelvin. The simplest representation of a ship, Figure 12.4(a), is to place a moving pressure field near the bow in order to simulate the bow wave system, together with a moving suction field near the stern to represent the stern wave system. In this model the bow pressure field will create a crest near the bow, observation showing that this occurs at about $\lambda/4$ from the bow, whilst the suction field will introduce a wave trough at the stern: both of these wave systems have a wavelength $\lambda = 2\pi V^2/g$. Figure 12.4(b) shows a photograph of the comparable wave system generated by a twin-screw passenger ferry.

The divergent component of the wave system derived from the bow and the stern generally do not exhibit any

Ship Resistance and Propulsion

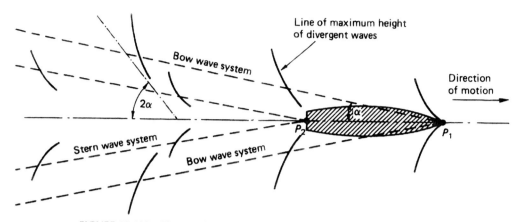

FIGURE 12.4(a) Simple ship wave pattern representation by two pressure points.

FIGURE 12.4(b) Photograph of the wave pattern developed by a ferry. *Source Unknown.*

strong interference characteristics. This is not the case, however, with the transverse wave systems created by the vessel, since these can show a strong interference behavior. Consequently, if the bow and stern wave systems interact such that they are in phase a reinforcement of the transverse wave patterns occurs at the stern and large waves are formed in that region. For such a reinforcement to take place, Figure 12.5(a), the distance between the first crest at the bow and the stern must be an odd number of half-wavelengths as follows:

$$L - \frac{\lambda}{4} = k\frac{\lambda}{2} \quad \text{where } k = 1, 3, 5,\ldots, (2j+1)$$
$$\text{with } j = 0, 1, 2, 3,\ldots$$

From which

$$\frac{4}{2k+1} = \frac{\lambda}{L} = \frac{2\pi V^2}{gL} = 2\pi(F_n)^2$$

that is,

$$F_n = \sqrt{\frac{2}{\pi(2k+1)}} \quad (12.9)$$

For the converse case, when the bow and stern wave systems cancel each other, and hence produce a minimum wave-making resistance condition, the distance $L - \lambda/4$ must be an even number of half wave lengths (Figure 12.5(b)):

$$L - \frac{\lambda}{4} = k\frac{\lambda}{2} \quad \text{where } k = 2, 4, 6,\ldots, (2j)$$
$$\text{with } j = 0, 1, 2, 3,\ldots$$

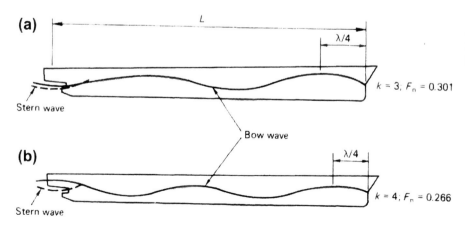

FIGURE 12.5 Wave reinforcement and cancellation at stern: (a) wave reinforcement at stern and (b) wave cancellation at stern.

Hence

$$F_n = \sqrt{\frac{2}{\pi(2k+1)}}$$

as before, but with k even in this case.

Consequently from equation (12.9), Table 12.1 can be derived, which for this particular model of wave action identifies the Froude numbers at which reinforcement (humps) and cancellation (hollows) occur in the wave-making resistance. Each of the conditions shown in Table 12.1 relates sequentially to maximum and minimum conditions in the wave-making resistance curves. The 'humps' occur because the wave profiles and hence the wave-making resistance are at their greatest in these conditions whilst the converse is true in the case of the 'hollows'. Figure 12.6 shows the general form of the wave-making resistance curve together with the schematic wave profiles associated with the various values of k. The hump associated with $k = 1$ is normally termed the 'main hump' since this is the most pronounced hump and occurs at the highest speed. The second hump, $k = 3$, is called the 'prismatic hump' since it is influenced considerably by the prismatic coefficient of the particular hull form.

The derivation of Figure 12.6 and Table 12.1 relies on the assumptions made in their formulation: for example, a single pressure and suction field, bow wave crest at $\lambda/4$; stern trough exactly at the stern, etc. Clearly, there is some latitude in all of these assumptions, and therefore the values of F_n at which the humps and hollows occur vary. In the case of warships the distance between the first crest of the bow wave and the trough of the stern wave has been shown to approximate well to $0.9L$, and therefore this could be used to re-derive equation (12.9). This would then derive slightly differing values of Froude numbers corresponding to the 'humps' and 'hollows'. Table 12.2 shows these differences, and it is clear that the greatest effect is formed at low values of k. Figure 12.6, for this and the other reasons

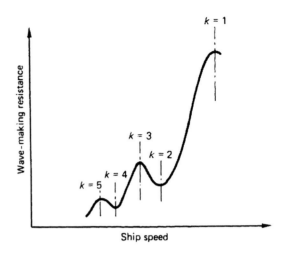

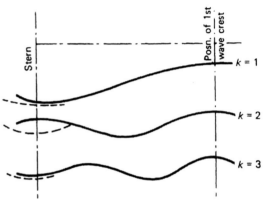

FIGURE 12.6 Form of wave-making resistance curve.

cited, is not unique but is shown here to provide awareness and guidance on wave-making resistance variations.

A better approximation to the wave form of a vessel can be made by considering the ship as a solid body rather than two point sources. Wigley initially used a simple parallel body with two pointed ends and showed that the resulting wave pattern along the body could be approximated by the sum of five separate disturbances of the surface (Figure 12.7). From this figure it is seen that a symmetrical disturbance corresponds to the application of Bernoulli's theorem with peaks at the bow and stern and a hollow, albeit with cusps at the start and finish of the parallel middle body, between them. Two wave forms starting with a crest are

TABLE 12.1 Froude Numbers Corresponding to Maxima and Minima in the Wave Making Resistance Component

K	F_n	Description
1	0.461	1st hump in R_w curve
2	0.357	1st hollow in R_w curve
3	0.301	2nd hump in R_w curve
4	0.266	2nd hollow in R_w curve
5	0.241	3rd hump in R_w curve
⋮	⋮	⋮

TABLE 12.2 Effect of Difference in Calculation Basis on Prediction of Hump and Hollow Froude Numbers

k	1	2	3	4	5
$L - \lambda/4$ basis	0.46	0.36	0.30	0.27	0.24
$0.9L$ basis	0.54	0.38	0.31	0.27	0.24

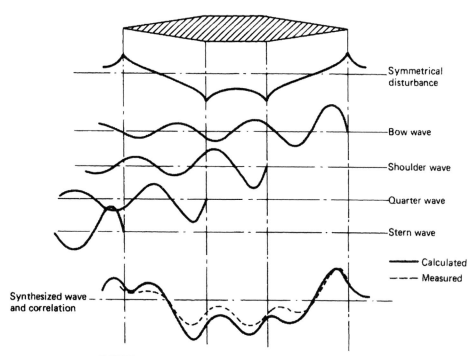

FIGURE 12.7 Components of wave systems for a simple body.

formed by the action of the bow and stern whilst a further two wave forms commencing with a trough originates from the shoulders of the parallel middle body. The sum of these five wave profiles is shown at the bottom of Figure 12.7 and compared with a measured profile which shows good general agreement. Since the wavelength λ varies with speed and the points at which the waves originate are fixed, it is easy to understand that the whole profile of the resultant wave form will change with speed length ratio.

This analysis was extended by Wigley for a more realistic hull form comprising a parallel middle body and two convex extremities. Figure 12.8 shows the results in terms of the same five components and the agreement with the observed wave form.

Considerations of this type lead to endeavoring to design a hull form to produce a minimum wave-making resistance using theoretical methods. The basis of these theories is developed from Kelvin's work on a traveling pressure source; however, the mathematical boundary conditions are difficult to satisfy with any degree of precision. Results of work based on these theories have been mixed in terms of their ability to represent the observed wave forms.

12.2.2 The Contribution of the Bulbous Bow

Bulbous bows are today commonplace in the design of ships. Although their origins are to be found before the turn of the last century, the first application appears to have been in 1912 by the US Navy. The general use in merchant applications appears to have waited until the late 1950s and early 1960s.

The basic theoretical work on their effectiveness was carried out by Wigley[6] in which he showed that if the bulb was nearly spherical in form, then the acceleration of the flow over the surface induces a low-pressure region which can extend towards the water surface. This low-pressure region then reacts with the bow pressure wave to cancel or reduce the effect of the bow wave. The effect of the bulbous bow, therefore, is to cause a reduction, in the majority of cases, of the effective power required to propel the vessel, the effective power P_E being defined as the product of the ship resistance and the ship speed at a particular condition in the absence of the propeller. Figure 12.9 shows a typical example of the effect of a bulbous bow from which it can be seen that a bulb is, in general, beneficial above a certain speed and gives a penalty at low speeds. This is due to the balance between the bow pressure wave reduction effect and increase in frictional resistance caused by the presence of the bulb on the hull.

The effects of the bulbous bow in changing the resistance and delivered power characteristics can be attributed to several causes. The principal of these are as follows:

1. The reduction of bow pressure wave due to the pressure field created by the bulb and the consequent reduction in wave-making resistance.
2. The influence of the upper part of the bulb and its intersection with the hull to introduce a downward flow component in the vicinity of the bow.

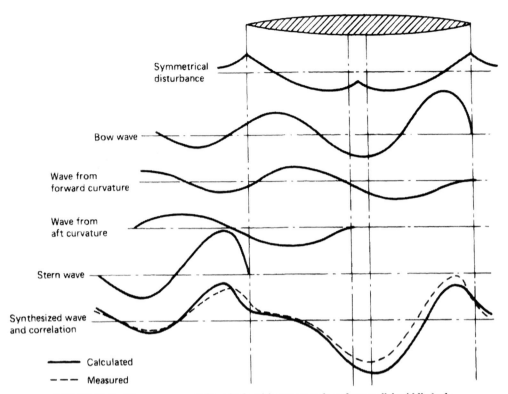

FIGURE 12.8 Wave components for a body with convex ends and a parallel middle body.

3. An increase in the frictional resistance caused by the surface area of the bulb.
4. A change in the propulsion efficiency induced by the effect of the bulb on the global hull flow field.
5. The change induced in the wave breaking resistance.

The shape of the bulb is particularly important in determining its beneficial effect. The optimum shape for a particular hull depends on the Froude number associated with its operating regime and bulbous bows tend to give good performance over a narrow range of ship speeds.

Consequently, they are most commonly found on vessels which operate at clearly defined speeds for much of their time. The actual bulb form, Figure 12.10, is defined in relation to a series of form characteristics as follows:

1. length of projection beyond the forward perpendicular;
2. cross-sectional area at the forward perpendicular (A_{BT});
3. height of the centroid of cross-section A_{BT} from the base line (h_B);
4. bulb section form and profile;
5. transition of the bulb into the hull.

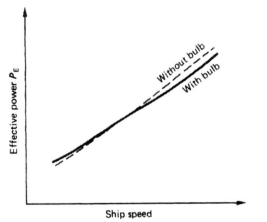

FIGURE 12.9 Influence of a bulbous bow of the effective power requirement.

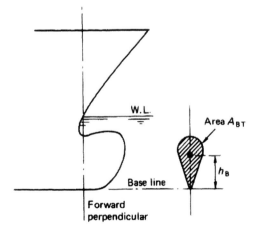

FIGURE 12.10 Bulbous bow definition.

With regard to section form many bulbs today are designed with non-circular forms so as to minimize the effects of slamming in poor weather. There is, however, still considerable work to be done in relating bulb form to power saving and much contemporary work is proceeding. For current design purposes reference can be made to the work of Inui,[7] Todd,[8] Yim[9] and Schneekluth.[10]

In addition to its hydrodynamic behavior the bulb also introduces a further complication into resistance calculations. Traditionally the length along the waterline has formed the basis of many resistance calculation procedures because it is basically the fundamental hydrodynamic dimension of the vessel. The bulbous bow, however, normally projects forward of the forward point of the definition of the waterline length and since the bulb has a fundamental influence on some of the resistance components, there is a case for redefining the basic hydrodynamic length parameter for resistance calculations.

Bulbous bows are only really effective over a limited range of draught conditions due to their interaction with the bow pressure wave. Consequently, when extreme changes in draught are required, such as with a tanker between loaded and ballast conditions, then cylindrical bow forms are contemplated: these being somewhat of a two-dimensional approximation to a conventional three-dimensional bulbous bow form.

12.2.3 Transom Immersion Resistance

In modern ships a transom stern is now normal practice. If at the design powering condition a portion of the transom is immersed, this leads to separation taking place as the flow from under the transom passes out beyond the hull (Figure 12.11). The resulting vorticity that takes place in the separated flow behind the transom leads to a pressure loss behind the hull which is taken into account in some analysis procedures.

The magnitude of this resistance is generally small and, of course, vanishes when the lower part of the transom is dry. Transom immersion resistance is largely a pressure resistance that is scale independent.

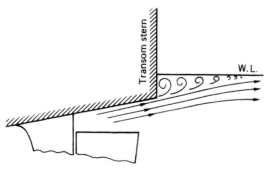

FIGURE 12.11 Flow around an immersed transom stern.

12.2.4 Viscous form Resistance

The total drag on a body immersed in a fluid and traveling at a particular speed is the sum of the skin friction components, which is equal to the integral of the shear stresses taken over the surface of the body, and the form drag, which is the axial component of the integral of the normal forces acting on the body.

In an inviscid fluid the flow along any streamline is governed by Bernoulli's equation and the flow around an arbitrary body is predictable in terms of the changes between pressure and velocity over the surface. In the case of Figure 12.12(a) this leads to the net axial force in the direction of motion being equal to zero since in the two-dimensional case shown in Figure 12.12(a),

$$\oint p \cos \theta \, ds = 0 \qquad (12.10)$$

When moving in a real fluid, a boundary layer is created over the surface of the body which, in the case of a ship, will be turbulent and is also likely to separate at some point in the afterbody. The presence of the boundary layer and its growth along the surface of the hull modifies the pressure distribution acting on the body from that of the potential or inviscid case. As a consequence, the left-hand side of equation (12.10) can no longer equal zero and the viscous form drag R_{VF} is defined for the three-dimensional case of a ship hull as

$$R_{VF} = \sum_{k=1}^{n} p_k \cos \theta_k \delta S_k \qquad (12.11)$$

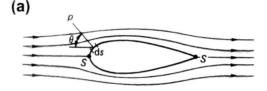

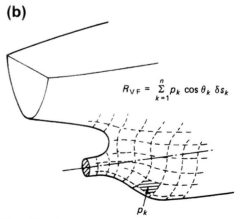

FIGURE 12.12 Viscous form resistance calculation: (a) inviscid flow case on an arbitrary body and (b) pressures acting on shell plate of a ship.

in Figure 12.12 the hull has been split into n elemental areas δS_k and the contribution of each normal pressure p_k acting on the area is summed in the direction of motion (Figure 12.12(b)).

Equation (12.11) is a complex equation to solve since it relies on the solution of the boundary layer over the vessel and this is a solution which can only be approached using significant computational resources for comparatively simple hull forms. As a consequence, for many practical purposes the viscous form resistance is normally accounted for using empirical or pseudo-empirical methods.

12.2.5 Naked Hull Skin Friction Resistance

The original data upon which to calculate the skin friction component of resistance was that provided by Froude in his plank experiments conducted at Torquay. This data, as discussed previously, was subject to error and in 1932 Schoenherr re-evaluated Froude's original data in association with other work in the light of the Prandtl–von Karman theory. This analysis resulted in an expression for the friction coefficient C_F as a function of Reynolds number R_n and the formulation of a skin friction line, applicable to smooth surfaces, of the following form:

$$\frac{0.242}{\sqrt{C_F}} = \log(R_n \cdot C_F) \quad (12.12)$$

This equation, known as the Schoenherr line, was adopted by the American Towing Tank Conference (ATTC) in 1947 and in order to make the relationship applicable to the hull surfaces of new ships an additional allowance of 0.0004 was added to the smooth surface values of C_F given by equation (12.12). By 1950 there was a variety of friction lines for smooth turbulent flows in existence and all, with the exception of Froude's work, were based on Reynolds number. Phillips-Birt[11] provides an interesting comparison of these friction formulations for a Reynolds number of 3.87×10^9 which is applicable to ships of the length of the former trans-Atlantic liner *Queen Mary* and is rather less than that for the large supertankers: in either case lying way beyond the range of direct experimental results. The comparison is shown in Table 12.3, from which it is seen that close agreement is seen to exist between most of the results except for the Froude and Schoenherr modified line. These last two friction lines, while giving comparable results, include a correlation allowance in their formulation. Indeed the magnitude of the correlation allowance is striking between the two Schoenherr formulations: the allowance is some 30 per cent of the basic value.

In the general application of the Schoenherr line some difficulty was experienced in the correlation of large and small model test data and wide disparities in the correlation factor C_A were found to exist upon the introduction of all

TABLE 12.3 Comparison of C_F Values for Different Friction Lines for a REYNOLDS NUMBER $R_n = 3.87 \times 10^9$ (Taken From Reference 11)

Friction Line	C_F
Gerbers	0.00134
Prandtl–Schlichting	0.00137
Kemph–Karham	0.00103
Telfer	0.00143
Lackenby	0.00140
Froude	0.00168
Schoenherr	0.00133
Schoenherr + 0.0004	0.00173

welded hulls. These shortcomings were recognized prior to the 1957 International Towing Tank Conference (ITTC) and a modified line was accepted. The ITTC–1957 line is expressed as

$$C_F = \frac{0.075}{(\log_{10} R_n - 2.0)^2} \quad (12.13)$$

and this formulation, which is in use with most ship model basins, is shown together with the Schoenherr line in Figure 12.13. It can be seen that the present ITTC line gives slightly higher values of C_F at the lower Reynolds numbers than the Schoenherr line whilst both lines merge towards the higher values of R_n.

The frictional resistance R_F derived from the use of either the ITTC or ATTC lines should be viewed as an instrument of the calculation process rather than producing a definitive magnitude of the skin friction associated with a particular ship. As a consequence, when using a Froude analysis based on these, or indeed any friction line data, it is necessary to introduce a correlation allowance into the

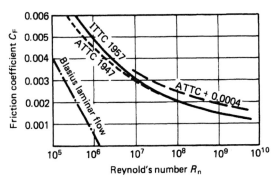

FIGURE 12.13 Comparison of ITTC (1957) and ATTC (1947) friction lines.

calculation procedure. This allowance is denoted by C_A and is defined as

$$C_A = C_{T(\text{measured})} - C_{T(\text{estimated})} \qquad (12.14)$$

In this equation, as in the previous equation, the resistance coefficients C_T, C_F, C_W and C_A are non-dimensional forms of the total, frictional, wave-making and correlation resistances, and are derived from the basic resistance summation

$$R_T = R_W + R_V$$

by dividing this equation throughout by $\frac{1}{2}\rho V_s^2 S, \frac{1}{2}\rho V_s^2 L^2$ or $\frac{1}{2}\rho V_s^2 \nabla^{2/3}$ according to convenience.

12.2.6 Appendage Skin Friction

The appendages of a ship such as the rudder, bilge keels, stabilizers, sea chest openings, duct head-box arrangements, transverse thruster orifices and so on introduce a skin friction resistance above that of the naked hull resistance.

At ship scale the flow over the appendages is turbulent, whereas at model scale it would normally be laminar unless artificially stimulated, which in itself may introduce a flow modeling problem. In addition, many of the hull appendages are working wholly within the boundary layer of the hull and since the model is run at Froude identity, not Reynolds identity, this again presents a problem. As a consequence the prediction of appendage resistance needs care if significant errors are to be avoided. The calculation of this aspect is further discussed in Section 12.3.

In addition to the skin friction component of appendage resistance, if the appendages are located on the vessel close to the surface then they will also contribute to the wave-making component since a lifting body close to a free surface, due to the pressure distribution around the body, will create a disturbance on the free surface. As a consequence, the total appendage resistance can be expressed as the sum of the skin friction and surface disturbance effects as follows:

$$R_{APP} = R_{APP(F)} + R_{APP(W)} \qquad (12.15)$$

where $R_{APP(F)}$ and $R_{APP(W)}$ are the frictional and wave-making components, respectively, of the appendages. In most cases of practical interest to the merchant marine $R_{APP(W)} \simeq 0$ and can be neglected. This is not the case, however, for some naval applications, such as where submarine hydrofoils are operating just submerged near the surface.

12.2.7 Viscous Resistance

Figure 12.2 defines the viscous resistance as being principally the sum of the form resistance, the naked hull skin friction and the appendage resistance. In the discussion on the viscous form resistance it was said that its calculation by analytical means was a complex matter and for many hulls of a complex shape this is difficult with any degree of accuracy at the present time.

Hughes[12] attempted to provide a better empirical foundation for the viscous resistance calculation by devising an approach which incorporated the viscous form resistance and the naked hull skin friction. To form a basis for this approach Hughes undertook a series of resistance tests using planks and pontoons for a range of Reynolds numbers up to a value of 3×10^8. From the results of this experimental study Hughes established that the frictional resistance coefficient C_F could be expressed as a unique inverse function of aspect ratio AR and, moreover, that this function was independent of Reynolds number. The function derived from this work had the form:

$$C_F = C_F|_{AR=\infty} \cdot f\left(\frac{1}{AR}\right)$$

in which the term $C_F|_{AR=\infty}$ is the frictional coefficient relating to a two-dimensional surface; that is, one having an infinite aspect ratio.

This function permitted Hughes to construct a two-dimensional friction line defining the frictional resistance of turbulent flow over a plane smooth surface. This took the form

$$C_F|_{AR=\infty} = \frac{0.066}{[\log_{10} R_n - 2.03]^2} \qquad (12.16)$$

Equation (12.16) quite naturally bears a close similarity to the ITTC−1957 line expressed by equation (12.13). The difference, however, is that the ITTC and ATTC lines contain some three-dimensional effects, whereas equation (12.16) is defined as a two-dimensional line. If it is plotted on the same curve as the ITTC line, it will be found that it lies just below the ITTC line for the full range of R_n and in the case of the ATTC line it also lies below it except for the very low Reynolds numbers.

Hughes proposed the calculation of the total resistance of a ship using the basic relationship

$$C_T = C_V + C_W$$

in which

$$C_V = C_F|_{AR=\infty} + C_{FORM},$$

there by giving the total resistance as

$$C_T = C_F|_{AR=\infty} + C_{FORM} + C_W \qquad (12.17)$$

in which C_{FORM} is a 'form' resistance coefficient which takes into account the viscous pressure resistance of the ship. In this approach the basic skin friction resistance coefficient can be determined from equation (12.16). To determine the form resistance, the ship model can be run at

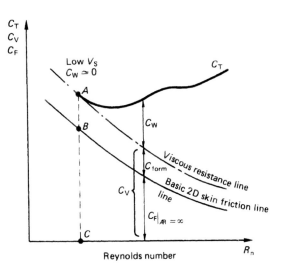

FIGURE 12.14 Hughes model of ship resistance.

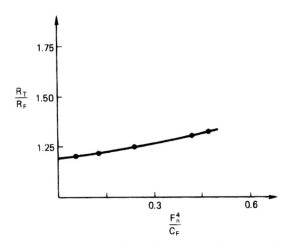

FIGURE 12.15 Determination of $(1 + k)$ using Prohaska method.

a very slow speed when the wave-making component is very small and can be neglected. When this occurs, that is to the left of point A in Figure 12.14, then the resistance curve defines the sum of the skin friction and form resistance components. At the point A, when the wave making resistance is negligible, the ratio

$$\frac{AC}{BC} = \frac{\text{viscous resistance}}{\text{skin friction resistance}}$$

$$= \frac{\text{skin friction resistance} + \text{viscous form resistance}}{\text{skin friction resistance}}$$

$$= 1 + \frac{\text{viscous form resistance}}{\text{skin friction resistance}}$$

and if $\quad k = \dfrac{\text{viscous form resistance}}{\text{skin friction resistance}}$

then $\dfrac{AC}{BC} = (1 + k)$

(12.18)

In equation (12.18), $(1 + k)$ is termed the form factor and it is assumed constant for both the ship and its model. Indeed the form factor is generally supposed to be independent of speed and scale in the resistance extrapolation method. In practical cases the determination of $(1 + k)$ is normally carried out using a variant of the Prohashka method by a plot of C_T against F_n^4 and extrapolating the curve to $F_n = 0$ (Figure 12.15). From this figure the form factor $(1 + k)$ is deduced from the relationship

$$1 + k = \lim_{F_n \to 0} \left(\frac{R}{R_F} \right)$$

This derivation of the form factor can be used in the resistance extrapolation only if scale-independent pressure resistance is absent; for example, there must be no immersion of the transom and slender appendages which are oriented to the direction of flow.

Although traditionally the form factor $(1 + k)$ is treated as a constant with varying Froude number the fundamental question remains as to whether it is valid to assume that the $(1 + k)$ value, determined at vanishing Froude number, is valid at high speed. This is of particular concern at speeds beyond the main resistance hump where the flow configuration around the hull is likely to be very different from that when $F_n = 0$, and, therefore, a Froude number dependency can be expected for $(1 + k)$. In addition a Reynolds dependency may also be expected since viscous effects are the basis of the $(1 + k)$ formulation. The Froude and Reynolds effects are, however, likely to mostly affect the high-speed performance and have a lesser influence on general craft.

The extrapolation from model- to full-scale using Hughes' method is shown in Figure 12.16(a). From this figure it is seen that the two-dimensional skin friction line, equation (12.16), is used as a basis and the viscous resistance is estimated by scaling the basic friction line by the form factor $(1 + k)$. This then acts as a basis for calculating the wave-making resistance from the measured total resistance on the model which is then equated to the ship condition along with the recalculated viscous resistance for the ship Reynolds number. The Froude approach (Figure 12.16(b)), is essentially the same, except that the frictional resistance is based on one of the Froude, ATTC (equation (12.12)) or ITTC (equation (12.13)) friction lines without a $(1 + k)$ factor. Clearly the magnitude of the calculated wave-making resistance, since it is measured as total resistance minus calculated frictional resistance, will vary according to the friction formulation used. This is also true of the correlation allowances as defined in equation (12.14) and, therefore, the magnitudes of these parameters should always be considered in the context of the approach and experimental facility used.

In practice both the Froude and Hughes approaches are used in model testing; the latter, however, is most

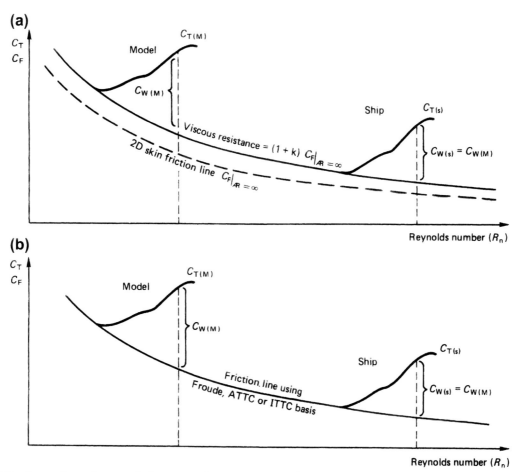

FIGURE 12.16 Comparison of extrapolation approaches: (a) extrapolation using Hughes approach and (b) extrapolation using Froude approach.

frequently used in association with the ITTC–1957 friction formulation rather than equation (12.16).

12.3 METHODS OF RESISTANCE EVALUATION

To evaluate the resistance of a ship the designer has several options available. These range, as shown in Figure 12.17, from what may be termed the traditional methods through to Computational Fluid Dynamics (CFD) methods. The choice of method depends not only on the capability available but also on the accuracy desired, the funds available and the degree to which the approach has been developed. Figure 12.17 identifies four basic classes of approach to the problem; the traditional and standard series, the regression-based procedures; the direct model test and the CFD approach. Clearly these are somewhat artificial distinctions and, consequently, break down on close scrutiny; they are, however, convenient classes for discussion purposes.

Unlike the CFD and direct model test approaches, the other methods are based on the traditional naval architectural parameters of hull form; for example, block coefficient, longitudinal center of buoyancy, prismatic coefficient, etc. These form parameters generally describe the hull form and have served the industry well in the past for resistance calculation purposes. However, as requirements become more exacting and hull forms become more complex these traditional parameters are less able to reflect the growth of the boundary layer and wave-making components. As a consequence an amount of research has been expended in the development of form parameters which will reflect the hull surface contours in a more equable way: in some methods this has extended to around 30 or 40 geometric form parameters.

12.3.1 Traditional and Standard Series Analysis Methods

A comprehensive treatment of these methods would require a book in itself and would also lie to one side of the main theme of this text. As a consequence an outline of four of the traditional methods starting with that of Taylor and passing through Ayer's analysis to the later methods of

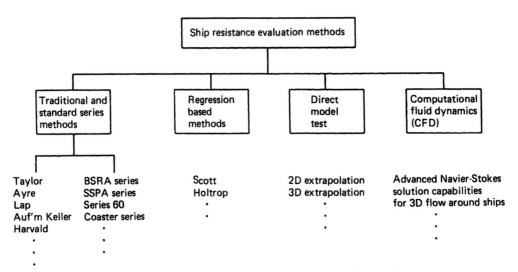

FIGURE 12.17 Ship resistance evaluation methods and examples.

Auf'm Keller and Harvald are presented in outline in order to illustrate the development of this class of methods.

Taylor's Method (1910–1943)

Admiral Taylor in 1910 published the results of model tests on a series of hull forms. This work has since been extended (Reference 13) to embrace a range of $V/\sqrt{L}$ from 0.3–2.0. The series comprised some 80 models in which results are published for beam to draught ratios of 2.25, 3.0 and 3.75 with five displacement length ratios. Eight prismatic coefficients were used spanning the range 0.48–0.80 and this tends to make the series useful for the faster and less full vessels.

The procedure is centered on the calculation of the residual resistance coefficients based on the data for each B/T value corresponding to the prismatic and $V/\sqrt{L}$ values of interest. The residual resistance component C_R is found by interpolation from the three B/T values corresponding to the point of interest. The frictional resistance component is calculated on a basis of Reynolds number and wetted surface area together with a hull roughness allowance. The result of this calculation is added to the interpolated residuary resistance coefficient to form the total resistance coefficient C_T from which the naked effective horsepower is derived for each of the chosen $V/\sqrt{L}$ values from the relation

$$\text{EHP}_n = A C_T V_S^3 \qquad (12.19)$$

where A is the wetted surface area.

Ayre's Method (1942)

Ayre[14] developed a method in 1927, again based on model test data, using a series of hull forms relating to colliers. His approach, which in former years achieved quite widespread use, centers on the calculation of a constant coefficient C_2 which is defined by equation (12.20)

$$\text{EHP} = \frac{\Delta^{0.64} V_S^3}{C_2} \qquad (12.20)$$

This relationship implies that in the case of full-sized vessels of identical forms and proportions, the EHP at corresponding speeds varies as $(\Delta^{0.64} V_S^3)$ and that C_2 is a constant at given values of $V/\sqrt{L}$. In this case the use of $\Delta^{0.64}$ is intended to avoid the necessity to treat the frictional and residual resistances separately for vessels of around 30 m.

The value of C_2 is estimated for a standard block coefficient. Corrections are then made to adjust the standard block coefficient to the actual value and corrections applied to cater for variations in the beam–draught ratio, the position of the l.c.b. and variations in length from the standard value used in the method's derivation.

Auf'm Keller Method

Auf'm Keller[15] extended the earlier work of Lap[16] in order to allow the derivation of resistance characteristics of large block coefficient, single-screw vessels. The method is based on the collated results from some 107 model test results for large single-screw vessels and the measurements were converted into five sets of residuary resistance values. Each of these sets is defined by a linear relationship between the longitudinal center of buoyancy and the prismatic coefficient. Figure 12.18 defines these sets, denoted by the letters A to E, and Figure 12.19 shows the residuary resistance coefficient for set A. As a consequence it is possible to interpolate between the sets for a particular l.c.b. versus C_P relationship.

The procedure adopted is shown in outline form by Figure 12.20 in which the correction for ζ_r and the ship

Chapter | 12 Ship Resistance and Propulsion

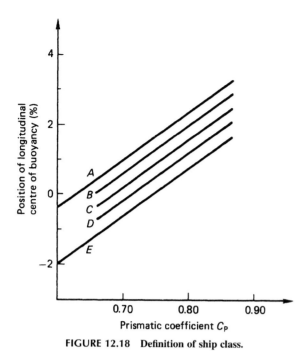

FIGURE 12.18 Definition of ship class.

model correlation C_A are given by equation (12.21) and Table 12.4, respectively:

$$\% \text{ change in } \zeta_r = 10.357[e^{1.129(6.5-L/B)} - 1] \quad (12.21)$$

As in the case of the previous two methods the influence of the bulbous bow is not taken into account but good experience can be achieved with the method within its area of application.

Harvald Method

The method proposed by Harvald[17] is essentially a preliminary power prediction method designed to obtain an estimate of the power required to propel a ship. The approach used is to define four principal parameters upon which to base the estimate; the four selected are:

1. the ship displacement (Δ),
2. the ship speed (V_s),
3. the block coefficient (C_b),
4. the length displacement ratio ($L/\nabla^{1/3}$).

By making such a choice all the other parameters that may influence the resistance characteristics need to be standardized: such as, hull form, B/T ratio, l.c.b., propeller diameter, etc. The method used by Harvald is to calculate the resistance of a standard form for a range of the four parameters cited above and then evaluate the shaft power using a Quasi-Propulsive Coefficient (QPC) based on the wake and thrust deduction method discussed in Chapter 5 and a propeller open water efficiency taken from the Wageningen B Series propellers. The result of this analysis led to the production of seven diagrams for a range of block coefficients from 0.55–0.85 in 0.05 intervals as shown in Figure 12.21. From these diagrams an estimate of the required power under trial conditions can be derived readily with the minimum of

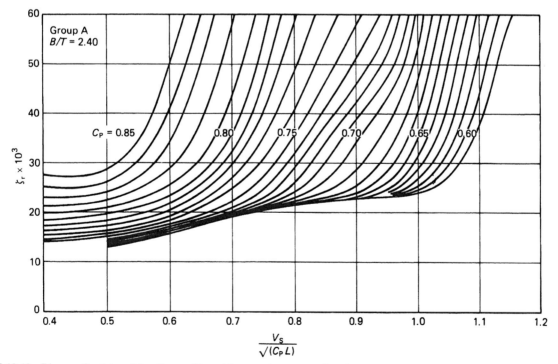

FIGURE 12.19 Diagram for determining the specific residuary resistance as a function of $V_s/\sqrt{(C_p L)}$ and C_p. *Reproduced with permission from Reference 15.*

in the form of model data and more particularly in model data relating to standard series hull forms. That is, those in which the geometric hull form variables have been varied in a systematic way. Much data has been collected over the years and Bowden[18] gives a useful guide to the extent of the data available for single-screw ocean-going ships between the years 1900 and 1969. Some of the more recent and important series and data are given in References 19–31. Unfortunately, there is little uniformity of presentation in the work as the results have been derived over a long period of time in many countries of the world. The designer therefore has to accept this state of affairs and account for this in his calculations. In addition hull form design has progressed considerably in recent years and little of these changes is reflected in the data cited in these references. Consequently, unless extreme care is exercised in the application of this data, significant errors can be introduced into the resistance estimation procedure.

In more recent times the Propulsion Committee[32] of the ITTC have been conducting a cooperative experimental program between tanks around the world. The data so far reported relates to the Wigley parabolic hull and the Series 60, $C_b = 0.60$ hull forms.

12.3.2 Regression-Based Methods

Ship resistance prediction based on statistical regression methods has been a subject of some interest for a number of years. Early work by Scott in the 1970s[33,34] resulted in methods for predicting the trial performance of single- and twin-screw merchant ships.

The theme of statistical prediction was then taken up by Holtrop[35–39] in a series of papers. These papers trace the development of a power prediction method based on the regression analysis of random model and full-scale test data together with, most recently, the published results of the Series 64 high-speed displacement hull terms. In this latest version the regression analysis is now based on the results of some 334 model tests and the results are analyzed on the basis of the ship resistance equation:

$$R_T = R_F(1 + k_1) + R_{APP} + R_W + R_B + R_{TR} + R_A \quad (12.22)$$

In this equation the frictional resistance R_F is calculated according to the ITTC–1957 friction formulation, equation (12.13), and the hull form factor $(1 + k_1)$ is based on a regression equation. It is expressed as a function of afterbody form, breadth, draught, length along the waterline, length of run, displacement, prismatic coefficient:

$$\begin{aligned}(1 + k_1) = {}& 0.93 + 0.487118(1 + 0.011 C_{\text{stern}}) \\ & \times (B/L)^{1.06806}(T/L)^{0.46106} \\ & \times (L_{\text{WL}}/L_R)^{0.121563}(L_{\text{WL}}^3/\nabla)^{0.36486} \\ & \times (1 - C_P)^{-0.604247}\end{aligned} \quad (12.23)$$

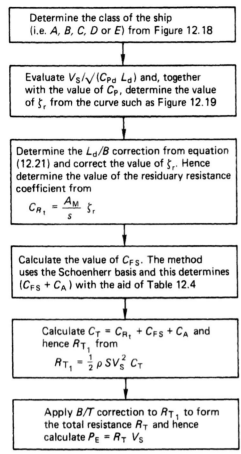

FIGURE 12.20 Auf'm Keller resistance calculation.

effort. However, when using a method of this type it is important to make allowances for deviations of the actual form from those upon which the diagrams are based.

Standard Series Data

In addition to the more formalized methods of analysis there is a great wealth of data available to the designer and analyst

TABLE 12.4 Values of C_A used in Auf'm Keller Method (taken From Reference 15)

Length of Vessel (m)	Ship Model Correlation Allowance
50–150	0.0004 → 0.00035
150–210	0.0002
210–260	0.0001
260–300	0
300–350	−0.0001
350–450	−0.00025

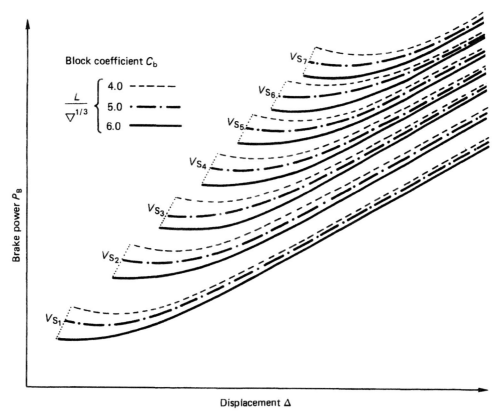

FIGURE 12.21 Harvald estimation diagram for ship power.

in which the length of run L_R, if unknown, is defined by a separate relationship as follows:

$$L_R = L_{WL}\left(1 - C_P + \frac{0.06 C_P \text{ l.c.b.}}{4C_p - 1}\right)$$

The sternshape parameter C_{stern} in equation (12.23) is defined in relatively coarse steps for different hull forms, as shown in Table 12.5.

The appendage resistance according to the Holtrop approach is evaluated from the equation

$$R_{APP} = \frac{1}{2}\rho V_S^2 C_F (1 + k_2)_{equv} \sum S_{APP} + R_{BT} \quad (12.24)$$

where the frictional coefficient C_F of the ship is again determined by the ITTC-1957 line and S_{APP} is the wetted area of the appendages of the vessel. To determine the equivalent $(1 + k_2)$ value of the appendages, denoted by $(1 + k_2)_{equv}$, appeal is made to the relationship

$$(1 + k_2)_{equv} = \frac{\sum (1 + k_2) S_{APP}}{\sum S_{APP}} \quad (12.25)$$

The values of the appendage form factors are tentatively defined by Holtrop as shown in Table 12.6.

In cases where bow thrusters are fitted to the vessel their influence can be taken into account by the term R_{BT} in equation (12.24) as follows:

$$R_{BT} = \pi \rho V_S^2 d_T C_{BTO}$$

in which d_T is the diameter of the bow thruster and the coefficient C_{BTO} lies in the range 0.003–0.012. If the thruster is located in the cylindrical part of the bulbous bow, then $C_{BTO} \rightarrow 0.003$.

The prediction of the wave-making component of resistance has proved difficult and in the last version of Holtrop's method[39] a three-banded approach is proposed to overcome the difficulty of finding a general regression formula. The ranges proposed are based on the Froude number F_n and are as follows:

Range 1: $F_n > 0.55$
Range 2: $F_n < 0.4$
Range 3: $0.4 < F_n < 0.55$

TABLE 12.5 C_{stern} Parameters According to Holtrop

Afterbody Form	Cstern
Pram with gondola	−25
V-shaped sections	−10
Normal section ship	0
U-shaped sections with Hogner stern	10

TABLE 12.6 Tentative Appendage Form Factors $(1 + k_2)$

Appendage Type	$(1 + k_2)$
Rudder behind skeg	1.5–2.0
Rudder behind stern	1.3–1.5
Twin-screw balanced rudders	2.8
Shaft brackets	3.0
Skeg	1.5–2.0
Strut bossings	3.0
Hull bossings	2.0
Shafts	2.0–4.0
Stabilizer fins	2.8
Dome	2.7
Bilge keels	1.4

within which the general form of the regression equations for wave-making resistance in ranges 1 and 2 is

$$R_W = K_1 K_2 K_3 \nabla \rho g \exp[K_4 F_n^{K_6} + K_5 \cos(K_7/F_n^2)] \quad (12.26)$$

The coefficients K_1, K_2, K_3, K_4, K_5, K_6 and K_7 are defined by Holtrop in Reference 39 and it is of interest to note that the coefficient K_2 determines the influence of the bulbous bow on the wave resistance. Furthermore, the difference in the coefficients of equation (12.26) between ranges 1 and 2 above lie in the coefficients K_1 and K_4. To accommodate the intermediate range, range 3, a more or less arbitrary interpolation formula is used of the form

$$R_W = R_W\big|_{F_n=0.4} + \frac{(10 F_n - 4)}{1.5} \times \left[R_W\big|_{F_n=0.55} - R_W\big|_{F_n=0.4} \right] \quad (12.27)$$

The remaining terms in equation (12.22) relate to the additional pressure resistance of the bulbous bow near the surface R_B and the immersed part of the transom R_{TR}. These are defined by relatively simple regression formulae. With regard to the model–ship correlation resistance the most recent analysis has shown the formulation in Reference 38 to predict a value some 9–10 per cent high; however, for practical purposes that formulation is still recommended by Holtrop:

$$R_A = \frac{1}{2} \rho V_S^2 S C_A$$

where

$$C_A = 0.006(L_{WL} + 100)^{-0.16} - 0.005205 \\ + 0.003 \sqrt{(L_{WL}/7.5)} C_B^4 K_2 (0.04 - c_4) \quad (12.28)$$

in which $c_4 = T_F/L_{WL}$ when $T_F/L_{WL} \leq 0.04$
and $c_4 = 0.04$ when $T_F/L_{WL} > 0.04$

where T_F is the forward draught of the vessel and S is the wetted surface area of the vessel.

K_2, which also appears in equation (12.26) and determines the influence of the bulbous bow on the wave resistance is given by

$$K_2 = \exp[-1.89\sqrt{C_3}]$$

where

$$c_3 = \frac{0.56(A_{BT})^{1.5}}{BT(0.31\sqrt{A_{BT}} + T_F - h_B)}$$

in which A_{BT} is the transverse area of the bulbous bow and h_B is the position of the center of the transverse area A_{BT} above the keel line with an upper limit of $0.6 T_F$ (see Figure 12.10).

Equation (12.28) is based on a mean apparent amplitude hull roughness $k_S = 150$ μm. In cases where the roughness may be larger than this use can be made of the ITTC–1978 formulation, which gives the effect of the increase in roughness ΔC_A as

$$\Delta C_A = (0.105 k_S^{1/3} - 0.005579)/L^{1/3} \quad (12.29)$$

The Holtrop method provides a most useful estimation tool for the designer. However, like many analysis procedures of this type it relies to a very large extent on traditional naval architectural geometric parameters. As these parameters cannot fully act as a basis for representing the hull curvature and its effect on the flow around the vessel, there is a natural limitation on the accuracy of the approach without using more complex hull definition parameters.

12.3.3 Direct Model Test

Model testing of a ship in the design stage is an important part of the design process and one that, in a great many instances, is either not explored fully or undertaken. In the author's view this is a false economy, bearing in mind the relatively small cost of model testing as compared to the cost of the ship and the potential costs that can be incurred in design modification to rectify a problem or the through-life costs of a poor performance optimization.

General Procedure for Model Tests

While the detailed procedures for model testing differ from one establishment to another the underlying general procedure is similar. Here the general concepts are discussed, but for a more detailed account reference can be made to Phillips-Birt[11] or to the ongoing ITTC proceedings. With regard to resistance and propulsion testing there are a limited number of experiments that are of interest: the resistance test, the open water propeller test, the propulsion test and the flow visualization test. The measurement of the wake field was discussed in Chapter 5.

Resistance Tests

In the resistance test the ship model is towed by the carriage and the total longitudinal force acting on the model is measured for various speeds (Figure 12.22). The breadth and depth of the towing tank essentially govern the size of the model that can be used. Todd's original criterion that the immersed cross-section of the vessel should not exceed 1 per cent of the tank's cross-sectional area was placed in doubt after the famous *Lucy Ashton* experiment. This showed that to avoid boundary interference from the tank walls and bottom this proportion should be reduced to the order of 0.4 per cent.

The model, constructed from paraffin wax, wood or glass-reinforced plastic, has to be manufactured to a high degree of finish and turbulence simulators placed at the bow of the model in order to stimulate the transition from a laminar into a turbulent boundary layer over the hull. The model is positioned under the carriage and towed in such a way that it is free to heave and pitch, and ballasted to the required draught and trim.

In general there are two kinds of resistance tests: the naked hull and the appended resistance test. If appendages are present local turbulence tripping is applied in order to prevent the occurrence of uncontrolled laminar flow over

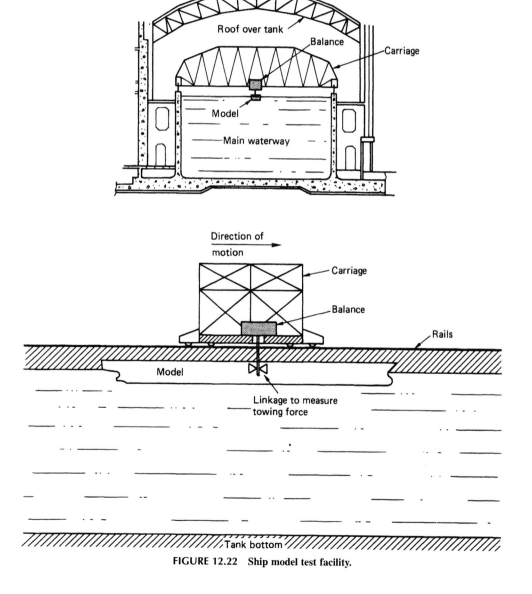

FIGURE 12.22 Ship model test facility.

the appendages. Furthermore, the propeller should be replaced by a streamlined cone to prevent flow separation in this area.

The resistance extrapolation process follows Froude's hypothesis and the similarity law is followed. As such the scaling of the residual, or wave-making component, follows the similarity law

$$R_{W_{ship}} = R_{W_{model}} \lambda^3 (\rho_S/\rho_M)$$

provided that $V_S = V_M\sqrt{\lambda}$, where $\lambda = L_S/L_M$.

In general, the resistance is scaled according to the relationship

$$R_s = [R_M - R_{F_M}(1+k)]\lambda^3 \left(\frac{\rho_S}{\rho_M}\right) + R_{F_S}(1+k) + R_A$$
$$= [R_M - F_D]\lambda^3 \left(\frac{\rho_S}{\rho_M}\right) \quad (12.30)$$

where

$$F_D = \frac{1}{2}\rho_M V_M^2 S_M (1+k)(C_{F_M} - C_{F_S}) - \frac{\rho_M}{\rho_S} R_A/\lambda^3$$

that is,

$$F_D = \frac{1}{2}\rho_M V_M^2 S_M [(1+k)(C_{F_M} - C_{F_S}) - C_A] \quad (12.31)$$

The term F_D is known as both the scale effect correction on resistance and the friction correction force. The term R_A in equation (12.30) is the resistance component, which is intended to allow for the following factors: hull roughness; appendages on the ship but not present during the model experiment; still air drag of the ship and any other additional resistance components acting on the ship but not on the model. As such its non-dimensional form C_A is the incremental resistance coefficient for ship—model correlation.

When $(1+k)$ in equation (12.30) is put to unity, the extrapolation process is referred to as a two-dimensional approach since the frictional resistance is then taken as that given by the appropriate line, Froude flat plate data, ATTC or ITTC−1957, etc.

The effective power (P_E) is derived from the resistance test by the relationship

$$P_E = R_s V_s \quad (12.32)$$

Open Water Tests

The open water test is carried out on either a stock or actual model of the propeller to derive its open water characteristics in order to estimate the ship's propulsion coefficients. The propeller model is fitted on a horizontal driveway shaft and is moved through the water at an immersion of the shaft axis which is frequently equal to the diameter of the propeller (Figure 12.23).

The loading of the propeller is normally carried out by adjusting the speed of advance and keeping the model revolutions constant. However, when limitations in the measuring range, such as a J-value close to zero or a high carriage speed needed for a high J-value, are reached the rate of revolutions is also varied. The measured thrust values are corrected for the resistance of the hub and

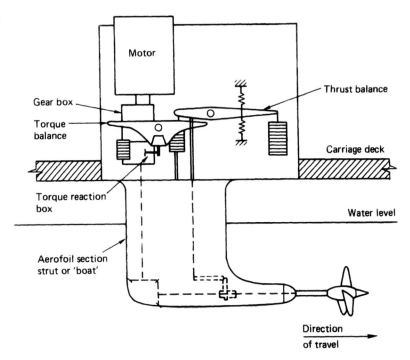

FIGURE 12.23 Propeller open water test using towing tank carriage.

streamlined cap, this correction being determined experimentally in a test using a hub only without the propeller.

The measured torque and corrected thrust are expressed as non-dimensional coefficients K_{QO} and K_{TO} in the normal way (see Chapter 6); the suffix O being used in this case to denote the open rather than the behind condition. The open water efficiency and the advance coefficient are then expressed as

$$\eta_0 = \frac{J}{2\pi} \frac{K_{TO}}{K_{QO}}$$

and

$$J = \frac{V_c}{nD}$$

where V_c is the carriage speed.

Unless explicitly stated it should not be assumed that the propeller open water characteristics have been corrected for scale effects. The data from these tests are normally plotted on a conventional open water diagram together with a tabulation of the data.

Propulsion Tests

In the propulsion test the model is prepared in much the same way as for the resistance experiment and turbulence stimulation on the hull and appendages is again applied. For this test, however, the model is fitted with the propeller used in the open water test together with an appropriate drive motor and dynamometer. During the test the model is free to heave and pitch as in the case of the resistance test.

In the propulsion test the propeller thrust T_M, the propeller torque Q_M and the longitudinal towing force F acting on the model are recorded for each tested combination of model speed V_M and propeller revolutions n_M.

Propulsion tests are carried out in two parts. The first comprises a load variation test at one or sometimes more than one constant speed whilst the other comprises a speed variation test at constant apparent advance coefficient or at the self-propulsion point of the ship. The ship self-propulsion point being defined when the towing force (F) on the carriage is equal to the scale effect correction on viscous resistance (F_D), equation (12.31).

The required thrust T_S and self-propulsion point of the ship is determined from the model test using the equation:

$$T_S = \left[T_M + (F_D - F)\frac{\partial T_M}{\partial F}\right] \lambda^3 \frac{\rho_S}{\rho_M} \quad (12.33)$$

In equation (12.33) the derivative $\partial T_M/\partial F$ is determined from the load variation tests which form the first part of the propulsion test. In a similar way the local variation test can be interpolated to establish the required torque and propeller rotational speed at self-propulsion for the ship.

In the extrapolation of the propulsion test to full-scale the scale effects on resistance (F_D), on the wake field and on the propeller characteristics need to be considered. At some very high speeds the effects of cavitation also need to be taken into account. This can be done by analysis or through the use of specialized facilities.

Flow Visualization Tests

Various methods exist to study the flow around the hull of a ship. One such method is to apply stripes of an especially formulated paint to the model surface, the stripes being applied vertical to the base line at different longitudinal locations. The model is then towed at Froude identity and the paint will smear into streaks along the hull surface in the direction of the flow lines.

In cases where the wall shear stresses are insufficient tufts can be used to visualize the flow over the hull. In general, woolen threads of about 5 cm in length will be fitted onto small needles driven into the hull surface. The tufts will be at a distance of between 1 and 2 cm from the hull surface and the observation made either visually or by using an underwater television camera. Interaction phenomenon between the propeller and ship's hull can also be studied in this way by observing the behavior of the tufts with and without the running propeller.

Model Test Facilities

Many model test facilities exist around the world, almost all of which possess a ship model towing tank. Some of the model facilities available are listed in Table 12.7; this, however, is by no means an exhaustive list of facilities and is included here to give an idea of the range of facilities available.

Two-Dimensional Extrapolation Method

This, as discussed previously, is based on Froude's original method without the use of a form factor. Hence the full-scale resistance is determined from

$$R_S = (R_M - F_D)\lambda^3 \left(\frac{\rho_S}{\rho_M}\right)$$

where

$$F_D = \frac{1}{2}\rho_M V_M^2 S_M (C_{F_M} - C_{F_S} - C_A)$$

and when Froude's friction data is used the value of C_A is set to zero; however, this is not the case if the ATTC−1947 or ITTC−1957 line is used.

When the results of the propulsion test are either interpolated for the condition when the towing force (F) is equal to F_D or when F_D is actually applied in the self-propulsion test the corresponding model condition is termed the 'self-

TABLE 12.7 Examples of Towing Tank Facilities Around the World (Reproduced with permission from Reference 55)

Facilities	Length (m)	Width (m)	Depth (m)	Maximum Carriage Speed (m/s)
European Facilities				
Qinetiq Haslar (UK)	164	6.1	2.4	7.5
	270	12.0	5.5	12.0
Experimental and Electronic Lab.	76	3.7	1.7	9.1
	188	2.4	1.3	13.1
B.H.C. Cowes (UK)	197	4.6	1.7	15.2
MARIN Wageningen (NL)	100	24.5	2.5	4.5
	216	15.7	1.25	5
	220	4.0	4.0	15/30
	252	10.5	5.5	9
MARIN Depressurized Facility, Ede (NL)	240	18.0	8.0	4
Danish Ship Research Laboratories	240	12.0	6.0	14
Ship Research Institute of Norway (NSFI)	27	2.5	1.0	2.6
	175	10.5	5.5	8.0
SSPA. Göteborg, Sweden	260	10.0	5.0	14.0
Bassin d'Essais de Carènes, Paris	155	8.0	2.0	5
	220	13.0	4.0	10
VWS Germany	120	8.0	1.1	4.2
	250	8.0	4.8	20
H.S.V. Hamburg Germany	30	6.0	1.2	0.0023–1.9
	80	4.0	0.7	3.6
	80	5.0	3.0	3.6
	300	18.0	6.0	8.0
B.I.Z. Yugoslavia	37.5	3.0	2.5	3
	23	12.5	6.2	8
	293	5.0	3.5	12
North American Facilities				
NSRDC Bethesda USA	845	15.6	6.7	10
	905	6.4	3.0–4.8	30
NRC, Marine Dynamics and Ship Laboratory, Canada	137	7.6	3.0	8
Far East Facilities				
Meguro Model Basin, Japan	98	3.5	2.25	7
	235	12.5	7.25	10
	340	6.0	3.0	20
Ship Research Institute, Mitaka Japan	20	8.0	0–1.5	2
	50	8.0	4.5	2.5
	140	7.5	0–3.5	6
	375	18.0	8.5	15
KIMM — Korea	223	16.0	7.0	
Hyundai — Korea	232	14.0	6.0	

propulsion point of the ship'. The direct scaling of the model data at this condition gives the condition generally termed the 'tank condition'. This is as follows:

$$\left.\begin{array}{l} P_{DS} = P_{DM}\lambda^{3.5}\left(\dfrac{\rho_S}{\rho_M}\right) \\ T_S = T_M\lambda^3\left(\dfrac{\rho_S}{\rho_M}\right) \\ n_S = n_M/\sqrt{\lambda} \\ V_S = V_M/\sqrt{\lambda} \\ R_S = (R_M - F_D)\lambda^3\left(\dfrac{\rho_S}{\rho_M}\right) \end{array}\right\} \quad (12.34)$$

The power and propeller revolutions determined from the tank condition as given by equation (12.34) require to be converted into a trial prediction for the vessel. In the case of the power trial prediction this needs to be based on an allowance factor derived from the results of trials of comparable ships of the same size or, alternatively, on the results of statistical surveys. The power trial allowance factor is normally defined as the ratio of the shaft power measured on trial to the power delivered to the propeller in the tank condition.

The full-scale propeller revolutions prediction is based on the relationship between the delivered power and the propeller revolutions derived from the tank condition. The power predicted for the trial condition is then used in this relationship to devise the corresponding propeller revolutions. This propeller speed is corrected for the over- or underloading effect and often corresponds to around a 0.5 per cent decrease of rpm for a 10 per cent increase of power. The final stage in the propeller revolutions prediction is to account for the scale effects in the wake and propeller blade friction. For the trial condition these scale effects are of the order of

$\dfrac{1}{2}\sqrt{\lambda}$ per cent for single-screw vessels

1–2 per cent for twin-screw vessels

The allowance for the service condition on rotational speed is of the order 1 per cent.

Three-Dimensional Extrapolation Method

The three-dimensional extrapolation method is based on the form factor concept. Accordingly, the resistance is scaled under the assumption that the viscous resistance of the ship and its model is proportional to the frictional resistance of a flat plate of the same length and wetted surface area when towed at the same speed, the proportionality factor being $(1+k)$ as discussed in Section 12.2. In addition it is assumed that the pressure resistance due to wave generation, stable separation and induced drag from non-streamlined or misaligned appendages follows the Froude similarity law.

The form factor $(1+k)$ is determined for each hull from low-speed resistance or propulsion measurements when the wave resistance components are negligible. In the case of the resistance measurement of form factor then this is based on the Prohaska derivation:

$$(1+k) = \lim_{F_n \to 0}\left(\dfrac{R}{R_F}\right)$$

In the case of the propulsion test acting as a basis for the $(1+k)$ determination then this relationship takes the form

$$(1+k) = \lim_{F_n \to 0}\left[\dfrac{F - T/(\partial T/\partial F)}{(F|_{T=0}/R)R_F}\right]$$

The low-speed measurement of the $(1+k)$ factor can only be validly accomplished if scale-independent pressure resistance is absent. This implies, for example, that there is no immersed transom. In this way the form factor is maintained independent of speed and scale in the extrapolation method.

In the three-dimensional method the scale effect on the resistance is taken as

$$F_D = \dfrac{1}{2}\rho_M V_M^2 S_M[(1+k)(C_{F_M} - C_{F_S}) - C_A]$$

in which the form factor is normally taken relative to the ITTC–1957 line and C_A is the ship–model correlation coefficient. The value of C_A is generally based on an empirical relationship and additional allowances are applied to this factor to account for extreme hull forms at partial draughts, appendages not present on the model, 'contract' conditions, hull roughnesses different from the standard of 150 μm, extreme superstructures or specific experience with previous ships.

In the three-dimensional procedure the measured relationship between the thrust coefficient K_T and the apparent advance coefficient is corrected for wake scale effects and for the scale effects on propeller blade friction. At model-scale the model thrust coefficient is defined as

$$K_{TM} = f(F_n, J)_M$$

whereas at ship-scale this is

$$K_{TS} = f\left(F_n, J\left(\dfrac{1 - w_{TS}}{1 - w_{TM}}\right) + \Delta K_T\right)$$

According to the ITTC–1987 version of the manual for the use of the 1978 performance reduction method, the relationship between the ship and model Taylor wake fractions can be defined as

$$w_{TS} = (t + 0.04) + (w_{TM} - t - 0.04)$$
$$\times \dfrac{(1+k)C_{FS} + \Delta C_F}{(1+k)C_{FM}}$$

where the number 0.04 is included to take account of the rudder effect and ΔC_F is the roughness allowance given by

$$\Delta C_F = \left[105\left(\frac{k_s}{L_{WL}}\right)^{1/3} - 0.64\right] \times 10^{-3}$$

The measured relationship between the thrust and torque coefficient is corrected for the effects of friction over the blades such that $K_{TS} = K_{TM} + \Delta K_T$

and

$$K_{QS} = K_{QM} + \Delta K_Q$$

where the factors ΔK_T and ΔK_Q are determined from the ITTC procedure discussed in Chapter 6.

The load of the full-scale propeller is obtained from the relationship

$$\frac{K_T}{J^2} = \frac{S}{2D^2}\frac{C_{TS}}{(1-t)(1-w_{TS})^2}$$

and with K_T/J^2 as the input value, the full-scale advance coefficient J_{TS} and torque coefficient K_{QTS} are read off from the full-scale propeller characteristics and the following parameters calculated:

$$\begin{aligned}n_S &= \frac{(1-w_{TS})V_S}{J_{TS}D}\\ P_{DS} &= 2\pi\rho D^5 n_S^3\frac{K_{QTS}}{\eta_R}\times 10^{-3}\\ T_S &= \frac{K_T}{J^2}J_{TS}^2\rho D^4 n_S^2\\ Q_S &= \frac{K_{QTS}}{\eta_R}\rho D^5 n_S^2\end{aligned} \quad (12.35)$$

The required shaft power P_S is found from the delivered power P_{DS} using the shafting mechanical efficiency η_S as

$$P_S = P_{DS}/\eta_S$$

12.3.4 Computational Fluid Dynamics

The analysis of ship forms to predict total resistance using the CFD approach is now an important and maturing subject and considerable research effort is being devoted to the topic.

With regard to the wave-making part of the total resistance, provided that the viscous effects are neglected, then the potential flow can be defined by the imposition of boundary conditions at the hull and free surface. The hull conditions are taken into account by placing a distribution of source panels over its surface. The problem comes in satisfying the free surface boundary conditions which ought to be applied at the actual free surface and which, of course, are unknown at the start of the calculation. A solution to this problem was developed by Dawson[40] and is one method in the class of 'slow-ship' theories. With this method the exact free surface condition is replaced by an approximate one that can be applied at a fixed location such as the undisturbed water surface. In such a case a suitable part of the undisturbed free surface is covered with source panels and the source strengths determined so as to satisfy the boundary conditions. Figure 12.24 shows the wave pattern calculated (Reference 41) using a variation of the Dawson approach for a Wigley hull at a Froude number of 0.40.

Free surface models in the CFD process pose problems for integrated solutions for the total resistance estimation. However, methods based on the transportation of species concentration show promise for an integrated CFD solution. These transport models are then solved additionally to the Navier–Stokes equations within the computational code. A typical example of one such model is:

$$\partial/\partial t\int_V C_i dv + \int_S C_i \mathbf{v}.\mathbf{n}\,ds = 0$$
$$\text{with } \rho = \sum\rho_i C_i \text{ and } \mu = \sum\mu_i C_i$$

and where C_i is the transport species concentration in a particular grid domain.

In the case of the viscous resistance the flow field is often considered in terms of three distinct regions: a potential or, more correctly, nearly potential zone, a boundary layer zone for much of the forward part of the hull and a thick boundary layer zone towards the stern of the ship (Figure 12.25). Analysis by CFD procedures has matured significantly in the last few years and in many cases yields good quantitative estimates of frictional resistance. It also enables the designer to gain valuable insights into the flow field around the ship, particularly in the afterbody region where unpleasant vorticity and separation effects may manifest themselves. In these computationally based analyses turbulence modeling has been problematic and while reasonable estimates of the frictional resistance have been made for fine form ships using k-Ω and k-Ω SST models, deployment of the more computationally intensive Reynolds stress models have

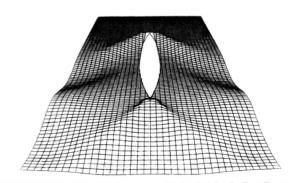

FIGURE 12.24 Calculated wave profile for Wigley hull at $F_n = 0.4$. *Courtesy MARIN.*

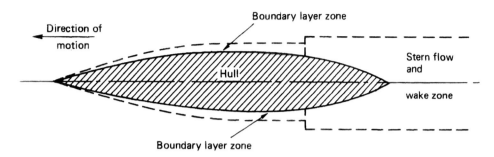

FIGURE 12.25 Zones for CFD analysis.

improved the accuracy of the prediction for the finer hull forms. Moreover, these more advanced models have extended the range of applicability in terms of quantitative estimates of resistance to full-form ships. Such developments, therefore, help to relieve concerns as to where the frictional resistance solution starts to diverge significantly from the true value for a given hull form.

When considering the propulsion aspects of a ship's design the use of a combination of model testing and analysis centered on CFD coupled with sound design experience is advisable. Moreover, notwithstanding the advances that have been made with the mathematical modeling processes, they should not at present replace the conventional model testing procedures for which much correlation data exists: rather they should be used to complement the design approach by allowing the designer to gain insights into the flow dynamics and develop remedial measures before the hull is constructed.

12.4 PROPULSIVE COEFFICIENTS

The propulsive coefficients of the ship performance form the essential link between the effective power required to drive the vessel, obtained from the product of resistance and ship speed, and the power delivered from the engine to the propeller.

The power absorbed by and delivered to the propeller P_D in order to propel the ship at a given speed V_S is

$$P_D = 2\pi nQ \quad (12.36)$$

where n and Q are the rotational speed and torque at the propeller respectively. The torque required to drive the propeller Q can be expressed for a propeller working behind the vessel as

$$Q = K_{Qb}\rho n^2 D^5 \quad (12.37)$$

where K_{Qb} is the torque coefficient of the propeller when working in the wake field behind the vessel at a mean advance coefficient J. By combining equations (12.36) and (12.37) the delivered power can be expressed as

$$P_D = 2\pi K_{Qb}\rho n^3 D^5 \quad (12.38)$$

If the propeller were operating in open water at the same mean advance coefficient J the open water torque coefficient K_{Qo} would be found to vary slightly from that measured behind the ship model. As such the ratio K_{Qo}/K_{Qb} is known as the relative rotative efficiency η_r

$$\eta_r = \frac{K_{Qo}}{K_{Qb}} \quad (12.39)$$

this being the definition stated earlier in Chapter 6.

Hence, equation (12.38) can then be expressed in terms of the relative rotative efficiency as

$$P_D = 2\pi \frac{K_{Qo}}{\eta_r}\rho n^3 D^5 \quad (12.40)$$

Now the effective power P_E is defined as

$$P_E = RV_s$$
$$= P_D QPC$$

where the QPC is termed the quasi-propulsive coefficient.

Hence, from the above and in association with equation (12.40),

$$RV_S = P_D QPC$$
$$= 2\pi \frac{K_{Qo}}{\eta_r}\rho n^3 D^5 QPC$$

which implies that

$$QPC = \frac{RV_S \eta_r}{2\pi K_{Qo}\rho n^3 D^5}$$

Recalling that the resistance of the vessel R can be expressed in terms of the propeller thrust T as $R = T(1 - t)$, where t is the thrust deduction factor as will be explained later. Also from Chapter 5 the ship speed V_s can be defined in terms of the mean speed of advance V_a as $V_a =$

$V_s(1 - w_t)$, where w_t is the mean Taylor wake fraction. Furthermore, since the open water thrust coefficient K_{To} is expressed as $T_o = K_{To}\rho n^2 D^4$, with T_o being the open water propeller thrust at the mean advance coefficient J,

$$\frac{T_o}{K_{To}} = \rho n^2 D^4$$

and the QPC can be expressed from the above as

$$\text{QPC} = \frac{T_o(1-t)V_a K_{To}\eta_r}{(1-w_t)2\pi K_{Qo}nDT_o}$$

which reduces to

$$\text{QPC} = \left(\frac{1-t}{1-w_t}\right)\eta_0\eta_r$$

since, from equation (6.8),

$$\eta_0 = \frac{J}{2\pi}\frac{K_{To}}{K_{Qo}}$$

The quantity $(1-t)/(1-w_t)$ is termed the hull efficiency η_h and hence the QPC is defined as

$$\text{QPC} = \eta_h\eta_0\eta_r \qquad (12.41)$$

or, in terms of the effective and delivered powers,

$$P_E = P_D \text{QPC}$$

that is,

$$P_E = P_D\eta_h\eta_0\eta_r \qquad (12.42)$$

12.4.1 Relative Rotative Efficiency

The relative rotative efficiency (η_r), as defined by equation (12.39), accounts for the differences in torque absorption characteristics of a propeller when operating at similar conditions in a mixed wake and open water flows. In many cases the value of η_r lies close to unity and is generally within the range

$$0.96 \leq \eta_r \leq 1.04$$

In relatively few cases it lies outside this range. Holtrop[39] gives the following statistical relationships for its estimation:

$$\left.\begin{aligned}
&\text{For conventional stern single-screw ships :}\\
&\eta_r = 0.9922 - 0.05908\,(A_E/A_0)\\
&\quad + 0.07424(C_P - 0.0225\,\text{l.c.b.})\\
&\text{For twin-screw ships :}\\
&\eta_r = 0.9737 + 0.111\,(C_P - 0.0225\,\text{l.c.b.})\\
&\quad - 0.06325\,P/D
\end{aligned}\right\}$$

If resistance and propulsion model tests are performed, then the relative rotative efficiency is determined at model-scale from the measurements of thrust T_m and torque Q_m with the propeller operating behind the model. Using the non-dimensional thrust coefficient K_{Tm} as input data the values of J and K_{Qo} are read off from the open water curve of the model propeller used in the propulsion test. The torque coefficient of the propeller working behind the model is derived from

$$K_{Qb} = \frac{Q_M}{\rho n^2 D^5}$$

Hence the relative rotative efficiency is calculated as

$$\eta_r = \frac{K_{Qo}}{K_{Qb}}$$

The relative rotative efficiency is assumed to be scale independent.

12.4.2 Thrust Deduction Factor

When water flows around the hull of a ship which is being towed and does not have a propeller installed, a certain pressure field is set up which is dependent on the hull form. If the same ship is now fitted with a propeller and is propelled at the same speed the pressure field around the hull changes due to the action of the propeller. The propeller increases the velocities of the flow over the hull surface and hence reduces the local pressure field over the after part of the hull surface. This has the effect of increasing, or augmenting, the resistance of the vessel from that which was measured in the towed resistance case and this change can be expressed as

$$T = R(1 + a_r) \qquad (12.44)$$

where T is the required propeller thrust and a_r is the resistance augmentation factor. An alternative way of expressing equation (12.44) is to consider the deduction in propeller effective thrust which is caused by the change in pressure field around the hull. In this case the relationship

$$R = T(1 - t) \qquad (12.45)$$

applies, in which t is the thrust deduction factor. The correspondence between the thrust deduction factor and the resistance augmentation factor can be derived from equations (12.44) and (12.45) as being

$$a_r = \left(\frac{t}{1-t}\right)$$

If a resistance and propulsion model test has been performed, then the thrust deduction factor can be readily calculated from the relationship defined in the ITTC–1987 proceedings

$$t = \frac{T_M + F_D - R_c}{T_M}$$

in which T_M and F_D are defined previously and R_c is the resistance corrected for differences in temperature between the resistance and propulsion tests:

$$R_c = \frac{(1+k)C_{FMC} + C_R}{(1+k)C_{FM} + C_R} R_{TM}$$

where C_{FMC} is the frictional resistance coefficient at the temperature of the self-propulsion test.

In the absence of model tests an estimate of the thrust deduction factor can also be obtained from the work of Holtrop[39] and Harvald.[17] In the Holtrop approach the following regression-based formulas are given:

For single-screw ships :

$$\left.\begin{array}{l} t = \dfrac{0.25014(B/L)^{0.28956}(\sqrt{(B/T)}/D)^{0.2624}}{(1 - C_P + 0.0225 \text{ l.c.b.})^{0.01762}} \\ \quad + 0.0015 C_{stern} \\[6pt] \text{For twin-screw ships :} \\ t = 0.325 C_B - 0.1888 D/\sqrt{(BT)} \end{array}\right\} \quad (12.46)$$

In equation (12.46) the value of the parameter C_{stern} is found from Table 12.5.

The alternative to this approach is that of Harvald for the calculation of the thrust deduction factor. This assumes that it comprises three separate components as follows:

$$t = t_1 + t_2 + t_3 \qquad (12.47)$$

in which t_1, t_2 and t_3 are basic values derived from hull form parameters, a hull form correction and a propeller diameter correction, respectively. The values of these parameters for single-screw ships are reproduced in Figure 12.26.

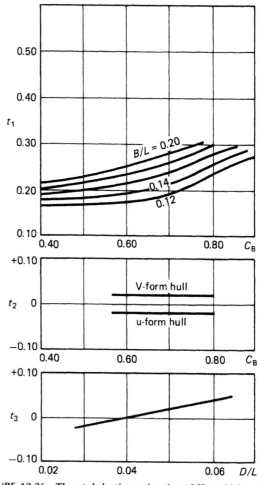

FIGURE 12.26 Thrust deduction estimation of Harvald for single-screw ships. *Reproduced with permission from Reference 17.*

12.4.3 Hull Efficiency

The hull efficiency can readily be determined once the thrust deduction and mean wake fraction are known. However, because of the pronounced scale effect of the wake fraction there is a difference between the full-scale ship and model values. In general, because the ship wake fraction is smaller than the corresponding model value, due to Reynolds effects, the full-scale efficiency will also be smaller.

12.4.4 Quasi-Propulsive Coefficient

It can be deduced from equation (12.41) that the value of the QPC is dependent upon the ship speed, pressure field around the hull, the wake field presented to the propeller and the intimate details of the propeller design, such as diameter, rate of rotation, radial load distribution, amount of cavitation on the blade surfaces, etc. As a consequence, the QPC should be calculated from the three component efficiencies given in equation (12.41) and not globally estimated.

Of interest when considering general trends is the effect that propeller diameter can have on the QPC; as the diameter increases, assuming the rotational speed is permitted to fall to its optimum value, the propeller efficiency will increase and hence for a given hull form the QPC will tend to rise. In this instance the effect of propeller efficiency dominates over the hull and relative rotative efficiency effects.

12.5 THE INFLUENCE OF ROUGH WATER

The discussion so far has centered on the resistance and propulsion of vessels in calm water or ideal conditions. Clearly the effect of bad weather is either to slow the vessel down for a given power absorption or, conversely, an additional input of power to the propeller in order to maintain the same ship speed.

To gain some general idea of the effect of weather on ship performance appeal can be made to the NSMB Trial Allowances 1976.[42] These allowances were based on the trial results of 378 vessels and formed an extension to the 1965 and 1969 diagrams. Figure 12.27 shows the allowances for ships with a trial displacement between 1000 and 320 000 tonnes based on the Froude extrapolation method and coefficients. Analysis of the data upon which this diagram was based showed that the most significant variables were the displacement, Beaufort wind force, model scale and the length between perpendiculars. As a consequence a regression formula was suggested as follows:

$$\text{trial allowance} = 5.75 - 0.793\Delta^{1/3} + 12.3B_n \\ + (0.0129L_{PP} - 1.864B_n)\lambda^{1/3} \quad (12.48)$$

where B_n and λ are the Beaufort number and the model scale, respectively.

Apart from global indicators and correction factors such as Figure 12.27 or equation (12.48) considerable work has been undertaken in recent years to establish methods by which the added resistance due to weather can be calculated for a particular hull form. Latterly, particular attention has been paid to the effects of diffraction in short waves which is a particularly difficult area to consider analytically.

In general, estimation methods range from those which work on databases for standard series hull forms whose main parameters have been systematically varied, to those where the calculation is approached from fundamental considerations. In its most simplified form the added resistance calculation is of the form

$$R_{TW} = R_{TC}(1 + \Delta_R) \quad (12.49)$$

where R_{TW} and R_{TC} are the resistances of the vessel in waves and calm water, respectively, and Δ_R is the added resistance coefficient based on the ship form parameters, speed and irregular sea state. Typical of results of calculation procedures of this type are the results shown in Figure 12.28 for a container ship operating in different significant wave heights (H_S) and a range of heading angles from directly ahead ($\theta = 0°$) to directly astern ($\theta = 180°$).

Shintani and Inoue[43] have established charts for estimating the added resistance in waves of ships based on a study of the Series 60 models. This data takes into account various values of C_B, B/T, L/B and l.c.b. position and allows interpolation to the required value for a particular design. In this work the compiled results have been empirically corrected by comparison with model test data in order to enhance the prediction process.

In general the majority of the practical estimation methods are based in some way on model test data:

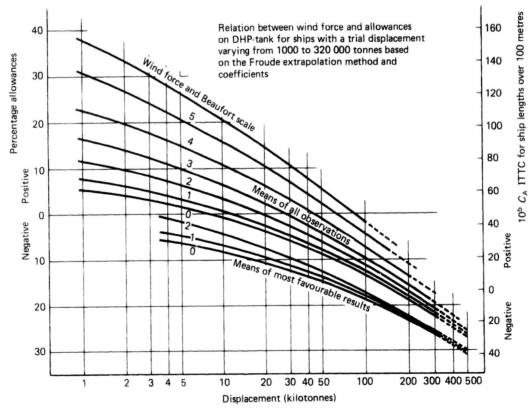

FIGURE 12.27 NSMB 1976 trial allowances. *Reproduced with permission from Reference 42.*

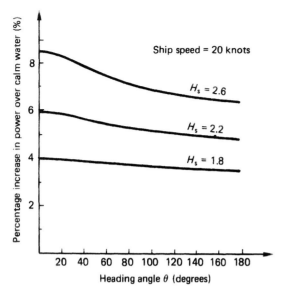

FIGURE 12.28 Estimated power increase to maintain ship speed in different sea states for a container ship.

either for deriving regression equations or empirical correction factors.

In the case of using theoretical methods to estimate the added resistance and power requirements in waves, methods based on linear potential theory tend to under-predict the added resistance when compared to equivalent model tests. In recent years some non-linear analysis methods have appeared which indicate that if the water surface due to the complete non-linear flow is used as the steady wave surface profile then the accuracy of the added resistance calculation can be improved significantly (References 56 and 57). Although CFD analyses are relatively limited, those published so far show encouraging results when compared to measured results, for example Reference 58.

In the context of added resistance numerical computations have suggested that the form of the bow above the calm water surface can have a significant influence on the added resistance in waves. Such findings have also been confirmed experimentally and have shown that a blunt-bow ship could have its added resistance reduced by as much as 20–30 per cent while having minimal influence on the calm water resistance.

12.6 RESTRICTED WATER EFFECTS

Restricted water effects derive essentially from two sources. These are first a limited amount of water under the keel, and second, a limitation in the width of water each side of the vessel which may or may not be in association with a depth restriction.

In order to assess the effects of restricted water operation, these being particularly complex to define mathematically, the ITTC[32] have cited typical influencing parameters. These are as follows:

1. An influence exists on the wave resistance for values of the Froude depth number F_{nh} in excess of 0.7. The Froude depth number is given by

$$F_{nh} = \frac{V}{\sqrt{(gh)}}$$

where h is the water depth of the channel.
2. The flow around the hull is influenced by the channel boundaries if the water depth to draught ratio (h/T) is less than 4. This effect is independent of the Froude depth number effect.
3. There is an influence of the bow wave reflection from the lateral boundary on the stern flow if either the water width to beam ratio (W/B) is less than 4 or the water width to length ratio (W/L) is less than unity.
4. If the ratio of the area of the channel cross-section to that of the mid-ship section (A_c/A_M) is less than 15, then a general restriction of the waterway will start to occur.

In the case of the last ratio it is necessary to specify at least two of the following parameters: width of water, water depth or the shape of the canal section, because a single parameter cannot identify unconditionally a restriction on the water flow.

The most obvious sign of a ship entering into shallow water is an increase in the height of the wave system, Figure 12.4(b), in addition to a change in the ship's vibration characteristics. As a consequence of the increase in the height of the wave system the assumption of small wave height, and consequently small wave slopes, cannot be used for restricted water analysis. This, therefore, implies a limitation to the use of linearized wave theory for this purpose; as a consequence higher-order theoretical methods need to be sought. Currently several researchers are working in this field and endeavoring to enhance the correlation between theory and experiment.

Barrass[44] suggests the depth–draught ratio at which shallow water just begins to have an effect is given by the equation

$$h/T = 4.96 + 52.68(1 - C_w)^2$$

in which the C_w is the water-plane coefficient. Alternatively, Schneekluth[45] provides a set of curves based on Lackenby's work (Figure 12.29) to enable the estimation of the speed loss of a vessel from deep to shallow water. The curves are plotted on a basis of the square of Froude depth number to the ratio $\sqrt{A_M}/h$. Beyond data of this type there is little else available with which to readily estimate the added resistance in shallow water beyond recourse to numerical methods.

One further effect of shallow water is the phenomenon of ship squat. This is caused by a venturi effect between the

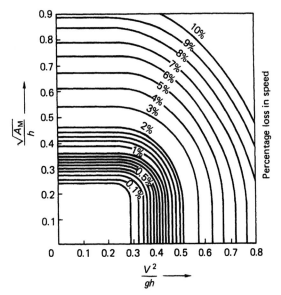

FIGURE 12.29 Loss of speed in transfer from deep to shallow water. *Reproduced from Reference 45.*

bottom of the vessel and the bottom of the seaway which causes a reduction of pressure to occur. This reduction of pressure then induces the ship to increase its draught in order to maintain equilibrium. Barras developed a relationship for ship squat by analyzing the results from different ships and model tests with block coefficients in the range 0.5−0.9 for both open water and restricted channel conditions. In his analysis the restricted channel conditions were defined in terms of h/T ratios in the range 1.1−1.5. For the conditions of unrestricted water in the lateral direction such that the effective width of the waterway in which the ship is traveling must be greater than $[7.7 + 45(1 - C_w)^2]B$, the squat is given by

$$S_{max} = (C_b(A_M/A_C)^{2/3} V_s^{2.08})/30 \text{ for } F_{nh} \leq 0.7$$

12.7 HIGH-SPEED HULL FORM RESISTANCE

For a conventional displacement hull the coefficient of wave-making resistance increases with Froude number, based on waterline length, until a value of $F_n \simeq 0.5$ is reached. Beyond this point it tends to reduce such that at high Froude numbers, in excess of 1.5, the wave-making resistance becomes a small component of the total resistance. The viscous resistance, however, increases due to its dependence on the square of the ship speed: this is despite the value of C_F reducing with Froude number. As a consequence of this rise in viscous resistance a conventional displacement hull requires excessive power at high-speed and other hull forms and modes of support need to be introduced. Such forms are the planing hull, the hydrofoil and the hovercraft.

The underlying principles of high-speed planing craft resistance and propulsion have been treated by several authors: for example, DuCane[46] and Clayton and Bishop.[47] These authors not only examine high-speed displacement and planing craft but also hydrofoils and hovercraft. As a consequence, for the details of their hydrodynamic principles of motion reference can be made to these works.

The forces acting on a planing hull are shown by Figure 12.30 in which the forces shown as W, F_p, F_n, F_s and T are defined as follows:

W is the weight of the craft;
F_p is the net force resulting from the variation of pressure over the wetted surface of the hull;
F_h is the hydrostatic force acting at the center of pressure on the hull;
F_s is the net skin friction force acting on the hull;
T is the thrust of the propulsor.

By the suitable resolution of these forces and noting that for efficient planing the planing angle should be small, it can be shown that the total resistance comprises three components:

$$R_T = R_I + R_{WV} + R_{FS} \qquad (12.50)$$

where

R_I is the induced resistance or drag derived from the inclination of F_p from the vertical due to the trim angle of the craft;
R_{WV} derives from the wave-making and viscous pressure resistance;
R_{FS} is the skin friction resistance.

At high-speed the wave-making resistance becomes small; however, the vessel encounters an induced drag component which is in contrast to the case for conventional displacement hulls operating at normal speeds.

To estimate the resistance properties of high-speed displacement and planing craft use can be made of either standard series data or specific model test results.

12.7.1 Standard Series Data

A considerable amount of data is available by which an estimate of the resistance and propulsion characteristics can

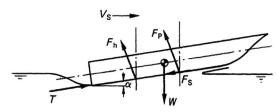

FIGURE 12.30 Forces experienced by a planing craft.

TABLE 12.8 Published Data for Displacement and Planing Craft

Standard Series Data	
Displacement Data	Planing Data
Norstrom Series (1936)	Series 50 (1949)
de Groot Series (1955)	
Marwood and Silverleaf (1960)	Series 62 (1963)
Series 63 (1963)	Series 65 (1974)
Series 64 (1965)	
SSPA Series (1968)	
NPL Series (1984)	
NSMB Series (1984)	
Robson Naval Combatants (1988)	

be made. Table 12.8 identifies some of the data published in the open literature for this purpose.

In addition to basic test data of this type various regression-based analyses are available to help the designer in predicting the resistance characteristics of these craft; for example, van Oortmerssen[48] and Mercier and Savitsky.[49] Additionally, Savitsky and Ward Brown[50] offer procedures for the rough water evaluation of planing hulls.

12.7.2 Model Test Data

In specific cases model test data is derived for a particular hull form. In deriving this data the principles for model testing outlined in Reference 51 and the various ITTC proceedings should be adhered to in order to achieve valid test results.

Multi-Hull Resistance

The wave resistance of a multi-hull vessel is commonly approximated by considering the waves generated by each hull of the vessel acting in isolation to then be superimposed on each other (References 59 and 60). If this approach is followed then an expression for the wave resistance for a pair of non-staggered identical hulls takes the form

$$R_W = 0.5\pi\rho V_s^2 \int |A(\theta)|_{SH}^2 \cdot F(\theta) \cdot \cos^3\theta \, d\theta$$

where $f |A(\theta)|_{SH}^2$ refers to the amplitude function for the side hull and $F(\theta)$ is a hull interference function which is dependent on the hull separation, ship length and Froude number. However, it is important to phase the waves generated by each hull correctly if their transverse components are to be cancelled. This cancellation effect is a function of the Froude number and the longitudinal relative positions of the hulls. Moreover, the cancellation effect of the transverse waves will only be beneficial for a range of Froude numbers around that for which the cancellation is designed to occur.

Approximations of the type do not, however, take into account that the waves generated by one hull will be incident upon another hull, whereupon they will be diffracted by that hull. These diffracted waves comprise a reflected and transmitted wave which implies that the total wave system of the multi-hull ship is not a superposition of the waves generated by each hull in isolation. In this context it is the divergent waves at the Kelvin angle that are responsible for the major part of the interaction. Three-dimensional Rankine panel methods are helpful for calculating the wave patterns around multi-hull ships and when this is done for catamarans it is seen that in some cases relatively large wave elevations occur between the catamaran hulls in the after regions of the ship.

A regression-based procedure was developed (Reference 61) to assess the wave resistance of hard chine catamarans within the range:

$10 \leq L/B \leq 20$
$1.5 \leq B/T \leq 2.5$
$0.4 \leq C_b \leq 0.6$
$6.6 \leq L/\nabla^{1/3} \leq 12.6$

Within this procedure the coefficient of wave-making resistance C_w is given by

$$C_W = \exp(\alpha)(L/B)^{\beta_1}(B/T)^{\beta_2}C_b^{\beta_3}(S/L)^{\beta_4}$$

where the coefficients a, β_1, β_2, β_3 and β_4 are functions of Froude number and s is the spacing between the two hulls.

In this procedure two interference factors are introduced following the formulation of Reference 62, one relating to the wave resistance term (τ) and the other a body interference effect expressed as a modified factor $(1 + \beta_k) = 1.42$) as established by Reference 63. This permits the total resistance coefficient to be expressed as

$$C_T = 2(1 + \beta_k)C_F + \tau C_w$$

Subsequently, an optimization scheme has been developed (Reference 64) for hard chine catamaran hull form basic designs based on the earlier work of Reference 61.

In the case of trimarans and pentamarans the relative dispositions of the outriggers relative to the main hull can have a profound effect on the ship's resistance characteristics. As such, attention needs to be paid to developing the correct longitudinal and thwartship positioning of the outriggers for the intended operating Froude number.

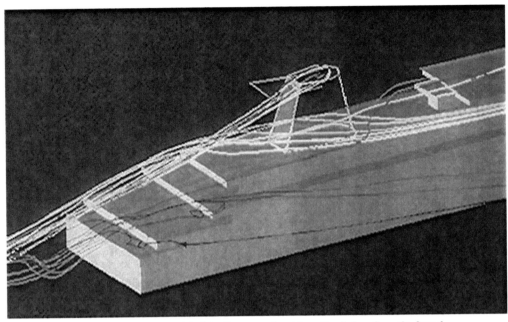

FIGURE 12.31 CFD model of the flow around a ship and emissions from the funnel.

12.8 AIR RESISTANCE

The prediction of the air resistance of a ship can be evaluated in a variety of ways ranging from the extremely simple to undertaking a complex series of model tests in a wind tunnel or in using CFD methods.

At its simplest the still air resistance can be estimated as proposed by Holtrop[52] who followed the simple approach incorporated in the ITCC−1978 method as follows:

$$R_{AIR} = \frac{1}{2}\rho_a V_S^2 A_T C_{air} \qquad (12.51)$$

in which V_S is the ship speed, A_T is the transverse area of the ship and C_{air} is the air resistance coefficient, taken as 0.8 for normal ships and superstructures. The density of air ρ_a is normally taken as 1.23 kg/m^3.

For more advanced analytical studies appeal can be made to the works of van Berlekom[53] and Gould.[54] The approach favored by Gould is to determine the natural wind profile on a power law basis and select a reference height for the wind speed. The yawing moment center is then defined relative to the bow and the lateral and frontal elevations of the hull and superstructure are subdivided into so-called 'universal elements'. Additionally, the effective wind speed and directions are determined from which the Cartesian forces together with the yawing moment can then be evaluated.

The determination of the air resistance from wind tunnel measurement would generally only be undertaken in exceptional cases and would most probably be associated with flow visualization studies too; for example, in the design of suitable locations for helicopter platforms, ensuring exhaust gases and particulate matter do not fall on passenger deck areas as well as to check that abnormal wind flows and vortex patterns do not occur in unwanted locations. When such tests are contemplated then most commonly they would be undertaken in openjet wind tunnels and using smoke for flow visualization purposes. However, for most commercial applications the cost of undertaking wind tunnel tests cannot be justified since air resistance is by far the smallest of the resistance components.

An alternative to wind tunnel testing is through the use of CFD methods. These computations, given the possession of an adequate model mesh generation capability, are relatively easy and quick to undertake; albeit they use a significant computation resource to model the ship to a required level of accuracy. Figure 12.31 shows the results of one such computation which was addressing a perceived exhaust gas problem.

REFERENCES AND FURTHER READING

1. Brown DK. *A Century of Naval Construction*. Conway; 1983.
2. Froude W. *The papers of William Froude*. Trans RINA; 1955.
3. Kelvin L. *On deep water two-dimensional waves produced by any given initiating disturbance*. Proc Roy Soc (Edin.) 1904;**25**: 185−96.
4. Kelvin L. *On the front and rear of a free procession of waves in deep water*. Proc Roy Soc (Edin.) 1904;**25**:311−27.
5. Kelvin L. *Deep water ship waves*. Proc Roy Soc (Edin.) 1904;**25**:562−87.

6. Wigley WCS. *The theory of the bulbous bow and its practical application.* Trans NECIES 1936;**52**.
7. Inui T. *Wavemaking resistance of ships.* Trans SNAME 1962;**70**.
8. Todd FH. *Resistance and propulsion.* In: Principles of Naval Architecture. SNAME, [Chapter 7].
9. Yim B. *A simple design theory and method for bulbous bows of ships.* J Ship Res 1974;**18**.
10. Schneekluth H. *Ship Design for Efficiency and Economy.* London: Butterworths; 1987.
11. Phillips-Birt D. *Ship Model Testing.* Leonard-Hill; 1970.
12. Hughes G. *Friction and form resistance in turbulent flow and a proposed formulation for use in model and ship correlation.* Trans RINA 1954;**96**.
13. Taylor DW. *The Speed and Power of Ships.* Washington: US Govt Printing Office; 1943.
14. Ayre, Sir Amos L. *Approximating EHP — Revision of data given in papers of 1927 and 1948.* Trans NECIES.
15. Auf'm Keller WH. *Extended diagrams for determining the resistance and required power for singlescrew ships.* ISP 1973;**20**.
16. Lap AJW. *Diagrams for determining the resistance of single screw ships.* ISP 1954;**1**(4).
17. Harvald Sv Aa. *Estimation of power of ships.* ISP 1978;**25**(283).
18. Bowden BS. *A Survey of Published Data for Single-Screw Ocean Going Ships.* NPL Ship Report 139; May 1970.
19. Dawson J. *Resistance and propulsion of singlescrew coasters. I, II, III and IV.* Trans IESS, Paper Nos 1168, 1187, 1207 and 1247 in 1953, 1954, 1956, 1959, respectively.
20. Moor DI. *The © of some $0.80C_b$ forms.* Trans RINA; 1960.
21. *The DTMB Series 60 — Presented in a set of papers by various authors 1951, 1953, 1954, 1956, 1957 and 1960.* Trans SNAME.
22. Moor DI. *The effective horse power of single screw ships — Average modern attainment with particular reference to variations of C_b and LCB.* Trans RINA; 1960.
23. Manen JD Van, et al. *Scale effect experiments on the victory ships and models. I, II, III and IV in three papers.* Trans RINA; 1955. 1958 and 1961.
24. Moor DI, Parker MN, Pattullo RN. *The BSRA methodical series — An overall presentation — Geometry of forms and variations of resistance with block coefficient and longitudinal centre of buoyancy.* Trans RINA 1961;**103**.
25. Moor DI. *Resistance, propulsion and motions of high speed single screw cargo liners.* Trans NECIES;**82**:1965—66.
26. Lackenby H, Parker MN. *The BSRA methodical series — An overall presentation — Variation of resistance with breadth—draught ratio and length—displacement ratio.* Trans RINA 1966;**108**.
27. Moor DI, Pattullo RNM. *The Effective Horsepower of Twin-Screw Ships — Best Modern Attainment for Ferries and Passenger Liners with Particular Reference to Variations of Block Coefficient.* BSRA Report No. 192; 1968.
28. *The SSPA Standard Series published in a set of papers from SSPA 1948—1959 and summarized in 1969.*
29. Moor DI. *Standards of ship performance.* Trans IESS 1973;**117**.
30. Moor DI. *Resistance and propulsion qualities of some modern single screw trawler and bulk carrier forms.* Trans RINA; 1974.
31. Pattullo RNM. *The resistance and propulsion qualities of a series of stern trawlers — Variations of longitudinal position of centre of buoyancy, breadth, draught and block coefficient.* Trans RINA; 1974.
32. *18th ITTC Conference, Japan,* 1987.
33. Scott JR. *A method of predicting trial performance of single screw merchant ships.* Trans RINA; 1972.
34. Scott JR. *A method of predicting trial performance of twin screw merchant ships.* Trans RINA; 1973.
35. Holtrop J. *A statistical analysis of performance test results.* ISP February 1977;**24**.
36. Holtrop J. *Statistical data for the extrapolation of model performance tests.* ISP 1978.
37. Holtrop J, Mennen GGJ. *A statistical power prediction method.* ISP October 1978;**25**.
38. Holtrop J, Mennen GGJ. *An approximate power prediction method.* ISP July 1982;**29**.
39. Holtrop JA. *Statistical re-analysis of resistance and propulsion data.* ISP November 1988;**31**.
40. Dawson CW. *A practical computer method for solving ship—wave problems.* Berkeley: 2nd Int. Conf. on Numerical Ship Hydrodynamics; 1977.
41. MARIN Report No. 26, MARIN, Wageningen, September 1986.
42. Jong HJ de, Fransen HP. *N.S.M.B. trial allowances 1976.* ISP 1976.
43. Shintani A, Inoue R. *Influence of hull form characteristics on propulsive performance in waves.* SMWP; December 1984.
44. Barrass CB. *Ship-handling problems in shallow water.* MER; November 1979.
45. Schneekluth H. *Ship Design for Efficiency and Economy.* London: Butterworths; 1987.
46. Cane Pdu. *High Speed Craft.* David and Charles; 1974.
47. Clayton BR, Bishop RED. *Mechanics of Marine Vehicles.* London: Spon; 1982.
48. Oortmerssen G. van. *A power prediction method and its application to small ships.* ISP 1971;**18**(207).
49. Mercier JA, Savitsky D. *Resistance of Transom—Stern Craft in the Pre-Planing Regime. Davidson Laboratory.* Stevens Institute of Tech., Rep. No. 1667; 1973.
50. Savitsky D, Ward Brown P. *Procedures for hydro-dynamic evaluation of planing hulls in smooth and rough waters.* Maritime Technology October 1976;**13**(4).
51. *Status of Hydrodynamic Technology as Related to Model Tests of High Speed Marine Vehicles.* DTNSRDC Rep. No. 81/026; 1981.
52. Holtrop J. *A statistical resistance prediction method with a speed dependent form factor.* SMSSH '88, Varna; October 1988.
53. Berlekom WB van. *Wind forces on modern ship forms — Effects on performance.* Trans NECIES 1981.
54. Gould RWF. *The Estimation of Wind Loads on Ship Superstructures.* Mar. Tech. Monograph No. 8, Trans RINA; 1982.
55. Clayton BR, Bishop RED. *Mechanics of Marine Vehicles.* London: E. & F.N. Spon; 1982.
56. Raven HC. *A solution method for the non-linear ship wave resistance problem.* PhD Thesis, TU Delft; 1996.
57. Hermans AJ. *Added resistance by means of timedomain models in seakeeping.* Cortona, Italy: 19th WWWFB; 2004.
58. Orihara H, Miyata H. *Evaluation of added resistance in regular incident waves by computational fluid dynamics motion simulation using an overlapping grid system.* J Marine Sci and Technol 2003;**8**(2):47—60.

59. Tuck EO, Lazauskas L. Optimum spacing of a family of multi-hulls. *Ship Technol Res* 1998;**45**:180–95.
60. Day S, Clelland D, Nixon E. Experimental and numerical investigation of 'Arrow' Trimarans. *Proc. FAST 2003* 2003; vol. III (Session D2).
61. Pham XP, Kantimahanthi K, Sahoo PK. *Wave resistance prediction of hard chine catamarans using regression analysis*. Hamburg: 2nd Int. Euro-Conference on High Performance Marine Vehicles; 2001.
62. Day AH, Doctors LJ, Armstrong N. *Concept evaluation for large, very high speed vessels*. 5th Int. Conf. on Fast Sea Transportation. *Fast*; 1997.
63. Insel M, Molland AF. An investigation into the resistance components of high speed displacement catamarans. *Trans RINA*; April 1991.
64. Anantha Subramanian V, Dhinesh G, Deepti JM. Resistance optimisation of hard chine high speed catamarans. *J Ocean Technol* Summer 2006;**I**(I).

Chapter 13

Thrust Augmentation Devices

Chapter Outline

13.1 Devices Before the Propeller 334
    13.1.1 Wake Equalizing Duct 334
    13.1.2 Asymmetric Stern 334
    13.1.3 Grothues Spoilers 335
    13.1.4 Stern Tunnels, Semi- or Partial Ducts 336
    13.1.5 Reaction Fins 336
    13.1.6 Mitsui Integrated Ducted Propulsion Unit 336
    13.1.7 Hitachi Zosen Nozzle 337
    13.1.8 Mewis Duct 337
13.2 Devices at the Propeller 337
    13.2.1 Increased Diameter/Low rpm Propellers 337
    13.2.2 Grim Vane Wheel 338
    13.2.3 Propellers with End-Plates 339
    13.2.4 Kappel Propellers 340
    13.2.5 Propeller Cone Fins 340
13.3 Devices Behind the Propeller 341
    13.3.1 Rudder-Bulb Fins Systems 341
    13.3.2 Additional Thrusting Fins 341
13.4 Combinations of Systems 342
References and Further Reading 342

The 1980s saw a proliferation of hydrodynamically based energy-saving devices enter the marine market. Similarly, today there is renewed interest in enhancing ship performance due both to the increasing price of fuel and because of international measures that are being undertaken for environmental reasons.

For discussion purposes energy-saving devices based on hydrodynamic principles can be considered as operating in three basic zones of the hull. Some are located before the propeller, some at the propeller station and some after the propeller and Figure 13.1 defines these three stages as Zones I, II and III respectively. Clearly some devices transcend these boundaries; however, these zones are useful to broadly group the various devices.

In Zone I the thrust augmentation device is reacting with the final stages of the growth of the boundary layer over the stern of the ship. This is to gain some direct benefit or to present the propeller with a more advantageous flow regime: in some cases perhaps both. Devices in Zones II and III are working within both the hull wake field and modifications to that wake field caused by the slipstream of the propeller. In this way they are attempting to recover

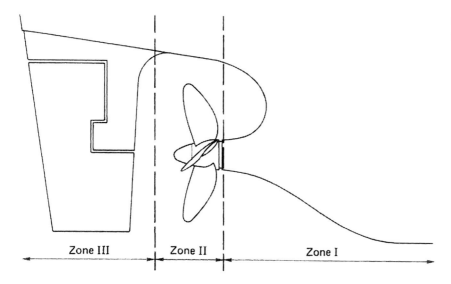

FIGURE 13.1 Zones for classification of energy-saving devices.

TABLE 13.1 Zones of Operation of Energy-Saving Devices

Energy-Saving Device	Zone(s) of Operation
Wake equalizing duct	I
Asymmetric stern	I
Grothues spoilers	I
Stern tunnels, semi- or partial ducts	I
Mewis duct	I
Reaction fins	I/II
Mitsui integrated ducted propellers	I/II
Hitachi Zosen nozzle	I/II
Increased diameter/low rpm propellers	II
Grim vane wheels	II
Propellers with end-plates	II
Propeller boss fins	II/III
Additional thrusting fins	III
Rudder-bulb fins	III

energy which would otherwise be lost. Table 13.1 identifies some of the principal thrust augmentation devices and attempts to categorize them into their principal operating zones.

To consider the influence of an energy-saving device its effect on the various components of the Quasi-Propulsive Coefficient (QPC) needs to be examined. From Chapter 12 it will be recalled that this coefficient is defined by

$$\text{QPC} = \eta_o \eta_H \eta_R \qquad (13.1)$$

where $P_E = P_D \cdot \text{QPC}$. Consequently, each of the devices identified in Table 13.1 will be considered briefly and, in doing so, an outline explanation given of their modes of operation together with some idea of the changes that may occur in the components of equation (13.1).

13.1 DEVICES BEFORE THE PROPELLER

Within Zone I it is seen from Table 13.1 that the wake equalizing duct, asymmetric stern, Grothues spoilers, stern tunnels of various forms and the Mewis duct operate on the flow in this region. Additionally, reaction fins, the Mitsui integrated ducted propellers, the Hitachi Zozen nozzle and to some extent the Mewis duct operate at the boundary of Zones I and II. Figure 13.2 illustrates many of these devices in outline form.

13.1.1 Wake Equalizing Duct

The wake equalizing duct[1,2] was proposed by Schneekluth and aims to improve the overall propulsive efficiency of the ship by reducing the amount of separation over the afterbody of the vessel. It endeavors to achieve this by helping to establish a more uniform inflow into the propeller by accelerating the flow in the upper part of the propeller disc and by attempting to minimize the tangential velocity components in the wake field. Additionally, it is claimed that a larger diameter propeller may be fitted in some cases since the wake field is made more uniform and hence is likely to give rise to smaller pressure impulses transmitted to the hull. As a consequence it might be expected that the mean wake fraction and thrust deduction may be reduced, the latter probably more so, thereby giving rise to moderate increase in hull and propeller open water efficiency components of the QPC. There is little reason to expect that the relative rotative efficiency component will change significantly in this or any of the other devices listed in Table 13.1. In general it can be expected that the power savings with a wake improvement duct will depend on the extent of the flow separation and non-uniformity of the wake field.

This device was first introduced in 1984 and since that time many ducts have been built. Moreover, this device lends itself to retrofitting on vessels; however, the designs need to be produced by experienced personnel and preferably with the aid of model tests at as large a scale as possible, although scale effects are uncertain. With the increasing quantitative abilities of CFD methods then the scale effect issues may be attenuated since the alignment to the flow of the duct can be computed at full scale.

13.1.2 Asymmetric Stern

The asymmetric stern (References 3 and 4) was patented in Germany by Nonnecke and is directed towards reducing separation in the afterbody of a vessel when the flow is influenced by the action of the propeller. However, efficiency gains have occurred where separation has not been noticed at model-scale and, accordingly, the disparity in Reynolds number between model- and full-scale must not be overlooked when considering this type of device in the model tank. Model tests show that this concept can mainly be expected to influence the hull efficiency by causing a significant reduction in the thrust deduction factor coupled with a slight reduction in the mean wake fraction. In this way the increase in hull efficiency is translated into an increase in the QPC for the vessel. Clearly, the asymmetry in the hull also has an effect on the swirl of the flow into the propeller.

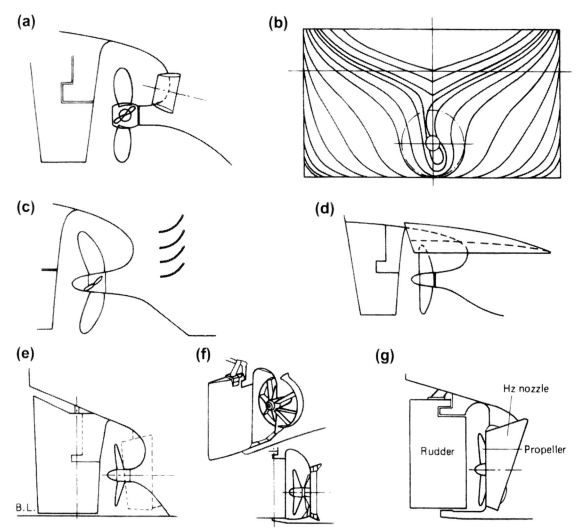

FIGURE 13.2 Zone I and Zone I/II devices: (a) wake equalizing duct; (b) asymmetric stern—body plan; (c) Grothues spoilers; (d) stern tunnel; (e) Mitsui integrated ducted propeller; (f) reaction fins and (g) Hitachi Zosen nozzle.

While such a concept could be fitted to an existing ship this would entail a major hull modification exercise and, therefore, it is probably most suitable for a new building. The design of an asymmetric stern must be done in connection with model tests in order to gain an idea of the extent of any separation present, subject to the reservations expressed previously and the flow configuration at the stern. As in the case of the previous device, CFD methods could provide qualitative and quantitative insights into the design of asymmetric sterns.

During the period 1982—87 some thirty vessels were built or were in the process of construction utilizing the concept of the asymmetric stern.

13.1.3 Grothues Spoilers

The Grothues spoilers (Reference 5) comprise a hydrodynamic fin system fitted to the stern of a vessel immediately ahead of the propeller: as a consequence this option is only applicable to single-screw vessels. The mode of action of the fins is to prevent cross-flow in the vicinity of the hull from reinforcing the bilge vortex and, thereby, impairing the consequent tendency towards energy loss. Each fin is curved with the intention that the leading edge of the fin aligns with the local flow directions within the boundary layer over the stern of the vessel while the trailing edge of the fin is parallel to the shaft line over the whole span. Consequently, the fin system comprises a plurality of spoilers that are capable of diverting the downward cross-flow over the hull surface to a horizontal flow through the propeller.

The spoilers in general can be expected to cause a reduction in hull resistance together with an increase of propeller efficiency induced by the homogenizing effect of the fins on the wake field. In addition to suppressing the effects of the bilge vortices, thereby giving less hull resistance, it is also possible that the fins, by changing the direction for the flow, contribute a component of thrust in the forward direction to overcome resistance. As a consequence, a probable effect of the spoiler system will be a reduction in hull effective power (P_E) together with an increase in hull and propeller open water efficiencies.

Since the spoiler system endeavors to inhibit the bilge vortex formation, it can be expected to perform best on moderately to significantly U-shaped hull forms. The spoiler needs careful design both in terms of its hydrodynamic design, preferably with the aid of model tests and CFD studies, as well as in the mechanical design to ensure the correct strength margin to prevent failure of the fin or hull structure through vibratory response or seaway motion.

13.1.4 Stern Tunnels, Semi- or Partial Ducts

These appendages exist in many forms and have been applied over a considerable number of years for one reason or another. Their use has not always been focused on propulsive efficiency improvement and originally they were more frequently used to attenuate propeller-induced vibration problems. They did this by attempting to reduce the wake peak effect of pronounced V-form hulls (Reference 6). Indeed, today this is still perhaps one of their principal roles.

When deployed with the purpose of improving propulsive efficiency, their aim is frequently to facilitate the accommodation of large-diameter, slow rpm propellers and to ensure that the propeller is kept sufficiently immersed in the ship's ballast draught condition. In these cases their design should be based on model flow visualization studies since detrimental influences on the ship speed have otherwise been known to result: a loss of the order of one knot due to poor tunnel design has not been unknown. Since the primary role of these devices is in the reduction of separation, then their principal influences are likely to be in the reduction in ship effective power, thrust deduction and wake fraction. As a consequence the hull efficiency and propeller open water efficiency could be expected to reflect these changes.

Bilge vortex fins, Figure 23.5(a), are fitted to the surface of the hull upstream of the propeller. In contrast to the stern tunnel concept discussed earlier the role of the bilge vortex fin is to inhibit the cross-flows on the hull surface which stimulate the formation of bilge vortices and hence give rise to energy losses and sources of vibration.

13.1.5 Reaction Fins

The reaction fin (References 7 and 8) normally comprises some six radially located fins which are reinforced by a slim ring nozzle circumscribing them. The device is placed immediately in front of the propeller as shown in Figure 13.2. The diameter of the nozzle ring, which has an aerofoil profiled section, is normally of the order of 10 per cent greater than the propeller diameter. The radial fins have a uniform aerofoil section profile along their length; however, the inflow angles are different for each radial station.

The design of the reaction fin is commonly based on the nominal wake field measurement at model-scale or a computational flow study and aims chiefly at creating a pre-swirl of the flow into the propeller. The pre-swirl created by the reaction fin needs to be sufficiently strong so that rotational flow aft of the propeller is prevented from occurring. If the reaction fin is fitted to an existing vessel then, due to this pre-swirl initiation, a decrease in propeller rpm will be found to occur: this is normally of the order of 2–3 rpm. As a consequence it is also necessary to adjust the propeller design to prevent it from becoming too 'stiff' in the ship's later life. The fitting of a reaction fin which has a thin duct ring integral within its design does not appear from either model- or full-scale tests to cause a deterioration in the cavitation or induced vibration behavior of the propeller (Reference 8).

Other types of reaction blading have been designed which are of a lesser diameter than the propeller and do not have a ring around the periphery of the fins. If the design of the reaction fin blades has not been completely sympathetic to the inflow from the hull, these have been known to induce thin bands of erosion on the propeller blades due to the tip vortex generated by the reaction fin. As such, care has to be exercised in the design of such fins in order to achieve an acceptable radial loading distribution over their span in order to prevent this situation from happening.

A further effect which can accrue from the application of the reaction fin in the mixed wake behind a hull is the production of a thrust on the fins. This tends to have greatest effect when the fins are placed in regions of the wake field having transverse velocity components.

As a consequence the introduction of the reaction fin can be expected to increase the magnitude of the mean wake field in which the propeller operates, this will both increase hull efficiency but also, to some extent, reduce propeller open water efficiency. At the same time it can also be expected that the reaction fin will decrease the rotational losses and gain some benefit, in certain applications, from a positive thrust on the fins. In view of this and the proximity of the fin to the propeller, care also needs to be exercised in the strength aspect of the reaction fin design.

Systems of asymmetric reaction fins have also been developed. These have different numbers of fins located on the port and starboard sides of the ship and do not have a supporting circumferential ring.

13.1.6 Mitsui Integrated Ducted Propulsion Unit

In essence the Mitsui Integrated Ducted Propulsion (MIDP) system (Reference 9) comprises a slightly non-axisymmetric duct which is located immediately ahead of the propeller. With systems of this type the interactions between the hull, duct and propeller are extremely complex and, as a consequence, they cannot be considered in isolation.

Mitsui, in their development of the concept, carried out extensive model tests. In these tests the effects of varying the axial location of the duct, duct entrance configuration and duct chord profiles all featured. From these tests the propulsive efficiency of the system is shown to be intimately related to the longitudinal location of the duct. Furthermore, the non-axisymmetric units, having larger chords at the top, appear to perform better than their axisymmetric counterparts.

Up to the present time a considerable number of these units have been manufactured and installed on relatively full-form vessels which range in size from 43 000—450 000 dwt.

13.1.7 Hitachi Zosen Nozzle

Although developed separately, the Hitachi Zosen system (Reference 10) closely resembles the MIDP system except that the degree of asymmetry in the nozzle appears far greater.

Kitazawa et al.[11] made an extensive study of propeller—hull interaction effects and essentially concluded that:

1. The resistance of an axisymmetric body increases after fitting a duct due to the pressure at the afterbody. However, the required propeller thrust decreased because the duct thrust is larger than the change in resistance.
2. For a given propeller thrust and rpm, the duct thrust increased significantly when placed behind a body.
3. The total propulsive efficiency of the vessel increases and of the components which comprise this total efficiency the relative rotative efficiency remains constant, the open water efficiency increases and the hull efficiency decreases.

As in the previous case, several ships have been fitted with this system, some new buildings and some retrofits, and the vessels so fitted tend towards having high block coefficients.

13.1.8 Mewis Duct

The Mewis duct, Figure 13.3, was originally developed within the context of smaller container ships and bulk carriers which have speeds less than around 20 knots and thrust coefficients greater than unity. However, it is planned to extend the concept to faster ships in the future. The device is intended to enhance three aspects of the flow into the propeller. First, it aims to bring uniformity to the propeller inflow by the action of the duct which is positioned ahead of the propeller station. In this context the duct, which has a smaller diameter than the propeller, is not

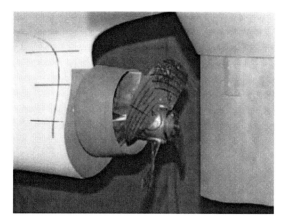

FIGURE 13.3 Mewis duct arrangement during model test. *Reference 12.*

located so as to be co-axial with the shaft center line: it is supported by a number of pre-swirl fins and positioned eccentrically with respect to the shaft center line. Its second effect through the use of the pre-swirl fins, located towards the aft end of the duct, is to reduce the rotational losses in the slipstream of the propeller. The third aim is to induce higher loading in the inner radii of the propeller and thereby improve the propulsion efficiency.

The underlying reasons for the eccentricity of the duct with respect to the shaft center line are so that the wake equality is distributed equitably in the upper part of the propeller disc and to distribute the flow of the duct through a wider range towards the propeller. The design process for these devices combines the advantages of model testing and computational fluid dynamics together with the experience of the designer. Based on model tests with the device for fuller form ships it was predicted that power reductions of the order of 6—7 per cent were achievable.[12]

13.2 DEVICES AT THE PROPELLER

The devices in Zone II are those which essentially operate at the propeller station. As such they include increased diameter/low rpm propellers, Grim vane wheels, Tip Vortex Free (TVF) propellers, Keppel propellers and propeller cone fins.

13.2.1 Increased Diameter/Low rpm Propellers

It can be simply demonstrated with the aid of a $B_p-\delta$ chart that, for a given propulsion problem, the propeller open water efficiency can be increased by reducing the rpm and allowing the diameter of the propeller to increase freely. As a consequence, propeller design should always take account of this within the constraints of the design problem.

FIGURE 13.4 Grim vane wheel general arrangement.

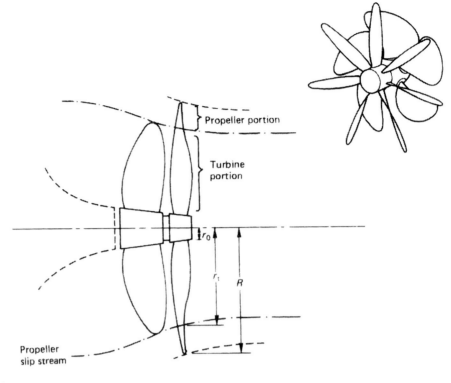

The constraints, however, which limit this design option are the available space within the propeller aperture, insufficient immersion, high propeller-induced surface pressures on the hull and, in extreme cases, the weight of the resulting propeller. Whilst the latter can normally be accommodated by suitable stern bearing design the former constraints generally act as the limiting criteria for this concept.

The principal effects of these propellers are to be found in increased open water efficiency; because of the increased diameter, however, the mean wake fraction decreases slightly, which has a reducing effect on the hull efficiency. The net effect, nevertheless, is generally an overall enhancement of the quasi-propulsive coefficient.

13.2.2 Grim Vane Wheel

The Grim vane wheel (References 13–15) derives its name from its inventor, Professor Grim, and is a freely rotating device which is installed behind the propeller. In the greater majority of cases the vane wheel is sited on a stub shaft bolted to the ship's tail shaft; however, there have been proposals to locate the stub shaft on the rudder horn.

The diameter of the vane wheel is larger than that of the propeller and its function is to extract energy from the propeller slipstream, which would otherwise be lost, and convert this into an additional propulsive thrust. To achieve this, the inner parts of the vane wheel blades act as a turbine while the outer part acts as propeller, Figure 13.4. The design basis of the vane wheel is, therefore, centered on satisfying the following two relationships:

$$\int_{r_t}^{R} \frac{dQ}{dr} dr + \int_{r_0}^{r_1} \frac{dQ}{dr} dr = 0 \quad \text{(ignoring bearing friction)}$$

$$\int_{r_t}^{R} \frac{dT}{dr} dr + \int_{r_0}^{r_1} \frac{dT}{dr} dr > 0$$

where dT and dQ are the elemental thrusts and torques acting on the blade section and R, r_t and r_0 are the vane wheel tip radius, transition (between propeller and turbine parts) radius and the boss radius respectively.

In less formal terms the propeller and turbine partial torques must balance, ignoring the small frictional component, and the net effect of the propeller and turbine portion axial forces must be greater than zero.

Vane wheels in general have rather more blades than the propeller, typically greater than six, and rotate at a somewhat lower speed, which is of the order of 30–50 per cent of the propeller rpm. Consequently, the blade passing frequencies in addition to the blade natural frequencies need careful consideration as blade fatigue failure may result if this is not taken into account.

Figure 13.5 shows the velocity diagrams relating to the inner and outer portions of the vanes. The in-flow

Thrust Augmentation Devices

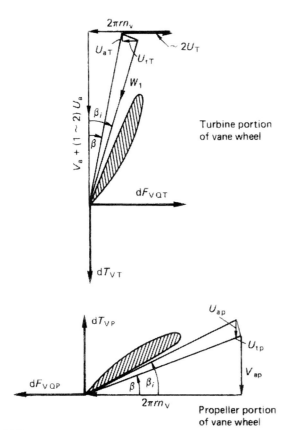

FIGURE 13.5 Velocity diagram of propeller–vane wheel combination.

velocities into the vane wheel are defined by the induced velocities in the slipstream created by the propeller and, therefore, the in-flow conditions to the vane wheel are derived from the propeller calculation. In view of the axial separation of the propeller and vane wheel these velocities need to be corrected for this effect in design: the extent of the correction will, however, depend on the type of mathematical model used in the propeller design process. The blade design of the vane wheel, together with a rotational speed and blade number optimization, is then usually undertaken either on a blade element or lifting line basis. From these analyses the resulting blade loadings and radial stress distribution in the vane wheel blade can be readily determined. The vane wheel diameter is determined primarily from the geometric constraints of the ship.

In the design process, if model tests are undertaken, care is needed when interpreting the results since differential scale effects between the propeller and vane wheel can manifest themselves. Calculation of the performance of the Grim vane wheel is, therefore, an essential feature of the design process.

When considering the application of a Grim vane wheel to a ship the greatest advantage can be gained in cases where the rotational energy losses are high, thereby giving a greater potential for conversion of this component of the slipstream energy. As a consequence, it is to be expected that single-screw vessels will provide a greater potential for energy saving than those having a high-speed twin-screw form. In real terms the increase in propulsion efficiency is governed by the value of C_T for the parent propeller. In the author's experience, the improvement in propulsive efficiency can be as low as 2 per cent or 3 per cent for high-speed, low-wake fraction vessels to something of the order of 13 per cent for full-form, single-screw ships.

13.2.3 Propellers with End-Plates

The underlying reason for the introduction of blade end-plate technology is to give the designer a greater freedom in the choice of the distribution of circulation over the propeller blades. Although the basic concept has been known for many years, it was Perez Gomez who developed the concept into a practical proposition in the mid-1970s in the form of the TVF (Tip Vortex Free) propeller. In essence the idea of the TVF propeller arose from endeavoring to simplify the concept of the ring propeller, particularly in trying to reduce the high viscous resistance of the ring, while at the same time trying to obtain a circulation at the blade tips. In its TVF form the propeller tip plates were effectively tangential to the cylindrical sections and were not considered to contribute to the thrust development beyond permitting the circulation to be non-zero at the blade tips.

The early TVF propellers were designed to work in association with a duct such that the propeller was located at the aft end of the duct. Such an arrangement allowed the flow into the propeller to be controlled so as to create shock-free entry of the incident flow onto the tip plates.

Subsequent to the introduction of the TVF propeller ongoing research gave rise to the present generation of CTL (Contracted and Loaded Tip) propellers. Figure 13.6 shows, by way of illustration, an example of a CLT propeller fitted to a tanker. Unlike the TVF propeller, which if it did not work behind a duct the blade tip plates would not be aligned to the contracting propeller slip-stream flow and would therefore give rise to considerable drag forces at large propeller radii, the CLT propeller has tip plates which are intended to be aligned to the direction of the flow through the propeller disc. To achieve this objective two propeller theories were developed by the designers: New Momentum

FIGURE 13.6 CLT Propeller fitted to a tanker with a wake equalizing duct mounted on the ship ahead of the CLT Propeller. *Courtesy Sistemar*

FIGURE 13.7 Kappel propeller. *Source unknown.*

Theory and New Cascades Theory (Reference 17). To date, a number of TVF and CTL propellers have been fitted to ships and an extensive literature has been published by the designers of the systems. Reference 16−18 are examples of this information; indeed, the latter reference provides a much fuller reference list. Theoretical development of the concept has also been provided by Klaren and Sparenberg[19] and de Jong.[20]

Dyne[21] conducted an investigation into propellers with end-plates which confirmed the calculated efficiency gains as well as model test predictions conducted by Anderson and Schwanecke[22] and also by de Jong et al.[23] However, Dyne was unable to explain why so many full-scale trials reported gains in excess of 10 per cent; nevertheless there is some evidence to suggest that with CLT propellers there is some propulsion benefit to be gained.

13.2.4 Kappel Propellers

The Kappel propeller endeavors to introduce higher propulsive efficiency by deploying a blade design which has modified blade tips which continuously and smoothly curve towards the suction side of the blade, Figure 13.7. This concept is based on similar considerations to winglets found on aircraft wings and has been applied to both fixed and controllable pitch propellers.

The design process is such that the propeller blades and their winglets addition are designed as a single integral curved blade (References 24 and 25). Friesch et al.[26] described a series of model- and full-scale trial measurements on a Kappel propeller In this program it was demonstrated that for a product tanker the propulsive efficiency was higher in the case of the Kappel propeller than for a conventional propeller. Moreover, it was shown that the frictional component and scale effect of the Kappel propeller were larger than for the conventional propeller and a new surface strip method was produced in order to scale the frictional forces over the blade.

13.2.5 Propeller Cone Fins

The concept of fitting fins to the cone of a propeller, located behind the blades, was proposed by Ouchi et al.[27] with the aim of enhancing the efficiency of the screw propeller by reducing the energy loss associated with the propeller hub vortex. In principle a number of small fins having a flat plate form and with a height of the order of 10 per cent of the propeller blade span are fitted at a given pitch angle to the cone of the propeller. The number of fins corresponds to the propeller blade number.

The role of the fins is to weaken the strength of the hub vortex and in so doing recover kinetic energy from the rotating flow around the propeller cone. In this way the fins contribute to an increase in propeller efficiency.

Much model testing of this concept has been undertaken and flow measurements in the propeller wake have been made using laser Doppler methods. From such tests it is clear that the fins have an influence on the hub vortex and that at model-scale there is a beneficial influence on the open water efficiency of the parent MAU standard series propeller. Clearly scale effects between model- and full-scale manifest themselves; however, the inventors claim that the analysis of full-scale trial results from several ships show a beneficial improvement in propulsion efficiency when using these fins.

Chapter | 13 Thrust Augmentation Devices

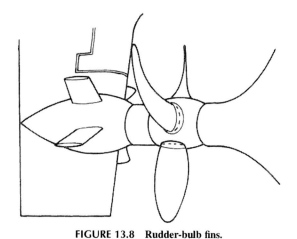

FIGURE 13.8 Rudder-bulb fins.

rudder close behind the propeller boss. The system appears in two versions, one with just a bulb and the other comprising a set of four fins, in an X-shape, protruding normally from the hub and extending to about $0.9R$ as indicated in Figure 13.8.

When applied without the fins it is not dissimilar to the Costa bulb, which was first applied in the 1950s to some ships (Reference 28). This system aimed to prevent flow separation and excessive vorticity behind the hub by effectively extending the propeller boss. When the fins are fitted to the system they produce a lift force since they are operating in the helical slipstream of the propeller and, therefore, receive flow at incidence. A component of this lift force then acts in the forward direction to produce an augmentation to the thrust force. The design of the fins needs to be based on fatigue considerations since the fins are working within the flow variations caused by the vortex sheets emanating from the propeller.

13.3 DEVICES BEHIND THE PROPELLER

Zone III devices, as implied by Figure 13.1, operate behind the propeller and consequently operate within the slipstream of the propeller. Rudder-bulb fins and additional thrusting fins fall into this category.

13.3.1 Rudder-Bulb Fins Systems

This system, developed by Kawasaki Heavy Industries, comprises a large bulb, having a diameter of some 30–40 per cent of the propeller diameter, which is placed on the

13.3.2 Additional Thrusting Fins

The additional thrusting fins (References 29 and 30) were developed and patented by Ishikawajima–Harima Heavy Industries. The system essentially comprises two fins, placed horizontally in the thwart-ship directions on the rudder and in line with or slightly above the propeller axis. Figure 13.9 shows this system in schematic form. The chord length of the fins is of the order of half that of the

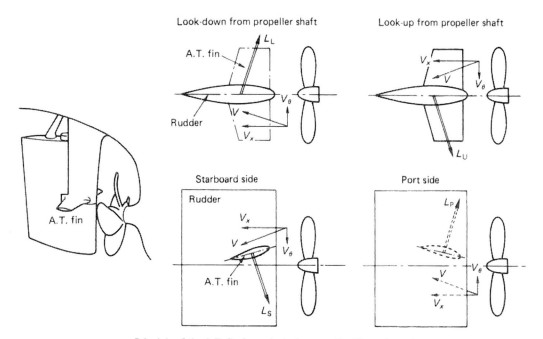

Principle of the A.T. fin for a clockwise propeller. Thrust is produced as propeller axial components of lift L_L, L_U act on the rudder with additional thrust produced as propeller axial components of lift L_S, L_P act on the fin.

FIGURE 13.9 Additional thrusting fins. *Reproduced with permission from Reference 30.*

rudder and the span is about 40 per cent of the propeller diameter.

The design of the fins is directed towards optimizing their lift–drag ratio whilst operating in the slipstream of the propeller and hence use is made of cambered aerofoil sections of variable incidence. The principle of operation can be seen from Figure 13.9 by examining the four positions in the propeller disc: top and bottom dead center and port and starboard athwart ships. At the top dead center position it can be seen that the flow, which comprises an axial component V_x and a tangential component V_θ, is incident on the rudder and consequently produces a horizontal force on the rudder: a component of which is directed in the forward direction. Similarly, with the conditions at the bottom dead center position. In the case of fins that are set normally to the rudder and at an incidence relative to the propeller shaft center line a similar situation also occurs. Then, by adjusting the incidence of the fins with respect to the hydrodynamic pitch angles of the propeller slipstream, the magnitudes of the lift forces can be made sufficient to overcome the drag of the fins and a positive contribution to the propulsion thrust can be developed (Figure 13.9).

When applying this system at full-scale attention has to be paid to the system of steady and non-steady forces acting on the fins: for example, the added mass, slamming forces, lift, drag and weight. These factors have important consequences for the rudder strength.

13.4 COMBINATIONS OF SYSTEMS

In many cases the question is posed as to whether the various energy-saving devices are compatible with each other so as to enable a cumulative benefit to be gained from fitting several devices to a ship. The general answer to this question is no, because some devices remove the flow regimes upon which others work. Nevertheless, if devices depend on different regions of the flow field around the ship and are mutually independent they can be used in combination in order to gain an enhanced benefit.

REFERENCES AND FURTHER READING

1. Schneekluth, H. *Wake Equalising Duct*. The Naval Architect.
2. Beek Tvan, Verbeek R, Schneiders CC. *Saving Fuel Cost – Improving Propulsive Efficiency with Special Devices and Designs*. Lips BV Paper; August 1986.
3. Collatz G. *The asymmetric afterbody, model tests and full scale experiences*. El Pardo: Int. Symp. on Ship Hydrodynamics and Energy Saving; September 1983.
4. Collatz G. *Non-symmetrical afterbodies*. Wageningen: MARIN Workshop on Developments in Hull Form Designs; October 1985.
5. Grothues-Spork K. *Bilge vortex control devices and their benefits for propulsion*. Int Shipbuilding Prog 1988;**35**(402).
6. Carlton JS, Bantham I. *Full scale experience relating to the propeller and its environment*. Propellers 78 Symp, Trans SNAME; 1978.
7. Takekuma K, et al. *Development of reaction fin as a device for improvement of propulsion performance of high block coefficient ships*. Soc Nav Arch Jpn December 1981;**150**.
8. Takekuma K. *Evaluations of various types of nozzle propeller and reaction fin as the devices for the improvement of propulsive performance of high block coefficient ships*. SNAME, New York: Shipboard Energy Conservation Symp; September 1980.
9. Narita H, et al. *Development and full scale experiences of a novel integrated duct propeller*. New York: SNAME Annual Meeting; November 1981.
10. Kitazawa T, Hikino M, Fujimoto T, Ueda K. *Increase in the Propulsive Efficiency of a Ship with the Installation of a Nozzle Immediately Forward of the Propeller*. Hitachi Zosen Annual Research Report; 1983.
11. Kitazawa T, et al. *Increase of the propulsive efficiency of a ship by a nozzle installed just in front of a propeller*. Kansai Soc Nav Arch Jpn; November 1981.
12. Mewis, F. *A novel power-saving device for full form vessels*. First Int. Symp. on Marine Propulsors, SMP'09, Trondheim, Norway; June 2009.
13. Grim O. *Propeller and vane wheel, second George Weinblum Memorial Lecture*. J Ship Res December 1980;**24**(4).
14. Grim O. *Propeller and vane wheel, contrarotating propeller*. 8th School, Gothenburg: WEGEMT; August 1983.
15. Kubo H, Nagase M, Itadani Y, Omori J, Yoshioka M. *Free rotating propeller installed on ship*. Bull MESJ March 1988;**16**(1).
16. Ruiz-Fornells R, et al. *Full scale results of first TVF propellers*. Madrid: Int. Symp. on Ship Hydrodynamics and Energy Saving; 1983.
17. Perez Gomez G, Gonzalez-Adalid J. *Detailed Design of Ship Propellers*. Madrid: FEIN; 1997.
18. Sistemar. *Fairplay Advertising Supplement*. Fairplay February 1992;**6**.
19. Klaren L, Sparenberg. JA. *On optimum screw propellers with end plates, inhomogeneous inflow*. J Ship Res December 1981;**25**(4).
20. de Jong K. *On the design of optimum ship screw propellers, including propellers with end plates*. Hamburg: Int. Symp. on Propulsors and Cavitation; June 1992.
21. Dyne G. *On the principles of propellers with end-plates*. Int J Maritime Eng, Trans RINA 2005;**147**(Part A3).
22. Andersen P, Schwanecke H. *Design of model tests of tip fin propellers*. Trans RINA 1992;**134**:315–28.
23. De Jong K, Sparenberg J, Falcao de Campos J, van Gent W. *Model testing of an optimally designed propeller with two-sided shifted end plates on the blades*. 19th Symp. on Naval Hydrodynamics, NRC; 1994.
24. Andersen SV, Andersen P. *Hydrodynamical design of propellers with unconventional geometry*. Trans RINA; 1986.
25. Andersen P, Friesch J, Kappel JJ, Lundegaard L, Patience G. *Development of a marine propeller with nonplanar lifting surfaces*. Marine Technol July 2005;**42**(3):144–58.
26. Friesch J, Anderson P, Kappel JJ. *Model/full scale correlation investigation for a new marine propeller*. NAV2003 Conf. Italy: Palermo; 2003.
27. Ouchi KB, Kono Y, Shiotsu T, Koizuka H. *PBCF (propeller boss cap fins)*. J Soc Nav Arch Jpn 1988;**163**.
28. Gregor O. *Etudes récentes du Bulbe de Propulsion Costa*. Assoc. Technique Maritime et Aéronautiques; 1983.
29. Ishida S. *Study on recovery of rotational energy in propeller slipstream*. IHI Eng Rev July 1983;**16**.
30. Mori M, IHI AT. *Fin – 1st report, its principle and development*. IHI Eng Rev January 1984;**17**.

Chapter 14

Transverse Thrusters

Chapter Outline

14.1 Transverse Thrusters 343
    14.1.1 Performance Characterization 346
    14.1.2 Unit Design 347
14.2 Steerable Internal Duct Thrusters 350
References and Further Reading 352

Many vessels depend for their effectiveness on possessing a good maneuvering capability in confined waters. Figure 14.1 illustrates this in the case of ferries maneuvering in confined waters and berthing stern first into link spans: in some cases, in order to maintain schedules, under poor weather conditions. In addition to the specific case of the ferry, many other vessels also require an enhanced maneuvering capability. To satisfy these requirements several methods of providing a directional thrusting capability are available to the naval architect. One of these is the provision of transverse fixed tunnel propulsion units and another is steerable internal duct thrusters, although of course some propriety designs transcend these two boundaries. While the former type is more common, the various options available and their comparative merits have to be carefully considered at the vessel's design stage. The principal types of units are shown schematically in Figure 14.2.

14.1 TRANSVERSE THRUSTERS

Transverse fixed tunnel thrusters essentially comprise an impeller mounted inside a tunnel which is aligned athwart the vessel and the essential features of the system are illustrated in Figure 14.3. It is important to emphasize that the system must be considered as an entity; that is impeller, tunnel, position in the hull, drive unit fairings, tunnel openings and the protective grid all need to be evaluated as a complete concept if the unit is to satisfy any form of optimization criteria. Incorrect, or at best misleading results, will be derived if the individual components are considered in isolation or, alternatively, some are neglected in the analysis. Although Figures 14.2 and 14.3 generally show a transverse propulsion unit located in the bow of the vessel, and in this position the unit is termed a bow thruster, such units can and are also located at the stern of the ship. The bow location is, however, the more common and for large vessels and where enhanced maneuverability in less than ideal situations, such as in the case of a ferry, is

FIGURE 14.1 Ro/Ro ships maneuvering in ferry terminal.

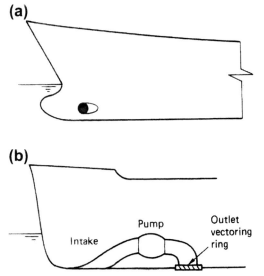

FIGURE 14.2 Types of thruster units: (a) transverse propulsion unit, and (b) steerable, internal duct thrusters.

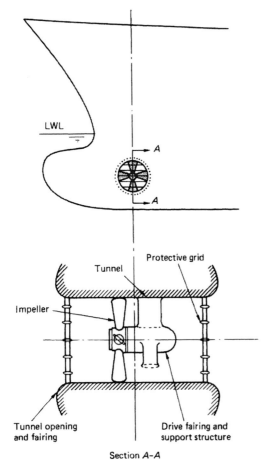

FIGURE 14.3 Transverse propulsion unit — general arrangement in hull.

required they are often fitted in pairs: for some larger ships such as cruise liners more are fitted. The decision as to whether to fit one or two units to a vessel is normally governed by the power or thrust requirement and the available draught.

The design process for a transverse propulsion unit located in a given vessel has two principal components: first, to establish the thrust, or alternatively the power, required for the unit to provide an effective maneuvering capability and, second, how best to design the unit to give the required thrust in terms of the unit's geometry. To satisfy the demand many manufacturers have elected to provide standard ranges of units covering, for example, a power range of 150 to around 4000 hp and then select the most appropriate unit from the range for the particular application. Other manufacturers, who perhaps tend today to be in the minority, design a particular unit for a given application.

In order to determine the size of a transverse propulsion unit for a given application two basic philosophical approaches can be adopted. In both cases the vessel is considered to be stationary with regard to forward ahead speed. The first approach is to perform a fairly rigorous calculation or undertake model tests, perhaps a combination of both, to determine the resistance of the hull in lateral and rotational motion. Such an exercise would also probably be undertaken for a range of anticipated currents. Additionally the wind resistance of the vessel would also be evaluated either by calculation, typically using a method such as Reference 1 or by model tests in a wind tunnel. The various wind and hydrodynamic forces on the vessel could then be resolved to determine the required thrust at a particular point on the ship to provide the required motion. A method of this type, whilst attempting to establish the loading from first principles, suffers particularly from correlation problems, scale effects and not least the cost of undertaking the exercise. As a consequence, although this method is adopted sometimes, more particularly with the azimuthing thruster design problem, it is more usual to use the second design approach for the majority of vessels.

This alternative approach either uses a pseudo-empirical formulation of the ship maneuvering problem coupled with experience of existing vessels of similar type or is based on one of the emerging ship maneuvering simulation capabilities. In essence, the first of these approaches attempts to establish a global approximation to the relationship between turning time, required thrust and wind speed for a particular class of vessel. An approximation to the turning motion of a ship then can be represented by equation (14.1) assuming the vessel rotates about a point as seen in Figure 14.4:

$$J_P \left(\frac{d^2\theta}{dt^2} \right) = M_H + M_W + M_P \quad (14.1)$$

where M_H and M_W are the hydrodynamic and wind moments, respectively, and M_P is the moment produced by the thruster about some convenient turning axis. J_p is the polar moment of inertia of the ship and $d^2\theta/dt^2$ the angular acceleration. However, by assuming a constant turning rate the left-hand side of equation (14.1) can be put to zero, thereby removing difficulties with the polar inertia term. That is:

$$M_H + M_W + M_P = 0 \quad (14.2)$$

In pursuing this pseudo-empirical approach it can be argued that the hydrodynamic moment is largely a function of $(d\theta/dt)^2$ and the wind moment is a function of the maximum wind moment times $\sin 2\theta$. The thruster moment is simply the thrust times the distance from the point of rotation and, assuming a constant power input to the unit, is a constant k_3. Hence, equation (14.2) can be rewritten as

$$k_1 \left(\frac{d\theta}{dt} \right)^2 + k_2 \sin 2\theta + k_3 = 0 \quad (14.3)$$

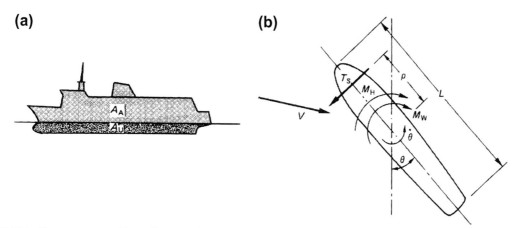

FIGURE 14.4 Transverse propulsion unit nomenclature: (a) surface area definition, and (b) force, moment and velocity definition.

where the coefficients k_1 and k_2 depend on the water and air densities (ρ_W and ρ_A), the underwater and above water areas (A_U and A_A), the vessel's length (L), wind speed (V), etc., as follows:

$$\left. \begin{array}{l} k_1 = 0.5\rho_w A_U L^3\ C_{MW} \\ \text{and} \\ k_2 = 0.5\rho_A A_A L V^2\ C_{MA|max} \end{array} \right\}$$

in which C_{MW} and $C_{MA|max}$ are the water and maximum air moment coefficients, respectively.

Consequently equation (14.3) can be rewritten as

$$\left(\frac{d\theta}{dt}\right) = -\frac{[(k_2\ \sin\ 2\theta + k_3)]^{0.5}}{k_1}$$

from which the time to turn through 90° can be estimated as follows:

$$t_{90} = \int_0^{\pi/2} \left(\frac{dt}{d\theta}\right) d\theta \qquad (14.4)$$

Several authors have considered this type of relation for transverse propulsion unit sizing. One such approach (Reference 2) uses a form of equation (14.4) to derive a set of approximate turning times for three classes of vessel in terms of the turning time for a quarter of a turn as a function of thruster power and with wind speed as a parameter. The relationship used in this case is

$$t_{90} = \left[\frac{0.308 C_{MW}\rho_W A_U L_{pp}^2}{kT_s - 0.5 C_{MA}\rho_A A_A V^2}\right]^{0.5}$$

in which k is the distance of the thruster from the point of rotation non-dimensionalized by ship length between perpendiculars, T_s is the propulsion unit thrust, and C_{MA} is a mean wind resistive moment coefficient. The vessels considered by Reference 2 are ferries, cargo liners and tankers or bulk carriers and Figure 14.5 reproduces the

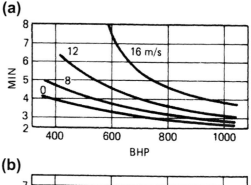

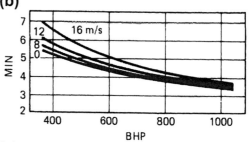

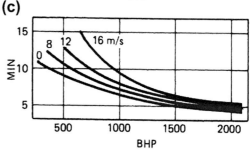

FIGURE 14.5 Average relationship between turning time and power of unit[2]: (a) Ro/Ro and ferries; (b) cargo ships; and (c) tankers and bulk carriers. *Reproduced with permission from Reference 2.*

results of that prediction. Implicit in this type of prediction is, of course, the coefficient of performance of the unit which relates the unit thrust T to the brake horsepower of the motor; however, coefficients should not introduce large variations between units of similar types; that is controllable pitch, constant speed units. Whilst curves such as those shown in Figure 14.5 can only give a rough estimate of turning capability they are useful for estimation purposes. With ships having such widely differing forms, one with another, due account has to be taken in the sizing procedure of the relative amounts of the vessel exposed to the wind and to the water. An alternative approach is to consider the thrust per unit area of underwater or above water surface of the vessel. Table 14.1 shows typical ranges of these parameters, compiled from References 3 and 4.

When interpreting Table 14.1 one should be guided by the larger resulting thrust derived from the coefficients. This is particularly true of the latest generation of Ro/Ro and Ro/Pax ferries in which considerable wind exposure is an inherent design feature; furthermore, in the case of tankers and bulk carriers the assessment of thruster size by the above water area is not a good basis for the calculation.

In the case of the alternative approach, that of using a dynamic simulation capability, this lends itself to a much wider class of simulation scenarios. This is because in many of these techniques not only can the transverse propulsion unit's characteristics be taken into account but a whole range of other contributing factors. Typically these might include the thrusts of the propellers, perhaps in opposite directions (reversing thrusts); the depth of water and proximity of the quay; water currents and wind speeds. Indeed, all of these factors are significant in deciding upon the maneuvering equipment necessary for a particular ship. To achieve this level of simulation the full equations of motion of the ship must be considered and then solved, albeit with some empirical data derived either from the analysis of other full-scale trials, model test data or the results of computational fluid dynamic studies. Many of these more advanced ship maneuvering simulation capabilities have reached a stage where reasonably reliable predictions can be developed with relatively modest computational facilities.

The question of an acceptable turning rate is always a subjective issue and depends on the purpose for which the vessel is intended and the conditions under which it is expected to operate. Consequently, there is no unique

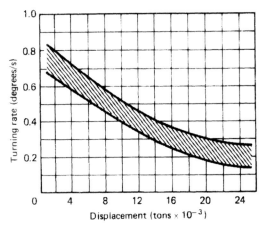

FIGURE 14.6 Band of rotation rates versus displacement at zero ship speed. *Reproduced with permission from Reference 11.*

answer to this problem; Hawkins et al.[5] made an extensive study of several types of maneuvering propulsion devices for the US Maritime Administration and Figure 14.6 presents curves based on their work showing measured turning rates as a function of displacement. The band shown in the figure represents turning rates which have been considered satisfactory in past installations.

14.1.1 Performance Characterization

The usual measure of propeller performance defined by the open water efficiency (η_0) and given by equation (6.2) decreases to zero as the advance coefficient J tends to zero. However, at this condition thrust is still produced and as a consequence another measure of performance is needed to compare the thrust produced with the power supplied.

Several such parameters have been widely used in both marine and aeronautical applications; in the latter case to characterize the performance of helicopter rotors and VTOL aircraft. The most widely used are the static merit coefficient (C) and the Bendemann static thrust factor (ζ) which are defined by the following relationships:

$$\left. \begin{array}{l} C = \dfrac{0.00182 T^{3/2}}{\text{SHP}\sqrt{\rho \pi D^2/4}} = \dfrac{K_T^{3/2}}{\pi^{3/2} K_Q} \\[2ex] \xi = \dfrac{T}{P_s^{2/3} D^{2/3} (\rho \pi / 2)^{1/3}} = \dfrac{K_T}{K_Q^{2/3} [\pi (2)^{1/3}]} \end{array} \right\} \quad (14.5)$$

In these equations the following nomenclature applies:

T is the total lateral thrust, taken as being equal to the vessel's reactive force (i.e., the impeller plus the induced force on the vessel).
SHP is the shaft horsepower.
P_s is the shaft power in consistent units.
D is the tunnel diameter.
ρ is the mass density of the fluid.

TABLE 14.1 Guide to Thrust Per Unit Area Requirements

Ship Type	T/A_U (kp/m²)	T/A_A (kp/m²)
Ro/Ro and ferries	10–14	4–7
Cargo, ships, tugs	6–10	4–8
Tankers/bulk carriers	4–7	14–16
Special craft (i.e. dredgers, pilot vessels, etc.)	10–12	5–8

and K_T and K_Q are the usual thrust and torque coefficient definitions.

Both of the expressions given in equation (14.5) are derived from momentum theory and can be shown to attain ideal, non-viscous maximum values for $C = \sqrt{2}$ and $\zeta = 1.0$ for normal, non-ducted propellers. In the case of a ducted propeller with no duct diffusion these coefficients become $C = 2$ and $\zeta = \sqrt[3]{2}$.

Clearly it is possible to express the coefficient of merit (C) in terms of the Bendemann factor (ζ) and from equation (14.5) it can easily be shown that:

$$C = \zeta^{3/2}\sqrt{2} \qquad (14.6)$$

14.1.2 Unit Design

Having determined the required size of the unit it is necessary to configure the geometry of the unit to provide the maximum possible thrust. The fundamental decision is to determine whether the unit will be a controllable pitch, constant speed machine or a fixed pitch, variable speed unit: the former type being perhaps the most common amongst larger vessels. In the case of controllable pitch impeller units the blades are designed as constant pitch angle blades to enable a nominal equality of thrust to be achieved in either direction for a given operational pitch angle. The term *nominal equality of thrust* is used to signify that equality of thrust is not achieved in practice due to the position of the impeller with respect to the pod and its position in the tunnel. The blades of the controllable pitch units are frequently termed *flat-plate blades* on account of their shape; although this is not strictly the case since they have an aerofoil cross-sectional shape when viewed normally to their cylindrical sections. In the alternative case of the fixed pitch unit the radial distribution of pitch angle over the blade can be allowed to vary in order to develop a suitable hydrodynamic flow regime over the blades, reversal of thrust in this case being achieved by a reversal of rotation of the impeller. Nevertheless, in some instances *flat-plate blades* are also used with these types of unit.

In both the controllable and fixed pitch cases the blade sections are symmetrical about their nose–tail lines; that is, the blades do not possess camber. Furthermore, the fixed pitch blade sections need to be bisymmetrical since both edges of the blade have to act as the leading edge for approximately equal times, whereas for the controllable pitch unit a standard National Advisory Committee for Aeronautics (NACA) or other non-cambered aerofoil section is appropriate.

Transverse propulsion units are a source of noise and vibration largely resulting from the onset of cavitation and the turbulent flow of the water through the tunnel within the vessel. The issue of noise emission is considered in Chapter 10; however, in order to design a unit which would be able to perform reliably and not cause undue nuisance, this being particularly important in passenger vessels, it is generally considered that the blade tip speed should be kept within the band of 30–34 m/s.

Blade design can be achieved by use of either model test data or by theoretical methods. Taniguchi et al.[6] undertook a series of model tests on a set of six transverse thrusters. These models had an impeller diameter of 200 mm; two had elliptic blade forms whilst the remaining four were of the Kaplan type and it is this latter type that is of most interest in controllable pitch transverse propulsion unit design. Taniguchi et al.'s model tests considered Kaplan blade designs having expanded area ratios of 0.300, 0.450 and 0.600 in association with a blade number of four. Additionally, there was also a version having a 0.3375 expanded area ratio with three blades. Each of the blades for these units was designed with a NACA 16-section thickness form in association with a non-dimensional hub diameter of 0.400 and a capability to vary the pitch ratio between 0 and 1.3. Using these models Taniguchi et al. evaluated the effects of changes in various design parameters on performance and Figure 14.7 shows a selection of these results. These highlight the effects of variations in expanded area ratio and pitch ratio; the effects of blade number and boss ratio. This latter test was carried out with the elliptic blade form unit and the results show that there is little difference between the efficiency (η) of the elliptic and Kaplan blade forms with the exception that the Kaplan form performs marginally better at all pitch settings. In Figure 14.7 K_T and K_Q are the conventional thrust and torque coefficients, respectively, and C_F represents the force measured on the simple block hull body containing the tunnel (Figure 14.8). In these experiments the efficiency of the unit is defined by

$$\eta = \frac{1}{K_Q}\left(\frac{K_T + C_F}{\pi}\right)^{3/2} \qquad (14.7)$$

The influence of cavitation on these types of blades forms can be seen in results from a different series of flat-plate blades shown in Figure 14.9. These results which relate to a blade area ratio of 0.5 and a blade number of four show how the breakdown of the thrust and torque characteristics occurs with reducing cavitation number for a series of pitch ratios. However, when comparing the results of these tests with those of Taniguchi et al., it should be noted that the test configurations between the results shown in Figures 14.7 and 14.9 are somewhat different.

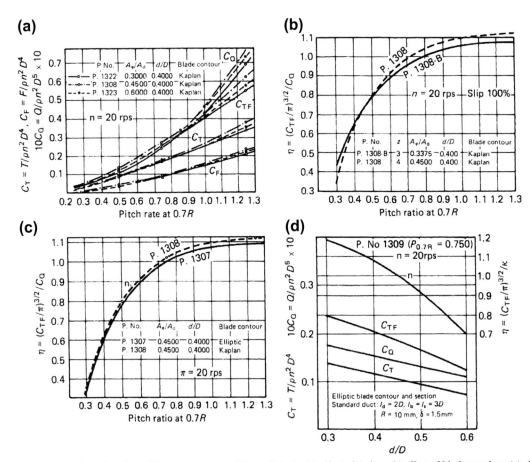

FIGURE 14.7 Examples of test data from CP transverse propulsion unit tests: (a) effect of A_C/A_D; (b) effect of blade number; (c) effect of blade form; and (d) effect of hub diameter. *Reproduced with permission from Reference 6.*

With regard to theoretical bases for impeller design several methods exist. These range from empirically based approaches, such as that by van Manen and Superine[7] to advanced computational procedures of the type discussed in Chapter 8. Indeed, in recent years the use of boundary element and computational fluid dynamics methods has been particularly helpful in understanding the flow configurations that exist in transverse propulsion units.

The impeller design process is only one aspect of the system design. The position of the impeller in the hull presents an equally important design consideration. Taniguchi et al.[6] in their extensive model test study examined this problem using simple block models of the hull. In these models the vertical position of the tunnel relative to the base line, the tunnel length and the effects of frame slope in the way of the tunnel opening could be investigated individually or in combination. Figure 14.10 shows the effects of these changes at model scale. From these results it can be seen that these parameters exert an important influence on the overall thrust performance of the unit. As a consequence it is seen that care needs to be exercised in determining the location of the unit in the hull so as to avoid any unnecessary hydrodynamic losses and also to maximize the turning moment of the unit on the ship system. It will be seen from Figure 14.10 that these two issues are partially conflicting and, therefore, an element of compromise has to be introduced within the design process.

Transverse propulsion units are at their most effective when the vessel is stationary in the water with respect to normal ahead speed and tend to lose effectiveness as the vessel increases its ahead, or alternatively astern, speed. English[8] demonstrated this effect by means of model tests from which it can be seen that the side thruster loses a significant amount of its effect with ship speeds, or conversely current speeds, of the order of 2—3 knots. The cause of this fall-off in net thrusting performance is due to the interaction between the fluid forming the jet issuing from the thruster tunnel and the flow over the hull surface, due principally to the translational motion of the hull but also in part to the rotational motion. Figure 14.11 shows the effect in diagrammatic form where it is seen that this interaction causes a reduced pressure region to occur downstream of the tunnel on the jet efflux side and

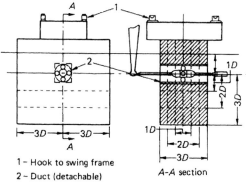

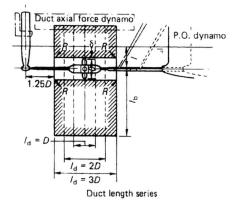

FIGURE 14.8 Taniguchi et al.'s simplified hull form arrangement. *Reproduced with permission from Reference 6.*

The tunnel openings need to be faired to some degree in order to prevent any undue thrust losses from the unit and also to minimize the hull resistance penalty resulting from the discontinuity in the hull surface. However, the type of fairing required to enhance the thrust performance of the transverse propulsion unit is not the same as that required to minimize hull resistance during normal ahead operation and, consequently, a measure of compromise is required in the design process within this context. This highlights the compromise necessary in designing the opening fairings to suit the nominally zero speed thrusting condition as well as minimizing the hull aperture resistance at service speeds. Indeed, Holtrop et al. made some regression-based estimates of the aperture size on ship resistance as discussed in Chapter 12. This compromise can normally be achieved provided that the ship's service speed is below around 20–24 knots; however, if a service speed at the top end of this range or above is contemplated then consideration should be given to the fitting of tunnel orifice doors. Figure 14.13 shows such a case in which doors have been provided to minimize the hull frictional resistance. Furthermore, it can be seen that the hinges on the doors are aligned such that the axis about which door opening occurs approximately aligns with the flow streamlines generated over the bulbous bow, which, in this case, lies to the left of the picture. Such an alignment can be particularly useful in minimizing the ship's frictional resistance should a door actuating mechanism fail and the door cannot be closed after use.

A further advantage of doors fitted to the ends of thruster tunnels is that the turbulent noise generated by the water passing over the tunnel orifice is considerably reduced at service ship speeds when compared to normal thruster openings. Additionally, within the context of noise generation, the traditional shape of controllable pitch transverse propulsion unit blades has been trapezoidal, when viewed in plan form as seen in Figure 14.9, and this is not conducive to quietness of operation. The application of moderate skew to the controllable pitch impeller blades of thruster units, Figure 14.14, has, by helping to control the effects of cavitation, given a further degree of control in minimizing the noise generated by these units. This can be particularly beneficial to passenger ship operation or other ship types where the accommodation is located in the vicinity of the thruster units and where the vessel may be required to maneuver in harbor while people are still sleeping. Furthermore, careful attention to the hydrodynamic fairing of the pod strut and body and to the changes of section that occur within the tunnel space make a significant difference to the noise generation potential of the unit.

this can extend for a considerable way downstream. This induces a suction force on that side of the hull which reduces the effect of the impeller thrust and alters the effective center of action of the force system acting on the vessel (Reference 9) Considerations of this type led some designers (Reference 10) to introduce a venting tube, parallel to the axis of the tunnel, in order to induce a flow from one side of the hull to the other. Figure 14.12 illustrates the effects of fitting such a device to two different types of vessel (Reference 10). More recent research and full-scale practice, however, has shown that the fall off in the effectiveness of bow thrusters with increasing ship speed is attenuated when very large units are employed; typically of the order of 3 MW and above.

Wall effects are important when considering the performance of a transverse propulsion unit. A low-pressure region can be created between the hull surface and the jetty wall when in the presence of a jet from a bow thruster unit. This induces suction between the wall and the hull to occur which, in the case of an idealized flat-plate at about three jet diameters from the wall, experiences suction of the order of the jet thrust. Such a magnitude, however, decays rapidly with increasing separation distance such that at about six jet diameters the suction is only about 10 per cent of the jet thrust.

FIGURE 14.9 Effect of cavitation on K_T and K_Q for a Kaplan blade form.

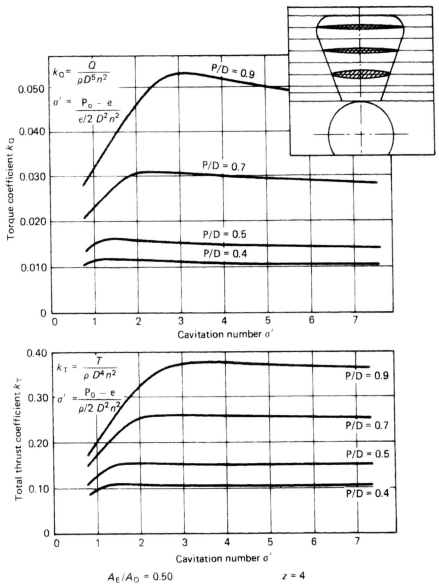

14.2 STEERABLE INTERNAL DUCT THRUSTERS

These types of thruster, sometimes erroneously referred to as pump jets, are particularly useful for navigating a ship at slow speed, as well as for the more conventional docking maneuvers. In the case of research ships, for example, when undertaking acoustic trials of one kind or another it is sometimes helpful not to be dependent upon the main propellers to drive the ship. This is because although the propellers may have been designed to be subcavitating at the speeds of interest for scientific measurements, there will still be turbulence noise generated by the flow over the propulsor and its supporting arrangements. Consequently, to have a propulsor driving the ship in the sense of a *tractor* at the bow of the ship may be useful since the noise and disturbance of propulsion can be largely removed from the scientific measurement positions or towed arrays.

In this type of thruster the water enters the system through an intake located usually in the vicinity of the ship's bow and then passes through an impeller-driven pump from where the water is exhausted to the outlet. At the outlet, located on the bottom of the ship, the efflux from the pump passes through a vectoring ring comprising a cascade of horizontally aligned deflector vanes which impart a change of direction to the water flow. The cascade can be rotated to any desired direction in the horizontal plane, generally through the

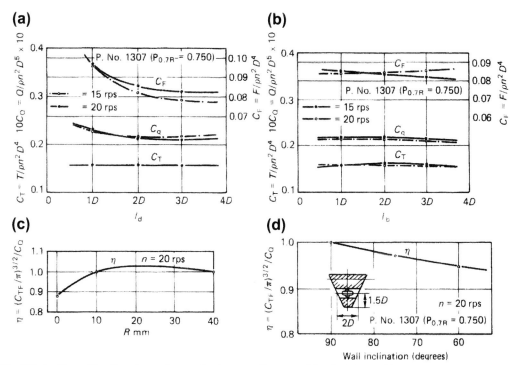

FIGURE 14.10 Effects of tunnel location, frame shape and entrance radius on model scale: (a) tunnel length series; (b) bottom immersion series; (c) tunnel entrance shape series; and (d) hull frame inclination. *Reproduced with permission from Reference 6.*

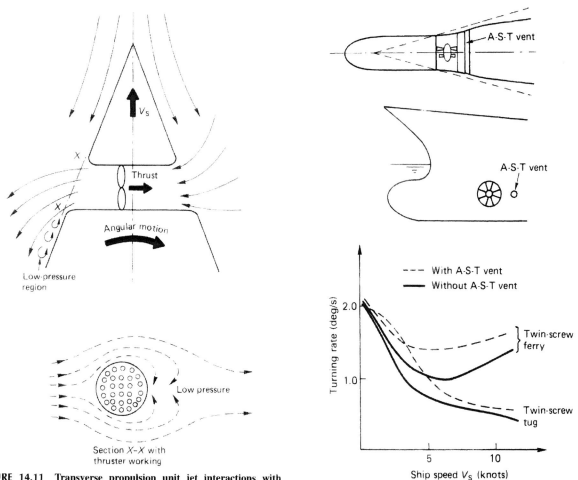

FIGURE 14.11 Transverse propulsion unit jet interactions with forward ship speed.

FIGURE 14.12 Effect of AST vent. *Reference 10.*

FIGURE 14.13 Doors fitted to a set of transverse propulsion unit openings.

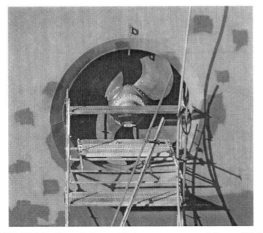

FIGURE 14.14 Use of skew with transverse propulsion unit blades.

full 360°, and resulting from the change in direction of the flow velocities a thrust force can be generated in the desired direction. When free running it is often possible, assuming that the unit has been sized properly and the ship is not too large, to drive the vessel at speeds of the order of five knots or so. In the case of berthing, then due to the azimuthing capability of the thruster unit, an additional directional degree of control is afforded.

REFERENCES AND FURTHER READING

1. Gould RWF. *The Estimation of Wind Loads on Ship Superstructures. Maritime Technology Monograph No. 8.* RINA; 1982.
2. *Steering Propellers with Controllable Pitch.* Ka-Me-Wa; 1968.
3. *Steering Propellers.* Ka-Me-Wa; 1976.
4. Lips Transverse Tunnel Thrusters. *Lips Publ.*
5. Hawkins S, et al. *The Use of Maneuvering Propulsion Devices on Merchant Ships. Robert Taggart, Inc.* Report RT-8518, Contract MA-3293; January 1965.
6. Taniguchi K, Watanabe K, Kasai H. Investigations into fundamental characteristics and operating performances of side thrusters. *Mitsubishi Tech Bulletin*; 1966.
7. Manen JD van, Superine A. Design of screw propellers in nozzles. *ISP* March 1959;**6**(55).
8. English JW. The design and performance of lateral thrust units for ships — Hydrodynamic considerations. *Trans RINA* 1963; **103**(3).
9. Chislett MS, Bjorheden O. *Influence of Ship Speed on the Effectiveness of a Lateral Thrust Unit. Hydroog Aerodynamisk Laboratorium.* Rep. No. Hy-8 Lyngby; 1966.
10. *Modern Lateral Thrusters with Increased Performance — The Anti-suction Tunnel — A-S-7.* Schottel Nederland b.v; 1985.
11. Beveridge JL. Design and performance of bow thrusters. *Marine Technol*; October 1972.

Chapter 15

Azimuthing and Podded Propulsors

Chapter Outline
15.1 Azimuthing Thrusters 353
15.2 Podded Propulsors 356
  15.2.1 Steady-State Running 356
  15.2.2 Turning Maneuvers 358
  15.2.3 Crash Stop Maneuvers 359
  15.2.4 Podded Propulsors in Waves 360
  15.2.5 General and Harbor Maneuvers 360
  15.2.6 Specific Podded Propulsor Configurations 361
References and Further Reading 362

The general class of azimuthing propulsors includes both azimuthing thrusters and podded propulsors. Before considering these systems in greater detail and to avoid confusion it is important to be clear on the definition of a podded propulsor as distinct from other forms of propulsion and azimuthing thrusters in particular. A podded propulsor is defined as a propulsion or maneuvering device which is external to the ship's hull and houses a propeller powering capability. This distinguishes them from azimuthing thrusters which have their propulsor powering machinery located within the ship's hull and commonly drive the propeller through a system of shafting and spiral bevel gearing.

15.1 AZIMUTHING THRUSTERS

Azimuthing thrusters have, as a class of propulsion units, gained importance in recent years due to the increasing demand for dynamic positioning capabilities and directional thrust requirements. These units fall into two distinct classes: the first is where a propeller is mounted on a rotatable pod beneath the ship and the second is the Voith Schneider or Kirsten–Boeing propulsion concept; this latter concept was considered in Chapter 2. With regard to the former class where a propeller is mounted on a pod beneath the ship, Figure 15.1 illustrates the basic features of the system. It can be seen that there are two basic types of unit: the pusher unit shown in Figure 15.1(a) and the tractor unit shown in Figure 15.1(b). Frequently, azimuthing units are fitted with ducted propellers having ducts of the Wageningen 19A form. This is because for many dynamically positioning applications it is necessary to maintain station against tide or wind forces and at low advance speeds this type of ducted propeller has a greater thrusting capability. For other applications, such as canal barge propulsion, the non-ducted propeller may have the advantage and is commonly used.

The resulting thrust from an azimuthing thruster is the sum of three components:

$$T = T_P + T_D + T_G \qquad (15.1)$$

where T_P, T_D and T_G are the component thrusts from the propeller, duct and the pod, respectively, and T is net unit thrust. As with any other propulsion device, the effective thrust acting on the ship is the net thrust adjusted by the augment of resistance (thrust deduction factor) induced by the unit on the vessel.

These types of unit experience a complex system of forces and moments which are strongly dependent on the

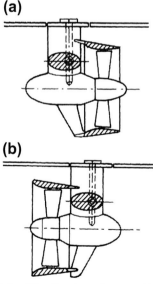

FIGURE 15.1 Azimuthing thruster unit types: (a) pusher unit and (b) tractor unit.

relative alignment of the unit to the incident flow as seen in Figure 15.2. The principal forces and moments which occur are

F_x the longitudinal force in the propeller shaft direction.
F_y the transverse force perpendicular to the propeller shaft.
Q the propeller torque.
M_z the steering or turning moment of the unit.

All of these forces and moments are dependent both upon the inflow incidence angle δ and the magnitude of the inflow velocity V_a. In general, however, six components of loading $\{F_x, F_y, F_z, M_x, M_y$ and $M_z\}$ will be present.

For design purposes two specific sets of model test data are commonly used. The first and most comprehensive (Reference 1) reports a set of test data conducted in both the cavitation tunnel and also a towing tank. This test data considers three blade forms mounted inside a duct of the Wageningen 19A form, two with *flat-plate blades*, as defined in Chapter 14, and the other with an elliptical outline and cambered aerofoil sections, in each of the tractor and pusher configurations. The model propeller diameters are 250 mm and the only difference between the two planar blade forms is the radial thickness distribution. All of the blade forms have a blade area ratio of 0.55.

The analysis of the model test results shows, as might be expected, that the open water efficiency of the propeller with the cambered sections is considerably better than for the propeller with flat-plate blades. This latter blade form, however, while losing on efficiency, has the advantage of a nominal equality of thrust in each direction, as in the case of the transverse propulsion units discussed in Chapter 14. It also prevents the otherwise cambered sections from working at negative angles of incidence which may, under certain off-design conditions, create cavitation and noise issues. Noise levels on the pusher unit were found to be some 10–20 dB higher than on the tractor unit and the propeller with cambered sections gave the lowest noise levels of the propellers for the conditions tested.

The pusher unit was shown to have a slightly better efficiency than the tractor version. However, due to the uniform inflow conditions experienced by the tractor unit in the zero azimuthing angle, less cyclic variation in cavitation pattern was observed when compared to the pusher unit where the propeller is operating in the wake of the gear housing. With regard to turning moment (M_z) the tractor unit showed a much higher moment than the pusher unit under equivalent conditions.

Figure 15.3 shows a typical set of characteristic curves for a propeller unit of this type. The conventional open water curves are shown in Figure 15.3(a) from which the general behavior of the components of equation (15.1) can be seen. Most notable here is the negative behavior of the pod thrust components, indicating that this is a drag. The corresponding diagram, Figure 15.3(b), shows how the various force and moment coefficients change with the angle β. This angle and the force and moment coefficients are consistent with the definitions given in Chapter 6. In this diagram the component relating to the lateral force F_y is plotted to half scale and, consequently, for large angles of β this force can be dominant.

The second source of data is from Oosteveldt.[2] This data is rather more limited than that published by Minsaas and Lehn[1] and relates to open water data using the K_a 4–55 propeller in a 19A duct form. Data of the type shown in Figure 15.3(b) is given for this single propeller duct combination; however, the force components are not broken down so as to be able to differentiate the pod drag.

The model testing of azimuthing units can present particular scale problems if it is intended to model experimentally the performance, for example, of an offshore structure. In these cases the model propeller size can become very small and this introduces hydrodynamic scale effect issues. In addition, if more than one unit is fitted, mutual interaction problems can exist between the units. Consequently, care needs to be exercised in the design of such experiments, the analysis and use to which they are put as well as a proper identification of the various thruster interactions that are made.

In an attempt to increase the propulsion efficiency of azimuthing units, contra-rotating propeller versions have been placed on the market by certain manufacturers. Similarly, tandem units are also available.

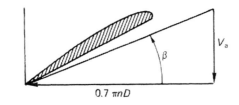

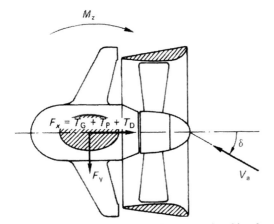

FIGURE 15.2 Forces and moments acting on an azimuthing thruster in uniform flow.

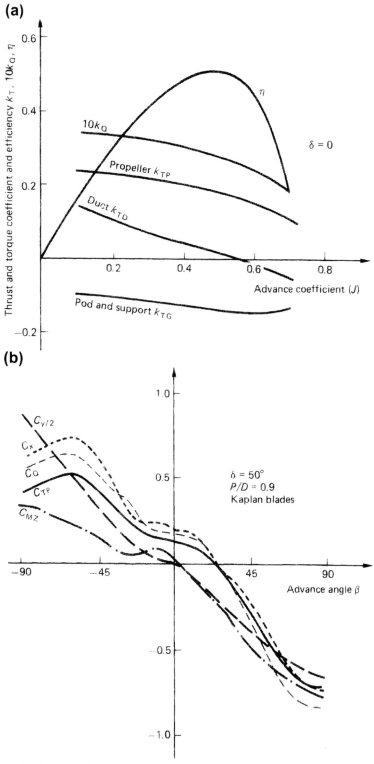

FIGURE 15.3 Azimuthing thruster characteristic curves: (a) open water curves with ahead advance and (b) open water curves for ahead and astern advance with $\delta = 50°$. *Reference 1.*

15.2 PODDED PROPULSORS

Podded propulsors, in their current form, were introduced into the marine industry in the 1990s. They derive from the concept of azimuthing thrusters which have been in common use for many years, the first application being in 1878. Indeed, many of the early design principles for podded propulsors were derived from azimuthing thruster practice. However, the demand from the marine industry for the growth in podded propulsor size occurred very rapidly during the latter half of the 1990s with units rising during that period from a few megawatts in size to the largest which are currently in excess of 20 MW. Their principal applications in the early years were for the propulsion of ice breakers and then cruise ships, but subsequently they have found application with Ro/Pax ferries, tankers, cable layers, naval vessels and research ships. Much of this rapid expansion was fuelled by claims for enhanced propulsive efficiency and ship maneuverability: the latter attribute having been clearly demonstrated.

In outline terms the mechanical system of a podded propulsor has normally comprised a short propulsion shaft on which an electric motor is mounted and supported on a system of rolling element radial and thrust bearings. Some contemplation within the industry is being given to changing from rolling element thrust bearings to a twin bearing arrangement comprising a conventional thrust bearing and a separate journal bearing: thereby splitting the duty of the single rolling element thrust bearing of reacting the propulsor thrust and shaft radial load over two bearings, each having a specific duty. The motor is likely to be either an a.c. machine or, in some cases of smaller units, a permanent magnet machine. Also mounted on the shaft line may be an exciter and shaft brake, together with an appropriate sealing system. The electrical power to drive the motor, some control functions and monitoring equipment are supplied by an arrangement of electrical cables and leads. These are connected to the inboard ship system by a slip-ring assembly located in the vicinity of the pod's slewing ring bearing at the interface between the propulsor and the ship's hull. The podded propulsor's internal machinery is supported within a structure comprising a nominally axisymmetric body suspended below the hull by an aerofoil-shaped fin. The propellers fitted to these units are currently of a fixed pitch design and are frequently of a built-up configuration in that the blades are detachable from the boss. As in the case of azimuthing thrusters, podded propulsors can be either tractor or pusher units and some designs have a system of tandem propellers mounted to the shaft: one propeller mounted at each end of the propulsor body.

While each manufacturer has variants about these basic forms, Figure 15.4 shows a typical schematic layout for

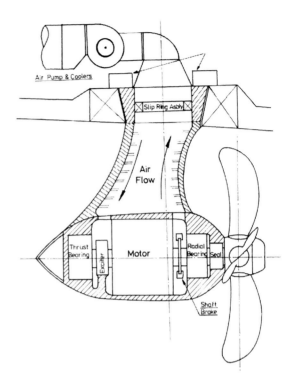

FIGURE 15.4 General arrangement of a podded propulsor.

a tractor unit, this being the most common form at the present time.

15.2.1 Steady-State Running

Assuming that the podded propulsor is of the tractor type, in its twin-screw propulsion configuration it will operate in relatively clear water, which will be disturbed principally by the boundary layer development over the hull. This is in contrast to a conventional twin-screw propulsion arrangement in which the incident propulsor wake field is disturbed by the shafting and its supporting brackets or, alternatively, for a pusher pod configuration which operates in the boundary layer and velocity field generated by the pod body and strut. Consequently, the wake field presented to the propeller of a tractor podded propulsor, in the absence of any separation induced by the effects of poor hull design, should be rather better for the ahead free running mode of operation than would be the case for a conventional twin-screw ship.

Notwithstanding the benefits of an improved wake field, the siting of the propulsors in relation to the hull and their attitude relative to the ship's buttocks and waterlines needs to be considered with care. If this is not adequately achieved then propulsion efficiency penalties may be incurred because the propulsion efficiency has been found at model scale to be sensitive to relatively small changes in propulsor location with respect to the hull. The optimum pod azimuth angle for ahead free running has to be derived from detailed consideration of the flow streamlines over the afterbody of the ship, particularly if a range of operating conditions is anticipated for the ship. Similarly with the tilt angle; however, this may be approximated for initial design purposes as being half the angle of the ship's buttocks relative to the baseline at the propulsor station. Table 15.1 illustrates a typical example of this sensitivity to pod attitude, in this case relating to the relative attitude of the propulsor for a cruise ship.

The computation of the propeller thrust and torque at or close to the zero azimuthing position can be satisfactorily accomplished using classical hydrodynamic lifting line, lifting surface or boundary element methods. Similarly, estimates can be made of the other forces and moments about the propeller's Cartesian reference frame. However, full-scale trial measurements conducted some years ago on cruise ships' propellers with conventional A-bracket shafting arrangements suggested that, within the then current state of development of the propeller computational codes, a greater error bound should be allowed for when extending the calculation of these loadings in the other Cartesian directions.

When undertaking maneuvers including turns and stopping, both at sea and when in harbor, as well as when operating in poor weather, model tests have indicated that the hydrodynamic loadings can significantly increase (References 3 and 4). Moreover, the predictions of these loadings do not at present lend themselves to assessment by the normal classical methods of analysis but must be estimated from model-or full-scale data. Similarly, Reynolds Averaged Navier–Stokes (RANS) codes are currently not at the required state of development to confidently make quantitative predictions of the loads; nevertheless they can give useful qualitative insights into the flow behavior and the various interactions involved.

The loadings developed by the pod are complex since the axisymmetric body and a part of the fin, or strut, need to be analyzed within the helicoidal propeller slipstream for a tractor unit. The remainder of the strut lies in a predominantly translational flow field and for analysis purposes has to be treated as such. Furthermore, the interaction between the propeller and pod body is complex and this also needs to be taken into account. A different flow regime clearly exists in the analysis of pusher units since the propeller then operates in the wake of the strut, and pod body and the propeller-pod body interaction effects are significant. Notwithstanding these complexities it is possible to make useful quantitative approximations using earlier empirical data, provided a proper distinction is made between those parts of the propulsor which are subjected to translational flow and those which will operate within the propeller slipstream. In this context the earlier work of Gutsch[5] for inclined propellers can be put to good use provided that appropriate corrections are made. Alternatively, systematic model test data, albeit in a limited form, is now beginning to emerge in the technical literature, for example that contained in Reference 6.

Yakovlev[7] has found that reasonable accuracy for loading estimations can be obtained by a combination of analytical methods and empirical relationships. In this methodology the computation of the propulsor characteristics at large flow angles or extreme advance coefficients utilizes the Rayleigh approach to address the influence of separation. Additionally, the analysis of model test data results in the development of coefficients which are then utilized in the procedure to give an empirically based calculation procedure.

The forces and moments in the three Cartesian directions need to be quantitatively estimated as accurately as possible, either by model test or by calculation for the full range of different operating conditions, since without such an assessment the reactive loads on the bearings cannot be properly estimated. Indeed, if these loading estimates are inadequate, then the necessary fatigue evaluations that are undertaken for the bearing materials will prove unreliable and this may then contribute to premature bearing failure. Figure 15.5, by way of example, illustrates a typical variation in propeller blade thrust generated at two different azimuthing angles, 15° and 35°, as a propeller rotates through one revolution. This should be contrasted with the nearly constant thrust and torque signature produced at a zero azimuthing angle.

Notwithstanding the implied reliance on empirical data from model tests, since full-scale data is difficult to obtain for podded propulsors, the scale effects relating to the pod-ship and pod-propeller interaction mechanisms are significant (Reference 3). Therefore, when measurements are made in a model facility the experiment must be carefully designed in order to minimize these effects. However, research effort still needs to be expended in refining the analysis of scale effects in order to gain a fuller understanding of their influence both in terms of the propeller loading and also for ship propulsion studies[8] Nevertheless, the ITTC, Appendix A of Reference 8, have developed a draft procedure and guidance for the extrapolation of podded propulsor model tests. A general problem is the

TABLE 15.1 Typical Change in Power Requirement with Pod Attitude Angle

Pod tilt angle (°)	2	4	6
Increase in P_D (%)	0	1.3	1.7
Azimuth angle (°)	−2	0	+2
Increase in P_D (%)	0.8	0	1.6

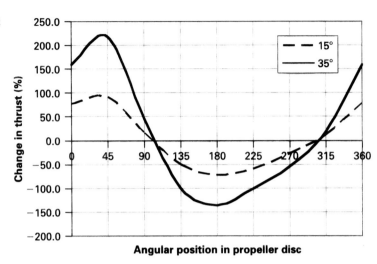

FIGURE 15.5 Typical propeller blade-induced thrust fluctuations for different pod azimuthing angles.

treatment of scaling of the pod housing drag where currently a number of methods exist. Sasaki et al.[9] have shown that a considerable scatter exists between the various methods that have been proposed. However, this scatter does not necessarily imply a similar scatter in the final power prediction for the ship.

Computational fluid dynamic RANS methods potentially offer a means whereby scale effects can be considered free from the constraints imposed by model testing and institutional practices. Chicherin et al.[10] have endeavored to draw conclusions from studies using RANS codes. First they considered that the numerical analyses do not support the application of the conventional appendage scaling procedures for full-scale pod housing drag estimates. Second, the form factor concept is inappropriate for pod housing drag scaling and finally, the most suitable extrapolation parameter is the non-dimensional resistance coefficient used as a correction to the drag of the complete pod unit.

When testing podded propulsor configurations at model scale it is important that care is taken to correct the results for the presence of the gap effect between the propeller and the pod body. Corrections of this type have been known to be necessary for similar configurations with ducted propellers, pump-jets and azimuthing thrusters for many years and their importance lies in the correct estimation of thrust. Similarly, there is also a gap effect between the top of the strut and the lower end-plate of the model test setup. Within present knowledge this is thought to be relatively small.

The pod body design is important in terms of minimizing boundary layer separation and vorticity development. Islam et al.[11-16] in a series of papers have made a useful model test and analytical examination of the influence of the various pod body geometric features. In particular, the radius of the axisymmetric pod body needs to be minimized, but this is dependent on the electro-dynamic design of the propulsion motor, which is, in turn, both motor speed and length dependent. Furthermore, the motor speed, along with the unit's speed of advance and power absorption, governs the propeller efficiency. In terms of future development, high-temperature superconducting motor research offers a potential for significant reductions in motor diameter and hence, if realized, will facilitate pod body drag reductions.

Full-scale experience shows that for twin-screw propulsion systems, podded propulsors, when operating close to the zero azimuthing position, generally have a superior cavitation performance when compared to conventional propulsion alternatives. This implies that the propeller radiated hull surface pressures will be significantly reduced. Typically for a non-ice classed, tractor podded propulsion system operating on a well-designed cruise ship hull form, the blade rate harmonic hull surface pressures can be maintained at around 0.5 kPa, with the higher blade rate harmonics generally being insignificant. Such a finding is compatible with the expected enhanced wake field in which the podded propeller operates. However, while not generally reaching these low levels, it should be recalled that the radiated hull surface pressures for conventionally designed ship afterbodies with shaft lines and A-brackets have improved significantly in recent years, typically returning values in the region of 1 kPa. Notwithstanding this benefit, it has been found that broadband excitation can have a tendency to manifest itself more frequently with podded propulsor propeller ships. When significant azimuthing angles are encountered or large incidence buttock flows are encountered then the effects of cavitation may be rather more significant and solutions to this may involve aspects of the propeller blades and the pod body.

15.2.2 Turning Maneuvers

When turning at speed in calm open sea conditions a complex flow regime is generated in the vicinity of the propeller which significantly alters the inflow velocity field. For a twin-screw ship undertaking a turn the resultant forces and moments generated by the propellers located on the port and starboard sides of the hull are different. This difference depends upon whether the propeller is on the inside or outside of the turn and on the extent of the influence of the ship's skeg on the transverse components of the global flow field in way of the propellers. Figure 15.6 shows an example of these differences, measured at model scale on the starboard propeller, during turns to port and starboard when operating at constant shaft rotational speed. Analogous variations are seen for the other force and moment components generated by the propeller in these types of maneuver.

By implication Figure 15.6 underlines the importance of implementing a proper speed control regime for podded propulsion systems. It can be seen that if the shaft speed were not reduced during the turn to port the starboard motor would be in danger of being overloaded if the shaft speed at the beginning of the maneuver was close to the normal service rating. Furthermore, in this context accelerations and decelerations of the ship are also important. In this latter case the rate of change of shaft speed during such a maneuver influences considerably the loadings generated by the propeller.

The thrust, torque, lateral forces and moments also vary significantly throughout a turning maneuver. A typical result measured from the model test programs discussed in Reference 17 and related to the work of Ball and Carlton[3,4] is shown in Figure 15.7.

Figure 15.7 relates to a maneuver which changes the heading of the ship though 180°. It can be seen that relative to the steady values recorded on the approach to the turn, as soon as the propulsors change their azimuthing angle the torque and thrust increase. Similarly, the fluctuating shaft bending moment measured on the shaft increases in amplitude and then decays to an extent during the turn but then maintains steady amplitude. In contrast the thrust and torque maintain their enhanced amplitude throughout the turn and then when corrective helm is applied to return the ship on a reciprocal course these parameters then decay back to their normal ahead values. However, this is not the case for the bending moment amplitudes which upon applying corrective helm then sharply increase before decaying back to their pre-maneuver values. When analyzing this data during constant shaft speed turns over a number of similar tests it is seen that while the thrust and torque increase their pre-maneuver levels by between 10 and 50 per cent, the maximum bending amplitudes are amplified by factors of between four and six times their original free running ahead values. In the case of zig-zag maneuvers carried out under the same operating conditions analogous characteristics are found.

These types of maneuvering examples, because of their potential to develop high bearing loadings, suggest that careful thought should be applied to sea trial maneuvering programs. Ships driven by podded propulsors generally exhibit a better maneuvering performance when compared to equivalent conventionally driven ships. The implications of this enhanced performance are discussed in Reference 17 and, in particular, the advisability of employing an equivalence principle for ship maneuverability between podded propulsor and conventionally driven ships is discussed so as to minimize the risk of shaft bearing overload in the podded propulsors. Such a principle essentially suggests that if the maneuvering of a conventionally driven ship is satisfactory then it should not be necessary to effectively demonstrate that a similar ship fitted with a podded propulsor has better maneuvering capabilities since this is already known. Consequently, the sea trials program should be adjusted to define turning rates and other conditions appropriate to the ship and podded propulsor configuration.

15.2.3 Crash Stop Maneuvers

In the case of linearly executed stopping maneuvers (Reference 3) exploratory model tests have indicated that if a crash stop maneuver is executed with the podded propulsors in a fore and aft orientation the bending moments generated can be limited to values consistent with the normal free running service speed. Notwithstanding this, the thrust and torque loadings change significantly during the maneuver. If, however, the pods were permitted to take up a toe-out attitude then the shaft forces and bending moments could be expected to significantly increase. In the case, for example, of a 25° toe-out maneuver at constant shaft speed, the ratio of induced bending moment during the maneuver to the free running bending moment at the start of the maneuver could be as high as 11, with similar ratios being developed in the other in-plane loadings.

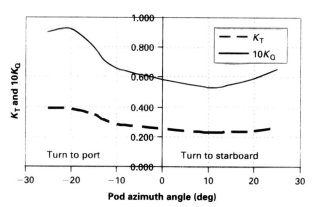

FIGURE 15.6 Variation in thrust and torque coefficient when turning at constant shaft speed.

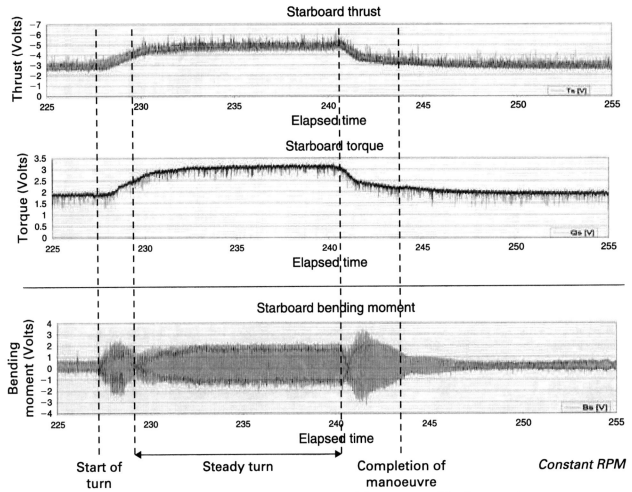

FIGURE 15.7 Typical fluctuations in propeller-induced thrust, torque and bending moment.

15.2.4 Podded Propulsors in Waves

The effects of poor weather have also been similarly explored, from which the relative motions between the propeller and the seaway have been seen to increase the loadings which have to be reacted to by the shaft bearings. For the model configuration tested (Reference 3) the shaft bending moments in irregular waves were found to increase by up to a factor of 1.8 over the free running condition when encountering significant wave heights of 7 m at constant shaft speed. In those tests the sea conditions which gave rise to the greatest increase in shaft forces and moments were those encountered in head quartering seas. Clearly, however, in the general case these loading factors will be ship motion and seaway dependent.

15.2.5 General and Harbor Maneuvers

Different loading regimes occur during low-speed harbor maneuvering. It has been known for many years that if a number of azimuthing thrusters are deployed on the bottom of a marine structure in a dynamic positioning mode, when particular relative azimuthing angles of the thrusters occur, they will mutually interfere with each other. In some cases if this mutual interference was severe, mechanical damage to the thruster shaft line components could result.

By simulating the underwater stern of a typical cruise ship with a deployment of propulsors and varying their relative azimuthing angles (Reference 4) a number of good and bad operating conditions were identified for a twin-screw ship. These may be summarized as follows:

i. If the pods are at arbitrary azimuthing angles and the angle of one of the pods is chosen such that its efflux passes into the propeller disc of the other then high fluctuating shaft bending moments and radial forces can be expected to occur on the latter pod. The magnitude of these shaft bending moments at model scale has been measured to reach values of up to 10 times the normal free running values at low ship speed and constant rotational speed. In contrast the thrust and torque forces appear to be relatively unaffected. At full scale when interference has been encountered between podded propulsors, vibration levels up to 116 mm/s have been recorded at the tops of the pods in the vicinity of the slewing ring.

ii. If both pods are positioned such that they are thrusting in approximately the same thwart-ship line, then the trailing pod will suffer significant fluctuating loads. The maximum loadings will be experienced when the trailing pod is slightly off the common transverse axis: whether this relative azimuthing angle is forward or aft of the thwart-ship line will depend upon the direction of rotation of the propulsors.

iii. It has been found that a benign harbor maneuvering condition is when both podded propulsors are in a toe-out condition. At this condition the mutual interference with respect to shaft loads is minimal.

iv. The control methodology of podded propulsors when undertaking dynamic positioning maneuvers requires careful consideration if unnecessary, and in some cases harmful, azimuthing activity is to be avoided.

It was also observed that the interference signatures created in conditions (i) and (ii) exhibited a finely tuned characteristic with respect to relative azimuthing angle.

In the case of a quadruple-screw ship, a poor operating condition was found to be when the podded propulsors on one side of the ship are both operating and are positioned such that the forward unit is in the fore and aft direction and the astern one is transverse. In this case the efflux from the forward propeller is attracted towards the transversely oriented propeller which then suffers strong fluctuating loads. This is because it is operating obliquely in the helical flow field generated by the fore and aft aligned propulsor. Similarly, when both propellers on one side of the ship are aligned in the fore and aft direction the efflux from the forward propeller, although relatively attenuated, is attracted towards the propeller in the astern location. Consequently, some benefit in minimizing these slight oblique flow characteristics can be achieved by azimuthing the astern propeller towards the location of the forward propeller.

The severity of mutual propulsor interference is load dependent in that at high speeds the effects are potentially considerably more harmful than at low speeds. Under many harbor maneuvering conditions, particularly in benign weather conditions, propulsor speeds are low and consequently the propeller's effluxes possess low energy. Due to energy dissipation this implies that the potential of the efflux to damage an adjacent unit does not extend too far from the location of a propulsor. Under these conditions it is therefore unlikely that significant adverse loadings will be encountered if, for example, one propeller of a twin-screw ship is aligned fore-aft to give longitudinal control while the other is placed in a transverse alignment to control sideways movement.

15.2.6 Specific Podded Propulsor Configurations

A number of configurations for podded propulsors have been developed in recent years and these have included units with tandem propellers, rim-driven propulsors and contra-rotating versions. In recent times a pump-jet variant has also been proposed (Reference 18).

In the case of contra-rotating podded propulsors these tend to be hybrid designs which deploy a conventional propeller and stern bearing arrangement for the ship with a tractor azimuthing podded propulsor located immediately astern of the conventional propeller; Figure 2.9. In this way, when moving ahead on a straight course the efficiency advantages of contra-rotating propeller can be gained without the complexities entailed in the mechanical shafting arrangements. Moreover, such an arrangement has the added benefit of being able to distribute the power between the two propellers, either equally or favoring perhaps the conventional propeller, and in so doing gives a rather better cavitation environment for the absorption of the total propulsion power. These types of arrangements are potentially attractive to relatively fast ships such as Ro/Pax ships. Under turning conditions, since the podded propulsor acts as the rudder it has to be ensured that the bearings in the podded propulsor can withstand the level of excitation generated from the periodic loadings generated by the leading propeller as well as the additional loads induced by the turning maneuver. Bushkovsky et al.[19] examined the mutual propeller interaction comprising the periodic forces and the crash stop behavior of these types of configuration. In the case of the periodic forces these were shown to be complex because of each propeller's induced velocities in the disc of the other and, furthermore, the induced velocities were both spatially and temporally dependent. This placed considerable demands on the computational procedures by which the induced velocities were calculated and overcome, in this case by coupling the vortex sheets of the propeller blades and making an analytical estimate of the viscous wake behind the propeller. From this analysis it was shown, perhaps rather unsurprisingly, that the periodic forces have a wider spectrum of harmonics than would be the case for a conventional single propeller. With regard to cavitation, it was considered important to avoid the podded propulsor of the contra-rotating pair interacting with the conventional propeller's hub vortex and blade generated cavitation when the podded propulsor turns for steering purposes.

As such, this situation needs to be carefully considered at the design stage.

Rim-driven podded propulsors comprise a multiple blade row propeller with a permanent magnet, radial flux motor rotor located on the tips of the propellers which then interacts with the motor's stator which is sited within a duct circumscribing the propeller. It is claimed that this arrangement yields the required thrust with a smaller pod when compared to the normal arrangement; it develops a higher efficiency, develops reduced unsteady hull surface pressures; and has improved cavitation performance. The concept has been model tested (Reference 20) and a 1.6 MW demonstrator unit was destined for sea trials during 2006. Such an arrangement is not dissimilar to an electromagnetic tip-driven propeller designed and tested by Abu Sharkh et al.[21] This 250 mm diameter, four-bladed Kaplan type propeller was tested over a wide range of advance and rotational speeds, with differing duct geometries and a limited variation of stator angles. It was also tested in sea water. The unit was benchmarked against a Ka 4–70 propeller in a No. 37 duct and it was found that at bollard pull conditions the thrust was about 20 per cent lower than the Wageningen series propeller and that the K_T values reduced more rapidly as the advance speed increased. This discrepancy was attributed to the additional drag of the propeller ring and the thicker duct.

REFERENCES AND FURTHER READING

1. Minsaas KJ, Lehn E. *Hydrodynamical Characteristics of Thrusters*. NSFI Report R-69.78; 1978.
2. Oosterveld MWC. *Ducted propeller characteristics*. Ducted Propeller Symp., Paper No. 5. Trans RINA; 1973.
3. Ball WE, Carlton JS. *Podded propulsor shaft loads from free-running model experiments in calm-water and waves*. J Maritime Eng, Trans RINA 2006;**148**(Part A4).
4. Ball WE, Carlton JS. *Podded propulsor shaft loads from model experiments for berthing manoeuvres*. J Maritime Eng, Trans RINA 2006;**148**(Part A4).
5. Gutsch F. *Untersuchung von Schiffsschrauben in Schräger Anstromung*. Schiffbanforschung 1964;**3**.
6. Frolova I, Kaprantsev S, Pustoshny A, Veikonheimo T. *Development of the propeller series for Azipod Compact*. Brest: T-Pod Conf.; 2006.
7. Yakolev A. *Calculation of propulsion pod characteristics in off-design operating conditions*. Trondheim, Norway: First Int. Symp. on Marine Propulsion, smp'09; June 2009.
8. *Report of the Podded Propulsor Committee*. Edinburgh: Proc. 25th ITTC; 2005.
9. Sasaki N, Laapio J, Fagerstrom B, Juurma K, Wilkman G. *Full scale performance of double acting tankers Mastera & Tempera*. University of Newcastle, UK: 1st T-Pod Conf; 2004, 155–172.
10. Chicherin IA, Lobatchev MP, Pustoshny AV, Sanchez-Caja A. *On a propulsion prediction procedure for ships with podded propulsors using RANS-code analysis*. University of Newcastle, UK: 1st T-Pod Conf; 2004, 223–236.
11. Islam MF. *Numerical investigation on effects of hub taper angle and pod-strut geometry on propulsive performance of pusher propeller configurations*. MSc Thesis. Canada: Memorial University of Newfoundland; 2004.
12. Islam M, Veitch B, Bose N, Liu P. *Cavitation characteristics of pushing and pulling podded propellers with different hub taper angles*. Halifax, NS: 7th MHSC; 2005.
13. Islam MF, Molloy S, Veitch B, Bose N. *Numerical and experimental studies of pod-strut geometry on podded propulsor performance*. Canada: CFD 2005, St John's, Nfdl; 2005.
14. Islam M, Veitch B, Bose N, Liu P. *Numerical study of effects of hub taper angle on the performance of propellers designed for podded propulsion systems*. Maritime Technology; 2005.
15. Islam MF, Molloy S, He M, Veitch B, Bose N, Liu P. *Hydrodynamic study of podded propulsors with systematically varied geometry*. Brest: T-Pod Conf.; 2006.
16. Islam MF, Veitch B, Bose N, Liu P. *Effects of hub taper angle on the performance of podded propulsors*. Marine Technology 2006;**43**(1):1–10.
17. Carlton JS, Rattenbury N. *Aspects of the hydro-mechanical interaction in relation to podded propulsor loads*. Brest: T-Pod Conf.; 2006.
18. Bellevre D, Copeaux P, Gaudin C. *The pump jet pod: An ideal means to propel large military and merchant ships*. Brest: 2nd T-Pod Conf.; 2006.
19. Bushkovsky VA, Frolova IG, Kaprantsev SV, Pustoshny AV, Vasiljev AV, Jakovlev AJ, Veikonheimo T. *On the design of a shafted propeller plus electric thruster contra-rotating propulsion complex*. University of Newcastle, UK: 1st T-Pod Conf; 2004. 247–261.
20. Lea M, Thompson D, Blarecom B, van, Eaton J, Richards J, Friesch J. *Scale model testing of a commercial rim-driven propulsor pod*. Trans SNAME Annual Mtg; 2002.
21. Abu Sharkh SM, Turnock SR, Hughes AW. *Design and performance of an electric tip-driven thruster*. Proc I Mech E 2003; **217**(Part M).

Chapter 16

Waterjet Propulsion

Chapter Outline

- 16.1 Basic Principle of Waterjet Propulsion — 364
- 16.2 Impeller Types — 366
- 16.3 Maneuvering Aspects of Waterjets — 367
- 16.4 Waterjet Component Design — 367
  - 16.4.1 Tunnel, Inlet and Supporting Structures — 368
  - 16.4.2 Impeller — 369
  - 16.4.3 Stator Blading — 370
  - 16.4.4 Nozzles, Steering Nozzles and Reversing Buckets — 371
- References and Further Reading — 371

The concept of waterjet propulsion dates back to 1661 when Toogood and Hays first proposed this form of propulsion. Its use was initially confined principally to small high-speed pleasure craft and work boat situations where high maneuverability was required with perhaps also a draught limitation. It is only comparatively recently that the waterjet has been considered for large high-speed craft and as a consequence the sizes of the units increased considerably during the last few years.

The principal reason for the comparatively infrequent early use of the waterjet in comparison to the screw propeller was that the propeller was generally considered to be a simpler, lighter and more efficient propulsor. However, there were situations where the propeller failed to give a satisfactory propulsion solution, particularly in relation to the commercial demand for higher-speed craft, and with the introduction of more efficient pumps this created two main reasons for their rapid growth in utilization.

As was seen from Figure 2.13 the waterjet has three main components: an inlet ducting, a pump and an outlet or nozzle. This rather simplified diagram can be enhanced as shown in Figure 16.1 which shows, albeit in schematic form, a typical waterjet in rather more detail. From the figure it is seen that the basic system comprises an inlet duct which is faired into the hull in the most convenient way for the vessel concerned. From this inlet duct the water then passes through the impeller, which may take a variety of forms: most usually this is a mixed or axial flow device comprising a number of blades ranging from four to eight. The next phase in the passage of the water through the unit is normally to pass through a stator ring which has the dual function of straightening the flow and also acting as a support for the hub body. Stator rings are likely to comprise some 7–13 blades, but it should be noted that not all designs utilize this feature. With some designs of waterjet the nozzle is steerable while in others deflector plates are used to control the direction of the flow and hence impart steering forces to the vessel through the change in direction of the momentum of the waterjet. The final feature of the system is the reversing bucket.

The reversing bucket is a mechanically or hydraulically actuated device which can be lowered over the waterjet exit so as to produce a retarding force on the vessel, again through a change in momentum. In some designs the bucket is designed so that it can, as well as providing a total braking capability, 'spill' part of the jet so that a fine control can be exerted over the propulsion force generated by the unit.

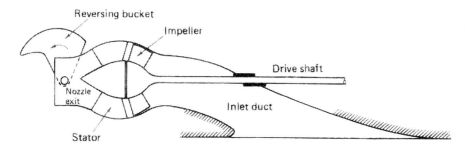

FIGURE 16.1 Typical waterjet general arrangement.

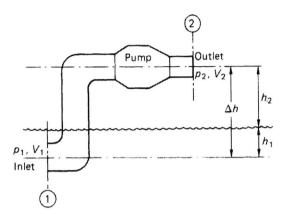

FIGURE 16.2 Idealized waterjet arrangement.

16.1 BASIC PRINCIPLE OF WATERJET PROPULSION

As a basis for considering the underlying principles of waterjet propulsion reference should be made to Figure 16.2 which shows an idealized waterjet system. Using this diagram, suppose that the water enters the system with the velocity V_1 and leaves with a different velocity V_2 by means of a nozzle of area A_2. The mass flow of the water through the waterjet is then given by

$$\dot{m} = \rho A_2 V_2$$

where ρ is the density of the water.

Hence, the increase in the rate of change of momentum of the water passing through the waterjet is given by $\rho A_2 V_2 (V_2 - V_1)$ and since force is equal to the rate of change of momentum, the thrust produced by the system is

$$T = \rho A_2 V_2 (V_2 - V_1)$$

and the propulsion power P_T is given by

$$P_T = T V_S = \dot{m} V_S (V_2 - V_1) \quad (16.1)$$

where V_S is the speed of the vessel and $\dot{m}$ is the mass flow rate of the water $\rho A_2 V_2$.

Now in order to derive a useful expression for the power required to drive the waterjet system it is necessary to appeal to the general energy equation of fluid mechanics and to apply this between the inlet and outlet of the unit. Hence it can be written for the system,

$$\frac{p_1}{\rho g} + \frac{V_1^2}{2g} + H_p = \frac{p_2}{\rho g} + \frac{V_2^2}{2g} + \Delta h + h_{\text{loss}} \quad (16.2)$$

where

H_P is the head associated with the energy supplied to the system (i.e. the pump head).

Δh is the difference in static head between the inlet and outlet of the waterjet (i.e. $\Delta h = h_1 + h_2$).

h_{loss} is the losses associated with the flow through the system and the pump losses.

In the case of the difference in static head between the inlet and outlet of the waterjet system it should be noted that this will be a variable between start-up and sailing conditions. This is particularly true for hydrofoils which are propelled by waterjets, and of which Figure 16.2 is particularly representative in the cruising condition. With regard to the loss term, h_{loss}, this is associated with frictional and eddy shedding losses which occur around bends in the ducting, in way of inlet grillages and the various obstructions throughout the system which may impede the flow during its passage through the unit.

Returning to equation (16.2) and for practical purposes assuming that p_2 is constant above the waterline since the altitudes involved and their effect on ambient pressure are small, equation (16.2) can be rewritten as:

$$H_p = \frac{V_2^2 - V_1^2}{2g} + h_2 + h_{\text{loss}} \quad (16.3)$$

since $p_1 = p_2 + h_1 \rho g$.

The power transferred to the water by the pump (P_{pump}) can be expressed in terms of energy per unit time as $\dot{m} g H_p$, which from equation (16.3) leads to the expression

$$P_{\text{pump}} = \dot{m} \left[\frac{1}{2}(V_2^2 - V_1^2) + g(h_2 + h_{\text{loss}}) \right] \quad (16.4)$$

Hence, the equivalent open water efficiency of a waterjet unit can be defined from equations (16.1) and (16.4) as being the ratio of the thrust horsepower to the delivered horsepower as follows:

$$\eta_o = \frac{V_s(V_2 - V_1)}{\left[\frac{1}{2}(V_2^2 - V_1^2) + g(h_2 + h_{\text{loss}}) \right]} \quad (16.5)$$

The loss term h_{loss} in equation (16.5) is the sum of two independent losses; those defined as internal losses h_D and those relating to the pump head loss h_p. Therefore h_{loss} can be written as

$$h_{\text{loss}} = h_D + h_P \quad (16.6)$$

The internal losses are primarily dependent on the waterjet configuration and comprise the intake losses h_{DI}, the diffuser head losses h_{DD} and the skin friction losses h_{DSF}. Thus, expressing this formally,

$$h_D = h_{DI} + h_{DD} + h_{DSF} \quad (16.7)$$

The intake losses are in themselves the sum of the losses arising from the intake guard, the guide vanes and the various bends. All of these losses are principally a function of the intake velocity V_1 and can consequently be expressed in the form:

$$h_{DI} = k \frac{V_1^2}{2g}$$

where the coefficient $k,^1$, is the sum of two other factors k_1 and k_2 which represent the losses due to the guard and guide vanes, and the losses due to the bends respectively. Typically values for k_1 and k_2 are 0.10 and 0.015 respectively.

The diffuser head loss can be estimated from normal hydraulic methods, from which an expression for h_{DD} can be obtained as,

$$h_{DD} = (1 - \eta_D)(1 - \varepsilon^2)\frac{V_1^2}{2g}$$

in which η_D is the diffuser efficiency, of the order of 90 per cent in normal circumstances, and ε is ratio of the entrance and exit areas of the diffuser.

The final term in equation (16.7), h_{DSF}, which defines the skin friction losses can be estimated from calculating the wetted surface areas of the intake, ducting, diffusers, supporting struts and vanes and the nozzle in association with their respective frictional coefficients.

If then the sum of the internal losses h_D, as defined in equation (16.7), is then represented in terms of a single loss coefficient this can take the form:

$$h_D = k_D \frac{(V_S + \Delta V)^2}{2g}$$

where $\Delta V = (V_2 - V_1)$, and van Walree[1] suggests that the value of k_D would normally lie in the range $0.04 < k_D < 0.10$.

The pump head loss term h_p of equation (16.6) is related solely to the pump configuration and its associated losses. This head loss can be expressed in terms of the pump head H and the efficiency of the pump η_p as

$$h_p = H\left(\frac{1 - \eta_p}{\eta_p}\right)$$

and for a modern well-designed axial or mixed flow pump the value of η_p should be of the order of 0.90.

By analogy with propellers the pump efficiency can be expressed as

$$\eta_p = \frac{\phi}{2\pi}\frac{\psi}{K_Q} \quad (16.8)$$

where ϕ and ψ are the flow and energy transfer coefficients defined by

$$\phi = \frac{Q}{ND^3} \quad \psi = \frac{gH}{N^2D^2}$$

and K_Q is the normal torque coefficient of propeller technology.

While the value of η_p is clearly higher for a waterjet than a propeller, this is not the basis upon which the comparison should be made. A proper comparison can only be made in terms of the corresponding quasi-propulsive coefficients which for the propeller include the hull and relative rotative efficiency and for the waterjet equation (16.5) together with the appropriate hull coefficient embracing the effect of the waterjet.

In a waterjet propulsion system the hull and waterjet mutually interfere with each other. The naked hull resistance is modified due to a distortion of the flow over the ship's afterbody which at high speeds may also introduce a change in trim, thereby influencing the resistance characteristics further. Similarly, the waterjet's performance is altered by the distortions in the hull flow because the vessel's boundary layer is ingested into the waterjet intake system and, consequently, differs from the normal free stream assumptions of waterjet theory. To address this problem a parametric model for the description of the overall powering behavior of a waterjet–hull configuration was developed by van Terwisga[12] which permits the separate identification of the interaction terms. Moreover, as part of this work an experimental procedure to determine these interaction effects was validated.

Numerical methods, in particular computational fluid dynamics, have now reached a state of development where the flow characteristics in waterjet systems can be determined at least in a qualitative manner but also in some cases quantitatively. Bulten[13] has explored the flow through a waterjet using Reynolds Averaged Navier–Stokes codes in which the turbulence modeling was done with the $k-\varepsilon$ two equation model. This approach was satisfactorily validated experimentally for the flow through the inlet and mixed flow pump regions. Moreover, within the investigation it was found that the magnitude of the radial rotor–stator interaction force depended on the flow rate through the pump under uniform inflow conditions. However, when operating with a non-uniform velocity field entering the pump an additional mean component of radial force was found to exist whose magnitude and direction are dependent on both the flow rate and the level of non-uniformity in the flow. This is considered to be due to the variation in angle of attack during one revolution causing an unbalanced torque on the impeller blades.

The deployment of computational fluid dynamics to the problems associated with waterjet flows is becoming increasingly widespread. The technique is being applied to thrust breakdown and cavitation as well as the design integration issues, (References 14–16), with some success. Kinnas et al.[17] have extended their ducted propeller boundary element code by developing a potential flow computational method to predict the performance of a cavitating waterjet. This was then validated against the results from the axial-flow pump designed by Lavis et al.[18] but the computed torque values were significantly lower than the experimental values. This was attributed to the modeling of the gap flow and the simplicity of the viscous methodology used in the method.

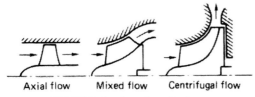

FIGURE 16.3 Pump impeller types.

16.2 IMPELLER TYPES

When designing a waterjet to perform a given duty, the most appropriate type of pump for the intended duty must be established. The choice of pumps will lie between a centrifugal, mixed flow, axial flow pump and inducer. Figure 16.3 categorizes the first three types of turbomachine while Figure 16.4 shows an inducer that has been laid out for inspection.

For a given hydraulic turbomachine there is a unique relationship between the unit's efficiency and the flow coefficient, assuming that both Reynolds and cavitation effects are negligible: this is analogous to the propeller efficiency curves. For a pump unit the efficiency versus flow coefficient curve will take the form shown in Figure 16.5.

Additionally, other performance coefficients can be determined from dimensional analysis and in this context the energy transfer coefficient ψ is particularly important. From Figure 16.5 it will be seen that as the flow coefficient is increased the efficiency tends to rise and then reach a maximum value after which it will fall-off rapidly. The optimum efficiency point can be used to identify a unique value of the flow coefficient.

Additionally, a corresponding value of ψ can be uniquely determined. In pump technology it is customary to define the specific speed N_s of a machine from the values of ϕ and ψ which correspond to the maximum efficiency point as being

$$N_S = \left.\frac{\phi^{1/2}}{\psi^{3/4}}\right|_{\eta_{max}}$$

which reduces to

$$N_S = \frac{NQ^{1/2}}{(gH)^{3/4}}$$

or, more typically, in its dimensional form

$$N_S = \frac{NQ^{1/2}}{H^{3/4}} \quad (16.9)$$

Because of the dependence of the specific speed on the maximum efficiency point on the pump characteristic curve, this parameter is of considerable importance in

FIGURE 16.4 Typical inducer design.

selecting the type of turbomachine required for the given duty. To change the maximum efficiency point with respect to the flow coefficient, as shown in Figure 16.5, requires that the pump geometry must change: as a consequence the maximum efficiency condition replaces the geometric similarity condition. Furthermore, each of the different classes of machine shown in Figures 16.3 and 16.4 have their optimum efficiencies defined within a fairly narrow band of specific speed. In general, the physical size of the impeller, for a given duty defined by the flow and head required, varies with specific speed. Hence, the higher the specific speed the more likely it is that an axial flow machine will be specified which is smaller physically than its centrifugal counterpart.

As a consequence, since high specific speed implies a smaller machine it is desirable to select the highest specific speed, consistent with good efficiency, for a particular application.

The centrifugal pump exhibits low flow—high pressure characteristics while the converse is generally true for the axial flow machine. Table 16.1 indicates the general ranges

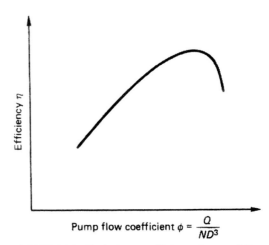

FIGURE 16.5 Typical pump efficiency characteristic.

Waterjet Propulsion

TABLE 16.1 Typical Specific Speed Ranges of Various Pump Types

Pump Type	Approximate N_s (rad)
Centrifugal pump	Below 1.2
Mixed flow pump	1.2–3.0
Axial flow pump	3.0–7.0
Inducers	Above 7.0

that the various pump classes suitable for waterjet design would be expected to operate within.

While the centrifugal and axial flow pumps were the original types of machine used for waterjets the mixed flow type, which is a derivative of the centrifugal machine, rapidly established itself. This was because it provided a smaller diameter unit than the centrifugal pump and offered an easier conversion of pump head to kinetic energy.

The machines shown in Figure 16.3 are well known and their theoretical background is well defined in many textbooks, for example References 2 and 3.

The inducer, Figure 16.4, was developed originally in response to the need for large liquid fuel pumps for rocket propulsion. From the figure it is seen that the waterjet inducer in this case comprised four full blades in the initial stage with a further four partial blades which allow the suction stage to be shortened to some extent. These large blades are then followed by a row of short blades which produce about 60 per cent of the head rise through the machine.

16.3 MANEUVERING ASPECTS OF WATERJETS

The waterjet principle lends itself particularly well to a propulsion system with integral steering capabilities. The majority of waterjet units are fitted with either a steerable nozzle or deflectors of one form or another in order to provide a directional control of the jet. The steering capability in each of these cases is produced by the reaction to the change in momentum of the jet (Figure 16.6(a)). The angle through which the jet can be directed is of course a variable depending on the manufacturer's particular design; however, it would generally be expected to be of the order of mn; 30°.

With regard to the stopping or retarding force capabilities of waterjets, these are normally achieved with the aid of a reversing bucket with the stopping force being produced by change of momentum principles. The reversing bucket design can be of the simple form shown in Figure 16.1 or, alternatively, of a more sophisticated form which allows a 'spilling' of the jet flow in order to give a fine control to the braking forces (Figure 16.6(b)). With this latter type of system the resultant thrust can also be continuously varied from zero to maximum at any power setting for the prime mover.

16.4 WATERJET COMPONENT DESIGN

The literature on the design and analysis of waterjet propulsion systems is extensive. The references at the end of this chapter give several examples from which further works can be traced. This section, however, is not so much concerned with the underlying mathematical and theoretical engineering principles, which were dealt with in

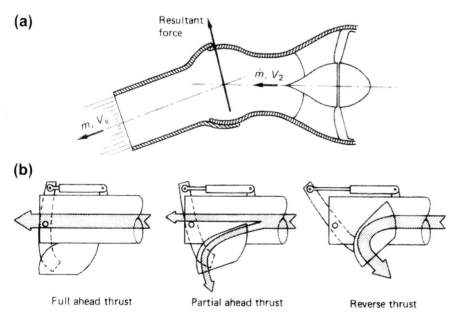

FIGURE 16.6 (a) Principle of waterjet steering capability and (b) waterjet thrust control mechanism.

TABLE 16.2 Waterjet Design Matrix

IVR Design/Condition	1	2	3	4	5
Off-design condition 1	*	*	*	*	*
Off-design condition 2	*	*	*	*	*
⋮	⋮	⋮	⋮	⋮	⋮
Off-design condition N	*	*	*	*	*

Section 16.1 but with the practical design aspects of the various components.

In terms of a general design approach to the problem, the fundamental parameter is the inlet velocity ratio, IVR. This parameter is defined as:

$$\text{IVR} = \frac{V_1}{V_s} \qquad (16.10)$$

where V_1 is the water inlet velocity and V_s is the craft velocity.

This parameter in effect controls the flow rate through the waterjet together with the velocity ratio, the pump head, the overall efficiency and in addition the inception of cavitation at the intake lips.

The basic design procedure is to consider a range of IVR values at the craft design speed from which a pump design, delivered power and efficiency can be calculated. Then for each of these pump designs the off-design performance can be considered and again a set of delivered powers, efficiencies and cavitation conditions can be considered. From the resulting matrix of values, Table 16.2, the designer can then select the most suitable combination of results to suit the craft conditions; typically the cruise and hump speeds. With this choice made, another iteration of the above procedure can be undertaken, if necessary, or alternatively the designer may progress to the detailed design of the unit. In this latter case the question then arises as to whether the waterjet should have a variable area water intake so as to allow some variation in the IVR. Depending on the answer, a further iteration of Table 16.2 may be necessary. With regard to the details of the calculation for a given IVR value, this would take the form shown in Figure 16.7. In practical terms this outline design procedure can be used to design a unique waterjet unit or, alternatively, to select the closest model from a predefined range for a particular duty.

With respect to the design of the various components of the waterjet unit, several aspects, in addition to strength, need to be taken into consideration. To detail the most important of these, it is necessary to consider each component separately.

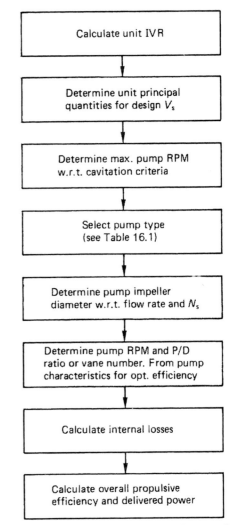

FIGURE 16.7 Outline waterjet calculation procedure.

16.4.1 Tunnel, Inlet and Supporting Structures

The inlet to the tunnel, in order to protect the various internal waterjet components, is frequently fitted with an inlet guard to prevent the ingress of large objects. Clearly the smaller the mesh of the guard, the better it is at its job of protection. However, the design of the guard must strike a balance between undue efficiency loss because of the flow restriction and viscous losses, the size of the object allowed to pass and the guard's susceptibility to clog with weed and other flow restricting matter. For small tunnels a guard may be unnecessary or indeed undesirable since a compromise between the above constraints may prove unviable, but for the large tunnels this is not the case. In this latter event the strength of guard needs careful attention since the flow velocities can be high.

The profile of the tunnel needs to be designed so that it will provide a smooth uptake of water over the range of vessel operating trims and, therefore, avoid any significant separation of the flow or cavitation at the tunnel intake.

In some waterjet applications, typically hydrofoil applications where the water flow has to pass up the foil legs, it is necessary to introduce guide vanes into the tunnel in order to assist the water flow around bends in the tunnel. The strength of these guide vanes needs careful attention, both from steady and fluctuating loadings, and if they form an integral part of the bend by being, for example a cast component, then adequate root fillets need to be provided. The guide vanes also need to be carefully aligned to the flow and the leading and trailing edges of the vanes should be faired so as not to cause undue separation or cavitation. Guide vanes, where fitted, need to be inspected periodically for fracture or impending failure during service. Therefore, some suitable means for inspection needs to be provided for either direct visual inspection or indirectly through the use of a boroscope.

Within the tunnel the dimensions are sometimes such that the drive shaft for the pump needs support from the tunnel walls. In such cases the supports, which should normally number three arranged at 120° spacing if there is danger of shaft lateral vibration, must be aligned to the flow and have an aerofoil section to minimize the flow disturbance of the incident flow into the pump and, additionally, the probability of cavitation erosion on the strut. The form and character of the wake field immediately ahead of the impeller is generally unknown. Some model tests have been undertaken in the past (References 3 and 4) and an example is shown in Figure 16.8. However, the aim should be to provide the pump with as small a variation in the flow field as possible in order to minimize the fluctuating blade loading. This can be done only by scrupulous attention to detail in the upstream tunnel design and with the aid of RANS computational codes.

The integrity of the tunnel wall both in intact and failure modes of operation is essential. If the wall fails this can lead to extensive flooding of the compartment in which the waterjet is contained and hence, by implication, have ship safety implications. There is, therefore, a need for attention to the detail of the tunnel design; for example, in adequately radiusing any penetrations or flanged connections and in terms of producing an adequate stress analysis of the tunnel both in the global and detailed senses. In addition to considering the waterjet as an integral component, the tunnel must be adequately supported, framed and fully integrated into the hull structure, taking due account of the different nature, response and interactions of the various materials used; for example, GRP, steel and aluminum.

16.4.2 Impeller

The hydrodynamic design of the pump impeller follows the general lines of Figure 16.7 and Table 16.2 since the

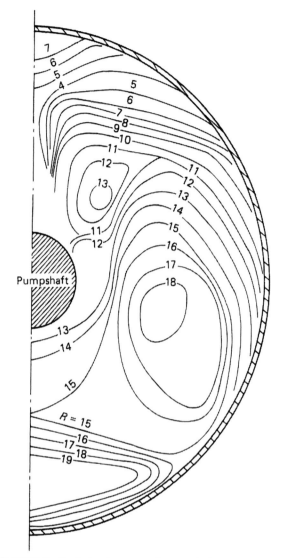

FIGURE 16.8 Typical wake survey of a waterjet inlet just upstream of the pump impeller. *Reproduced with permission from Reference 5.*

detailed features and form of the impeller are defined in this way. The detailed calculation of impeller components in terms of their strength and integrity has to be based on the maximum rated power of the machinery. Hence the mean loads on the blading need to be predicted on this basis and from there the stress analysis of the various components can be undertaken. One method of undertaking this prediction is to use an adaptation of the cantilever beam technique described in Chapter 19. However, it must be remembered that the impeller blades, certainly for mixed flow pumps, axial pumps and inducers, have in general a low aspect ratio, and therefore cantilever beam analysis has an inherent difficulty in coping with this analytical situation. Due allowance has therefore to be made for this in the eventual fatigue analysis of the blades and in determining the appropriate factor of safety. Some guidance can be obtained, however, by undertaking finite element calculations on certain classes of blade and from these determining the likely level of inaccuracy in the cantilever beam result. Alternatively, given the relative ease of finite element computations this is a more satisfactory method of calculating the stress distributions in the blades and hub and should be used if possible.

Computational fluid dynamics RANS codes are capable of estimating the likely flow velocity field in way of the impeller station; however, in the absence of wake field data, unless this is also estimated by calculation, a true estimate of the fluctuating stresses is difficult to achieve in practice. A realistic estimate is nevertheless required and this has to be based, in the former case, on a consideration of the upstream obstructions and their effect on flow into the impeller. Once this estimate is complete it can be used in association with the mean stress and an estimate for the residual stress (see Chapter 19) to undertake a fatigue evaluation of the design using the Soderberg or modified Goodman approaches.

The blades of the impeller must be provided with adequate fillets at the root. Such fillets need to be designed with care in order to provide the required degree of stress relief; in this context the elliptical or compound fillet design is generally preferred, although a single radius will suffice, provided its radius is greater than the blade thickness, but will not be as effective as the compound design. In addition if, as in some designs, the blades are bolted onto the pump hub, then extreme attention to the detail of the bolting arrangements and the resulting stresses in the palms and hub body is required since this is a serious potential source of failure.

The blade section design of the impeller demands considerable attention. Pump impellers work at high rotational speeds in comparison to the majority of propellers and, consequently, while their overall design is based on acceptable cavitation development and the control of its harmful effects, lack of attention to the section design detail can completely destroy this overall premise. As a consequence in all but centrifugal pumps, the blades require aerofoil forms to assist in controlling the cavitation properties of the pump. The blading can be either of the cambered or non-cambered type according to the head required to be developed by the pump. Furthermore, adequate control must be exercised over the manufacture of the blading in order to ensure that the manufacturing accuracy of the blade profiles is adequate for their proper cavitation performance.

The blade tip clearances, as with ducted propellers, need to be kept to a minimum to prevent undue hydrodynamic losses. However, this need must be balanced by the conflicting requirement to provide adequate clearance to cater for any transient vibration behavior of the rotating mechanism, axial shaft movement or differential thermal expansion of various components of the transmission system.

To guard against blade failure by vibration, the natural frequency of the blading should be calculated by a suitable means (see Chapter 21). The results of this calculation then need to be shown to lie outside of the primary operating ranges of the pump unit. In making this calculation the appropriate allowance needs to be made for the effects of the water on the blades rather than simply undertaking a calculation based on the assumption of the blades being in air.

Since the pump impellers work at high speed there is clearly a need for them to be balanced. In many cases, where the resulting couple is likely to be small, it is sufficient to limit the balancing operation to a static procedure conducted to an appropriate standard; typically the ISO standard. However, if it is considered that the 'out-of-balance' couple is likely to be significant in terms of the shafting system, then dynamic balancing must be implemented.

Because of the nature of the pump impeller and its inherent susceptibility to damage, provision needs to be made for this component to be inspected during service, preferably without dismantling the whole unit. In the case of the impeller, it is clearly preferable that the inspection is visual; however, if this is not practical for whatever reason, then boroscope inspection will suffice but this will not be as satisfactory an option as the provision of a direct visual capability.

A fundamental starting requirement for a pump impeller is that it should be self-priming. That is, it should, when the vessel is at rest in the water, have sufficient water to be able to start and effectively develop the required head. If the self-priming condition cannot be satisfied, then this is likely to involve an expensive priming capability which, in turn, may have important safety implications.

16.4.3 Stator Blading

Not all waterjet units are designed with a stator blade stage; however, where they are then the design has to consider both the maximum continuous power-free running condition and also stopping maneuvers, since these will introduce a back pressure on to the unit. The effects of steering maneuvers generally produce less severe conditions than stopping maneuvers.

As with the impeller blading many of the same principles apply to the blade design with regard to their section form, loading and strength. However, in the case of the stator blades a root fillet should be introduced at both ends of the blade and the effects on the tunnel strength of the reactive forces at the blade–tunnel interaction need to be considered. Furthermore, the natural frequency of the stator blades needs to be shown to lie outside the range of anticipated rotor blade passing and flow frequencies.

16.4.4 Nozzles, Steering Nozzles and Reversing Buckets

The nozzle design and its fixing or actuating arrangements, whether it be a steering nozzle or otherwise, need to be designed in the full knowledge of the forces acting on it during the various modes of operation of the waterjet unit. In particular, these relate to the pressure distribution along the nozzle internal surface and also the reactive forces produced by the rate of change of momentum of the fluid in the case of a steering nozzle.

The design of the bucket and its supporting and actuating mechanism have critical implications for the vessel's safety since it is the only means of stopping the craft quickly. As a consequence it regularly experiences significant transient loadings and therefore requires a careful assessment of its mechanical integrity. This, however, is far from easy to achieve from direct calculation at the present time and a measure of design experience based on previous installations is required for a proposed new unit. Model tests may be of assistance but questions of scaling and representation need careful attention.

Since the bucket and also the nozzle are normally exposed at the stern of the vessel the influence on these components of external loadings must also be considered. These loadings would normally comprise those from the impact of the sea in various weather conditions; collision with harbour walls and other vessels and fouling with buoys. Consequently, care needs to be taken either to ensure that the unit can withstand these interferences or, alternatively, that a suitable level of protection is provided.

REFERENCES AND FURTHER READING

1. Walree F van. *The powering characteristics of hydrofoil craft*. 6th Int. High Speed Surface Craft Conf; January 1988.
2. Dixon SL. *Fluid Mechanics and Thermodynamics of Turbomachinery*. Oxford: Pergamon Press; 1982.
3. Sayers AT. *Hydraulic and Compressible Flow Turbomachines*. McGraw-Hill; 1990.
4. Holden K, et al. *On development and experience of waterjet propulsion systems*. 2nd Int. Cong. of Int. L. Maritime Ass. of the East Mediterranean, Trieste; 1981.
5. Haglund K, et al. *Design and testing of a high performance waterjet propulsion unit*. Proc. 2nd Symp. on Small Fast Warships and Security Vessels. Trans RINA, Paper No. 17, London; May 1982.
6. Johnson Jr VE. *Waterjet propulsion for high speed hydrofoil craft*. J Aircraft March/April 1966;**3**(3).
7. Barham HL. *Application of waterjet propulsion to high performance boats*; June 1976.
8. Brandau JH. *Performance of waterjet propulsion systems – A review of the state of the art*. J Hydronaut April 1968;**2**(2).
9. Alder RS, Denny SB. *Full Scale Evaluation of Waterjet Pump Impellers*. Bethesda: DTNSRDC; September 1977. Report No. SPD-718-02.
10. Svensson R. *Experience with waterjet propulsion in the power range up to 10 000 kW*. SNAME Power Boat Symp; February 1985.
11. Venturini, G. *Waterjet propulsion in high speed surface craft*. High Speed Surface Craft.
12. Terwisga TJC van. *Waterjet–hull interaction*. PhD Thesis. Technical University of Delft; 1997.
13. Bulten NWH. *Numerical analysis of a waterjet propulsion system*. PhD Thesis. University of Eindhoven; 2006.
14. Delaney K, Donnelly M, Ebert M, Fry D. *Use of RANS for waterjet analysis of a high-speed sealift concept vessel*. Trondheim, Norway: First Int. Symp. on Marine Propulsors Symp'09; June 2009.
15. Rhee B, Coleman R. *Computation of viscous flow for the joint high speed sealift ship with axial flow waterjets*. Trondheim, Norway: First Int. Symp. on Marine Propulsors Symp'09; June 2009.
16. Schroeder S, Kim S-E, Jasak H. *Toward predicting performance of an axial flow waterjet including the effects of cavitation and thrust breakdown*. Trondheim, Norway: First Int. Symp. on Marine Propulsors Symp'09; June 2009.
17. Kinnas SA, Lee H, Michael T, Sun H. *Prediction of a cavitating waterjet propulsor performance using a boundary element method*. Ann Arbor, USA: Proc. 9th Int. Conf. on Numerical Ship Hydrodynamics; 2007.
18. Lavis DR, Forstell BG, Purnell JG. *Compact waterjets for high-speed ships*. Ships and Offshore Structures 2007;**2**(2):115–25.

Chapter 17

Full-Scale Trials

Chapter Outline

17.1 Power Absorption Measurements and Trials 373
   17.1.1 Techniques of Measurement 376
   17.1.2 Methods of Analysis 378
17.2 Bollard Pull Trials 379
   17.2.1 Trial Location and Conditions 380
   17.2.2 Measurements Required 381
17.3 Propeller-Induced Hull Surface Pressure Measurements 381
17.4 Cavitation Observations 382
References and Further Reading 383

Model tests and full-scale measurements are two equally important sources of data for the study of ship model correlation. Full-scale data, however, derives further importance in providing both a basis for the demonstration of a ship's contractual requirements and also in defining an experimental database from which the solution to some in-service problems can be developed. As a consequence the accuracy of the trials data in all of these cases is of the utmost importance and calls for precision in measurement, as far as is realistic under sea trial conditions, and a consistency of approach.

Full-scale trials on ships which relate specifically to the propeller fall broadly into three classes. The first relates to power absorption, the second is the measurement of propeller-induced vibration and noise on the vessel, while the third relates to the observation of cavitation on the propeller and rudder. In addition, there are several other specific measurements and trials, such as bollard pull estimation, that can be conducted and have to be soundly based in order to be meaningful.

17.1 POWER ABSORPTION MEASUREMENTS AND TRIALS

Measurements to define the power absorption characteristics of a vessel fall broadly into two categories. These are full-ship speed trials as would normally be conducted for the demonstration of contractual conditions and the less comprehensive power absorbed versus shaft speed characteristic. This latter trial is, however, merely a subset of the former. While accepting the second type of trial is a useful diagnostic tool its limitation in preparing propeller design remedial action must be recognized, since it ignores the ship speed component of the design triumvirate of power, revolutions and ship speed. As such, to prevent repetition, the discussion will center on the former type of trial.

A full-speed trial can be conducted either on a measured distance as specified on a maritime chart or by the use of electronic navigational position of fixing systems such as GPS (Global Positioning System). With the exception of the ship speed measurement procedure, many of the measurement requirements are common to both types of trial so as to obtain a valid result. The basic requirements of these trials are as follows:

1. *Measured distance trial area.* The area selected for the trial should not be one where the effects of tide are large since this will introduce large corrections into the trial analysis procedure. Furthermore, if the direction of flow is oblique to the trial course, this may lead to difficulty in course keeping in strong tides leading to further errors in the measured speeds.

In addition to the requirement for reasonable tidal activity, it is necessary to ensure that there is both sufficient water depth at the intended time of the trial and adequate space to conduct approach runs both from geographical and marine traffic density considerations. With regard to water depth a value of $3\sqrt{BT}$ or $2.75\, V_s^2/g$, whichever is the greater, is recommended by the ITTC.[1] Similarly, the length of the approach run to the start of the measured distance, or the start of a GPS measurement, must be adequate to allow the vessel to reach a uniform state of motion after the various course changes that will occur between one run and another. It is difficult to specify these distances precisely, but for guidance purposes a distance of 25 and 40 ship lengths have been suggested[1] for a high-speed cargo liner and a 65 000–100 000 dwt tanker, respectively.

2. *Measured distance course.* Where these are used, the measured distance course should be of the standard form shown in Figure 17.1 and the ship when sailing the measured distance must be navigated parallel to the line between the distance posts on the coast. Additionally, the direction of the turn after completing the measured course should normally be away from the coast in order to take advantage of any deep water and also enhance navigational safety. Upon completion of the measured distance, the angle of turn in preparation for the return run should comprise gradual rudder movements which should be limited to around 15° in bringing the vessel back on to a reciprocal course. More abrupt turning procedures, such as the Williamson Turn which was designed for life-saving purposes, are unacceptable since they disturb too greatly the dynamic equilibrium of the ship.
3. *Vessel condition.* The condition of the ship should be checked prior to the trials to ensure that both the hull and propeller are in a clean state. The inspection should in all cases be done in a dry dock and only exceptionally by an in-water survey. Moreover, the cleanliness of the underwater surfaces should be checked in this way as close to the trial date as possible, but not at a greater time interval than two weeks. This is because significant levels of slime and biological growth can occur in very short periods of time, given the correct conditions of biological infestation, light and temperature (see Chapter 24). Where possible at the time of cleaning an observation and measurement of the topography of the hull and propeller surfaces should take place.
4. *Weather conditions.* To avoid undue corrections to the trial results, the trials should be run in sea states of preferably less than force 2–3 on the Beaufort scale and in low swell conditions. This clearly cannot always be met in view of the constraints on time and location.
5. *Number of trial runs.* The number of runs should comprise at least four double runs; that is, consecutive traverses of the measured distance in each direction. The nominal power of each of the double runs and their total span is largely dependent on the purpose for which the trial is being conducted. However, it is suggested that at least two double runs at full power should be considered.

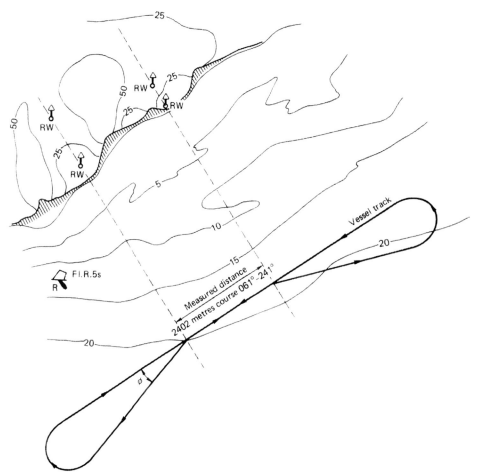

FIGURE 17.1 Typical measured distance course.

6. *Trial procedure.* The trials should be under the overall control of a 'trials master' on whom the responsibility of the trial should rest. It is he who should make certain that all those responsible for the safe navigation and control of the vessel understand clearly what is required at all times within the trial period.

When a new power setting is required this should be set immediately upon leaving the measured distance or on completion of the GPS run at the previous measurement condition. All adjustments should then be completed prior to the vessel turning in order to make a new approach to the measurement area on a reciprocal course. Under no circumstances must the engine or propeller pitch, in the case of a controllable pitch propeller, then be altered until the full set of double runs at that condition has been completed. When the vessel is turning the propeller revolutions will tend to decrease: this is perfectly normal and these will recover themselves when the vessel straightens course on the approach to the measured distance. If the engine or propeller controls are altered during a double run, for whatever reason, then the results should be discarded and that part of the trial recommenced. The use of rudder adjustments to maintain course at the time of the measurement should be kept to the absolute minimum consistent with the prevailing weather conditions. If this is not done it will introduce additional resistance components to the vessel and may invalidate the trial. Prior to the trial commencing it is essential that all instrumentation to be used is properly calibrated and 'zero values' taken. A repeat set of 'zeros' should also be undertaken upon completion of the trial and for some instrumentation it is desirable to take intermediate 'zero' readings. If time is not allowed for collecting this reference data then the value of the measurement will be degraded and in some cases will call the accuracy of the entire trial into question.

7. *Measurements required.* In order to demonstrate that the vessel has achieved a certain hydrodynamic performance, or to provide data from which a solution to some propulsion problem can be developed, it is necessary to measure an adequate data set. This data should comprise:
   a. Draught and trim.
   b. Ambient conditions.
   c. Ship motions.
   d. Machinery measurements.
   e. Ship speed and course.

More specifically these various headings embrace the following aspects:

a. *Draught and static trim.* This should ideally be measured both before and after the trials for all vessels: in the case of small high-speed craft, however, a 'before' and 'after' measurement is an essential requirement. This measurement for small vessels should be taken immediately prior to and after finishing the measured distance runs, since the fuel weight and, indeed, personnel weight can often form a significant component of the total craft weight. In addition, for small craft not subject to survey during building, or in cases of doubt, the draught marks should be checked by a competent person. Indeed, in a significant number of cases, in the author's experience, these marks are found to be in error and in some cases there is need to put temporary draught marks on to the hull where none exist. This can be relatively easily achieved in most cases with some form of waterproof tape.

b. *Ambient conditions.* These include measurements of the prevailing weather and sea conditions at the time of and during the trial. For the atmospheric conditions these should include: air temperature, wind speed and direction, atmospheric pressure and both relative humidity and visibility in order that the analyst of the trial result might have a true picture of the trial conditions.

In the case of sea conditions, the measurements need to include: sea temperature, sea state, an estimate of the swell height and direction. It is important to distinguish between sea state and swell since the sea state largely defines the local surface conditions, whereas the swell defines the underlying perturbation which may originate from some remote sea area. In most cases, these observations are made by reference to experienced personnel on board the vessel: for example, the senior navigational officers. When more detailed analysis is required, typically for research purposes or difficult contractual situations, then wave buoys may be used for these measurements.
It is the author's practice, under normal circumstances, to require that ambient conditions are recorded on either a half-hourly or hourly basis, since for protracted trials the weather in many areas can change significantly in a comparatively short space of time.

c. *Ship motions.* Some record should be kept of the ship motions in terms of pitch and roll magnitude and period during the trials; this record should be kept on a 'run by run' basis. The normal ship's equipment is usually sufficient for this purpose. In addition, for craft which change their trim considerably during high-speed runs, the dynamic running trim should be recorded by means of a suitable inclinometer placed in the fore and aft direction on the vessel. The instrument used for this purpose needs to have a measure of damping inherent in it otherwise it will be difficult to read during the trials due to sea-induced transients occurring during the trial runs.

d. *Machinery measurements.* From the viewpoint of the ship resistance and propulsion the principal measurements required are shaft horsepower, propeller revolutions and propeller pitch in the case of a controllable pitch propeller. Shaft axial thrust measurements, between the thrust block and the propeller, are also extremely useful, and in some cases essential, but notoriously difficult to measure for trial analysis purposes.

Naturally, during a contractor's or acceptance trial many other engine measurements will be taken that are of value in quantifying the machinery performance. However, as a means of support to the resistance and propulsion analysis additional data such as engine exhaust temperatures, turbo-charger speeds, temperatures and pressures, fuel rack setting, etc. should always be obtained where possible.

e. *Ship speed and course.* Clearly in any propulsion trial this is an essential ingredient in the measurement program; without this measurement the trilogy of parameters necessary to define propeller performance cannot be established. It can be measured using a variety of methods such as by the conventional measured distance or by means of a position fixing navigation system such as GPS. Coincident with the speed measurement, the course steered, together with any deviations, should be noted by both magnetic and gyrocompass instruments: the dates when these instruments were last 'swung' or calibrated also need to be ascertained.

17.1.1 Techniques of Measurement

There are many techniques available for the measurement of the various propulsion parameters; some of the more common methods are outlined here for guidance purposes:

a. *Propeller pitch angle.* In the case of a controllable pitch propeller the pitch should be read from the Oil Distribution (OD) box scale, or its equivalent, and this value interpreted via a valid calibration into blade pitch angle at the propeller. On no account should a bridge or engine room consul indicator be taken as any more than an approximate guide, unless this has a proven calibration attached to it. This is because zeroing and other adjustments are often made to electrical dials during the life of the vessel and blade pitch angle changes of the order of 1° make significant alterations to the power absorbed. If possible, before the sea trial and when the ship is in dry dock, it is useful to check the OD box indicator scale against the manufacturer's blade angular markings situated between the blade palms and the hub body.

b. *Propeller shaft power.* Shaft power measurements can be measured in one of two ways; these being classified as either permanent or temporary for the purposes of the trial. Permanent methods principally involve the use of a torsion meter fitted to the vessel in which a value of the torque being transmitted is read either directly or in terms of a coefficient which needs scaling by a calibration factor. While such instruments, in whatever state of calibration, are frequently sufficient for measurements between one power setting and the next, if they are intended for a quantitative scientific measurement then their calibration should be validated immediately prior to the trial.

Temporary methods of measurement normally involve the use of strain gauge procedures configured to form a Wheatstone bridge circuit. The most fundamental of these measurements is to place a strain gauge bridge comprising four strain gauges as shown in Figure 17.2(a) onto the shaft. This bridge is activated by a battery pack or power induction loop and the strain signals are sent via a radio telemetry transmitter to a stationary receiver. Several variants of this strain gauge bridge system are in use; for example, where the two halves of the Wheatstone bridge are placed diametrically on opposite sides of the shaft. This can be very helpful in cases where space is limited or the effects of shear in the shaft are significant. The size of the strain gauges does not need to be particularly small since when used on industrial shafting beneficial effects can accrue from the averaging of the strain signal that takes place over the strain gauge length: typically the gauge length is of the order of 5 mm or so. Calibration of the system can be effected by means of high-quality standard resistances shunted across the arms of the bridge to simulate the torsional strain in the shaft when under load. The reader is referred to Reference 4 for a detailed account of strain measurement techniques.

c. *Propeller revolutions.* Most vessels are fitted with shaft speed instruments and in general these are reasonably accurate. Notwithstanding this generalization, the calibration should always be checked prior to a trial. When more specialized trials requiring greater accuracy or where independence is needed, then inductive proximity or optical techniques may be used. In ship vibration studies, Section 17.3, a separate shaft speed measurement technique of this type is considered essential for vibration order reference purposes. This is also true for some forms of cavitation observation trials.

d. *Propeller thrust measurements.* The measurement of propeller thrust on a long-term basis is notoriously difficult to undertake from a measurement reliability and stability point of view. Several techniques exist for permanent installation; however, for trial purposes the experimenter is well advised to check the calibration of these measurements.

If a short-term installation is required, then a strain gauge technique is probably the most reliable at the present time and also the easiest to use. When applying strain gauge techniques to marine propeller thrust

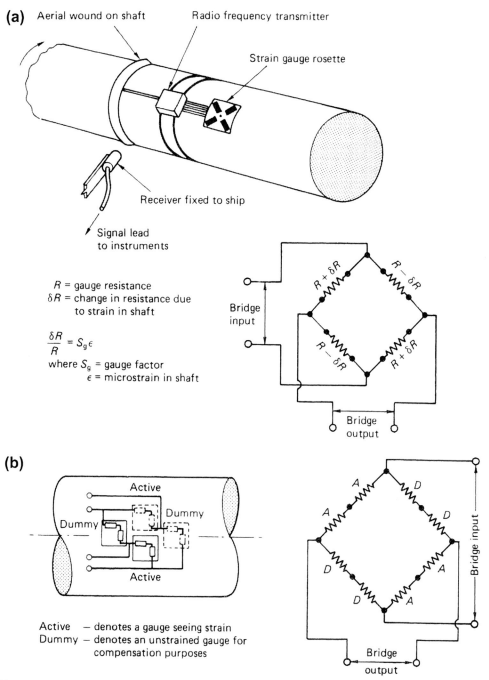

FIGURE 17.2 Measurement of thrust and torque by strain gauge methods: (a) measurement of shaft torque and (b) Hylarides bridge for thrust measurement.

measurement, the problem is that the axial strain on the vessel's intermediate shaft is generally of an order of magnitude less than the torsional shear strain. This difference in strain magnitudes can lead to cross-interference problems if the measurement installation is not carefully and accurately completed.

A useful experimental technique for axial strain measurement, and hence thrust determination, is the Hylarides bridge, Figure 17.2(b). This bridge system has the benefit, through its system of eight strain gauges, four of which have a null-strain compensating function, of alleviating the worst effects of strain cross sensitivity. The strain signal is transmitted from the rotating shaft in much the same way as for the torsional strain signals. An alternative procedure to strain gauge methods is to measure the thrust in terms of the axial deflection of the intermediate shaft using a ring gauge coupled with rods fitted parallel to the shaft.[5]

Thrust measurements on trials, for the reasons cited above as well as the additional cost involved, are relatively infrequently recorded. However, the information that these measurements provide completes the necessary propulsion information required to undertake a complete and rigorous propulsion factor analysis of the vessel.

e. *Ship speed measurement.* The traditional measurement of ship speed on a measured mile requires the use of at least three independent observers timing the vessel over the measured distance with the aid of stop watches. The stop watches need to be of high quality with a measurement resolution of the order of one hundredth of a second. On trials the observers should be left to measure the time that the ship takes to travel along the measured distance independently and without any prompting from one of their number or, alternatively, from an independent observer as to when the vessel is 'on' or 'off' the mile. Indeed, the author has found it advisable for observers to be sufficiently far apart so that they do no hear, and therefore become subject to influence by, the activating clicks of each other's stop watches.

The alternative form of speed measurement is to use a navigational position fixing system in which a specified distance can be traversed and the time recorded in a not dissimilar way to that used for the classical measured mile trials. When using this type of equipment, however, it must be first ascertained that the system is working correctly and is free of any interference or astronomical aberration at the time of the trial.

The accuracy of current day GPS systems is very high and, therefore, the distance between two way points can be quite accurately determined and with the time similarly accurately recorded a good speed estimate can be determined. This speed, however, is the ship's speed over the ground and not through the water, which is the parameter normally required for sea trial analysis purposes. With this type of trial the greatest source of error is usually the vagaries of the sea or river currents which can change significantly in terms of speed and direction over a short distance. Consequently, a strategy depending on the sea area in which the trial is to be conducted has to be formulated before the trial commences. This has to take into account the length of run, which should not significantly exceed that of classical measured distances; the direction of the run with respect to the prevailing sea conditions and the underlying swell; the timing of the trial with respect to tidal activity; the number of runs to be undertaken in order to minimize any errors, including the need for double runs which in the author's opinion should always be undertaken.

17.1.2 Methods of Analysis

The machinery measurements are relatively easy to analyze and the quantities derived from the measurement are normally readily deduced; for example, K_T and K_Q. In the case of the torque coefficient K_Q it should be remembered that it is normal to measure shaft power and not delivered power. Hence an allowance for the transmission efficiency needs to be made: typically this will lie between 0.98 and unity for most vessels assuming the measurement is made aft of any shaft-driven auxiliaries. If this is not the case, then the appropriate allowances will need to be made based on the power absorption of these auxiliaries, which will need to be known accurately.

Ship speed analysis presents perhaps the greatest problem. If the mean speed is taken between a consecutive pair of runs over a measured distance this implicitly assumes the tidal variation is linear. In many instances this is a reasonable assumption provided the time between runs is short compared to the prevailing tidal change. If more than a single double run is made, then a mean of means can be taken, as shown in the example illustrated in Table 17.1, where each run was made at a regular time interval.

Table 17.1 shows what may be termed as the standard textbook way of performing the analysis. In practice, however, regular time intervals between runs seldom, if ever, occur. Notwithstanding this, it is still valid to use a mean-value analysis between any two runs forming a consecutive pair on a measured distance, provided that all parties to the trial analysis are happy with the use of a linear approximation for the tide over the time interval concerned. However, the trials analyst, whatever method is used, is well advised to check the tidal assumptions with the predictions for the sea area in which the trial was carried out, both in terms of magnitudes and tidal flows.

TABLE 17.1 Example of a Mean of Means Analysis

Time	Measured Speed (knots)	Mean 1	Mean 2	Mean 3
14:00	25.18			
		24.70		
14:45	24.22		24.71	
				24.72
			24.72	
15:30	25.22		24.73	
			24.74	
16:15	24.27			

Mean ship speed = 24.72 knots

If a higher-order tidal model is felt desirable, then several methods are available. It is most common, however, to use either a polynomial or sinusoidal approximation depending on the circumstances prevailing and, of course, the analyst's own preferences. In the polynomial expression, the standard technique is to adopt a quadratic approximation for the speed of the tide (v_t):

$$v_t = a_0 + a_1 t + a_2 t^2 \quad (17.1)$$

where t is the time measured from the initial run on the measured distances. This tidal speed is then used in the analysis procedure by assuming the measured ship speed V_m represents the ship speed V_s in the absence of the tide plus the speed of the tide v_t:

$$V_m = V_s + v_t \quad (17.2)$$

Hence, by taking any set of four consecutive runs on the measured distance, either at regular or irregular time intervals, a set of four linear equations is formed from equations (17.1) and (17.2):

$$\begin{bmatrix} V_{m1} \\ V_{m2} \\ V_{m3} \\ V_{m4} \end{bmatrix} = \begin{bmatrix} 1 & 1 & t_1 & t_1^2 \\ 1 & -1 & -t_2 & -t_2^2 \\ 1 & 1 & t_3 & t_3^2 \\ 1 & -1 & -t_4 & -t_4^2 \end{bmatrix} \begin{bmatrix} V_s \\ a_0 \\ a_1 \\ a_2 \end{bmatrix} \quad (17.3)$$

The difference in signs in equation (17.3) merely indicates that the tide is either with or against the ship. If equation (17.3) is applied to the example of Table 17.1, then the resulting mean ship speed is given by the vector

$$\begin{bmatrix} 25.18 \\ 24.22 \\ 25.22 \\ 24.27 \end{bmatrix} = \begin{bmatrix} 1 & 1 & 0 & 0 \\ 1 & -1 & -0.75 & -0.5625 \\ 1 & 1 & 1.5 & 2.25 \\ 1 & -1 & -2.25 & -5.0625 \end{bmatrix} \begin{bmatrix} V_s \\ a_0 \\ a_1 \\ a_2 \end{bmatrix}$$

from which: $V_s = 24.721$ knots, $a_0 = 0.4587$, $a_1 = 0.08667$, $a_2 = -0.03999$, giving the equation of the tide as

$$V_t = 0.4587 + 0.08667 t - 0.03999 t^2 \text{ knots}$$

where t is measured in hours.

Such a tidal model is sufficient provided the measurement is not conducted over a protracted period of time. A quadratic function, being a second-order polynomial, only has one turning point, and therefore cannot adequately represent the periodic nature of the tide. Figure 17.3 illustrates this point, from which it can be seen that the above tidal model cannot predict accurately the tidal effect after about two hours from the start of the trial described in Table 17.1. However, it would not be the first time that the author has seen such a model, based on four such points as those in Table 17.1, used to predict the tidal effect at a much later

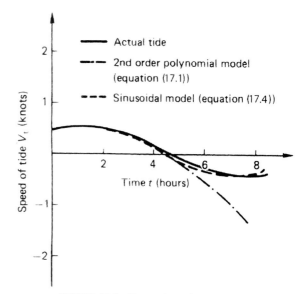

FIGURE 17.3 Comparison of tidal models.

time and then this prediction used to justify a trial speed! Such a second-order representation will prove inadequate if the trial is conducted over a period greater than about four or five hours, even in the absence of extrapolation, due to the form of the polynomial and the tidal characteristics.

As a consequence of these problems one can either use a higher-order polynomial, requiring rather more double runs to be conducted on the measured distance, or use a sinusoidal model of the general form

$$v_t = a \sin(\omega t + \phi) \quad (17.4)$$

An equation of this form can, by judicious construction of the coefficients, be made to approximate the true physics of the tidal motion. Figure 17.3, however, shows its use in its simplest form of constant coefficients.

17.2 BOLLARD PULL TRIALS

In general the bollard pull trial is conducted to satisfy a contractual requirement and, as such, would normally make use of the vessel's own instrumentation with the exception of a calibrated load cell which is introduced into the vessel's tethering line system. For those cases where either the instrumentation fitted to the vessel is insufficient or where an independent certification is required, temporary instrumentation would be fitted and the relevant parts of the discussion of the previous section would normally apply.

Some authorities identify three different definitions of bollard pull for certification purposes. These are:

1. *Maximum bollard pull,* which is the maximum average of the recorded tension in the towing wire over a period of one minute at a suitable trial location. As such, this

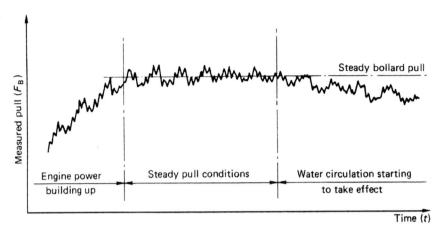

FIGURE 17.4 Measured bollard pull time history.

would normally correspond to the maximum engine output.
2. *Steady bollard pull*, which is the continuously maintained tension in the towing wire which is achievable over a period of five minutes at a suitable trial location.
3. *Effective bollard pull*, which is the bollard pull that the vessel can achieve in an open seaway. Since this is not ascertainable in a normal trial location it is normally characterized as a certain percentage of the steady bollard pull. This is frequently taken as 78 per cent after making due allowance for the weather conditions.

The general bollard pull characteristic of a vessel is outlined in Figure 17.4. From the figure it is seen that there is a general rise in bollard pull in the initial stages of the trial as the engine speed is increased. The pull then remains sensibly constant for a period of time, after which a decay in level is then frequently observed as water recirculation through the propeller starts to build up. A vibratory component of thrust superimposed on the mean trend of the bollard pull signature normally occurs and this frequently has a cyclic variation which appears to relate to the rudder movements required to keep the vessel on station. Additionally, higher-frequency thrust components, having a period of around an order of magnitude less than those rudder-induced variations, will also be observed which correspond to the natural period of oscillation of the vessel on the end of the cable.

17.2.1 Trial Location and Conditions

Bollard pull trials should be conducted at a location which provides a sufficient extent of deep and unobstructed water together with a suitable anchorage point on the shore. The extent of water required is largely governed by the recirculation effects into the propeller and the attempt to try and minimize these as far as practicable. The reason for this concern is that at the bollard pull condition the advance coefficient $J = 0$; however, if water circulation, in either the vertical or horizontal planes, becomes significant, then the effective value of J increases and an inspection of any open water propeller characteristic curve will show that under these conditions the value of the propeller thrust coefficient will fall off. Each trial location should clearly be treated on its merits with due regard to the vessel and its installed power. The trial location should preferably be that shown in Figure 17.5(a), which has clear water all around the vessel: the alternative location indicated in Figure 17.5(b) cannot be considered satisfactory since it encourages recirculation. When undergoing bollard pull trials it is normal for the vessel to 'range around' to a limited extent: this requires consideration when determining the trial location as does accommodating the emergency situation of the vessel breaking free and the consequent need for clear water ahead. For general guidance purposes the following conditions should be sought in attempting to achieve the best possible bollard pull trial location:

1. The stern of the vessel should not be closer than two ship lengths from the shore and in general the greater this distance, the better.
2. The vessel should have at least one ship length of water clear of the shore on each beam.
3. The water depth under the keel of the ship should not be less than twice the draught at the stern with a minimum depth of 10m.
4. Current and tidal effects should ideally be zero, and consequently a dock location is preferable from this viewpoint. If these flow disturbance effects are unavoidable, then the trial should be conducted at the 'top of the tide' with an ambient water speed not exceeding 0.5 m/s.
5. The wind conditions should not exceed force three or four and the sea or river should be calm with no swell or waves.

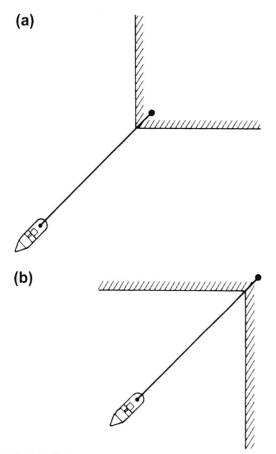

FIGURE 17.5 Bollard pull trial location: (a) good location for trial and (b) poor location for trial.

17.2.2 Measurements Required

The required measurements, in addition to the ambient weather and sea conditions, are the bollard pull, the engine power, propeller speed and the propeller pitch if the ship is fitted with a controllable pitch propeller. In addition, a record of the rudder movements and vessel position with time should be kept, as should the angle of yaw of the vessel so that these measurements can be synchronized with the recorded bollard pull signature. In general terms, the discussion in Section 17.1 applies as far as the measurement methods are concerned. The load cell which could be either mechanically or electrically based must be calibrated and should form part of the tow line at the shore end of the tethering system. The reason for locating the gauge at the shore end is that it is generally easier to protect the gauge in this location and it is not influenced by external effects such as friction on the towing horse of a tug. Ideally, the trial results from all of the various measurements should be continuously and simultaneously measured against time which will then enable them to be fully considered for analysis purposes.

17.3 PROPELLER-INDUCED HULL SURFACE PRESSURE MEASUREMENTS

The propeller-induced hull surface pressures are of interest because of the vibration that they excite in the ship. It is normal practice for these measurements to use pressure transducers inserted flush into the hull surface at appropriate locations above the propeller. The signal from these transducers is then recorded together with other measured ship parameters, as desired, but in all cases in association with propeller shaft speed and a reference mark on the shaft: in some instances these are combined. The pressure recorded by the transducers is the apparent propeller-induced pressure $p'(t)$, which is given by:

$$p'(t) = p_H(t) + p_V(t) \quad (17.5)$$

where $p_H(t)$ is the true propeller-induced pressure on the hull surface and $p_V(t)$ is the hull-induced pressure caused by its own vibratory behavior.

In order, therefore, to correct the measured result for the pressure induced by the vibration of the hull, it is necessary to measure the vibration of the hull surface in the vicinity of the pressure transducer. From the recorded vibration signals, the local motion of the hull can be determined from which a first-order correction, $p_V(t)$, can be determined on the basis of a vibrating plate in an infinite medium. As such, the propeller-induced pressures on the hull surface can be derived from equation (17.5) by rewriting it as

$$p_H(t) = p'(t) - p_V(t) \quad (17.5a)$$

It is beneficial to have as many pressure transducers as possible distributed in a matrix over the hull surface, above and in the vicinity of the propeller, in order to be able to define the magnitudes and relative phases of the pressure signature at the various locations. Then from these results estimate the total force transmitted to the hull; however, this is seldom practicable in commercial trials. As a minimum requirement for sea trial purposes the number of pressure transducers and their associated vibration transducers should not fall below about five to seven. A suitable distribution of the transducers, based around a vertical measurement reference plane passing through the mid-chord positions of the $0.8R$ blade sections and in the case of a right-handed propeller when viewed from above, is shown in Figure 17.6; in the case of a left-handed propeller then the distribution would be a mirror image of that shown. Figure 17.6 relates to a ship with a conventional transom stern. However, if the ship has a cruiser type stern which protrudes well beyond the conventional location of a transom it would be prudent to include further transducers well aft of the propeller station in order to capture any untoward activity from the behavior of the tip vortex. Typically, transducers might be placed at distances up to $2D$ aft from the propeller plane in-line with the expected principal activity of the systems of tip vortices.

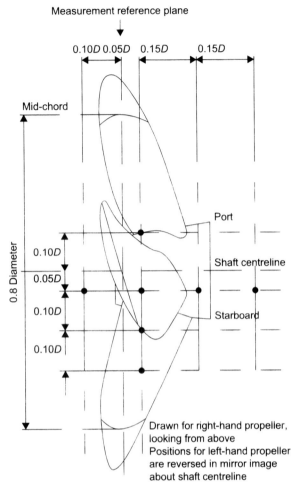

FIGURE 17.6 A typical distribution of pressure transducers.

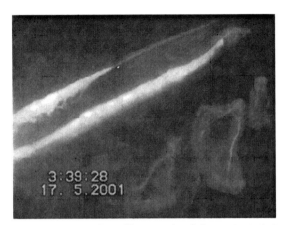

FIGURE 17.7 Image taken with conventional observation techniques.

In this context, the parameter D is the propeller diameter. With regard to frequency response, for most situations, pressure transducers with a response of up to 5 kHz will be found adequate.

17.4 CAVITATION OBSERVATIONS

The traditional approach to the full-scale observation of cavitation on propeller blades has developed from still cameras and flash units linked to the shaft rotational position, through cine-cameras with stroboscopic lighting, high-speed cine-cameras to the present day use of conventional video camcorders used with natural daylight and fast shutter speeds. Stroboscopic techniques have usually required the correct positioning of the camera with respect to the lighting source, generally this being as orthogonal as possible, and operation at night in order to achieve the correct photographic discrimination. Where the water clarity is sufficient, then sharp pictures can generally be achieved as seen in Figure 17.7. Stroboscopic sources can also be used with a trigger, normally placed on the intermediate shaft, in order for the light source to be kept in phase with the shaft at positions pre-defined for observation. This, in turn, implies that relatively long time frames elapse between successive images, although it is normally possible to observe the same blade on successive revolutions. Nevertheless, temporal changes in the flow velocities do occur between successive revolutions and this can complicate interpretation of the resulting images.

Natural daylight is, however, preferable and this can now be utilized due to improvements in video camera technology, both in frame speed and the ability to accommodate poorer lighting conditions. This enables the time-series recording of dynamic cavity events, rather than ensemble averages. Moreover, the fact that the trials are undertaken under daylight conditions makes them far simpler to implement. With this type of observational photography there is a need to place a series of windows in the ship's hull in the appropriate positions to observe the phenomenological behavior of the cavitation. Figure 17.8 shows a drawing of such a window and unless the ship can be ballasted sufficiently so as to expose the area of hull above the propeller, the fitting process demands that the ship is placed in dry dock.

An alternative procedure which does not generally require the ship to be dry-docked is to use a borescope system inserted onto small penetrations in the hull at locations above the propeller location. This system of propeller observation, pioneered by Fitzsimmons[6], normally requires a set of M20 tapped holes to be placed in the hull and the majority of ships can be adequately ballasted to reduce the water static head on the outside of the hull plating sufficiently to prevent a serious ingress of water during fitting. Experience with the borescope system has been good and although the image reproduction is not as good as that obtained from using the conventional window-based system, it is frequently sufficient for practical investigation purposes. Moreover, the system most frequently uses the same penetrations used for pressure transducers and swing prism borescopes have the ability to

Full-Scale Trials

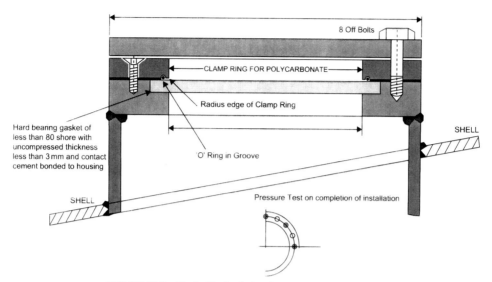

FIGURE 17.8 Typical hull window for cavitation observation.

be rotated through the full 360°, which gives a distinct viewing advantage over the conventional observation methods. This procedure, therefore, enables good insights into the cavitating behavior of rudders, A-brackets and propellers, and, moreover, gives the ability to discriminate between the various cavitation types. Figure 17.9 shows an example of the image quality obtained on a small warship. The figure shows a view of a blade tip together with two vortex structures on the suction side of the blade: one emanating from the leading edge. Since their introduction to cavitation observation trials, the pace of development of digital video and borescope optical technology has been rapid. This has in turn permitted the enhancement of image capture and the replacement of the original low light cameras with selectable shutter speed by more powerful low light, high-speed digital capabilities. The newer systems, which also utilize lower-energy loss borescope optics, are now able to have a capability of between 200 and 1500 frames per second. Moreover, simultaneously with the increase in digital video technology has been the development of software which permits the synchronization of time-series data with the video images. Consequently, it becomes possible to display measured hull surface pressure or vibration data correctly phased with the video recordings.

Fitzsimmons has explored the use of acoustic emission methods to predict the onset of cavitation erosion. This approach has been applied to a number of fixed pitch, interference fitted propellers. The measurement method, which has been used in conjunction with dual, high-speed video systems, comprises acoustic emission sensors, signal conditioning and telemetry fitted onto the propeller shaft just forward of the aft-peak tank bulkhead. In this approach the acoustic path passes from the cavitation sources in the outer regions of the blades, along the blades and into the propeller boss in preparation for transmission up the tail shaft to the measurement location. Using this approach tentative acoustic thresholds have been determined during particular trials.

REFERENCES AND FURTHER READING

1. *ITTC guide for measured mile trials*. Rome: Rep. of Perf. Comm., Proc. 12th ITTC; 1969.
2. *BSRA Code of Procedure for Measured-Mile Trials*. BSRA; 1977.
3. *Norwegian Standard Association*. Oslo; 1985.
4. Dally JW, Riley WF. *Experimental Stress Analysis*. McGraw-Hill; 1978.
5. Hylarides S. *Thrust measurement by strain gauges without the influence of torque*. Shipping World and Shipbuilder; December 1974.
6. Fitzsimmons PA. *Observations of cavitation on propellers*. Trans., Lloyd's Register Technology Days 2011; Lloyd's Register, London; 2011.

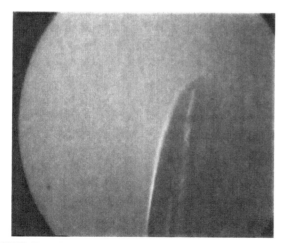

FIGURE 17.9 Propeller and leading edge image from the borescope system.

Chapter 18

Propeller Materials

Chapter Outline

- 18.1 General Properties of Propeller Materials — 385
- 18.2 Specific Properties of Propeller Materials — 388
  - 18.2.1 High-Tensile Brass — 389
  - 18.2.2 Aluminum Bronzes — 390
  - 18.2.3 Stainless Steels — 391
  - 18.2.4 Cast Iron — 392
- 18.2.5 Cast Steel — 392
- 18.2.6 Carbon-Based Composites — 392
- 18.3 Mechanical Properties — 393
- 18.4 Test Procedures — 394
- References and Further Reading — 396

The materials from which propellers are made today can predominantly be classed as members of the bronzes or stainless steels. The once popular material of cast iron has now virtually disappeared, even for the production of spare propellers, in favor of materials with better mechanical and cavitation-resistant properties. Figure 18.1 introduces the better-known materials that have been in use for the manufacture of all types of propellers ranging from the large commercial vessels and warships through to pleasure runabouts and model test propellers.

The relative popularity of the two principal materials has changed over the years[1] in that in the early 1960s the use of high-tensile brass accounted for some 64 per cent of all of the propellers produced with manganese—aluminum bronze and nickel—aluminum bronze accounting for comparatively small proportions: 12 per cent and 19 per cent, respectively. However, by the mid- to late 1980s nickel—aluminum bronze had gained an almost complete dominance over the other materials, accounting for some 82 per cent of the propellers classed by Lloyd's Register during that period. This trend has continued to the present time. High-tensile brass, sometimes referred to as manganese bronze, which in the early 1960s was the major material used for propeller manufacture, now accounts for less than 7 per cent of the propellers produced and manganese—aluminum bronze below 8 per cent. The stainless steels became relatively popular during the period from the mid-1960s through to the mid-1970s, but then progressively lost favor to the copper-based materials. Today they appear to account for some 3 per cent of the propeller manufacturing materials and are most commonly used for ice class propellers.

18.1 GENERAL PROPERTIES OF PROPELLER MATERIALS

Pure copper, which has a face-centered-cubic structure as shown in Figure 18.2, has a good corrosion resistance. It is a particularly ductile material having an elongation of around 60 per cent in its soft condition together with a tensile strength of the order of 215 N/mm². Therefore, when considered in terms of its tensile strength properties it is a relatively weak material in its pure form. But since plastic deformation of metallic crystals normally results from the slipping of close-packed planes over each other in close-packed directions, the high ductility of copper is explained on the basis of its face-centered-cubic structure.

By combining copper with quantities of other materials to form a copper-based alloy the properties of the resulting material can be designed to give an appropriate blend of high ductility and good corrosion resistance, coupled with reasonable strength and stiffness characteristics. One such alloy is the copper—zinc alloy, which contains up to about 45 per cent zinc frequently in association with small amounts of other elements. Such copper—zinc alloys, where zinc has the close-packed hexagonal structure, are collectively known as the brasses and a phase diagram for these materials is shown in Figure 18.3(a). In their α phase, containing up to about 37 per cent zinc, the brasses are noted principally for their high ductility which reaches a maximum for a 30 per cent zinc composition. If higher levels of zinc are used, in the region of 40—45 per cent, then the resulting structure is seen from Figure 18.3(a) to be of a duplex form. In the β' phase, which exhibits an ordered structure, the material is found to be hard and brittle, which

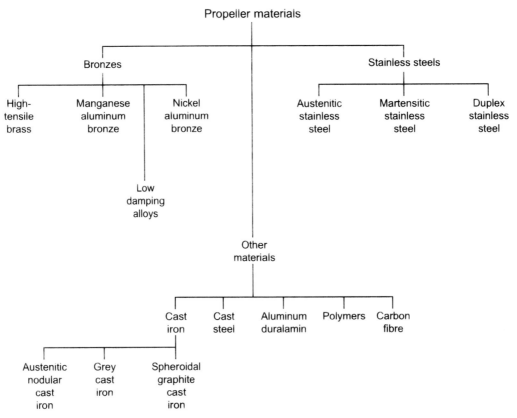

FIGURE 18.1 Family of propeller materials.

is in contrast to the β phase where it has a disordered solid solution and has a particularly malleable characteristic. From the figure it is readily seen that when a brass having a 40 per cent zinc composition is heated to around 700°C, the alloy becomes completely β in structure. A second important alloy composition is the copper–nickel system whose phase diagram is shown in Figure 18.3(b). Nickel, like copper, is a face-centered-cubic structured element and has similar atomic dimensions and chemical properties to copper, and so these two elements form a substitutional solid solution when combined in all proportions. The resulting material is tough, ductile, reasonably strong and has good corrosion resistance.

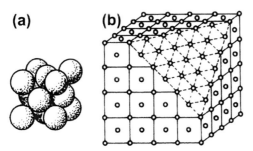

FIGURE 18.2 Face-centered-cubic structure of copper: (a) cell unit and (b) arrangement of atoms on the (111) close-packed plane.

The properties required of a propeller material will depend to a very large extent on the duty and service conditions of the ship to which the propeller is being fitted. However, the most desirable set of properties which it should possess are as follows:

1. High corrosion fatigue resistance in sea water.
2. High resistance to cavitation erosion.
3. Good resistance to general corrosion.
4. High resistance to impingement attack and crevice corrosion.
5. High strength to weight ratio.
6. Good repair characteristics including weldability and freedom from subsequent cracking.
7. Good casting characteristics.

The majority of propellers are made by casting. However, cast metal is not homogeneous throughout and the larger the castings, the more the differences between various parts of the casting are accentuated. These differences in properties are due to differences in the rates of cooling in the various parts of the casting, assuming that the liquid metal is initially of uniform temperature and composition. For example, the rate of cooling of the metal at a blade tip, which may be of the order of 15 mm in thickness, will be much faster than that at the boss, which may be 1000 mm thick for the same propeller.

Propeller Materials

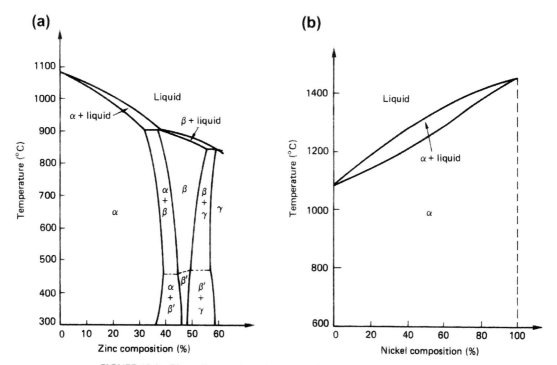

FIGURE 18.3 Phase diagram for: (a) copper–zinc and (b) copper–nickel alloy.

In general, the faster the cooling rate the smaller the crystal or grain size of the material will be. The slower the cooling rate, the more nearly equilibrium conditions will be reached; consequently, at the center of the boss of a large propeller the structure of the alloy tends to approach the conditions defined by the phase diagram. The difference in the microstructure, therefore, between the metal in the blade tip and in or near the boss region can be considerable depending upon the level of control exercised and the type of alloy being cast. This difference assumes importance in propeller technology because for conventional low skew propellers the maximum stress in service is normally incurred in thick sections of the casting at the blade root.

Apart from differences between the thin and thick parts of the same casting, there are also differences through the section thickness. The difference in through-section properties arises because the metal at the skin of the casting is the first to freeze, since the metal here is chilled by contact with the mold. Consequently, the cooling rate is fast and hence the grain size of the material is the smallest close to the surface. However, towards the center of the section there is a slower cooling rate and the metal at this location is last to freeze. Therefore, because alloys solidify over a range of temperatures, the structural composition can be expected to vary in the metal between that which is the first and last to freeze. Additionally, other material which is not in solution when in the liquid state, such as slag or other impurities, is pushed, while still liquid, towards the center as the dendrites of the solidifying metal grow from the sides. Furthermore, in a casting which is not adequately fed by liquid metal, there may not be sufficient metal to fill the space in the center of the casting and unsoundness due to shrinkage will result. Consequently, the poorest properties can be expected near the center of a thick casting. Figure 18.4 shows the expected variation in grain size through a propeller root section.

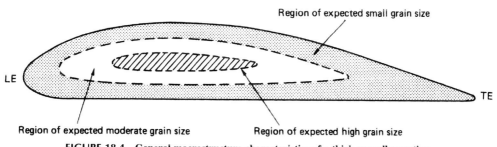

FIGURE 18.4 General macrostructure characteristics of a thick propeller section.

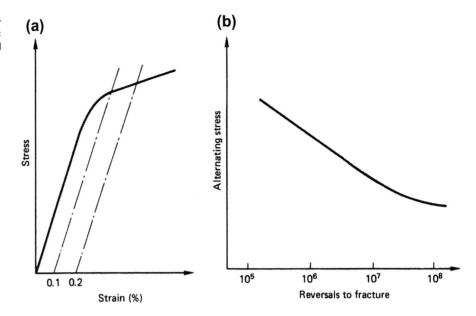

FIGURE 18.5 Mechanical characteristics of propeller materials: (a) stress–strain relationship and (b) fatigue resistance.

Because of the differences in the material properties that can take place in a propeller casting, it is reasonable to expect that the through-thickness mechanical properties of the material will also show considerable variation. This is indeed the case and care needs to be taken in selecting the location, size and test requirements of material specimens in order to gain representative mechanical properties for use in design: these aspects are discussed more fully in Section 18.4. In more general terms, however, the stress–strain relationship for bronze materials takes the form of Figure 18.5(a). Since these materials, like the stainless steels, do not possess a clearly defined yield point, as in the case of a low carbon steel, consequently, the stress–strain curve is characterized in terms of 0.1 per cent and 0.2 per cent proof stresses. The more important mechanical characteristic, the fatigue resistance curve, is shown in Figure 18.5(b). In the case of propeller design it is important to consider data at least to 10^8 cycles, preferably more, if realistic mechanical fatigue properties are to be derived. For example, a ship having a propeller rotating at 120 rpm and operating for 250 days per year will accumulate on each blade 8.6×10^8 first-order stress cycles over a twenty-year life. It is, however, instructive to consider the way in which cyclic fatigue life builds up on a propeller blade: Table 18.1 demonstrates this accumulation for the example cited above.

A moment's consideration of Table 18.1 in relation to Figure 18.5(b) gives a measure of support to the fatigue failure 'rule of thumb' for a propeller which implies that 'If it lasts more than two or three years, then it is probably unlikely to suffer a fatigue failure from normal service loadings causing the loss of a blade.'

When considering the fatigue characteristics of a material it is important to consider the relationship shown in Figure 18.5(b) in relation to the amount of tensile stress acting on the material in question. The effect of tensile stress on the fatigue resistance of bronzes has been quite extensively investigated; for example Webb et al.[2] From these studies it has been shown that the effect of the tensile stress is considerable, as shown in Figure 18.6.

The chemical composition of the metal is of importance in determining the mechanical properties of the material. Langham and Webb[3] show, for example, how the effects of changes to the manganese and aluminum contents of Cu–Mn–Al alloys influence the mechanical strength of the material (Figure 18.7).

18.2 SPECIFIC PROPERTIES OF PROPELLER MATERIALS

The preceding discussion has considered in general terms the properties and influences on copper-based materials starting from the nature of pure copper and finishing with

TABLE 18.1 Build-Up of First-Order Fatigue Cycles on a Blade of Propeller

Time	1st Hour	1st Day	1st Month	1st Year	2nd Year	10th Year	20th Year
Number of first-order fatigue cycles	7.2×10^3	1.7×10^5	3.6×10^6	4.3×10^7	8.6×10^7	4.3×10^8	8.6×10^8

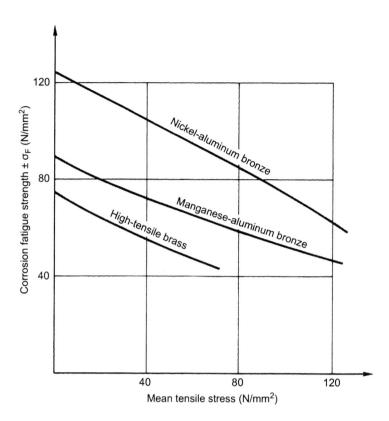

FIGURE 18.6 Effect of mean tensile stress on corrosion fatigue properties.

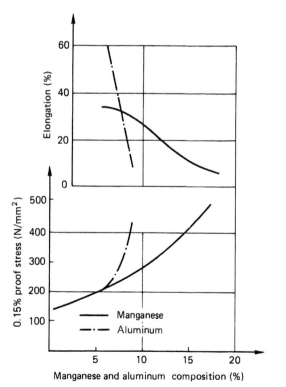

FIGURE 18.7 Typical effect of chemical composition on mechanical properties of a copper–manganese–aluminum alloy.

the basic characteristics of realistic propeller alloys. From this basis, consideration now turns to an overview of the specific properties of the more common propeller materials shown in Figure 18.1. To develop a greater insight into propeller materials, particularly from the metallurgical viewpoint, the reader is referred to References 4 and 5: indeed, these references have formed a basis for the present discussion.

18.2.1 High-Tensile Brass

These alloys are frequently referred to as 'manganese bronze'; however, this is a misnomer as they are essentially alloys of copper and zinc and are, therefore, brasses rather than bronzes. Furthermore, although a small amount of manganese is usually present, this is not an essential constituent of these alloys.

High-tensile brasses have the advantage of being able to be melted very easily and cast without too much difficulty. Care, however, has to be exercised in melting the alloy for the manufacture of very large propellers, since any contamination with hydrogen gas leads to unsoundness in the propeller casting. The composition of these alloys varies considerably, but they are essentially based on a 60 per cent copper, 40 per cent zinc brass formulation together with additions of aluminum, tin, iron, manganese and sometimes nickel. Aluminum is a strengthening addition which also helps to improve the corrosion resistance and is generally present in proportions of between 0.5 and 2 per cent; this, however, is sometimes increased to around 3 per cent in order to produce a stronger alloy. If tin is omitted from the material, then the alloys corrode rapidly by the process of dezincification so that the surface appearance of the material remains unchanged except for some degree of coppering.

The high-tensile brasses basically comprise two separate phases; however, when dezincification occurs, the beta phase in the structure is initially replaced by copper. While dezincification may occur when in fast flowing sea water, it most readily occurs under stagnant conditions, particularly where there are crevices in the material. To give reasonable resistance against this form of attack a tin content of at least 0.2 per cent has to be incorporated in the alloy and the higher the tin content, the greater the resistance against this type of corrosion. High tin contents, however, lead to difficulties in propeller casting and the alloys become more sensitive to stress corrosion cracking. Tin contents, therefore, seldom exceed 0.8 per cent and never exceed 1.5 per cent.

The mechanical properties obtained from castings are very dependent upon the grain size: the maximum strength being obtained with fine-grained material. Iron is an essential constituent to produce grain refinement in the alloy and is present, in the absence of high aluminum or nickel contents, at levels of the order of 0.7–1.2 per cent. In cases where the aluminum or nickel contents are high, higher iron contents are necessary in order to achieve the requisite degree of grain refinement; however, little benefit is gained by increasing the iron content above 1.2 per cent. Manganese appears to have a generally beneficial but non-critical influence on the alloy properties and about 1 per cent is usually present in the material. Nickel is not harmful, but at the same time does not appear to introduce any worthwhile benefits which could not be obtained more economically by increasing the aluminum content.

The copper and zinc contents are adjusted to give the best balance of properties: these are obtained when the microstructure of the alloy contains about 40 per cent of the softer more ductile alpha phase and 60 per cent of the harder, less ductile, beta phase. The relative proportions of these two phases also have a controlling influence on the tensile properties and fatigue strength of the alloy. If the zinc content is raised to too high a level, the alloy will contain none of the alpha phase (Figure 18.3(a)) and in that condition it will be very susceptible to stress corrosion in sea water. Therefore, if a high-tensile stress is continuously experienced in the material during immersion in sea water, then spontaneous cracking can occur. This susceptibility to stress corrosion exists even when some of the alpha phase is present in the alloy; however, sensitivity to stress corrosion cracking is believed to decrease as the alpha content is increased: an alpha content of 25 per cent is generally regarded as a minimum in a material used for propeller manufacture. As in the case of the simple Cu–Zn alloy discussed in Section 18.1, when the two-phase alloy is heated the alpha particles gradually dissolve into the beta phase until, at temperatures of around 550°C, the alloy consists entirely of the beta phase. If the metal is allowed to cool slowly to room temperature, the alpha particles precipitate out once more and a structure similar to the original is recovered. Alternatively, if the cooling is rapid the alpha phase does not precipitate fully and, with very fast cooling, a completely beta structure can be retained down to room temperature. Similar structures are frequently produced in areas adjacent to welds where residual internal stresses can be of a very high order and this combination of high residual stress and undesirable microstructure, in terms of low alpha phase, has frequently led to stress corrosion cracking in high-tensile brass propellers. It is of the utmost importance, therefore, that welds in these materials be stress relieved by heat treatment.

Stress relief can be effected by heating the material at temperatures in the range of 350–550°C as referenced in a later chapter when discussing maintenance and repair; the higher temperatures allowing full precipitation of the alpha phase. Residual stresses can be reduced by localized heating of the surface of a high-tensile brass propeller and this should be undertaken wherever possible.

High-tensile brass is an easy material to machine and can be bent or worked at any temperature. When it is heated above 600°C it consists entirely of the beta phase and is quite soft and ductile, facilitating any straightening repairs which may be necessary.

18.2.2 Aluminum Bronzes

For discussion purposes it is possible to classify aluminum bronzes into three types:

1. Those containing more than 4 per cent of nickel and very little manganese.
2. Those containing in excess of 8 per cent of manganese.
3. Those containing very little nickel or manganese.

The majority of large aluminum bronze propellers are manufactured using either the first or second types of alloy, which are normally known by the names of nickel–aluminum bronze, and manganese–aluminum bronze, respectively. The latter of the three alloys has low impact strength and poor corrosion resistance.

The first of the manganese–aluminum bronzes was patented around 1950 and had a composition of some 12 per cent manganese, 8 per cent aluminum, 3 per cent iron and 2 per cent nickel. These manganese and aluminum contents were selected at that time to give a phase structure comprising about a 60–70 per cent alpha content. Whilst some alloys containing 6–9 per cent of manganese have been used for propeller manufacture it is found that an increase in manganese content above 10 per cent results in a general improvement in mechanical properties. The presence of the manganese of around 6–10 per cent concentration inhibits a decomposition of the beta phase into a brittle eutectoid mixture containing a hard gamma phase which would otherwise occur in heavy cast sections. Indeed, some alloys contain up to about 15 per cent manganese.

All manganese–aluminum alloys have similar microstructures and, therefore, somewhat similar characteristics. Their structures are similar to those of the high-tensile brasses; however, the structure tends to be finer and the proportion of the alpha phase is higher with the manganese–aluminum bronzes. In keeping with the high-tensile brasses they have no critical temperature range in which they lose ductility and are susceptible, but less sensitive, to stress corrosion in sea water in the presence of high internal stresses. This lower sensitivity to stress corrosion is probably due to their lower beta content. The same precautions that are recommended for the high-tensile brasses regarding stress relief after welding should be applied, although the risk of cracking is less in manganese–aluminum alloys if these precautions are not taken.

The nickel–aluminum bronze alloys usually contain some 9–9.5 per cent aluminum with nickel and iron contents each in excess of 4 per cent: this level of nickel is required to obtain the best corrosion resistance. In BS 1400-AB2, lead is normally permitted up to a level of 0.05 per cent, except where welding is to be carried out when it should be limited to a maximum of 0.01 per cent. Manufacturers, however, can experience difficulty in maintaining the lead at levels as low as 0.01 per cent because of its tramp persistence in secondary metals. Although published work dealing with the effect of lead on the weldability of nickel–aluminum bronze is sparse, it is generally considered that its presence should not be detrimental to weldability if maintained below 0.03 per cent.

The microstructure of nickel–aluminum bronze is quite different from that of high-tensile brass. It comprises a matrix of the alpha phase in which small globules and plates of a hard constituent are distributed which is frequently designated a kappa phase. Figure 18.8 shows a typical nickel–aluminum bronze microstructure at a magnification of 200. At ambient temperature the alloy is tough and ductile but as the temperature is raised it becomes less ductile and tough with elongation values at about 400°C that are only about a quarter of those at room temperature. The ductility is recovered, however, at higher temperatures and bent propeller blades can be straightened at temperatures in excess of 700°C. Furthermore, at temperatures above 800°C nickel–aluminum bronze

FIGURE 18.8 Microstructure of nickel–aluminum bronze (x200).

becomes quite malleable and ductile, allowing repairs to be made with relative ease.

The nickel–aluminum bronzes have a considerably higher proof stress than the high-tensile brasses as well as having somewhat higher impact strength. The corrosion fatigue resistance in sea water is approximately double that of high-tensile brasses and this permits the use of higher design stresses and hence reduced section thicknesses of the propeller blades. Nickel–aluminum bronze is also found to be more resistant to cavitation erosion than high-tensile brass by a factor of two or three, and it is also much more resistant to the impingement type of corrosion, often referred to as wastage, which removes metal from the leading edges and the tips of propeller blades.

An in-depth treatment of the nickel–aluminum bronzes is given in Reference 6.

18.2.3 Stainless Steels

Two principal types of stainless steel have been used for propeller manufacture. These are the 13 per cent chromium martensitic and the 18 per cent chromium, 8 per cent nickel, 3 per cent molybdenum austenitic stainless steel. The former is perhaps the more widely used; however, its use has generally been confined to small propellers and component parts of controllable pitch propellers. The main advantage of austenitic stainless steel is to be found in its toughness, which enables it to withstand impact damage, and its good repairability.

Both types of stainless steel have a good resistance to impingement corrosion, but tend to suffer under crevice corrosion conditions. Their resistance to corrosion fatigue in sea water and also to cavitation erosion is generally lower than those of the aluminum bronzes. In recent years, stainless steels with more than 20 per cent chromium and about 5 per cent nickel with microstructures containing roughly equal proportions of austenitic and ferrite phases have been designed for propeller manufacture. These materials have better resistance to corrosion fatigue in sea water than either the martensitic or austenitic types.

Much work has been undertaken, particularly in Japan, on the development of stainless steels for marine propellers. In essence, the main thrust of this development work has been to make stainless steels more competitive to nickel—aluminum bronze in terms of their relative corrosion fatigue strength. Currently, many stainless steels have reduced allowable corrosion fatigue strength when compared to nickel—aluminum bronze of around 27 per cent (Reference 7): such a reduction translates to an increased blade section thickness requirement of the order of 17 per cent for a zero-raked propeller. Kawazoe et al.[8] discuss the development of a stainless steel and the results of laboratory and full-scale trials on a number of vessels of differing types. The chemical composition for this stainless steel is nominally 18 per cent chromium; 5—6 per cent nickel; 1—2 per cent molybdenum with manganese less than 3 per cent and cobalt and silicon less than 1.5 per cent. Indications are that this material, based on Wöhler rotating beam tests, can develop fatigue strengths of the order of 255 N/mm^2 at 10^8 reversals.

18.2.4 Cast Iron

Ordinary flake graphite cast iron has in the past been used mainly for spare propellers that are carried on board a ship for emergency purposes. This material has a very poor resistance to corrosion, particularly of the impingement type, and the life of a cast iron propeller must be regarded as potentially very short. Because the resistance to corrosion is adversely affected by removal of the cast skin, it is not normal to grind the blades to close dimensional control. Additionally, since much heavier section thicknesses are required for strength purposes this makes these propellers far less efficient. Cast iron is brittle and this renders it susceptible to breakage on impact with an underwater object and, moreover, only very minor repairs can be affected.

The enhanced ductility of spheroidal graphite cast iron when compared with grey iron makes it a more attractive material for propeller usage. It is, however, subject to rapid corrosion and erosion and the use of heavy section thicknesses is still necessary.

Austenitic nodular cast iron has been used for the manufacture of small propellers. It contains around 20—22 per cent nickel and 2.5 per cent chromium and the microstructure has an austenitic matrix with graphite in spheroidal form. Its resistance to impingement attack and corrosion approaches that of high-tensile brass, but its impact strength and resistance to cavitation erosion are rather lower.

18.2.5 Cast Steel

Low alloy and plain carbon cast steels are occasionally used for the manufacture of spare propellers. While the tensile properties are reasonable, the resistance to corrosion and erosion in sea water is much inferior to that of the copper-based alloys. Cathodic protection is essential when using this material for propellers.

18.2.6 Carbon-Based Composites

In recent years carbon-based composites as a propeller material have made an entry into the commercial craft and yacht market although they have been used in specialized naval vehicles such as some submarines for many years. Apart from useful acoustic properties for these propellers there is a further advantage in their light weight when compared to conventional propeller materials. For example, in the case of a 130 tonne finished weight container ship propeller, the weight would reduce if the blades were manufactured from carbon fiber materials to something of the order of 50 tonnes.

There is a distinction between carbon and graphite. While graphite is composed of the carbon element, the term graphite should only be applied to those carbons with a perfect hexagonal structure. This ideal is, however, rarely achieved in manufacturing practice and manufactured graphite tends to be a heterogeneous agglomeration of near perfect crystallites intermingled with less well-ordered areas. Nevertheless the term graphite has become reasonably well accepted as embracing carbons which approach a near perfect crystal structure. Graphite, however, has one serious drawback in that it is mechanically weak: typically the tensile strength of high-grade polycrystalline graphite at room temperature is of the order of 35 MPa compared with 670 MPa for nickel—aluminum bronze.

To understand the properties of carbon in a solid form appeal has to be made to the hexagonal crystal structure of graphite. The atoms are arranged in planar basal layers and within each of the layers the atoms are hexagonally close packed with an interatomic distance of 0.14 nm. They are covalently bonded with sp^2 hybridization and the bond strength is 522 kJ/mol. Between the basal layers there are van der Waal bonds which have strengths of only 17 kJ/mol and these arise from delocalized electrons. The layers are separated by a distance of 0.3354 nm but due to alternate layers being in the atomic register the repeat unit distance is double this distance. Clearly such a structure gives rise to a highly anisotropic behavior of the crystal. For example, the modulus of elasticity in the basal plane is 1050 GPa while that in the orthogonal plane is 35 GPa. Such behavior therefore has a profound influence on the design of the fiber and matrix composition of the material.

Carbon fibers with diameter of the order of 5—10 μm are not handled individually in the manufacture of propeller blades. The manufacturers of carbon fibers normally supply

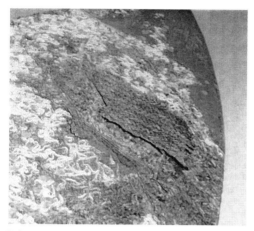

FIGURE 18.9 Delamination in a composite propeller.

18.3 MECHANICAL PROPERTIES

The form of the general stress–strain curve for the copper-based alloys and the stainless steels was shown in Figure 18.5(a). Table 18.2 shows typical comparative properties of the more common materials used for propeller manufacture as determined by separately cast test pieces.

These properties are important from the general stress analysis viewpoint. In determining the allowable stress, however, it is the fatigue properties that are of most importance for a ship's ahead operation. In Table 18.1 it was shown that 10^9 cycles can be attained in a matter of twenty years or so for, say, a large bulk carrier, and correspondingly sooner for many smaller vessels. However, if fatigue tests on these materials were carried out to that number of cycles and over a sufficiently large number of specimens for the results to become meaningful, the necessary design data would take an inordinately long time to collect. Consequently, it is more usual to conduct tests up to 10^8 reversals and although it can be argued that this data tends to be suspect when extrapolated to 10^9 cycles, the criteria of assessment are normally based on the lower number of reversals for marine propellers.

Throughout the development of propeller materials many tests using the Wöhler fatigue testing procedure have been made for the various materials. However, these tests have several limitations in this context since they do not readily permit the superimposition of mean loads on the specimen and the stress gradients across the test specimen tend to be large. For these reasons and, furthermore, since the exposed areas of the test piece tend to be small, the

the fibers in bundles, termed tows, and these might contain somewhere between 1000 and 12 000 individual fibers. In the case of untwisted tows these can be spread out to generate a unidirectional tape, alternatively in order to maintain the coherence of the tow it can be twisted. The tows can then be woven into a number of different weave patterns, some typical of those seen in the textile industry while others may be linear tapes or interwoven into chevron patterns.

While the use of composite materials holds out significant potential for propulsor design care has to be taken in the way the fiber weaves are laid up to form the propeller blade. If this is not done correctly, then delamination can occur as seen in Figure 18.9.

TABLE 18.2 Typical Comparative Material Properties

	Material	Modulus of Elasticity (kgf/cm^2)	0.15% Proof Stress (kgf/mm^2)	Tensile Strength (kgf/mm^2)	Brinell Hardness Number	Specific Gravity	Elongation (%)
Copper-based alloys	High-tensile brass	1.05×10^6	19	45–60	120–165	8.25	28
	High-manganese alloys	1.20×10^6	30	66–72	160–210	7.45	27
	Nickel–aluminum alloys	1.25×10^6	27.5	66–71	160–190	7.6	25
Stainless steels	13% chromium	2.0×10^6	45.5	69.5	220	7.7	20
	Austenitic	1.9×10^6	17	50.5	130	7.9	50
	Ferritic–austenitic	1.8×10^6	55	80	260	7.9	18
Cast iron	Grey cast iron	1.1×10^6	–	23.5	200	7.2	–
	Austenitic SG	1.1×10^6	–	44	150	7.3	25
Polymers	Nylon	0.008×10^6	1.1	4.7	–	–	35
	Fiberglass	0.14×10^6	–	20	–	–	1.5

results from these tests are used primarily for qualitative analysis purposes. In order to overcome these difficulties and, thereby, provide quantitative fatigue data for use in propeller design, fatigue testing machines such as the one shown by Figure 18.10 have been designed. Machines of this type are normally able to test material specimens of the order of 75 mm in diameter. In addition to applying a fluctuating component of stress, a mean stress can also be superimposed by using hollow specimens which permit pre-stressing by means of a suitable linkage. In order to simulate the corrosive environment a 3 per cent sodium chloride solution is normally sprayed onto the specimen. The use of this solution to simulate sea water is generally considered preferable for testing purposes since the properties of sea water are found to vary considerably with time. Therefore, unless the sea water is continuously replaced, it decays to such an extent that it becomes unrepresentative of itself.

Testing machines of the type shown in Figure 18.10 have been used extensively in order to examine the behavior of various propeller materials. Figure 18.6, by way of demonstration of these researches, shows the comparative behavior of three copper alloys based on a fatigue life of 10^8 cycles as determined by the authors of Reference 2. From these test results the superior corrosion fatigue properties of the nickel—aluminum bronzes become evident. However, in establishing these results care is necessary in controlling the solidification and cooling rates of the test specimens after pouring in order that they can correctly simulate castings of a significantly greater weight. In the case of Figure 18.6 a simulation of a casting weight of around four tonnes was attempted.

Casting size has long been known to influence the material properties as witnessed by the sometimes significant differences between test bar results and the mechanical properties of the propeller blade when destructively tested. For these reasons controllable pitch propeller blades are generally believed to have superior mechanical properties to monoblock propellers of an equivalent size. Many attempts have been made to correlate this effect by using a variety of parameters. Webb et al.,[2] from their research some years ago, suggested a relationship based on blade weight referred to a basis of a casting weight of ten ton (10.16 tonnes). The relationship proposed is as follows:

$$\sigma_w = \sigma_{10}\left(0.70 + \frac{30}{w + 90}\right) \quad (18.1)$$

where σ_w is the estimated fatigue strength at zero mean stress of a propeller weighing w ton in relation to that for a ten ton propeller, σ_{10}. Values of σ_{10} for high-tensile brass, manganese—aluminum bronze and nickel—aluminum bronze were proposed as being 6.8, 9.2 and 11.8 kgf/mm^2 respectively.

The approach by Meyne and Rauch,[9] in which tensile strength and proof stress are plotted to a base of propeller weight divided by the product of blade number and the area of one blade, gives encouraging results. In this approach, which is effectively defining a pseudo-blade thickness correlation parameter, the analysis procedure is taken a stage further than the simpler methods using blade weight alone. The correlation of elongation with casting size is, however, still far from resolved.

Later work by Wenschot[10] in examining some hundred or so nickel—aluminum bronze propellers undertook mechanical and corrosion fatigue studies on material taken from the thickest parts of propeller castings. From the various castings included in this study, section thicknesses varied from 25–450 mm and the analysis resulted in a relationship between cast section thickness and fluctuating stress amplitude for zero mean stress of the form

$$\sigma_a = 160.5 - 24.4\log(t) \quad (18.2)$$

where the cast section thickness (t) is measured in mm and the corrosion fatigue strength in sea water (σ_a) is based on 10^8 reversals.

18.4 TEST PROCEDURES

Because of the variations in mechanical properties within one casting it is practically impossible to cast a test bar, or test bars, which will represent the actual properties of the material in all parts of a casting. The best that can be done

FIGURE 18.10 Corrosion fatigue testing machine for propeller materials.

is to cast a number of test bars in separate molds using the same molten metal as for the casting. The mold for the test bar should be correctly designed so that the part from which test pieces are to be cut is properly fed to obtain sound metal. Test bars or coupons which are cast in an integral way with the casting may give an indication of the properties in the casting near to the position of the test bar provided the section thickness is similar. However, there is always the danger with cast-on bars that they are not properly fed and their properties may be inferior to those of the adjacent metal. Alternatively, there may be overriding requirements, such as the assurance that a test bar is cast from the same metal it is supposed to represent, that insist that it should be integral with the casting.

The usual geometric form of test bar for propeller alloys is the keel bar casting, having a circular cross-section of diameter 25 mm and a feeder head along its full length. From such a bar it is not possible to machine a tensile test piece with a parallel portion much more than 15 mm in diameter and, clearly, a test piece of this type will not represent the properties of the thick sections of a large propeller. It is nevertheless quite satisfactory for sorting out a poor cast of metal from a number of casts. When examining the microstructure of high-tensile brass, the test specimen should be cut from the test bar to ensure that it has been cooled at a standard rate for comparison of the amount of alpha and beta phases present.

Most fatigue testing is carried out on rotating beam Wöhler machines, using a round specimen held in a chuck with a load applied in bending as a cantilever. In this way a complete reversal of the stress is applied to the specimen with each revolution of the test piece. Other types of fatigue tests employ rectangular specimens with the load applied in the plane of bending and others apply the fluctuating stress axially in tension and compression by a pulsating load. For testing in air using a rotating beam machine, the specimen is usually about 10–15 mm in diameter and good reproducibility of results is obtained for most wrought materials with this method. When cast materials are to be tested the results show more scatter, but still give an indication of the fatigue limit which can be expected on a comparative basis. Clearly, the larger the test piece of a cast material, the more useful the results will be in representing the fatigue properties of large castings.

The evaluation of the resistance of a material to a fluctuating stress in a corrosive environment is a much more difficult proposition: since the conditions involve corrosion, short-time tests are of little value. Because there is no stress, however low, which will not induce failure if the corrosive conditions are maintained for long enough, the longer the time of the test the better. Clearly some time limit must be defined before testing starts and for this purpose a year is a good criterion; however, from the practical standpoint of getting results for design purposes shorter periods must be permitted.

As the corrosion fatigue test relies on the stress and corrosion acting together, due account must be taken that the stress acts through the material whereas corrosion acts over the surface area. The ratio of the area of the cross-section of the test piece to its diameter is therefore important, and the smaller the diameter of the specimen, the greater will be the effect of the corrosion parameter while maintaining a constant stress on the specimen.

Since rotating beam fatigue tests in air use a specimen of about 10 mm in diameter, this size has frequently been pursued for corrosion fatigue testing. Work in Japan has, however, shown a 30 per cent reduction in corrosion fatigue resistance when the specimen size was increased from 25 to 250 mm diameter. This work was carried out over a short time frame and is therefore not a realistic appraisal of corrosion fatigue resistance. However, it does show the effect of specimen size on the fatigue resistance of cast copper alloys.

When large specimens are used the contact with the environment becomes difficult to arrange with a rotating specimen. Machines have, therefore, been devised (Figure 18.10) to apply reversed bending on a static specimen by a rotating out-of-balance load developed through counterweights attached to the specimen. These machines can test specimens of 76 mm in diameter exposing about 225 cm^2 to the corrosive medium. It is also recognized that a copper alloy propeller as cast contains significant internal stresses. In corrosion fatigue testing, therefore, it is useful to be able to apply a mean stress to the test piece so that a fluctuating stress can be superimposed on it. The large 76 mm specimens referred to above are consequently made hollow and a screwed insert within the bore enables a tensile stress of known magnitude to be applied during the cyclic fatigue test. Failures in these tests have fractures very similar to those on propellers in service which have failed.

It will be appreciated that with all the variables contingent on corrosion fatigue testing, the test results on a particular material can have a great deal of scatter, as shown in Figure 18.11. Each spot on the figure is the result of the failure of a nickel–aluminum bronze test bar, all cast

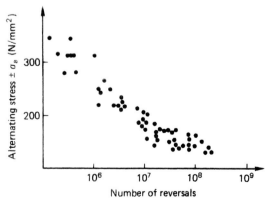

FIGURE 18.11 Typical scatter of corrosion fatigue tests on a nickel–aluminum bronze alloy.

to the same specification, and scatter of this type is not unusual for such tests on cast material. Clearly, any attempt to extrapolate the curve beyond 2×10^8 reversals in such a case is unwise.

REFERENCES AND FURTHER READING

1. Carlton JS. Propeller service experience. *7th Lips Symposium*; 1989.
2. Webb AWO, Eames CFW, Tuffrey A. *Factors affecting design stresses in marine propellers*. Propellers '75 Symposium, Trans SNAME 1975.
3. Langham MA, Webb AWO. *The new high strength copper–manganese–aluminium alloys – Their development, properties and applications*. Detroit: Proc. Int. Foundry Congress; 1962.
4. Webb AWO. *High Strength Propeller Alloys*. SMM Technical Paper No. 5, May 1965.
5. Webb AWO, Capper H. *Propellers* (ISBN 0 900976 42 X). In: Materials for Marine Machinery. I.Mar.E; 1976 [Chapter 8].
6. Meigh H. *Cast and Wrought Aluminium Bronzes: Properties, Processes and Structure* (ISBN 978 1 861250 62 9). Copper Development Association; August 2009.
7. *Rules and Regulations for the Classification of Ships*. Part 5, [Chapter 7]. Lloyd's Register; January 1991.
8. Kawazoe T, Matsuo S, Sasajima T, Daikoku T, Nishikido S. *Development of Mitsubishi corrosion resistance steel (MCRS) for marine propeller*. 4th Int Symp on Mar Eng Trans MESJ 1990.
9. Meyne KJ, Rauch O. *Some results of propeller material investigations*. Propellers '78 Symposium, Trans SNAME 1978.
10. Wenschot P. *The properties of Ni–Al bronze sand cast ship propellers in relation to section thickness*. JSP June 1987;**34**.

Chapter 19

Propeller Blade Strength

Chapter Outline

19.1 Cantilever Beam Method 397
19.2 Numerical Blade Stress Computational Methods 402
19.3 Detailed Strength Design Considerations 405
19.4 Propeller Backing Stresses 408
19.5 Blade Root Fillet Design 408
19.6 Residual Blade Stresses 409
19.7 Allowable Design Stresses 410
19.8 Full-Scale Blade Strain Measurement 413
References and Further Reading 414

The techniques of propeller stressing remained in essence unchanged throughout the development of screw propulsion until the early 1970s. Traditionally the cantilever beam method has been the instrument of stress calculation and formed the cornerstone of commercial propeller stressing practice. However this method has, in many instances, been superseded by finite element methods which lend themselves to a more detailed stress analysis of the propeller blade.

The cantilever beam method was originally proposed by Admiral Taylor in the early years of the last century and since that time a steady development of the method can be traced.[1-7] Currently several expositions of the method have been made in the technical literature, all of which, although developing the same basic theme, have differing degrees of superficial emphasis. The version published by Sinclair,[8] based on the earlier work of Burrill,[5] is typical of the cantilever beam methods.

19.1 CANTILEVER BEAM METHOD

The cantilever beam method relies on being able to represent the radial distributions of thrust and torque force loading, as shown in Figure 19.1, by equivalent loads, F_T and F_Q, at the center of action of these distributions. Having accepted this transformation, the method proceeds to evaluate the stress at the point of maximum thickness on a reference blade section by means of estimating each of the components in the equation

$$\sigma = \sigma_T + \sigma_Q + \sigma_{CBM} + \sigma_{CF} + \sigma_\perp \quad (19.1)$$

where

σ_T is the stress component due to thrust action.
σ_Q is the stress component due to torque action.
σ_{CBM} is the stress component due to centrifugal bending.
σ_{CF} is the stress component due to direct centrifugal force.
$\sigma_\perp$ is the stress component due to out of plane stress components.

Using the definitions of Figure 19.1 the bending moment due to hydrodynamic action (M_H) on a helical section of radius (r_0) is given by

$$M_H = F_T a \cos\theta + F_Q b \sin\theta$$

in which F_T and F_Q are the integrated means of the thrust and torque force distributions and a and b define their respective centers of action.

The mechanical loadings on a particular section of a propeller blade are a function of the mass of the blade outboard of the section considered and the relative position of its center of gravity with respect to the neutral axis of the section being stressed. Hence, a system of forces and moments is produced, which can be approximated, for all practical purposes in conventional non-skewed propeller forms, to a direct centrifugal loading together with a centrifugal bending moment acting about the plane of minimum section inertia. In the case of conventional low skew propeller designs the centrifugal loadings can be readily calculated as indicated by Figure 19.2 and in general it will be found that they give rise to much smaller stresses than do their hydrodynamic counterparts; the exception to this is the case of small high-speed propellers.

The total bending moment (M) acting on the blade section due to the combined effects of hydrodynamic and centrifugal action is therefore given by:

$$M = M_H + M_C \quad (19.2)$$

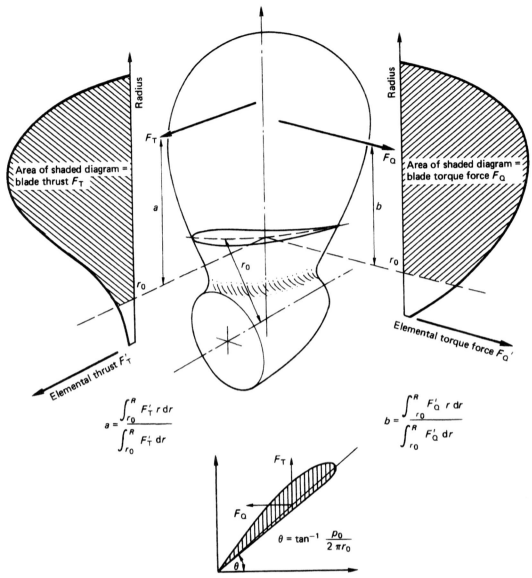

FIGURE 19.1 Basis of the cantilever beam method of blade stressing.

the centrifugal component (M_C) being the product of the centrifugal force by that part of the blade beyond the stress radius at which the stress is being calculated and the distance perpendicular to the neutral axis of the line of this force vector.

Hence from equations (19.1) and (19.2) the maximum tensile stress exerted by the blade on the section under consideration is given by:

$$\sigma = \frac{M}{Z} + \frac{F_C}{A} \qquad (19.3)$$

where F_C is the centrifugal force exerted by the blade on the section. The term M/Z embraces the first three terms of equation (19.1), the term F_C/A is the fourth term of equation (19.1), while the final term $\sigma_\perp$ is considered negligible for most practical purposes. The calculation of the section area and modulus are readily undertaken from the information contained on the propeller drawing. The procedure, in its most fundamental form, being basically to plot the helical section profile according to the information on the propeller drawing and then, if undertaking the calculation by hand, to divide the section chord into ten equally spaced intervals, see Chapter 11. The appropriate values of the local section thickness (t) and the pressure face ordinate (y_p) can then be interpolated and integrated numerically according to the following formulae:

$$A = \int_0^C t \, dc \qquad (19.4)$$

Propeller Blade Strength

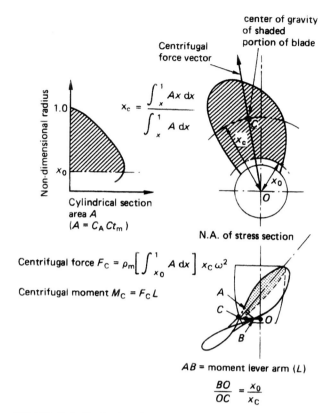

FIGURE 19.2 Derivation of mechanical blade loading components.

and for the section tensile modulus

$$Z_m = \frac{2\int_0^C [3y_p(y_p+t)+t^2]t\,dc \cdot \int_0^C t\,dc}{3\int_0^C (2y_p+t)t\,dc} - \frac{1}{2}\int_0^C (2y_p+t)t\,dc \quad (19.5)$$

It will be noted that the final form of the blade stress equation (19.3) ignores the components of stress resulting from bending in planes other than about the plane of minimum inertia. This simplification has been shown to be valid for all practical non-highly skewed propeller blade forms and therefore is almost universally used by the propeller industry for conventional propeller blade stressing purposes.

Clearly the cantilever beam method provides a simple and readily applicable method of estimating the maximum tensile, or alternatively maximum compressive, stress on any given blade section. To illustrate the details of this method, a worked example appears in Table 19.1. This example considers the evaluation of the mean value of the maximum tensile stress at the $0.25R$ section of a propeller blade and it can be seen that the calculation is conveniently divided into six steps. The first two are devoted principally to the collection of the necessary data required prior to performing the calculations. The propeller section data is given at a variety of chordal stations, depending upon the manufacturer's preference; consequently, it is usually necessary to obtain values by interpolation at intermediate stations in order to satisfy the requirements of the numerical integration method. It has been found by experience that for hand calculations a conventional Simpson's rule integration procedure over eleven ordinates is perfectly adequate for calculating section areas and moduli and the appropriate stages for this calculation are outlined by steps (3) and (4). Having evaluated the section properties the calculation proceeds as shown in the remainder of the Table 19.1.

A method of this type depends for its ease and generality of application upon being able to substitute values for the lengths of the moment arms a and b without recourse to a detailed analysis of the blade radial loading distribution. Again, experience has shown that this can be satisfactorily done providing that the propeller type is adequately taken into account. Typically, for a conventional, optimally loaded fixed pitch propeller, the moment arms a and b would be of the order of $0.70R$ and $0.66R$, respectively, whereas for the corresponding controllable pitch propeller they would be marginally higher. Similar considerations also apply to the position of the blade centroid.

Cantilever beam analysis provides a useful means of examining the relative importance of the various blade stress components delineated in equation (19.1). Table 19.2 shows typical magnitudes of these components expressed as percentages of the total stress for a variety of ship types and although variations will naturally occur within a given ship group, several important trends can be noted from such a comparison. It becomes apparent from the table that the thrust component accounts for the greatest part of the total stress for each class of vessel and that the direct centrifugal components, although comparatively small for the larger propellers, assume a greater significance for the smaller and higher-speed propellers. However, probably most striking is the effect of propeller rake as shown by the two propellers designed for the same fast cargo vessel. These propellers, although designed for the same powering conditions, clearly demonstrate a potential advantage of employing a reasonable degree of forward rake, since this effect leads to a compressive stress on the blade face. Consequently, this effect can allow the use of slightly thinner blade sections which is advantageous from blade hydrodynamic considerations, although if carried too far may lead to casting problems. Nevertheless, although the use of forward rake is desirable and indeed relatively commonly used, its magnitude is normally limited by propeller–hull interaction considerations; typically by classification society clearance limitations or from

TABLE 19.1 Blade stress computation using the cantilever beam method

Calculation of the Maximum Tensile Stress Acting on a Helical Section of a Propeller Blade by the Cantilever Beam Method

(1) **Stress Basis**

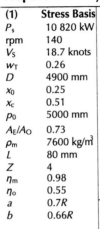

P_s	10 820 kW
rpm	140
V_S	18.7 knots
w_T	0.26
D	4900 mm
x_0	0.25
x_c	0.51
p_0	5000 mm
A_E/A_O	0.73
ρ_m	7600 kg/m³
L	80 mm
Z	4
η_m	0.98
η_o	0.55
a	0.7R
b	0.66R

(2) Stress section data

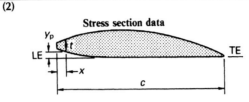

$C = 1000$ mm $r_0 = x_0 R = 0.25 \times (4900/2) = 612.5$ mm

x	0	55	110	220	330	497	665	832	1000	mm
y_p	55.0	39.0	25.0	6.5	0	0	0	0	0	mm
t	30.2	92.5	143.5	211.0	238.5	226.5	187.5	114.0	114.0	mm

(3) **Interpolated Section Data**

x	0	100	200	300	400	500	600	700	800	900	1000	mm
y_p	55	26	10	1	0	0	0	0	0	0	0	mm
t	30	195	195	232	236	226	205	173	128	73	15	mm

Chordal increment $\Delta C = \dfrac{C}{10} = 100$ mm

(4) **Evaluation of Section Properties**

Column	1	2	3	4	5	6	7	8	9
Ordinate	x	y_p	t	Simpson's mult.	$t \times$ S.M.	$(2y_p+t)t$	$(2y_p+t)t \times$ S.M	$[3y_p(y_p+t)+t^2]t$	$[3y_p(y_p+t)+t^2]t \times$ S.M.
1	0	55	30	½	15	4200	2 100	447 750	223 875
2	100	26	136	2	272	25 568	51 136	4 233 952	8 467 904
3	200	10	195	1	195	41 925	41 925	8 614 125	8 614 125
4	300	1	232	2	464	54 288	108 576	12 649 336	25 298 672
5	400	0	236	1	236	55 696	55 696	13 144 256	13 144 256
6	500	0	226	2	452	51 076	102 152	11 543 176	23 086 352
7	600	0	205	1	205	42 025	42 025	8 615 125	8 615 125
8	700	0	173	2	346	29 929	59 858	5 177 717	10 355 434
9	800	0	128	1	128	16 384	16 384	2 097 152	2 097 152
10	900	0	73	2	146	5 329	10 638	389 017	778 034
11	1000	0	15	½	7	225	112	3 375	1687
Total	–	–	–	–	2466	–	490 622	–	100 682 617

$$A = \int_0^C t\, dc = \frac{2 \times \Sigma\, \text{Col.5} \times \Delta C}{3} = \frac{2 \times 2466 \times 100}{3} = 164\,433\, mm^2$$

$$\int_0^C (2y_p + t)t\, dc = \frac{2 \times \Sigma\, \text{Col.7} \times \Delta C}{3} = \frac{2 \times 490622 \times 100}{3} = 32\,708\,167\, mm^3$$

$$\int_0^C [3y_p(y_p + t) + t^2]t\, dc = \frac{2 \times \Sigma\, \text{Col.9} \times \Delta C}{3} = \frac{2 \times 100\,682\,617 \times 100}{3} = 6\,712\,174\,433\, mm^3$$

From equation (19.5)

$$Z_m = \frac{2 \times 6\,712\,177\,733 \times 164\,400}{3 \times 32\,708\,133} - \frac{1}{2} \times 32\,708\,133 = 6\,141\,934\, mm^3$$

TABLE 19.1 Blade stress computation using the cantilever beam method—cont'd

(5) Blade centrifugal force
  (a) Calculate blade mass by either
  Evaluating Columns (1)–(5) of the previous step for each defined helical section, thereby obtaining the radial distribution of section area (A). The blade mass (m) is then calculated from

  $$m = \rho_m \int_{r_0}^{R} A\, dr$$

  N.B. (The position of the blade centroid (x_c) can also be calculated in an analogous way as shown in Figure 11.1)
  or
  by use of approximation

  $$m = 0.75 \times \text{mean radial thickness above stress section} \times \left(\frac{\text{total surface area}}{\text{number of blades}}\right) \times \text{density}$$

  viz.

  $$m = 0.75 \times 0.110 \times \left(\frac{0.73 \times \pi (4.90)^2}{4 \times 4}\right) \times 7600 \text{ kg}$$

  $m = 2158$ kg

  (b) Centrifugal force is given by
  $$F_c = 2\pi^2 m x_c D n^2$$
  $$F_c = 2\pi^2 \times 2158 \times 0.51 \times 4.90 \times \left(\frac{140}{60}\right)^2 \text{ N}$$

  that is $F_c = 580$ kN

(6) Calculation of section maximum tensile stress

  $$\text{Section pitch angel } \theta = \tan^{-1}\left(\frac{p_0}{\pi x_0 D}\right)$$

  $$= \tan^{-1}\left(\frac{5000}{\pi \times 0.25 \times 4900}\right) = 52.41°$$

  Propeller speed of advance $V_a = V_s (1 - w_T)$
  $= 18.7 \times (1 - 0.26) = 13.8$ knots
  $= 13.8 \times 0.515 = 7.10$ m/s

  (a) Component due to propeller thrust:

  $$\sigma_T = \frac{P_s \times \eta_m \times \eta_o \times (a - r_0) \times \cos\theta}{V_a \times Z \times Z_m}$$

  $$= \frac{(10\,820 \times 10^3) \times 0.98 \times 0.55 \times (0.7 - 0.25) \times 2450 \times \cos(52.41)}{7.10 \times 4 \times 6\,141\,934} = 22.40 \text{ MPa}$$

  (b) Component due to propeller torque:

  $$\sigma_Q = \frac{P_s \times \eta_m \times (b - r_0) \times \sin\theta}{2\pi \times n \times b \times Z \times Z_m}$$

  $$= \frac{(10820 \times 10^3) \times 0.98 \times (0.66 - 0.25) \times \sin(52.14) \times 10^3}{2\pi \times \left(\frac{140}{60}\right) \times 0.66 \times 4 \times 6\,141\,931} = 14.49 \text{ MPa}$$

  (c) Component due to centrifugal bending moment:

  $$\sigma_{CBM} = \frac{F_c \times L}{Z_m} = \frac{580\,000 \times 80}{6141934} = 7.55 \text{ MPa}$$

  (d) Component due to centrifugal force:

  $$\sigma_{CF} = \frac{F_c}{A} \times \frac{580\,000}{164\,433} = \underline{3.52 \text{ MPa}}$$

  TOTAL = 47.96 MPa

Maximum tensile stress acting on section (σ) = 47.96 MPa

TABLE 19.2 Breakdown of the Total Maximum Root Tensile Stress for a Set of Four Different Vessels

	Ship Type				
		Fast Cargo Vessel			
Component of Stress	Bulk Carrier	5° Astern rake	15° Forward rake	Twin-Screw Ferry	High-Speed Craft
Thrust	72%	58%	71%	54%	51%
Torque	23%	33%	41%	36%	35%
Centrifugal bending	1%	5%	−17%	2%	3%
Centrifugal force	4%	4%	5%	8%	11%
Total	100%	100%	100%	100%	100%

propeller-induced hull surface pressure calculations and studies.

In addition to providing a procedure for calculating the maximum stress at a given reference section, the cantilever beam method is frequently used to determine radial maximum stress distributions by successively applying the procedure described by Table 19.1 at discrete radii over the blade span. If such a procedure is adopted, then the resulting blade stress distributions have the form shown by Figure 19.3 and where typical bands of radial stress distribution for both linear and non-linear thickness distributions

can be seen. The non-linear distribution is the most commonly employed because although it encourages higher blade stresses, it permits a lower blade weight and also the use of thinner blade sections, which are advantageous from both the hydrodynamic efficiency and cavitation inception viewpoints. The linear distribution is frequently employed in towing and trawling situations in order to give an added margin against failure. These distributions, however, are frequently adopted in the case of many smaller propellers for the sake of simplicity in manufacture. However, this latter design philosophy can sometimes be mistakenly employed, since many small high-speed patrol craft have presented considerably more difficult hydrodynamic design problems than the largest bulk carrier.

Although the cantilever beam method provides a basis for propeller stress assessment, it does have certain disadvantages. These become apparent when the calculation of the chordal stress distribution is attempted since it has been found that the method tends to give erroneous results away from the maximum thickness location. This is partly due to assumptions made about the profile of the neutral axis in the helical sections since the method, as practically applied, assumes a neutral axis approximately parallel to the nose—tail line of the section. However, the behavior of propeller blades tends to indicate that a curved line through the blade section would perhaps be more representative of the neutral axis when used in conjunction with this procedure. Complementary reservations are also expressed since the analysis method is based on helical sections, whereas observations of blade failures tend to show that propellers break along 'straight' sections as typified by the failure shown in Figure 19.4.

19.2 NUMERICAL BLADE STRESS COMPUTATIONAL METHODS

To overcome these fairly fundamental problems with the cantilever beam approach which manifest themselves when

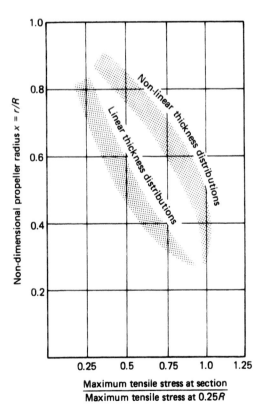

FIGURE 19.3 Comparative relationship between thickness and radial stress distribution.

Propeller Blade Strength

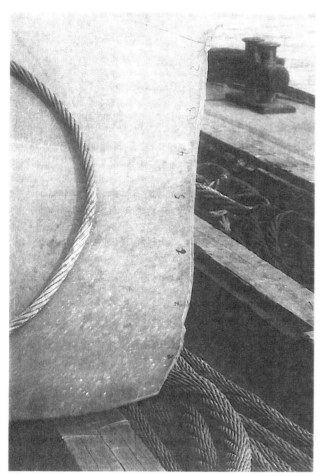

FIGURE 19.4 Propeller blade failure.

formulation. In each of the approaches the elements naturally require the normal considerations of aspect ratio and of near-orthogonality at the element corners that are normally associated with finite elements. Figure 19.5 shows some discretizations for a range of biased skew propellers: clearly, in the extreme tip regions the conditions of near-orthogonality are sometimes difficult to satisfy completely and compromises have to be made.

The finite element method is of particular importance for the stressing of highly skewed propellers, which form a significant subset of the propellers that are produced today, since the presence of large amounts of skew significantly influence the distribution of stress over the blades. Figure 19.6, taken from Carlton[14] shows the distributions of blade stress for a range of balanced and biased skew variants of the same blade configuration in comparison to a non-skewed version. In each case the blade thickness distribution remained unchanged and for ease of comparison, the iso-stress contour lines in this figure are drawn at 20 MPa intervals on each of the

more advanced studies are attempted, research efforts led in the first instance to the development of methods based upon shell theory.[9,10] However, as computers became capable of handling more extensive computations and data, work concentrated on the finite element approach using plate elements initially and then progressed towards isoparametric and superparametric solid elements. Typical of these latter methods are the approaches developed by Ma and Atkinson.[11–13] The principal advantage of these methods over cantilever beam methods is that they evaluate the stresses and strains over a much greater region of the blade than can the simpler methods, assuming, of course, that it is possible to define the hydrodynamic blade loadings accurately. Furthermore, unlike cantilever beam methods, which essentially produce a criterion of stress, finite element techniques develop blade stress distributions which can be correlated more readily with model and full-scale measurements.

In order to evaluate blade stress distributions by finite element methods, the propeller blade geometry is discretized into some sixty or seventy thick-shell finite elements; in some approaches more elements can be required, depending upon the element type and their

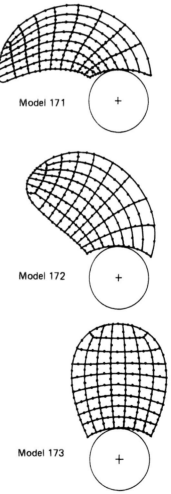

FIGURE 19.5 Finite element discretization.

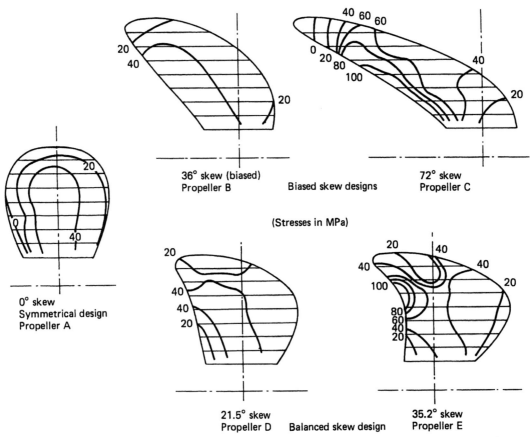

FIGURE 19.6 Distribution of maximum principal stress about a series of blades having different skew designs.

expanded blade outlines. From this figure it is immediately obvious that the effect of skew, whether of the balanced or biased type, is to redistribute the stress field on each blade so as to increase the stresses near the trailing edge. In particular, both highly skewed propellers C and E give trailing edge stresses of similar magnitudes and also relatively high stresses, of the order of the root stress for a symmetrical design, on the leading edge. This is not the case for the symmetrical or low-skew designs. The highly stressed region on propeller E is also seen to be rather more concentrated than that on propeller C. Furthermore, the tendency for the tip stresses on the blade face, which are of a low tensile or compressive nature in the symmetrical and biased skew designs, does not so clearly manifest itself in the balanced design. The accuracy of the tip stress prediction is, however, limited by the finite element representation.

An important feature also noted, although not directly shown in Figure 19.6, is that small changes in the trailing edge curvature can cause a marked change in the trailing edge stress distribution. For example, if the blade surface area was reduced by, say, 5–10 per cent but the leading edge profile kept constant, thus effectively increasing the skew or blade curvature, this would significantly increase the trailing edge stresses.

The orientation and nature of the stress field which exists on the propeller blade is an important consideration from many aspects. For the traditional low-skewed designs of propeller, the orientation of the maximum principal stresses is generally considered to be approximately in the radial direction for the greater part of the blade away from the tips and the leading and trailing edges at the root. Furthermore, the chordal stress components are generally considered to be less than about 25 per cent of the maximum radial stress. Analysis of the results obtained from propeller studies shows that these ideas, although requiring some modification, can to a very large extent be generalized to highly skewed designs. It is seen that the orientation of the maximum principal stresses normally lies within a band 30° either side of the radial direction. With regard to the magnitudes of the chordal stresses, it is also generally found that these rise to between 30 and 40 per cent of the maximum radial stress in the case of the highly skewed designs. As might be expected in the case of biased skew designs the magnitude of the chordal component of stress tends to achieve a maximum nearer to the trailing edge than for the other propellers at the root section.

Blade deflection, although not of primary importance for the strength integrity of the blade, is important for hydrodynamic considerations of the section angle of attack and camber distribution. For conventional and balanced skew propellers the deflection characteristics seem to be predominantly influenced by a linear displacement of the section together with a slight rotation. In the case of a biased skew propeller, however, the rotational and translational components of the blade deflection are considerably magnified. These changes effectively reduce the section angle of attack and owing to the non-linear values of the deflection curve, the section camber is reduced as a result of the 'lifting' of leading and trailing edges. In the case shown for propeller C in Figure 19.6, the rotational component approximated to a reduction in pitch of the section of the order of 0.5° relative to its unloaded condition. The problem of the hydroelastic response of propeller blades is an important one, and this has been addressed by Atkinson and Glover.[15]

When undertaking finite element studies the choice of element type has a direct bearing on the validity of the analysis. It is insufficient to simply use arbitrary formulations for the blades: use needs to be made of elements which can readily accept all of the loadings conventionally met in blade analysis problems. This can be readily illustrated by considering comparative studies: for example, those undertaken by the ITCC[16] in which the results derived from finite element computations from six organizations, using some seven different finite element formulations of the problem, were compared to experimental results at model scale. The propeller chosen for the study was a 254 mm diameter, 72° biased skew design: propeller C of Figure 19.6 taken from Reference 14. The model had been experimentally subjected to point loading at the $0.7R$ and 50 per cent chordal location and was instrumented with four sets of strain gauge rosettes located in the root section of the blade on the pressure side at $0.3R$. Figure 19.7 shows the results obtained from the subsequent ITTC correlation exercise based on finite element computations undertaken by the contributing organizations. Also shown in the figure is the result of a cantilever beam calculation for the same loading condition. It can be seen that although the general trend of the measured result tends to be followed by the various finite element computations, there is a considerable scatter in terms of the magnitudes achieved between the various methods employed. As a consequence, Figure 19.7 underlines the need for a proper validation of the finite element methods. Such validation can only be undertaken by a correlation between a theoretical method with either a model or full-scale test, since while the trends may be predicted by a non-validated procedure as seen by the figure; it is the actual stress magnitudes which are important for fatigue assessment purposes. Furthermore, Figure 19.7 amply demonstrates that the cantilever beam method does not realistically predict the magnitudes of the loadings experienced in the root section of highly skewed blades.

The discussion so far has concentrated on the use of isotropic materials. If anisotropic materials, such as carbon fiber-based composites, are used then the finite element modeling process needs to take particular account of the lay-up of the carbon fibers since, as discussed in Chapter 18, these have directionally dependent properties. The directional properties of these types of material require that proper consideration is given to the induced blade deflections, in both the radial and chordal directions, as load is taken up by the propeller. Kane and Smith[32] discuss the design of a prototype composite propeller for a full-scale trimaran research ship.

19.3 DETAILED STRENGTH DESIGN CONSIDERATIONS

The detailed design of propeller thickness distributions tends to be a matter of individual choice between the propeller manufacturers, based largely on a compromise between strength, hydrodynamic and manufacturing considerations. Additionally, in the case of the majority of vessels there is also a requirement for the propeller blade thickness to meet the requirements of one of the classification societies. In the case of Lloyd's Register, as indeed with most of the other classification societies, these rules are based upon the cantilever beam method of analysis and are essentially based on equation (19.3). The techniques of propeller blade stressing discussed in Sections 19.1 and 19.2 are applied to all types of propeller and it is, therefore, relevant to consider briefly the special characteristics of particular types of propeller in relation to the conventional fixed pitch propeller upon which the discussion has so far centered:

1. *Ducted propellers.* As ducted propellers, in common with transverse propulsion unit propellers, tend to have rather more heavily loaded blade outer sections than conventional propellers, the effective centers of action of the hydrodynamic loading tend to act at slightly larger radii. However, since a proportion of the total thrust is taken by the duct, the appropriate adjustment must be made for this in the stress calculation. Additionally, the duct can also have an attenuating influence over the wake field, which to some extent improves the fluctuating load acting on the blades.
2. *Tip-unloaded propellers.* Noise-reduced or tip-unloaded propellers, which have largely evolved from naval practice and contemporary thinking on reducing propeller-induced hull surface pressures on merchant vessels, tend to concentrate the blade loading nearer the root sections as shown by Figure 19.8. This characteristic tends to reduce the effective centers of action of the

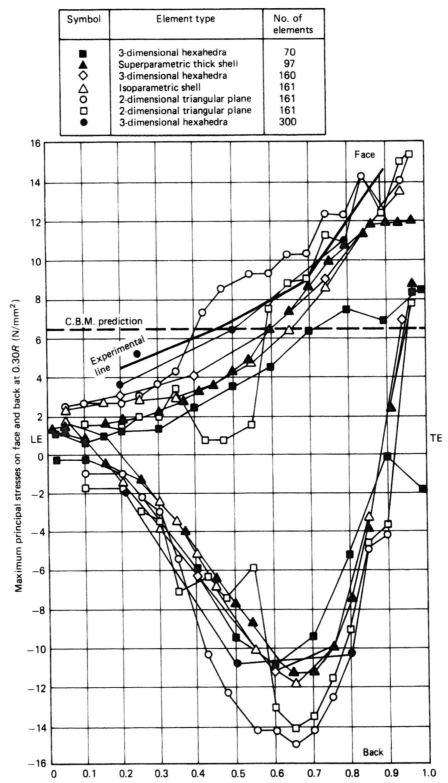

FIGURE 19.7 Correlation of different finite element calculation methods with experiment. *Reproduced with permission from Reference 16.*

Propeller Blade Strength

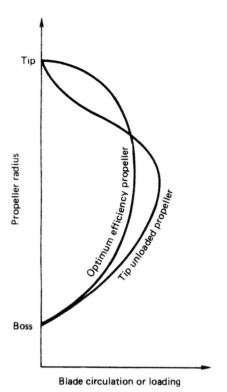

FIGURE 19.8 Comparison between tip unloaded and optimum efficiency radial loadings.

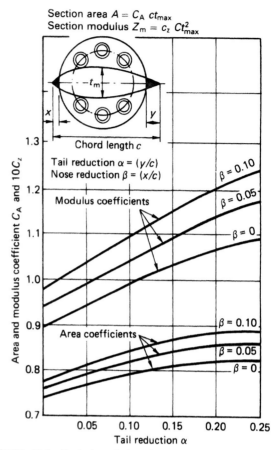

FIGURE 19.9 Typical variations in root section properties for controllable pitch propellers.

hydrodynamic loading coupled with the slightly lower propulsive efficiency for these propellers.

3. *Controllable pitch propellers.* Controllable pitch propellers tend to present a more difficult situation in contrast to fixed pitch propellers due to the problems of locating the blade onto the palm. The designers of hub mechanisms prefer to use the smallest diameter blade palms in order to maximize the hub strength, and, conversely, the propeller designer prefers to use a larger palm in order to give the greatest flexibility to the blade root design. These conflicting requirements inevitably lead to a compromise, which frequently results in the root sections of the blade being allowed to 'overhang' the palm. This feature, although introducing certain discontinuities into the design, is not altogether undesirable, since it allows both the root section modulus and area to be increased, as seen for a typical controllable pitch propeller root section profile in Figure 19.9. Additionally, in many designs of controllable pitch propeller the blade bolting arrangements are such as to place a further limitation on the maximum section thickness. It therefore becomes necessary on occasion, although undesirable, for the blade bolt holes to significantly penetrate the root fillets in order to fit the blade onto the palm.

The modes of operation of a controllable pitch propeller are varied, almost by definition, as discussed in Reference 17. Generally, however, from the stressing point of view these off-design operating conditions remain unconsidered unless prolonged working in any given mode is indicated.

Ice class requirements can also present additional problems for controllable pitch propellers. Since the blades have restricted root chord lengths, the additional ice class thickness requirements in some instances result in root section thickness to chord ratios in excess of 0.35. From the hydrodynamic viewpoint this gives both poor efficiency and greater susceptibility to cavitation erosion.

4. *High-speed propellers.* High-speed propellers generally have better in-flow conditions than their larger and slower-running counterparts, although poorly designed shafting support brackets are sometimes troublesome. Consequently, high wake-induced cyclic loads are not usually a problem unless and shafting is highly inclined. Centrifugal stresses, as seen in Table 19.2, tend to take on a greater significance due to the higher rotational speeds, and therefore greater attention needs to be paid to the calculation of the mechanical loading components.

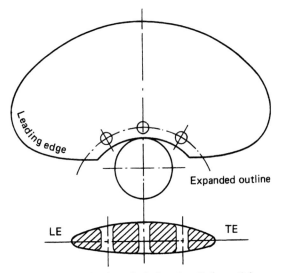

FIGURE 19.10 A method of root cavitation relief.

Naturally these propellers, if configured in either a ducted or controllable pitch form, can adopt some of the characteristics of the previously discussed classes. One feature which may be introduced occasionally in attempts to control root cavitation erosion is a system of holes bored through the blade along the root section as sketched in Figure 19.10. While the purpose of these holes is to relieve the root section pressure distributions and to modify the flow behind the fixed cavitation at the point of its break-up, their presence necessitates a careful review of the root section thicknesses. Additionally, large blending radii need to be specified so as to merge the holes into the blade surface in as fair a way as possible.

19.4 PROPELLER BACKING STRESSES

When a propeller undergoes a transient maneuver considerable changes occur both in the magnitudes and distribution of the blade stress levels. Figure 19.11 shows typical changes in the stress measured on the blades of a single-screw coaster undergoing a stopping maneuver. This vessel was fitted with a conventional non-highly skewed fixed pitch propeller.

Experience with fixed pitch, highly skewed propellers when undertaking emergency stopping maneuvers has led to the bending of the blade tips in certain cases (Reference 18). This type of bending, which frequently occurs in the vicinity of a line drawn between about $0.8R$ on the leading edge to a point at about $0.60R$ on the trailing edge, is thought to be due to two principal causes, see Figure 23.10. The first of these causes is due to simple mechanical overload of the blade tips from the quasi-steady hydrodynamic loads causing stresses leading to the plastic deformation of the material; the second is from the transient vibratory stresses, of the type shown in Figure 19.11, which occur during the maneuver. These latter stresses are not wholly predictable within the current state of theoretical analysis but an estimate needs to be made of the severity of the stresses during backing maneuvers.

As part of the design process a fixed pitch, highly skewed propeller should always be checked for overload against the material proof stress capability based on the quasi-steady mean hydrodynamic stresses using a suitable hydrodynamic criterion. Most commonly this criterion is the bollard pull astern condition since at present this is thought to be the most representative idealization of the worst condition the propeller is likely to experience during a transient maneuver. Clearly, the backing stress estimations need to be based on a lifting surface hydrodynamic model which is used together with a finite element analysis. However, it must be recognized that hydrodynamic codes, when used for backing stress calculations, are operating far from their originally intended purpose: as a consequence, the analysis must be viewed in this context.

In contrast to fixed pitch, highly skewed propellers, their controllable pitch variants do not suffer with the same tendency towards blade tip bending when operating astern. This is because in the case of a controllable pitch propeller the leading edge normally functions as the leading edge during these types of maneuver and, therefore, the trailing edge is protected from high loading. Consequently, for controllable pitch propellers it is normal to consider them only in the ahead operating condition for the strength analysis which is based on the normal fatigue considerations.

19.5 BLADE ROOT FILLET DESIGN

So far consideration has centered only upon the blade stresses without any account being taken for the root fillets where the blade meets either the propeller boss or blade palm. The root fillet geometry is complex since it is required to change, for conventional propellers, in a continuous manner from a maximum cross-sectional area in the mid-chord regions of the blade to comparatively small values at the leading and trailing edges. Notwithstanding the complexities of the geometry, the choice of root fillet radius is of extreme importance. For conventional propeller types, if a single radius configuration is to be deployed, it is considered that the fillet radius should not be less than the blade thickness at $0.25R$.

The use of a single radius at the root of the blade always introduces a stress concentration. However, the introduction of a compound radius fillet reduces these concentrations considerably. Therefore, the use of fillet profiles of the type described by Baud and Tum and Bautz are desirable which, for most marine propeller applications, can be approximated to two single radii having common tangents. Typically, such a representation may be achieved by using

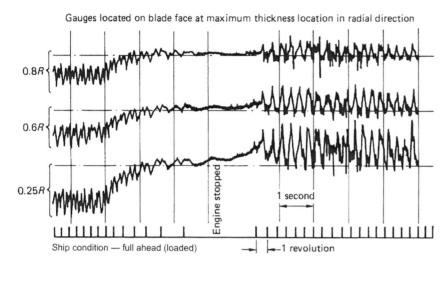

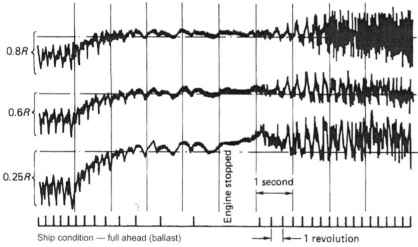

FIGURE 19.11 Typical crash stop maneuvers measured on a single-screw coaster from full ahead.

radii of magnitudes $3t$ and $t/3$, having common tangents with each other and with the blade and boss, respectively. In these expressions t is the blade thickness at the section of interest: typically at 0.25R.

The results of the blade surface stress distribution for symmetrical and balanced skew designs imply that the full size of the fillet should be maintained at least over the middle 50 per cent of the root chord length. In the case of extreme biased skew designs, there is a sound case for continuing the full-fillet configuration to the trailing edge of the blade in order to minimize the influence of the stress concentration factor in the highly stressed trailing edge regions.

19.6 RESIDUAL BLADE STRESSES

The steady and fluctuating design stresses produced by the propeller when rotating in the wake field developed by the ship represent only one aspect of the total blade stress distribution. Residual stresses, which are introduced during manufacture or during repair, represent the complementary considerations.

Full-scale experience relating to residual stresses is limited to a comparatively few studies. Webb et al.[19] is typical of these studies in which measurements have been made for propellers subjected to local heating. These measurements related to high-tensile brass and manganese—aluminum bronze propellers which had been subjected to heating subsequent to manufacture. In these cases, residual stresses of the order of 155 and 185 MPa were measured by the trepanning technique of residual stress measurement. Little published information exists, however, for the level or nature of residual stress in new or unrepaired castings. Clearly this is principally due to the semi-destructive nature of the measurement procedure involved in determining a residual stress field for these types of casting.

From investigations undertaken by Lloyd's Register[14] into the causes of propeller failures, due other than by poor repair or local heating of the boss, have shown that residual surface stresses measured in blades adjacent to the failed blade can attain significant magnitudes. The technique used for those measurements was that of bonding purpose-designed strain gauge rosettes to the surface of the blade and then incrementally milling a carefully aligned hole through the center of the three rosette configuration. At each increment of depth within the hole, a measurement of the relaxed strain recorded by each gauge of the rosette was made. This method, used in association with a correctly designed milling guide, is relatively easy to apply and also has been shown to give reliable results in the laboratory on specially designed calibration test specimens. An example of the results gained using this procedure is given by Figure 19.12 for a five-blade, nickel–aluminum bronze, forward-raked propeller having an approximate finished weight of approximately fourteen tonnes. From the figure it is seen that the measured residual stresses in this case are of a significant magnitude and tensile in nature over much of the blade. Indeed, the magnitudes in this case reach tensile values of between two and three times the normally accepted design stress levels. Furthermore, it can also be seen that the principal stresses at a given measurement point are of similar magnitudes. This implies the introduction of a strong bi-axial characteristic into the stress field on the blade surface, which under pure design considerations, in the absence of residual stress, would be expected to be of a predominantly radial nature. Analysis of the through-thickness characteristics of the relieved strain for the same propeller blade also suggests that the residual stresses posses a strong through-thickness variation, having high stresses on the blade surface which then decay fairly rapidly within the first 1–2 mm below the surface.

To extrapolate the results of a particular residual stress measurement to other propellers would clearly be unwise. Nevertheless, since these stresses play an important part in the fatigue assessment of a propeller, the designer should be aware that they can obtain high magnitudes. However, full-scale experience in terms of the number of propeller failures would suggest that either residual stresses are not normally this high or there are significant safety margins in design procedures. The magnitudes of residual stress, although unclear in their precise origins, are strongly influenced by the thermal history of the casting, material of manufacture and the type or nature of the finishing operation. Furthermore, it is also known from measurements that large variations can exist between measurements made at equivalent positions on consecutive blades of the same propeller.

19.7 ALLOWABLE DESIGN STRESSES

The strength design of a propeller in the ahead condition must be based on a fatigue analysis; it is insufficient and

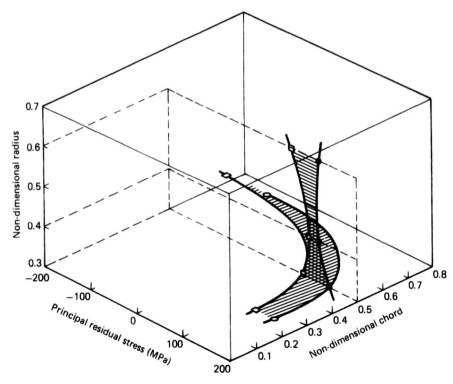

FIGURE 19.12 Measured residual stresses on a propeller blade.

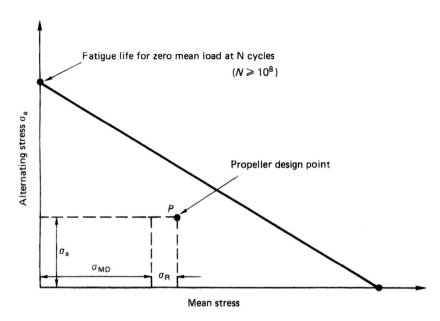

FIGURE 19.13 Propeller fatigue analysis.

inaccurate to base designs on simple tensile strength or yield stress criteria. In order to relate the blade stresses, both steady state and fluctuating, to a design criteria some form of fatigue analysis is essential. The most obvious choices are the modified Goodman and Soderberg approaches of classical mechanical fatigue analysis. In these approaches the mean stress is plotted on the abscissa and the fluctuating stress on the ordinate (Figure 19.13). To evaluate the acceptability of the particular design a linear relationship is plotted between the fatigue life at zero mean load and some point on the abscissa. While the fatigue life should always relate to 10^8 cycles or greater, as discussed in Chapter 18, the intersection point on the abscissa to which the linear relationship should be drawn is less well defined. In general engineering practice the ultimate tensile strength is the basis of the modified Goodman approach, which is generally considered a satisfactory basis for analysis. However, there is a body of experimental material data (Reference 19) which suggests that the most appropriate point of intersection may well be in the region of the 0.15 per cent proof stress. If this is valid then the more conservative Soderberg approach is probably the more correct for marine propellers.

As is seen in Figure 19.13, the magnitude of the alternating stress σ_a is a single component dependent largely on the fluctuating velocities in the wake field in which the propeller blade is operating. The steady-state stress component is the sum of two components σ_{MD} and σ_R, where these respectively relate to the mean design component, as determined from either cantilever beam or finite element studies, and the level of residual stress considered appropriate.

The comparison of the design stresses with the fatigue characteristics of the propeller material is a complex procedure. Figure 19.14 demonstrates this in outline terms as part of the overall propeller design process: this is discussed more fully in Chapter 22. From Figure 19.14 it is apparent that the design mean (σ_{MD}) and alternating (σ_a) stresses derive directly from the hydrodynamic analyses of the blade working in the wake field. Hence, these parameters are directly related to the blade design and the environment in which the propeller is operating. The residual stress allowance σ_R is a function of the casting size, propeller material and manufacturing technique. The magnitude of this stress allowance is therefore very largely indeterminate in the general sense. However, in the absence of any other information or indications to the contrary it would be prudent to allow a value of between 15 and 25 per cent of the 0.15 per cent proof stress to account for the residual stress σ_R.

The propeller fatigue characteristics are clearly dependent on the choice of material (see Figure 18.6); however these basic characteristics need to be modified to account for casting size and other environmental factors, as discussed in Chapter 18. Having, therefore, defined the various parameters in Figure 19.13, a judgment based on normal engineering principles can be made as to whether the apparent factor of safety is appropriate. In propeller technology it is unlikely that a factor of safety of less than 1.5 would be considered acceptable for the ahead operating condition.

Casting quality has a profound influence on the life of a propeller in service. The defects found in copper alloy propellers are generally one of two kinds. First, they may be attributable to porosity in the form of small holes resulting

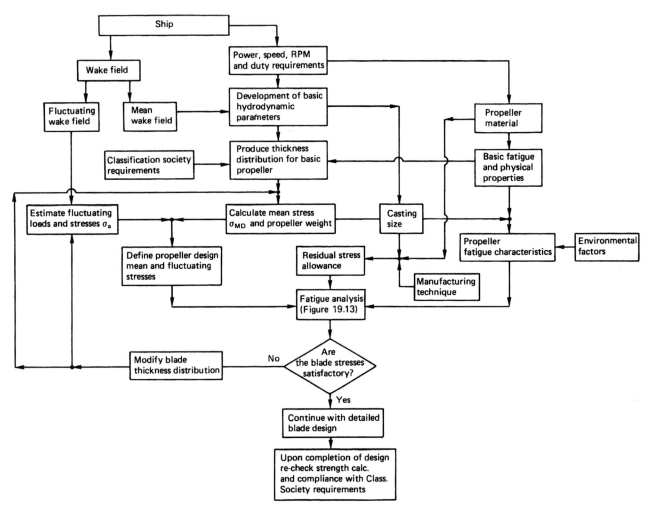

FIGURE 19.14 Propeller strength analysis design procedure.

from either the releasing of excess gases or shrinkage due to solidification. Alternatively, the defects can be oxide inclusions in the form of films of alumina which are formed during the pouring stage of propeller manufacture and have a tendency to collect near the skin of the casting. The location of a defect is obviously critical. For conventional, low-skew propellers, defects on the suction face are of less concern than those located on the pressure face in the midchord region of the inner part of the blade: particularly, although not exclusively, close to the run-out of the fillet radii. Alternatively, in the case of highly skewed propellers casting defects in the trailing edge region of the blade are of critical importance in view of the location of the stress concentrations within the blade. Considerations of this type lead to the concept of acceptable defect criteria for marine propellers, which in turn introduces the subject of fracture mechanics.

The visual characteristics, as sketched by Figure 19.15 and seen in Figure 23.9, of a propeller blade which has failed by fatigue action, are generally similar for all propellers: although in some cases the beach marks are more clearly visible than in others. Beach marks are usually formed at points of crack arrest such as when a ship is moored alongside and discharging cargo. They are not in themselves proof of fatigue action although they can be a good indicator; the proof comes from the observation of striations on the material fracture surface when viewed under a scanning electron microscope and, in certain cases, with high power optical microscopes. Attempts at correlating the relative geometric form of beach marks during

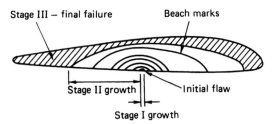

FIGURE 19.15 Visual characteristics of fatigue failure.

Chapter | 19 Propeller Blade Strength

TABLE 19.3 Material Constants for Crack Propagation Equation

Material	c	m	Mean Stress (kgf/mm²)	Condition
Mn–Al bronze	6.6×10^{-11}	3.7	7.0	Sea water at 4 Hz
Ni–Al bronze	4.97×10^{-13}	4.7	0	Simulated sea water at 2.5 Hz
	3.37×10^{-14}	5.2	0	Simulated sea water at 5 Hz

(Threshold value $= 25$ kgf/mm$^{3/2}$)

crack growth have been made based on observations of failed propellers. The advantages of obtaining such a relationship are that the aspect ratio of the crack can be directly related to the stress intensity factor, which may then be used in conjunction with fracture toughness information to assess acceptable defects and crack propagation rates. Work by Roren et al.[20] and Tokuda et al.[21] has derived coefficients, Table 19.3, for the Stage II Paris Law crack propagation equation:

$$\frac{da_c}{dN} = c(\Delta k)^m \qquad (19.6)$$

This data was derived from sample test specimens cut from failed propeller blades. The tests for the manganese–aluminum alloy used center slotted type specimens while those for the nickel–aluminum alloy were defined as being of the wedge opening load type.

Notwithstanding the encouraging work that has been done in the field of acceptable defects and on Stage II crack propagation, for example References 22 and 23, in which the crack moves from its initiation phase, Stage I, through to eventual rapid failure at the end of Stage III, much work remains to be done in understanding fully the mechanisms of the Stage I crack growth for propeller materials.

19.8 FULL-SCALE BLADE STRAIN MEASUREMENT

By comparison with the amount of theoretical work undertaken on the subject of propeller blade and boss stresses, there have been few full-scale measurement exercises. The reason for this comparative dearth of full-scale data has undoubtedly been due to the difficulties hitherto encountered in instrumenting the chosen ship. Early attempts at these measurements required that the ship's tail shaft was hollow bored in order to conduct the signal wires from the strain gauges located on the propeller blades through to a system of slip rings inside the vessel. Figure 19.16(a) shows in schematic form this arrangement. Despite the obvious disadvantages of this method some

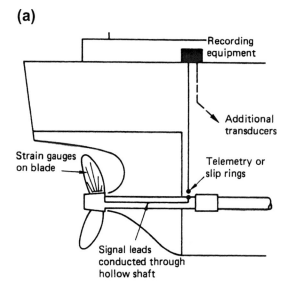

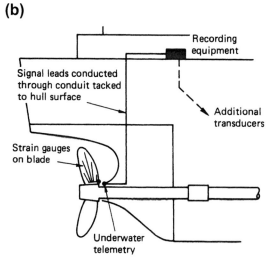

FIGURE 19.16 Full-scale blade strain measurement techniques: (a) hollow bored shaft method and (b) underwater telemetry method.

notable full-scale studies have been conducted (References 24–31) and these have formed the nucleus of full-scale data in the publicly available literature.

In recent years the use of underwater telemetry techniques has been explored as an alternative form of measurement.[14] The use of telemetry methods has obvious advantages in that the signal can be transmitted at radio frequencies across a suitable water gap and consequently avoids the requirement to bore the tail shaft. The most usual procedure is to attach the transmitter to the forward face of the propeller boss, under the rope guard, and transmit the signals to a receiver located on the stern seal carrier, as seen in Figure 19.16(b). Having bridged the rotating to stationary interface in this way, the signal leads can then be conducted over the hull surface, protected by conduit tack welded to the hull plating, to a convenient location for the recording instruments to be stationed.

With regard to the conduct of blade strain measurement trials, the general principles of ship speed trials discussed in Chapter 17 should be adhered to, including the requirements for the measurement of ship speed as this is an important parameter in the blade stress determination.

REFERENCES AND FURTHER READING

1. Hecking J. *Strength of propellers*. Mar. Eng. & Shipping Age; October 1921.
2. Rosingh WHCE. *Design and strength calculations for heavily loaded propellers*. Schip en Werf; 1937.
3. Romson JA. *Propeller strength calculation*. Mar. Eng. & Nav. Arch. 1952;**75**.
4. Morgan WB. *An Approximate Method of Obtaining Stress in a Propeller Blade*. DTMB Report No. 1954;**919**.
5. Burrill LC. *A short note on the stressing of propellers*. The Shipbuilder and Mar. Eng. Builder; August 1959.
6. Weber R. *Festigkeits berechnung für Schiffspropeller*. Schiffbautechnik 1961;**II**(8).
7. Schoenherr KE. *Formulation of propeller blade strength*. Trans. SNAME; April 1963.
8. Sinclair, L. *Propeller blade strength*. Trans. IESS, 199, 1975–76.
9. Cohen, JW. *On Stress Calculations in Helicoidal Shells and Propeller Blades*. TNO Report No. 215.
10. Conolly JE. *Strength of propellers*. Trans. RINA 1961;**103**.
11. Atkinson P. *The prediction of marine propeller distortion and stresses using a superparametric thick shell finite element model*. Trans. RINA 1973;**115**.
12. Ma JH. *Stresses in marine propellers*. J. Ship Res. 1974;**18**(4).
13. Atkinson P. *A practical stress analysis procedure for marine propellers using curved finite elements*. Propellers '75 Symp. Trans. SNAME; 1975.
14. Carlton JS. *Marine propeller blade stresses*. Trans. I.Mar.EST.; 1984.
15. Atkinson P, Glover EJ. *Propeller hydroelastic effects*. Propellers '88 Symp. Trans. SNAME; 1988.
16. *Report of Propulsor Committee*. Kobe: 18th ITTC Conf; 1987.
17. Hawdon L, Carlton JS, Leathard FI. *The analysis of controllable pitch propellers at off-design conditions*. Trans. I. Mar. Est. 1976;**88**.
18. Blake WK, Meyne K, Kerwin JE, Weithedorf E, Frisch J. *Design of APL C-10 propeller with full scale measurements and observations under service conditions*. Trans. SNAME; 1990.
19. Webb AWO, Eames CFW, Tuffrey A. *Factors affecting design stresses in marine propellers*. Propellers '75 Symp. Trans. SNAME; 1975.
20. Roren E, Solumsmoen O, Tenge P, Sontvedt T. *Marine propeller blades — Allowable stresses, cumulative damage and significance of sharp surface defects*. 2nd Lips Propeller Symp; 1973.
21. Tokuda S, Okuyamu Y, Inoue H, Denoh S. *Fatigue failure in marine propeller blades*. Propellers. '78 Symp. Trans. SNAME; 1978.
22. Denoh S, Morimoto K, Nakano I, Moriya K. *Fatigue strength in marine propeller blades. I. Fatigue strength of propeller materials*. J. MESJ 1979;**15**(4).
23. Nakano I, Oka M, Gotoh N, Moriya K, Denoh S, Saski Y. *Fatigue strength in marine propeller blades. II. Allowable flaw size in propeller blades*. J. MESJ 1979;**15**(5).
24. Wereldssma R. *Stress measurements on a propeller blade of a 42 000 ton tanker on full scale*. ISP; 1964.
25. Chirila JV. *Dynamische Beanspruchung von Propellerflugeln auf dem Motorschiff 'Pekari'*. Schiff und Hafen 1970;**3**.
26. Dashnaw FJ, Everett-Reed F. *Propeller strain measurements and vibration measurements on the SS Michigan*. Marine Technology; October 1971.
27. Keil HG, Blaurock JJ, Weitendorf EA. *Stresses in the blades of a cargo ship propeller*. J. Hydronaut. January 1972;**1**.
28. Chirila JV. *Investigation of stresses at the propeller blade of a container ship*. Shipping World and Shipbuilder; November 1971.
29. Watanabe K, Hieda N, Sasajima T, Matsuo T. Propeller Stress Measurements on the Container Ship 'Hakone Maru'; 1972;.
30. Ueta Y, Maebashi M, Shiode K, Jakesawa S, Takai M. *Stress measurements on a propeller blade*. Bull. MESJ March 1974;**2**.
31. Sontvedt T, Johansson P, Murgass B, Vaage B. Loads and response of large ducted propeller systems. Symp. on Ducted Propeller. Trans. RINA; 1973.
32. Kane C, Smith J. *Composite blades in marine propulsors*. Proc. Int. Conf. on Advanced Marine Materials: Tech. and Applications. London: Trans. RINA; 2003.

Chapter 20

Propeller Manufacture

Chapter Outline
20.1 Traditional Manufacturing Method 415
20.2 Changes to the Traditional Technique of Manufacture 419
References and Further Reading 420

Propeller manufacture is an extensive subject embracing the arts and sciences of foundry techniques relating to the casting of large quantities of metal, sometimes of the order of 200 tonnes, together with the many engineering machine shop skills to create an artifact of complex and precise geometric form. The skill of propeller manufacture lies both in interpreting the hydrodynamic design into physical reality and in ensuring that the manufacturing process does not introduce defects which could bring about the premature failure of the propeller.

Propeller manufacture relies on two basic techniques: the use of full solid patterns for multiple uses or the construction of a unique mold which will be broken up after the casting is complete. Which technique is used is a techno-economic question depending on the type of propeller, the number to be produced, the finishing technique and the size of the propeller. However, to gain an understanding of the manufacturing process in generalized terms the traditional method of manufacture will be described before outlining a range of variants to this process.

20.1 TRADITIONAL MANUFACTURING METHOD

Originally propellers were of a simple shape and made in either cast iron or steel. These early propellers were usually cast in the engine builder's own foundry and were fitted to the vessel in a largely 'as-cast' condition, except for some necessary fettling and machining of the bore. Today the materials, as was seen in Chapter 18, have largely changed to the bronzes and propellers are manufactured to a high standard of surface finish and dimensional accuracy in foundries and workshops devoted solely to the manufacture of propellers.

Each propeller is nominally of a different design and as a consequence it is relatively rare for a propeller manufacturer to receive orders for a significant number of propellers to the same design: particularly for large propellers. The traditional method of manufacture reflects this situation and is based on the production of a mold for each propeller that is to be manufactured.

In some propeller foundries the propellers will be cast in large pits sunken into the floor, while in others the mold will be built onto the actual floor of the foundry. There is no general procedure for this and each manufacturer works out an individual technique which takes into account safety, versatility, space available and costs of production. Therefore the manufacturing process will be found to vary from one manufacturer to another in matters of detail, but the general theme of manufacture follows a similar pattern and it is that underlying theme which will be outlined here.

A mold for each propeller is constructed in two halves: the bed, the upper surface of which defines the pressure side or pitch face of the blade, and the top, the lower surface of which defines the suction surface or back of the propeller blades. As a consequence, propellers are generally cast 'face downwards' in the mold.

The traditional mold material is a pure washed silica sand, having an average grading of between 20 and 50 mesh, and is mixed with controlled amounts of ordinary Portland cement and water using the Randupson process.[1]

A typical Randupson sand mixing and reclamation plant is shown in Figure 20.1. From this figure it can be seen that previous molds, once they have been broken into manageable proportions and the reinforcing rods have been broken out, are passed through a crusher and then mechanically transported to a series of vibrating sieving screens. The first of these screens is sufficient to reject lumps and foreign matter such as nails, while the latter stages of the sieves pass only grains and dust. From the vibrating screens the reconditioned sand passes into a hopper which is adjacent to two other hoppers, one containing new sand and the other cement. The mixing mill is then fed in the required amounts from each of these

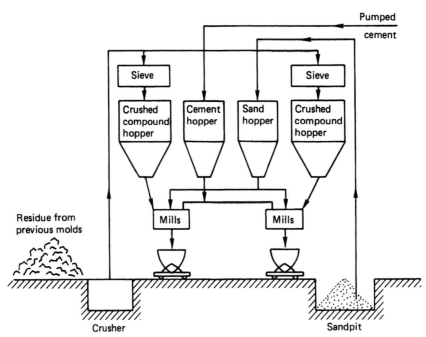

FIGURE 20.1 Randupson sand plant.

hoppers and after dry mixing for a period of time, of the order of 2–5 minutes, water is then added in carefully controlled amounts. The amount of water depends upon the moisture content of the new sand and also the workshop's humidity. Wet mixing is then continued for a period of time, whereupon the contents are discharged into a portable skip for transport to the molding site elsewhere in the foundry. In many cases, for economic reasons, previous mold material is used to construct those parts of the mold which are not directly in contact with the molten metal of the new propeller; however, a new sand mixture should be used for those parts of the mold which are in contact with the molten metal.

Prior to the Randupson process being introduced, loam was almost universally used for molds. However, extensive artificial drying is necessary with this material and, therefore, this represents a disadvantage in addition to its lower strength properties.

The first stage in the manufacture is to construct the bed of the mold around the shaft center line, which is defined as being vertical relative to the shop floor. Using this line as the basic reference datum, the angular spacing of the directrices of each of the blades is carefully marked out on the shop floor and the approximate shape of each blade defined about the blade directrices. Based on this approximate shape, a wooden shuttering is then erected to form a box into which the mold material is rammed together with suitable reinforcing rods. Having formed the body of the bed of the mold in this way the pitch face of the propeller is formed by a technique known as 'strickling'. This process uses a striking board fixed to a long arm at one end and which has a roller at the other. The arm is free to rotate about and slide vertically up and down a spindle which has been erected vertically on the shaft center line of the propeller: the roller, at the other end of the striking board, runs on a pitch rail. Figure 20.2 shows this arrangement in schematic form. The pitch rail defines a portion of a helix, centered on the shaft center line and constructed at a suitable radius which is greater in magnitude than the propeller tip radius. The slope or pitch angle of the helical rail is appropriate to the required pitch and radius of the propeller under construction. The striking board is then pushed up the rail to generate a true helicoidal surface. To cater for a non-uniform pitch distribution the

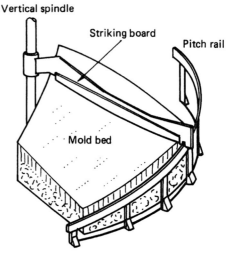

FIGURE 20.2 Sweeping the mold bed.

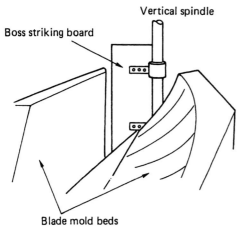

FIGURE 20.3 Sweeping the propeller boss.

maximum pitch is either first swept and the surface corrected for other radii by rubbing down to templates, or in some cases it is possible to use articulated striking boards and multiple pitch rails. Whichever method is used, the resulting surface is then sleeked by hand. To cater for propeller rake the striking board is set to the appropriate angle relative to the rotating arm.

To form the outside profile of the propeller boss, assuming it is a fixed pitch propeller, another striking board, seen schematically in Figure 20.3, is rotated about the shaft center line.

The next stage in the construction is to construct the blade form on the mold bed by means of patterns. These patterns are accurately cut from either thin wooden sheet or metal, most usually the former material, and represent the designed cylindrical section profiles together with appropriate contraction and machining allowances. When the bed of the mold has dried and is hard these patterns, which define the helical sections of the blade, are carefully positioned at the appropriate radii and fixed vertically such that they lie along circumferential paths; Figure 20.4. The space

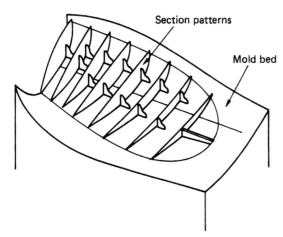

FIGURE 20.4 Location of section patterns.

between these patterns is then packed with a sand and cement mixture, whereupon the resulting surface is again carefully sleeked to form an upper surface and edge contour of the blade.

Once this second sand–cement mixture forming the blade has dried a reinforcing iron grillage is placed over the blade, at a height of some 50–70 mm above it, and wooden shuttering is then positioned to form a box for the construction of the top half of the mold. Another sand–cement mixture is then rammed into the box against the blade pattern to form the top of the mold in a similar way to the procedure that was adopted for the bed. At a convenient height above the blade pattern the top of the mold is leveled off and allowed to dry. When thoroughly dry, the top of the mold is parted and the top is lifted off by means of lifting hooks which are attached to the reinforcing frame. In this open state the sand and wood patterns are completely removed from the mold and the formation of the root fillets generally then takes place by rubbing down the sharp edges of both the bed and top of the mold to the designed fillet form with the aid of templates.

This method of construction is then applied to each blade in turn and when complete the mold surfaces are cleaned and dressed to a high state of finish. The mold tops are then fitted back onto their beds and secured by means of mechanical ties and braces to prevent relative slippage or bursting during the pouring process.

Prior to pouring the metal, the mold is heated for several hours by blowing hot air through the boss aperture and out through vents which have been incorporated in the mold near the blade tips. When a predetermined temperature is reached at the outlet vents, typically of around 110–120°C, it is then fairly certain that the surfaces of the mold are free from moisture and a sufficient level of pre-heat of the mold has been achieved. A uniformity of pre-heat throughout the mold is essential and this is achieved by means of suitable ducts.

The final stage in the propeller mold construction, as distinct from the casting feeder system, is the securing of the head ring on top of the mold such that it is concentric with the shaft axis; Figure 20.5. At this time the core is also inserted and will form a basic hole for the shaft and lightning chamber if one is needed. The core is, of course, fitted to be concentric with the shaft center line and has been frequently constructed of alternate layers of foam and straw rope on a former that, for example, may be constructed from a perforated iron cylinder. This construction gives a measure of flexibility to the core so that it does not offer serious resistance when the casting cools and contracts around it.

During the construction of the mold a runner system is also built into it to enable the molten metal to be fed into the mold in a controlled and proper way. Figure 20.5 shows a typical runner system. From this figure it can be seen that

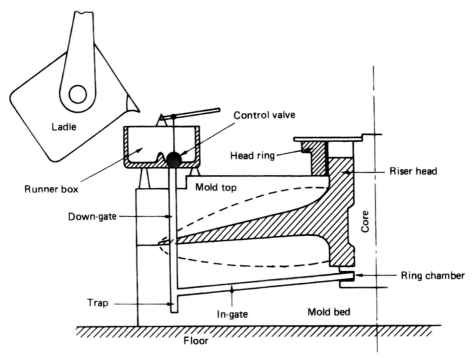

FIGURE 20.5 Runner system for a typical mold.

the molten metal is poured into a runner box, which is fitted with a simple control valve to govern the flow of metal into the mold. For large castings there may well be two or more of these runner systems fed from different ladles simultaneously and providing metal to different points at the base of the casting. From the runner box the metal passes into a vertical down-gate which was built into the mold at the time of construction. These runners are made of pre-cast sand pipe sections and it is found that the exact shape and dimensions of the runners are very important in order to get an efficient flow into the casting which minimizes turbulence and oxide formation. From the down-gate an in-gate is constructed from either a cylindrical or rectangular section. The in-gate is built to run up an inclined line from just above the dirt trap at the bottom of the down-gate to the bottom of the boss. The entry into the bottom boss is normally flared and made tangentially so as to avoid a direct impact of the metal flow onto the core. It has been found that the construction of a chamber below the box enhances the pouring of the casting by eliminating much of the initial turbulence and allows the molten metal to rise gently into the casting.

Transport of the metal from the furnaces is by one or more ladles, depending on the size of the cast. The metal is poured into the ladles at a slightly higher temperature than is required for casting in order to allow for cooling during the transportation process to the mold. Accordingly, upon arrival at the mold the temperature of the molten metal is checked using optical pyrometers and the casting operation is delayed until the correct temperature is reached.

Notwithstanding the primary need to minimize turbulence during casting operations, the propeller mold needs to be filled through the runner system as quickly as possible. Langham[2] quotes pouring rates of up to 8 tonnes/min although this is a variable depending on the size of the propeller. Throughout the pouring operation, the rising surface of the metal that is accessible within the mold is skimmed in order to prevent oxide being trapped in the blades or in the core. In general, a large feeding head in the form of an extension to the boss at the forward end is needed for two reasons: first, to allow for the contraction of the metal during cooling and second, to provide a reservoir of heat to help provide uniform cooling and directional solidification.

The mold is filled to a predetermined head during the initial casting and after casting the surface is skimmed and covered with an insulating compound. During the cooling process the casting is 'topped up' with small additions of molten metal at certain intervals. Exothermic materials are regularly used to assist in the feeding process. These applications are usually in the form of direct applications of powder to the surface: the amount of powder and the interval between applications are dependent upon the size of the cast.

The cooling of the mold takes place over a number of days, depending on the size of the casting. When ready, the mold is dismantled and the cast propeller lifted. The first process after lifting the casting is fettling which involves the removal of all extraneous riser and venting appendages. The general dimensions of the casting are checked next and

Propeller Manufacture

various datum lines are established by means of measurement. Following the satisfactory completion of this process the propeller is bolted to a large horizontal boring machine with its shaft axis horizontal. The first process is then to remove the riser head after which the taper bore of the propeller is machined to suit the appropriate plug gauge, or template, by means of a concentric boring bar placed through the cored hole of the casting. After this the forward and aft faces of the boss are machined by means of facing arms attached to the boring bar and during this process the various features found on these faces are also finished.

The next stage in production is the lining out of the blades in order to determine the amount of material to be removed. In former times this amount was considerable and required lengthy periods of pneumatic chiseling to be undertaken to develop the basis of the required blade surfaces. To provide control to this process datum grooves or spot drillings were cut into the blades and the pneumatic chisels were used to remove excess metal. This made parts of propeller foundries extremely noisy and unpleasant places in which to work but today with the greater use of precision casting techniques, less metal needs to be removed.

In certain areas of the blade, templates are deployed and eventually the whole propeller is ground and polished using high-speed portable grinders. Finally, the blade edges, leading and trailing, are contoured to the designed shape: again in some cases with the aid of templates. The whole propeller, in the case of large propellers, is then statically balanced as the final stage of manufacture.

The traditional manufacturing method described is based on the use of a sand—cement pattern constructed around thin wooden or metal section patterns and both Langham[2] and Tector[3] describe this process more fully. A good photographic record of the traditional manufacturing technique is seen in Reference 4.

The alternative to the traditional procedure is to use solid patterns. In the case of large fixed pitched propellers this technique is rarely used on account of cost; however, for controllable pitch propellers or the once popular built-up propeller, their use is more common since the extra cost of building a solid pattern is justified when an increased number of blades to the same design are required. In these cases the manufacturing process is analogous to that described but making use of the solid patterns for individual blades is often achieved in conjunction with the use of separate casting boxes.

20.2 CHANGES TO THE TRADITIONAL TECHNIQUE OF MANUFACTURE

Modern foundry techniques permit the use of finer casting allowances through the use of precision casting methods. This provides the benefit of requiring considerably less material to be removed during the manufacturing process and hence enables production costs to be potentially reduced. Such techniques, however, need to be used in controlled environments so that an increased incidence of surface imperfections is not encountered.

When considering casting tolerances it is tempting, in order to reduce manufacturing costs, to reduce these tolerances to the smallest possible value. This, however, may lead to unwelcome imperfections remaining on the finished blade surfaces which are directly attributable to the manufacturing process. These may then show up as defects in the subsequent non-destructive examination processes and will need to be removed, hopefully without strength integrity questions being raised.

The choice of charge stock for the furnace is also an important consideration. Traditionally this has been done from prismatic bars of the different charge materials being placed into the furnace. However, for economic reasons small pieces of scrap material from other industries are sometimes used to charge the furnace. If these scrap materials have a surface area to volume ratio which is relatively large then there will be an increased probability of oxide formations taking place within the melting process which, in turn, may lead to an increased incidence in casting defects.

The machining or mechanical working processes of propeller blades subsequent to casting have undergone major changes in many manufacturers' works since the advent of numerically controlled machine tools. Early usage of automated machinery required the use of a solid pattern which acted as the master blade from which the machine could work a new casting. However, the advances in geometry handling using computer-based techniques, together with interfaces to multi-axis machines, enable computer-assisted manufacturing machines to be used for propeller manufacture. Most of the large manufacturers have introduced these methods and since about 1970 machines ranging from three-axis numerically controlled gantry units to nine-axis machines have been installed. In many cases fully automated flexible manufacturing propeller blade machining cells have been supplied to many manufacturers.

Many manufacturers are today using an integrated design, manufacturing and inspection concept.[5] Such methods start with the preliminary design of the blade based on polynomial representations of methodical propeller series data and cavitation criteria to which pitch and thickness distributions are added in order to define the overall power absorption characteristics. Once these are approved the design proceeds by means of theoretical hydrodynamic and finite element methods to produce a fully detailed propeller design tailored to the particular ship. When the design is completed, the blade geometry is filed within a computer and carried forward into the NC blade milling process and final geometric inspection.

While NC machines coupled to CAD/CAM facilities clearly provide a means of enhancing the manufacturing process from both machining and inspection viewpoints, many smaller propellers are manufactured using approximations to the traditional methods. In either case, a highly satisfactory propeller is likely to result provided the appropriate tolerance specification procedures are deployed. The decision as to which manufacturing process to use in a particular case to satisfy the required design tolerance requirements is largely one of the scale of production and economics.

REFERENCES AND FURTHER READING

1. Rowe FW. *The Randupson Sand-Mixing Process.* London: Institute of British Foundrymen; London 1938.
2. Langham JM. *The manufacture of marine propellers with particular reference to the foundry.* Proc. 21st Inst. Foundry Congress, Florence, September 1954; also SMM Technical Paper No. 1; February 1964.
3. Tector FJ. *The manufacture of ships propellers.* The Marine Engineer and Naval Architect; April/May, 1951.
4. *Lips Drunen. Anniversary Publication.* Lips, Drunen; 1974.
5. Navarra P, Rivara G, Rollando G, Grossi L. *Propeller blade manufacturing — Fincantieri advanced technology of production by CAM-procedure.* Nav. '88-WEMT '88 Symp., Trieste; October 1988.

Chapter 21

Propeller Blade Vibration

Chapter Outline

21.1	Flat-Plate Blade Vibration in Air	421	21.5 Finite Element Analysis	426
21.2	Vibration of Propeller Blades in Air	422	21.6 Propeller Blade Damping	427
21.3	The Effect of Immersion in Water	422	21.7 Propeller Singing	428
21.4	Simple Estimation Methods	425	References and Further Reading	429

The modes of vibration of a propeller blade, beyond the fundamental and first torsional and flexural modes, are extremely complex. This complexity arises from the non-symmetrical outline of the blade, the variable thickness distribution both chordally and radially and the twist of the blade caused by changes in the radial distribution of pitch angle. In addition, the effect of the water in which the propeller is immersed causes both a reduction in modal frequency and a modified mode shape when compared with the corresponding characteristics in air. To introduce the problem of blade vibration it is easiest to consider the vibration of a symmetrical flat blade form in air because, in this way, many of the practical complexities are eliminated for a first consideration of the problem.

21.1 FLAT-PLATE BLADE VIBRATION IN AIR

Some experiments with a flat-plate propeller blade form by Grinstead are cited by Burrill[1] as a basis for understanding the basic composition of the modal forms of vibrating blades. These tests were conducted on a symmetrical blade having an elliptical form and a constant thickness in the chordal and radial direction. The blade was cantilevered at one end of its major axis and the various modes of vibration in air were excited by bowing with the aid of a rotating disc. In these experiments the nodes in the various modes of vibration were traced by means of sand patterns. The blade used for this work was small in propeller terms since it had a span of 131.32 mm and at maximum chord length the minor axis of the ellipse was 86.11 mm; the thickness of the plate was 13.59 mm. As a consequence, the frequencies of the various modal forms are considerably higher than would be expected from a full-size propeller blade. The modal forms established by Grinstead are shown in Figure 21.1 for the first ten modes beyond the fundamental. The fundamental mode was a simple flexural cantilevered mode with its node coincident with the blade root. The various modal forms, other than the fundamental, are identified in Table 21.1 together with the measured frequencies.

The results of the frequencies can be plotted as shown in Figure 21.2, from which it can be seen that the pure flexural frequencies have the lowest frequencies with the one- and two-node torsionally based frequencies having progressively higher frequencies. In the cases of the two-node torsional and one-node flexural modes, the experiment could not distinguish the pure mode shapes owing to its proximity to the three-node flexural mode. The results of these model tests show that in addition to the pure modes, cross and diaphragm modes arise if the frequencies for the secondary lateral modes are close to the natural flexural modes.

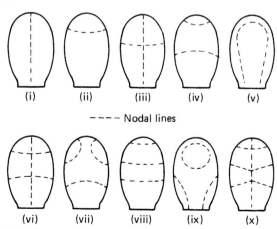

FIGURE 21.1 Mode shapes for an elliptical, flat-plate blade.

TABLE 21.1 Modes of Vibration of Blade Shown in Figure 21.1 (Compiled From Reference 1)

Mode Number	Mode Form	Frequency (Hz)
0	Fundamental mode	73
i	One-node torsional mode	249
ii	One-node flexural mode	415
iii	One-node torsional and one-node flexural mode	889
iv	Two-node flexural mode	1135
v	Two-node torsional or 'hoop' mode	1365
vi	Two-node flexural and one-node torsional mode	1819
vii	First cross-coupled mode	2155
viii	Three-node flexural mode	2202
ix	Second cross-coupled	2418
x	Cross-coupled, three-node flexural and one-node torsional	3009

21.2 VIBRATION OF PROPELLER BLADES IN AIR

Having identified the major vibration characteristics associated with a flat plate approximation to a propeller blade, the actual vibratory characteristics of a propeller blade may now be considered more easily. Burrill[1,2] conducted a series of model and full-scale experiments on propellers and Figure 21.3 shows the results of one set of vibratory tests on a propeller in air. The propeller chosen was a four-bladed, 1320 mm diameter propeller having a mean pitch ratio of 0.65 and a blade area ratio of 0.524 and the propeller was of a conventional design for the period. The tests were performed in a 6.4 m square tank and the blades vibrated by means of a vibrator, acting through a universal ball-joint clip, capable of exciting a range of frequencies from around 20 to 2000 Hz. The similarities in modal forms between Figures 21.1 and 21.3 are immediately apparent, although the torsional modes are seen to undergo some changes. The fundamental frequency of the propeller shown in Figure 21.3 was 160 Hz.

The effects of blade area can to some extent be seen by comparing the results shown by Figure 21.3 with the work of Hughes.[3] Hughes examined the response of a series of blade forms; however, for these purposes a four-bladed propeller having a blade area ratio of 0.85 is of interest. Figure 21.4 shows the results from that series of experiments and when comparing these results to those of Figure 21.3 a far more complex pattern of modal forms is observed. In particular, the importance of the blade 'edge nodes' is apparent in the case of the higher blade area ratio propeller. This latter propeller used by Hughes had circular back sections and therefore possessed symmetry about the directrix.

Blade form clearly has an important influence on the modal shapes of the vibrating blade. Some years ago Carlton and Filcek, in unpublished work, examined the effects of blade form on the vibration characteristics of controllable pitch propeller blades. Figure 21.5 shows the differences in the vibration patterns derived from this work for two different blades: one highly skewed and the other of conventional form but both having aerofoil sections. The propellers had diameters in the region 3.0–3.5 m. The propeller with a symmetrical blade outline, Figure 21.5(b), clearly shows analogous modal forms with those derived by Hughes for the smaller and simpler design (see Figure 21.4). In the case of the highly skewed blade, the pattern is somewhat more complex, although the presence of distinct flexural modes is clearly apparent. In both cases the presence of edge modes is apparent: more so with the symmetrical design.

21.3 THE EFFECT OF IMMERSION IN WATER

The principal effect of immersing the propeller in water is to cause a reduction in the frequency at which a particular mode of vibration occurs. The reduction factor is not a constant value for all modes of vibration and appears to be larger for the lower modes than for the higher modes. To investigate this effect in global terms a frequency reduction ratio Λ can be defined as:

$$\Lambda = \frac{\text{frequency of mode in water}}{\text{frequency of mode in air}} \quad (21.1)$$

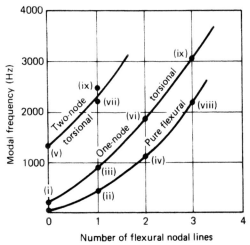

FIGURE 21.2 Modal frequencies of flat-plate blade.

Propeller Blade Vibration

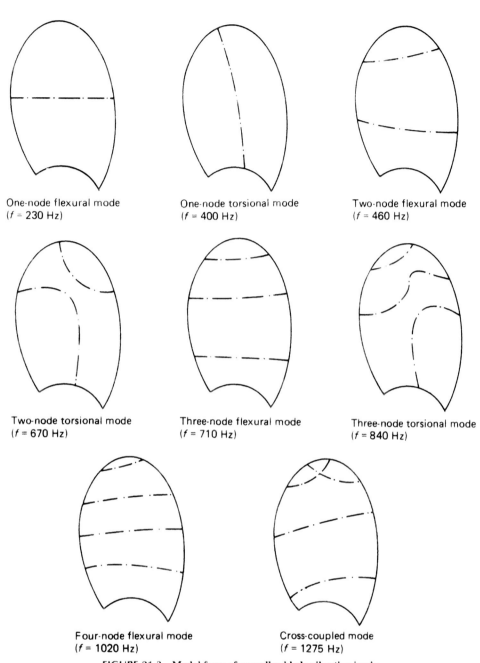

FIGURE 21.3 Modal form of propeller blade vibration in air.

Burrill[1] investigated this relationship for the propeller whose vibratory characteristics in air are shown by Figure 21.3. The results of his investigation are shown in Table 21.2 for both the flexural and torsional modes of vibration. From the table it can be seen that for this particular propeller the value of Λ increases with the modal number for each of the flexural and torsional modes. For the higher blade area ratio propeller results of Hughes, shown in Figure 21.4, Table 21.3 shows the corresponding trends. Again, from this table the general trend of increasing values of Λ with increasing complexity of the modal form is clearly seen. Hughes also investigated the effect of pitch on response frequency by comparing the characteristics of pitched and flat-plate blades. He found that for most modes, with the exception of the first torsional mode, the pitched blade had a frequency of around 10 per cent higher than the flat-plate blade and in the case of the first torsional mode the increase in frequency was of the order of 60 per cent. The influence of water immersion on these variations was negligible. In this study it was also shown that for a series of other blade forms, having broadly similar dimensions

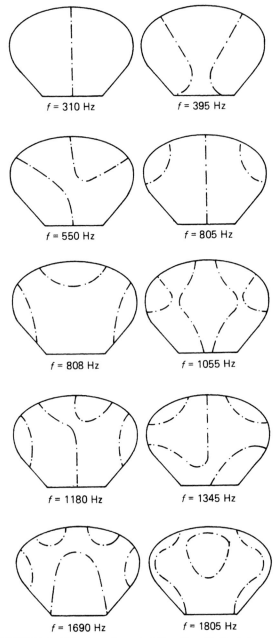

FIGURE 21.4 Vibratory characteristics of a wide-bladed propeller in air.

$$f = \frac{1}{2\pi}\sqrt{\frac{k}{m}} \qquad (21.2)$$

By assuming that the stiffness remains unchanged, then by combining equations (21.1) and (21.2) we have

$$\Lambda = \sqrt{\frac{\text{equivalent mass of the blade}}{\text{equivalent mass of the blade } + \text{ added mass due to water}}}$$

The effect of the modal frequency on the value of Λ can be explained by considering the decrease in virtual inertia due to the increased cross-flow induced by the motion of adjacent blade areas which are vibrating out of phase with

(a)

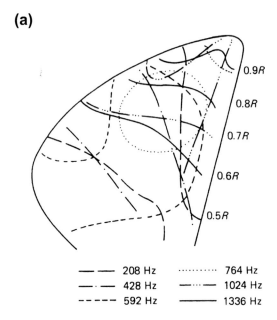

(b)

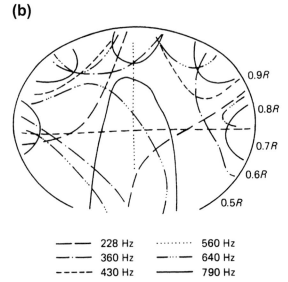

FIGURE 21.5 Vibration characteristics of two controllable pitch propeller blades.

except for blade area and outline, a reasonable correlation existed between the value of Λ and the frequency of vibration in air. It is likely, however, that such a correlation would not be generally applicable.

The influence of immersing a blade in water is chiefly to introduce an added mass term due to the water which is set in motion by the blade. If a blade is considered as a single degree of freedom system at each of the critical frequencies, then the following relation holds from simple mathematical analysis for undamped motion,

Chapter 21 Propeller Blade Vibration

TABLE 21.2 The Effect on Modal Frequency of Immersion in Water for a Four-Bladed Propeller with a BAR of 0.524 and $P/D = 0.65$ (Compiled From Reference 1)

Flexural Vibration Modes	Frequency (Hz) In Air	In Water	Λ
Fundamental	160	100	0.625
One-node mode	230	161	0.700
Two-node mode	460	375	0.815
Three-node mode	710	625	0.880
Four-node mode	1020	1000	0.980
Torsional vibration modes			
One-node mode	400	265	0.662
Two-node mode	670	490	0.731
Three-node mode	840	–	–

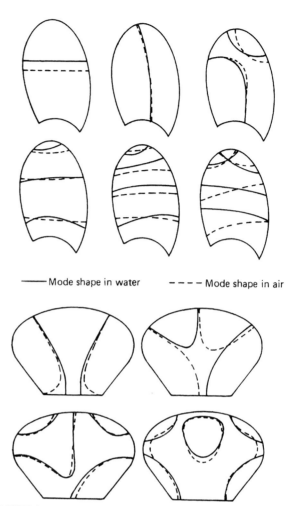

—— Mode shape in water   - - - - Mode shape in air

FIGURE 21.6 Mode shapes in air and water for the two different propeller forms.

each other: the greater the number of modal lines, the greater is this effect.

With respect to the effect of immersing the propeller blades on the mode shapes, Figure 21.6 shows that this is generally small in the examples taken from Burrill's and Hughes' work. While the basic mode shape is preserved it is seen that there is sometimes a shift in position of the modal line on the blade.

21.4 SIMPLE ESTIMATION METHODS

The estimation of the modal forms and their associated frequencies is clearly a complex matter and one that lies outside the scope of simple estimation methods. As a consequence, estimation techniques are normally confined to the determination of the fundamental flexural mode of vibration in air and a correction Λ, as identified in equation (21.1), is then applied to account for the immersion of the blade in water.

In the case of the associated problem of turbine and compressor blading, several solution procedures have been developed over the years. These methods, which rely in varying degrees on the mathematical formulation of the elasticity problem, are designed generally for comparatively high aspect ratio blades and, as such, are not always suitable for direct application to the propeller blade

TABLE 21.3 The Effect of Modal Frequency of Immersion for a Four-Bladed Propeller with a BAR of 0.85 and a $P/D = 1.0$ (Compiled From Reference 1)

Mode Shape	Frequency (Hz) In Air	In Water	Λ
i	310	200	0.645
ii	395	280	0.709
iii	550	395	0.718
iv	805	605	0.751
v	808	650	0.804
vi	1055	810	0.768
vii	1180	910	0.771
viii	1345	1055	0.784
ix	1690	1330	0.786
x	1805	1435	0.795

problem. In the case of a propeller blade, the method proposed by Baker[4] still finds fairly widespread use as an initial estimation technique for non-highly skewed propellers. The method, while giving a reasonable approximation to the fundamental frequency, also has the advantage of being simple to use and does not require the application of numerical computational analysis. According to Baker the fundamental frequency of a propeller blade in air approximates in inch-pound-second units, to

$$f_{\text{air}} = \frac{0.305}{(R - r_h)^2} \left[\left(\frac{gE}{\rho_m}\right) \left(\frac{\bar{t}}{\bar{c}}\right) c_h t_h \right]^{1/2} \quad (21.3)$$

where

$\bar{c}$ is the blade mean chord length.
c_h is the blade chord at the root section.
$\bar{t}$ is the blade mean thickness.
t_h is the blade thickness at the root section.
R is the tip radius.
r_h is the root radius.
E is Young's modulus of elasticity.
ρ_m is the material density.
g is the acceleration due to gravity.

Equation (21.3) is based on classical analysis procedures which are then used in association with the results of experimental studies conducted on flat-plate blades. The series of propellers, numbering seven in total, had a diameter of 305 mm and two blades; each propeller had differences in section form ranging from circular back to aerofoil sections and blade outlines from symmetrical to the moderate skew forms of the day.

To estimate the fundamental frequency in water, equations (21.1) and (21.3) are combined as follows:

$$f_{\text{water}} = \frac{0.305\Lambda}{(R - r_h)^2} \left[\frac{gE}{\rho_m} \left(\frac{\bar{t}}{\bar{c}}\right) c_h t_h \right]^{1/2} \quad (21.4)$$

where the value of Λ would normally take a value within the range of 0.62–0.64.

Baker also attempted an estimation formula for the primary torsional frequency of vibration which he estimated to have an accuracy of ± 5 per cent based on the tests and model forms used. This relationship in air is,

$$f_{t\,\text{air}} = \frac{0.92}{(R - r_h)} \left(\frac{t_{0.5}}{c_{0.5}}\right) \left(\frac{c_n}{\bar{c}}\right) \sqrt{\frac{gG}{\rho_m}} \quad (21.5)$$

in which $c_{0.5}$ and $t_{0.5}$ are the chord length and thickness at $0.5\,R$ respectively and G is the modulus of rigidity of the material. To estimate the torsional frequency f_t in water, it is necessary to introduce the appropriate value of Λ into equation (21.5) as was the case with equation (21.4).

In general terms, equations of the type (21.4) and (21.5) are useful for estimating purposes at the design stage of a propeller or in troubleshooting exercises. They provide an approximation to the basic vibration characteristics of the propeller blade; however, for more detailed examinations it is necessary to employ finite element based studies which enable then further exploration of the blade vibration problem.

21.5 FINITE ELEMENT ANALYSIS

The finite element technique offers a method which has the potential to define the blade natural frequencies and mode shapes with greater accuracy than by the use of the simple estimation formulae. Nevertheless, the finite element method when applied to propeller blade vibration problems relies on being able to satisfactorily model the blade in terms of the type and geometric form of the elements used. Additionally, there is also the issue of adequately representing the effect of the water in which the propeller is immersed.

In the first instance, the choice of element type is governed by the nature of the problem and reasonable correlation has been derived from the use of quadrilateral plate or isoparametric elements. The latter is particularly useful when blade rotation is included. With regard to the geometric form of the elements, the requirements of the particular elements with regard to aspect ratio and included angle at the element corners must clearly be adhered to if erroneous results are to be avoided.

Figure 21.7 shows an example of a blade discretization taken from Holden.[5] The conditions at the blade root require some consideration in order to achieve realistic conditions. Some authorities suggest that a fully built-in condition at the root is unrepresentative and that some relaxation of that condition needs to be made. Clearly the amount that can be done to meet this criticism depends upon the flexibility of the finite element capability being used.

With regard to the fluid effect on the blade, appeal can first be made to the two-dimensional analysis for laminae since the effect of blade thickness is likely to be small. For the case of a lamina, three motions can be identified which are of interest: translation motion of the lamina, rotational motion about the lamina axis and transverse or chordwise flexure. In the case of translational motion, if this is normal to the plane of the lamina, then the added mass of water per unit length is $\pi \rho b^2$ for a chord length $c = 2b$. Consequently, in the case of oblique motion at an angle θ to the plane of the lamina, then the added mass per unit length is given by:

$$m_{at} = \pi \rho b^2 \sin^2 \theta \quad (21.6)$$

For rotational motion about the blade axis the effective added moment of inertia of the blade section per unit length of the section is

$$I_{ar} = \frac{\pi}{8} \rho b^4 \quad (21.7)$$

Propeller Blade Vibration

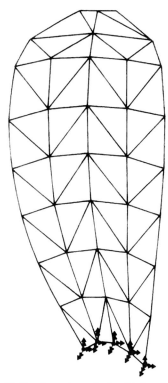

FIGURE 21.7 Finite element mesh for blade vibration analysis. *Reproduced with permission from Reference 5.*

In the case of a segmental section Lockwood Taylor[6] suggested that for transverse or chord-wise flexure the ratio of fluid to blade inertia can be approximated by

$$\frac{I_{af}}{I_f} = 1.2 \left(\frac{\rho}{\rho_m}\right)\left(\frac{b}{t}\right) \qquad (21.8)$$

where ρ_m and t are the blade material density and thickness, respectively.

Lockwood Taylor also suggested that equations (21.6) and (21.8) can be used directly for a propeller blade provided the blade is of sufficiently large aspect ratio. For wider blades of more common interest to propeller designers, three-dimensional corrections must be applied, such as those outlined by Lindholm et al.[7]

In the various pseudo-empirical approaches to the prediction of blade vibratory characteristics the blade is assumed to be stationary and so the effect of centrifugal stiffening is not considered. This situation is, however, partially redressed in numerical approaches to the problem. For conventional propeller designs the centrifugal stiffening effect is not thought to be significant; however, this may not necessarily be the case for very high rotational speed applications.

21.6 PROPELLER BLADE DAMPING

As with all mechanical structures the propeller blade material exerts a degree of damping on the vibration characteristics exhibited on the blades. Holden[5] investigated this relationship for a series of three propellers, one model and two full scale. The results obtained are shown in Table 21.4 for free oscillations at natural frequency. In Table 21.4 the damping factor is defined by

$$\zeta = \frac{1}{2\pi n} \ln\left(\frac{a_1}{a_n}\right) \qquad (21.9)$$

and was calculated from the variation in amplitude (a) of strain gauge measurements at $0.6R$ over 20 oscillations; that is, $n = 20$ in equation (21.9).

Under forced oscillations the damping factors were found to increase slightly to 0.0100 and 0.0328 for air and water, respectively.

While propeller materials in general exhibit low damping for most commercial applications, it is possible to use material with a very high damping if the blade design demands a high level of suppression of the vibration characteristics. These high damping alloys, an example of which

TABLE 21.4 Damping Factors for Three Propeller Blades Measured Experimentally (Compiled From Reference 1)

Propeller Dimensions	Material	Natural Frequency in Air (Hz)	Damping Factor ζ In Air	Damping Factor ζ In Water
Model propeller: Diameter = 1770 mm; $Z = 6$; $A_F/A_o = 0.595$	Al	118.5	0.0044	0.0405
Full-scale propeller: Diameter = 8850 mm; $Z = 6$; $A_F/A_o = 0.595$	Cu–Al–Ni	20.8	–	0.0073
Full-scale propeller: Diameter = 2050 mm; $Z = 3$; $A_F/A_o = 0.40$	Cu–Al–Ni	95.0	0.0044	0.0060

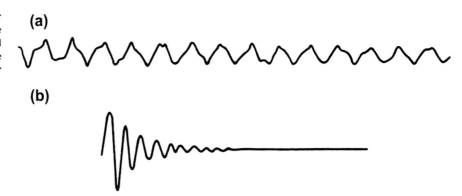

FIGURE 21.8 Comparison of propeller alloy damping properties: (a) free vibration signature of three-bladed propeller in Table 21.4 and (b) example of a high damping alloy for propeller manufacture.

is given in Reference 8, have damping characteristics as shown in Figure 21.8(b) which can be compared to that for the three-bladed propeller of Table 21.4 shown in Figure 21.8(a).

21.7 PROPELLER SINGING

Singing is a troublesome phenomenon that affects some propellers and its incidence for a particular design is unpredictable within the bounds of present analysis capabilities. It is quite likely, and indeed known, that two propellers can be manufactured to the same design and one propeller will sing while the other will not.

Singing may take many forms, ranging from a deep grunting noise through to a high-pitched warbling noise such as might be expected from an incorrectly set turning operation on a lathe. The deeper 'grunting' noise is most commonly associated with larger vessels such as bulk carriers and, in general terms, the faster rotating and smaller the propeller, the higher the singing frequency will be. The noise may be intermittent or may have an apparent period of about once per revolution: most frequently the latter. Furthermore, it is unlikely that singing will occur throughout the whole range of propeller loading but will occur only within certain specific revolution ranges. The classic example in this respect is of some controllable pitch propellers which, when working at slightly reduced pitch settings, will sing for a short period of time. Similarly, in the case of fixed pitch propellers when slowing down relatively quickly. Both of these examples occur when the blade tips are relatively lightly loaded and there is some indication that propeller designs which have significant pitch reduction at their blade tips may be more at risk of incurring the singing phenomenon at operational speeds.

The phenomenon of propeller singing has inspired many researchers to investigate the problem; much of this work being done in the 1930s and 1940s, for example References 9–15. Singing is generally believed to be caused by a vortex shedding mechanism in the turbulent and separated part of the boundary layer on the blade surface exciting the higher-mode frequencies of the blade and particularly those associated with blade edge modes. As a consequence, it is currently not possible to predict the conditions for the onset of singing in a propeller design procedure or indeed whether a particular design will be susceptible to the singing phenomenon. In addition to the theoretical complexity of the problem, the practical evidence from propellers manufactured to the same design and specification where one sings whilst the others do not, leads to the conclusion that small changes in dimensional tolerances are sufficient, given the appropriate circumstances, to induce singing.

Although prediction of singing inception is not possible, the cure of the phenomenon is normally not difficult; indeed some manufacturers incorporate the cure as a standard feature of their design whilst others prefer not to take this measure so as not to weaken the trailing edges and tip region of the blade. The most commonly used cure is to introduce a chamfer to the trailing edge of the blade and to ensure that the knuckle of the chamfer and trailing edge wedge, points a, b and c in Figure 21.9, are sharp. The purpose of this edge form is to deliberately disrupt the boundary layer growth in the trailing edge region and thereby alleviate the effects of the vortex shedding mechanism. Van Lammeren, in the discussion to Reference 1, suggests that the dimensions of an anti-singing edge can be calculated from

$$x = [20 + 5(D-2)]|_{\max=30} \text{ mm} \quad (21.10)$$

$$y = 0.1x \text{ mm}$$

where D is the propeller diameter in meters and where the other parameters are defined in Figure 21.9. The anti-singing edge is normally defined between the geometric tip of the propeller and a radial location of around $0.4R$ on the trailing edge, after which point it is then faired into the normal edge detail. Anti-singing edges of this type are applied to the suction surface of the blade; however, there are some anti-singing edge forms which are applied to both

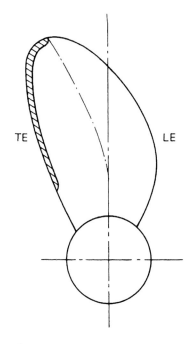

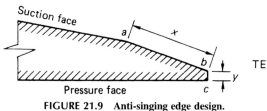

FIGURE 21.9 Anti-singing edge design.

sides of the blade at the trailing edge. These latter forms are less frequently used since the flow on the suction face of the blade, because it separates earlier, is the most likely cause of the singing problem. Edge forms of the type shown in Figure 21.9 do not cause any particular power absorption problems to arise; this is because the anti-singing edge operates wholly within the separated flow in the wake of the blade section and, therefore, does not operate in the manner of the power absorption modifications discussed in Chapter 23.

It has been found that on occasions with highly skewed propellers it is necessary to extend the anti-singing edge forward by a small amount from the geometric tip onto the leading edge of the blade in order to cure a singing problem. This extension, however, should be done with caution so as not to introduce unwanted cavitation problems due to the sharpened leading edge which results. When this extension of the anti-singing edge has been found necessary the cure of the singing problem has been completely satisfactory.

In certain other cases of small high-speed propellers it has also been found that some do not respond to the normal treatment for singing propellers. In such cases, Reference 16, it has been found necessary to create a system of notches along the trailing edge region of the propeller blades in order to effect a cure.

REFERENCES AND FURTHER READING

1. Burrill LC. *Underwater propeller vibration tests.* Trans NECIES *1949*;**65**.
2. Burrill LC. *Marine propeller blade vibrations: full scale tests.* Trans NECIES 1946;**62**.
3. Hughes WL. *Propeller blade vibrations.* Trans NECIES 1949;**65**.
4. Baker GS. *Vibration patterns of propeller blades.* Trans NECIES 1940;**57**.
5. Holden K. *Vibrations of marine propeller blades.* Norwegian Mar Res 1974;**3**.
6. Taylor Lockwood J. *Propeller blade vibrations.* Trans RINA; 1945.
7. Lindholm, Kana, Chu, Abramson. *Elastic vibration characteristics of cantilever plates in water.* J Ship Res; June 1965.
8. *Sonoston High Damping Capacity Alloy.* SMM Technical Brief No. 7.
9. *Report of the 'singing' propeller committee.* Trans RINA; 1936.
10. Conn JFC. *Marine propeller blade vibration.* Trans IESS; 1939.
11. Shannon JF, Arnold RN. *Statistical and experimental investigations on the singing propeller problem.* Trans IESS 1939;(996).
12. Davis AW. *Characteristics of silent propellers.* Trans IESS 1939;(1005).
13. Kerr W, Shannon JF, Arnold RN. *The problem of the singing propeller.* Trans I Mech E 1940;**143**.
14. Hughes G. *Influence of the shape of blade section on singing of propellers.* Trans IESS 1941;**85**.
15. Hughes G. *On singing propellers.* Trans RINA 1945;**87**.
16. Kruppa S, Henschke EH, Gutsche F. *Zickzackförmige Propellerkanten verhindern Singen.* Schiffbautechnik 1965;**157**.

Chapter 22

Propeller Design

Chapter Outline

- 22.1 The Design and Analysis Loop — 431
- 22.2 Design Constraints — 433
- 22.3 The Energy Efficiency Design Index — 433
- 22.4 The Choice of Propeller Type — 435
- 22.5 The Propeller Design Basis — 438
- 22.6 The Use of Standard Series Data in Design — 442
  - 22.6.1 The Determination of Diameter — 442
  - 22.6.2 Determination of Mean Pitch Ratio — 443
  - 22.6.3 Determination of Open Water Efficiency — 443
  - 22.6.4 To Find the rpm of a Propeller to give the Required P_D or P_E — 443
  - 22.6.5 Determination of Propeller Thrust at given Conditions — 445
  - 22.6.6 Exploration of the Effects of Cavitation — 445
- 22.7 Design Considerations — 445
  - 22.7.1 Direction of Rotation — 445
  - 22.7.2 Blade Number — 448
  - 22.7.3 Diameter, Pitch–Diameter Ratio and Rotational Speed — 448
  - 22.7.4 Blade Area Ratio — 448
  - 22.7.5 Section Form — 449
  - 22.7.6 Cavitation — 449
  - 22.7.7 Skew — 449
  - 22.7.8 Hub Form — 449
  - 22.7.9 Shaft Inclination — 450
  - 22.7.10 Duct Form — 450
  - 22.7.11 The Balance Between Propulsion Efficiency and Cavitation Effects — 450
  - 22.7.12 Propeller Tip Considerations — 451
  - 22.7.13 Propellers Operating in Partial Hull Tunnels — 451
  - 22.7.14 Composite Propeller Blades — 452
  - 22.7.15 The Propeller Basic Design Process — 452
- 22.8 The Design Process — 452
- References and Further Reading — 458

While each of the previous chapters has considered specific aspects of the propeller in some detail, this chapter draws together the various threads of the subject into the propeller design process so that the subject can be considered as a whole. The finished propeller depends for its success on the satisfactory integration of several scientific disciplines: these are hydrodynamics, stress analysis, metallurgy and manufacturing technology, with supportive inputs from mathematics, dynamics and thermodynamics. Indeed, within the design process it is not uncommon to find that several conflicting requirements develop from these disciplines. The test for the designer is to be found in how satisfactorily these conflicts can be resolved to develop a design that lies within an acceptable and optimal solution set. Moreover, it may be inferred that in propeller technology, as in all other aspects of engineering design, there is no single unique solution for a particular propulsion problem.

22.1 THE DESIGN AND ANALYSIS LOOP

The phases of the propeller design process can be summarized in the somewhat abstract terms of design textbooks as shown in Figure 22.1. From the figure, it is seen that the creation of the artefact commences with the definition of the problem and this implies that a sufficient and unambiguous specification for the propulsion problem has been produced. This design specification must include the complete definition of the inputs and required outputs, including any permissible deviations from these definitions, as well as any constraints that may be placed on the design.

Following the design definition phase, the process moves to the synthesis phase where the basic propeller design is formulated using the various capabilities that are at the designer's disposal. To provide a notionally optimal solution, the synthesis phase cannot exist in isolation and has to be conducted with the analysis and optimization phases in an interactive loop. This iterative approach is needed so as to refine the design to that required: that is, a design that complies with the original specification and also has an optimal property about it. However, the design loop must be flexible enough should an unresolvable conflict arise with the original definition of the design problem, to allow for an appeal to be made to change the definition of the design problem. In some cases it is also likely that this appeal process may lead to the

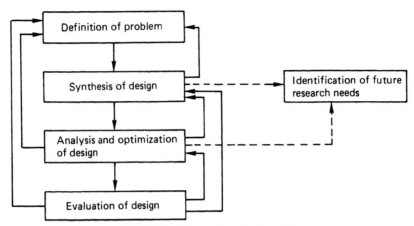

FIGURE 22.1 Phases of engineering design.

identification of areas for longer-term research to enhance future design solutions.

Design is an interactive process in which several steps are negotiated and where results are evaluated, after which it may be necessary to return to an earlier phase of the design procedure. Consequently, we may synthesize several individual components of the propeller design, then analyze and optimize them and finally return to see what effect this has on the remaining parts of the system. The analysis process may also include model testing in either a towing tank, cavitation tunnel or other facility. When the design loop of synthesis, analysis and optimization is complete, the process then passes on to the evaluation phase. This phase is the final proof of the design concept from which its success is determined, since it usually involves the testing of a prototype in the wider engineering context. However, in propeller design the luxury of a prototype is rare, since the propeller is normally a unit volume production item. Hence, the evaluation stage is frequently the sea trial phase within the ship-building program. Nevertheless, when a design does not perform as expected, it is normal, as in the generalized design process, to return to an earlier phase of the process to explore the reasons for failure and propose remedial action.

These general design ideas, although abstract, are nevertheless useful and are directly applicable to the propeller design process. How then are they applied?

Since in general a propeller can only be designed to satisfy a single design point, this involves a unique specification of the power absorbed, shaft rotational speed, ship speed and a mean radial wake field: the controllable pitch propeller being the partial exception to this rule when it would be normal to consider two or more design points. Although there is a unique design point, in general the propeller operates within a variable circumferential wake field and may also be required to work at off-design power absorption conditions. Indeed, in some instances the sea trial condition is an off-design point. Therefore, in addition to the synthesis phase of Figure 22.1, which might be conducted using a mean radial wake distribution, there is the analysis and optimization phase to study the effects of the propeller operating, for example, in the full wake field variations or at off-design conditions.

In the case of a propeller design, the conceptual and abstract design approach shown in Figure 22.1 can be considered in the following way. The definition of the problem is principally the specification of the propeller design point, or points in the case of controllable pitch propellers or ships with significant changes in operating condition, together with the constraints which are applicable to that particular design or to the vessel to which it is to be fitted. The resulting specification should be a jointly agreed document into which the owner, shipbuilder, engine builder and propeller designer have contributed: to do otherwise can lead to a grossly inadequate or unreasonable specification being developed. Following the creation of the design specification the synthesis of the design can commence. This will normally be based on a propeller type agreed during the specification stage, because it is very likely that some preliminary propeller design studies will have been conducted at that time. The blade number may also have been chosen so as to avoid global or local natural frequencies of the ship's structure and probably also the maximum propeller diameter that can be accommodated. Consequently, during the synthesis phase the basic design concept will be worked up into a detailed design proposal typically using, for advanced designs, a wake adapted lifting line with lifting surface correction capability. The choice of method, however, will depend on the designer's own capability and the data available, and may, for small vessels, be an adaptation of a standard series propeller model test data and this may work in a perfectly satisfactory manner from the cost-effectiveness point of view.

The design that results from the synthesis phase, assuming the former of the two synthesis approaches have been adopted, will then pass into the analysis and

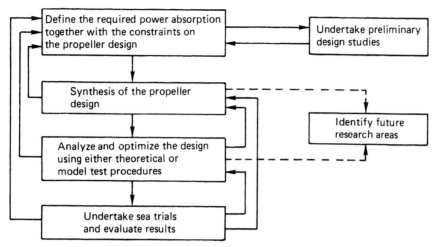

FIGURE 22.2 The phases of propeller design.

optimization phase. This phase may contain elements of both theoretical analysis and model testing. The theoretical analysis will vary, depending upon the designer's capabilities and the perceived cost benefit of this stage, from adaptations of Burrill's vortex analysis procedure through to unsteady lifting surface, vortex lattice or boundary element capabilities (Chapter 8). With regard to model testing in this phase, this may embrace a range of towing tank studies for resistance and propulsion purposes through to cavitation tunnel studies for determination of cavitation characteristics and noise prediction. Although today our understanding of the various phenomena has progressed considerably from that of say twenty or thirty years ago, there are still many areas where that understanding is far from complete. The important lesson in propeller technology, therefore, is to appreciate that each of the analysis techniques, theoretical or model testing, gives a partial answer to an aspect of the design problem. As a consequence, the basis of undertaking a good analysis and optimization phase is not simply to take the results of the various analyses at face value. Rather it is to examine them in the light of previous experience and knowledge of the techniques' various strengths and weaknesses so as to form a balanced view of the likely performance of the proposed propellers: however, this is only the essence of good engineering practice.

Figure 22.2 translates the more abstract concept of the phases of engineering design, shown in the previous figure, into a propeller-related design concept in the light of the foregoing discussion.

22.2 DESIGN CONSTRAINTS

The constraints on propeller design may take many forms. Each places a restriction on the designer and in many cases if more than one constraint is imposed then this places a restriction on the upper bound of performance that can be achieved in any one area. For example, if a single constraint is imposed, requiring the most efficient propeller for a given rotational speed, then the designer will most likely choose the optimum propeller diameter and blade loading with the smallest blade area ratio, consistent with any blade cavitation erosion criteria, in order to maximize efficiency. If then a second constraint is imposed, requiring the radiated pressures on the hull surface not to exceed a certain value, then the designer will start to alter the blade loading distribution and adjust other design parameters in order to control cavitation. Therefore, since the diameter, blade area and blade loading are no longer optimized, this will cause a reduction in efficiency but enhance the hull pressure situation. Although this is a somewhat simplified example, it adequately illustrates the point. Consequently, it is important that all concerned with a ship's design consider the various constraints with the full knowledge of their implications and the realization that the setting of unnecessary or overstrict constraints will most likely lead to degradation in the propeller's overall performance.

22.3 THE ENERGY EFFICIENCY DESIGN INDEX

The maritime shipping industry contributes some 3–4 per cent of the world's CO_2 production. Within this context the Energy Efficiency Design Index (EEDI) is a developing ship design parameter which seeks to govern the CO_2 production of ships in relation to their usefulness to society. It is one of three initiatives being developed by IMO under the auspices of the MEPC Sub-Committee: the others being the Energy Efficiency Operational Index (EEOI) and the Ship Energy Efficiency Management Plan (SEEMP). The EEDI parameter is however the most important of the three since it governs the ship design philosophy and when implemented would have to be verified against defined criteria by an independent organization in order to obtain certification.

Although it is frequently regarded as an imperfect parameter, the EEDI in its simplest terms can be regarded as the ratio between the carbon dioxide production potential of the ship and its benefit to society. In this context the CO_2 production potential of the ship is defined as comprising four components:

- The carbon dioxide that is directly attributable to the ship's propulsion machinery.
- The carbon dioxide arising from the auxiliary and hotel power loads of the ship.
- The reduction of carbon dioxide due to energy efficiency technologies. For example, heat recovery systems.
- The reduction of carbon dioxide due to the incorporation of innovative energy efficiency technologies in the design. Typically, these might include the introduction of sails or kites or other hydrodynamic devices aimed at enhancing the propulsive efficiency of the ship.

The benefit to society of the ship is seen as a function of the cargo carrying capacity of the ship and its speed. Within the formulation of the EEDI definition there are a number of correction terms that have been introduced either for specialized ship types or specific design features.

To define the reference lines, or criteria of performance, for the particular ship types to which the Index is to be applied, parametric studies have been undertaken and these have included variations in size for the ship types being considered. At the present time it is envisaged that the Index will be applied to the design of ships above 400 grt and will include tankers, gas carriers, container ships, cargo ships and refrigerated cargo ships. These ships will require an International Energy Efficiency Certificate (IEEC) once the EEDI procedure becomes implemented by IMO. However, certain ship types, whatever their size, are for the time being excluded from Index compliance and these are diesel-electric and turbine driven ships; fishing vessels; offshore and service vessels.

Full implementation is expected to be achieved within a phased process, not dissimilar to the MEPC Annex VI requirements for NO_x and SO_x emissions. Within this phased process a factor (x) will be applied to the reference EEDI versus ship capacity relationship derived from the parametric studies and this factor will vary with respect to implementation year group intervals; currently set at 2013–17, 2018–22 and 2023–27. In each of these periods the value of the parameter (x) is intended to be set to a prescribed value such that it reduces the required Index value in successive steps. Consequently, within this context the required EEDI will be given by the relationship:

$$\text{Required EEDI} = (1 - x)a.Y^b$$

where the values of a and b relate to each ship type and will be included in the regulations. The Actual EEDI, calculated from the proposed ship design, must then be shown to be less than or equal to the Required EEDI such that,

$$\text{Actual EEDI} \leq \text{Required EEDI}$$

The computation of the Actual EEDI for a specific ship design is achieved through the use of the following relationship which embraces the four CO_2 potentially producing components in the numerator while in the denominator is the product of ship speed and capacity. It will also be seen that in both the numerator and denominator there are number correction factors included which adjust the value of the Index for particular circumstances. The Actual EEDI is then given by:

$$\text{Actual EEDI} = \frac{\prod_{j=1}^{M} f_j \left(\sum_{i=1}^{nME} P_{ME(i)} C_{FME(i)} SFC_{ME(i)} \right) + (P_{AE} C_{FAE} SFC_{AE}) + \left(\left(\prod_{j=1}^{M} f_j \sum_{i=1}^{nPTI} P_{PT(i)} - \sum_{i=1}^{neff} f_{eff(i)} P_{AEeff(i)} \right) C_{FAE} SFC_{AE} \right) - \left(\sum_{i=1}^{neff} f_{eff(i)} P_{eff(i)} C_{FME} SFC_{ME} \right)}{f_i \text{Capacity } V_{ref} f_w}$$

where:

Capacity is the ship's capacity measured in deadweight or gross tonnage at the summer load line. In the case of container ships this is taken as 65 per cent of the deadweight. [tonnes]

C_{FAE} is the carbon factor for the auxiliary engine fuel. [g_{CO_2}/g_{fuel}]

C_{FME} is the carbon factor for the main engine fuel. [g_{CO_2}/g_{fuel}]

EEDI is the Actual Energy Efficiency Design Index for the ship. [g_{CO_2}/tonne.nm]

f_{eff} is a correction factor for the availability of innovative technologies.

f_i is a correction factor for the capacity of ships with technical or regulatory limitations in capacity.

f_j is a correction factor for ships having specific design features: for example, an ice breaker.

f_w is a correction factor for speed reduction due to representative sea conditions.

M is the number of propulsion shafts possessed by the ship.

n_{eff} is the number of innovative technologies contained within the design.

n_{ME} is the number of main engines installed in the ship.
n_{PTI} is the number of power take-in systems.
P_{AE} is the ship's auxiliary power requirements under normal sea-going conditions. [kW]
P_{EAeff} is the auxiliary power reduction due to the use of innovative technologies. [kW]
P_{eff} is taken as 75 per cent of the installed power for each innovative technology that contributes to the ship's propulsion. [kW]
P_{PTI} is taken as 75 per cent of the installed power for each power take-in system. For example, propulsion shaft motors. [kW]
SFC_{AE} is the specific fuel consumption for the auxiliary engines as given by the NO_x certification. [g/kWh]
SFC_{ME} is the specific fuel consumption for the main engines as given by the NO_x certification. [g/kWh]
V_{ref} is the ship speed under ideal sea conditions when the propeller is absorbing 75 per cent of the main propulsion engine(s) MCR when the ship is sailing in deep water.

Consideration of this equation suggests a number of ways that compliance with the EEDI requirements might be achieved and also options for reducing the value of the Index for a given ship. These are:

- The installation of engines with less power and, thereby, the adoption of a lower ship speed.
- To incorporate a range of energy efficient technologies in order to minimize the fuel consumption for a given power absorption.
- The use of renewable or innovative energy reduction technologies so as to minimize the CO_2 production.
- To employ low carbon fuels and in so doing produce less CO_2 than would otherwise have been the case with conventional fuels.
- To increase the deadweight of the ship by changes to or enhancements to the design.

If the option to install engines of a lower power rating into the ship was adopted, this would be a relatively simple way to reduce the value of EEDI. Such an option, however, begs the question as to whether the ship would then have sufficient power to navigate safely in poor weather conditions or, alternatively, maneuver satisfactorily in restricted channels or harbors under the full range of tidal and weather conditions that might be encountered.

With regard to energy efficient technologies there are a range as discussed in Chapter 13. Clearly, the deployment of these technologies in specific cases, as well as others like air lubrication of the hull, will be dependent on the hull form and speed of the proposed ship as well as what other devices have been fitted either upstream or downstream of the proposed device.

22.4 THE CHOICE OF PROPELLER TYPE

The choice of propeller type for a particular propulsion application can be a result of considering any number of factors. These factors may, for example, be the pursuit of maximum efficiency, noise reduction, ease of maneuverability, cost of installation and so on. Each ship and its application has to be considered on its own merits taking into account the items listed in Table 22.1.

In terms of optimum open water efficiency van Manen[1] developed a comparison for a variety of propeller types based on the results of systematic series data. In addition to the propeller data from experiments at MARIN he also included data relating to fully cavitating and vertical axis propellers (References 2 and 3): the resulting comparison is shown in Figure 22.3. The figure shows the highest obtainable open water efficiency for the different types of propeller as a function of the power coefficient B_p. As may be seen from the legend at the top of the figure the lightly loaded propellers of fast ships lie towards the left-hand side of the diagram while the more heavily loaded propellers of the tankers, bulk carriers, trawlers and the towing vessels lie to the right-hand side of the figure. Such a diagram is able to give a quick indication of the type of propeller that is likely to give the best efficiency for a given type of ship. As is seen from the diagram the accelerating duct becomes a more attractive proposition at higher values of B_p whereas the contra-rotating and conventional propellers tend to be most efficient at the lower values of B_p.

In cases where cavitation is a dominant factor in the propeller design, such as in high-speed craft, Tachmindji et al.[4] developed a useful basic design diagram to determine the applicability of different propeller types with respect to the cavitating conditions of these types of craft. This diagram is reproduced in Figure 22.4 from which it is seen that it comprises a series of regions which define the applicability of different types of propeller. In the top right-hand region are to be found the conventional propellers fitted to most merchant vessels, while in the bottom right-hand region are the conditions where super-cavitating

TABLE 22.1 Factors Affecting Choice of Propulsor

Role of Vessel
Special requirements
Initial installation costs
Running costs
Maintenance requirements
Service availability
Legislative requirements

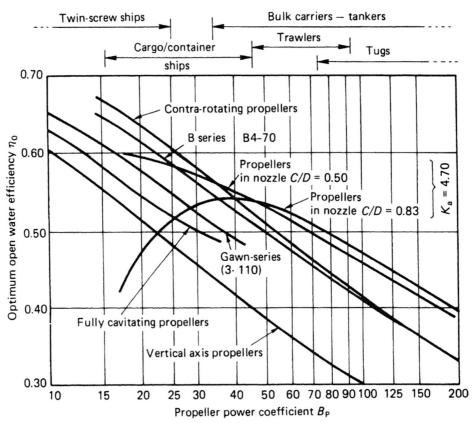

FIGURE 22.3 Typical optimum open water efficiencies for different propeller types. *Reproduced with permission from Reference 1.*

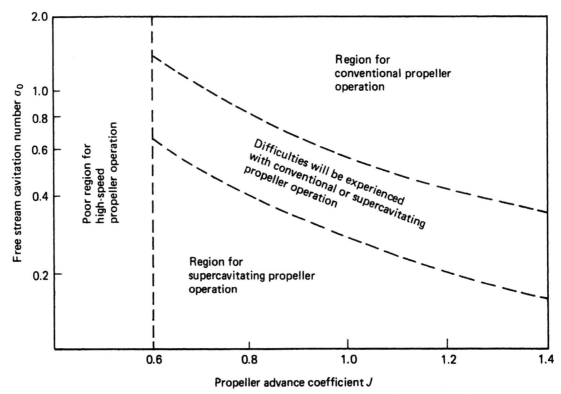

FIGURE 22.4 The effect of cavitation number on propeller type for high-speed propellers.

propellers are likely to give the best efficiencies. Propellers that fall towards the left-hand side of the diagram are expected to give low efficiency for any type of propeller and since low advance coefficient implies high B_p the correspondence between the Figures 22.4 and 22.3 can be seen.

The choice between fixed pitch propellers and controllable pitch propellers has been a long contested debate between the proponents of the various systems. In Chapter 2 it was shown that the controllable pitch propellers have gained a significant share of the Ro/Ro, ferry, fishing, offshore and tug markets. This is because there is either a demand for high levels of maneuverability or a duality of operation that can best be satisfied with a controllable pitch propeller rather than through reduction gearing. Alternatively, there may be a need for constant speed shaft driven auxiliaries. For classes of vessel which do not have these specialized requirements, then the mechanically simpler fixed pitch propeller provides a satisfactory propulsion solution. With regard to reliability of operation, as might be expected the controllable pitch propeller has a higher failure rate due to its increased mechanical complexity. By way of example Table 22.2 details the failure rates for both fixed pitch and controllable pitch propellers over a period of about a quarter of a century (Reference 5). In either case, however, it is seen that the propeller has achieved the status of being a very reliable marine component.

The controllable pitch propeller does have the advantage of permitting constant shaft speed operation of the propeller. Although this generally establishes a more onerous set of cavitation conditions, it does readily allow the use of shaft-driven generators should the economics of the ship operation dictate that this is advantageous. In addition there is some evidence to suggest (Reference 6) that the NO_x exhaust emissions can be reduced on a volumetric basis at intermediate engine powers when working at constant shaft speed. Figure 22.5 shows this trend from which a reduction in the NO emissions, which form about 90 per cent of the total NO_x component, can be seen at constant speed operation for a range of fuel qualities. Such data, however, needs to be interpreted in the context of mass emission for particular ship applications.

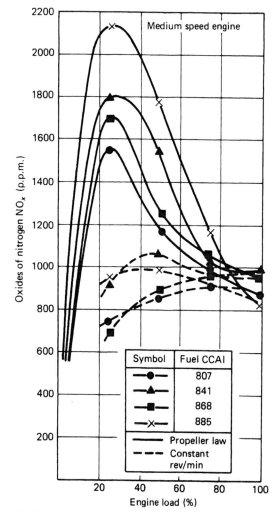

FIGURE 22.5 Influence of engine operating conditions and fuel CCAI number.

In cases where maneuverability or directional control is important, the controllable pitch propeller, steerable duct, azimuthing or podded propeller and the cycloidal propeller can offer various levels of solution to the problem, depending on the specific requirements.

In summary, Table 22.3 lists some of the important features of the principal propeller types.

TABLE 22.2 Change in the Propeller Defect Incidence with Time for Propellers in the Range $5000 < BHP < 10000$ (1960—1989)

	1960—64	1965—69	1970—74	1975—79	1980—84	1985—89
Fixed pitch propeller	0.018	0.044	0.067	0.066	0.065	0.044
Controllable pitch propellers	0.080	0.161	0.128	0.157	0.106	0.079

Defects recorded in defect incidence per year per unit.

TABLE 22.3 Some Important Characteristics of Propeller Types

Propeller Type	Characteristics
Fixed pitch propellers	Ease of manufacture Design for a single condition (i.e. design point) Blade root dictates boss length No restriction on blade area or shape Rotational speed varies with power absorbed Relatively small hub size
Controllable pitch propellers	Can accommodate multiple operating conditions Constant or variable shaft speed operation Restriction on blade area to maintain blade reversibility Blade root is restricted by palm dimensions Increased mechanical complexity Larger hub size, governed by spindle torque requirements
Ducted propellers	Can accommodate fixed and controllable pitch propellers Duct form should be simple to facilitate manufacture Enhanced thrust at low ship speed Duct form can be either accelerating or decelerating Accelerating ducts tend to distribute thrust equally between duct and propeller at bollard pull Ducts can be made steerable
Azimuthing units	Good directional control of thrust Increased mechanical complexity Can employ either ducted or non-ducted propellers of either fixed or controllable pitch type
Cycloidal propellers	Good directional control of thrust Avoids need for rudder on vessel Increased mechanical complexity
Contra-rotating propellers	Provides ability to cancel torque reaction Enhanced propulsive efficiency in appropriate conditions Increased mechanical complexity Can be used with fixed shaft lines or azimuthing units

22.5 THE PROPELLER DESIGN BASIS

The term 'propeller design basis' refers to the power absorbed, shaft rotational speed and ship speed that are chosen to act as the basis for the design of the principal propeller geometric features. Defining this basis is an extremely important matter even for the controllable pitch propellers, since in this latter case the design helical sections will only be absolutely correct for one pitch setting. This discussion, however, will largely concentrate on the fixed pitch propeller since for this type of propeller the correct choice of the design basis is critical to the performance of the ship.

The selection of the design basis begins with a consideration of the mission profile for the ship. Each vessel has a characteristic mission profile which is determined by the operator to meet the commercial needs of the particular service under the economic conditions prevailing. It must also be recognized that the mission profile of a particular ship may change throughout its life, depending on a variety of circumstances. When this occurs it may be economically wise to change the propeller design, as witnessed by the slow steaming of the large tankers after the oil crisis of the early 1970s and also in the last few years, to enhance the

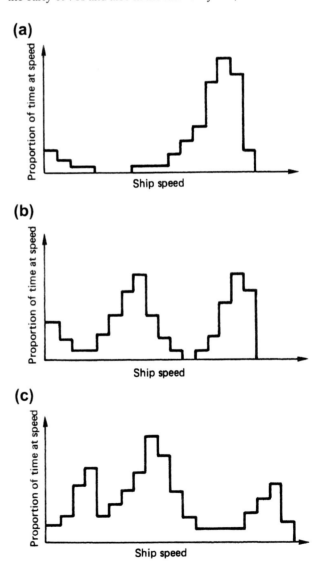

FIGURE 22.6 Examples of ship mission profiles: (a) container ship; (b) Ro/Ro passenger ferry and (c) warship.

ship's efficiency at the new operating conditions dictated by the market. The mission profile is determined by several factors, but is governed chiefly by the vessel type and its intended trade pattern; Figure 22.6 shows three examples relating to a container ship, a Ro/Ro ferry and a warship. The wide divergence in the form of these curves amply illustrates that the design basis for a particular vessel must be chosen with care such that the propeller can be designed to give the best overall performance in the required operating regions. To achieve this may well require several preliminary design studies to be undertaken in order to establish the best combination of diameter, pitch distribution, blade area and section forms to satisfy the operational constraints of the ships.

In addition to satisfying the mission profile requirements it is also necessary that the propeller and engine characteristics match, not only when the ship is new but also after it has been in service for some years. Because the diesel engine at the present time is used for the greater majority of propulsion plants, this will form the primary basis for the discussion. The diesel engine has a general characteristic of the type shown in Figure 22.7 with a propeller demand curve superimposed on it which is shown, in this instance, to pass through the Maximum Continuous Rating (MCR) of the engine. It should not, however, be assumed that in the general case the propeller demand curve must pass through the MCR point of the engine. The propeller demand curve is frequently represented by the so-called 'propeller law', which is a cubic relationship between power and rpm. This, however, is an approximation, since the propeller demand is dependent on the various hull resistance and propulsion components and, consequently, has a more complex functional relationship. In practice, however, the cubic approximation is generally valid over limited power ranges. If the pitch of the propeller has been selected incorrectly, then the propeller will be either over-pitched (stiff), curve A, or underpitched (easy), curve B. In either case, the maximum power of the engine will not be realized, since in the case of over-pitching the maximum power attainable will be X at a reduced rpm, this being governed by the engine torque limit. In the alternative under-pitching case, the maximum power attainable will be Y at 100 per cent rpm, since the engine speed limit will be the governing factor. In addition to purely geometric propeller features, a number of other factors influence the power absorption characteristics. Typical of these are sea conditions, wind strength, hull condition in terms of roughness and fouling, and, of course, displacement. It is generally true that increased severity of any of these factors requires an increase in power to drive the ship at the same speed. This has the effect of moving the power demand curve of the propeller (Figure 22.7) to the left in the direction of curve A. As a consequence, if the propeller is designed to operate at the MCR condition when the ship's hull is clean and in a light displacement with favorable weather, such as might be found on a trial condition, then the ship will not be able to develop full power in its subsequent service when the draughts are deeper and the hull fouls or when the weather deteriorates. Under these conditions the engine torque limit will restrict the brake horsepower developed by the engine.

Clearly this is not a desirable situation and a method for overcoming these needs to be sought. This is most commonly achieved by designing the propeller to operate at a few revolutions fast when the vessel is new, so that by mid-docking cycle the revolutions will have fallen to the

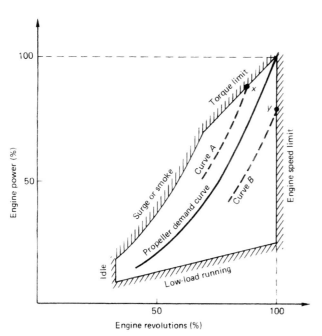

FIGURE 22.7 Engine characteristic curve.

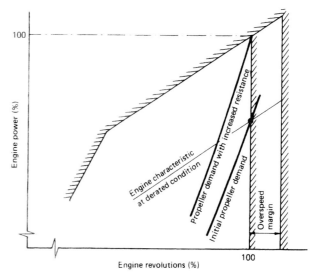

FIGURE 22.8 Change in propeller demand due to weather, draught changes and fouling.

desired value. Additionally, when significant changes of draught occur between the sea trial and the operating conditions, appropriate allowances need to be made for this effect. Figure 22.8 illustrates one such scenario, in which the propeller has been designed so that in the most favorable circumstances, such as the trial condition, the engine is effectively working at a derated condition. Consequently, the ship will not attain its maximum speed because the engine will reach its maximum speed before reaching its maximum power. Therefore in poorer weather or when the vessel fouls or works at a deeper draught, the propeller characteristic moves to the left so that the maximum power becomes available. Should it be required on trials to demonstrate the vessel's full-speed capability then engine manufacturers often allow an overspeed margin with a restriction on the time the engine can operate at this condition. This concept of the difference in performance of the vessel when on trial and in service introduces the term 'sea margin', which is imposed by the prudent owner to ensure the vessel has sufficient power available in service and throughout the docking cycle.

In practice the propeller designer will use a derated engine power as the basis for the propeller design. This is to prevent excessive maintenance and warranty costs in keeping the engine at peak performance throughout its life. Hence the propeller is normally based on a Normal Continuous Rating (NCR) of between 85 per cent and 90 per cent of the MCR conditions: sometimes this condition is called a continuous service rating (CSR). Figure 22.9 shows a typical propeller design point for a vessel working with a shaft generator. For this ship an NCR of 85 per cent of the MCR was chosen and the power of the shaft generator P_G deducted from the NCR. This formed the propeller design power. The rotational speed for the propeller design was then fixed such that the power absorbed by the propeller in service, together with the generator power when in operation, could absorb the MCR of the engine at 100 per cent rpm. This was done by deducting the power

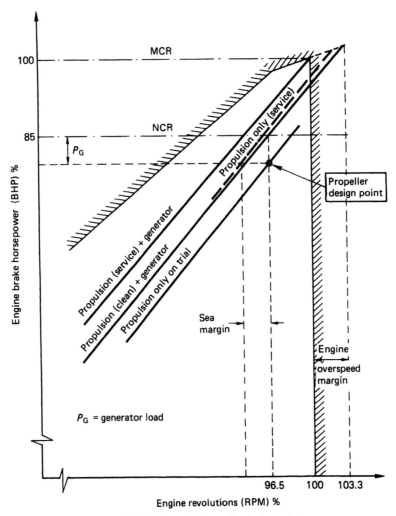

FIGURE 22.9 Typical propeller design point.

required by the generator from the combined service propeller and generator demand curve to arrive at the service propulsion only curve and then applying the sea margin which enables the propeller to run fast on trial. In this way the design power and revolutions basis became fixed.

In the case of a propeller intended for a towing duty, the superimposition of the propeller and engine characteristics presents an extreme example of the relationship between curve A and the propeller demand curve shown in Figure 22.7. In this case, however, curve A is moved far to the left because of the added resistance to the vessel caused by the tow. These situations normally require correction by the use of a gearbox in the case of the fixed pitch propeller, or by the use of a controllable pitch propeller.

The controllable pitch propeller presents an interesting extension to the fixed pitch performance maps shown in Figures 22.7 to 22.9. A typical example is shown in Figure 22.10, in which the controllable pitch propeller characteristic is superimposed on an engine characteristic. The propeller demand curve through the design point clearly does not pass through the minimum specific fuel consumption region of the engine maps: this is much the same as for the fixed pitch propeller. However, with the controllable pitch propeller it is possible to adjust the blade pitch at partial load condition to move towards this region.

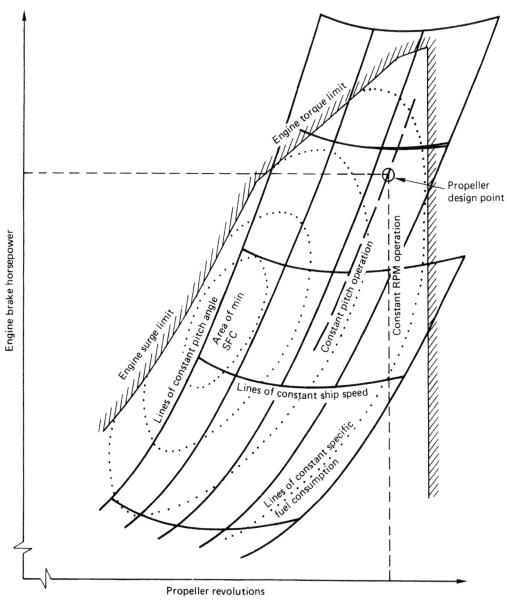

FIGURE 22.10 Controllable pitch propeller characteristic curve superimposed on a typical engine mapping.

However, when this is done it can be seen that the propeller mapping may come very close to the engine surge limit which is not a desirable feature. Nevertheless, the controllable pitch propeller pitch–rpm relationship, frequently termed the 'combinator diagram' can be programmed to give an optimal overall efficiency for the vessel.

In general, in any shaft line three power definitions are assumed to exist, these being the brake horsepower, the shaft horsepower and the delivered horsepower. The following definitions generally apply:

Brake power (P_B) The power delivered at the engine coupling or flywheel.

Shaft power (P_S) The power available at the output coupling of the gearbox, if fitted. If no gearbox is fitted then $P_S = P_B$. If a shaft-driven generator is fitted on the line shaft, then two shaft powers exist; one before the generator P_{SI} and one aft of the generator $P_{SA} = P_{SI} - P_G$. In this latter case some bearing losses may also be taken into account.

Delivered power (P_D) The power available at the propeller after the bearing losses have been deducted and any power take-off provisions made.

In design terms, where no shaft generators exist to absorb power it is normally assumed that P_D is between 98 and 99 per cent of the value of P_S depending on the length of the line shafting and the number of bearings. When a gearbox is installed, then P_S usually lies between 96 and 98 per cent of the value of P_B, depending on the gearbox type.

22.6 THE USE OF STANDARD SERIES DATA IN DESIGN

Standard series data is one of the most valuable tools that the designer has for undertaking preliminary design and feasibility study purposes. Design charts, or in many cases today regression formulae, based on standard series data, can be used to explore the principal dimensions of a propeller and their effect on performance and cavitation prior to the deployment of more detailed design or analysis techniques. In a number of cases, however, propellers are designed solely on the basis of standard series data, the only modification being to the section thickness distribution for strength purposes. This practice, which is common for small general duty propellers, is also seen to a limited extent on the larger merchant propellers.

When using design charts, however, the user should be careful of the unfairness that exists between some of the early charts and, therefore, should always, where possible, use a cross-plotting technique between these earlier charts for different blade area ratios. The unfairness arose at a time when scale effects were less well understood than they are today and in several of the model test series this has now been eradicated by recalculating the measured results to a common Reynolds number basis.

Some examples of the use of standard series data are given below. In each of these cases, which are aimed to illustrate the use of the various design charts, the hand calculation procedure has been adopted. This is quite deliberate since if the basis of the procedure is understood, then the computer-based calculations and spreadsheets will be more readily accepted and able to be critically reviewed. The examples shown are clearly not exhaustive, but serve to demonstrate the underlying use of standard series data.

22.6.1 The Determination of Diameter

To determine the propeller diameter D for a propeller when absorbing a certain delivered power P_D and operating at a rotational speed N and in association with a ship speed V_s.

It is first necessary to determine a mean design Taylor wake fraction (w_T) from experience, published data or model test results. The mean speed of advance V_a can then be determined as $V_a = (1 - w_T)V_s$. This then enables the power coefficient B_p to be determined as follows:

$$B_p = \frac{P_D^{1/2} N}{V_a^{2.5}}$$

which is then entered into the appropriate design chart as seen in Figure 22.11(a). The value of the parameter δ_{opt} is then read off from the appropriate 'constant δ line' at the point of intersection of this line and the maximum efficiency line at the required B_p value. From this value the optimum diameter D_{opt} can be calculated from the equation

$$D_{opt} = \frac{\delta_{opt} V_a}{N} \qquad (22.1)$$

If undertaking this process manually it should be repeated for a range of blade area ratios in order to interpolate for the required blade area ratio. In general, optimum diameter will decrease for increasing blade area ratio, see insert to Figure 22.11(a).

Several designers have produced regression equations for calculating the optimum diameter. One such example, produced by van Gunsteren[7] and based on the Wageningen B series, is particularly useful and is given here as:

$$\delta_{opt} = 100 \left[\frac{B_p^3}{(155.3 + 75.11 B_p^{0.5} + 36.76 B_p)} \right]^{0.2}$$

$$\times \left[0.9365 + \frac{1.49}{Z} - \left(\frac{2.101}{Z} - 0.1478 \right)^2 \times \frac{A_E}{A_O} \right]$$

(22.2)

Chapter 22 Propeller Design

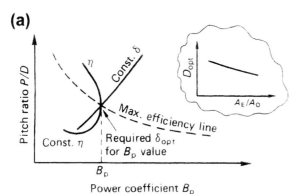
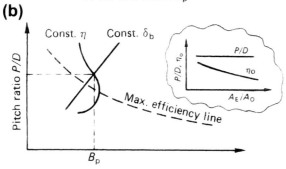
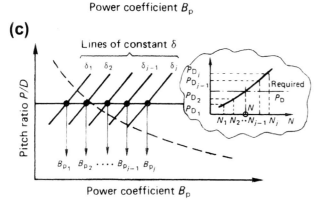

FIGURE 22.11 Examples of use of standard series data in the $B_p \sim \delta$ from: (a) diameter determination; (b) pitch ratio and open water efficiency determination and (c) power absorption analysis of a propeller.

where B_p is calculated in British units of British horsepower, rpm and knots; Z is the number of blades and A_E/A_O is the expanded area ratio.

Having calculated the optimum diameter in either of these ways, it then needs to be translated to a behind hull diameter D_b to establish the diameter for the propeller when working under the influence of the ship rather than in open water. Section 22.7 discusses this aspect of design.

22.6.2 Determination of Mean Pitch Ratio

Assuming that the propeller B_p value together with its constituent quantities and the behind hull diameter D_b are known, then evaluate the mean pitch ratio of the standard series equivalent propeller.

The behind hull value of delta (δ_b) is calculated as

$$\delta_b = \frac{ND_b}{V_a} \qquad (22.3)$$

from which this value together with the power coefficient B_p is entered onto the $B_p-\delta$ chart, as shown in Figure 22.11(b). From this chart the equivalent pitch ratio (P/D) can be read off directly. As in the case of propeller diameter this process should be repeated for a range of blade area ratios in order to interpolate for the required blade area ratio. It will be found, however, that P/D is relatively insensitive to blade area ratio under normal circumstances.

In the case of the Wageningen B series all of the propellers have constant pitch with the exception of the four-blade series, where there is a reduction of pitch towards the root (see Chapter 6). In this latter case, the P/D value derived from the chart needs to be reduced by 1.5 per cent in order to arrive at the mean pitch.

22.6.3 Determination of Open Water Efficiency

This is derived at the time of the mean pitch determination when the appropriate value of η_o can be read off from the appropriate constant efficiency curve corresponding to the value of B_p and δ_b derived from equation (22.3).

22.6.4 To Find the rpm of a Propeller to give the Required P_D or P_E

In this example, which is valuable in power absorption studies, a propeller would be defined in terms of its diameter, pitch ratio and blade area ratio and the problem is to define the rpm to give a particular delivered power P_D or, by implication, P_E. In addition it is necessary to specify the speed of advance V_a either as a known value or as an initial value to converge in an iterative loop.

The procedure is to form a series of rpm values, N_j (where $j = 1, \ldots, k$), from which a corresponding set of δ_j can be produced. Then, by using the $B_p-\delta$ chart in association with the known P/D values, a set of B_{pj} values can be produced, as seen in Figure 22.11(c). From these values the delivered powers P_{Dj} can be calculated, corresponding to the initial set of N_j, and the required rpm can be deduced by interpolation to correspond to the particular value of P_D required. The value of P_D is, however, associated with the blade area ratio of the chart, and consequently this procedure needs to be repeated for a range of A_E/A_O values to allow the unique value of P_D to be determined for the actual A_E/A_O of the propeller.

By implication this can be extended to the production of the effective power to correlate with the initial value of V_a chosen. To accomplish this the open water efficiency needs to be read off at the same time as the range of B_{pj} values to form a set of η_{oj} values. Then the efficiency η_o can be calculated to correspond with the required value of P_D in order to calculate the effective power P_E as,

$$P_E = \eta_o \eta_H \eta_r P_D$$

Figure 22.12 demonstrates the algorithm for this calculation, which is typical of many similar procedures

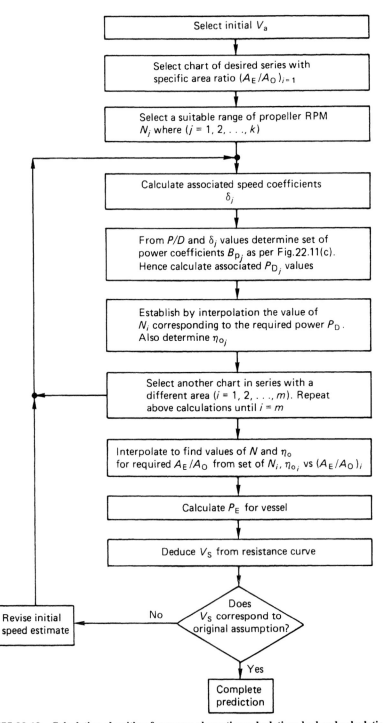

FIGURE 22.12 Calculation algorithm for power absorption calculations by hand calculation.

that can be based on standard series analysis to solve particular problems.

22.6.5 Determination of Propeller Thrust at given Conditions

The estimation of propeller thrust for a general free running condition is a trivial matter once the open water efficiency η_o has been determined from a B_p–δ diagram and the delivered power and speed of advance are known. In this case the thrust becomes

$$T = \frac{P_D \eta_o}{V_a} \quad (22.4)$$

However, at many operating conditions, such as towing or the extreme example of zero ship speed, the determination of η_o is difficult or impossible since, when V_a is small then $B_p \to \infty$ and, therefore, the B_p–δ chart cannot be used. In the case when $V_a = 0$ the open water efficiency η_o loses significance because it is the ratio of thrust power to the delivered power and the thrust power is zero because $V_a = 0$: additionally, equation (22.4) is meaningless since V_a is zero. As a consequence, a new method has to be sought.

Use can be made either of the standard K_T–J propeller characteristics or alternatively of the μ–σ diagram. In the case of the K_T–J curve, if the pitch ratio, rpm and V_a are known, then the advance coefficient J can be determined and the appropriate value of K_T read off directly and the thrust determined. Alternatively, the μ–σ approach can be adopted as shown in Figure 22.13.

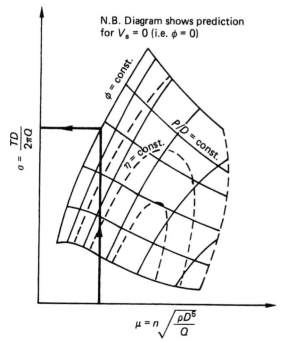

FIGURE 22.13 The use of the μ–σ chart in thrust prediction.

22.6.6 Exploration of the Effects of Cavitation

In all propellers the effects of cavitation are important. In the case of general merchant propellers some standard series give guidance on cavitation in the global sense: for example, the KCD series of propellers where generalized face and back cavitation limits are given. The problem of cavitation for merchant ship propellers, while addressed early in the design process, is nevertheless generally given more detailed assessment in later stages of the design in terms of the pressure distribution, cavitation growth, extent, type and decay.

In the case of high-speed propellers the effects of cavitation need particular consideration at the earliest stage in the design process, along with pitch ratio, diameter and, by implication, the choice of propeller rpm. Many of the high-speed propeller model test series include the effects of cavitation by effectively repeating the model tests at a range of free stream cavitation numbers based on advance velocity. Typical in this respect is the KCA series. From propeller series of this kind the influence of cavitation on the propeller design can be explored, for example, by taking a series of charts for different blade area ratios and plotting for a given advance coefficient K_Q against the values of σ tested to show the effect of blade area against thrust or torque breakdown for a given value of cavitation number. Figure 22.14 demonstrates this approach. In the design process for high-speed propellers several analogous design studies need to be undertaken to explore the effects different diameters, pitch ratios and blade areas have on the cavitation properties of the propeller.

22.7 DESIGN CONSIDERATIONS

The design process of a propeller should not simply be a mechanical process of going through a series of steps such as those defined in the previous section. Like any design it is a creative process of resolving the various constraints to produce an optimal solution. An eminent propeller designer once said 'It is very difficult to produce a bad propeller design but it is equally difficult to produce a first class design.' These words are very true and should be engraved on every designer's heart.

22.7.1 Direction of Rotation

The direction of rotation of the propeller has important consequences for maneuvering and also has cavitation and efficiency considerations for twin-screw vessels. In terms of maneuvering, for a single-screw vessel the influence on maneuvering is entirely determined by the 'paddle wheel effect'. When the vessel is stationary and the propeller started, the propeller will move the afterbody of the ship in the direction of rotation: that is in the sense of a paddle or

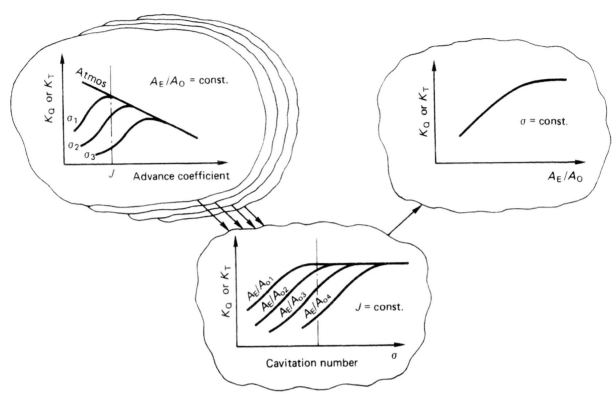

FIGURE 22.14 The use of high-speed standard series data to explore the effects of cavitation.

road wheel moving relative to the ground. Thus with a fixed pitch propeller, this direction of initial movement will change with the direction of rotation, that is left or right-handed and with ahead or astern thrust. In the case of a controllable pitch propeller the movement will tend to be unidirectional.

In the case of twin-screw vessels, certain differences become apparent. In addition to the paddle wheel effect other forces due to the pressure differential on the hull and shaft eccentricity come into effect. The pressure differential on either side of the hull, caused by reversing thrusts of the propellers, produces a lateral force and turning moment, Figure 22.15, which remains largely unchanged for fixed and controllable pitch propellers and direction of rotation. The magnitude of this thrust is of course a variable depending on the underwater hull form and in the case of some gondola hull forms it is practically non-existent. However, in the general case of maneuvering van Gunsteren[7] undertook an analysis between propeller rotational directions and fixed and controllable pitch propellers

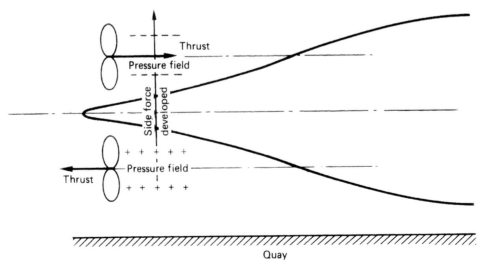

FIGURE 22.15 Side force developed by reversing thrusts of propellers on a twin-screw vessel due to pressure field in hull.

Chapter 22 Propeller Design

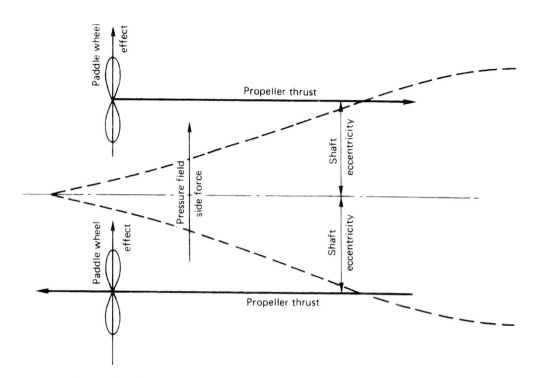

Induced turning moment = propeller thrust eccentricity + paddle wheel effect + pressure field side force

FIGURE 22.16 Induced turning moment components.

to produce a ranking of the magnitude of the turning moment produced. This analysis took into account shaft eccentricity, the axial pressure field and the paddle wheel effect (Figure 22.16) based on full-scale measurements (Reference 8) for frigates. The results of his analysis are shown in Table 22.4 for maneuvering with two propellers giving equal thrusts and in Table 22.5 for maneuvering on a single propeller.

While the magnitudes in Table 22.4 and 22.5 relate to particular trials, they do give guidance on the effects of propeller rotation on maneuverability. The negative signs were introduced to indicate a turning moment contrary to nautical intuition. From the twin-screw maneuverability point of view it can be deduced that fixed pitch propellers are best when outward turning; however, no such clear-cut conclusion exists for the controllable pitch propeller.

TABLE 22.4 Turning Moment Ranking of Two Propellers Producing Equal Thrusts (Compiled From Reference 7)

Twin-Screw Installation (Reverse Thrusts)	Turning Moment Ranking
F.p.p.; inward turning	−2.1
F.p.p.; outward turning	10.1
C.p.p.; inward turning	3.3
C.p.p.; outward turning	4.6

C.p.p.: controllable pitch propeller.
F.p.p.: fixed pitch propeller.

TABLE 22.5 Turning Moment Ranking of One Propeller Operating on a Twin-Screw Installation (Compiled From Reference 7)

Twin-Screw Installation (Single Propeller Operation)	Direction of Thrust	Turning Moment Ranking
F.p.p.; inward turning	Forward	−1.2
F.p.p.; inward turning	Astern	−1.1
F.p.p.; outward turning	Forward	5.6
F.p.p.; outward turning	Astern	4.5
C.p.p.; inward turning	Forward	−1.2
C.p.p.; inward turning	Astern	4.5
C.p.p.; outward turning	Forward	5.6
C.p.p.; outward turning	Astern	−1.1

C.p.p.: controllable pitch propeller.
F.p.p.: fixed pitch propeller.

In the context of propeller efficiency, it has been found for twin-screw ships that the rotation present in the wake field, due to the flow around the ship, at the propeller disc can lead to a gain in propeller efficiency when the direction of rotation of the propeller is opposite to the direction of rotation in the wake field. However, if concern over cavitation extent is present, then this can partially be helped by considering the propeller rotation in relation to the wake rotation. If the problem exists for a twin-screw ship at the blade tip, then the blades should turn in the opposite sense to the rotation in the wake, while if the concern is at root then the propellers should rotate in the same sense as the wake rotation. As a consequence the dangers of blade tip and tip vortex cavitation need to be carefully considered against the possibility of root cavitation.

22.7.2 Blade Number

The number of blades is primarily determined by the need to avoid harmful resonant frequencies of the ship structure and the machinery. However, as blade number increases for a given design the extent of the suction side sheet cavity generally tends to decrease. This is in contrast to the situation at the blade root where cavitation problems can be enhanced by choosing a high blade number, since the interblade clearances become less in this case close to the root.

In addition to resonant excitation and cavitation considerations, it is also found that both propeller efficiency and optimum propeller diameter increase as blade number reduces. As a consequence of this latter effect, it will be found, in cases where a limiting propeller diameter is selected, that propeller rotational speed will be dependent on blade number to some extent.

The cyclical variations in thrust and torque forces generated by the propeller are also dependent on blade number and this dependence was discussed in Table 11.13.

22.7.3 Diameter, Pitch—Diameter Ratio and Rotational Speed

The choice of these parameters is generally made on the basis of optimum efficiency. When the delivered horsepower is held constant, efficiency is only moderately influenced by small deviations in the diameter, P/D and revolutions. However, the effect of these parameters on the cavitation behavior of the propeller is extremely important and consequently needs careful exploration at the preliminary design stage. For example, it is likely the propellers of high-powered or fast ships should have an effective pitch diameter ratio larger than the optimum value determined on the basis of optimum efficiency. Furthermore, it is generally true that a low rotational speed of the propeller is a particularly effective means of retarding the development of cavitation over the suction faces of the blades.

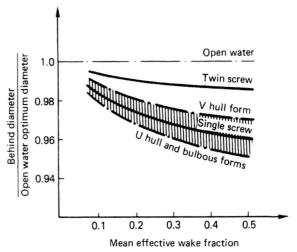

FIGURE 22.17 Correction to optimum open water diameter. *Reference 9.*

In Section 22.6.1 the optimum diameter calculation was discussed. For an actual propeller working behind a ship the diameter usually needs to be reduced from the optimum value predicted from the standard series data. Traditionally this was done by reducing the optimum diameter by 5 per cent and 3 per cent, for single- and twin-screw vessels, respectively. This correction is necessary because the resultant propulsion efficiency of the vessel is a function of both the open water propeller efficiency and the propeller—hull interaction effects. Hawdon et al.[9] conducted a study into the effect of the character of the wake field on the optimum diameter. From this study they derived a relationship of the form shown in Figure 22.17; however, the authors note that, in addition to the mean effective wake, it is necessary to take into account the radial wake distribution as implied by the distinction between different hull forms.

22.7.4 Blade Area Ratio

In general, the required expanded area ratio when the propeller is operating in a wake field is larger than that required to simply avoid cavitation at shock-free angles of attack. Furthermore, a larger variation in the section angle of attack can to some extent be supported by increasing the expanded area ratio of the propeller. There is also a margin, actual or implied, to accommodate the uncertainties surrounding the onset of material erosion for cavitation effects. Notwithstanding the advantages of increasing blade area, it must be remembered that this leads to an increased section drag and hence a loss in efficiency of the propeller (see Figure 22.11(b)).

In the case of a controllable pitch propeller there is a limit to the extent of the blade area due to the requirement of blade—blade passing in order to obtain reversibility of the blades.

22.7.5 Section Form

With regard to section form, the most desirable thickness distribution from the cavitation viewpoint is an elliptic form. This, however, is not very practical in the context of section drag and in practice the National Advisory Committee for Aeronautics (NACA) 16, 65 and 66 (modified) forms are the most utilized. In the case of mean lines, the NACA $a = 1.0$ is not generally considered a good form since the effect of viscosity on lift for this camber line is large and it is doubtful whether the load distribution can be achieved in practice. The most favored camber lines form would seem to be the NACA $a = 0.8$ or 0.8 (modified), although a number of organizations use proprietary section forms.

22.7.6 Cavitation

Sheet cavitation is generally caused by the suction peaks in the way of the leading edge being too high whilst bubble cavitation tends to be induced by too high section cambers being used in the mid-chord region of the blade.

The choice of section pitch and the associated camber line should aim to minimize or eradicate the possibility of face cavitation, although this form of cavitation is not as aggressive as previously thought. Hence the section form and its associated angle of attack needs to be designed so that it can accommodate the full range of negative incidence.

There are few propellers in service that do not cavitate at some point around the propeller disc. The focus of design should be to accept that cavitation will occur but to minimize its effects, both in terms of the erosive and pressure impulse effects.

The initial blade design can be undertaken using one of the basic estimation procedures, notably the Burrill cavitation chart or the Keller formula (see Chapter 9). These methods usually give a reasonable first approximation for the blade area ratio required for a particular application. The full propeller design process needs to incorporate procedures to design the radial distribution of chord length and camber in association with cavitation criteria rather than through the use of standard outlines.

If the blade area, or more specifically the section chord length (c), is unduly restricted, then in order to generate the same lift from the section, this being a function of the product cc_l, the lift coefficient (c_l) must increase. This generally implies a larger angle of attack or camber, which in turn leads to higher suction pressures, and hence a greater susceptibility towards cavitation. Therefore, to minimize the extent of cavitation, the variations in the angle of attack around the propeller disc should only give rise to lift coefficients in the region of shock-free flow entry for the section if this is possible.

In general terms the extent of sheet cavitation, particularly with high-powered fast ships, tends to be minimized when the blade section thickness is chosen to be sufficiently high so as to fall just below the inception of bubble cavitation on the blades. The selection of the blade camber and pitch should normally be such that the attitude of the resultant section can satisfactorily accommodate the negative incidence range that the section has to meet in practice while the radial distribution of chord length needs to be selected in association with the variations of the in-flow angle.

Tip vortex cavitation is best controlled by adjustment of the radial distribution of blade loading near the tip. The radial distribution of bound circulation at the blade tip can lay within the range:

$$0 \leq \frac{d\Gamma}{dr} < \infty \qquad (22.5)$$

Hence, the closer ($d\Gamma/dr$) is to zero, the greater will be the control of the tip vortex strength. In addition to the control exerted by ($d\Gamma/dr$) further control can be exerted by choosing the highest number of propeller blades, since this means that the total load is distributed over a greater number of blades. While using the rate of change of bound circulation near the tip as a control for the tip vortex strength, it must be remembered that moving away from the optimum value of ($d\Gamma/dr$) will induce a loss of propeller efficiency.

22.7.7 Skew

The use of skew has been shown to be effective in reducing both shaft vibratory forces and hull pressure-induced vibration (see Figure 23.6). The effectiveness of a blade skew distribution for retarding cavitation development depends to a very large extent on the matching of the propeller skew with the skew of the maximum or minimum in-flow angles in the radial sense.

Moderate skew of the type used with most marine propellers may also induce the phenomenon known to aerodynamicists of leading edge vorticity in relation to swept wings. To produce these vortices the flow at the leading edge separates, due to the effective sweep of the blade plan form in the flow field, and then flows on downstream over the propeller blade surface and into the propeller wake.

22.7.8 Hub Form

It is clearly advantageous for the propeller hub to be as small as possible consistent with its strength and the flexibility it gives to the blade root section design.

In addition to the hub diameter consideration, the form of the hub is of considerable importance. A convergent hub form is normally quite satisfactory for slow merchant vessels; however, for higher-speed ships and fast patrol

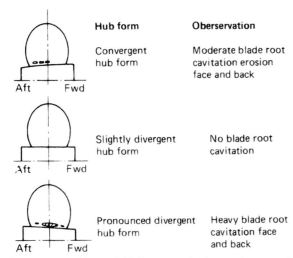

FIGURE 22.18 Observed blade root cavitation erosion on a fast patrol craft propeller.

vessels or warships experience suggests that a slightly divergent hub form is best from the point of view of reducing the risk of root erosion problems. In the case of a fast patrol craft van Gunsteren and Pronk[10] experimented with different hub profiles, and the results are shown in Figure 22.18. The convergent hub enlarges the flow disc area between the hub and the edge of the slipstream, which has only minimal contraction, from forward to aft and, therefore, decelerates the flow which results in a positive pressure gradient. This may introduce flow separation that promotes cavitation. The strongly divergent hub accelerates the flow and therefore reduces the pressure, which again promotes cavitation.

In addition to the use of a slightly divergent hub form, where appropriate, the use of a parallel or divergent cone (Figure 22.19) can assist greatly in reducing the strength of the root vortices and their erosive effects on the rudder.

22.7.9 Shaft Inclination

When the propeller shafts are inclined to any significant degree relative to the incident flow field direction, this will give rise to a cyclic variation in the advance angle of the flow entering the propeller. The amplitude of this variation is given by

$$\Delta\beta = \frac{\sin\phi}{1 + \left(\frac{\pi x}{J}\right)} \quad (22.6)$$

where ϕ is the inclination of the shaft relative to the flow and β is the advance angle at the particular radius. It should be noted that the value of ϕ will vary between a static and dynamic trim condition and this variation can in some cases be quite significant. In addition to the consequences for cavitation at the root sections, since equation (22.6) gives a larger value for $\Delta\beta$ at the root than at the tip, shaft

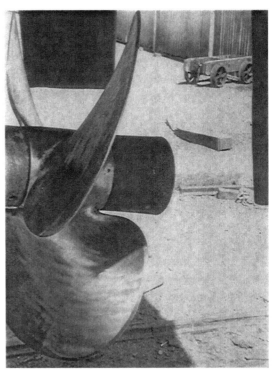

FIGURE 22.19 Truncated fairwater cone fitted to a high-speed patrol vessel.

inclination can induce significant lateral forces, eccentrically disposed, at the propeller station together with large turning moments which have to be reacted by the shaft and its bearings as discussed in Chapter 6.

22.7.10 Duct Form

When a ducted propeller is selected the choice of duct form needs to be made. When selecting a duct form for normal commercial purposes it is necessary to ensure that it is both hydrodynamically reasonable and practical and easy to manufacture. For many commercial purposes a duct form of the Wageningen 19a type will suffice when a predominately unidirectional accelerating duct form is required. However, if an improved astern performance is required, then a duct based on the Wageningen No. 37 form usually provides an acceptable compromise between ahead and astern operation.

The uses of decelerating duct forms are comparatively rare outside of naval practice and generally operate at rather higher B_p values than the conventional accelerating duct form.

22.7.11 The Balance Between Propulsion Efficiency and Cavitation Effects

The importance of attaining a balance between the achievements of maximum propulsion efficiency and attaining an acceptable cavitation performance has been noted on a number of occasions during this discussion. These references have mostly been in the context of the design point for the propeller and, by implication, for ships with a relatively narrow operational spectrum. Important as this is, for ships which operate under very variable conditions this balance has then to be maintained across a much wider spectrum of operating conditions.

A typical example of such a situation might be a cruise ship and Figure 22.20 illustrates the problems that can occur if this balance is not maintained and the design specification is incorrectly developed. In this case a high maximum ship speed was required and the builder offered a premium for achieving a maximum speed above the contract speed with the given power installation. As a consequence the ship was designed to achieve as high a speed as possible since no mention in the contract had been made of the importance of acceptable vibratory performance at lower ship speeds. The result was a pleasing performance from the controllable pitch propellers at the ship's maximum contract speed from both an efficiency and cavitation viewpoint. However, when the ship operated on legs of the cruise schedule which called for lower speeds, complex cavitating tip and leading edge vortex structures were developed by the propeller which then gave rise to broadband excitation of the hull structure at these lower operational speeds. In Figure 22.20 it can be seen that the resulting excitation levels in the restaurant, when evaluated in accordance with the ISO 6954 (2000) Code, were rather higher at 8 knots than was the case with the higher design speed of 27 knots.

The example, therefore, underlines the importance of attaining the correct balance of performance characteristics across the operating spectrum of the ship and, moreover, of defining a design specification having due regard of the way in which it is intended to operate the ship.

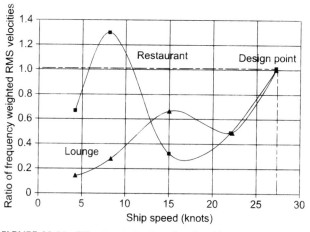

FIGURE 22.20 Vibratory behavior of cruise ship whose propellers had not been designed for use across the operating spectrum.

22.7.12 Propeller Tip Considerations

There are many factors which can be deployed in the design of the propeller tip in order to influence the behavioral characteristics of the propeller: particularly in relation to noise and cavitation. Apart from increasing the strength of the blade tips for ships such as dredgers or which regularly take the ground, one of the primary aims in designing the blade tip is to influence the characteristics of the tip vortex as well as minimizing any unwelcome interactions between super-cavitating tip sheet cavities and the tip vortex. Moreover, in some designs there is the desire to increase the tip loading by the use of end-plates. However, among the more important influencing parameters are the chordal and radial profile of thickness; camber design; section length; the use of tip plates or winglets and tip skew and rake.

Vonk et al.[27] examined the influence of tip rake on propeller efficiency and cavitation behavior through a series of computational fluid dynamic studies. They suggested that the cavitation characteristics in the mid-chord areas, where bubble cavitation can arise, and in the tip region can be enhanced by the use of aft tip rake. Conversely, they also concluded that forward tip rake, although not generally helpful to the cavitation characteristics, has a greater potential for improving the propeller efficiency. Notwithstanding this by utilizing the cavitation benefits of aft tip rake, then for the same set of cavitation criteria the design can be adjusted to yield a greater efficiency in the design balancing process. Dand[28] examined the behavior of a forward raked propeller. In these types of propeller there is a tendency to generate a pre-swirl which is in the opposite direction to the rotation of the tip vortex and this tends to disperse the vortices from the tip region. Indeed, there is a large measure of similarity between the conclusions derived from this work and that of Vonk et al.

22.7.13 Propellers Operating in Partial Hull Tunnels

Where a ship's draught may be restricted for operational reasons, designing the hull form so as to have partial tunnels sometimes benefits. Figure 22.21 shows typical configurations for both a single- and twin-screw ship.

As has been discussed previously, in general, the slowest turning, largest diameter propeller is likely to return the highest propulsive efficiency: additionally, slow rotational speed can also have cavitation benefits. However, where operating draughts are restricted propeller immersion can be a dominating factor in the propeller design. This is not only from a reduced cavitation number perspective but also from the ever attendant possibility of air-drawing into the propeller disc. To counteract these effects the

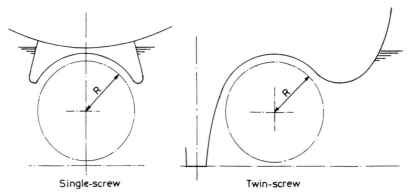

FIGURE 22.21 Examples of partial tunnels.

designing of partial ducts into the hull-form permits the largest propeller diameter, slowest turning propeller to be installed in a flow field which also frequently can have attenuated ship boundary layer influences and minimal risk of air-drawing taking place. Such arrangements have been fitted to single- and multi-screw ships and if correctly designed may enhance not only the propeller efficiency but also the hull efficiency to a limited extent.

An alternative reason for the employment of partial tunnels is to be found in the case of lifeboats, where for reasons of giving a measure of a protection to people in the water, the propeller is located within a tunnel. However, in these cases the tunnel is normally rather more encasing than that shown in Figure 22.21.

22.7.14 Composite Propeller Blades

Although all propellers are subject to hydroelastic effects, the isotropic behavior of the conventional propeller metals tends generally to reduce these effects to negligible proportions except for highly biased skew or some specialized designs. The anisotropic behavior of carbon fiber composites allows the designer extra degrees of freedom in exploiting the potential advantages of hydroelasticity.

Because carbon fiber material is normally supplied in tapes with the fiber having specific orientations, typically $0°$, $\pm 45°$, $0-90°$, etc., the primary strength of the fibers also corresponds to these directions. Consequently, the lay-up of the fibers and the way they are combined in the matrix will give different deflection properties in each of the radial and chordal directions of the blade. Therefore, the blade can be designed to deflect in ways which are beneficial from a power absorption or cavitation inception and control perspective as the rotational speed is increased. This implies that the design process must be fully hydroelastic in the sense that a finite element procedure, capable of accommodating composite material lay-up processes, and a hydrodynamic analysis code are integrated into a convergent solution capability. Composite propellers, in their larger sizes, commonly have a metallic boss with the composite blades keyed into the boss using a number of proprietary configurations. An additional feature with composite propellers is that it is likely that radiated noise emissions can be reduced significantly; perhaps of the order of 5 dB in certain cases.

Currently, with certain naval exceptions, composite propellers have only been produced in relatively small sizes: the biggest to date probably being for the experimental trimaran *Triton*. Nevertheless, in addition to their potential hydrodynamic advantages there is also a weight advantage since composite blades are much lighter than those made from conventional materials.

22.7.15 The Propeller Basic Design Process

In order to outline the overall basic design process for a propeller an example for the design of a small coastal ferry has been chosen and the resulting EXCEL spreadsheet for one operating condition is shown in Table 22.6 Within the overall design process many such spreadsheets are developed and cross-plotted in order to arrive at the final basic design. Moreover, such processes are integrated into other similar capabilities relating to hull resistance and propulsion analysis in order to achieve an integrated design.

22.8 THE DESIGN PROCESS

The level of detail to which the propeller design process is taken is almost as variable as the number of propeller designers in existence. The principal manufacturers all have detailed design capabilities, albeit based on different methods. While the computational capability of the designer plays a large part in the detail of the design process, the information available upon which to base the design is also an important factor: there is little value in using advanced and high-level computational techniques requiring detailed input when gross assumptions have to be

TABLE 22.6 Typical Basic Propeller Design Calculation

PROPELLER BASIC DESIGN
Program E/BD1
Twin Screw Passenger Ferry
T=2.88 m
10th July 2005

Ship speed (Vs)	15.5	kts	Immersion to CL	2.2 m	
Delivered power	836	kW	Height of stern wave	1.2 m	
	1137	hp(m)	Total immersion	3.4 m	
Revolutions	300	rpm			
Wake fraction	0.112		Static head (p0-e)	19.47 lbf/in^2	
Speed of advance	13.764	kts	Dynamic head (qt)	38.82 lbf/in^2	
Diameter	1.980	m	Cavitation number	0.501	
Blade number	4		Tc (Burrill)	0.183	
$W_{max}-W_{min}$	0.3		d/R	1.110	
W_{max}	0.35		Non-cavitating po	1.3 kPa (N.B. for (d/r) ≤ 2)	
Tip clearance	1.0	m	Cavitating pc	2.5 kPa (d/r) ≤ 1	1.0 kPa (d/r) >
Ship displacement	1118	m^3	Blade rate hull pressure	2.8 kPa (d/r) ≤	1 1.6 kPa (d/r)

Wageningen B4 analysis

BAR	0.40	0.55	0.70	0.85	
P/D	0.841	0.859	0.869	0.875	
η0	0.679	0.693	0.694	0.683	
Thrust	8.15	8.32	8.33	8.20	tonnes
Ap	1.7	1.7	1.7	1.7	m^2
Ad	1.9	1.9	2.0	1.9	m^2
Ae/A0	0.616	0.632	0.634	0.625	

Basic propeller

Propeller Ae/A0	0.627	Kq	0.0192
P/D mean	0.869	Kt	0.123
η0	0.675	J	0.678
Thrust	4.9 tonnes		

Expanded blade area (0.1667R)   1.93 m^2

r/R	wc	Chord (mm)	pc	Pitch (mm)	tc	Thick (mm)	Ca	Area (mm^2)
1.0000	0.000	0	1.026	1765	0.055	4.0	1.000	0
0.9375	0.702	410	1.037	1784	0.107	7.7	0.769	2432
0.8750	0.920	538	1.043	1795	0.167	12.0	0.732	4735
0.7500	1.143	668	1.040	1789	0.293	21.1	0.714	10067
0.6250	1.206	705	1.030	1772	0.433	31.2	0.712	15653
0.5000	1.176	688	1.008	1734	0.588	42.3	0.709	20640
0.3750	1.072	627	0.971	1671	0.770	55.4	0.704	24464
0.2500	0.886	518	0.909	1564	1.000	72.0	0.695	25923
0.1667	0.712	416	0.832	1432	1.190	85.7	0.688	24541

CSR/MCR powering ratio	0.850			
Developed power (stressing)	984	kW		
Revolutions (stressing)	316.7	rpm		
Ship speed (stressing)	16.4	kts		
Density of material	0.271	lb/in^3	Area coefficient	0.690
Blade thickness @ 0.25R	72	mm ←	Modulus coefficient	0.110
Mean blade thickness	31	mm	Section area	25737 mm^2
Mass of blade	83	kg	Section modulus	295412.517 mm
Position of blade centre of gravity	0.51	R		
Centrifugal force	46	kN	Pitch angle @ 0.25R	45.16°
Centrifugal lever	100	mm		
Position of centre of thrust	0.7	R		
Position of centre of torque	0.66	R		
Blade number	4			
Bending stress due to thrust	23.6	MPa		
Bending stress due to torque	11.1	MPa		
Centrifugal bending stress	15.5	MPa		
Direct centrifugal stress	1.8	MPa		
Tensile stress @ 0.25R	51.9	MPa		
Allowable tensile stress	49	MPa		
New thickness @ 0.25R	74.1	mm		

(Continued)

TABLE 22.6 Typical Basic Propeller Design Calculation—cont'd

r/R	Area	SM	Area × SM	Lever	A.SM.L	Lever	A.SM.L^2
1.0000	0	0.5	0	0	0	0	0
0.8750	4735	2	9469	1	9469	1	9469
0.7500	10067	1	10067	2	20133	2	40266
0.6250	15653	2	31305	3	93915	3	281746
0.5000	20640	1	20640	4	82558	4	330233
0.3750	24464	2	48928	5	244640	5	1223202
0.2500	25923	0.5	12962	6	77770	6	466617

Crown of boss radius	0.23 R		Shaft power basis	1004 kW
Volume of blade to 0.25R	0.0110 m^3		Shaft RPM basis	300 rpm
Volume 0.25R to cob	0.0006 m^3		Blade chord @ cob	494 mm
Volume of fillets	0.0003 m^3		Blade pitch @ cob	1532 mm
Blade volume	0.0119 m^3		Boss length	368 mm
Blade weight	90	kgf	Shaft tensile strength	600 N/mm^2
Weight of blade to 0.25R	83	kgf	Fitting factor (k)	1.22
Propeller weight	1062 kgf		Tail shaft diameter	198 mm
	1.1 tonnes		cob diameter	455.4 mm

Centroid of blade beyond 0.25R	0.505 R	Shaft taper ratio	1 in 30			
MI blade beyond 0.25R about tip	0.0030 m^5	Boss posn	Fwd		Cob	Aft
MI blade to cob about tip	0.0035 m^5	r/R	0.24		0.23	0.19
MI of blade about tip	0.0035 m^5	Ext. radius	239.1		227.7	191.3 mm
MI of blade about shaft center-line	27 kg.m ^2 (Wk^2)	Int. radius	98.9		92.7	86.6 mm
Radius of gyration of blade	0.557 R	cob/tailshaft dia. ratio			2.30	
		Fwd boss dia/tail shaft dia			2.42	
		Volume of boss			0.094 m^2	
		MI of boss			11 kg.m^2 (Wk^)2	

Dry moment of inertia of propeller about shaft CL	120 kg.m^2 (Wk2)
	0.1 tonne.m^2 (Wk2)
Radius of gyration of propeller	0.340 R

made concerning the basis of the design. Figures 22.22 and 22.23 show two extreme examples of the design processes used in propeller technology.

Leaving to one side the design of propellers which are standard 'off the shelf' designs such as may be found on outboard motor boats, the design process shown in Figure 22.22 represents the most basic form of propeller design that could be considered acceptable by any competent designer. Such a design process might be expected to be applied to a small fishing boat or large workboat, where little is known of the in-flow into the propeller. It is not unknown, however, to see standard series propellers applied to much larger vessels of 100 000 tonnes deadweight and above. Such occurrences are, however, comparatively rare and more advanced design processes normally need to be used for these vessels. The design of high-speed propellers can also present a complex design problem. In the calculation of such propellers the second box, which identifies the calculation of the blade dimensions, may involve a considerable amount of chart work with standard series data: this is particularly true if unfavourable cavitation conditions are encountered.

Blade stresses should always, in the author's view, be calculated as a separate entity by the designer, using as a minimum the cantilever beam technique followed by a fatigue estimate based on the material's properties. The use of classification society minimum thicknesses should always be used as a check to see that the design satisfies these conditions. However, they should not be used as the sole design tool as they are generalized minimum standards of strength.

Since many standard series propellers are of the flat face type an increase in thickness gives an implied increase in camber which will increase the propeller blade effective pitch. After the propeller has been adjusted for strength the design needs to be analyzed for power absorption using the methods in Chapters 3 and 6 in order to derive the appropriate blade pitch distribution. During the design process the question of design tolerances needs to be addressed whatever level of design is used, otherwise significant departures between design and practice will occur.

While Figure 22.22 shows the simplest form of design method and such processes are used to design perfectly satisfactory propellers for many vessels, more complex

Propeller Design

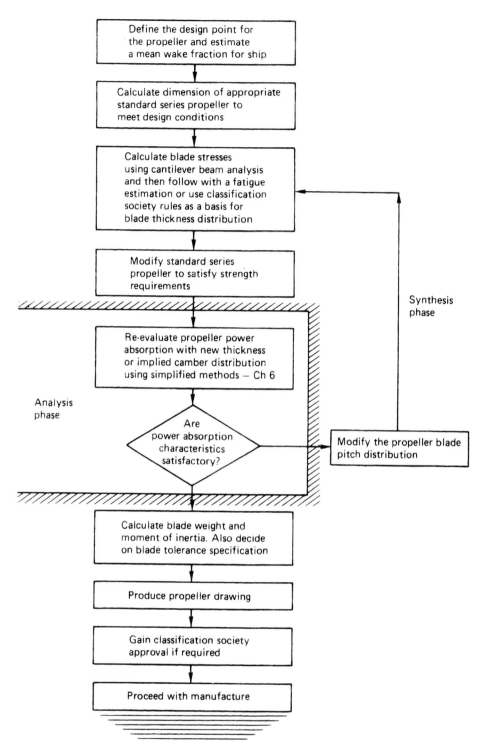

FIGURE 22.22 Example of a simplified design procedure.

design procedures become necessary when increasing constraints are placed on the design and increasing amounts of basic information are available upon which to base the design. Variants of the design process shown in Figure 22.22 normally increase in complexity when a mean circumferential wake distribution is substituted for the mean wake fraction. This then enables the propeller to become wake adapted through the use of lifting line or higher-order design methods and the analysis phase may then embody a blade element, lifting line, lifting surface or boundary element based analysis procedure for different angular positions in the propeller disc: this pre-supposes

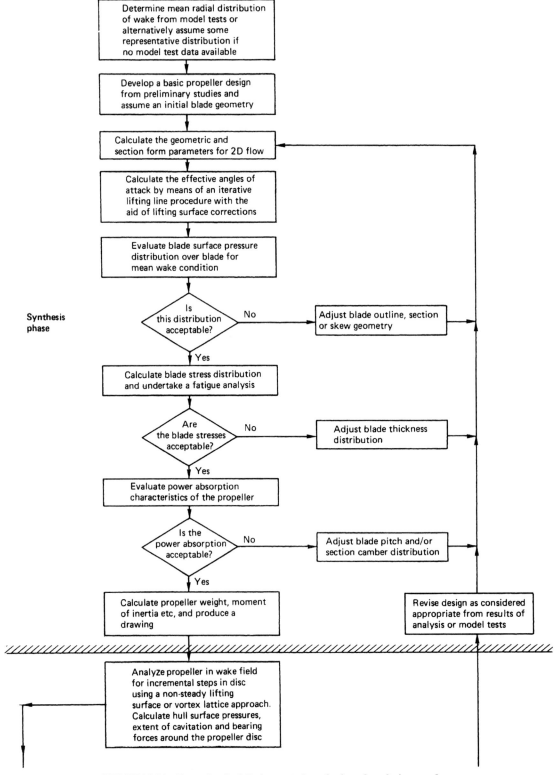

FIGURE 22.23 Example of a fully integrated synthesis and analysis procedure.

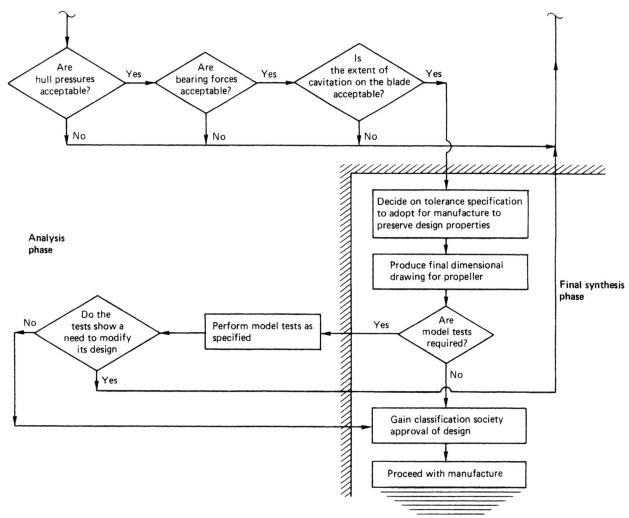

FIGURE 22.23 (*Continued*).

that model wake data is available rather than the mean radial wake distribution being estimated from the procedures discussed in Chapter 5. As the complexity of the design procedure increases, the process outlined in Figure 22.23 is approached and this embodies most of the advanced design and analysis techniques available today. Each designer, however, will use different theoretical methods and his correlation with full-scale experience will be dependent on the methods used. This underlines the reason why it may be dangerous and unjust to criticize a designer for not using the most up-to-date theoretical methods, since the extent of his theoretical to full-scale correlation database may outweigh any advantages gained by use of more up-to-date methods.

Theoretical design methods, and analysis methods too for that matter, will only take the designer so far in the design process. Knowledge is lacking in many detailed aspects of propeller design and nowhere is this truer than in defining the flow at the blade—boss interface of all propellers. In such cases careful assumptions regarding the assumed blade loading at the root have to be made in the context of the anticipated severity of the in-flow conditions: this may dictate that a zero circulation or some other condition, determined from experience, is an appropriate assumption. In either case, the actual circulation which occurs on the blade will remain unknown due to the nature of the complex three-dimensional flow regime in this region of the blade. Another classic example is the definition of the geometric and flow conditions that cause singing, although in this case a remedy is well known from normal propeller types but this has not worked in all circumstances.

It will, however, be noted that each of the design processes shown in Figures 22.22 and 22.23 contain the elements of synthesis and analysis phases shown in Figure 22.2. Much has been written on the subject of propeller design and analysis by many practitioners of the subject. The references here comprise some of this information, however,

References 11–26 contain further information specifically related to propeller design and analysis.

The traditional approach to the detailed design of propeller blades has been that the propeller blade sections are designed for the mean inflow conditions around the circumference at a set of specific radii in the propeller disc. During this process the design is then balanced against the various constraints and velocity excursions relating to that particular design. Kinnas et al. in References 29 and 30 have explored an alternative approach of optimizing the design for the actual flow conditions without the necessity of employing circumferential flow averaging processes. Their method uses a B-spline representation of the blade and determines the blade performance characteristics via second order Taylor expansions of the thrust, torque, cavitation extent and volume in the region of the solution using the MPUF-3A code. However, to converge to an optimum solution using this procedure a considerable amount of computer time is required which tends to limit the method's general applicability. To overcome these problems they developed an approach in which the optimum blade geometry that was being sought is found from a set of geometries which have been scaled from a basic geometry. In this alternative procedure the blade performance is computed from the MPUF-3A code for selected geometries within the set of geometries derived from the basic propeller and then interpolation curves are used to establish continuous analytical functions of performance. These functions are then used within an optimization procedure to establish the required final optimum blade geometry. Deng[31] presents the detail of the optimization method used in this procedure.

REFERENCES AND FURTHER READING

1. Manen JD van. *The choice of the propeller.* Marine Technology; April 1966.
2. Tachmindji AJ, Morgan WB. *The design and estimated performance of a series of supercavitating propellers.* Washington: 2nd Symp. on Navel Hydrodynamics; 1958.
3. Manen JD van. *Ergebnisse systematischer versuche mit propellern mit annahernd senkrecht slehender achse.* Jahrbuch STG; 1963.
4. Taschmindji AJ, Morgan WB, Miller ML, Hecher R. *The Design and Performance of Supercavitating Propellers*, DTMB Rep Bethesda, Washington, DC: C-807; February 1957.
5. Carlton JS. *Propeller service experience.* 7th Lips Symposium; 1989.
6. Carlton JS. *Marine diesel engine exhaust emissions when operating with variable quality fuel and under service conditions.* Trans I Mar E, IMAS 90 Conf; May 1990.
7. Gunsteren, LA van. *Design and performance of controllable pitch propellers.* Trans SNAME, New York Sect. 1972.
8. Voorde CB van de. *Effect of Inward and Outward Turning Propellers on Manoeuvring Alongside for Ships of the Frigate Class.* NSMB Report No. 68-015-BT; 1968.
9. Hawdon L, Patience G, Clayton JA. *The effect of the wake distribution on the optimum diameter of marine propellers.* Trans NECIES, NEC 100 Conf; 1984.
10. Gunsteren L.A. van, Pronk C. *Propeller design concepts.* ISP July 1973;**20**(227).
11. Burrill LC. *Considérations sur le Diamétre Optimum des Hélices*, Assoc Tech. Paris: Maritime et Aéronautique; 1955.
12. Burrill LC. *Progress in marine theory, design and research in Great Britain.* Appl Mech Convention, Trans I Mech E; 1964.
13. Hannan TE. *Strategy for Propeller Design.* Thos Reed Publ; 1971.
14. Hannan TE. *Principles and Design of the Marine Screw Propeller.*
15. Cummings RA, et al. *Highly skewed propellers.* Trans SNAME; November 1972.
16. Sinclair L, Emerson A. *Calculation, experiment and experience in merchant ship propeller design.* IMAS 73 Conf Trans I Mar E; 1973.
17. Boswell RJ, Cox GG. *Design and model evaluation of a highly skewed propeller for a cargo ship.* Mar Tech; January 1974.
18. Parsons CG. *Turning the screw.* Marine Week September 1975;**12**.
19. Oossanen P van. *Trade-Offs in the Design of Sub-Cavitating Propellers.* NSMB Publ No. 491; 1976.
20. Dyne G. *A note on the design of wake-adapted propellers.* J Ship Res December 1980;**24**(4).
21. Hawdon L, Patience G. *Propeller design for economy.* London: 3rd Int. Marine Propulsion Conf; 1981.
22. Handler JB, et al. *Large diameter propellers of reduced weight.* Trans SNAME 1982;**90**.
23. Holden K, Kvinge T. *On application of skew propellers to increase propulsion efficiency.* 5th Lips Propeller Symposium; 1983.
24. Beek T van, Verbeek R. *Design aspects of efficient marine propellers.* Jahrbuch der Schiffbantechnischen Gesellschaft; 1983.
25. Norton JA, Elliot KW. *Current practices and future trends in marine propeller design and manufacture.* Mar Tech April 1988;**25**(2).
26. Klintorp H. *Integrated design for efficient operation.* Motor Ship; January 1989.
27. Vonk KP, Terwisga Tvan, Ligtelijn JTh. *Tip rake: Improved propeller efficiency and cavitation behaviour.* London: WMTC Conf; 2006.
28. Dand J. *Improving cavitation performance with new blade sections for marine propellers.* J Int Shipbuilding Progress 2004;**51**.
29. Mishima S, Kinnas SA. *Application of a numerical optimisation technique to the design of cavitating hydrofoil sections.* J Ship Res 1997;**40**(2):93–107.
30. Kinnas SA, Griffin PE, Choi J-K, Kosal EM. *Automated design of propulsor blades for highspeed ocean vehicle applications.* Trans SNAME 1998;**106**:213–40.
31. Deng Y. *Performance database interpolation and constrained nonlinear optimisation applied to propulsor blade design.* Master's Thesis, Department of Civil Engineering, The University of Texas at Austin; 2005.

Chapter 23

Operational Problems

Chapter Outline

23.1 Performance Related Problems 459
   23.1.1 Power Absorption Problems 459
   23.1.2 Blade Erosion 461
   23.1.3 Noise and Vibration 462
23.2 Propeller Integrity Related Problems 465
   23.2.1 Blade Failure 465
   23.2.2 Previous Repair Failures 466
   23.2.3 Casting Integrity 467
23.3 Impact or Grounding 467
References and Further Reading 467

When a propeller enters service, despite the best endeavors of the designers and manufacturers, problems in performance may from time to time arise. Equally, during the service life of the ship a range of problems may also be encountered. Figure 23.1 outlines some of the more common problems that can be encountered during the lifetime of the propeller.

Figure 23.1 essentially draws the distinction between accidental damage, due to impact or grounding of the blades, and the other types of problem which relate to the performance and integrity of the propeller in the 'as designed' condition. Each of these issues needs to be treated differently and, as a consequence, invites separate consideration.

23.1 PERFORMANCE RELATED PROBLEMS

Problems related to propeller performance can in a many instances be traced to a lack of knowledge of the wake field in which the propeller is operating during the design process. When a ship has had the benefit of model testing prior to construction, a model nominal wake field is very likely to have been measured. This allows the designer to understand in a qualitative sense the characteristics of the wake field in which the propeller is to operate. As discussed in Chapters 5 and 22, the designer needs to transform the model nominal wake field into a ship effective velocity distribution before it can be used for quantitative design purposes. Although computational fluid dynamics has begun to address this problem errors may, however, develop in the definition of the effective wake field. In the case where the ship has not been model tested, the designer has less information with which to work. In these cases reliance on heuristic knowledge of other similar ships and the way they performed is demanded as well as making empirically based estimates of the type discussed in Chapter 5.

Clearly not all performance problems are traceable to lack of knowledge about the wake field. Other causes, such as poor tolerance specification, poor specification of design criteria, incorrect design and manufacture and so on may also contribute to poor performance of the propeller. Figure 23.1 identified three principal headings under performance related problems and these are considered individually.

23.1.1 Power Absorption Problems

Such problems are normally identified by observing that the engine will not produce the required power at either the NCR or MCR conditions. In such conditions the engine attains the required power at either too low or too high an engine speed and this condition will also be reflected in the engine exhaust temperatures. Indeed, if the engine speed rises too much the engine may not be able to develop the necessary rated power. Alternatively, in vessels where a torsion meter is fitted, and assuming it has been calibrated properly and maintained its calibration, the condition becomes obvious. Another class of power absorption problem is seen in twin-screw vessels, where a power imbalance may sometimes be noted between the port and starboard shaft systems.

In the first case, and assuming the vessel is new and the hull is in a clean state, it is likely that the cause will be found in the choice of pitch for the propeller. Notwithstanding this, before attempting to change the effective pitch of the propeller, the blade manufacturing tolerances should be checked against specification and also consideration should be given to ensure that the level of tolerances specified for the vessel were adequate; see Chapter 25. The effective pitch of

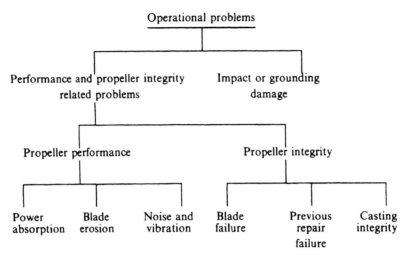

FIGURE 23.1 Common operational problems.

the propeller can be changed in one of two ways: either by reducing the diameter of the propeller, frequently termed 'cropping' the blades, or by modifying the blade section form to change the pitch of the blades. If the change required is small, then one or either of these methods will be found to be satisfactory; however, for situations requiring larger changes a combination of the two methods should be undertaken in order to preserve the efficiency of the propeller at its highest level; References 1 to 3 discuss these effects and show the type of modification that can be achieved. Figure 23.2, taken from Reference 2, shows the effects of modifying the blades by 'cropping' and pitch reduction.

In the case where the vessel is not new and a 'stiffening' of the propeller is seen to take place over time, assuming the original propeller is still in place and is clean and undamaged, the most likely cause of the problem is the roughening or fouling of the hull. Clearly in such cases the hull should be cleaned, but if the problem still persists and the engine cannot, over a docking cycle, work within the original design rpm band, then an 'easing' of the propeller pitch may prove desirable since, despite shot blasting and repainting of the hull, an old ship tends to roughen and increase in resistance.

When an imbalance between shaft powers in a twin-screw installation occurs this may be due either to one or both propellers being out of specification or to the chosen manufacturing tolerance being too relaxed and allowing a significant change of effective pitch to occur between the propellers. Although in general there are always differences between the propellers of twin-screw ships these are normally too small to cause concern; however, there are cases where one propeller tends to be at one end of the tolerance band whilst the other is at the opposite end. If this difference in power absorption is too large, then the tolerance specification needs to be tightened and the appropriate geometric changes made.

Two further examples of power absorption problems are to be found in deficiencies in the bollard pull characteristics of tugs, anchor handlers, trawlers or similar ships and in the thrust breakdown on propellers. In the first case, that of a lack of bollard pull, and assuming that the propeller design is satisfactory, the most common cause of a deficiency is that insufficient clear water has been allowed around the propeller at the bollard pull trial location. If the

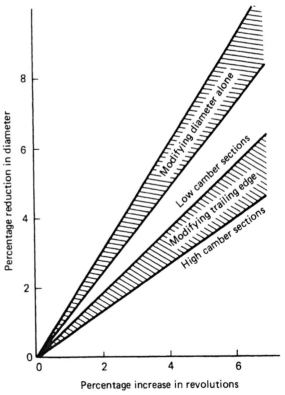

FIGURE 23.2 The effect of changes in propeller diameter and pitch on performance. *Reproduced with permission from Reference 2.*

water around the propeller is restricted in width and depth, water circulation tends to take place. This effectively increases the propeller advance coefficient, sometimes quite considerably, and causes a reduction in thrust generated by the propeller. Chapter 17 discusses the conditions desirable for conducting bollard pull trials to ensure a realistic propulsor performance. Furthermore, when considering bollard pull conditions, it is important not to confuse the terms 'propeller thrust' and 'bollard pull'.

Thrust breakdown of propellers due to cavitation is a condition rarely seen today, since it is normally caused by a grossly inadequate blade area being specified for the propeller. Figure 6.4 shows this effect in terms of a K_T versus J diagram in which at low J the K_T characteristic can be seen to fall off rapidly due to the effects of extensive cavitation. For conventional propeller types, that is not super-cavitating or surface piercing designs, this occurs when significant back sheet cavitation occurs; typically of the order of 30—40 per cent and above of the total area of the backs of the blades. In machinery terms this condition manifests itself as the shaft revolutions increasing very quickly at the higher rpm of the speed spectrum without a corresponding increase in vessel speed. The cure normally involves a redesign of the propeller; however, it is important to ensure that the cause is thrust breakdown due to cavitation and not air-drawing. In this latter case if high thrusts are being developed which induce high suction pressures when the immersion is low relative to the free surface then air-drawing from the surface is possible. The symptoms described above apply to both conditions. Air can be drawn into propellers by a variety of routes: for example, down an 'A' or 'P' bracket. To differentiate between the two causes can sometimes be difficult and requires consideration of the propeller design in terms of whether it is likely to cavitate sufficiently to cause thrust breakdown as well as observing the noise emanating from the propeller. In addition, air-drawing often leads to a *snatching* characteristic in small boats and vessels.

23.1.2 Blade Erosion

The avoidance of the harmful effects of cavitation on the marine propeller blade again hinges on being able to predict accurately the effective wake field of the vessel, since it is the wake field that forms the basis of the incident flow into the propeller and, hence, affects the distribution of loading over the propeller blade surfaces.

The gross effects of cavitation caused by the lack of provision of sufficient blade surface area are now comparatively infrequently seen, since designers can generally predict, in global terms, the amount of blade surface required for a given propulsion application. More common, however, are the localized effects of cavitation due either to variations in the local angles of attack

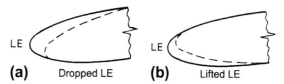

FIGURE 23.3 Traditional LE modifications to alleviate local cavitation problems.

encountered by the propeller at some point during its passage around the propeller disc or to the use of too high cambers for a particular application. Localized cavitation caused by deviations in incidence angles from those anticipated during design can frequently be alleviated by the traditional method of either 'lifting' or 'dropping' the leading edge; Figure 23.3. Alternatively, reprofiling the leading edge in terms of its radius and blending this change into the rest of the section may make the blade section more tolerant to the changes in angle of attack that it experiences. The effects of the use of too high a camber in a particular situation are more difficult to deal with, since this frequently involves attempting to generate a new section profile from the existing section form while preserving the strength integrity of that particular blade section. In such a process this inevitably leads to the loss of some blade chord length and the consequent effect on blade strength.

The incidence of root cavitation and its associated erosive effect is due largely to the difficulty of calculating the flow regime in this area and often an insufficient control of dimensional accuracy during manufacture in this region of the propeller. Such problems when encountered can be difficult to solve. Whilst a certain amount can be done to alleviate a root cavitation erosion problem by section modification, interference with the mechanical strength of the propeller blade in this region is always of concern and is, therefore, uppermost in determining the extent of the modification that can be implemented. Notwithstanding this, the intractability of certain root cavitation problems has led designers, on occasions, to the somewhat aggressive measure of drilling comparatively large holes in the blade root, from the pressure surface through to the suction face of the blade, in an attempt to alleviate the problem (Figure 19.10). Such measures, however, are not to be recommended except as a very last resort.

Much can be done at the design stage to alleviate potential cavitation problems in the blade root area by making the correct choice of hub profile, since comparatively small changes from mildly convergent hub forms to a divergent hub form will have a significant effect on the resulting root cavitation inception properties of the blade in this region. Van Gunsteren and Pronk[4] outlined this effect some years ago, but it is one that is often, in the author's experience, found to be ignored and is the source of many root cavitation problems. Figure 22.18 shows the changes caused by the use of convergent, divergent and parallel

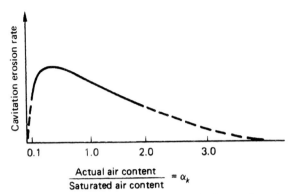

FIGURE 23.4 Effect of air content on cavitation erosion rate.

forms. In cases where a strongly convergent hull form is used, this can on occasion lead to a very strong root vortex being formed which collapses on the rudder. This, in turn, may lead to erosion on the rudder and this effect has been noted on vessels as diverse in size and duty as container ships and pilot cutters. The cure for this is often to change the form of the cone on the propeller to either a divergent or parallel form as shown in Figure 22.18.

An effective treatment of root cavitation problems can sometimes be achieved by means of injecting air into the root sections of the propeller blade at a station just immediately ahead of the propeller. Figure 23.4 illustrates, in generalized terms, how the effect of air content in the water influences the cavitation erosion rate. The form of the curve can be explained on the basis that when no dissolved air is present in the water or boundary crevices the tensile strength of the fluid is large and therefore inhibits the inception of cavitation. However, as air is introduced, this provides a basis for nuclei to form which in turn will lead to greater levels of cavitation being experienced until an air content is reached in which further nuclei seeding does not materially increase the cavitation development. Beyond this point, which lies somewhere in the range $0.1 < \alpha_k < 0.8$, if the amount of air introduced into the system is increased, then the presence of this excess air will essentially have a cushioning effect on the rapid collapse of the cavitation bubbles which would otherwise lead to the erosion mechanism: both with respect to the formation of pressure waves during the rapid bubble collapse and microjets from the collapsing bubbles directed at the blade surface. If the air content is increased significantly beyond saturation, as shown in Figure 23.4, then the erosion rate has been observed by various experimenters to reduce significantly. In the context of cavitation erosion in the propeller blade root, the author has used this technique with considerable effect on a number of high-speed vessels. However, care needs to be exercised, particularly with the smaller propellers, in choosing the amount of air for the particular application so as to prevent a fall-off in thrust performance of the propeller by effectively reducing the density of the fluid. Notwithstanding this, by the correct choice of air mass flow and injecting it in the correct place a significant erosion problem, for example of the order of 4–5 mm incurred over a period of some twelve hours, has been reduced to zero over a similar trial period by deploying the air injection technique.

Another method which is frequently debated as a means to protect a blade from erosion is the use of protective epoxy type coatings. Opinion is divided as to the usefulness of this approach; however, several new and improved materials are coming on to the market. In order that a coating has a fair chance of survival in the hostile environment on the propeller surface it should be applied to the propeller under strict conditions of cleanliness and environmental control: this implies at the very least enclosing it within an 'environmentally controlled' tent in a dock bottom, but most preferably in a workshop. When undertaking this type of coating practice care should be exercised in preserving the coating intact during the operational life so as to avoid the possibility of an electrical cell developing with the consequent effects of propeller blade corrosion occurring.

23.1.3 Noise and Vibration

The vibratory behavior of a ship in general takes one of two forms: it may be either resonant or forced in character. In the case of a resonant behavior of some part of the ship structure this can manifest itself on either a single component, such as the vibration of a bridge wing or an appendage, or of a major structure such as the entire deckhouse or the superstructure. As a consequence, for resonant vibration problems there are two principal alternatives to affect a cure. First, the resonant frequency of the offending component can be changed by structural modification and this is generally easier and a less costly option in the case of smaller components. Alternatively, the propeller blade number can be changed to alter the blade rate frequency or some multiple of it. This latter approach is usually the most convenient option where resonances at frequencies of nZ ($n = 1, 2, 3, \ldots$) in large structural components are encountered or in certain types of torsional vibration problems. Nevertheless, when changing the number of blades in an attempt to solve resonant vibration problems, while the effects of blade rate excitation are predictable by calculation for the natural frequency characteristics of the particular structural member and are normally of primary concern, the magnitude of the propeller-induced excitation at blade rate harmonics will also change — not always downwards. Care, therefore, needs to be taken to ensure that this will not cause further problems of either a forced or a resonant nature.

In the case of forced vibration, this, from a propeller viewpoint, is often caused by the harmonic pressures

Operational Problems

generated by the variations in the cavitation dynamics on the propeller blades as the blades rotate through the propeller disc. In Chapter 11 it was shown that the pressure (p_c) induced at some distance from a fluctuating cavity volume is related by the following function:

$$p_c \propto \left(\frac{\partial^2 V}{\partial t^2}\right)$$

where V is the cavity volume and t is time. As a consequence, if a forced vibration problem of this type is to be attacked at source, then a method must be found to reduce the cavity volume and the rate of its structural variation. In practical terms this means a change either to the blade geometry, which in turn frequently implies a change to the skew and radial load distributions along the blade, or to the in-flow conditions into the propeller.

In the latter case this usually implies the fitting of some appendages to the hull, which may take a variety of forms, in order to control a known or anticipated undesirable feature of the wake field. Typical examples of these appendages were shown in Figure 13.2. The particular device used depends on the type of wake feature which needs attention: for example in Figure 23.5, device (a) is normally used to control bilge vortex formation while that of (b) attempts to modify wake peak of a highly 'V' formed hull. The fitting of such devices as these needs considerable skill and should not be attempted on an ad hoc basis unless one is prepared to accept a high risk of failure. As a consequence, their choice and fitting needs a considerable reliance on past experience coupled with the results of model tests and computational studies. The model test results are, however, only a guide for the designer because the ship model is run at Froude identity and hence considerable Reynolds dissimilarity exists in the region where the particular device is to be located. While recognizing these problems, it must be noted that there are many cases where devices of the types shown in Figure 23.5 have been used with success without incurring significant speed penalties or fatigue fractures.

The alternative approach to a forced vibration problem of this type is to modify the propeller blade in such a way so as to relieve the particular condition which is being experienced by the vessel. The designer's choices in this case are many; for example, vary the radial load distribution along the blade, change the skew of the blades, increase the blade chord lengths or adjust the relationship governing the proportion of the section lift generated from angle of attack and section camber. Generally, the use of increasing amounts of skew together with the consequent changes in the other parameters alleviates a hull pressure problem but in the case of bearing forces, the skew distribution should be matched to the harmonic content of the wake field if undesirable results are to be avoided. The particular technique, or more frequently combination of techniques, used depends on which of the many types of cavitation-related vibration or noise problem requires solution.

The radial distribution of loading near the blade tip has an important influence on the strength of the tip vortex, as discussed in Chapter 9. Consequently, it is frequently desirable to limit the rate of change of load in the tip region of the propeller as shown in Figure 19.8. The penalty for doing this, however, can be a loss in propulsive efficiency, because the design has then deviated from the optimum radial loading. The strength of the tip vortex needs to be carefully controlled since the collapse of this vortex can, if the correct circumstances prevail, give rise to excessive noise and, in some circumstances, high levels of vibration in the aftbody of the ship due to the pressure waves of the cavitation collapse mechanism being transmitted through the water and onto the hull surface. The control of the strength of this vortex can most realistically be achieved by attention to the radial distribution of blade loading near the tip. However, should the phenomenon of vortex—sheet cavitations interaction occur, as discussed in Chapter 9, then this can be a source of significant excitation on the hull. With highly skewed propellers, particularly at off-design operating conditions, vortex interactions can take place. Typically, these can occur between the tip and leading edge vortices and may give rise to significant broadband excitation of the ship's structure. Furthermore, the presence of a strong tip vortex which impinges on the rudder has been known in many cases to cause significant cavitation erosion on either the rudder or the rudder horn. However, much still needs to be learnt about the behavior of the tip vortex in terms of its prediction from theoretical methods and its scaling from model tests.

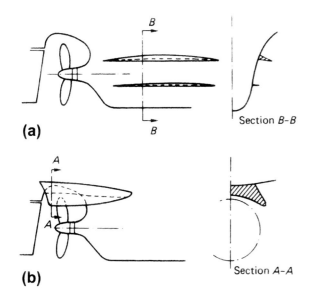

FIGURE 23.5 Fin arrangements commonly used in flow correction problems: (a) type of fin normally associated with 'U' form hulls and (b) type of fin normally associated with 'V' form hulls.

The highly skewed propeller has been particularly successful in overcoming certain classes of vibration and noise problem in both its 'biased' and 'balanced' forms, although the balanced blade form has become pre-eminent in recent times. Bjorheden[5] discusses their use, particularly with reference to controllable pitch propeller applications, and Figure 23.6(a) shows one example of the reduction in first-and second-order hull pressure induced vibration on a Ro/Ro vessel resulting from the change to a highly skewed form from a conventional design. Also from this figure it can be seen that in this particular case the third and fourth harmonics increased slightly and this underlines the importance of acknowledging, whilst not being able to predict theoretically, that changes to the higher harmonics will inevitably occur by changing the propeller form. The alternative figure, Figure 23.6(b), taken from Carlton and Bantham,[6] shows the use of a highly skewed propeller in reducing the axial vibratory characteristics on a fishing trawler. In this context it must be recalled, from Chapter 11, that propeller–ship interaction is manifested in terms of both hull surface pressures and mechanical excitations of the line shafting and its supports. The highly skewed propeller form is particularly useful in solving problems where the cavitation growth and collapse rate and cavity structure are considered to be the cause of the problem. Set against these advantages are a tendency towards increased manufacturing costs and the advisability of undertaking wake field tests if the design is to be properly optimized: this latter aspect is, of course, a general point, but is particularly true in the context of highly skewed propellers.

Detailed blade geometry changes, other than the skew or radial load distribution, are normally used where cavity structural changes or extent are required to be made. These can be particularly effective in many cases and they can frequently be carried out on an existing propeller or, alternatively, be incorporated into a new propeller of similar generic form. The use of analysis procedures based on unsteady lifting surface or boundary element methods is of considerable assistance in determining the effect of changes in blade geometry and in-flow on first and second blade order excitation frequencies. At higher orders reliance has to be made on the results of experimental cavitation studies, although here questions of the adequacy of the flow field simulation, cavitation scaling and model geometry need to be carefully considered. It is insufficient to model only the axial flow field.

A particular form of cavitation which can, on occasions, be troublesome in ship vibration terms is Propeller–Hull Vortex (PHV) cavitation. The formation of this type of cavitation was discussed in Chapter 9, as were the conditions favorable to its formation. In those cases where PHV cavitation occurs a small amount of erosion can often, but not always, be observed on the hull plating in the region above the propeller and the vibration signature will be intermittent in character, as seen in Figure 23.7. The noise

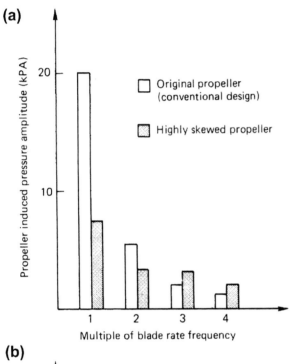
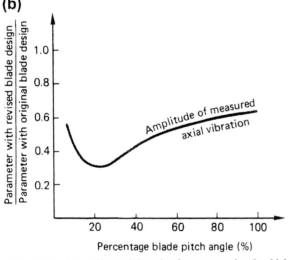

FIGURE 23.6 Some effects of changing from conventional to highly skewed propeller designs: (a) propeller-induced hull pressures recorded on a Ro/Ro vessel and (b) influence of highly skewed propeller on axial shaft vibration.

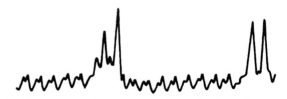

FIGURE 23.7 Typical hull pressure fluctuation indicating possible presence of PHV cavitation.

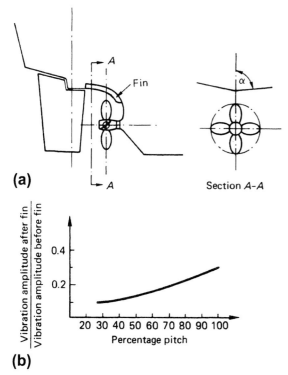

FIGURE 23.8 PHV cavitation problems and their solution.

generated, again of an intermittent nature, sounds much like a series of single sharp blows with a scaling hammer on the hull surface above the propeller. The cure for this type of cavitation is simple and effective for a non-ducted propeller: it comprises the fitting of a single vertical fin above the propeller as shown in Figure 23.8(a). This fin prevents the formation of the vortex motion necessary to the formation of the cavity discussed in Chapter 9. By way of example, the effects of fitting such a fin on a coaster are shown in Figure 23.8(b) from which it can be seen that a marked reduction in vibration level can be observed.

In the case of a ducted propulsion system it is suggested (Reference 7) that the formation of PHV cavitation can be prevented by fitting an appendage between the hull and the duct which is aimed at accelerating the flow into the upper part of the duct.

Propeller noise for the greater majority of merchant ships is intimately connected with the cavitation behavior of the propellers and as such the noise control problem, to some extent, reduces to a cavitation control problem. In dealing with these problems it is important, however, to differentiate between the sympathetic 'chattering' of loose fittings to vibration and the true level of noise originating from the propeller or other machinery components. In the case of research vessels, for example, or in many naval applications where noise emissions interfere with the operation of the ship then it is necessary to consider in detail the structure of the flow around the propeller blade sections, see Chapter 10; notwithstanding the unknowns concerning the nature of the inflow to the propeller at full scale.

Propeller singing and its cure was discussed in Chapter 21. In practical terms it manifests itself as a periodic noise which ranges from a low-frequency grunt to a high-pitch periodic tonal noise. The low-frequency *grunt* tends to be associated with larger ships with slower turning propellers while the higher tonal noises arise from smaller high-speed propellers.

23.2 PROPELLER INTEGRITY RELATED PROBLEMS

In Figure 23.1 three areas of propeller integrity problems were identified, although in reality these often interact. However, for convenience of discussion these will be treated separately.

23.2.1 Blade Failure

The over-stressing of blades in relation to the generally expected properties of materials at the design stage is an extremely rare occurrence in merchant vessels which have been designed to meet classification society requirements. These rules govern the strength requirements of marine propellers in relation to the absorption of the full machinery power and any special operational regimes that the vessel is required to undertake.

Depending upon the integrity of the casting and the size and distribution of defects within the casting, the material properties vary continuously over the surface and throughout the blade. The factors of safety incorporated in the design procedure attempt to take this into account; however, the defect geometry, location and proximity to other defects cause stress raisers which can induce a propagating fatigue crack in the blade. In the majority of cases a propeller blade fails by fatigue action, a typical example of which is shown in Figure 23.9. In the majority of these cases the *beach marks* seen in Figure 23.9 are clearly visible and represent points of crack arrest during the fatigue crack growth. When looking at a failure the Stage I area around the defect is normally visible with the Stage II area containing the *beach marks* forming the major part of the failure surface. The final rupture area, Stage III, caused by mechanical overload of the material forms a band around the edge of the fatigue failure morphology. By undertaking simple fracture mechanics estimations in relation to a failed propeller's lifetime, it can be deduced that the crack spends about 90 per cent of its life in Stage I growing to a size where Stage II propagation according to the Paris law can take place. Therefore, it is unlikely that a crack will be observed by inspection other than when it is in its Stage I phase and is small. Stage III is effectively an

FIGURE 23.9 Typical fatigue failure of a propeller blade.

instantaneous failure. When a blade fails in this way, since the failure is normally between $0.6R$ and the root, the propeller, or blade in the case of a controllable pitch propeller, is unsuitable for repair. In these cases the spare propeller should be fitted as soon as possible, and in the intervening time the vessel should be run at reduced speed. The required reduction in speed can be determined practically on the vessel at the time of failure but that empirical determination should be checked by calculation of the out-of-balance forces in relation to the hydrodynamic and mechanical loading of the stern bearing. If, as is sometimes the case, the spare propeller is in another part of the world it may be necessary to run for some time in this failed condition. When this situation occurs and a significant portion of the blade is lost, then the opposite blade, in the case of an even-bladed propeller, should be suitably cropped and for an odd-bladed propeller the opposite pair of blades partially cropped. This cropping action, although altering the power—rpm relationship of the propeller and increasing the thrust loading per blade and hence the tendency towards cavitation, helps to protect the stern tube bearing from damage from the out-of-balance force generated by the failure. If a spare propeller does not exist, then the propeller should be approximately balanced in the manner described above until a new propeller can be produced. For lesser damage, in which smaller parts of the blade are lost, drastic cropping actions are normally unnecessary since the propeller may be able to be repaired; nevertheless, the effect of the damage should always be considered in relation to its effect on the lubrication film in the stern tube bearing.

If a propeller fails in fatigue, the underlying cause should always be sought. This is because the reason for the failure will have an influence on whether redesign is necessary or whether a repeat propeller can be ordered to the same design.

In Chapter 19, the effect of backing or emergency stopping on highly skewed, fixed pitch propellers was discussed. In a limited number of cases this leads to bending, which is normally found in the region of the blade shown by Figure 23.10. While in many cases the blade could be straightened, the plastic behavior would recur the next time the offending astern maneuver was undertaken. As a consequence blade redesign is normally necessary to either thicken the blade or adjust the blade shape, perhaps a combination of both.

23.2.2 Previous Repair Failures

As discussed in Chapter 26 propeller repairs need to be conducted strictly in accordance with the manufacturer's recommendations and classification society requirements. Otherwise failure of the repair will very likely result and other undesirable features such as stress corrosion cracking may occur. In these circumstances the failure of a local

Operational Problems

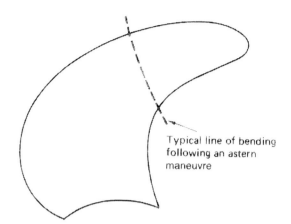

FIGURE 23.10 Typical location of bending following an astern maneuver with a highly skewed, fixed pitch propeller.

repair can act as the origin of a blade fatigue failure and actually cause failure in a very short space of time.

23.2.3 Casting Integrity

It is practically impossible to produce a propeller casting without defects and as a consequence potential sites for fatigue crack initiation. In the majority of cases the defects are of no consequence to the long-term integrity of the propeller. The question of defining an acceptable defect size has occupied several research workers in recent years, but as yet no generally accepted criteria has evolved.

In several cases it is possible and perfectly valid to repair a casting by welding; however, as in the case of service repairs the manufacturer's recommendations and the rules of the various classification societies must be rigidly adhered to when undertaking this type of repair.

With castings it is unlikely that their inherent defect state will deteriorate in service, except in the case of the joining up of closely packed defect sites under the action of a tensile stress field. Hence the casting integrity is defined at the time of manufacture. Where a number of relatively small defect sites occur in close proximity and it is considered inadvisable to leave them in the casting, then given due consideration to the cavitation and strength constraints of the propeller, it may be possible to gently fair them out by grinding the blade surface. Such action, however, has to be carried out advisedly and with care. In Chapter 20 some discussion is offered on reasons why such situations might occur.

23.3 IMPACT OR GROUNDING

Propellers by virtue of their position and mode of operation are likely to suffer impact or grounding damage during their life. This sometimes results in a complete or partial blade failure due to overload or, more likely, in blade bending or the tearing of small pieces from the blade edges. In by far the majority of cases these damages can be rectified by repair, again with the caveat of the use of repair specialists and under the jurisdictions of the appropriate classification society.

REFERENCES AND FURTHER READING

1. *Modifying the propeller for optimum efficiency.* Shipping World and Shipbuilder; April 1972.
2. Patience G. *Modifying propeller characteristics for better efficiency in ageing ships.* The Motor Ship; January 1973.
3. *Edge Modification of Propeller.* Kobe Steel Ltd; 1979.
4. Gunsteren LA van, Pronk, C. *Propeller design concepts.* 2nd Lips Symp; 1973.
5. Bjorheden O. *Vibration performance of highly skewed CP propellers.* Symp on Propeller Induced Vibration, Trans RINA; 1979.
6. Carlton JS, Bantham I. *Full scale experience relating to the propeller and its environment.* Propellers '78 Symp., Trans. SNAME; 1978.
7. Kooij J Van der, Berg, W Van den. *Influence of Hull Inclination and Hull-Duct Clearance on Performance*, Cavitation and Hull Excitation of a Ducted Propeller: Part II. NSMB, the Netherlands.

Chapter 24

Service Performance and Analysis

Chapter Outline
- 24.1 Effects of Weather — 469
- 24.2 Hull Roughness and Fouling — 469
- 24.3 Hull Drag Reduction — 478
- 24.4 Propeller Roughness and Fouling — 478
- 24.5 Generalized Equations for the Roughness-Induced Power Penalties in Ship Operation — 482
- 24.6 Monitoring of Ship Performance — 485
- References and Further Reading — 492

In general the performance of a ship in service is different from that obtained on trial. Apart from differences in loading conditions and for which due correction should be made, these differences arise principally from engine deterioration, the weather, fouling and surface deterioration of the hull and propeller.

The subject of service performance quite naturally, therefore, can be divided into four component parts for discussion purposes as follows:

1. Effects of weather — both sea and wind.
2. Hull roughness and fouling.
3. Propeller roughness and fouling.
4. The monitoring of ship performance.

As such, the discussion in this chapter will essentially fall into these four categories.

24.1 EFFECTS OF WEATHER

The influence of the weather, both in terms of wind and sea conditions, is an important factor in ship performance analysis. The analytical aspects of the prediction of the effects of wind and sea state were discussed in Chapter 12, and need not be reiterated here. In the case of the service data returned from the ship for analysis purposes it is insufficient to simply record wind speed and sea state according to the Beaufort scale. With regard to wind, it is important to record both its speed and direction, since both of these parameters clearly influence the drag forces experienced by the vessel. The recording of sea conditions is somewhat more complex, since most commonly the actual sea state will contain both a swell component and a local surface disturbance and these two attributes are not related. For example, if a sea is not fully developed, then the apparent Beaufort number will not be representative of the conditions actually prevailing at the time. In the case of the underlying swell components these may have had their origin in meteorological disturbances many miles distant, whereas surface waves are more likely to reflect local disturbances and geographical features in the vicinity of the ship. Nevertheless, both swell and surface disturbance effects and their direction relative to the ship's heading need to be taken into account if a realistic evaluation is to be made of the weather effects in the analysis procedures. In making these comments it is fully recognized that, in the absence of instrumented data, the resulting data will contain a subjective observational error bound on the part of the deck officer. However, an experienced estimate of the conditions is essential to good analysis practice.

24.2 HULL ROUGHNESS AND FOULING

The surface texture or hull roughness of a vessel is a continuously changing parameter which has a comparatively significant effect on ship performance. This effect derives from the way in which the roughness and texture of the hull surface influences the boundary layer and its growth over the hull. Hence, the effect of hull roughness can be considered as an addition to the frictional component of resistance of the hull. By way of example, for new ships Table 24.1 shows typical comparative proportions of frictional resistance (C_F) to total resistance (C_T) at design speed for a series of ship types. From this table it is clearly seen that the frictional components (C_F) play a large role for almost all types of vessel but the larger full-form vessels have the greatest proportion of their resistance accounted for by the frictional components.

The roughness of a hull can be considered to be the sum of two separate components as follows:

Hull surface roughness = Permanent roughness
+ Temporary roughness

TABLE 24.1 Typical Proportions of Frictional to Total Resistance for a Range of Ship Types

Ship Type		C_F/C_T
ULCC – 516 893 dwt	(loaded)	0.85
Crude carrier – 140 803 dwt	(loaded)	0.78
	(ballast)	0.63
Product tanker – 50801 dwt	(loaded)	0.67
Refrigerated cargo ship – 8500 dwt		0.53
Container ship – 37 000 dwt		0.62
Ro/Ro ferry		0.55
Cruise liner		0.66
Offshore tug supply vessel		0.38

in which the permanent roughness refers to the amount of unevenness in the steel plates and the temporary roughness is that caused by the amount and composition of marine fouling.

Permanent roughness derives from the topographical condition of the hull plates and the condition of the painted surface directly due either to the application or the drying of the paint on the hull. Admittedly, when considering the full implication of the word permanent, this latter component of permanent roughness can to some extent be rectified by the shot blasting of the hull surface to remove the accumulation of previous paint deposits so that the underlying steel surface topology is restored to that of the underlying hull material. The condition of the hull plates embraces the bowing of the ship's plates, weld seams and the condition of the steel surface. The bowing of the plates, colloquially frequently referred to as the *hungry horse* appearance, has a comparatively small effect on resistance: generally not greater than about 1 per cent. Similarly, the welded seams also have a small contribution: for example, a VLCC or container ships might incur a penalty of the order 3/4 per cent and so a decision has to be made, dependent upon the absolute magnitude of the penalty and the cost of grinding, whether it may be cost effective to remove these by grinding the surface of the weld seams. By far the greatest influence on resistance is to be found in the local surface topography of the steel plates. This topography is governed by a wide range of variables:

1. Corrosion.
2. Mechanical damage.
3. Deterioration of the paint film.
4. A build-up of old coatings.
5. Rough coating caused by poor application.
6. Cold flow resulting from too short a drying time prior to immersion.
7. Scoring of the paint film resulting from scrubbing to remove fouling.
8. Poor cleaning prior to repainting, etc.

Consequently, it can be seen that the permanent roughness, which is permanent in the sense of providing the basic surface after building or dry-docking during service, cannot be eliminated by subsequent paint coatings. Therefore, to enhance the situation in terms of local surface topography, complete removal of the old coating is necessary to restore the hull surface.

In contrast, temporary roughness, which refers to the temporal changes in the hull surface during a given time period, can be removed or reduced by the removal of the fouling organisms or subsequent coating treatment. It is caused in a variety of different ways: for example, the porosity of leached-out anti-fouling; the flaking of the current coating caused by internal stresses; corrosion caused by the complete breakdown of the coating system and by marine fouling. While permanent roughness can be responsible for an annual increment of, say, 30–60 μm in roughness perhaps, the effects of marine fouling on performance can be considerably more and may be responsible, given the right circumstances, for 30–40 per cent increases in fuel consumption in a relatively short time if due anti-fouling provision has not been made.

The sequence of marine fouling commences with slime, comprising bacteria and diatoms, which then progresses to algae and in turn on to animal foulers such as barnacles, culminating in the climax community. Within this cycle Christie[1] describes the colonization by marine bacteria on a non-toxic surface as being immediate, their numbers reaching several hundred in a few minutes, several thousand within a few hours and several millions within two to three days. Diatoms tend to appear within the first two or three days and then grow rapidly, reaching peak numbers within the first fortnight. Depending on the prevailing local conditions this early diatom growth may be overtaken by fouling algae.

The mixture of bacteria, diatoms and algae in this early stage of surface colonization is recognized as the primary slime film. The particular fouling community which will eventually establish itself on the surface is known as the climax community and is particularly dependent on the localized environment. In conditions of good illumination this community may be dominated by green algae, or by barnacles or mussels: these forms are often observed on static structures such as pier piles or drilling rigs.

The vast numbers and diversity of organisms comprising the primary slime film results in the inevitable formation of slime on every submerged marine surface, whether it is toxic or non-toxic. The adaptability of the bacteria is such that these organisms are found in nature colonizing habitats varying in temperature from below 0° to 75°C. The adaptability of diatoms is similarly impressive; they can be found in all aquatic environments from fresh water to hyper-saline conditions and are even found growing on the undersides of ice floes. These life cycles and the adaptability of the various organisms combine to produce a particularly difficult control problem.

Severe difficulty of fouling control is not, however, restricted to micro-fouling. For example, it has been seen in recent times that the emergence of oceanic, stalked barnacles is a serious problem fouling VLCCs working between the Persian Gulf and Northern Europe. This group of barnacles is distinguished from the more familiar 'acorn' barnacles in both habitat and structure.

Whereas acorn barnacles are found in coastal waters, characteristically attached directly to fixed objects such as rocks, buoys, ships, pilings and sometimes to other organisms, such as crabs, lobsters and shellfish, stalked barnacles are usually found far from land attached to flotsam or to larger animals such as whales, turtles and sea snakes by means of a long, fleshy stalk. The species, the most important of which is the Conchoderma, is recognized as a problem for large slow-moving vessels, and considerable research dealing with their life cycle and habits has been undertaken. The conclusions of this work indicate that VLCCs may become fouled with Conchoderma while under way in the open ocean. The results of the shipboard studies suggest that vessels traveling between the Gulf and Northern Europe are most likely to become fouled in the Atlantic Ocean between the Canary Islands and South Africa and particularly in an area between 17°S and 34°S. Adult Conchoderma, however, have been reported to be in every ocean in the world, and so there are no areas of warm ocean where vessels can be considered immune from attack.

The fouling of underwater surfaces is clearly dependent on a variety of parameters such as ship type, speed, trading pattern, fouling pattern, dry-dock interval, basic roughness and so on. To assist in quantifying some of these characteristics Evans and Svensen[2] produced a general classification of ports with respect to their fouling or cleaning characteristics; Table 24.2 reproduces this classification.

A number of solutions of varying effectiveness were developed from the early days of sailing ships to combat the problems of hull fouling. These embraced resin and pitch coatings to lime, arsenic and mercury compounds and, until it was banned in the United States of America, dichlorodithenyl trichloroethane (DDT). More recently the organotin compounds were developed during the 1960s to satisfy an industry need to alleviate the hull fouling characteristics, exemplified by Figure 24.21, in which the ship's analysis effective wake fraction exhibited a saw-tooth characteristic over successive dry-docking intervals. So successful were these developments that by the 1970s most

TABLE 24.2 Port Classification According to Reference 2

Clean Ports	Fouling Ports		Cleaning Ports	
	Light	Heavy	Non-Scouring	Scouring
Most UK ports	Alexandria	Freetown	Bremen	Calcutta
Auckland	Bombay	Macassar	Brisbane	Shanghai
Cape Town	Colombo	Mauritius	Buenos Aires	Yangtze Ports
Chittagong	Madras	Rio de Janeiro	E. London	
Halifax	Mombasa	Scurabaya	Hamburg	
Melbourne	Negapatam	Lagos	Hudson Ports	
Valparaiso	Karadii		La Plata	
Wellington	Pernambuco		St Lawrence Ports	
Sydney*	Santos		Manchester	
	Singapore			
	Suez			
	Tuticorin			
	Yokohama			

*Variable conditions.

seagoing ships used tributylin (TBT) based hull coatings in order to suppress the hull fouling. Figure 24.22 shows a typical example of the results obtained.

Subsequently, environmental studies began to show that relatively high concentrations of TBT were apparent in the water and sediments and, moreover, these had the effect of killing sealife, other than those attached to the hull surfaces, together with the potential to enter the food chain. Indeed, it was also noticed that residual TBT had unwelcome immune response, genetic and neurotoxic effects as well as giving rise to sex change characteristics in some 72 marine species. Based on this evidence in November 1999 the plenary session of the 21st Assembly of the IMO adopted resolution A.895 (21), developed during MEPC 42, which urged the MEPC 'to work towards the expeditious development of a global legally-binding instrument to address the harmful effects of anti-fouling systems used on ships'. Following the subsequent work of the MEPC in October 2001 the IMO adopted a new *International Convention on the Control of Harmful Anti-fouling Systems on Ships*. This both prohibited the use of harmful organotin compounds in anti-fouling paints and established a mechanism to prevent future uses of other harmful substances in ship anti-fouling systems.

When considering the properties of a good anti-fouling system, since their use is an important economic issue for ship operators, it is probable that the following set of properties constitute such a system:

1. Have broad spectrum activity.
2. Have low mammalian toxicity.
3. Have no bio-accumulation in the food chain.
4. Possess low water solubility.
5. Be not persistent in the environment.
6. Be compatible with the paint coating raw materials.
7. Be able to exhibit a favorable price-performance characteristic.

In order to develop a TBT-free anti-fouling system the coating might comprise a seawater soluble matrix which contains biologically active ingredients which are tin-free and dispersed throughout the matrix, but not necessarily chemically bonded to it. The principle of operation, therefore, is that as the matrix dissolves in the sea water successive layers of biocide are revealed and leached out at a controlled rate, thereby enabling a predictable performance to be achieved as outlined in Figure 24.1.

A number of tin-free anti-fouling alternatives have been developed.[25,26] These include natural biocides which are either considered harmless to the natural environment or substances produced in nature and prevent or hinder the fouling process. These might include active metabolites such as ceratinamine and mauritiamine. A common alternative to the organotin biocide is copper in the form of cuprous oxide which is added to the paint matrix. Copper, by itself, is susceptible to diatom and algae fouling and therefore has to be supplemented with other compounds within the coating matrix. Additionally, the life of conventional copper coatings is relatively short, typically less than a docking interval, but self-polishing copper-based coating systems have been produced which have lives more in keeping with the traditional hull painting intervals.

It is also known that enzymes can stop the ability of bacteria to stick to surfaces and hydrophilic coatings, where the organisms cannot maintain a grip on the surface, may also prove useful. In the case of the silicone-based elastomeric coatings, which are not thought to have toxic effects, these prevent marine life from adhering to the hull surface by virtue of the coating's surface properties, provided that ship speed is maintained above a critical value, typically in the region of 17–18 knots and the ship does not spend long periods stationary in port. Moreover, some advantage in terms of a reduced turbulent flow wall shear stress is also possible.

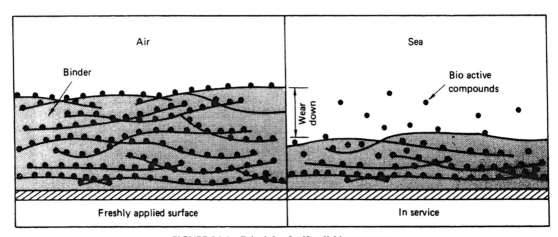

FIGURE 24.1 Principle of self-polishing process.

Among other options are the creation of an electrical charge between the hull and sea water and also prickly coating systems have been suggested. However, there are some potential disadvantages with these options which would need to be satisfactorily overcome; such as the increased risk of corrosion and higher energy consumption in the first case and in the latter option increased frictional resistance.

The wear-off rate or polishing rate of anti-foulings is not always completely uniform, since it depends on both the turbulence structure of the flow and the local friction coefficient. The flow structure and turbulence intensities and distribution within the boundary layer change with increasing ship speed, which gives a thinner lamina sublayer, and consequently a hydrodynamically rougher surface, since more of the roughness peaks penetrate the sublayer at higher ship speeds. A further consequence of the reduced lamina sublayer at high speed is that the diffusion length for the chemically active ingredients is shorter, which leads to a faster chemical reaction, and therefore faster renewal, at the surface. In addition to the ship speed considerations, the hull permanent roughness is also of importance. While this will not, in general, affect the polishing rate of the coating, it will be found that in the region of the peaks the anti-fouling will polish through more quickly since the coating surface will be worked harder by the increased shear stresses and turbulent vortices. Figure 24.2 shows this effect in schematic form. Additionally, the average polishing rate for the coating is likely to be the same for a rough or smooth hull; however, the standard deviation on the distribution curve for polishing rate will give a much bigger spread for rough hulls. Figure 24.3 illustrates this effect by showing the results of model experiments (Reference 3) for both a smooth and rough surface, 50 and 500 μm, respectively. Consequently, it will be seen that the paint coating needs to be matched carefully to the operating and general conditions of the vessel.

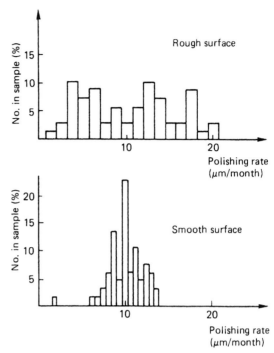

FIGURE 24.3 Influence of roughness on polishing rate. *Reference 3.*

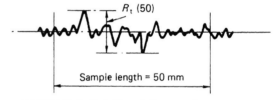

FIGURE 24.4 Definition of $R_{t(50)}$ roughness measure.

The standard measure of hull roughness that has been adopted within the marine industry is $R_{t(50)}$. This is a measure of the maximum peak-to-valley height over 50 mm lengths of the hull surface, as shown in Figure 24.4.

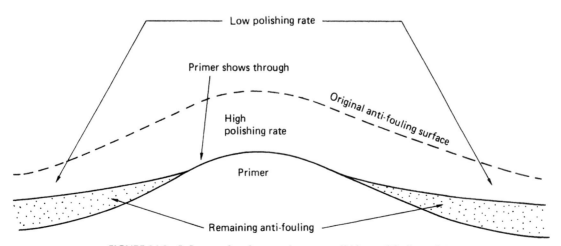

FIGURE 24.2 Influence of surface roughness on polishing anti-fouling paints.

When undertaking a survey of a hull, several values of $R_{t(50)}$ will be determined at and around a particular location on the hull and these are combined to give a mean hull roughness (MHR) at that location defined by

$$\text{MHR} = \frac{1}{n}\sum_{i=1}^{n} h_i \qquad (24.1)$$

where h_i are the individual $R_{t(50)}$ values measured at that location.

The Average Hull Roughness (AHR) is an attempt to combine the individual MHR values into a single parameter defining the hull conditions at a particular time. Typically the vessel may have been divided up into a number of equal areas, perhaps 100, and a value of MHR determined for each area. These MHR values are then combined in the same way as equation (24.1) to give the AHR for the vessel:

$$\text{AHR for vessel} = \frac{\sum_{j=1}^{m} w_j (\text{MHR})_j}{\sum_{j=1}^{m} w_j} \qquad (24.2)$$

where w_j is a weight function depending on the location of the patch on the hull surface. For many purposes w_j is put equal to unity for all j values; however, by defining the relation in the general way some flexibility is given to providing an opportunity for weighting important areas of the hull with respect to hull roughness. Most notable here are the regions in the fore part of the vessel.

Townsin et al.[4] suggest that if a full hull roughness survey is made, the AHR will be statistically correct using $w_j = 1$ in equation (24.2). However, should some stations be left out for reasons of access or some other reason, then the AHR can be obtained as follows:

AHR for vessel = (MHR of sides)

   × fraction of the sides covered

   + (MHR of flats)

   × fraction of the flats covered

   + (MHR of boot topping)

   × fraction of the boot topping covered

(24.3)

Much debate has centered on the use of a simple parameter such as $R_{t(50)}$ in representing non-homogenous surfaces. The arguments against this parameter suggest that the lack of data defining the surface in terms of its texture is serious and has led to the development of replica-based criteria for predicting power loss resulting from hull roughness (Reference 5). With this method, the surface of the actual ship is compared to those reproduced on replica cards, which themselves have been cast from other ships in service and the surfaces tested in a water tunnel to determine their drag. When a particular card has been chosen as being representative of a particular hull surface, a calculation of power penalty is made by use of diagrams relating the principal ship particulars; these diagrams having been constructed from a theoretical analysis procedure.

There is unfortunately limited data to be found that gives a statistical analysis and correlation with measured roughness functions for typical hull surfaces. Amongst the tests carried out, Musker,[12] Johannson[13] and Walderhaug,[14] feature as well-known examples. In the case of Musker, for example, he found that the measured roughness function for a set of five surfaces did not show a good correlation with $R_{t(50)}$ and used a combination of statistical parameters to improve the correlation. The parameters used in his study were:

1. the standard deviation (σ_r);
2. the average slope (S_p);
3. the skewness of the height distribution (S_k);
4. the Kurtosis of the distribution (K_u);

and he combined them into an 'equivalent height' (h') which correlated with the measured roughness function using a filtered profile with a 2 mm long wavelength cut off. The relationship used was

$$h' = \sigma_r(1 + aS_p)(1 + bS_k K_u) \qquad (24.4)$$

With regard to $R_{t(50)}$ as a parameter, Townsin[6] concludes that for rough surfaces, including surface-damaged and deteriorated anti-fouling coatings — in excess of around 250 μm AHR, it is an unreliable parameter to correlate with added drag. However, for new and relatively smooth hulls it appears to correlate well with other available measures of roughness function and so can form a basis to assess power penalties for ships.

It is found that the majority of new vessels have AHR of the order of 90–130 μm provided that they have been finished in a careful and proper manner. McKelvie[7] notes, however, that values for new vessels of 200–250 μm have not been uncommon in the period preceding 1981. The way in which this value increases with time is a variable depending on the type of coating used. To illustrate this Figure 24.5 shows a typical scenario (Reference 8) for a vessel in the first eight years of its life. In the figure it will be seen that the initial roughness AHR increased after four years to a value of around 250 μm using traditional anti-fouling coatings (Point A on the diagram). If the vessel is shot blasted, it can be assumed that the initial hull roughness could be reinstated since an insignificant amount of corrosion should have taken place. If, after cleaning, the vessel is treated with a reactivatable or SPA, after a further period of four years in service the increase in roughness would be small. Alternatively, if the vessel had been treated with traditional anti-fouling, as in the previous four-year period, then a similar increase in roughness would be noted. As illustrated in the diagram, the rate of increasing roughness depends on the coating system employed and the

Service Performance and Analysis

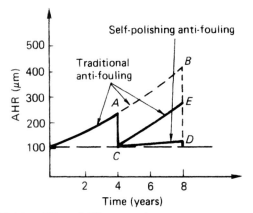

FIGURE 24.5 Effect of different coatings on hull roughness. *Reproduced from Reference 8.*

TABLE 24.3 Typical Annual Hull Roughness Increments

Coating Type	Annual Increase in Roughness (μm/year)
Self-polishing paints	10–30
Traditional coating	40–60

figures shown in Table 24.3 will give some general indication of the probable increases.

Clearly, significant deviations can occur in these roughening rates in individual circumstances for a wide variety of reasons. Figure 24.6, which is taken from Townsin et al.[9] shows the scatter that can be obtained over a sample of some 86 surveys conducted over the two-year period 1984–85.

Assuming that the AHR can be evaluated, this value then has to be converted into a power penalty if it is to be of any practical significance beyond being purely an arbitrary measure of paint quality. Lackenby[10] proposed an early approximation that for every 25 μm increase in roughness an increase in fuel consumption of around 2.5 per cent could be anticipated.

More recently Bowden and Davison[11] proposed the relationship

$$\frac{\Delta P_1 - \Delta P_2}{P} \times 100\% = 5.8\left[(k_1)^{1/3} - (k_2)^{1/3}\right] \quad (24.5)$$

where k_1 and k_2 are the AHR for the rough and smooth ship, respectively, and ΔP_1 and ΔP_2 are the power increments associated with these conditions, P is the maximum continuous power rating of the vessel.

The relationship was adopted by the 1978 International Towing Tank Conference (ITTC) as the basis for the formulation of power penalties and appeared in those proceedings in the form:

$$\Delta C_F \times 10^3 = 105 \left(\frac{k_s}{L}\right)^{1/3} - 0.64 \quad (24.6)$$

in which k_s is the mean apparent amplitude of the surface roughness over a 50 mm wavelength and L is the ship length. With equation (24.6) a restriction in length of 400 m

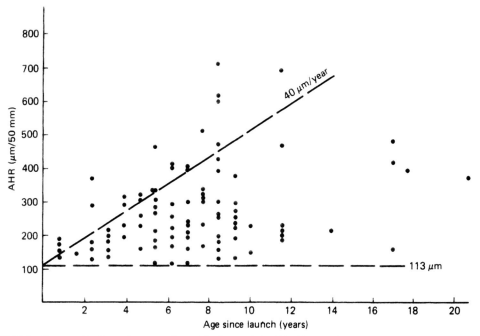

FIGURE 24.6 Survey of hull roughness conducted during period 1984–85. *Reproduced with permission from Reference 9.*

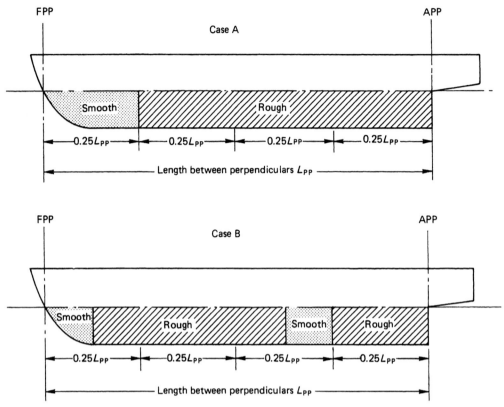

FIGURE 24.7 Hull smoothing regimes considered by Kauczynski and Walderhaug. *Reproduced from Reference 15.*

was applied and it is suitable for resistance extrapolation using a form factor method and the 1957 ITTC friction line. It assumes a standard roughness of 150 μm.

Townsin[6] has recently produced a modified expression for the calculation of ΔC_F based on the AHR parameter and is applicable to new and relatively smooth vessels:

$$\Delta C_F \times 10^3 = 44 \left[\left(\frac{\text{AHR}}{L} \right)^{1/3} - 10(R_n)^{-1/3} \right] + 0.125 \tag{24.7}$$

The effects of the distribution of roughness on the skin friction of ships were explored by Kauczynski and Walderhaug.[15] They showed that the bow region was the most important part of the hull with respect to the increase in resistance due to roughness. However, the length of the significant part of this portion of the hull decreases as the block coefficient increases. In the case of vessels with higher block coefficients, of the order of 0.7–0.8, the afterbody also plays a significant role. Figure 24.7, based on Reference 15, illustrates this point by considering two smoothing regimes for a vessel. In case A, a smooth strip equal to 25 per cent of L_{WL} was fixed to the bow, whereas in case B the smooth area was divided into two equal portions, both with a length equal to 12.5 per cent L_{WL}. In both cases the smoothed areas were equal. Calculations showed that the reductions in C_F compared to the whole rough surface were 0.105×10^{-3} and 0.119×10^{-3} for cases A and B, respectively, thus showing an advantage for the smoothing regimes of case B. In order to compute the value of C_F corresponding to paint roughness, Kauczynski and Walderhaug based their calculations on a conformal mapping technique for describing the hull form and used a momentum integral method for the calculation of the three-dimensional turbulent boundary layer characteristics. The results of these calculations for five hull forms of the Series 60 models with block coefficients between 0.60 and 0.80 have shown that the increase of frictional resistance due to roughness ΔC_F is a function of block coefficient, Reynolds number, $R_{t(50)}$ and $R_{t(1)}$. A regression procedure was applied by the authors to these results in order to give a readily applicable approximation of the form

$$\Delta C_F = a_0 + a_i \bar{k}_B^{*1/i} + b_j \Delta C_B^{*j} \frac{\bar{k}_B^*}{L_{WL}} \bar{k}_1^* R_n^* \tag{24.8}$$

where $i, j = 1, 2, 3$ and

$$\bar{k}_B^* = \frac{\bar{k}_B / L_{WL}}{3.32 \times 10^9}; \quad R_n^* = \frac{R_n}{2.7 \times 10^6}; \quad \bar{k}_1^* = \frac{k_1}{105}$$

with

$$\Delta C_B^* = \frac{C_B - 0.6}{0.2}$$

TABLE 24.4 Values of Coefficient $f_{i,p}$ (taken from Reference 15)

p	$f_{0,p} \times 10^3$	$f_{1,p} \times 10^3$	$f_{2,p} \times 10^3$	$f_{3,p} \times 10^3$
1	−0.05695	−0.08235	−0.48093	0.43460
2	−0.25473	−0.73105	1.01946	−1.37640
3	−0.18337	−2.01563	1.31724	−0.11176
4	0.38401	0.79786	2.02432	−2.30461
5	−0.27985	0.27460	−2.56908	2.26801
6	0.12397	0.47117	1.30053	−1.43575
7	1.95506	−10.87320	35.18020	−24.04790
8	−4.89111	17.57430	−63.66010	49.25690
9	1.70315	−5.44915	19.77400	−15.92990
10	0.72533	−2.50564	9.33041	−7.31122
11	−0.07676	−0.74104	0.62533	−0.06440
12	−2.93232	9.38549	−36.49980	29.61980
13	1.88597	−0.39504	6.04098	−11.67790
14	6.04607	−23.66800	88.78930	−64.38880
15	−5.02286	17.10700	−64.93670	49.93150
16	0.07829	0.00438	0.56607	−0.56425
17	0.04596	0.25232	−0.09525	−0.39173
18	3.04651	−10.75950	42.36420	−31.51610
19	−7.47250	23.29540	−93.06020	72.79150
20	4.26166	−13.00580	51.38600	−40.80970

In order to derive the coefficients a_0 and a_i in equation (24.8) a further polynomial expression has been derived as follows:

$$a_i = \sum_{n=1}^{4} \sum_{m=1}^{5} fi, p (\overline{k}_1^*)^{n-1} (R_n^*)^{m-1}$$

where $i = 0, 1, 2, 3$

$$P = m + 5(n-1)$$

The coefficients $f_{i,p}$ are given by Table 24.4 for all values of $p = 1, 2, 3 \ldots 20$. The coefficients b_j are given by Table 24.5.

The calculation procedure is subject to the constraints imposed by the model series and the conditions examined. Thus $\overline{k}_{1_{max}}$, $(\overline{k}_B/L_{WL})_{max}$, $R_{n_{max}}$ and $C_{B_{min}}$ are defined as 105 μm, 3.32×10^9, 2.7×10^6 and 0.6, respectively. The method described has been examined in comparison with others, notably those by Hohansson, Townsin and Bowden, for a 16 knot, 350 m tanker, and the results are shown in Figure 24.8. The range of values predicted for $(\overline{k}_B/\overline{k}_1)$ in the range 4−8, typical values for painted surfaces, embrace the result from Bowden's formula. Nevertheless, Bowden's formula does not consider the effects of R_n and C_B and consequently in other examples differences may occur. With regard to Townsin and Johansson's formulas, close agreement is also seen in the region where $\overline{K}_1$ is of the order of 30 μm.

TABLE 24.5 Coefficients b_j (taken from Reference 15)

j	$bj \times 10^3$
1	−0.09440
2	0.01126
3	0.13756

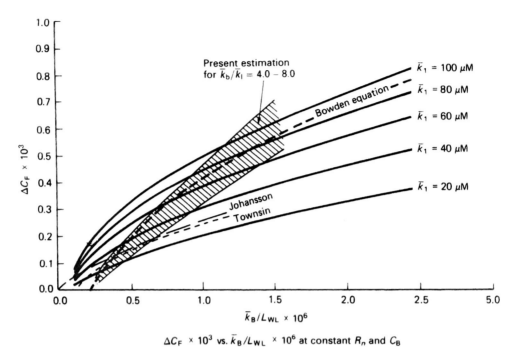

FIGURE 24.8 Comparison of roughness ΔC_F values. *Reproduced with permission from Reference 15.*

Walderhaug[14] suggests an approximation to the procedure outlined above, which has the form

$$\Delta C_F \times 10^3 \simeq 0.5 \left(\frac{k_E \times 10^6}{L}\right)^{0.2} \times \left[1 + \left(\frac{C_B - 0.75}{0.7}\right)^2\right] \times \left[\ln\left(1 + \frac{u_\tau k_E}{v}\right)\right]^{0.7}$$

(24.9)

where the effective roughness k_E is given by

$$k_E = \left(\frac{\bar{k}}{\lambda}\right)_1 (R_{t(50)} - K_A)$$

with the roughness to wavelength ratio $(\bar{k}/\lambda)_1$ at $= 1$ mm and the admissible roughness k_A given by

$$k_A = \frac{fv}{V} (\ln R_n)^{1.2}$$

with $f = 2.5$ for painted surfaces and the friction velocity

$$u_\tau = \frac{V}{(\ln R_n)^{1.2}}$$

24.3 HULL DRAG REDUCTION

Methods involving the injection of small quantities of long-chain polymers into the turbulent boundary layer surrounding a hull form, such as polyethylene oxide, were shown in the 1960s to significantly reduce resistance, provided the molecular weight and concentration were chosen correctly. Experiments conducted at that time suggested that the reduction in drag was linked to changes in the structure of the turbulence by the addition of the long-chain polymers. Frenkiel *et al.*[22] and Berman[23] discuss these effects in detail.

Those methods which relied on the injection of chemical substances into the sea, however, are unlikely to be environmentally acceptable today. Nevertheless, current research is focusing on a range of methods involving boundary layer fluid injection and manipulation. These methods embrace the injection of low-pressure air, either in the formation of air bubble interfaces between the hull and the sea water or through the provision of an air cushion trapped by an especially developed hull form. Some attention is also devoted to the injection of non-toxic or environmentally friendly fluids into the hull boundary layer.

24.4 PROPELLER ROUGHNESS AND FOULING

Propeller roughness is a complementary problem to that of hull roughness and one which is no less important. As in the hull roughness case, propeller roughness arises from a variety of causes, chief of which are marine growth, impingement attack, corrosion, cavitation erosion, poor maintenance and contact damage.

The marine growth found on propellers is similar to that observed on hulls except that the longer weed strands tend to get worn off. Notwithstanding this, weed having a length of the order of 10—20 mm is not uncommon on the minor regions of the blade, as indeed are stalked barnacles which are frequently found alive on the blades after a vessel has docked subsequent to undertaking a considerable journey. Marine fouling of this type increases the power absorption of the propeller, which for a fixed pitch propeller will result in a reduction of service rotational speed.

Impingement attack resulting from the passage of the water and the abrasive particles held in suspension over the blade surfaces normally affects the blades in the leading edge region and particularly in the outer radii of the blade where the velocities are highest. This results in a comparatively widespread area of fairly shallow depth surface roughness: similar to corrosion of either the chemical or electrochemical kind. Furthermore, with both corrosive and impingement roughness the severity of the attack tends to be increased with the turbulence levels in the boundary layer of the section. Consequently, subsequent to an initial attack, increased rates of surface degradation could be expected with time.

Cavitation erosion is normally, but not always, confined to localized areas of the blade. It can vary from a comparatively slight and relatively stable surface deterioration of a few millimeters in depth to a very rapid deterioration of the surface reaching depths of the order of the section thickness in a few days. Fortunately, the later scenario is comparatively rare. Cavitation damage, however, presents a highly irregular surface, as seen in Figures 24.9 and 26.1, which will have an influence on the drag characteristics of the blade sections. Blade-to-blade differences are likely to occur in the erosion patterns caused by cavitation and also, to some extent, with the forms of roughness. These differences will influence the individual drag characteristics of the sections.

Both poor maintenance and contact damage influence the surface roughness. In the former case perhaps by the use of too coarse grinding discs and incorrect attention to the edge forms of the blade and in the latter case, by gross deformation leading both to a propeller drag increase and also to other secondary problems; for example, cavitation damage. With regard to the frequency of propeller polishing there is a consensus of opinion between many authorities that it should be undertaken in accordance with the saying 'little and often' by experienced and specialized personnel. Furthermore, the pursuit of super-fine finishes to blades is generally not worth the expenditure, since these high polishes are often degraded significantly during transport or in contact with ambient conditions.

The effects of surface roughness on aerofoil characteristics have been known for a considerable period of time. These effects are principally confined to the drag coefficient and a typical example taken from Reference 16 is seen

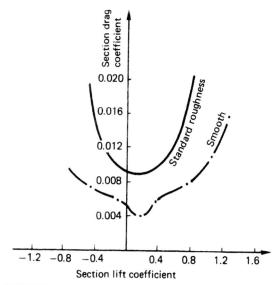

FIGURE 24.10 Effect of roughness on NACA 65–209 profile.

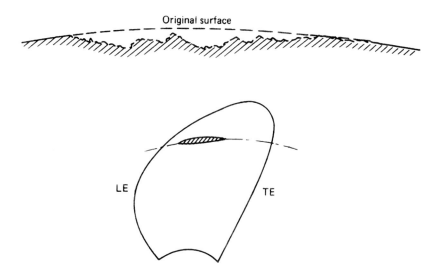

FIGURE 24.9 Typical cavitation damage profile.

in Figure 24.10 for a National Advisory Committee for Aeronautics (NACA) 65-209 profile.

The effect on section lift is small since the lift coefficient is some 20–30 times greater than the drag coefficient and studies conducted by the ITTC showed that the influence of roughness on the lift coefficient can be characterized by the relationship

$$\Delta C_L = -1.1 \Delta C_D \qquad (24.10)$$

Results such as those shown in Figure 24.10 are based on a uniform distribution of sand grain roughness over the section surface. In practice, however, this is far from the case, and this implies that a multiparameter statistical representation of the propeller surface embracing both profile and texture might be more appropriate than a single parameter such as the maximum peak-to-valley height. Grigson[17] shows two surfaces to illustrate these points (Figure 24.11) which have approximately the same roughness amplitudes but quite different textures. In general propeller surface roughness is of the Colebrook–White type and can be characterized in terms of the mean apparent amplitude and a surface texture parameter.

The topography of a surface can be reduced into three component terms: roughness, waviness and form errors, as shown in Figure 24.12. Clearly, the definition of which category any particular characteristic lies in is related to the wavelength of the characteristic. The International Standards Organization (ISO) has used two standards in the past; these are the peak-to-valley average (PVA) and the center-line average (CLA or R_a). The definitions of these terms are as follows:

Peak-to-valley average (PVA). This is the sum of the average height of the peaks and the average depth of the valleys. It does not equate to the R_t parameter, since this latter term implies the maximum rather than the average value.

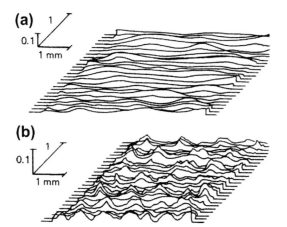

FIGURE 24.11 **Example of two different textures having approximately the same roughness amplitude.** *Reproduced with permission from Reference 17.*

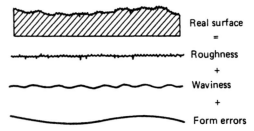

FIGURE 24.12 Reduction of surface profile into components.

Center-line average (CLA or R_a). This is the average deviation of the profile about the mean line and is given by the relation

$$R_a = \frac{1}{l} \int_{x=0}^{l} |y(x)| dx \qquad (24.11)$$

where l is the length of the line over which the roughness distribution $y(x)$ is measured.

There is unfortunately very little correspondence between the values derived from a PVA or CLA analysis. Some idea of the range of correspondence can be deduced from Figure 24.13, taken from Reference 18, for mathematically defined forms. The authors of Reference 18 suggest a value of the order of 3.5 when converting from CLA to PVA for propeller surfaces. The difference between these two measurement parameters is important when comparing the 1966 and 1981 ISO surface finish requirements for propellers, since the former was expressed in terms of PVA whilst the latter was in CLA. Sherrington and Smith[19] discuss the wider aspects of characterizing the surface topography of engineering surfaces. For reference purposes Table 24.6 itemizes the ISO surface finish requirements for Class 'S' and Class '1' propellers.

Several methods of surface roughness assessment exist and these range from stylus-based instruments through to the Rubert comparator gauge. For the stylus-based instruments it has been generally found that a wavelength cut-off value of the order of 2.5 mm gives satisfactory values for the whole range of propellers. The stylus-based instrument will give a direct measure of the surface profile which is in contrast to the comparator gauge method. In this latter method the surface of the blades at particular points are 'matched' to the nearest surface on the reference gauge. The 'Rubert' gauge, which is perhaps the most commonly used, comprises six individual surfaces tabulated A through to F as seen in Figure 24.14. These surfaces have been the subject of extensive measurement exercises by a number of authorities. Townsin et al.[20] undertook a series of studies to determine the value of Muskers' apparent height h' from both his original definition and a series of approximations. The values derived for the apparent roughness together with the maximum peak-to-valley amplitude R_t (2.5) quoted by

Type	Shape	PVA (μm)	PVA/CLA
Sinusoidal		8	$\pi =$ 3.142
Triangular		10	4
Occasional triangular protrusions	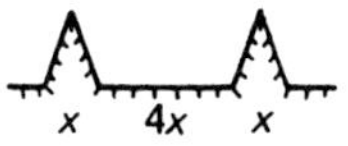	15.5	6.2
Occasional triangular indentations	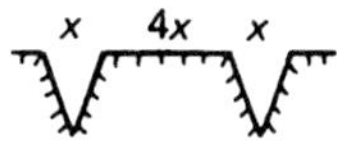	15.5	6.2
Occasional triangular protrusions and indentations	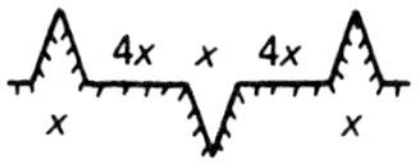	50	20
Parabolic indentations (mathematical scratches)		9.7	3.9
Occasional parabolic indentations (scratches)	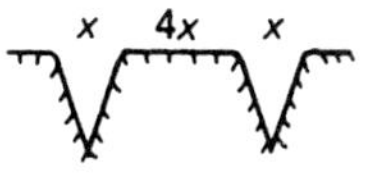	22.5	9
Parabolic cusps		9.7	3.9
Occasional parabolic cusps	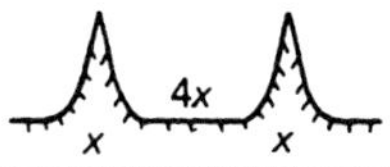	22.5	9

CLA = centre line average (the average deviation from the mean line)
PVA = peak-to-valley average (the average height of the peaks plus the average depth of the valleys)

FIGURE 24.13 Comparison between CLA and PVA measurements of roughness for a constant CLA value of 2.5 μm R_a. *Reproduced with permission from Reference 18.*

the manufacturers of the Rubert gauge are given in Table 24.7. Also shown in this table is the approximation to h' derived from the relation

$$h' \simeq 0.014 R_a^2 (2.5) P_c \qquad (24.12)$$

TABLE 24.6 ISO Surface Finish Requirements

Specification		Class 'S'	Class 'I'	Units
ISO R484	1966	3	9	μm (PVA)
ISO R484/1 ⎫ ISO R484/2 ⎭	1981	3	6	μm (Ra)

where P_c is the peak count per unit length and is used as a texture parameter.

When measuring the roughness of a propeller surface it is not sufficient to take a single measurement or observation on a blade. This is because the roughness will vary over a blade and different parts of the blade will be more significant than others: principally the outer sections since the flow velocities are higher. Furthermore, differences will exist from blade to blade. To overcome this problem a matrix of elements should be superimposed on the suction and pressure surfaces, as shown in Figure 24.15. In each of the twelve regions defined by the matrix on each surface of the blade several roughness measurements should be taken in the direction of the flow and widely spaced apart. A minimum of three measurements is recommended in each patch from which a mean value can be assessed (Reference 20).

analysis is to consider the power delivered to the propeller in order to propel a ship at a given speed V_s through the water:

$$P_D = \frac{RV_s}{\text{QPC}}$$

where R is the resistance of the ship at the speed V_s and the QPC is the quasi-propulsive coefficient given by

$$\text{QPC} = \eta_H \eta_r \eta_0 = \eta_H \eta_r \frac{K_T}{K_Q} \frac{J}{2\pi}$$

Consequently, the basic relationship for the delivered power P_D can be re-expressed as follows:

$$P_D = \frac{\pi \rho S V_s^3 C_T K_Q}{K_T J \eta_H \eta_r} \qquad (24.13)$$

by writing the ship resistance R as $\frac{1}{2} S V_s^2 C_T$ equation (24.13) can be linearized by taking logarithms and differentiating the resulting equation to give

$$\frac{dP_D}{P_D} = \frac{d\rho}{\rho} + \frac{dS}{S} + \frac{3dV_s}{V_s} + \frac{dC_T}{C_T} + \frac{dK_Q}{K_Q} - \frac{dK_T}{K_T} - \frac{dJ}{J} - \frac{d\eta_H}{\eta_H} - \frac{d\eta_r}{\eta_r}$$

In this equation it can be assumed for all practical purposes that the density (ρ), the wetted surface area (S) and the relative rotative efficiency (η_r) are unaffected by increases in roughness of the order normally expected in ships in service. As a consequence these terms can be neglected in the above equation to give

$$\frac{dP_D}{P_D} = \frac{3dV_s}{V_s} + \frac{dC_T}{C_T} + \frac{dK_Q}{K_Q} - \frac{dK_T}{K_T} - \frac{dJ}{J} - \frac{d\eta_H}{\eta_H}$$

In addition, since roughness, as distinct from biological fouling, is likely to cause only relatively small changes in the power curve, these can then be approximated to linear

FIGURE 24.14 The Rubert gauge.

24.5 GENERALIZED EQUATIONS FOR THE ROUGHNESS-INDUCED POWER PENALTIES IN SHIP OPERATION

Townsin et al.[20] established a valuable and practical basis upon which to analyze the effects of roughness on the hull and propeller of a ship. In this analysis they established a set of generalized equations, the derivations of which form the basis of this section. The starting point for their

TABLE 24.7 Rubert Gauge Surface Parameters

Rubert Surface	h' Equation (23.4) (μm)	h' (Approximation) Equation (23.12) (μm)	R_t (2.5) (μm)	R_a (2.5) (μm)
A	1.32	1.1	6.7	0.65
B	3.4	5.4	14.2	1.92
C	14.8	17.3	31.7	4.70
D	49.2	61	50.8	8.24
E	160	133	97.2	16.6
F	252	311	153.6	29.9

Note: a and b in equation (24.4) taken as 0.5 and 0.2, respectively.

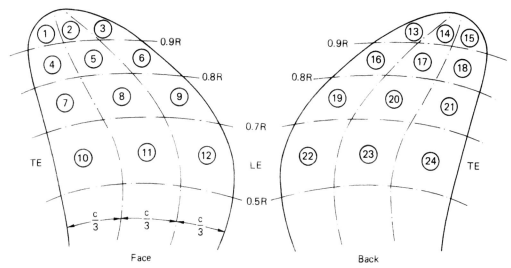

FIGURE 24.15 Definition of patches for recording propeller roughness.

functions. Consequently, the differentials can be considered in terms of finite differences:

$$\frac{\Delta P_D}{P_D} = \frac{3\Delta V_s}{V_s} + \frac{\Delta C_T}{C_T} + \frac{\Delta K_Q}{K_Q} - \frac{\Delta K_T}{K_T} - \frac{\Delta J}{J} - \frac{\Delta \eta_H}{\eta_H} \quad (24.14)$$

This equation clearly has elements relating to both the propeller and the hull, and can be used to determine the power penalty for propulsion at constant ship speed V_s:

$$\frac{\Delta P_D}{P_D} = \frac{\Delta C_T}{C_T} + \frac{\Delta K_Q}{K_Q} - \frac{\Delta J}{J} - \frac{\Delta K_T}{K_T} - \frac{\Delta \eta_H}{\eta_H} \quad (24.15)$$

Clearly, it will simplify matters considerably if equation (24.15) can be decoupled into hull and propeller components and, therefore, treated separately. This can be done subject to certain simplifications in the following way.

The terms $\Delta K_T/K_T$ and $\Delta K_Q/K_Q$ can be divided into two components; one due to propeller roughness and one due to the change in operating point assuming the propeller remained smooth:

$$\left.\begin{array}{l}\dfrac{\Delta K_Q}{K_Q} = \left(\dfrac{\Delta K_Q}{K_Q}\right)_R + \left(\dfrac{\Delta K_Q}{K_Q}\right)_J \\ \dfrac{\Delta K_T}{K_T} = \left(\dfrac{\Delta K_T}{K_T}\right)_R + \left(\dfrac{\Delta K_T}{K_T}\right)_J\end{array}\right\} \quad (24.16)$$

where the suffixes R and J denote propeller roughness and operating point, respectively. This distinction is shown in Figure 24.16 for the torque coefficient characteristics. The relative changes to the propeller characteristic due to roughness alone can be estimated from Lerb's theory of equivalent profiles.

Considering the second term in each of equations (24.16), since for a smooth propeller

$$\Delta K_Q = \frac{dK_Q}{dJ}\Delta J$$

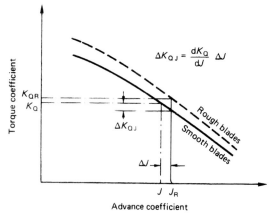

FIGURE 24.16 Effect of change of operating advance on propeller torque characteristics with rough and smooth blades.

and similarly for ΔK_T, we write for the change in operating point terms in equations (24.16):

$$\left.\begin{array}{l}\left(\dfrac{\Delta K_Q}{K_Q}\right)_J = \dfrac{J}{K_Q}\left(\dfrac{dK_Q}{dJ}\right)\left(\dfrac{\Delta J}{J}\right) \\ \left(\dfrac{\Delta K_T}{K_T}\right)_J = \dfrac{J}{K_T}\left(\dfrac{dK_T}{dJ}\right)\left(\dfrac{\Delta J}{J}\right)\end{array}\right\} \quad (24.17)$$

Now the term ΔJ is the difference between the rough and smooth or original operating points, as seen in Figure 24.16:

$$\Delta J = J_R - J$$

that is,

$$\Delta J = \frac{V_s}{D}\left[\frac{(1-w_{TR})}{N_R} - \frac{(1-w_T)}{N}\right]$$

since V_s is assumed constant from equation (24.15). Hence by referring to the original operating point

$$\frac{\Delta J}{J} = \left(\frac{1-w_{TR}}{1-w_T}\right)\frac{N}{N_R} - 1 \quad (24.18)$$

Furthermore, since $\Delta K_T = (K_{TR} - K_T)$,

$$\frac{\Delta K_T}{K_T} = \left(\frac{K_{TR}}{K_T} - 1\right)$$

that is,

$$\frac{\Delta K_T}{K_T} = \frac{T_R}{T}\left(\frac{N}{N_R}\right)^2 - 1 \qquad (24.19)$$

and by assuming an identity of thrust deduction between the rough and original smooth condition for a given ship speed, this implies

$$\frac{T_R}{T} = \frac{R_R}{R} = \frac{C_{TR}}{C_T} = \frac{C_T + \Delta C_T}{C_T} = \left[1 + \frac{\Delta C_T}{C_T}\right]$$

Hence, substituting this relationship into equation (24.19), eliminating propeller revolutions between equations (24.18) and (24.19) and noting that C_T is wholly viscous so that $\Delta C_T = \Delta C_V$, we obtain

$$\frac{\Delta J}{J} = \left(\frac{1 - w_{TR}}{1 - w_T}\right)\left[\frac{1 + (\Delta K_T/K_T)}{1 + (\Delta C_V/C_T)}\right]^{1/2} - 1 \qquad (24.20)$$

By applying the binomial theorem to equation (24.20) and since $\Delta K_T/K_T$ and $\Delta C_V/C_T$ are small,

$$\frac{\Delta J}{J} = \left(\frac{1 - w_{TR}}{1 - w_T}\right)\left[1 + \frac{1}{2}\left(\frac{\Delta K_T}{K_T}\right) - \left(\frac{\Delta C_V}{C_T}\right)\right]$$

Hence, from equations (24.16) and (24.17) and substituting these into the above, an explicit relationship can be found for the term $\Delta J/J$ as follows:

$$\frac{\Delta J}{J} = \frac{\left(\frac{1 - w_{TR}}{1 - w_T}\right)\left[1 + \frac{1}{2}\left(\frac{\Delta K_T}{K_T}\right)_R - \frac{\Delta C_V}{C_T}\right] - 1}{1 - \frac{1}{2}\left(\frac{1 - w_{TR}}{1 - w_T}\right)\frac{J}{K_T}\left(\frac{dK_T}{dJ}\right)}$$

In this equation the terms CV and w_T relate to the hull roughness, excluding any propeller-induced wake considerations, and the term $(\Delta K_T/K_T)_R$ relates to the propeller roughness. Separating these terms out, we have

$$\frac{\Delta J}{J} = \frac{\left(\frac{1 - w_{TR}}{1 - w_T}\right)\left[1 - \frac{\Delta C_V}{C_T}\right] - 1}{1 - \frac{1}{2}\left(\frac{1 - w_{TR}}{1 - w_T}\right)\frac{J}{K_T}\left(\frac{dK_T}{dJ}\right)}$$

$$+ \frac{\frac{1}{2}\left[\frac{\Delta K_T}{K_T}\right]_R}{\left(\frac{1 - w_T}{1 - w_{TR}}\right) - \frac{1}{2}\frac{J}{K_T}\frac{dK_T}{dJ}} \qquad (24.21)$$

The first term in equation (24.21) is a function of hull roughness only and is the relative change in advance coefficient due to hull roughness only $(\Delta J/J)_H$. The second term is a function of both propeller and hull roughness; this can, however, be reduced to a propeller roughness function by assuming that

$$\left(\frac{1 - w_r}{1 - w_{TR}}\right) \cong 1$$

when the change in propeller roughness can be approximated by the function

$$\left(\frac{\Delta J}{J}\right)_R \cong \frac{\frac{1}{2}\left(\frac{\Delta K_T}{K_T}\right)_R}{1 - \frac{1}{2}\left(\frac{J}{K_T}\right)\left(\frac{dK_T}{dJ}\right)}$$

Consequently, the total change in advance coefficient, equation (24.21), can be decoupled into the sum of independent changes in hull and propeller roughness:

$$\frac{\Delta J}{J} \simeq \left(\frac{\Delta J}{J}\right)_{\text{Hull rough}} + \left(\frac{\Delta J}{J}\right)_{\text{Prop. rough}}$$

This immediately allows the power penalty $\Delta P_D/P_D$, expressed by equation (24.15), to be decoupled into the following:

$$\left(\frac{\Delta P_D}{P_D}\right)_{\text{Prop. rough}} = \left(\frac{\Delta K_Q}{K_Q}\right)_{\text{Prop. rough}}$$
$$- \left(\frac{\Delta J}{J}\right)_{\text{Prop. rough}} \qquad (24.22)$$
$$- \left(\frac{\Delta K_T}{K_T}\right)_{\text{Prop. rough}}$$

and

$$\left(\frac{\Delta P_D}{P_D}\right)_{\text{Hull rough}} = \frac{\Delta C_V}{C_T} - \frac{\Delta \eta_H}{\eta_H} + \left(\frac{\Delta K_Q}{K_Q}\right)_{\text{Hull. rough}}$$
$$- \left(\frac{\Delta J}{J}\right)_{\text{Hull rough}}$$

But, since propellers generally work in a region of the propeller curve, where the ratio, over small changes, of K_T/K_Q is relatively constant, this latter equation reduces to

$$\left(\frac{\Delta P_D}{P_D}\right)_{\text{Hull rough}} = \frac{\Delta C_V}{C_T} - \frac{\Delta \eta_H}{\eta_H} - \left(\frac{\Delta J}{J}\right)_{\text{Hull rough}}$$
$$(24.23)$$

Equations (24.15), (24.22) and (24.23) form the generalized equations of roughness-induced power penalties in ship operation. These latter two equations can, however, be expanded to give more explicit relationship for the hull and propeller penalties.

In the case of equation (24.22) for the propeller penalty, the propeller roughness effects $(\Delta KQ/KQ)$ and $(\Delta K_T/K_T)$ can be estimated from Lerb's equivalent profile method, from which the following relationship can be derived in association with Burrill's analysis:

$$\left(\frac{\Delta P_D}{P_D}\right)_{\text{Prop. rough}} = \left[\frac{2.2}{(P/D)} - 1.1 + \left\{\frac{3.3(P/D) - 2J}{2.2(P/D) - J}\right\}\right.$$
$$\left. \times (0.45(P/D) + 1.1)\right]\left(\frac{\Delta c_D}{c_L}\right) 0.7$$
(24.24)

For the hull roughness contribution, Townsin et al.[20] show that by assuming a constant thrust deduction factor and employing the ITTC 1978 formula for wake scaling such that

$$w_{\text{TR}} = t + \left(w_{\text{T}} - t\right)\left[\frac{\Delta C_{\text{FS}} - \Delta C_{\text{FT}}}{(1+k)C_{\text{FS}} + \Delta C_{\text{FT}}} + 1\right]$$

then the full roughness power penalty becomes

$$\left(\frac{\Delta P_D}{P_D}\right)_{\text{Hull rough}}$$
$$= \left[1 - \left(\frac{w_{\text{r}} - t}{1 - w_{\text{TR}}}\right)\frac{C_{\text{T}}}{(1+k)C_{\text{FS}} + \Delta C_{\text{FT}}}\right] \quad (24.25)$$
$$\times \left(\frac{\Delta C_F}{C_T}\right) - \left(\frac{\Delta J}{J}\right)_{\text{Hull rough}}$$

where

P_D is the delivered power at the propeller,
C_T is the ship thrust coefficient,
C_F is the ship frictional coefficient,
C_{FS} is the smooth ship frictional coefficient,
J is the advance coefficient,
P is the propeller pitch,
D is the propeller diameter,

Δc_d is the change in reference section drag, coefficient,
c_l is the reference section lift coefficient,
ΔC_{FT} is the increment in ship skin friction, coefficient in trial condition.

24.6 MONITORING OF SHIP PERFORMANCE

The outline role of the ship service analysis is summarized in Figure 24.17. The data obtained from the ship should have two primary roles for the ship operator. The first is to develop a data bank of information from which standards of performance under varying operational and environmental conditions can be derived. The resulting standards of performance, derived from this data, then become the basis of operational and chartering decisions by providing a reference for a vessel's performance in various weather conditions and a reliable comparator against which the performance of sister or similar vessels can be measured. The second role for the data records is to enable the analysis of trends of either the hull or machinery to be undertaken, from which the identification of potential failure scenarios and maintenance decisions can be derived.

Table 24.8 identifies the most common set of parameters that are traditionally recorded to a greater or lesser extent by seagoing personnel in the ship's engine and bridge logbooks. It is this information which currently forms the database from which analysis can proceed. In the table the measurement of shaft power has been noted with an asterisk, this is to draw attention to the fact that this extremely important parameter is only recorded in relatively few cases, due to the lack of a torsion meter having

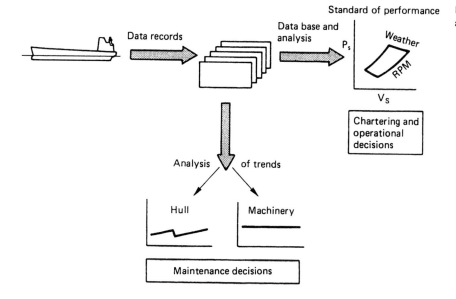

FIGURE 24.17 Role of ship service analysis.

TABLE 24.8 Traditionally Recorded Parameters in Ship Log Books

Deck Log

Ship draughts (fore and aft)

Time and distance traveled (over the ground)

Subjective description of the weather (wind, sea state, etc.)

Ambient air and sea water temperature

Ambient air pressure

General passage information

Engine Log

Cooling sea water temperature at inlet and outlet

Circulating fresh water cooling temperature and pressures for all engine components

Lubricating oil temperature and pressures

Fuel lever, load indicator and fuel pump settings

Engine/shaft revolution count

Turbocharger speed

Scavenge and injection pressures

Exhaust gas temperatures (before and after turbocharger)

Main engine fuel and lubricating oil temperatures

Bunker data

Generator and boiler performance data

Evaporator and boiler performance data

Torsion meter reading*

*denotes if fitted.

---

been fitted, and, as such this cannot be considered to be a commonly available parameter.

Traditionally, Admiralty coefficient (A_c) based methods have formed the foundation of many practical service performance analysis procedures used by shipowners and managers. If a simple plot of Admiralty coefficient against time is made, and Figure 24.18(a) shows a typical example of such a plot for a 140 000 tonnes dwt bulk carrier, it can be seen that it is difficult to interpret in any meaningful way due to the inherent scatter in this type of plot. One may, however, move a stage further with this type of study by analyzing the relationship between the Admiralty coefficient and the apparent slip (S_a) as seen in Figure 24.18(b). This figure shows a convergence in the data and invites the drawing of a trend line through the data. The form of these coefficients is given by the well-known relationships

$$A_c = \frac{\Delta^{2/3} V_s^3}{P_s}$$

and

$$S_a = 1 - 30.86 \left(\frac{V_s}{PN}\right) \text{ metric}$$

The data for this type of analysis is extracted from the ship's deck and engine room log abstracts and the resulting curves of A_c plotted against S_a would normally be approximated by a linear relationship over the range of interest. Furthermore, apparent slip can be correlated to the weather encountered by the vessel by converting the description of the sea state, as recorded by the ship's navigating officers, to wave height according to an approved scale for that purpose. The wave heights derived in this way can be modified to take account of their direction relative to the ship and, having established the wave height versus apparent slip lines for the propeller, the Admiralty coefficient or other similar variable can be plotted against the appropriate line using the recorded apparent slip from the log book. Methods such as these,

Service Performance and Analysis

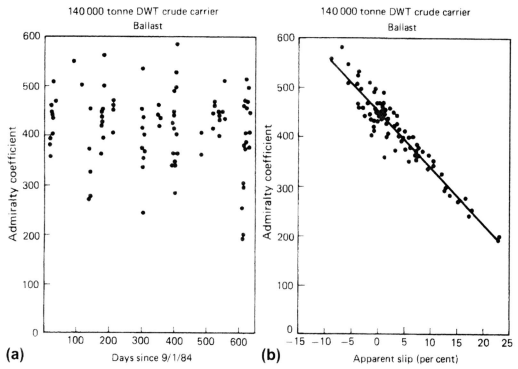

FIGURE 24.18 Common ship service procedures in use by the shipping community: (a) Admiralty coefficient versus time and (b) Admiralty coefficient versus apparent slip.

whilst providing a basis for analysis, can lead in some circumstances to misinterpretation. Furthermore, the Admiralty coefficient, although a useful criterion, is a somewhat 'blunt instrument' when used in this way since it fails to effectively distinguish between the engine and hull-related parameters. The same is also true for the alternative version of this equation, termed the fuel coefficient, in which the shaft horsepower (P_s) is replaced with the fuel consumption. This latter derivative of the Admiralty coefficient serves where the vessel is not fitted with a torsion meter.

Several coefficients of performance have been proposed based on various combinations of the parameters listed in Table 24.8. Whipps,[21] for example, attempts to split the overall performance of the vessel into two components — the responsibility of the engine room and the responsibility of the bridge watch-keepers. Accordingly, three coefficients of performance are proposed:

1. K_1 — nautical miles/tonne of fuel (overall performance).
2. K_2 — meters traveled/shp/h (navigational performance).
3. K_3 — grams of fuel/shp/h (engine performance).

Clearly, these coefficients require the continuous or frequent monitoring of the parameters concerned and the presentation of the coefficients of performance to the ship's staff on a continuous or regular basis. Experience with these and other similar monitoring techniques suggests that they do aid the ship's staff to enhance the performance of the vessel by making them aware of the economic consequences of their decisions at the time of their actions in terms that are readily understandable: this latter aspect being particularly important.

More recently Bazari[24] has considered the application of energy auditing to ship operation and design. This process is designed to undertake energy audits during a ship's operation either singly or across a fleet, particularly where there are a number of ships of the same design. From the results of these audits it then becomes possible to assess the potential for improvement in propulsion efficiency. This procedure involves three principal activities in the benchmarking or rating process shown in Figure 24.19. The three main activities include:

1. Selection of Key Performance Indicators (KPIs) and specifying their reference target values.
2. Data collection and assuring the data quality.
3. Estimating the KPIs, comparing these to the reference targets, estimating deviations and allocating a rating to the ship.

This analysis procedure is applicable to a range of ship types; for example, passenger ships, tankers and container ships. However, to carry out the process effectively it is essential to give consideration to all aspects of ship design, machinery procurement, ship operation, alternative technologies and fuels within the analysis process and to take a holistic view of the ship operation.

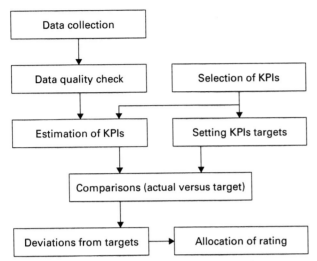

FIGURE 24.19 Outline of the benchmarking or rating process.

In the case of new ships significant reductions in the ship's overall fuel consumption are considered feasible using these auditing processes to make improvements to the ship design and use of energy-efficient machinery. While procuring a more energy-efficient ship may be slightly more expensive in the first instance, when fuel prices are high or show a general upward trend, the extra initial investment may well be recovered in the ship's operational account. Indeed it has been found that the majority of the effort within the auditing process, given that the hydrodynamic design process has been satisfactorily undertaken, is concentrated on the engineering systems; the use of energy-efficient machinery; optimization of hotel, HVAC and refrigeration systems and the wider use of shore services. These considerations need to be input at the conceptual design phase of the ship and reviewed at a pre-contract specification stage to ensure that energy efficiency is fully considered as part of the ship design process. Additionally, the constraints to be imposed by the EEDI requirements will also have an impact on this process.

When applied to ships that are in service the primary focus of the auditing processes should be on the reduction of fuel consumption. This can be achieved as outlined in Figure 24.19 using a combination of benchmarking, energy audits and performance monitoring. Within this process a systematic and holistic investigation needs to be undertaken which considers both technical and non-technical aspects of the operation. Furthermore, to obtain optimum results from the process it is often better if this is done by both an independent auditing practitioner and the ship's operator so that the two viewpoints are fully considered and agreed in the auditing process and as a result joint ownership of the result can then be achieved. This position has to be attained from a comprehensive level of data gathering and analysis combined with a shipboard energy survey. Moreover, in addition to the technical systems, the process should also take into account the operational profile of the ship and its main machinery together with any reference data from other similar ships.

In order to progress beyond the basic stages of performance monitoring it is necessary to attempt to address the steady-state ship powering equations:

$$P_s = \frac{RV_s}{\eta_0 \eta_r \eta_m} \left[\frac{1 - w_T}{1 - t} \right]$$

$$R = (1 - t)T$$

These equations clearly require knowledge of the measured shaft power and thrust, together with the ship and shaft speeds in association with the appropriate weather data. All of these parameters are potentially available, with the possible exception of shaft axial thrust. Thrust measurement has, in the past, proved notoriously difficult. In many instances this is due to the relative order of the magnitudes of the axial and torsional strain in the shaft and this measurement has generally only been attempted for specific measurement exercises under carefully controlled circumstances, using techniques such as the eight gauge Hylarides bridge (Chapter 17). When this measurement has been attempted on a continuous service basis the long-term stability of the measurement has frequently been a problem.

Consequently, it is generally possible to attempt only a partial solution to the steady-state powering equations defined above. To undertake this partial solution, the first essential task is to construct a propeller analysis model so as to determine the thrust, torque and hence efficiency characteristics with advanced coefficient. The method of constructing these characteristics can vary depending on the circumstances and the data available and, as such, can range from standard series open water curves to more detailed lifting line, vortex lattice techniques or boundary element methods. The resulting model of propeller action should, however, have the capability to accommodate allowances for propeller roughness and fouling, since this can, and does, influence the power absorption and efficiency characteristics to a marked extent. For analysis purposes it is clearly desirable to have as accurate a representation of the propeller characteristics as possible, especially if quantitative cost penalties are the required outcome of the exercise. However, if it is only performance trends that are required, then the absolute accuracy requirements can be relaxed somewhat since the rates of change of thrust and torque coefficients, dK_q/dJ and dK_t/dJ, are generally similar for similar types of propeller.

Figure 24.20 demonstrates an analysis algorithm. It can be seen that the initial objective, prior to developing standards of actual performance, is to develop two time series — one expressing the variation of effective wake fraction and the other expressing the specific fuel consumption with time. In the case of the effective wake fraction analysis this

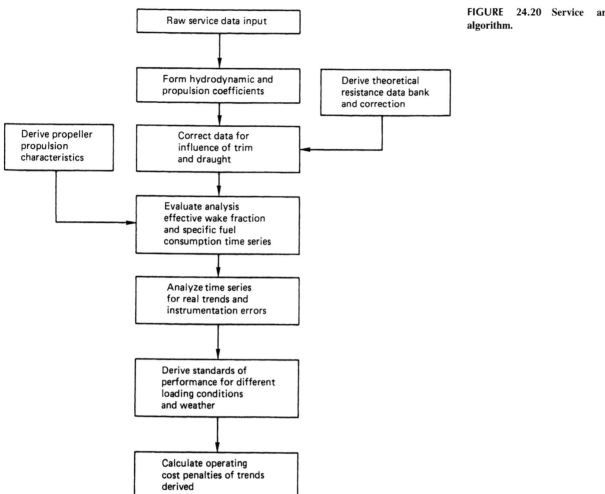

FIGURE 24.20 Service analysis algorithm.

is a measure, over a period of time, of the change in the condition of the underwater surfaces of the vessel because if either the hull or propeller surfaces deteriorate, the effective analysis wake fraction can be expected to reflect this change in particular ways. The second series, relating the specific fuel consumption to time, provides a global measure of engine performance. Should this latter parameter tend to deteriorate and it is shown by analysis that it is not a true trend, for example an instrument failure, then the search for the cause of the fault can be carried out using the other parameters listed in Table 24.8. Typically, these other parameters might be exhaust temperatures, turbocharger performance, bearing temperatures and so on.

The capabilities of this type of analysis can be seen in Figure 24.21, which relates to the voyage performance of a bulk carrier of some 50 years ago. The upper time series relates to the specific fuel consumption, from which it is apparent that, apart from the usual scatter, little deterioration takes place in this global engine characteristic over the time interval shown. The second series is that of the analysis effective wake fraction, from which it can be seen that a marked increase in the wake fraction occurs during each docking cycle and coincident with the dry-docking periods, when cleaning and repainting takes place, the wake fraction falls to a lower level. It is of interest to note that after each dry-docking the wake fraction never actually regained its former value, and consequently underlines the fact that hull deterioration has at least two principal components. The first is an irreversible increase with the age of the vessel and is the general deterioration of the hull condition with time, while the second is on a shorter time cycle and related to repairable hull deterioration and biological fouling (see Section 24.2).

Comparison of this analysis with that of a recent 140 000 dwt crude oil carrier (Figure 24.22) shows how the deterioration between docking cycles has been reduced despite the docking cycle having increased from the order of a year in Figure 24.21 to around three and a half years in this latter example. As might be expected, in Figure 24.22 there is still an upward trend in the wake fraction with time,

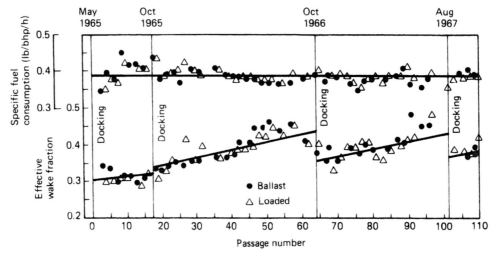

FIGURE 24.21 Service analysis for a bulk carrier (mid-1960s).

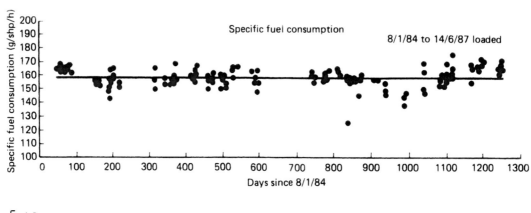

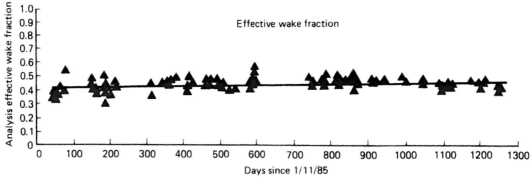

FIGURE 24.22 Service analysis for a 140 000 tonnes dwt crude carrier.

but nothing as significant as in the earlier case. The improvement in this case is almost entirely due to the use of modern paints and good propeller maintenance.

Instrumentation errors are always a potential source of concern in performance analysis methods. Such errors are generally in the form of instrument drift, leading to a progressive distortion of the reading, and these can generally be detected by the use of trend analysis techniques. Alternatively, if they are in the form of a gross distortion of the reading the principles of deductive logic can be applied.

On some ships today, as was the case in former times, a complete record in the ship's logs of all of the engine measured parameters and ship's operational entries are made relatively few times a day: typically one entry per day for the comprehensive set of data on many deep sea vessels assuming an automatic data logging system is not installed. Provided that the vessel is of simple design and is working on deep sea passages lasting a number of days, then this single entry practice, although not ideal from the analysis viewpoint, will probably be satisfactory for the building up of a profile of the vessel's operating characteristics over a period of time. In the alternative case, of a short sea route ferry for example, this once or twice per day level of recording is not appropriate since the vessel may make many passages in a day lasting for one or two hours.

Data logging by automatic or semi-automatic means clearly enhances this situation and leads to a much more accurate profile of ship operation in a significantly shorter-time frame. This is to some extent only an extension of the present procedures for alarm monitoring. Over a suitable period of time a databank of information can be accumulated for a particular ship or group of vessels. This databank enables the average criterion of performance for the ship to be derived. A typical example of such a criterion is shown in Figure 24.23 for a medium-sized container ship. This diagram, which is based on the actual ship measurements and corrected for trim, draught and fouling, relates the principal operational parameters of power, ship and shaft speed, and weather. Consequently, such data, when compiled for different trim draught and hull conditions can provide a reliable guide to performance for chartering purposes on any particular class of trade route.

Trim and draught have important influences on the performance of the vessel. Draught is clearly a variable determined by the cargo that is being carried. Trim, however, is a variable over which, for a great many vessels, some control can be exercised by the ship's crew. If this is done effectively and with due regard to weather conditions, then this can result in considerable savings in the transport efficiency of the vessel.

The traditional method of data collection was via the deck and engine room logs. In terms of current data processing capabilities, which involve both significant statistical trend analysis and detailed hydrodynamic analysis components, this method of data collection is far from ideal since, of necessity, it involves the translation of the data from one medium to another.

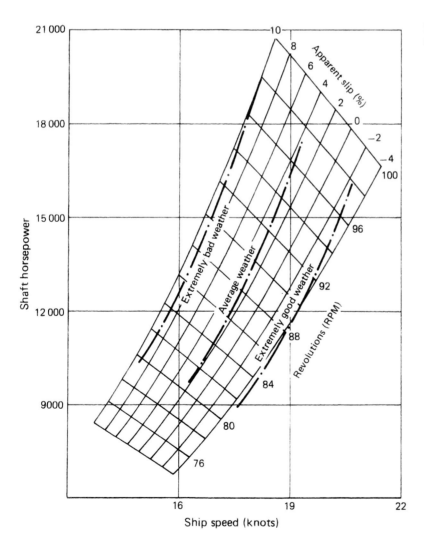

FIGURE 24.23 Typical power diagram for a container ship.

The immediate solution to this problem is to be found in the use of desktop or personal computers working in either the on-line or off-line mode. Such computers are ideal for many shipboard applications since, in addition to having large amounts of memory, they are small and user friendly, can readily be provided with custom-built software and are easily obtainable in most parts of the world. When used in the off-line mode, they are to some extent an extension of the traditional method of log entries, where instead of being written by hand in the book the data is typed directly into the computer for both storage on disc media and also for the production of the normal log sheet. This permits both easy transfer of the data to shore-based establishments for analysis and the undertaking of simple trend analysis studies on board. Such 'on board' analyses methods produce tangible benefits if conducted advisedly.

When a small computer is used in the 'on-line' mode the measured parameters are input directly from the transducers via a data acquisition system to the computer and its storage medium. In this way, a continuous or periodic scanning of the transducers can take place and the data, or representative samples of it, can be stored as well as providing data for a continuous statistical analysis. Such methods readily raise alarms when a data parameter moves outside a predetermined boundary in a similar way to conventional alarm handling.

This clearly is advantageous in terms of man hours but does tend to add to the degree of remoteness between operator and machine unless considerable attention has been paid to the 'user-friendliness' of the system. This ergonomic aspect of data presentation is particularly important if the system is to be accepted and used to its full potential by the ship's operating personnel. All too often poorly designed computer-based monitoring equipment is largely discarded either because it has been insufficiently ruggedly designed for marine use and is prone to failure and regular breakdown or, more frequently, because the engineering design has been adequately undertaken, but the data and information it presents is not in an easily assimilative form for the crew, and the manuals describing its operation contain too much specialized jargon which is unfamiliar to the operator. This underlines the importance of choosing a monitoring system which satisfies the company's commercial objectives as well as being compatible with the operator's actual and perceived requirements.

Small on-line systems of this type are the first step towards an integrated ship management system embracing the activities of the deck, engine room and catering departments. Such systems have made their appearance in many large shipping companies. With these systems, the vessel's operating and shore-based staff are required to assimilate this data, albeit presented in a much more generally comprehensible form than has previously been the case and use it in the context of the commercial constraints, classification society requirements and statutory regulations.

By way of example of the advanced models available or in the process of development, for a given trade route or operating pattern, the operational economics look to establish the most efficient routing and voyage planning for a ship so as to avoid the penalties of added resistance when encountering poor weather.

The assessment of the added resistance of a ship can be conveniently made from model tests or, alternatively, estimated from non-linear computational methods, Chapter 12, and such an assessment is made for a variety of sea conditions. Given, therefore, the knowledge of the ship's powering behavior in a variety of sea conditions and where the propeller design point is fixed with respect to the engine operating diagram, use can be made of weather forecast information to optimize the voyage plan. Such planning processes have been put to good use in passenger liner trades for voyage time-keeping purposes, but can also be used to minimize voyage costs in the sense of optimizing the voyage plan with respect to any number of voyage attributes. These attributes might be the ship performance characteristics, engine performance parameters, ship loading and so on which can then be relaxed with respect to the constraints acting on the voyage, for example, anticipated poor weather, port slots and the wider issues surrounding the transport chain of which the ship voyage is but one part. As such, it is possible to formulate a mathematical problem:

$$\text{Min } V_c = f(A_1, A_2, A_3, \ldots; C_1, C_2, C_3)$$

to minimize the voyage overall cost V_c against a set of voyage attributes A_n and constraints C_n. If the voyage attributes and constraints can be linearized then the solution to this exercise is relatively trivial in mathematical terms; however, for most practical situations the solution will exhibit at least some non-linear characteristics which then makes the cost minimization function a more complex problem to solve but, nevertheless, soluble in many cases using available numerical methods.

REFERENCES AND FURTHER READING

1. Christie AO. *Hull roughness and its control.* Shanghai, China: Marintech Conference; 1981.
2. Evans JP, Svensen TE. *Voyage simulation model and its application to the design and operation of ships.* Trans RINA; November 1987.
3. Bidstrup K, et al. *Self activating module systems, a new approach to polishing antifouling paints.* California: 21st Annual Marine and Offshore Coatings Conference; 1981.
4. Townsin RL, Byrne D, Svensen TE, Milne A. *Estimating the technical and economic penalties of hull and propeller roughness.* Trans SNAME; 1981.
5. *A hull roughness evaluation project.* Marine Propulsion; February 1987.

6. Townsin RL. *Developments in the circulation of rough underwater surface power penalties.* Genoa: Cetena 25th Anniversary Symp.; 1987.
7. McKelvie AN. *Hull preparation and paint application: their effect on smoothness.* Marine Eng Review; July 1981.
8. Naess E. *Reduction of drag resistance by surface roughness and marine fouling.* Norwegian Maritime Research 1980;(4).
9. Townsin RL, Byrne D, Svensen TE, Milne A. *Fuel economy due to improvements in ship hull surface conditions 1976–1986.* ISP; July 1986.
10. Lackenby H. *The resistance of ships with special reference to skin friction and hull surface conditions. 34th Thomas Lowe Grey Lecture.* Trans I Mech E; 1962.
11. Bowden BS, Davison NJ. *Resistance increments due to hull roughness associated with form factor extrapolation methods.* NMI Ship TM January 1974;**3800**.
12. Musker AJ. *Turbulent shear-flow near irregularly rough surfaces with particular reference to ship's hulls.* PhD Thesis. University of Liverpool; December 1977.
13. Johansson LE. *The local effect of hull roughness on skin friction. Calculations based on floating element data and three dimensional boundary layer theory.* Trans RINA 1985;**127**.
14. Walderhaug H. *Paint roughness effects on skin friction.* ISP June 1986;**33**(382).
15. Kauczynski W, Walderhaug H. *Effects of distributed roughness on the skin friction of ships.* ISP January 1987;**34**(389).
16. Abbott Ira H, Doenhoff AE von. *Theory of Wing Sections.* Dover, New York, 1959.
17. Grigson CWB. *Propeller roughness, its nature and its effect upon the drag coefficients of blades and ship power.* Trans RINA; 1981.
18. *Propeller Surface Roughness and Fuel Economy.* SMM Technical Brief No. 18.
19. Sherrington I, Smith EH. *Parameters for characterising the surface topography of engineering components.* Trans I Mech E; 1987.
20. Townsin RL, Spencer DS, Mosaad M, Patience G. *Rough propeller penalties.* Trans SNAME; 1985.
21. Whipps SL. *On-line ship performance monitoring system: Operational experience and design requirements.* Trans I Mar E (TM) 1985;**98**(8).
22. Frenkiel FN, Landahl MT, Lumley JL. *Structure of turbulence and drag reduction.* Washington, DC: IUTAM Symp.; June 1976.
23. Berman NS. *Drag reduction by polymers.* Ann Review Fluid Mech 1978;**10**:27–64.
24. Bazari Z. *Ship energy performance benchmarking/rating: Methodology and application.* London: Trans. WMTC Conf. (ICMES 2006); March 2006.
25 Swain G. *Redefining antifouling coatings.* IPCL-PMC; September 1999.
26. Chambers LD, Stokes KR, Walsh FC, Wood RJK. *Modern approaches to marine antifouling coatings.* Surface & Coatings Technology 2006;**201**.

Chapter 25

Propeller Tolerances and Inspection

Chapter Outline
25.1 Propeller Tolerances 495
25.2 Propeller Inspection 495
    25.2.1 Inspection During Manufacture and Initial Fitting 496
    25.2.2 Inspection During Service 499
References and Further Reading 500

Although part of the manufacturing process, the related subject of propeller tolerances and inspection deserves separate attention. This is because it is only by correctly specifying the tolerances and then checking these have been adhered to, that the intentions of the designer can be properly realized. Without proper attention to blade manufacturing tolerances many serious problems can be encountered in the service life of the propeller: for example, cavitation, power absorption, noise, fatigue failure, and so on.

25.1 PROPELLER TOLERANCES

For general propeller design work the ISO specifications usually serve as the criteria for assessment. References 1 and 2 define the requirements for propellers greater than 2.5 m and between 0.80 and 2.5 m respectively. In certain cases, such as naval propellers, the purchasers of the propeller may impose their own particular tolerance specifications and methods of assessment: for example, the US Navy standard drawing method.

In general tolerances are normally specified on the set of dimensions shown in Table 25.1. This is because they affect the performance of the propeller or adjacent components in some particular way. Table 25.1 deals only with geometric parameters, but a propeller should also be shown to meet both the required material chemical composition tolerances and the minimum mechanical properties. The latter are of course classification society requirements for those vessels built under survey. For all vessels, however, attention needs to be given to the actual characteristic of the material for strength and repair purposes and these are not generally represented by the cast test pieces.

Of the geometric properties quoted in Table 25.1, each has some bearing on performance. It is, therefore, essential to understand the ways in which they influence the various propeller operational characteristics if the correct tolerance is to be specified. Unfortunately, the importance of each characteristic requires particular consideration for a given design but, notwithstanding this, certain general conclusions can be drawn and these are shown in Table 25.2. In this table an attempt is made to distinguish between the primary and secondary effects of the various parameters specified in Table 25.1, but not necessarily in relation to the ISO requirements.

As a consequence of the various effects detailed in Table 25.2, the designer and purchaser of the propeller need to determine what level of tolerance is required such that the propeller will be fit for the purpose for which it is intended. Indeed, one could specify the most stringent tolerance for every propeller; this, however, would be extremely wasteful in terms of additional costs of manufacture. The ISO specification defines four levels of tolerance: Classes S, 1, 2 and 3, these being in descending order of stringency. Again there is some latitude in deciding the correct tolerance level for a particular ship, but as a rough guide Table 25.3 has been prepared.

Table 25.3 is generally self-explanatory and other ship types and the appropriate tolerance classes can be deduced from those given in the table. Of particular concern, however, are the small high-speed vessels such as patrol or chase boats. All too often, in the author's experience, the subject of blade tolerances is completely neglected with these vessels leading to a host of cavitation related problems. Such vessels, by virtue of their speed, both in terms of ship and shaft rotational speeds, in association with a low static pressure head, should generally qualify the propeller for a Class S or 1 tolerance notation — sometimes more.

25.2 PROPELLER INSPECTION

The inspection of propellers needs to be undertaken both during the manufacture of the propeller and also during its service life. In general the former is undertaken in the

TABLE 25.1 Normal Propeller Tolerances Specification Parameters

Diameter
Mean pitch
Local section pitch
Section thickness
General section form (camber)
Section chord length
Blade form and relative location
Leading edge form
Rake and axial position
Surface finish
Static balance

machines. Such machines, whether they be manually operated or part of a CAM system, are in general a variant of the drop height measurement system. With this system the measurements are either made on the cylindrical sections defining the propeller or, alternatively, at various points defining a matrix over the blade surface. In the former case direct comparison with the cylindrical design sections can be made, whereas with the latter, as this frequently requires an interpolation procedure to be invoked, may raise questions of the validity of the mathematical model defining the blade surface.

In its most fundamental form the classical cylindrical measurement requires that the propeller be mounted in a gravitational or other known plane with a vertical pole, relative to the plane of mounting, erected on the shaft center line. To this pole is fixed a rotating arm which is free to rotate in a plane parallel to the plane on which the propeller is mounted; Figure 25.1. From this arm the various radii can be marked on the blade surface and drop heights to the blade surface measured at known intervals along the chord length. By undertaking this exercise on both surfaces of the propeller the blade section shape can be compared to the original design section. Whilst this is the basis of the measurement system, many refinements aimed at improving accuracy have been incorporated by manufacturers, each having their own version of the system.

The area which causes most concern is the detail of the leading edge. In a great many cases this is checked with the aid of a template (References 1 and 2). However, there is

relatively controlled conditions of the manufacturer's works, whilst the latter frequently, but not always, takes place in a dock bottom. Both types of inspection are important: the former to ensure design compliance with the design and the latter to examine the propeller condition in service.

25.2.1 Inspection During Manufacture and Initial Fitting

In the case of the fixed pitch propeller the inspection procedure is carried out with the aid of purpose-built

TABLE 25.2 Principal Effects of the Various Propeller Geometric Variations

Parameter	Primary Effect	Secondary Effect
Diameter	Power absorption	–
Mean pitch	Power absorption	Cavitation extent
Local section pitch	Cavitation inception and extent	Power absorption
Section thickness	Cavitation inception, blade strength	Power absorption
General section form (camber)	Power absorption, cavitation inception	Blade strength
Section chord length	Cavitation inception	Blade strength power absorption
Blade form and relative location (excluding leading edge)	Generally small effects on cavitation inception and shaft vibratory forces at frequencies dependent on wake harmonics and blade irregularities	–
Leading edge form	Critical to cavitation inception	–
Rake and axial position	Minor mechanical vibratory forms	–
Surface finish	Blade section drag and hence power absorption	–
Static balance	Shaft vibratory loads	–

Chapter | 25 Propeller Tolerances and Inspection

TABLE 25.3 Typical Tolerances for Certain Ship Types

ISO Tolerance	Typical Ships Where Tolerance Might Apply
S	Naval vessels* (e.g. frigates, destroyers, submarines, etc.). High-speed craft with a speed greater than 25 knots; research vessels; certain special purpose merchant vessels where nose or vibration is of paramount importance (e.g. cruise vessels, high-grade ferries).
1	General merchant vessels; deep sea trawlers; tugs, ferries, naval auxiliaries.
2	Low-power, low-speed craft, typically inshore fishing vessels, work boats, etc.
3	As for Class 2.

*Naval vessels are often specified on an 'ISO S Class Plus' basis.

also considerable use of optical methods for leading and trailing edge inspection: this is particularly true in the case of model propeller manufacture for model testing purposes where fine contol over dimensional tolerances is essential.

The current trend in manufacturing tolerance checking is to progress towards the introduction of electronic and laser-based techniques. These developments embrace electronic pitchometers, numerically controlled geometric inspection through to fully integrated design, manufacturing and inspection capabilities, outlined in Chapter 20, and the use of laser measurement techniques which exhibit very fine accuracy in either their fixed or portable forms. In all cases, however, it is of fundamental importance to define a sufficient set of inspection points in order to fully define the actual blade surface adequately so as to act as a basis of comparison with the design specification.

In addition to the blade profile tolerances, in the case of fixed pitch propellers, rigorous inspection needs to be given to the bore of the boss. The fitting of the propeller to the shaft requires considerable attention and the requirements for this are governed by the classification societies. For keyed propellers a satisfactory fit between the propeller and the shaft should show a light overall marking of the cone surface of the shaft taper with a tendency towards heavier marking in way of the larger diameter of the cone face. When conducting these inspections the final fit to the cone should be made with the key in place. In some cases the propeller is offered up to a shaft mandrel in the manufacturer's works so that the proper degree of face contact can be developed as required by the classification society rules. In cases where hand fitting is required this must be done by scraping the bore of the propeller; it should never be done by filing of the shaft cone.

With regard to the axial push-up required, Table 25.4 gives some typical guidance values for a shaft having a cone taper of 1 in 12. In this table D_s is the diameter of the shaft at the top of the cone and the axial push-up is measured from a reliable and stable zero mark obtained from the initial bedding of the propeller to the shaft. In cases where hydraulic nuts are used, great care needs to be exercised to ensure that the hub is not overstressed in way of the keyway and that the appropriate classification rule requirements are adhered to.

In the case of the keyed propeller it is of the utmost importance to prevent the ingress of sea water into the cone and as a consequence inspection needs to be particularly rigorous in this area. When the sealing arrangement comprises a rubber ring completely enclosed in a recess in the propeller boss, ample provision must be made for the rubber to displace itself properly to form a good seal. Alternatively, if an oil gland is fitted the following points should be carefully considered:

1. To ensure that rubber rings for forming the seal between the flange of the oil gland sleeve and the propeller boss are of the correct size and properly supported in way of the propeller keyway.
2. The fair water cones protecting propeller nuts and the flanges of sleeves of oil glands should be machined

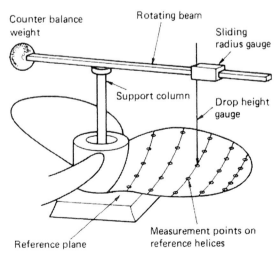

FIGURE 25.1 Mechanical pitch measurement principle.

TABLE 25.4 Typical Axial Push-up Values for Copper Alloy Propellers

Propeller Material	Axial Push-Ups
Aluminium bronage	$0.006 D_s$
High-tensile brass	$0.005 D_s$

smooth and fitted with efficient joints at their connection to propellers.

3. Drilling holes through the propeller boss should be discouraged but when these are essential to the design special attention needs to be paid to the efficient plugging of the holes.
4. The arrangement for locking all screwed components should be verified.
5. The propeller boss should be provided with adequate radius at the large end of the bore.
6. When the design of the oil gland attachment to the propeller is similar to that shown in Figure 25.2, it is good practice to subject the propeller boss to a low-pressure air test, checking all possible sources of leakage with a soapy water solution in order to prove tightness.

When a keyless propeller is fitted to the shaft the inspector should pay particular attention to ensuring that the design and approved interference fit is attained. As a prerequisite for this procedure the inspector needs to know the start point load to be applied and the axial 'push-up' required. These should normally be supplied at two temperatures, typically 0°C and 35°C, to allow interpolation between the two values to take place to cater for the actual fitting conditions. The inspector should carefully examine the final marking of the screw shaft cone fit; this should show a generally mottled pattern over the entire surface with harder marking at the large end of the cone.

Two basic techniques are employed to fit keyless propellers: the dry press-fit or the oil injection method. With both methods the propeller is pushed up the shaft cone by means of a hydraulic nut, but the fitting procedure differs somewhat between the two methods and the manufacturer's fitting procedure must be rigidly adhered to in all cases. As a final stage in the inspection it is essential that the propeller, both working and spare, be hard-stamped with information of the form detailed below on the outside of the boss away from any stress raisers or fillets:

1. Oil injection type of fitting
    i. Start point load (tonnes)
    ii. Axial push-up at 0°C (mm)
    iii. Axial push-up at 35°C (mm)
    iv. Identification mark on associated screw shaft
2. Press fit type of fitting (dry)
    i. Start point load (tonnes)
    ii. Push-up load 0°C (tonnes)
    iii. Push-up load 35°C (tonnes)
    iv. Axial push load 0°C (mm)
    v. Axial push load 35°C (mm)
    vi. Identification mark on associated screw shaft

With regard to the process of fitting the propeller to the shaft a good review of methods is given by Eames and Sinclair.[3]

Casting defects always occur in propellers. On occasions those outcropping on the surface of the propeller are concealed by the use of unauthorized local welding. When subsequently polished and on a newly manufactured propeller which has been kept in a workshop, the existence of small amounts of surface welding can be very difficult to spot with the naked eye. This type of welding process, when not fully authorized, is a dangerous practice from the propeller integrity point of view. Therefore, if any doubt exists about the processes that have been undertaken the propeller surfaces should be lightly etched with an appropriate solution. This, depending upon the etching solution used, will reveal the presence of any such actions in a longer or shorter time. However, if the propeller has been left outside of the manufacturer's compound and subjected to rain over a period of a week or so the acids in the rain will naturally etch the propeller surfaces and reveal any weld processes that have taken place. Similarly with propellers that have been in service, the action of the sea water has much the same effect and the welding history will be seen in dry dock.

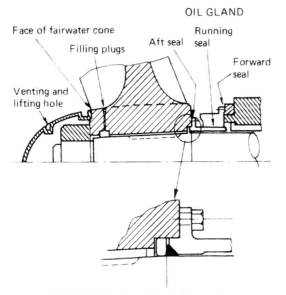

FIGURE 25.2 Propeller shaft assembly.

Propeller Tolerances and Inspection

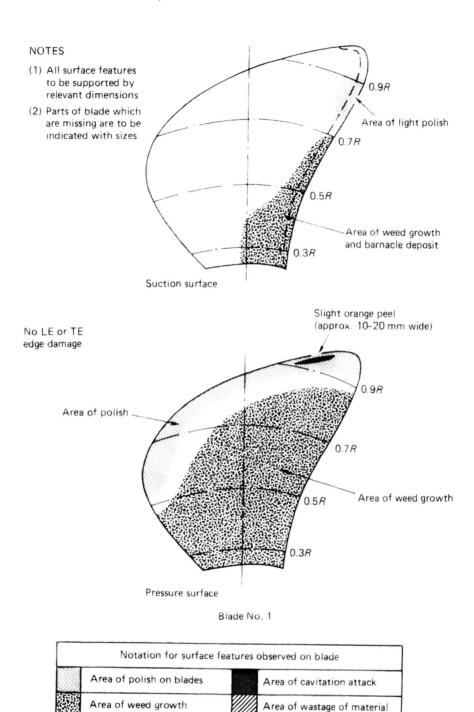

FIGURE 25.3 Typical blade inspection diagram.

25.2.2 Inspection During Service

In-service inspections are normally carried out for maintenance or survey reasons: to examine the propeller after suspected damage has take place or to check for fouling, or, alternatively, to form part of a survey of the ship. In the former case this may be carried out in the water by a diver if a superficial check is required, or in a dry dock for a more detailed survey. As a general comment on in-water and out-of-water inspections it should be remembered that a commercial diver is usually a highly trained person but is not normally a propeller technologist and, therefore, can give only generalized engineering reports. Furthermore, in many instances, typically in the North Sea area, while looking at one part of the propeller the diver may not be able to see the rest of the propeller due to the clarity of the water. Irrespective of water clarity, the in-water survey is greatly enhanced if the diver can talk, preferably with the aid of video techniques, to a propeller specialist while undertaking the survey, so that important details will not be missed and other less important features given undue weight.

When a propeller is operating any small cracks or defects tend to collect salt deposits. As a consequence, before conducting a blade surface inspection in a dry dock aimed at identifying cracks, the surface should be lightly cleaned to remove marine growth and then washed with a 10 per cent concentration of a sulphuric acid in water to dissolve the salts. If the propeller is removed from the shaft for this examination, then this is extremely helpful to the inspector and considerably increases the chances of finding small defects. It is, however, completely pointless to attempt an examination designed to look for small cracks in water.

To undertake a general propeller inspection it is a prerequisite to have an outline of the propeller which, although not needing to be absolutely correct in every geometric detail, must represent the main propeller features adequately. The outline should show both the face and back of the blades and have suitable cylindrical lines marked on it at say $0.9R$, $0.8R$, $0.6R$, $0.4R$ Without such a diagram serious misrepresentations of information can occur. In the author's experience the best blade outline to use for damage recording is the developed blade outline, since this tends to represent the blade most closely to the way an inspector observes it. In addition to a blade outline there should also be a consistent way of recording information to signify, for example, cavitation damage, bending, missing portions, marine growth, and so on. Figure 25.3 shows a typical diagram for inspection purposes and in addition to showing the location of the damages on the blade, typical dimensions of damage length, width and depth need to be recorded. These records need to be taken individually for each blade, on both the back and face of the propeller, so as to answer questions of the similarities of blade damage since a propeller may exhibit damage from different sources simultaneously.

In the case of a classification society inspection the propeller is normally required to be removed from the tail shaft at each screw shaft survey. On these occasions particular attention should be paid to the roots of the blades for signs of cracking.

If a new propeller is to be installed, the accuracy of fit on the shaft cone should be tested with and without the key in place. Identification marks stamped on the propeller should be reported for record purposes and if the new propeller is substantially different from the old one, it should be recalled that the existing approval of torsional vibration characteristics may be affected by significant changes in the propeller design.

It is particularly important to ensure that rubber rings between propeller bosses and the aft ends of liners are the correct size and so fitted that the shaft is protected from sea water. Failure in this respect is often found at the ends of keyways due to the fact that the top part of the key itself is not extended to provide a local bedding for the ring in way of the recess in the boss. In such cases it may be found practicable to weld an extension to the forward end of the key. It should also be recollected that water may enter the propeller boss at the aft end and attention should therefore be paid to this part of the assembly. Filling the recess between the aft end of the liner and the forward part of the propeller boss with grease, red lead or a similar substance is not in itself a satisfactory method of obtaining water-tightness. Sealing rings in connection with approved type oil glands should be similarly checked.

If an oil gland is fitted the various parts should be examined at each inspection and particular attention paid to the arrangement for preventing the ingress of water to the shaft cone. All oil glands, on reassembling, should be examined under pressure and shown to be tight.

If the ship has a controllable pitch propeller, the working parts and control gear should be opened up sufficiently to enable the inspector to be satisfied of their condition. In the case of directional propellers, at each docking the propeller and fastenings should be examined as far as practicable and the maneuvering of the propeller blades should be tested.

REFERENCES AND FURTHER READING

1. ISO 484/1. *Shipbuilding — Ship Screw Propellers — Manufacturing Tolerances — Part 1: Propellers of Diameter Greater than 2.50m*, 1981.
2. ISO 484/2. *Shipbuilding — Ship Screw Propellers — Manufacturing Tolerances — Part 2: Propellers of Diameter Between 0.80 and 2.50m Inclusive*, 1981.
3. Eames CFW, Sinclair L. *Methods of attaching a marine propeller to the tailshaft*. Trans IMarE 1980;**92**.

Chapter 26

Propeller Maintenance and Repair

Chapter Outline

- 26.1 Causes of Propeller Damage — 501
  - 26.1.1 Cavitation Erosion Damage — 501
  - 26.1.2 Maltreatment Damage — 501
  - 26.1.3 Mechanical Service Damage — 502
  - 26.1.4 Wastage Damage — 503
- 26.2 Propeller Repair — 503
  - 26.2.1 Blade Cracks — 503
  - 26.2.2 Boss Cracks — 504
  - 26.2.3 Repair of Defective Castings — 504
  - 26.2.4 Edge Damage to Blade — 504
  - 26.2.5 Erosion Damage — 504
  - 26.2.6 Maintenance of the Blade in Service — 505
  - 26.2.7 Replacement of Missing Blade Sections — 505
  - 26.2.8 Straightening of Distorted Blades — 505
- 26.3 Welding and the Extent of Weld Repairs — 505
- 26.4 Stress Relief — 507
- References and Further Reading — 508

The repair and maintenance of a propeller is of considerable importance if a propeller is to give a high-performance and reliable service throughout its life. Damage can arise from a number of causes and the propellers should be regularly examined for signs of this occurrence.

26.1 CAUSES OF PROPELLER DAMAGE

In general terms, propeller damage can be classified into one of four distinct types: cavitation erosion damage, maltreatment, mechanical service damage and wastage. Sometimes one type of damage gives rise to another; for example, mechanical damage then initiating cavitation erosion.

26.1.1 Cavitation Erosion Damage

Cavitation damage will occur in situations where the propeller is either working in a particularly onerous environment, in terms of immersion or inflow conditions, that cannot be accommodated by good design. Alternatively, poor design may also be a cause of damage. In either case cavitation is likely to be the primary source of damage. Conversely, cavitation damage can result from flow disturbances created by mechanical damage, such as a leading edge tear or bend, in which case the cavitation erosion is a secondary damage source. Figure 26.1 shows a typical example of cavitation erosion on a propeller blade and the mechanism by which this damage occurs is discussed in Chapter 9. A further example of cavitation damage is the phenomenon of trailing edge curl, which is also discussed in Chapter 9 and shown in Figure 9.34.

Cavitation erosion damage may either stabilize at some depth or continue to progress. If it progresses the time taken may vary considerably from case to case. In some cases it can develop relatively slowly or, alternatively, at an extremely fast rate: this might be as high as a few hours for complete penetration in some particularly aggressive instances. As a consequence, when cavitation damage is first noticed a check should be made to determine the rate at which the erosion attack is progressing. This is so that effective repair and remedial action can be planned and implemented.

26.1.2 Maltreatment Damage

Maltreatment damage may result from many causes; for example, incorrect handling, surface deterioration or severe heating of the boss or blades of a propeller.

Incorrect handling most commonly leads to blade edge or tip damage on the blades during transport. This damage should be avoided by the proper use of soft edge protectors, typically copper, rubber or lead, in order to prevent the propeller coming into direct contact with craneage slings. Additionally, these edge protectors should be augmented by the use of timbers or heavy tires in the appropriate fulcrum positions. If lifting eyebolts are fitted these should always be used.

Surface deterioration is most commonly the result of a loss of protective coating, paint splashes or other markings. Moreover, the protective coating applied to new propellers, which is normally a colorless varnish or self-hardening dewatering oil, should be completely removed

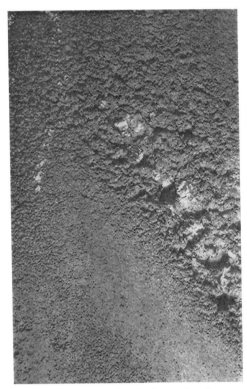

FIGURE 26.1 Typical cavitation erosion damage.

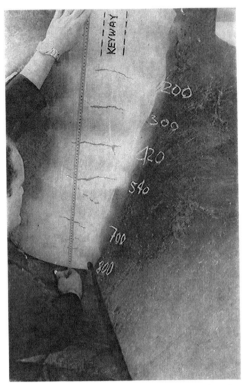

FIGURE 26.2 Boss cracking due to local heating abuse.

before the vessel enters service after fitting out. The purpose of the coating is to prevent fouling or other light damage such as paint splashing and impact damage. During painting operations the propeller should be covered at all stages during the operation so as to prevent splashes adhering to the surface as these can lead to cavitation, erosion or severe local pitting.

If heating is applied in an arbitrary manner to a propeller, this will create internal stresses in the material. These internal stresses occur when a local region of metal is heated and tries to expand, but is prevented from doing so by the surrounding cooler metal. As a consequence, compressive stresses build up until they are relieved eventually by the onset of plasticity. This continues until red heat is reached when internal stresses in the heated member can no longer exist. On cooling the heated area contracts and the incipient tensile stresses are relieved by plastic flow. However, as the cooling progresses the plastic flow becomes slow and at about 250°C it ceases, which then allows a tensile stress field to build up in the cooling metal. Typical causes of this local heating abuse are incorrect heating of the boss to aid removal and the repair of the blade by inappropriate methods. In the first instance stress corrosion cracks are frequently initiated by concentrated heat sources such as oxy-acetylene and oxy-propane processes being applied to the boss, which then produce high-tensile residual stresses in the manner just described. Although giving the appearance of a satisfactory propeller removal and replacement operation, this will lead to cracking some weeks or months after re-immersion in the sea water. Figure 26.2 shows a typical cracking pattern caused by local heating abuse. If heat is applied to the propeller boss this should be done with great care using either steam or low-temperature electric blankets positioned all over the boss surface. Alternatively, the use of a soft flame may be acceptable provided it is applied to each section of the boss and kept moving to prevent hot spots from occurring.

26.1.3 Mechanical Service Damage

This is perhaps the most common form of damage that a propeller blade encounters and it is normally caused by contact with floating debris, cables or chains. Figure 26.3 shows a propeller after having fouled the chains of a mooring buoy.

In most cases impact damage involves only minor damage to the blades, typically in the edge region. While such damage does not normally impair the strength integrity of the blade, advice should be sought as to whether there is the likelihood of secondary cavitation damage occurring and, if so, a temporary grinding repair may prove helpful and save further damage if the full repair cannot be effected immediately.

When a significant portion of the propeller is damaged and perhaps missing, or if the blade has suffered a fatigue

Propeller Maintenance and Repair

FIGURE 26.3 Propeller damage due to impact damage with mooring chains.

failure, then immediate attention is required. It may be, however, that a dry dock or the replacement propeller is some considerable distance away from the ship; in such cases a calculation of the out-of-balance forces should be made to establish the rotational speed at which the lubricant film can just be maintained in the stern bearing and a suitably de-rated shaft speed determined. Moreover, as discussed in Chapter 23, when a blade failure occurs cropping of the other blades may be considered in order to reduce the out-of-balance forces.

26.1.4 Wastage Damage

All propellers will exhibit some damage throughout their life due to wastage when compared to their new state: this is a quite normal process of corrosion.

For example, in the case of a high-tensile brass propeller it would be reasonable to expect a loss of the order of 0.05 mm per annum, assuming that the composition of the material was such that dezincification did not occur and that the propeller was lying at rest in still or slowly moving water. If, however, the propeller were operating normally with tip speeds in the region of 30 m/s, then this wastage might well rise to around 0.15 or 0.20 mm per annum and perhaps more in the case of ships like suction dredgers or trawlers. This effect leads to a general roughening of the propeller when compared to its new state.

Wastage is principally a process of corrosion, and it is well known that when small anodic areas are adjacent to large cathodic surfaces an electrochemical attack can take place. Moreover, this attack is accelerated when the surfaces are exposed to high-velocity turbulent flows. When abnormal wastage occurs on a propeller it is generally due to a corrosion process taking place and a typical example of this is often found during the fitting out stage of the vessel. At this time the propeller becomes a cathode in the propeller–hull electrolytic cell and a hard and strongly adherent coating of magnesium and calcium carbonate appears on the surfaces. When the vessel enters service this

cathodic chalk film becomes worn away in a patchy manner in the outer regions of the propeller blades. These areas and others where the establishment of the film has been delayed allow corrosion to proceed at an abnormally high rate, leading to localized depressions and pitting of the surface. This also increases the turbulence level and hence promotes the conditions to enhance the corrosion rates.

26.2 PROPELLER REPAIR

Propeller repair is a complex and extensive subject and the precise details of repair methods should be left to specialists in this field. The owner who, for whatever reason, allows non-experienced personnel or companies to repair a damaged propeller may well find that decision will eventually result in the premature loss of the propeller. The propeller is an extremely complex engineering artifact, manufactured from complex and advanced materials that operate in a hostile and corrosive environment. Its maintenance and repair therefore deserve considerable care and respect.

As a consequence of these remarks, the remainder of this section will set out some of the underlying principles of repair but leave the detail, which relates to each material specification, to the specialist repairer.

26.2.1 Blade Cracks

Experience has shown that all cracks in propeller blades are potentially dangerous and this is particularly true of cracks close to the leading edge. Cracks normally grow by fatigue action, but in cases where the propeller fouls some substantial object a notch can be created and then initiate rapid failure.

Where a crack is found on the leading edge of a propeller blade the crack should be ground out after any straightening has been carried out. During the grinding process care must be taken to ensure that the crack tip has been eradicated because if the tip remains it can act as a further initiation point for another crack. If the crack is very small it is generally best to fair the ground out portion into the existing blade form; however, if this is not possible, then recourse has to be made to a weld repair.

In the case of highly skewed propellers, if cracks are found in the vicinity of the trailing edges of the blades then this is potentially a very serious situation. As discussed previously, in Chapter 19 and shown in Figure 19.6, the highly skewed propeller may exhibit high stress concentrations along the trailing edge. Therefore, when cracks are observed in the trailing edge regions of highly skewed blades immediate advice should be sought on how to resolve the problem. Simply grinding the cracks out may not be a solution because if significant subsurface defects are found in this region of the blade, these may only accentuate the additional stress field induced during the re-profiling by grinding and, in this way, may induce further crack initiation sites. If cracks are noticed in the trailing edge region of controllable pitch propeller blades, the affected blades should be replaced with spares and remedial action on the cracked blades considered.

If a crack is found in the body of the blade then it should first be ground out to a significant depth into the section, typically rather more than half the thickness of the section at that point. The cross-sectional profile of the ground portion should be such as to give a 'V' form with an included angle of the order of $90°$. The repair must then be completed with a suitable welding technique. Following completion of the weld repair the surface of the weld must then be ground down to conform to the designed blade surface profile and the repair examined for lack of penetration or similar defects prior to initiating the stress-relieving operation.

If cracks are found close to the propeller boss, inboard of about $0.45R$, then welding processes leave high residual stress fields which in general can only be relieved by annealing the whole propeller. If the crack is small in this inner region, then consideration should be given to grinding the crack out and fairing the resulting depression into the blade form: the determinant in this situation is the strength of the blade.

26.2.2 Boss Cracks

Cracks in the propeller boss or hub of a controllable pitch propeller are always serious. They are, however, not normally found until they are too deep to repair. If they are sufficiently shallow to permit their complete removal by grinding, then this can be contemplated as a repair; however, before returning the propeller to service a strength evaluation should be made. When the boss is capable of repair in this way it would clearly be beneficial to apply some low-temperature stress relief.

26.2.3 Repair of Defective Castings

The propeller casting which is free from defect has yet to be made; however, most propellers are cast with an acceptable level of defects present. When a defect is considered to need repair caution must be exercised to ensure that this is done properly.

In general small surface defects such as pores of the order of 1 mm in diameter do not need rectification except where they occur in close-packed groups in highly stressed regions of the blade. However, when an unacceptable defect is found in a blade the defect should be ground out and gradually faired into the surrounding blade and a check made on the effects of this action on the blade strength. The integrity of the blade should then be demonstrated by a dye penetrant examination.

The repair of defects by welding should not be the first immediate response after noticing a defect on the blade surface. This is because welding can potentially introduce a greater problem than the original defect if the welding process is not properly carried out. Classification societies in general impose limits on the extent of weld repairs to propeller castings, as discussed in Section 26.3. Additionally, welds having an area of less than 5 cm^2 are not to be encouraged.

26.2.4 Edge Damage to Blade

Edge damage to propeller blades generally takes the form of local bending or tearing of the metal. Sometimes cracking also occurs at the time the damage occurs or afterwards, in which case the comments of 26.2.1 are applicable.

If the tearing is of a minor nature, the affected blade edges should be dressed back to the undamaged shape, which may result in some slight loss of blade area. If the loss of material is confined to around 10 mm in the chordal direction then there should be little effect on the propeller's performance, provided proper attention has been paid to reprofiling of the blade edge. Should the tearing be greater than that identified above, consideration should be given to replacement of that section of the blade by cutting the blade back and inserting a new piece of material. As discussed in Section 26.2.1, care needs to be exercised in the case of the trailing edge of highly skewed propellers.

Small distortions of the blade along the edge can normally be corrected by cold working with the use of clamps, but for greater distortions hot straightening techniques need to be deployed.

After repair of edge damage the edges need to be examined closely for signs of any remaining cracks that may give trouble in the future. The size of these cracks may be very small and difficult to see with the unaided eye and, therefore, a dye pennant inspection should be made.

26.2.5 Erosion Damage

Damage caused by cavitation erosion is generally repaired by welding. Consequently, to undertake repairs the propeller normally needs to be transferred to a protected area where good environmental control can be exercised. The area subjected to attack needs to be cut or ground out cleanly and then filled by welding. After welding the repaired surface must be ground to the blade profile and the repair examined for its integrity. Stress-relieving should then be carried out in accordance with the manufacturer's recommendations.

Whenever cavitation erosion damage is seen, the necessary modifications to the blade geometric form to prevent recurrence should be contemplated, assuming the erosion is not secondary to some other damage that the propeller has experienced. If feasible, these modifications should be implemented because if this is not done the erosion repair will most likely become a regular feature of the docking cycle.

26.2.6 Maintenance of the Blade in Service

During the life of a propeller it should be regularly checked to see that it remains in a clean and unfouled state. The implications of fouling on the propeller were discussed in Chapter 24.

If when inspected the blade surfaces are smooth, showing no roughness, chalking, fouling or wastage, then they should not be touched and left until the next inspection. If, however, any of these attributes are seen the blade should be lightly ground and polished by a competent organization: to allow this work to be done by others can lead to the abuse of the propeller and a consequent set of other in-service problems. To allow a propeller to continue in operation with local pits or more widespread roughness can, in time, give rise to the need for repair rather than simple and straightforward polishing.

Even when a propeller has been neglected, provided the local pits and depressions are not more than 1 mm or so deep, this can generally be rectified by light grinding and polishing. If the neglect has resulted in the propeller being in a worse state, then more serious attention is necessary and this is best left to a manufacturer because heavy grinding of the surface can lead to surface changes sufficient to affect propulsive efficiency and cavitation performance.

26.2.7 Replacement of Missing Blade Sections

If a section of a blade is partially torn off, lost by impact damage or cut away to remove a severely bent or damaged area this can often be replaced. The affected area is trimmed and dressed and a new blade piece, which has been especially cast for the purpose, is either welded or burnt on.

Before considering this action the extent of the repair has to be carefully considered. Such action is possible for a blade tip or parts of the edges, but it is not suitable for a complete blade replacement. The replacement of part of a blade is a major undertaking and, therefore, requires expert craftsmanship and advice.

26.2.8 Straightening of Distorted Blades

Apart from cold straightening methods for small areas of damage, as mentioned earlier when discussing edge damage, the method for straightening large amounts of bending damage is through the use of hot working techniques.

With these techniques, after having carefully assessed the extent of the damage, the back of the blade should be heated slowly in the area of damage and a significant area around it. The ideal method for this is to use a coke brazier which can be blown up to the appropriate temperature by forced draught fans. An alternative to the coke brazier would be a soft flame such as paraffin, coal gas or propane air burners. On no account should a hard flame be used, for example oxy-gas burners, because of the risk of local melting of the blade. During the heating process the top surface of the blade should be lagged to prevent heat loss.

When the blade has been heated to a suitable temperature depending on the material for working, about 150–250°C below the material melting point, the straightening process is best carried out using weights and levers. This process should not be hurried, but carried out very slowly avoiding the use of hammers as far as practicable. When the straightening process has been completed the propeller should be allowed to cool very slowly with the blade thermal lagging in place. With this process stress relief should not normally be necessary, but where doubt exists a full heat treatment should be applied.

Clearly, after such a process has been carried out the blade geometry must be checked, and this normally requires that the propeller is removed from the shaft. As a consequence, this type of blade straightening exercise would normally be carried out off the shaft: this is especially true if major straightening is required and needs the use of a hydraulic press.

In addition to the repair of damage, these methods have been deployed to adjust the pitch of a propeller which has not given the correct power absorption characteristics. In this case the blade needs to be carefully instrumented with thermocouples and the process is preferably undertaken with a number of flexible electrical heating mats which are easily controlled to give the correct temperature distribution: in these applications the leading and trailing edges heat up more quickly than the central body of the blade. When the blade has been heated to the hot forming temperature, it is twisted using a hydraulic ram placed near the tip of the blade. The forces involved are generally small and the twisting process takes only a short time.

26.3 WELDING AND THE EXTENT OF WELD REPAIRS

As a general rule all welding on a propeller should be done with metal arc processes using approved electrodes or wire filler. In certain circumstances gas welding can be used on high-tensile brass but not on other materials and most classification societies do not generally approve of gas welding.

The area to be welded needs to be both clean and dry and so it is preferable to conduct these operations under cover and in places relatively free from dust, moisture or draughts. If flux-coated electrodes are used they should be preheated for about one hour at a temperature of about 120°C. Where required by the material specification the area to be repaired should also be preheated to the desired temperature and the preheat maintained until the welding is complete. Welding should always be conducted in the downhand position and all slag must be chipped away from any undercuts or pockets between consecutive weld runs. Upon completion of the weld repair the area should be stress-relieved with the exception, in some cases, of the nickel–aluminum bronzes. With these materials classification societies (see, for example, Reference 1) require that stress relief be implemented where the weld has been carried out to the blade edge between $0.7R$ and the hub. Stress relief for the nickel–aluminum bronzes is also required where welding has been carried out between the bolt holes on controllable pitch propeller blades.

When a weld repair is complete the weld should then be ground smooth for visual examination and dye penetrant testing. If stress relief is to be employed, then a visual examination should be carried out prior to the stress relief and both a visual and a dye penetration examination carried out afterwards.

The extent of welding is governed by the requirements of the various classification societies. These requirements are based on the likely stress fields present in the blades and the potential consequences of welding in the various parts of the propeller. For example, the permitted areas are given in Table 26.1 and the Zones A, B and C are defined for conventional, highly skewed and vane wheel blade forms. The Zones A, B, C are defined for illustration purposes in Figure 26.4 taken from Reference 1: in practice, however, the current rules of the appropriate classification society must be used. In addition to the blade requirements the boss of the fixed pitch propellers is divided into three individual

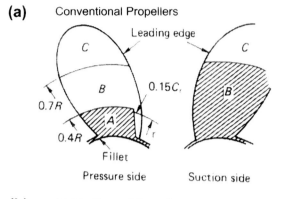

(a) Conventional Propellers

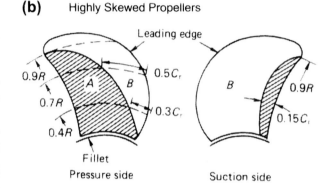

(b) Highly Skewed Propellers

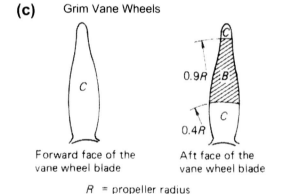

(c) Grim Vane Wheels

R = propeller radius
C_r = chord length at radius r

FIGURE 26.4 Blade welding severity zones.

TABLE 26.1 Permissible Areas of Welding as Defined by LR Rules 2006		
Severity Zone or Region	Maximum Individual Area of Repair	Maximum Total Area of Repairs
Zone A	Weld repair not generally permitted	
Zones B, C	60 cm² or 0.6% × S, whichever is the greater	200 cm² or 2% × S, whichever is the greater in combined Zones B and C, but not more than 100 cm² or 0.8% × S, whichever is the greater, in Zone B on the pressure side
Other regions	17 cm² or 1.5% × area of the region, whichever is the greater	50 cm² or 5% × area of the region, whichever is the greater

S: area of one side of a blade = $0.79\, D^2 A_D/Z$.
D: finished diameter of propeller.
A_D: developed area ratio.
Z: number of blades.
Note: When separately cast blades have integral journals, weld repairs are not generally permitted in the fillet radii or within 12 mm of the ends of the radii. When repairs are proposed in these locations, full particulars are to be submitted for special consideration.

Propeller Maintenance and Repair

TABLE 26.2 Other Regions of the Propeller as Defined by LR Rules

Fixed Pitch Propeller Bosses	Controllable Pitch Propeller and Built-Up Propeller Blades
The bore	The surfaces of the flange to the start of the fillet radius
The outer surfaces of the boss to the start of the fillet radius	The integrally cast journals
The forward and aft faces of the boss	

TABLE 26.3 Stress Relief Soaking Times Taken From LR Rules 2006

Stress Relief Temperature (°C)	Manganese Bronze and Nickel–Manganese Bronze		Nickel–Aluminum Bronze		Manganese–Aluminum Bronze	
	Hours per 25 mm of Thickness	Maximum Recommended Total Time (Hours)	Time per 25 mm of Thickness (Hours)	Maximum Recommended Times (Hours)	Time per 25 mm of Thickness (Hours)	Maximum Recommended Time (Hours)
350	5.0	15.0	–	–	–	–
400	1.0	5.0	–	–	–	–
450	0.5	2.0	5.0	15.0	5.0	15.0
500	0.25	1.0	1.0	5.0	1.0	5.0
550	0.25	0.5	0.5	2.0	0.5	2.0
600	–	–	0.25	1.0	0.25	1.0

regions while the palms of controllable pitch propeller blades are similarly divided into two other regions. These regions are given in Table 26.2 for the current edition of these regulations (Reference 1) and relate to the other regions defined in Table 26.1.

26.4 STRESS RELIEF

The stress relief of a propeller blade after a repair involving heat, particularly welding, is an important matter. The residual stresses induced by these heating operations are sufficient to lead to stress corrosion cracking in high-tensile brass and the manganese–aluminum bronzes. The operations of principal concern are welding, straightening and burning on of sections of the blades. In the case of hot straightening applied to a blade it is normally sufficient to lag the heated area with asbestos blankets so that it cools slowly: if the method of heating has been via a coke brazier then it is helpful to let the fire die down naturally under the blade.

In the case of the high-tensile brasses and manganese–aluminum bronzes a large area of the blade which embodies the repair should be heated to a predetermined temperature, around 550°C in the case of high-tensile brass and 650°C in the case of manganese–aluminum bronze. The heat should be applied slowly and uniformly such that the isothermals are generally straight across the blade. The blade should then be allowed to cool as slowly as possible by lagging the blade and protecting the area from draughts; during the entire stress relief process the control of the temperature needs to be carefully monitored. When the stress relief process is complete it is desirable to grind a portion of the blade surface and polish and etch it in order to demonstrate that a satisfactory microstructure has been produced.

For the nickel–aluminum bronzes the production of residual stresses can be minimized by ensuring that a large enough area of the blade is heated during the repair process. When the repair is complete this area should then be allowed to cool as slowly as possible so as to permit a slow plastic flow and relief of stress at the lower temperatures. During this process and particularly in the region of

300–400°C, the isothermals should be maintained as straight as possible across the blade.

Classification societies have strict procedures for stress relief and require that for all materials, except in certain cases for the nickel–aluminum bronzes, stress relief be carried out for weld repairs. Reference 1 gives the exception for nickel–aluminum bronze to be where the weld repair has not been carried out to the blade edge inboard of the $0.7R$ and where repairs have not been made between the bolt holes or flange of a separately cast blade. Table 26.3 illustrates soaking times for the various propeller materials, again taken from Reference 1.

REFERENCES AND FURTHER READING

1. *Rules for the manufacture, testing and certification of materials.* Lloyd's Register of Shipping; January 1994.
2. Sinclair L. *The economics and advisability of ships propellers care and maintenance.* Ships Gear Int Symp; July 1968.
3. Meyne KJ. *Propeller repairs.* Ship Repair; May 1977.

Bibliography

Bahgat F. *Marine Propellers*. Alexandria: Al Maaref. Est.; 1966.

Bavin, Zavadovski, Levkovski, Mishevich. *Propellers: Modern Methods of Analysis*. Idatelstvc Sudoost-royeniye, Leningrad.

Biles JH. *The Design and Construction of Ships*, Vol. 2. London: Griffin; 1923.

Brennen CE. *Fundamentals of Multiphase Flow*. Cambridge; 2005.

Breslin JP, Andersen P. *Hydrodynamics of Ship Propellers*. Cambridge; 1994.

Cane P du. *High Speed Small Craft*. Temple Press Books; 1951.

Faltinsen OM. *Hydrodynamics of High-Speed Marine Vehicles*. Cambridge; 2005.

Geer D. *Propeller Handbook*. Nautical; 1989.

Hannan TE. *Strategy for Propeller Design*. Reed; 1971.

Isay WH. *Kavitation*. Schiffahrts-Verlag, Hansa C. Schroedter; 1981.

Lammeren WPA van. *Resistance, Propulsion and Steering of Ships*. Haarlem, Holland: The Technical Publishing Company H. Stam; 1948.

Molland AF, Turnock SR. *Marine Rudders and Control Surfaces*. Butterworth-Heinemann; 2007.

Milne-Thompson LM. *Theoretical Hydrodynamics*. 5th edn. Macmillan; 1968.

Newman JN. *Marine Hydrodynamics*. MIT Press; 1977.

O'Brien TP. *The Design of Marine Screw Propellers*. Hutchinson; 1962.

Perez-Gómez G, Gonzalez-Adalid J. *Detailed Design of Ship Propellers*. FEIN; 1997.

Philips-Birt D. *Ship Model Testing*. Leonard Hill; 1970.

Principles of Naval Architecture. SNAME; 1988.

Rusetskii AA. *Hydrodynamics of Controllable Pitch Propellers*. Sudoostroyeniye; 1968.

Saffman PG. *Vortex Dynamics*. Cambridge; 1992.

Saunders HE. *Hydrodynamics in Ship Design*. SNAME; 1957.

Taggart R. *Marine Propulsion: Principles and Evolution*. Gulf; 1969.

Warnecke H-J. *Schiffsantriebe*. Koehler; 2005.

Young FR. *Cavitation*. McGraw-Hill; 1989.

Index

Note: Page numbers followed by f, indicate figures and t, indicate tables.

A

Absolute salinity, 48–49, *see also* Salinity
Accelerating duct, 14–15, 14f, 91, 110, 201, 435, 450
Added mass, inertia and damping, 275–279
Additional thrusting fins, 341–342, 341f
Adjustable pitch propeller, 7
Advance coefficient, 21, 70, 79–82, 87–88, 90–91, 102, 112, 115–117, 120–133, 136, 216–217, 221, 239, 243, 245–246, 286, 319, 322–324, 346, 357, 380, 435–437, 442, 445, 460–461, 484
Advance ratio, 178–179, 239
Aerofoil section, 38–39
   boundary layer growth over, 154–158
   characteristics, 139–141
   thickness, 39, 43t, 44t
Air resistance, 330–331
Allowable design stresses, 410–412
Aluminum bronzes, 390–391
American Towing Tank Conference (ATTC), 308–310, 319
Annular gap, 297
Appendage skin friction, 309
Area, of propeller, 35–38
   calculation algorithm for, 37f
   developed, 36
   expanded, 36–37
   projected, 35–36
   swept, 37–38
Asymmetric stern, 334–335, 335f
Auf'm Keller method, 312–313
Auxiliary propellers, 6
   characteristics of, 438t
Average hull roughness (AHR), 474–476
Axial flow pump, 365, 367
Axial push-up, 497, 498t
Ayre's method, 312
Azimuthing propellers, 7, 16, 16f
Azimuthing thrusters, 16

B

Backing stresses, 407
Balanced skew design, 33, 402–403, 408
Bearing forces, in propeller–ship interaction, 271–288
   forces and moments, 279–287
   induced by turning maneuvers, 287–288
Beaufort Wind Scale, 53, 54t
Behind-hull propeller characteristics, 133–134
Bernoulli's propeller wheel, 1, 2f
Bernoulli's theorem, 171, 304–305
Bevis–Gibson reversible propeller, 7
Biased skew design, 33, 402–404, 408, 452
Biot–Savart law, 143–144, 143f, 181, 189
   semi-infinite line vortex, 143
   semi-infinite regular helical vortex, 143–144, 144f
Blades
   area ratio, 36, 37t, 448
   composite propeller, 452
   cracks, 503–504
   damping, 427–428
   edge damage, 504
   element model, 161
   element theory, 171–172, 172f
   erosion, 461–462
   failure, 370, 401, 402f, 465–466, 502–503
   interference limits, 43
   number, 11–12
   reference lines, 30f
   root fillet design, 221, 369–370, 407–408
   thickness distribution, 42–43
   thickness fraction, 42–43
   weight, 272–273, 394, 401
Blade strength, 397–414
   allowable design stresses, 410–412
   blade root fillet design, 407–408
   cantilever beam method, 397–401
   design considerations, 404–407
   full-scale blade strain measurement, 412–413
   numerical blade stress computational methods, 401–404
   propeller backing stresses, 407
   residual blade stresses, 408–409
Blade vibration, 421–430
   in air, 422
   blade damping, 427–428
   finite element analysis, 426–427
   flat-plate blade vibration, in air, 421
   propeller singing, 428–429
   simple estimation methods, 425–426
   water immersion effect, 422–425
Bollard pull condition, 13–14, 79, 82–84, 91, 115, 362, 379–381, 407, 460–461
   effective, 380
   maximum, 379
   required measurements, 381
   steady, 380
   trial location and conditions, 380
Boss cracks, 504
Boss weight, 272–273
Boundary element methods, 161, 198–199
Boundary layer growth, over aerofoil, 154–158
Brake horsepower, 345–346, 439, 442
Brake power, 345
Bubble cavitation, 215–216, 216f
Bubble collapse, 212–214, 235–237
Built-up propellers, 11
Bulbous bow, 305–307, 316, 349
Burrill's analysis procedure, 174–180, 177f, 179f, 222–223, 223f

C

Calm water resistance, components of, 301–311
   appendage skin friction, 309
   bulbous bow, 305–307
   naked hull skin friction resistance, 308–309
   transom immersion resistance, 307
   viscous form resistance, 307–308
   viscous resistance, 309–311
   wave-making resistance, 302–305
Cantilever beam method, 397–401
Carbon-based composites, 392
Carbon fiber composites, 11
Cast iron, 11, 392
Cast steel, 392
Casting integrity, 467
Cavitating blade contribution, in hydrodynamic interaction, 289–291
Cavitation, 13, 13f, 51, 209–250
   basic physics of, 210–214
   bucket diagram, 224, 226f
   CFD prediction of, 245–247
   control of, 15
   defined, 209
   design considerations, 221–229
   effects on open water characteristics, 85
   effects and propulsion efficiency, balance between, 450–451
   erosion, 220–221, 230, 234–238, 243
   inception, 186f, 229–234
   -induced damage, 234–239

511

-induced thrust breakdown, 18
noise, 258–261
number, 13, 70–71, 79–80, 85, 99–102, 105–107, 210, 215–216, 221–228, 222t, 239, 261, 294–295
observations, 382–383
pressure data analysis, 243–245
propeller–rudder interaction, 293–297
testing of propellers, 239–243
types of, 214–221
vortices, 235–236
Center-line average (CLA), 480
Centrifugal pump, 21–22, 366–367, 370
Ceramic oxides, 26–27, 27t
Chimera technique, 203
Chlorinity and salinity, relationship between, 48
Chordal pitch, propellers with variable, 6
Chord length, 6, 36, 38–39, 88–89, 94–98, 110, 139–140, 154–155, 161, 185, 221, 272, 341, 406
Chord line, 38–39
Classification societies, 42, 399, 404–406, 454, 465–467, 492, 495, 497, 500, 504–508
Cloud cavitation, 216, 216f
Common Screw propeller, 6
Composite propeller blades, 452
Computational fluid dynamics (CFD), 293–294, 311, 322–323
analysis, 202–204
models, 161
prediction of cavitation, 245–247
Conchoderma, 471
Cone fins, 340
Conservation of energy, 164–166
Continuity equation, 164–165
Contra-rotating propellers, 2–4, 4f, 7, 7f, 16–17, 17f, 18f, 201–202
characteristics of, 438t
Controllable pitch propellers, 19–21, 20f, 200, 437
advantages of, 19
blade interference limits for, 43
characteristics of, 438t
design of, 19–20
Gutsche and Schroeder propeller series, 110
off-design section geometry, 43–46
open water characteristics of, 89–90
specialist types, 19
strength design considerations for, 406
vibration characteristics of, 424f
Conventional propeller geometry, 46
Crash stop maneuvers, 359
Cycloidal propellers, 22, 22f
characteristics of, 438t
Cylindrical blade section definition, 30f

D

Decelerating duct, 14–15, 14f, 91, 110, 201, 450
Defective castings, repair of, 502
Delivered horsepowerp, 81–82
Delivered power, 345

Detached Eddy Simulations (DES), 202–203
Differential Doppler, 74–75, 76f
multi-color, 75–77, 76f
Direction of rotation, 445–448
Direct Numerical Simulations (DNS), 202–203
Directrix, 30
Dissolved gases, in sea water, 50–51
Double integral approach, 244–245
Downwash, 158–159
Drag coefficient, 139
Drawing methods, of propeller, 38, 38f
Dry propeller inertia, 274
Dry weight, of propeller, 273
Duct form, 450
Ducted propellers, 13–15, 14f, 15f, 16f, 200–201
accelerating duct, 14–15, 14f, 91, 110, 201, 435, 450
characteristics of, 438t
decelerating duct, 14–15, 14f, 91, 110, 201, 450
Hannan slotted duct, 14f, 15
open water characteristics of, 90–91
pull–push duct, 14, 14f
strength design considerations for, 404

E

Eckhardt–Morgan design method, 183–187, 184f
Edge damage, to blade, 504
Effective pitch, 32, see also Pitch
Effective velocity, 65–66
Effective weight, of propeller, 279–285
Electrolysis, effects on cavitation inception, 230, 230f
Electromagnetic thrusters, 24–25
Energy Efficiency Design Index (EEDI), 433–435
Energy Efficiency Operational Index (EEOI), 433
Energy saving devices, see Thrust augmentation devices
Engineering design, phases of, 433f
Entrained nuclei model, 211
Erosion damage, 234–239
Euler model, 164
Extrapolation, 42–43, 88–89, 231, 310, 311f, 318, 358, 475–476
three-dimensional, 321–322
two-dimensional, 319–321

F

Face cavitation, 233–234
Face pitch, 32, see also Pitch
Fatigue failure, 338, 388, 411–412, 465–466
Fatigue testing machine, 393–394
Feathering float paddle, 23
Field point velocities, 142–144
Finite aspect ratio wing, 158
Finite element analysis, 426–427
Finite element discretization, 402f
Finite wing, 158–160
Fixed cavity, 214, 215f

Fixed pitch propellers, 11–13
characteristics of, 438t
design, 7
open water characteristics of, 89
operating regimes of, 13, 13f
types of, 11, 12f
Flexible blades, 8
Flat-plate blades, 347, 354
vibration, in air, 421
Flow spoilers, 297
Flow velocities, slipstream contraction and, 119–133
Flow visualization tests, of ship model, 319
Fouling
hull roughness and, 469–478
port classification for, 56t
Four-bladed propeller, 6–7
Fourier analysis, of wake field, 60
Frames of reference, 29–30
global, 29f
local, 29f
Froude's analysis procedure, 59–60, 300–301
Froude number, 239
Full-scale blade strain measurement, 412–413
Full-scale trials, 373–384
bollard pull trials, 379–381
cavitation observations, 382–383
power absorption measurements, 373–379
propeller-induced hull surface pressure measurements, 381–382

G

Gawn series, 99, 103t
Generator line, 30
rake, 33–35, see also Rake
Geometry, of propeller, 29–46
blade thickness distribution, 42–43
blade thickness fraction, 42–43
controllable pitch propeller off-design section geometry, 43–46
controllable pitch propellers, blade interference limits for, 43
conventional propeller geometry, 46
drawing methods, 38, 38f
frames of reference, 29–30, 29f
outlines and area, 35–38
pitch, 31–33, 31f, 32f, 33f
propeller reference lines, 30
rake, 33–35, 34f, 35f
section geometry, 38–42
skew, 33–35, 34f
Grand tunnel hydrodynamique (GTH), 242f
Grim vane wheel, 338–339
Grothues spoilers, 335–336
Guide vane, 15, 200, 369
Gutsche and Schroeder controllable pitch propeller series, 110, see also Controllable pitch propeller

H

Hannan slotted duct, 14f, 15
Harbor maneuvers, 360–361

Index

Harvald method, 313–314
Helmholtz's vortex theorems, 141, 143
Henry's law, 211
High-speed hull form resistance, 328–330
  model test data, 329–330
  standard series data, 329
High-speed propellers
  open water characteristics of, 91–93
  strength design considerations for, 406
High-tensile brass, 389–390
Highly skewed propeller, 11, 37–38, 40, 191, 203–204, 246, 259–260, 402–403, 407, 410–411, 429, 463–464, 503–504
Hitachi Zosen nozzle, 335f, 337
Hot-film anemometer, 73–74
Hot-wire anemometry, 73–74, 74f
Hub boss, 20
Hub form, 449–450
Hub piston system, 20f
Hub vortex cavitation, 216, 216f
Hub weight, 273
Hull drag reduction, 478
Hull efficiency, 325
Hull roughness, and fouling, 469–478
Hull surface
  influence on hydrodynamic interaction, 291
  pressure measurements, propeller-induced, 381–382
  pressures, predicting methods for, 291–293
Hydrodynamic interaction, 288–293
  cavitating blade contribution, 289–291
  components of, 289t
  hull surface influence on, 291
  hull surface pressures, predicting methods for, 291–293
  non-cavitating blade contribution, 289
Hydrodynamic pitch, 32, 185, see also Pitch
Hydrogen bubble technique, 230
Hydrophones, 264, 267
HYKAT, 241–243
Hylarides bridge, 377–378, 377f

I

Impact/grounding damage, 467
Impeller, 343–344, 369–370
  types of, 366–367
Increased diameter/low rpm propeller, 337–338
Inducers, 367
Inlet velocity ratio (IVR), 368
Inspection, of propellers, 495–500
  during manufacture and initial fitting, 496–498
  during service, 499–500
Integrity related problems, 465–467
  blade failure, 465–466
  casting integrity, 467
  previous repair failures, 466–467
International Towing Tank Conference (ITTC), 29–30, 33, 68, 87–88, 231, 243, 308–310, 314–319, 322, 325, 327–328, 404, 475–476, 480, 485

J

Japanese AU-series, 98–99, 99t, 100t–101t
JD–CPP series, 110–112

K

Kappel propeller, 340
KCA series, 99–102
KCD series, 106–107
Kelvin's circulation theorem, 145, 196
Keyed propeller, 497–498
Keless propeller, 498
Kirsten–Boeing propellers, 1, 22, 22f
k–ϵ model, 157, 202–204, 246–247
Kort nozzles, see Ducted Propellers
Kutta condition, 144–145, 196
Kutta–Joukowski theorem, 137

L

Lagrangian method, 165–166
Large Cavitation Channel (LCC), 243
Large Eddy Simulation (LES), 154, 166, 202–204
Laser-Doppler methods, 74
  differential, 74–75, 76f
  multi-color differential, 75–77, 76f
  reference beam method, 75, 76f
Left-handed propellers, 46
Lerbs analysis method, 180–183
  axial induced velocities, 181
  tangential induced velocities, 181
l'Hôpital's rule, 182–183
Lift coefficient, 139
Lifting line–lifting surface hybrid models, 194
Lifting line models, 161
Lifting line theory, 137, 138f
Lifting surface correction factors, 188–191
Lifting surface models, 162, 193–194
Lindgren series (Ma-series), 102, 105t
Line vortex, 141–142, 141f
Lorentz force, 25
Lowes's screw propeller, 5

M

Magnetohydrodynamic propulsion, 24–27, 25f, 26f
Main hump, 304
Maltreatment damage, 501–502
  incorrect handling, 13
  severe heating, 13
  surface deterioration, 13
Manganese, 11
Marine organisms, 55–56
MARIN series, 107, 110, 111t
Ma-series, see Lindgren series
Materials, 385–396
  general properties of, 385–388
  mechanical properties of, 393–394
  specific properties of, 388–393
  test procedures, 394–396
Maximum Continuous Rating (MCR), 439
Mean axial velocity, 60, 60f
Mean hull roughness (MHR), 473–474
Mean pitch ratio, determination of, 443
Mechanical service damage, 435–437
Meridian series, 107
Mewis duct, 337
Mitsui Integrated Ducted Propulsion (MIDP) unit, 336
Mixed flow pump, 365
Model testing, for ship resistance evaluation, 316–317
  facilities, 319
  flow visualization tests, 319
  open water tests, 318–319
  propulsion tests, 319
  resistance tests, 317–318
  three-dimensional extrapolation method, 321–322
  two-dimensional extrapolation method, 319–321
Models, of propeller action, 160–162
  blade element model, 161
  boundary element models, 161
  computational fluid dynamic models, 161
  lifting line models, 161
  lifting surface models, 162
  momentum model, 161
  surface vorticity model, 162
  vortex lattice models, 162
Models, of propeller action
  blade element theory, 171–172, 172f
  boundary element methods, 198–199
  computational fluid dynamics analysis, 202–204
  lifting surface models, 193–194
  momentum theory, 169–171
  vortex lattice methods, 194–198
Molecular attraction, 51
Moment coefficient, 139
Momentum equation, 164–165
Momentum model, 161
Momentum theory, 169–171
Mono-block propellers, 11
Multi-color differential Doppler, 75–77, 76f, see also Differential Doppler
Multi-hull resistance, 329–330
Multi-quadrant series data, 112–119

N

Naked hull skin friction resistance, 308–309
Napier's screw propeller, 5, 5f
NASA, see National Advisory Committee for Aeronautics
National Advisory Committee for Aeronautics (NACA), 38, 139, 151, 185
  series camper, 40, 41t–42t
Navier–Stokes equation, 154, 165–166
Newcomen steam engine, 1
Newtonian fluids, 49–50
Newton–Rader series, 102–106, 105t
Nickel–aluminum bronzes, 11
No-lift pitch, 32
Noise, 251–269
  nature of, 255–261
  prediction and control, 262–263
  radiated noise, measurement of, 263–264

in relation to marine mammals, 264–268
scaling relationships, 261
signature, 253–258, 260, 264–265, 268
transverse propulsion unit, 263
underwater sound, physics of, 251–255
Nominal equality of thrust, 347, 354
Nominal wake field, 61–62
  frictional, 61–62
  potential, 61
  wave-induced, 61–62
Non-cavitating blade contribution, in hydrodynamic interaction, 289
Non-cavitating propeller noise, 257–258
Non-Newtonian fluids, 49–50
Normal Continuous Rating (NCR), 440–441
Nose–tail pitch, 32, *see also* Pitch
Nozzles, 371
Nucleation models, 211
Nuclei density distribution, 51, 52f
Numerical blade stress computational methods, 402–405

O

Octave filter, 253
Off-design section geometry, 43–46
Oil gland, 497–498, 500
Open water characteristics
  cavitation effects on, 85
  of controllable pitch propellers, 89–90
  of ducted propellers, 90–91
  of fixed pitch propellers, 89
  general, 79–85
  of high-speed propellers, 91–93
Open water efficiency, determination of, 443
Open water tests, of ship model, 318–319
Operational problems, 459–468
  impact/grounding damage, 467
  performance-related problems, 459–465
  propeller integrity related problems, 465–467
Outlines, of propeller, 35–38, 37f
  developed, 36
  expanded, 36–37
  projected, 35–36
  swept, 37–38
Out-of-balance forces, 288, 465–466
Overlapping propellers, 17–18, 18f

P

Paddle propulsion, 5–6
Paddle wheels, 23–24, 24f
  float relative velocities, 23, 24f
Partial ducts, 336
Partial hull tunnels, propellers operating in, 451–452
Particle Image Velocimetry (PIV), 77
Peak-to-valley average (PVA), 480
Performance characteristics, 79–137
  behind-hull propeller characteristics, 133–134
  multi-quadrant series data, 112–119
  open water characteristics, 79–85, 89–93
  propeller scale effects, 87–89
  propeller ventilation, 134–136

slipstream contraction and flow velocities, 119–133
  standard series data, *see* Standard series data
Performance-related problems, 459–465
  blade erosion, 461–462
  noise and vibration, 462–465
  power absorption problems, 459–461
Permanent roughness, 469–470
Pinnate propellers, 20–21
Pirouette effect, 217
Pitch, 31–33, 32f, 33f
  angle, 376
  definition of, 31–32, 31f
  effective, 32
  face, 32
  hydrodynamic, 32
  lines, 32, 32f
  mean definition, 32–33, 33f
  nose–tail, 32
Pitch–diameter ratio, 448
Podded propulsors, 16, 17f, 356–362
  steady-state running, 356–358
  turning maneuvers, 358–359
  crash stop maneuvers, 359
  in waves, 360
  general and harbor maneuvers, 360–361
  configuration of, 361–362
Polymers, 11
Power absorption measurements, 373–379
  methods of analysis, 378–379
  techniques of, 376–378
Power absorption problems, 459–461
Power coefficient, 80
Power penalties, 482–485
Practical salinity, 48–49
Prandtl–von Karman theory, 308
Pressure distribution calculations, 149–154
Pressure transducers, 381–382
Prismatic hump, 304
Propeller damage, 431–433
  cavitation erosion damage, 433
  maltreatment damage, 433–435
  mechanical service damage, 435–437
  wastage damage, 433–435
Propeller design, 431–458
  and analysis loop, 431–433
  basis of, 438–442, 452
  blade area ratio, 448
  blade number, 448
  cavitation effects on, 445, 449
  composite propeller blades, 452
  constraints of, 433
  diameter, 442–443, 448
  direction of rotation, 445–448
  duct form, 450
  hub form, 449–450
  mean pitch ratio, determination of, 443
  open water efficiency, determination of, 443
  operating in partial hull tunnels, 451–452
  phases of, 433f
  pitch–diameter ratio, 448
  process, 452–458
  propeller thrust at given conditions, determination of, 445

propeller tip considerations, 451
propeller type, choice of, 435–437
propulsion efficiency and cavitation effects, balance between, 450–451
rotation speed, 443–444, 448
section form, 449
shaft inclination, 450
skew, use of, 449
standard series data, use of, 442–445
Propeller environment, 47–56
  dissolved gases, in sea water, 50–51
  marine organisms, 55–56
  salinity, 48–49
  silt, 55–56
  surface tension, 51
  vapor pressure, 50
  viscosity, 49–50
  water density, 47
  water temperature, 49
  weather, 51–55
Propeller hub, 182
Propeller–hull interaction, 219
Propeller-induced hull surface pressure measurements, 381–382
Propeller reference lines, 30
Propeller revolutions, 376
Propeller roughness, 478–481
Propeller–rudder interaction, 293–297
  full scale remedial measures, 297
  model testing, 296–297
  single-phase approach, 294–295
  two-phase approach, 295–296
Propeller scale effects, 87–89
Propeller–ship interaction, 293–297
  bearing forces and moments, 271–288
  hydrodynamic interaction, 288–293
Propeller thrust measurements, 376
Propeller ventilation, 134–136
Propeller singing, 428–429
Propellers with end-plates, 339–340
Propeller tolerances, 495
Propeller ventilation, 134–136
Propeller weight, 272–273
Propulsion efficiency and cavitation effects, balance between, 450–451
Propulsion systems, 11–28
  azimuthing propellers, 16, 16f
  contra-rotating propellers, 16–17, 17f, 18f
  controllable pitch propellers, 19–21, 20f
  cycloidal propellers, 22, 22f
  ducted propellers, 13–15, 14f, 15f, 16f
  fixed pitch propellers, 11–13, 12f, 13f
  magnetohydrodynamic propulsion, 24–27, 25f, 26f
  overlapping propellers, 17–18, 18f
  paddle wheels, 23–24, 24f
  podded propellers, 16, 17f
  surface piercing propellers, 21
  tandem propellers, 18–19, 19f
  waterjet propulsion, 21–22, 21f
  whale-tail propulsion, 27
Propulsion test, of ship model, 319
Propulsive coefficients, 323–325

Index

hull efficiency, 325
  quasi-propulsive coefficient, 325
  relative rotative efficiency, 324–325
  thrust detection factor, 325
Propulsor–Hull Vortex (PHV) cavitation, 216–217, 217f
Pull–push duct, 14, 14f
Pull–push rod system, 20f
Pump jet, 15, 15f

Q

Quasi-propulsive coefficient (QPC), 325

R

Radiated noise, measurement of, 263–264
Rake, 33–35
  generator line, 33–35
  skew induced, 33–35
  tip, 35f
  total, 34f
Rankine–Froude momentum theory, *see* Momentum theory
Rayleigh model, of bubble collapse, 212–213
Rayleigh–Plesset equation, 229, 245
Reaction fins, 336
Re-entrant jet model, 214–215
Reference beam method, 75, 76f
Reference frames, 29–30
  global, 29f
  local, 29f
Regression-based methods, in ship resistance evaluation, 314–316
Relative rotative efficiency, 324–325
Repair, of propeller, 503–505
  blade cracks, 503–504
  blade maintenance, in service, 505
  boss cracks, 504
  defective castings, 504
  distorted plates, straightening of, 505
  edge damage to blade, 504
  erosion damage, 504–505
  missing blade sections, replacement of, 505
Residual blade stresses, 408–409
Resistance tests, of ship model, 317–318
Restricted camber, propeller with, 7
Restricted water effects, 327–328
Reversing bucket, 371
Reynolds Averaged Navier Stokes (RANS) method, 62, 67, 69–70, 151, 154, 202–204, 245, 293, 357–358, 370
Reynolds number, 50, 80–82, 87–89, 139–140, 154–157, 197–198, 221, 223–224, 227, 230–233, 239, 301, 308, 310, 312
Reynolds stress model, 157, 202–203
Riabouchinsky cavity termination model, 215, 227, 228f
Right-handed propellers, 46
Rim-driven podded propulsors, 362
Root cavitation, 220–221
Rotating beam fatigue test, 395
Rotation speed, 443–444

Roughness-induced power penalties, generalized equation for, 482–485
Rough water, 326–327
Rubert gauge, 480–481
Rudder-bulb fins systems, 314–316

S

Salinity, 48–49
  relationship with chlorinity, 48
Scale effects, 87–89
Schoenherr line, 308
Scissor plates, 297
Screw propeller, early development of, 1–10
Sea margin, 439–441
Sea trials
  ambient conditions, 375
  draught and static trim, 375
  machinery measurements, 375–376
  measured distance course, 374
  measured distance trial area, 373
  number of trial runs, 374
  ship motions, 375
  ship speed and course, 376
  vessel condition, 374
  weather conditions, 374
Sea water
  dissolved gases in, 50–51
  surface tension of, 53t
Section form, 449
Section geometry, 38–42
Section washback, 42, 45f
Self-feathering propeller, 6
Self-noise, 255
Self-pitching propellers, 20–21
Semi-ducts, 336
Sequential solution technique, 229
Service analysis algorithm, 488–489
Service performance and analysis, 469–494
  hull drag reduction, 478
  hull roughness and fouling, 469–478
  propeller roughness and fouling, 478–481
  roughness-induced power penalties, generalized equation for, 482–485
  ship performance, monitoring of, 485–492
  weather effects, 469
Shaft horsepower, 346, 375–376, 442
Shaft incidence, test with propellers, 107–110
Shaft inclination, 450
Shaft power, 376
Shaft speed measurements, 376
Sheet cavitation, 215, 216f, 233–234
Ship Energy Efficiency Management Plan (SEEMP), 433
Ship performance monitoring, 485–492
  benchmarking/rating process, 488
  service analysis algorithm, 488–489
Ship resistance and propulsion, 299–332
  Froud's analysis procedure, 300–301
  calm water resistance, components of, 301–311
  resistance evaluation methods, 311–323
  propulsive coefficients, 323–325
  rough water influence, 326–327
  restricted water effects, 327–328

high-speed hull form resistance, 328–330
air resistance, 330–331
Ship resistance evaluation methods, 311–323
  Auf'm Keller method, 312–313
  Ayre's method, 312
  computational fluid dynamics, 322–323
  direct model test, 316–322
  Harvald method, 313–314
  regression-based methods, 314–316
  standard series data, 314
  Taylor's method, 312
Ship speed measurement, 373–379
Silt, 55–56
Simple estimation methods, 425–426
Skew, 33–35, 449
  angle, 33
  definition, 34f
  induced rake, 33–35, *see also* Rake
  types of, 33
Slipstream contraction and flow velocities, 119–133
Solid boundary factor, 289t, 291
Source vortex panel methods, 162–164, 163f
Specialist propulsors, methods for, 200–202
  contrarotating propellers, 201–202
  controllable pitch propellers, 200
  ducted propellers, 200–201
  super-cavitating propellers, 202
Spindle axis, 30
SSPA series, 107
Stationary crevice model, 211
Stacking line, *see* Generator line
Stainless steels, 391–392
  cladding, 297
Standard series data, 93–112
  Gawn series, 99, 103t
  Gutsche and Schroeder controllable pitch propeller series, 110
  high-speed hull form resistance, 329
  Japanese AU-series, 98–99, 99t, 100t
  JD–CPP series, 110–112
  KCA series, 99–102
  KCD series, 106–107
  Lindgren series (Ma-series), 102, 105t
  MARIN series, 107, 110, 111t
  Meridian series, 107
  Newton–Rader series, 102–106, 105t
  shaft incidence, test with propellers, 107–110
  SSPA series, 107
  use in propeller design, 442–445
  Wageningen B-screw series, 93–95, 94t, 97t
  Wageningen ducted propeller series, 110
Starting vortex, 145–146, 145f
Stationary crevice model, 211
Stator blading, 370
Steady-state running propulsors, 356–358
Steerable ducted propellers, 15, 16f
Steerable internal duct thrusters, 350–352
Steering nozzles, 371
Stern tunnels, 336
Stevens' formula, 53
Strain gauge technique, 376
Streak cavitation, 216

Stress relief, 507–508
Strickling, 416–417
Stylus-based instruments, 480–481
Super-cavitating propellers, 202
Superconducting effect, 26, 26f
Superconducting motors, for marine propulsion, 27
Surface deterioration, 13
Surface piercing propellers, 21
Surface roughness, 469–470, 475–476, 479–481
Surface tension, 51
  of fresh water, 53t
  molecular expansion of, 53f
  of sea water, 53t
Surface vorticity model, 162

T

Tandem propellers, 18–19, 19f
Taylor's method, 59, 185
Taylor's screw propeller, 5, 5f
Taylor wake fraction, 59
Temporary roughness, 469–470
Theodorsen method, 151
Theoretical development, of propellers, 172–174
Thin aerofoil theory, 146–149
Third-octave filters, 253
Three-dimensional extrapolation method, of ship model, 321–322
Thrust augmentation devices, 333–343
  combinations of systems, 342
  zone I devices, 334–337
  zone II devices, 337–340
  zone III devices, 341–342
Thrust breakdown, 18, 221–222, 365, 460–461
Thrust coefficient, 79–82
Thrust detection factor, 325
Thrust measurements, 376, 378
Thwaites' approximation, 155
Tidal models, 379
Tip-unloaded propellers, strength design considerations for, 404
Tip vortex cavitation, 216–219, 216f, 218f, 230
Torque coefficient, 79–82
Traditional manufacturing method, 415–419
Trailing edge curl, 238–239
Trailing edges, of aerofoil, 38–39
Traling vortex, 160f
Transom immersion resistance, 307
Transverse propulsion unit noise, 263
Transverse thrusters, 343–349
  performance characterization, 346–347
  unit design, 347–349
Traveling cavity, 214, 215f
Tunnels, 336
Turning maneuvers, 217–219, 358–359
  propeller forces and moments induced by, 287–288
Twisted rudders, 297
Two-dimensional extrapolation method, of ship model, 319–321
Two-bladed propeller, 6
  with perforated blades, 6

U

Underwater sound, physics of, 251–255

V

van der Waals' equation, 211
Vaporization, 210
Vapor pressure, 50
Velocity distribution, 223–224, 224f
Velocity ratio method, 59
Ventilated/partially submerged propellers, 21
Venturi effect, 328
Vertical axis propellers, *see* Cycloidal propellers
Viscosity, 49–50
Viscous form resistance, 307–308
Viscous–inviscid interactive method, 229
Viscous resistance, 309–311
Voith–Schneider propellers, 22, 22f
Vortex
  filaments, 141–142
  interaction, 219–220
  lattice methods, 162, 194–198
  sheets, 141–142

W

Wageningen 19A duct form, 15, 353–354, 450
Wageningen B-screw series, 93–95, 94t, 97t
Wageningen ducted propeller series, 110
Wake equalizing duct, 334
Wake field, 57–78
  characteristics of, 57–59
  definition, 59–60
  distribution, 57–59, 58f, 59f
  effective, 65–67, 66f
  Fourier analysis of, 60
  Froude method, 59–60
  measurement of, 72–77
  nominal wake field, 61–62
  parameters, estimation of, 62–65
  quality assessment, 70–72
  scaling of, 67–70
  Taylor's method, 59
  velocity ratio method, 59
  wake fraction, 60
Wake fraction, 59–60
Wake quality assessment, 70–72
Wastage damage, 433–435
Water
  density, 47
  immersion effect, 422–425
  phase diagram for, 210f
  rough, 326–327
  saturation temperature of, 210t
  temperature, 49
Waterjet propulsion, 21–22, 21f, 363–372
  basic principles of, 364–365
  component design, 367–371
  impeller types, 366–367
  maneuvering aspects of, 367
Wavelet technique, 244
Wave-making resistance, 302–305
Waves, podded propulsors in, 360
Weather, 51–55, 374
Weber number, 239
Welding, 505–507
Whale-tail propulsion, 27
Wind speed, 53f
World Meteorological Organization (WMO), 55
  sea state code, 55t

Z

Zones of operation, for propellers, 221, 222f